Fundamentals of Data Science

Trilogy

Experiment—Model—Learn

Jared M. Maruskin

Fundamentals of Data Science

Trilogy

Experiment—Model—Learn

Cayenne Canyon Press

Dr. Jared M. Maruskin

Published by Cayenne Canyon Press
San José, California

ISBN: 978-1-941043-01-1 (softcover)
ISBN: 978-1-941043-03-5 (hardcover)

10 9 8 7 6 5 4 3 2 1

More human than a human is our motto.
—Blade Runner

Preface

The goal of this project is to lay out a mathematical foundation of data science, from basic statistics through machine learning. In my experience, many data science books largely focus on data-science applications and programming packages, and not the nitty gritty of the underlying mathematical theory, which is largely buried across numerous graduate texts in statistics and machine learning. In addition to laying out the mathematics, I also complement the reading with numerous examples in Python, and strive (especially as this series progresses) to follow an object-oriented approach to programming.

Due to the encyclopedic scope of this project, I made certain necessary trade offs in order to make a condensed version of the topics presented throughout these pages, while covering as much material as possible in a reasonable amount of time. In particular, I favored clear explanation of the mathematical principles, supplemented with working Python code, over a parade of interesting examples, which I find to be commonly available in many other wonderful texts. Most chapters are themselves condensed versions of entire textbooks, which I refer to for additional discussion and depth throughout.

This work is intended for students of data science with a prerequisite of mathematical maturity, PhDs from the broader panoply of mathematical sciences with a weak statistics background who are interested into transitioning into the field, and as a reference source for practicing data scientists. I personally belonged to this second camp: my own PhD was in dynamical systems theory, with applications to aerospace engineering (tracking space debris; see Maruskin, *et al.* [2012]) and physics (geometric theory of the variational principles of nonholonomic systems; see Maruskin [2018]), whereas my knowledge of statistics, modeling, and machine learning, on the other hand, I picked up on the job after transitioning into the data science field. Since my understanding of many of these topics was superficial or tangential, I found myself one day wishing that I had a masters degree in statistics and machine learning and figured, *no worries, I used to be a professor, I can do that myself.* So I sat down and outlined my own personal curriculum and divided it among three volumes—statistics and experiment, statistical modeling, and machine learning—which form the basis of this series. In short: my goal was to lay a foundation of expertise in statistical modeling and machine learning for myself, while leaving a foundational path for others to follow.

I am very excited to present this trilogy, which covers the mathematical aspects of data science, from basic statistics, to hypothesis testing and experimentation, to statistical modeling and machine learning. This project has been a labor of love, and I've dedicated many weekends and early mornings to read, learn, solve and write the content of these pages. I hope the result bring benefit to your learning and understanding and proficiency of the topics as it has mine.

Coding examples throughout are written in Python. I have found that writing simple simulations is an effective tool in learning theory. It also serves as a tool to validate our theoretical formulas. Several times in the process of writing this text, I developed a formula, ran a simulation, and found that it didn't work the way I expected. This led me to discover my errors and, when the simulation finally worked, served to validate that I had arrived at the correct conclusions.

Python Distribution

I recommend obtaining the free Anaconda Python 3 distribution, available from `anaconda.com`. It comes with all of the scientific packages pre-installed. Anaconda is a bundle of applications, so you can use it to launch R-Studio, Jupyter Notebook, or Spyder. I write my code in Spyder. The Sypder interface gives you access to a code editor, an iPython console, and help docs, all in one screen. I find it extremely useful running code in the console while I develop source code in the editor. I have also have several friends who use PyCharm from `jetbrains.com` as an alternative.

Many of the examples throughout this text will make use of various packages. The import statements that are used throughout are collected in Code Block 1. I will further assume the reader to have familiarity with Python. For those starting out, however, I recommend McKinney [2017] as an excellent place to start. For additional references on Python for machine learning, featuring sci-kit learn and tensorflow, see Gèron [2019].

San José, California   *Jared M. Maruskin*
                                                              April 2022

```python
1   import numpy as np
2   import pandas as pd
3   import sklearn, scipy
4   import time, datetime
5   import matplotlib.pyplot as plt
6   import pymc3 as pm
7   from mpl_toolkits.mplot3d import Axes3D # For 3D
8   from pandas import DataFrame, Series
9   from abc import ABC, abstractmethod
10  from queue import Queue
11  from collections import deque
12  from collections import Counter
13  from scipy.special import comb
14  from scipy.special import beta as BETA
15  from scipy.special import gamma as GAMMA
16  from sklearn.preprocessing import OrdinalEncoder, OneHotEncoder,
        KBinsDiscretizer, KBinsDiscretizer, StandardScaler
17  from sklearn.impute import SimpleImputer
18  from sklearn.pipeline import Pipeline
19  from sklearn.compose import ColumnTransformer
20  from sklearn.model_selection import train_test_split
21  from sklearn.metrics import r2_score, roc_curve, auc,
        confusion_matrix
22  from sklearn.calibration import calibration_curve
23  from statsmodels.tsa.stattools import acf, pacf
24  from tqdm import tqdm
```

Code Block 1: Common package imports.

Contents

Part III Machine Learning

Part I

Inference and Experiment

1

Basecamp: Statistics

1.1 Probability

We begin with a brief review of a few of the main definitions in probability theory. Please see Capinksi and Kopp [2005], Ross [2012] or Wasserman [2004] for more details.

Definition 1.1. *The sample space, typically denoted Ω, of an experiment or random trial is the set of possible outcomes of that experiment. An event is a subset of the sample space. Each element $\omega \in \Omega$ is called an* outcome.

Probability theory deals with potential outcomes of whatever phenomenon we are trying to study; the sample space is the set of all possible outcomes. We shall consider the null set to be implicitly included in the sample space, so that the sample space is closed under the operation of complement ($\Omega^c = \emptyset$).

Example 1.1. A coin is flipped N times. By using the encoding

$$\text{``heads''} \longrightarrow 1$$
$$\text{``tails''} \longrightarrow 0,$$

we may represent the sample space by the set of all 2^N binary strings of length N. If $N \geq 2$, we may further define the event E as the set of all outcomes such that the first two coin flips result in a heads, i.e.,

$$E = \{s \in \mathbb{B}^N : s_0 = 1 \text{ and } s_1 = 1\},$$

where $\mathbb{B} = \{0, 1\}$ is the set of binary digits (bits). *Note*: we will enumerate components of vectors starting with zero, not one, to be consistent with Python. Hence, $x \in \mathbb{R}^n$ may be represented as $x = \langle x_0, x_1, \ldots, x_{n-1} \rangle$. Another example of an event is the outcome in which the ith coin flip results in heads:

$$E_i = \{s \in \mathbb{B}^N : s_i = 1\}, \qquad i = 0, 1, \ldots, N-1.$$

The intersection

$$\bigcap_{i=0}^{N-1} E_i = \{s \in \mathbb{B}^N : s_i = 1 \; \forall \; i \in \{0, \ldots, N-1\}\}$$

is the event in which *each* coin flip is a heads, whereas the union

$$\bigcup_{i=0}^{N-1} E_i = \{s \in \mathbb{B}^N : \exists \; i \in \{0, \ldots, N-1\}, \; s.t. \; s_i = 1\}$$

is the event in which there is *at least one* heads in the series of coin flips. ▷

We will further say that the events $E_1, E_2, \ldots$ are *disjoint* or *mutually exclusive* if $E_i \cap E_j = \emptyset$ whenever $i \neq j$. A *partition* of the sample space Ω is a set of disjoint sets $E_1, E_2, \ldots$ that cover the sample space, i.e., such that $\cup_{i=0}^{\infty} E_i = \Omega$. Finally, given an event E, we shall define its indicator function as

$$\mathbb{I}_E(x) = \begin{cases} 1 & \text{if } x \in E, \\ 0 & \text{if } x \notin E. \end{cases}$$

In general, we do not allow events to be *any* subset of the sample space. Rather, the set of possible events should satisfy some basic axioms.

Definition 1.2. *Let Ω be a nonempty set, and let $\mathcal{E}$ be a collection of subsets of Ω; i.e., $E \in \mathcal{E}$ means that $E \subset \Omega$. The set $\mathcal{E}$ is called an* algebra *if*

1. *the empty set $\emptyset \in \mathcal{E}$,*
2. *for any $E_1, E_2 \in \mathcal{E}$, the union $E_1 \cup E_2 \in \mathcal{E}$,*
3. *for any $E \in \mathcal{E}$, the complement $E^c \in \mathcal{E}$.*

An algebra is a σ-algebra if, in addition, it is closed under countable unions; i.e., if $E_1, E_2, \ldots \in \mathcal{E}$, then $\cup_{i=1}^{\infty} E_i \in \mathcal{E}$.

We are now ready to define a probability distribution over our sample space.

Definition 1.3. *A* probability space *is a triple $(\Omega, \mathcal{E}, \mathbb{P})$, where Ω is a nonempty set, called the* sample space, *$\mathcal{E}$ is a σ-algebra over Ω, and $\mathbb{P} : \mathcal{E} \to [0,1]$ is a function, known as a* probability measure *or* probability distribution, *that satisfies the following axioms:*

1. *$\mathbb{P}(E) \geq 0$ for every event $E \subset \Omega$,*
2. *$\mathbb{P}(\Omega) = 1$,*
3. *If $E_1, E_2, \ldots$ are disjoint, then*

$$\mathbb{P}\left(\bigcup_{i=0}^{\infty} E_i\right) = \sum_{i=0}^{\infty} \mathbb{P}(E_i).$$

If the sample space Ω is finite, the function

$$\mathbb{P}(E) = \frac{|E|}{|\Omega|}$$

constitutes a probability distribution over Ω.

Definition 1.4. *Given a sample space Ω and a probability distribution $\mathbb{P}$, two events E and F are* independent *if*

$$\mathbb{P}(EF) = \mathbb{P}(E)\mathbb{P}(F).$$

Here, we use the shorthand $E \cap F = EF$.

Definition 1.5. *Given a probability distribution $\mathbb{P}$ and an event F with $\mathbb{P}(F) > 0$, The* conditional probability *of E given F is defined by*

$$\mathbb{P}(E|F) = \frac{\mathbb{P}(EF)}{\mathbb{P}(F)}.$$

The intuition behind this definition is that we are restricting our world of possible outcomes, i.e., our sample space, to the subset F. We then define our conditional probability as the fraction of F that is occupied by E. As an exercise, the reader may show that the function $\mathbb{P}(E|F)$ satisfies the three probability axioms for the reduced sample space F.

1.2 Random Variables

Definition 1.6. *Given a probability space $(\Omega, \mathcal{E}, \mathbb{P})$, a* random variable X *is a mapping*

$$X : \Omega \to \mathbb{R}$$

that assigns a real number to each outcome, such that, for any real number c,

$$\{\omega : X(\omega) \le c\} \in \mathcal{E}.$$

A discrete *random variable is one that takes countably many values $\{x_0, x_1, \ldots\}$. A random variable that is not discrete is* continuous.

Example 1.2. Consider a single roll of two six-sided dice. The sample space is the set $\Omega = \{(i,j) : i, j = 1, \ldots, 6\}$. The random variable X is defined as the sum of the number on each die, i.e., $X = i + j$. This is a discrete random variable as its values take on only a countable (and finite) range: $X = 2, 3, 4, 5, 6, 7, 8, 9, 10, 11, 12$. ▷

Regardless of whether a random variable is continuous or discrete, we may define a related function that acts as a cumulative sum of probability over the values that the random variable takes.

Definition 1.7. *Given a random variable X, its associated* cumulative distribution function *(cdf) is the function $F_X : \mathbb{R} \to [0,1]$ given by*

$$F_X(x) = \mathbb{P}(X \le x).$$

Next, we may define a *probability mass function* for discrete random variables and a *probability density function* for continuous random variables, as follows. We will show in Section 1.2 that this distinction is a bit of an illusion, and that probability density functions can actually be defined for both continuous and discrete random variables, though this formalism is not a standard approach.

Definition 1.8. *Given a discrete random variable X, its associated* probability mass function (PMF) *f_X is the function defined by*

$$f_X(x) = \mathbb{P}(X = x).$$

Example 1.3. Continuing Example 1.2, the probability mass function may be expressed as

$$f_X(x) = \frac{|\Omega_{X=x}|}{|\Omega|},$$

where $\Omega_{X=x}$ is the subset of the sample space with $X = x$. Since the size of the sample space is 36, the probability mass function takes the following values:

$$f(2) = f(12) = \frac{1}{36}, \qquad f(3) = f(11) = \frac{2}{36}, \qquad f(4) = f(10) = \frac{3}{36},$$

$$f(5) = f(9) = \frac{4}{36}, \qquad f(6) = f(8) = \frac{5}{36}, \qquad f(7) = \frac{6}{36}.$$

One can additionally verify that the probability mass function is correctly normalized:

$$\sum_{x=2}^{12} f(x) = 1.$$

Finally, the associated cumulative distribution may be expressed as

$$F(x) = \sum_{t=2}^{x} f(t).$$

$\triangleright$

Definition 1.9. *Given a continuous random variable X, its associated* probability density function (PDF) *f_X is the function that satisfies $f_X(x) \ge 0$ for all x,*

$$\int_{-\infty}^{\infty} f_X(x)\, dx = 1,$$

and

$$\int_a^b f_X(x)\,dx = \mathbb{P}(a < X < b),$$

for all $a < b$.

Note that, for continuous random variables, the cumulative distribution function may be represented as

$$F_X(x) = \int_{-\infty}^x f(t)\,dt.$$

Example 1.4. A continuous random variable X is called a *uniform random variable* on the interval $[a, b]$, denoted $X \sim \text{Unif}(a, b)$, if its PDF may be expressed as

$$f(x) = \frac{1}{b - a}$$

for $a < x < b$. ▷

Definition 1.10. *The* expected value *of a function $g : \mathbb{R} \to \mathbb{R}$ of a random variable X is given by*

$$\mathbb{E}[g(X)] = \sum g(x_i)f(x_i), \tag{1.1}$$

if the random variable is discrete, and by

$$\mathbb{E}[g(X)] = \int g(x)f(x)\,dx \tag{1.2}$$

if the random variable is continuous. In particular, we say that the expected value *of a random variable X is given by $\mathbb{E}[X]$, and we use Equations (1.1) or (1.2) with $g(x)$ replaced with x for the case of a discrete or continuous random variable, respectively.*

Example 1.5. Let $X \sim \text{Unif}(a, b)$. Then the expected value of X is given by

$$\mathbb{E}[X] = \frac{1}{b - a}\int_a^b x\,dx = \frac{b^2 - a^2}{2(b - a)} = \frac{a + b}{2}.$$

▷

Proposition 1.1. *Given random variables X and Y and constants a and b,*

$$\mathbb{E}[aX + bY] = a\mathbb{E}[X] + b\mathbb{E}[Y].$$

This follows immediately from the linearity property of integrals and summations.

Definition 1.11. *The* variance *of a random variable X is defined by*

$$V(X) = \mathbb{E}\left[(X - \mathbb{E}[X])^2\right]. \tag{1.3}$$

Proposition 1.2. *Given random variables X and Y and constants a, b, and c,*

$$V(aX + bY + c) = a^2 V(X) + b^2 V(Y).$$

Proposition 1.3. *The variance of a random variable may be expressed as*

$$V(X) = \mathbb{E}[X^2] - \mathbb{E}[X]^2. \tag{1.4}$$

This follows by expanding the square and applying the linearity property of expectation as given by Proposition 1.1.

1.3 The Probability Density Function of a Discrete Random Variable

When considering discrete random variables, on invariably refers to their *probability mass functions*. However, we will do something surprising: we will show that discrete variables also have probability density functions. This will help us unify notation, so that we may consider both cases with a single stroke of the pen.

The "Dirac delta function," $\delta(t)$ is a generalized function that can be thought of as a singularity at the origin:

$$\delta(x) = \begin{cases} \infty & x = 0, \\ 0 & x \neq 0 \end{cases},$$

such that the resulting graph encloses unit area. One may think of the delta function as the limit of a family of standard normal distributions as $\sigma \to 0$. As a result, the delta function has the following properties:

$$\int_{-\infty}^{\infty} \delta(x)\,dx = 1 \qquad \text{and} \qquad \int_{-\infty}^{\infty} \delta(x) f(x)\,dx = f(0),$$

for any function $f(x)$ that is continuous at $x = 0$. We may further think of the delta function as the derivative of the Heaviside function, defined by

$$H(x) = \mathbb{I}[x \geq 0] = \begin{cases} 1 & \text{for } x \geq 0 \\ 0 & \text{for } x < 0 \end{cases},$$

where $\mathbb{I}$ is the *indicator function*, defined such that $\mathbb{I}[S] = 1$ if statement S is true, and 0 otherwise, for any logical expression S. Thus, integrating the delta function produces a "unit step" or a "unit response."

Now we can do something unexpected: write down the *probability density function* for a discrete random variable.

Proposition 1.4. *Given a discrete random variable X with probability mass function $p(x_i) = m_i$, defined on the countable set $\{x_0, x_1, \ldots\}$, its associated probability density function and cumulative distribution are given by*

$$f(x) = \sum_{i=0}^{\infty} m_i \delta(x - x_i) \qquad and \qquad F(x) = \sum_{i=0}^{\infty} m_i H(x - x_i). \qquad (1.5)$$

We may refer to the individual terms $m_i \delta(x - x_i)$ as point masses, *and the delta functions $\delta(x - x_i)$ as* unit point masses.

This is a very useful result, as now we may represent the expected value of a function by the formula

$$\mathbb{E}[g(X)] = \int_{-\infty}^{\infty} g(x) dF(x),$$

regardless whether the random variable X is discrete or continuous. This follows since, in the discrete case, we have

$$\mathbb{E}[g(X)] = \int_{-\infty}^{\infty} g(x) dF(x) = \sum_{i=0}^{\infty} \int_{-\infty}^{\infty} m_i g(x) \delta(x - x_i) \, dx = \sum_{i=0}^{\infty} g(x_i) f(x_i).$$

Using this notation, we may further express the expected value of a random variable X, as given by Definition 1.10, in the form

$$\mathbb{E}[X] = \int x \, dF(x) = \begin{cases} \sum x_i f(x_i) & \text{if } X \text{ is discrete} \\ \int x f(x) \, dx & \text{if } X \text{ is continuous} \end{cases}. \qquad (1.6)$$

Note 1.1. A more theoretical development of this framework involves measure theory. For our purposes, however, we find that using delta functions and Heaviside functions to be sufficient to unify the theory of discrete and continuous random variables into a common framework.   ▷

1.4 Multiple Random Variables

1.4.1 Joint and Marginal Distributions

Of course, it is possible to define more than one random variable on a given sample space.

Definition 1.12. *Given a discrete bivariate random vector (X, Y), the* joint probability mass function *(joint pmf) is the function $f : \mathbb{R}^2 \to \mathbb{R}$ defined by*

$$f_{X,Y}(x, y) = \mathbb{P}(X = x, Y = y).$$

When we are at no risk of confusion, we may drop the subscript and refer to this function simply as $f(x, y)$.

Example 1.6. Let us consider again the roll of two dice from Example 1.2. Suppose that the two dice are easily discernible; perhaps the first die is red and the second is blue. Let X represent the outcome of the red die, and Y the outcome of the blue die. Here, X and Y are both discrete random variables, which may take the values $\{1, 2, 3, 4, 5, 6\}$. The probability mass function is given by

$$f(x_i, y_j) = \frac{1}{36}, \qquad \text{for } x_i, y_j = 1, 2, 3, 4, 5, 6.$$

Notice that each die is *independent* of the other die. (We will define this more formally momentarily.) In fact, the probability mass function for X is simply $f_X(x) = 1/6$, and the probability mass function for Y is $f_Y(y) = 1/6$. This captures the intuitive meaning of *independence*, i.e.,

$$f_{X,Y}(x, y) = f_X(x) f_Y(y),$$

and *marginalization*, i.e.,

$$f_X(x) = \sum_{y=1}^{6} f_{X,Y}(x, y) \qquad \text{and} \qquad f_Y(y) = \sum_{x=1}^{6} f_{X,Y}(x, y).$$

$\triangleright$

Definition 1.13. *Given a continuous bivariate random vector (X, Y), a function $f : \mathbb{R}^2 \to \mathbb{R}$ is the* joint probability density function *(joint pdf) if, for every $A \subset \mathbb{R}^2$,*

$$\mathbb{P}((X, Y) \in A) = \iint_A f_{X,Y}(x, y) \, dx \, dy.$$

When there is no risk of confusion, we may drop the subscript and refer to this function as $f(x, y)$.

Let us take a step back. Since X and Y are both random variables over some sample space Ω, they must each have their own probability mass or probability density function. Our next definition shows us how to reconstruct these individual probability mass or density functions given only the joint distribution.

Definition 1.14. *Given a bivariate random vector (X, Y) and is joint probability mass or probability density function $f_{X,Y}(x, y)$. If the random variables are discrete, the* marginal probability mass functions *of X and Y are given by*

$$f_X(x) = \sum_{y} f_{X,Y}(x, y) \qquad \text{and} \qquad f_Y(y) = \sum_{x} f_{X,Y}(x, y). \qquad (1.7)$$

Similarly, if the random variables are continuous, we define the marginal probability density functions *of X and Y by the equations*

$$f_X(x) = \int f_{X,Y}(x,y)\,dy \qquad and \qquad f_Y(y) = \int f_{X,Y}(x,y)\,dx. \qquad (1.8)$$

We refer to the process of transforming the joint probability mass or probability density function to the univariate probability mass or probability density functions as marginalization.

1.4.2 Independence

Definition 1.15. *Given random variables X and Y, we say that the random variables X and Y are* independent, *denoted $X \perp\!\!\!\perp Y$, if the joint distribution can be factored as*

$$f(x,y) = f_X(x)f_Y(y). \qquad (1.9)$$

Similarly, given a multivariate random vector $(X_1, \ldots, X_n)$ with joint pmf or pdf $f(x_1, \ldots, x_n)$, we say that the random variables $X_1, \ldots, X_n$ are independent if

$$f(x_1, \ldots, x_n) = \prod_{i=1}^{n} f_{X_i}(x_i). \qquad (1.10)$$

Definition 1.16. *We say that a set of random variables $X_1, \ldots, X_n$ are* independent and identically distributed (or IID) *if they are independent and if each marginal distribution follows the same form, i.e., $F_{X_i}(x) = F_{X_j}(x) = F(x)$ for all $i, j = 1, \ldots, n$. In such a case, we may also say that the random vectors form a* random sample *of the distribution F, and we denote this $X_1, \ldots, X_n \sim F$.*

Example 1.7. Consider tossing a coin n times. Let $X_i \in \{0, 1\}$ represent the outcome of the ith coin toss. Then $X_1, \ldots, X_n$ are IID as they each share the same distribution

$$F(x) = \frac{1}{2}\left(H(x) + H(x-1)\right).$$

Alternatively, they share a common probability mass function: $f(0) = f(1) = 1/2$. Independence implies that the result of one coin toss does not influence another.   ▷

1.4.3 Covariance and Correlation

Definition 1.17. *Let X and Y be random variables with means μ_X and μ_Y and standard deviations σ_X and σ_Y. The* covariance *between X and Y is defined by*

$$\text{COV}(X, Y) = \mathbb{E}\left[(X - \mu_X)(Y - \mu_Y)\right]. \tag{1.11}$$

The correlation *between* X *and* Y *is defined by*

$$\rho = \rho(X, Y) = \frac{\text{COV}(X, Y)}{\sigma_X \sigma_Y}. \tag{1.12}$$

Finally, we say that the random variables X *and* Y *are* uncorrelated *if* $\text{COV}(X, Y) = 0$.

Proposition 1.5. *Given random variables* X *and* Y, *the covariance satisfies*

$$\text{COV}(X, Y) = \mathbb{E}(XY) - \mathbb{E}(X)\mathbb{E}(Y). \tag{1.13}$$

The correlation satisfies

$$-1 \leq \rho(X, Y) \leq 1. \tag{1.14}$$

Proposition 1.6. *Any two independent random variables* $X \perp\!\!\!\perp Y$ *are uncorrelated, i.e.,*

$$\mathbb{E}[XY] = \mathbb{E}[X]\mathbb{E}[Y].$$

The converse it not necessarily true.

Definition 1.18. *Let* $X = (X_1, \ldots, X_n) \in \mathbb{R}^n$ *be a random vector. Then its* variance–covariance matrix[1] *is defined by*

$$\mathbb{V}(X) = \mathbb{E}[(X - \mathbb{E}[X])(X - \mathbb{E}[X])^T]; \tag{1.15}$$

i.e., the matrix $\mathbb{V}(X)$ *whose* ijth *component is given by* $[\mathbb{V}(X)]_{ij} = \text{COV}(X_i, X_j)$.

Proposition 1.7. *Let* $X = (X_1, \ldots, X_n)$ *be a random vector. Then the variance of* X *satisfies*

$$\mathbb{V}(X) = \mathbb{E}[XX^T] - \mathbb{E}[X]\mathbb{E}[X]^T. \tag{1.16}$$

Proof. Expanding Equation (1.15), we have

$$\mathbb{V}(X) = \mathbb{E}\left[XX^T - X\mathbb{E}[X]^T - \mathbb{E}[X]X^T + \mathbb{E}[X]\mathbb{E}[X]^T\right].$$

Simplifying, while noting that $\mathbb{E}[X^T] = \mathbb{E}[X]^T$, yields the result.   □

1.4.4 Conditional Distributions and Expectation

In the case of two random variables, we previously saw how one might *marginalize* over one of the random variables in order to obtain the probability function of the other. This is tantamount to taking a probability-weighted average over all possible values of the random variable one is marginalizing over, thereby accounting for each of these possibilities. Next, we consider how one may *condition* over one of the random variables, by which we mean restricting the sample space to a subset consistent with a given value of one of the random variables. More formally, we have the following.

---
[1] Often referred to simply as the *covariance matrix* for X.

Definition 1.19. *Given a bivariate random vector* (X, Y) *with joint pmf or pdf* $f(x, y)$ *and associated marginal pmf or pdf* $f_X(x)$*, we define the conditional pmf or pdf of* X *given* $Y = y$ *as the function*

$$f_{X|Y}(x|y) = \frac{f_{X,Y}(x, y)}{f_Y(y)}, \qquad (1.17)$$

whenever $f_Y(y) \neq 0$.

Note 1.2. Observe that

$$f_{X|Y}(x|y) = \mathbb{P}(X = x|Y = y) = \frac{\mathbb{P}(X = x, Y = y)}{\mathbb{P}(Y = y)}.$$

Thus, the conditional distribution $f(X|Y = y)$ can be viewed as the resulting (re-normalized) probability distribution on the subset $\Omega_{Y=y} = \{s \in \Omega : Y(s) = y\}$ of the sample space Ω that satisfies $Y = y$. ▷

Note 1.3. If $f_Y(y) = 0$, this means that the subset $\Omega_{Y=y} \subset \Omega$ is either the empty set or occurs with probability zero, i.e., $\mathbb{P}(Y = y) = 0$. Thus, defining a probability distribution for X on this subset is ill defined since the outcome $Y = y$, according to our probability measure, cannot occur. ▷

Definition 1.20. *Given random variables* X *and* Y*, the* conditional expectation *of* X *given* $Y = y$ *is the expected value of* X *relative to* $f_{X|Y}(x|y)$*, i.e.,*

$$\mathbb{E}[X|Y = y] = \begin{cases} \sum x f_{X|Y}(x|y) & \text{discrete case} \\ \int x f_{X|Y}(x|y)\, dx & \text{continuous case} \end{cases}. \qquad (1.18)$$

If we allow the random variable Y to vary, the conditional expectation $\mathbb{E}[X|Y]$ may be thought of as a function of the random variable Y, and therefore a random variable itself. This motivates the following result.

Theorem 1.1. *Given two random variables* X *and* Y*, we have*

$$\mathbb{E}[X] = \mathbb{E}\left[\mathbb{E}[X|Y]\right]. \qquad (1.19)$$

The right-hand side of Equation (1.19) is interpreted as follows: The inner expectation, $\mathbb{E}[X|Y]$ is viewed as a random function of Y (i.e., a function of the random variable Y). Therefore the outer expectation is taken over the random variable Y, i.e.,

$$\mathbb{E}\left[\mathbb{E}[X|Y]\right] = \int \mathbb{E}[X|Y = y] f_Y(y)\, dy.$$

As we gather more and more samples from the distribution F, we can construct a histogram of our point estimator $\hat{\theta}$. This is (an approximation of) the sampling distribution.

In short: since a point estimator $\hat{\theta}_n$ is a function of a number of random variables, the point estimator itself is a random variable, and as such, it has its own distribution.

1.5 Sampling

1.5.1 Point Estimators and Sampling Distributions

Often we do not know the form of a distribution, but are instead given a *sample* of values from the distribution, and are tasked with using our data to draw statistical inferences. In particular, we are often interested in inferring the distribution that generated our data, i.e., given a sample $X_1, \ldots, X_n \sim F$, how do we infer the distribution F?

Definition 1.21. *Given a set of data $\mathcal{X} = \{X_1, \ldots, X_n\}$, a statistic is any function of the data.*

Definition 1.22. *Given some unobserved quantity of interest θ—it could be an unknown parameter of the distribution, the distribution itself, our some property of the distribution—a point estimator $\hat{\theta}_n$ is any statistic that estimates θ. Sometimes we may denote the point estimator as $\hat{\theta}_n$ to emphasize the number of data points in our sample.*

Suppose we have a data set $X_1, \ldots, X_n \sim F$ that was generated by some unknown distribution F, and we have an estimator $\hat{\theta}_n$ of some unknown quantity of interest θ. Of course we will want to know *how good* our estimator actually is. It is therefore natural to be interested in the *distribution* of $\hat{\theta}_n$.

Definition 1.23. *Given a point estimator $\hat{\theta}_n$ of some quantity of interest θ, the distribution of $\hat{\theta}_n$ is called the* sampling distribution, *and the standard deviation of $\hat{\theta}_n$ is called the* standard error:

$$\mathrm{se}(\hat{\theta}_n) = \sqrt{\mathbb{V}(\hat{\theta}_n)}.$$

But what do we mean by "the distribution of $\hat{\theta}_n$"? Isn't our point estimate $\hat{\theta}_n$ just a number?

As a statistic, our point estimator is (clearly) a function of the data, i.e., $\hat{\theta}_n = \hat{\theta}_n(X_1, \ldots, X_n)$. As such, in practice, we are only able to observe the single value of our statistic. However, let us do a little thought experiment. Suppose we are able to draw a second random sample from our distribution, say $X_{12}, X_{22}, \ldots, X_{n2} \sim F$. We can then obtain a second estimator $\hat{\theta}_n^2$ of the unknown quantity θ. But why stop at two? What if we were able to generate many many samples from our distribution, i.e., suppose

$$X_{1j}, \ldots, X_{nj} \sim F, \qquad \text{for } j = 1, \ldots, M,$$

for some large integer M. Then we could compute $\hat{\theta}_n^j = \hat{\theta}_n(X_{1j}, \ldots, X_{1n})$, for $j = 1, \ldots, M$. (Forgive the slight abuse of notation: by $\hat{\theta}_n^j$ we currently mean the point estimator from the jth sample—not the point estimator to the power j.)

1.5.2 Sample Mean and Sample Standard Deviation

The most common examples of statistical estimators are, of course, the sample mean and sample variance of a set of data.

Definition 1.24. *Given a set of random variables* $X_1, \ldots, X_n$, *we define the* sample mean *by*

$$\overline{X}_n = \frac{1}{n} \sum_{i=1}^{n} X_i. \tag{1.20}$$

Definition 1.25. *Given a set of random variables* $X_1, \ldots, X_n$, *we define the* sample variance *by*

$$S_n^2 = \frac{1}{n-1} \sum_{i=1}^{n} \left(X_i - \overline{X}_n \right)^2. \tag{1.21}$$

Naturally, the sample standard deviation *is* S_n.

If our data is an independent and identically distributed sample from some distribution F, the sample mean and sample variance are estimators of the mean and variance of the originating distribution, as we will see in our next theorem. But first, let us state an important lemma that will be useful later, when proving the theorem.

Lemma 1.1. *Let* X *be a random variable with expected value* $\mathbb{E}[X] = \mu$ *and variance* $\mathbb{V}(X) = \sigma^2$. *Then*

$$\mathbb{E}[X^2] = \sigma^2 + \mu^2. \tag{1.22}$$

Proof. This follows immediately from Proposition 1.3. □

Theorem 1.2. *Let* $X_1, \ldots, X_n \sim F$ *be* IID *and suppose* $\mathbb{E}[X_i] = \mu$ *and* $\mathbb{V}(X_i) = \sigma^2$. *Then*

$$\mathbb{E}[\overline{X}_n] = \mu, \qquad \mathbb{V}(\overline{X}_n) = \frac{\sigma^2}{n}, \qquad \mathbb{E}[S_n^2] = \sigma^2. \tag{1.23}$$

Proof. The first two parts of Equation (1.23) are easy. By Proposition 1.1, we have

$$\mathbb{E}[\overline{X}_n] = \mathbb{E}\left[\frac{1}{n} \sum_{i=1}^{n} X_i \right] = \frac{1}{n} \sum_{i=1}^{n} \mathbb{E}[X_i] = \frac{1}{n} \sum_{i=1}^{n} \mu = \mu.$$

Similarly, by Proposition 1.2, we have

$$\mathbb{V}(\overline{X}_n) = \frac{1}{n^2} \sum_{i=1}^{n} \mathbb{V}(X_i) = \frac{\sigma^2}{n}.$$

For the third part, let us consider the random variable

$$Y = \sum_{i=1}^{n}(X_i - \overline{X}_n)^2 = \sum_{i=1}^{n}\left(X_i^2 - 2X_i\overline{X}_n + \overline{X}_n^2\right).$$

(This is just the sample variance times the divisor $n - 1$.) Summing over the second and third term, we find this expression to be equivalent to

$$Y = \sum_{i=1}^{n}X_i^2 - n\overline{X}_n^2.$$

From Lemma 1.1, we have $\mathbb{E}[X_i^2] = \sigma^2 + \mu^2$, so that

$$\mathbb{E}\left[\sum_{i=1}^{n}X_i^2\right] = n\sigma^2 + n\mu^2.$$

We may also apply Lemma 1.1, along with the results of the first two parts of this theorem, to the random variable $\overline{X}_n^2$, thereby obtaining

$$\mathbb{E}\left[\overline{X}_n^2\right] = \frac{\sigma^2}{n} + \mu^2.$$

Combining the above three equations, we find

$$\mathbb{E}[Y] = n\sigma^2 + n\mu^2 - \sigma^2 - n\mu^2 = (n-1)\sigma^2,$$

thereby proving the result.   □

Example 1.8. In this example, we use built-in Python functions to generate samples from a given distribution. In particular, we will generate samples from a normal distribution with mean $\mu = 3$ and standard deviation $\sigma = 2$ (variance $\sigma^2 = 4$). We will pretend, however, that these values are unknown, and see how well we can infer them from the data.

First, let us compute the sample mean and sample variance from five samples of various size, as shown in Code Block 1.1.   Line 3 generates

```
np.random.seed(444)
for i in range(1, 6):
    x = np.random.normal(loc=3, scale=2, size=10**i)
    print('sample size:', 10**i, '- sample mean:',
          x.mean().round(3), '- sample variance:', x.var().round(3))
```

Code Block 1.1: Example 1.8

a random sample (from a normal distribution with $\mu = 3$, $\sigma = 2$) of a particular size. Line 4 then prints the sample size along with the sample mean and sample standard variance of the given sample. The output is displayed below.

```
sample size: 10 - sample mean: 3.108 - sample variance: 2.563
sample size: 100 - sample mean: 3.017 - sample variance: 3.969
sample size: 1000 - sample mean: 2.959 - sample variance: 4.261
sample size: 10000 - sample mean: 2.969 - sample variance: 3.951
sample size: 100000 - sample mean: 3.004 -sample variance: 4.032
```

We see that as the sample size n increases, our (single, random) point estimate of the mean and variance improves. It is important to note that the sample variance stabilizes as the sample size increases. For any sample size n, the sample variance is a point estimate of the population (true) variance, and the larger the sample size, the better this estimate is. Similarly, the larger the sample size, the better the sample mean approximates the population mean.

It is important to point out that $\overline{X}_n$ is not *just a number*—it is a random variable, as it depends on the data. If we seed the random number generator (line 2) with a different integer, it will produce different sample means and sample variances. And, in general, if we keep drawing random samples, of a fixed size, from a distribution, the sample mean (and sample variance) will be different for each one.

So how good of an approximation is the sample mean? This depends on the population variance and the sample size, according to the middle part of Equation (1.23). To see this in action, we must consider what is meant by: *the variance of the sample mean.*

Instead of considering a single sample of size n, what if we consider many samples of the same size and compute the sample mean for each one? We could then monitor the behavior as the sample size n increases. Let's

```
1   np.random.seed(444)
2   n_samples = 10000
3   for i in range(1, 6):
4       sample_means = np.zeros(n_samples)
5       for j in range(n_samples):
6           x = np.random.normal(loc=3, scale=2, size=10**i)
7           sample_means[j] = x.mean()
8       print('sample size:', 10**i, '- avgerage:',
            sample_means.mean().round(3), '- variance',
            sample_means.var().round(6))
```

Code Block 1.2: Example 1.8 (cont.)

walk through Code Block 1.2. The index i fixes the sample size $n = 10^i$. For a given sample size (i.e., for each i), we then initialize (line 4) an empty vector of length n_samples = 10,000, where n_samples is the number of

samples we will generate for each i. We then create an inner for-loop that
runs from $j = 1, \ldots,$ n_samples.

The output is recorded, below. (For brevity, "average" means the average, or sample mean, of the sample of sample means, and "variance" means
the sample variance of the sample of sample means. The code in Code
Block 1.2 should illuminate the meaning of this.)

```
sample size: 10 - average: 2.997 - variance: 0.410871
sample size: 100 - average: 3.003 - variance: 0.040004
sample size: 1000 - average: 3.001 - variance: 0.004055
sample size: 10000 - average: 3.0 - variance: 0.000395
sample size: 100000 - average: 3.0 - variance: 0.00004
```

In Code Block 1.2, we consider five different sample sizes, ranging from 10 to
100,000. For a given sample size, the sample mean $\overline{X}_n$ is a random variable.
To sample this random variable , say, 10,000 times, we generate 10,000
samples of size n from the original distribution and compute the sample
mean of each one. This results in a sample of 10,000 sample means $\overline{X}_n$, for
each sample size n. This gives us a feel for how good our estimator $\overline{X}_n$ is
at a given sample size. As expected, the variance of the sample of sample
means decreases like $1/n$, in agreement with Equation (1.23).

For a given sample size, we may further plot a histogram of the outputs
sample_means, which will aid in our visualization of the sample mean as a
random variable. We have done this for the cases $n = 10$ and $n = 100$ in
Figure 1.1. As the sample size increases, the range of values obtained for
$\overline{X}_n$ in accordance with Theorem 1.2. The code used to generate Figure 1.1
is given in Code Block 1.3.

In practice, of course, we typically only have a single sample from our
unknown distribution and, therefore, only a single sample mean. By generating many sample means from a known distribution, however, we gain an
insight into how far off any one particular sample mean might be from the
true mean. Moreover, we see how this error decreases as the sample size
increases.   ▷

Note 1.4. In Figure 1.1, we see that the distribution of the sample mean $\overline{X}_n$
is indistinguishable from a normal distribution, i.e., $\overline{X}_n \sim \mathcal{N}(\mu, \sigma^2/n)$. We
will see later that this is *exactly* true whenever the sample $X_1, \ldots, X_n$
arises from a normal distribution. We will further see, when we discuss the
central limit theorem, that the result is *approximately* true, regardless of
the originating distribution, as the sample size n becomes large. A typical
rule of thumb is that we may approximate the distribution of $\overline{X}_n$ with the
normal distribution $\mathcal{N}(\mu, \sigma^2/n)$ for $n > 30$.   ▷

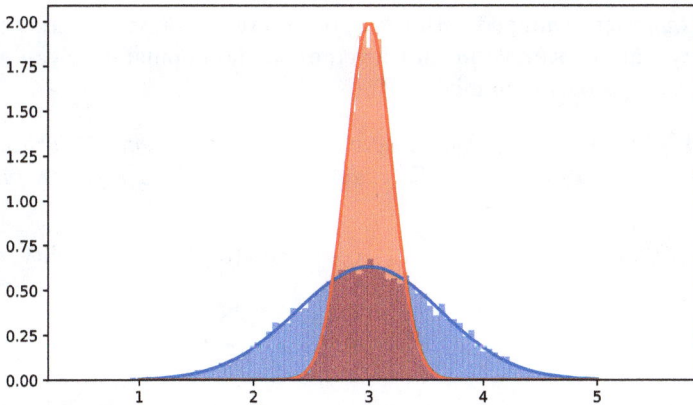

Fig. 1.1: Histogram of sample means for $n = 10$ (blue) and $n = 100$ (red). Solid curves are normal distributions with mean $\mu = 3$ and variance $\sigma^2 = 0.4$ (blue) and $\sigma^2 = 0.04$ (red).

```
plt.figure(figsize=(8, 9/2))
n_samples = 10000
sample_means_10 = np.zeros(n_samples)
sample_means_100 = np.zeros(n_samples)
for j in range(n_samples):
    x = np.random.normal(loc=3, scale=2, size=10)
    sample_means_10[j] = x.mean()
    x = np.random.normal(loc=3, scale=2, size=100)
    sample_means_100[j] = x.mean()
plt.hist(sample_means_10, color='b', bins=100, alpha=0.6,
    normed=True)
plt.hist(sample_means_100, color='r', bins=50, alpha=0.6,
    normed=True)
x = np.linspace(1, 5, num=100)
plt.plot(x, scipy.stats.norm.pdf(x, loc=3, scale=2/np.sqrt(10)),
    'b', linewidth=2)
plt.plot(x, scipy.stats.norm.pdf(x, loc=3, scale=2/np.sqrt(100)),
    'r', linewidth=2)
```

Code Block 1.3: To generate Figure 1.1

1.5.3 Law of Large Numbers

It turns out that, equipped with the appropriate notion of convergence, we can show that the sample mean converges to the population mean as the sample size approaches infinity.

Definition 1.26. *Let $X_1, X_2, \ldots$ be a sequence of random variables, with respective distributions $X_n \sim F_n$, and let $X \sim F$ be a separate random variable.*

1. *We say that the sequence $\{X_n\}$ converges to X in probability if, for every $\varepsilon > 0$, we have*

$$\lim_{n \to \infty} \mathbb{P}(|X_n - X| > \varepsilon) = 0;$$

2. *We say that the sequence $\{X_n\}$ converges to X in distribution if*

$$\lim_{n \to \infty} F_n(x) = F(x),$$

for the set of x at which the distribution F is continuous.

Theorem 1.3 (Law of Large Numbers). *The sample mean $\overline{X}_n$ converges in probability to the population mean $\mathbb{E}[X_i]$ as $n \to \infty$.*

For a proof, see Wasserman [2004]. The following useful theorem is also stated without proof.

Theorem 1.4 (Slutsky's Theorem). *Given two sequences $\{X_n\}_{n=1}^{\infty}$ and $\{Y_n\}_{n=1}^{\infty}$ of random variables with the properties that $X_n \to X$ in distribution and $Y_n \to c$ in probability, then*

(a) $X_n Y_n \to cX$ in distribution, and
(b) $X_n + Y_n \to X + c$ in distribution.

1.5.4 Bias

At first pass, it may appear strange that we use $n-1$ as opposed to n as the divisor in the definition of the sample variance, as given by Definition 1.25. This choice is made, however, so that this estimator is *unbiased*, a concept we shall introduce momentarily.

Definition 1.27. *Given a point estimator $\hat{\theta}_n$ of a quantity θ, estimated from an IID set of data $X_1, \ldots, X_n$, the bias of the estimator is defined by*

$$\text{bias}(\hat{\theta}_n) = \mathbb{E}_{\theta}[\hat{\theta}_n] - \theta. \tag{1.24}$$

Note 1.5. The expectation $\mathbb{E}_\theta$ is taken with respect to the distribution that generated the data, i.e.,

$$f(x_1, \ldots, x_n; \theta) = \prod_{i=1}^{n} f(x_i; \theta),$$

where θ is a fixed, often unknown, quantity. In other words, given the true distribution (which depends on θ), what is the expected value of the estimator over all possible n-point samples from the distribution? For the following discussion, we shall drop the subscript θ from $\mathbb{E}_\theta$, as there is no danger for ambiguity.   ▷

Note 1.6. From Theorem 1.2, it follows immediately that the sample mean and sample variance, as defined in Definitions 1.24 and 1.25, are unbiased estimators of the population mean and population variance, respectively. ▷

A useful measure of accuracy is the mean-squared error, defined as follows.

Definition 1.28. *The* mean-squared error *of a point estimator $\hat{\theta}_n$ is given by*

$$\mathrm{mse}(\hat{\theta}_n) = \mathbb{E}[(\theta - \hat{\theta}_n)^2]. \tag{1.25}$$

The definition of bias for a point estimator leads to our first encounter of a result that will be important in broader contexts later on during our discussion of machine learning.

Theorem 1.5 (Bias–Variance Tradeofff). *Given a point estimator $\hat{\theta}_n$ of an unknown, fixed quantity θ, the mean-squared error can be expressed in terms of the bias and variance of the estimator as follows:*

$$\mathrm{mse}(\hat{\theta}_n) = \mathrm{bias}(\hat{\theta}_n)^2 + \mathbb{V}(\hat{\theta}_n)^2. \tag{1.26}$$

Proof. We begin by expanding the quantity $(\theta - \hat{\theta}_n)^2$ as follows:

$$(\theta - \hat{\theta}_n)^2 = (\theta - \mathbb{E}[\hat{\theta}_n] + \mathbb{E}[\hat{\theta}_n] - \hat{\theta}_n)^2$$
$$= (\theta - \mathbb{E}[\hat{\theta}_n])^2 + (\mathbb{E}[\hat{\theta}_n] - \hat{\theta}_n)^2 - 2(\theta - \mathbb{E}[\hat{\theta}_n])(\hat{\theta}_n - \mathbb{E}[\hat{\theta}_n]).$$

Now, since θ and $\mathbb{E}[\hat{\theta}_n]$ are constants, we have

$$\mathbb{E}\left[(\theta - \mathbb{E}[\hat{\theta}_n])(\hat{\theta}_n - \mathbb{E}[\hat{\theta}_n])\right] = (\theta - \mathbb{E}[\hat{\theta}_n])\mathbb{E}\left[(\hat{\theta}_n - \mathbb{E}[\hat{\theta}_n])\right] = 0.$$

Thus, the mean-squared error may be expressed as

$$\mathbb{E}\left[(\theta - \hat{\theta}_n)^2\right] = (\theta - \mathbb{E}[\hat{\theta}_n])^2 + \mathbb{E}\left[(\hat{\theta}_n - \mathbb{E}[\hat{\theta}_n])^2\right],$$

and the result follows from the definitions of bias and variance.   □

Note 1.7. As a result of Theorem 1.5, the mean-squared error of an unbiased estimator is simply equal to its variance, i.e.,

$$\mathrm{mse}(\hat{\theta}_n) = \mathbb{E}[(\hat{\theta}_n - \theta)^2] = \mathbb{E}[(\hat{\theta}_n - \mathbb{E}[\hat{\theta}_n])^2] = \mathbb{V}(\hat{\theta}_n), \qquad \text{if } \mathbb{E}[\hat{\theta}_n] = \theta.$$

▷

1.6 The Empirical Distribution

In this section, we discuss estimating both the distribution function and certain functions of the distribution function from data. For a more details, see Wasserman [2004] and Wasserman [2006].

1.6.1 Estimating Distributions from Data

We next address how one may estimate a distribution given a sample of values generated by the distribution, without making any assumptions as to the underlying form of the distribution.

Definition 1.29. *Let* $X_1, \ldots, X_n$ *be a data set. The* empirical distribution function $\hat{F}_n$ *is the cumulative distribution function over* $\mathbb{R}$ *given by the function*

$$\hat{F}_n(x) = \frac{1}{n} \sum_{i=1}^{n} \mathbb{I}[X_i \leq x]. \tag{1.27}$$

Note 1.8. Suppose the data are IID from a common (oftentimes unknown) distribution, i.e., $X_1, \ldots, X_n \sim F$. In this case, the empirical distribution $\hat{F}_n$ is a point estimator of the original distribution F: it is a statistic, as it is a function of the observed data, and it seeks to estimate an underlying quantity of interest, namely the distribution itself.   ▷

Note 1.9. In the language of Section 1.3, the empirical distribution is simply the distribution obtained by placing an equally weighted *point mass* at each of the observed data points $X_1, \ldots, X_n$.   ▷

We can actually code Definition 1.29 in a single line, as shown in Code Block 1.4. In this code, the variable x is a scalar, whereas **data** is a vector

---

```
1  def emp_dist(x, data):
2      # x (float) point to evaluate empirical distribution
3      # data (array) observed data
4      # returns a float
5      return (data <= x).sum() / len(data)
```

Code Block 1.4: Empirical distribution (at a point)

(np.array). The logical evaluation **data <= x** returns a Boolean vector, i.e., a vector whose components are either **True** or **False**. Summing over this vector adds the number of **True** instances, thereby performing the calculation in Equation (1.27). The problem, however, is that if one were to pass a vector in for the first argument, the function would fail. We would

```
1   def emp_dist(x, data):
2       # x (float or array) point to evaluate empirical distribution
3       # data (array) observed data
4       # returns an array equal in size to x
5       data = data.reshape((len(data), 1))
6       if np.isscalar(x):
7           x = np.array([x])
8       x = x.reshape((len(x), 1))
9       return (data <= x.T).sum(axis=0) / len(data)
```

Code Block 1.5: Empirical distribution (vectorized)

therefore like to take this one step further, by *vectorizing* the function given in Code Block 1.4, as done in Code Block 1.5. By reshaping x and data as column vectors, and then taking the transpose of x, the operation data <= x.T suddenly makes sense, and may further be summed down the rows (axis=0). We invite the reader to dissect this code, playing with the outputs in iPython console, one line at a time. For an np.array object, the shape method is also useful: in particular, one may invoke the call x.shape() before and after the reshape call to see its effect.

Example 1.9. We can use the empirical distribution function from Code Block 1.5 to plot the empirical distribution from scikitlearn's Boston housing price dataset, as shown in Code Block 1.6. The result is shown in Figure 1.2. (We will explain the meaning of the dashed lines later, in Example 1.14.)

```
1   boston = sklearn.datasets.load_boston()
2   df = DataFrame(boston['data'], columns=boston['feature_names'])
3   df.head() # shows top 5 rows in DataFrame
4   print( boston['DESCR'] ) # prints description of data set
5   y = boston['target'] # target variable
6   x = np.linspace(0, 60)
7   F_hat = emp_dist(x, y)
8   plt.plot(x, F_hat)
```

Code Block 1.6: Computing the Empirical Distribution

▷

1.6.2 Bias and Variance of the Empirical Distribution

Since the empirical distribution $\hat{F}_n$, generated from a set of data $X_1, \ldots, X_n$, is a point estimator of the true (often unknown) distribution, it is natural

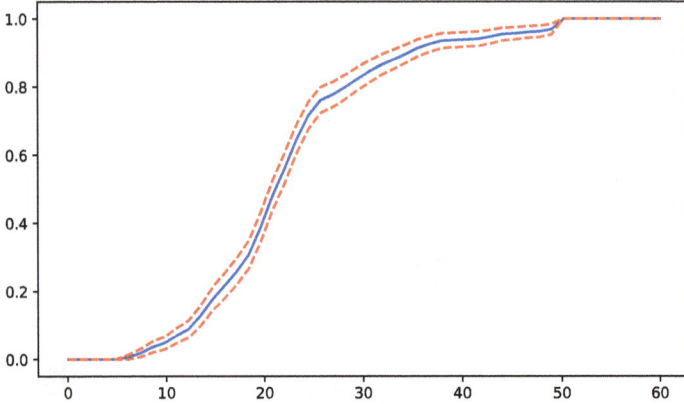

Fig. 1.2: Empirical Distribution of the Boston housing dataset.

to ask what its expected value and variance is, similar to what we achieved with Theorem 1.2.

Theorem 1.6. *Given a set of* IID *data* $X_1, \ldots, X_n \sim F$, *its empirical distribution* $\hat{F}_n$, *and a fixed point* $x \in \mathbb{R}$, *the expectation and variance are given by*

$$\mathbb{E}\left[\hat{F}_n(x)\right] = F(x), \tag{1.28}$$

$$\mathbb{V}\left(\hat{F}_n(x)\right) = \frac{F(x)(1 - F(x))}{n}. \tag{1.29}$$

Equation (1.28) implies that $\hat{F}$ is unbiased. Consequently, from Theorem 1.5, it follows that $\mathrm{mse}(\hat{F}_n(x)) = \mathbb{V}(\hat{F}_n(x))$. We will postpone the proof until our discussion of the binomial random variable.

1.6.3 Plug-in Estimators

One may be interested in many quantities associated with a probability distribution: its mean, variance, quantiles, etc. We next provide a definition of such quantities and then proceed to show how they may be estimated from the empirical distribution.

Definition 1.30. *A statistical functional is any real-valued function of a distribution.*

Example 1.10. The expected value and variance of a random variable X is a functional of its probability distribution F, as

$$\mu = \mu(F) = \int x \, dF \qquad \text{and} \qquad \sigma^2 = \sigma^2(F) = \int (x - \mu(F))^2 \, dF.$$

Hence, the mean and variance may be expressed as a functional of the distribution itself.   ▷

Definition 1.31. *Given a statistical functional* $\theta = T(F)$ *and an* IID *dataset* $X_1, \ldots, X_n \sim F$, *the* plug-in estimator *of* θ *is defined by*

$$\hat{\theta}_n = T(\hat{F}), \tag{1.30}$$

where $\hat{F}$ *is the empirical distribution generated by the dataset.*

A plug-in estimator is simply the point estimator of a given functional obtained by using the empirical distribution in lieu of the actual distribution, which is often unknown.

Proposition 1.8. *If* $T(F)$ *is a linear functional, i.e., a functional of the form* $\theta = T(F) = \int r(x) \, dF$, *and* $X_1, \ldots, X_n \sim F$ *is an* IID *sample from* F, *then the plug-in estimator is given by*

$$\hat{\theta}_n = \int r(x) \, d\hat{F}_n = \frac{1}{n} \sum_{i=1}^{n} r(x_i). \tag{1.31}$$

Proof. This result follows immediately from our discussion in Section 1.3 and the definition of the empirical distribution. Definition 1.29 is equivalent to the CDF generated by a point mass of strength $1/n$ placed at each datum:

$$\hat{F}_n(x) = \frac{1}{n} \sum_{i=1}^{n} \mathbb{I}[x_i \le x] = \frac{1}{n} \sum_{i=1}^{n} H(x - x_i).$$

Hence $\hat{f}_n(x) = \hat{F}'_n(x) = \frac{1}{n} \sum \delta(x - x_i)$, and

$$\hat{\theta} = \int r(x) \, d\hat{F}_n = \sum_{i=1}^{n} \int r(x) \delta(x - x_i) \, dx = \frac{1}{n} \sum_{i=1}^{n} r(x_i),$$

and the result follows.   □

Example 1.11. Recalling that the empirical distribution is the distribution that places a point mass of strength $1/n$ at each data point $X_1, \ldots, X_n$, we may express the plug-in estimator for the mean as

$$\hat{\mu}_n = \int x \, d\hat{F}_n = \frac{1}{n} \sum_{i=1}^{n} x_i,$$

which is equivalent to the sample mean. (The calculus of the preceding equation was discussed in Section 1.3 and Proposition 1.8.) As previously discussed, the standard error is $\text{se}(\hat{\mu}) = \sqrt{\mathbb{V}(\overline{X})} = \sigma/\sqrt{n}$. If $\hat{\sigma}$ is an estimate of σ, the corresponding estimated standard error is $\hat{\text{se}} = \hat{\sigma}/\sqrt{n}$.   ▷

Example 1.12. The plug-in estimator of the variance of a probability distribution is given by

$$\hat{\sigma}^2 = \int (x - \hat{\mu})^2 \, d\hat{F}_n = \frac{1}{n} \sum_{i=1}^{n} (x_i - \hat{\mu}_n)^2,$$

which is equivalent to the sample standard deviation S_n scaled by a factor of $(n-1)/n$, i.e., $\hat{\sigma}^2 = (n-1)/n S_n$. Unlike the sample standard deviation, the plug-in estimator of the variance is therefore biased.   ▷

Not all statistical functionals, however, are computed by integrating over a probability distribution. Case in point: quantiles.

Definition 1.32. *Given a distribution function F and a number $p \in (0, 1)$, the pth quantile, Q_p, is defined by*

$$Q_p = F^{-1}(p) = \inf\{x : F(x) \geq p\}. \qquad (1.32)$$

The pth quantile is clearly a statistical functional, as it is a function of the distribution function, i.e., $Q_p = Q_p(F)$. The inverse $F^{-1}(p)$ is defined using an infimum[2], to ensure its existence even when no x-value maps directly onto p.

Example 1.13. Consider the probability mass function that places equal weight $1/3$ on the values $x_1 = 1$, $x_2 = 2$, and $x_3 = 3$. The distribution function is

$$F(x) = \begin{cases} 0 & \text{for } x < 1 \\ 1/3 & \text{for } 1 \leq x < 2 \\ 2/3 & \text{for } 2 \leq x < 3 \\ 1 & \text{for } 3 \leq x \end{cases}.$$

Let's compute the 50th percentile ($p = 0.5$). The set $\{x : F(x) \geq 0.5\} = [2, \infty)$. Hence $Q_{0.5} = 2$. A similar calculation shows, for example, that $Q_{0.33} = 1$ and $Q_{0.34} = 2$.   ▷

Proposition 1.9. *Given a set of IID data $X_1, \ldots, X_n \sim F$, the plug-in estimator of the pth quantile Q_p is given by*

$$\hat{Q}_p = \hat{F}_n^{-1}(p) = \inf\{\hat{F}_n(x) \geq p\}. \qquad (1.33)$$

We call $\hat{Q}_p$ the pth sample quantile.

---

[2] The infimum is the *greatest lower bound* of a set. So the infimum of the set $(0, 1)$ is $\inf(0, 1) = 0$. An infimum is well defined, even if the minimum value does not exist (as is the case for the open interval $(0, 1)$). Similarly, the *supremum* of a set is its *least upper bound*; e.g., $\sup(0, 1) = 1$.

An interesting application of plug-in estimators is found by approximating the standard error of the cumulative distribution itself. Recall that the empirical distribution function $\hat{F}_n(x)$ is itself a point estimator, with expected value and variance given by Theorem 1.6. Since the standard error $\text{se}(\hat{F}_n(x)) = \sqrt{\mathbb{V}(\hat{F}_n(x))}$ is a function of the distribution, we may compute its plug-in estimator, as done in the following example.

Example 1.14. The plug-in estimator for the standard error of the empirical distribution $\hat{F}_n$ is obtained by replacing F with $\hat{F}_n$ in the right-hand side of Equation (1.29), yielding the expression

$$\hat{\text{se}}(\hat{F}_n(x)) = \sqrt{\frac{\hat{F}_n(x)(1 - \hat{F}_n(x))}{n}}.$$

We may use this to estimate the standard error of the empirical distribution from Example 1.9. We can then plot $\hat{F}_n(x) \pm 2\hat{\text{se}}$, as is done in Code Block 1.7. This results in a *confidence interval* for our empirical distribution function, as shown in Figure 1.2.

```
se = np.sqrt(F_hat * (1 - F_hat) / len(y))
F_up = np.fmin(1, F_hat + 2 * se)
F_dn = np.fmax(0, F_hat - 2 * se)
plt.plot(x, F_up, 'r--')
plt.plot(x, F_dn, 'r--')
```

Code Block 1.7: Computing the standard error (continued from Code Block 1.6).

Note that we cap the upper and lower bounds of our confidence interval at 100% and 0%, respectively, using numpy's built-in fmin and fmax functions. (In this case, the sample size is large enough so that these caps don't actually affect our bounds.) ▷

1.7 The Bootstrap

In this section, we build on our discussion of the empirical distribution and devise a method for computing the variance and the distribution of any statistic $T_n = g(X_1, \ldots, X_n)$. Typically we will not have access to the distribution F that actually generated the data.

1.7.1 Introduction

The *bootstrap* is a method for approximating the distribution of any test statistic T_n from a given sample of data. In particular, we are interested in computing the standard error of the test statistic along with a confidence interval (a set of values for T_n which we deem likely to coincide with reality).

The idea of the bootstrap is to replace the variance function $\mathbb{V}_F(T_n)$ with $\mathbb{V}_{\hat{F}}(T_n)$; i.e., when the distribution that generated the data is unknown, we approximate the variance of our statistic using the empirical distribution instead. In most cases, the empirical variance $\mathbb{V}_{\hat{F}}(T_n)$ must be approximated by simulation. Before we discuss this, however, it will be advantageous to perform the computation for a simple case, for which the empirical variance is known.

Example 1.15. Let us consider the familiar case of the sample mean $T_n = \overline{X}_n$. We know that the variance of the sample mean depends on the distribution that generated the data, since $\mathbb{V}_F(\overline{X}_n) = \sigma^2/n$. We saw in Example 1.12 that the plug-in estimator for the variance is given by $\hat{\sigma}^2 = 1/n \sum_{i=1}^{n}(X_i - \overline{X}_n)^2$. Hence, our estimate for the variance of the sample mean is $\mathbb{V}_{\hat{\sigma}}(\overline{X}_n) = \hat{\sigma}^2/n$.   ▷

The case of the sample mean is the exception, however, not the rule. In most cases, the empirical variance cannot be directly represented as a function of the data. This is where the bootstrap comes into play.

1.7.2 Bootstrap Samples

A statistic T_n is a function of the data. Given an IID sample $X_1, \ldots, X_n \sim F$, we may compute the corresponding point estimate of the statistic $T_n = g(X_1, \ldots, X_n)$. In order to approximate the distribution for T_n, we may instead take many samples from our original distribution F, compute the value of our statistic for each sample, and use the result as a proxy for the actual distribution of T_n.

In general, however, we do not know the form of the distribution F that generated our data. Moreover, we only have access to a single sample from that distribution $X_1, \ldots, X_n \sim F$. This sample is our data set. Our goal is to use these data to draw inferences about our statistic T_n, beyond the simple point estimate $T_n = g(X_1, \ldots, X_n)$.

If we had access to the distribution function F, we could, of course, produce an arbitrary number of samples, compute the value of T_n for each sample, thereby approximating the distribution of T_n. To resolve this, we instead use the empirical distribution function $\hat{F}_n$ that is obtained from our single, real sample, i.e., our actual data set. But how can we draw a random sample from our empirical distribution?

Definition 1.33. *Given a set of* IID *data* $X_1, \dots, X_n \sim F$, *a bootstrap sample of size* k *of the data is a set of* k *numbers* $X_1^*, \dots, X_k^*$ *that are drawn randomly, with replacement, from the original data set* $X_1, \dots, X_n$.

Since a bootstrap sample is drawn *with replacement* from the original data set, the bootstrap sample size k is not limited by the size of the data n. In practice, however, one typically considers a bootstrap sample to be the same size as the original data, i.e., with $k = n$.

Example 1.16. Consider the data set $\mathcal{D} = (R, G, B)$, representing the colors red, green, and blue. Our first draw from the set $\mathcal{D}$ will result in an R with a probability of $1/3$, G with a probability of $1/3$, and B with a probability of $1/3$, i.e.,

$$p(R) = p(G) = p(B) = 1/3.$$

Suppose we happen to draw a B. Since this draw occurs *with replacement*, it means that the probability distribution for our next draw is unchanged, i.e., we still have a $1/3$ probability of drawing a B. (Contrast this to a random draw *without replacement*: having first obtained a B, the probable outcome of our *second* draw would be $p(R) = p(G) = 1/2$, $p(B) = 0$, since we have already removed the B from the data set, and we are not "replacing" it.)

A bootstrap sample of size, say, $k = 10$ can be done in a single line of Python code:

```
np.random.choice(('R','G','B'), size=10, replace=True) .
```

This generated an output of

```
['B', 'G', 'B', 'B', 'R', 'G', 'G', 'B', 'B', 'R'].
```

(The output will be different each time, because it is generated randomly.)
▷

As one may have already suspected, the bootstrap sample is closely related to a sample from the empirical distribution.

Proposition 1.10. *A bootstrap sample* $X_1^*, \dots, X_k^*$ *obtained from a set of data constitutes a random sample* $X_1^*, \dots, X_k^* \sim \hat{F}_n$ *from the empirical distribution* $\hat{F}_n$ *generated by the data.*

Proof. Since the empirical distribution is equivalent to the distribution function of a set of equal-strength point masses, located at $X_1, \dots, X_n$, a random number $X \sim \hat{F}_n$ is equivalent to selecting one of the data points $X_1, \dots, X_n$ at random, i.e., with equal probability. Taking k IID samples from the empirical distribution $\hat{F}_n$ is therefore equivalent to drawing k values from the data set, with replacement, which is the definition of a bootstrap sample. □

1.7.3 Estimating the Variance of a Statistic

An immediate application of the bootstrap is estimating the variance of a statistic. The pseudocode is given in Algorithm 1.1.

Algorithm 1.1: Bootstrap Variance Estimation

Data: $X_1, \ldots, X_n \sim F$; a statistic $T_n = g(X_1, \ldots, X_n)$
Result: Estimate of the variance $v_{boot} \approx \mathbb{V}(T_n)$

1 **for** $b = 1$ **to** B **do**
2 $\quad$ Compute a bootstrap sample $X_{1,b}^*, \ldots, X_{n,b}^* \sim \hat{F}_n$
3 $\quad$ Compute $T_{n,b}^* = g(X_{1,b}^*, \ldots, X_{n,b}^*)$
4 **end**

5 Compute the sample mean $\overline{T}_n^* = \dfrac{1}{B} \sum_{b=1}^{B} T_{n,b}^*$

6 Output $v_{boot} = \dfrac{1}{B} \sum_{b=1}^{B} \left(T_{n,b}^* - \overline{T}_n^* \right)^2.$

Essentially, we perform many bootstrap samples from our data set, compute the value of the statistic T_n for each sample, storing the result, and use this collection of values as an approximate distribution for the statistic T_n. In particular, the variance of this vector of values yields an estimate for the variance of the statistic T_n.

In Python, functions are objects, and therefore may be passed in as arguments to other functions. Let's begin by defining a function for our statistic T_n. For example, if we were interested in computing the 90th quantile $Q_{0.9}$, we could code this using numpy's built-in quantile function, as in Code Block 1.8.

```
def t(x):
    # x is an array, set of data
    return np.quantile(x, 0.9)
```

Code Block 1.8: Function definition of our statistic T_n

For a given set of data x, the function t(x) outputs the point estimate of the quantile $Q_{0.9}$. We may pass this function, along with our observed data set, into a second function, bootstrap, defined in Code Block 1.9. The bootstrap function follows lines 1–4 of Algorithm 1.1. The bootstrap sample of the data x is computed by np.random.choice(x, size=len(x)). Passing the bootstrap sample into the function t produces a single simulated value for the statistic t. We do this many times and return the output as an array of values. To obtain the boostrap variance estimate, we simply compute the variance of this output array, as shown for the Boston housing

```
1   def bootstrap(x, t, B=10000):
2       # x is an array, set of data
3       # t = t(x) is a function: array --> scalar
4       # returns: array, sample from distribution for t
5       ts = np.zeros(B)
6       for b in range(B):
7           ts[b] = t( np.random.choice(x, size=len(x)) )
8       return ts
```

Code Block 1.9: Bootstrap sample of values for T_n

price data in Code Block 1.10. The variance output is `1.368`, which is the

```
1   y = data['target']
2   ts = bootstrap(y, t)
3   print(ts.var())
4   plt.hist(ts, bins=30, normed=True)
```

Code Block 1.10: Boostrap variance for $Q_{0.90}$ for the Boston house price data

bootstrap estimate for $\mathbb{V}(Q_{0.90})$. (The standard error is the square-root of this, so $\hat{se}(Q_{0.90}) = 1.17$.) But since we retained *all* of our samples, we can further plot a histogram of the data, as shown in Figure 1.3. By eyeballing this histogram, we can conclude that we expect the quantile $Q_{0.90}$ to take a value between 32 and 37. And, indeed, we find that

$$\texttt{ts.mean() + 2 * ts.std()} = 36.79$$
$$\texttt{ts.mean() - 2 * ts.std()} = 32.11\text{'}$$

confirming our earlier statement that $\hat{T}_n \pm 2\,\hat{se}(T_n)$ is normally a good approximation for a confidence interval.

1.8 Transformations of Random Variables

In this section, we shall consider functions of random variables. In general, since a random variable X is a function from our sample space into the set of real numbers, i.e., $X : \Omega \to \mathbb{R}$, one may apply a function $g : \mathbb{R} \to \mathbb{R}$ to a random variable to obtain a new random variable $Y = g(X)$. Such mappings are referred to as *transformations* of random variables. Our primary application for this section is in defining several more advanced distributions later on, so the ready can skip without loss of continuity. For further details, we direct the reader to, e.g., Casella and Berger [2002].

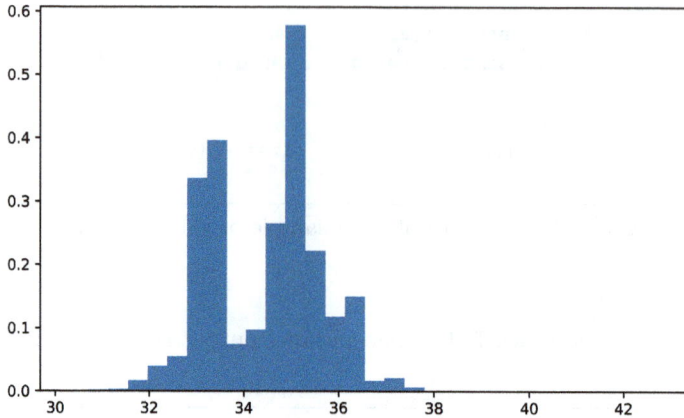

Fig. 1.3: Bootstrap sample of values for $Q_{0.90}$ for Boston housing data.

1.8.1 Univariate Transformations

If X is a random variable with CDF F_X, then any function $Y = g(X)$ is also a random variable. For any set $A \subset \mathbb{R}$, we have

$$\mathbb{P}(Y \in A) = \mathbb{P}(g(X) \in A) = \mathbb{P}(X \in g^{-1}(A)),$$

where we define the inverse mapping of the set A as $g^{-1}(A) = \{x \in \mathbb{R} : g(x) \in A\}$.

For convenience, let us define the following subsets of the real line:

$$\mathcal{X} = \{x \in \mathbb{R} : f_X(x) > 0\} \quad \text{and} \quad \mathcal{Y} = \{y \in \mathbb{R} : y = g(x) \text{ for some } x \in \mathcal{X}\}.$$

If X is a discrete random variable, then both $\mathcal{X}$ and $\mathcal{Y}$ are countable, and the probability mass function for Y is given by

$$f_Y(y) = \sum_{x \in g^{-1}(y)} f_X(x) \quad \text{for } y \in \mathcal{Y}.$$

In other words, the probability that $Y = y$ is simply the sum of the probabilities that $X = x$, over every x such that $g(x) = y$.

The case of a continuous random variable is not so neatly packaged, as it required invertibility of the function g. Often, one must deal with piece-wise invertibility.

Theorem 1.7. *Let X be a continuous random variable with pdf $f_X(x)$, and consider the transformation $Y = g(X)$. Suppose there is a partition $A_0, A_1, \ldots, A_k$ of $\mathcal{X}$ such that $\mathbb{P}(X \in A_0) = 0$ and f_X is continuous on A_i.*

Furthermore, let us define the subsets $\mathcal{Y}_1, \ldots, \mathcal{Y}_k \subset \mathbb{R}$ by the relations $\mathcal{Y}_i = \{y : y = g(x), \text{ for some } x \in A_i\}$. Finally, let us suppose that g is locally invertible on each A_i, i.e., suppose there exists functions $g_1(x), \ldots, g_k(x)$, defined on $A_1, \ldots, A_k$, respectively, such that

1. $g(x) = g_i(x)$, for $x \in A_i$,
2. $g_i(x)$ is monotone on A_i,
3. $g_i^{-1}(y)$ has a continuous derivative on $\mathcal{Y}_i$, for each $i = 1, \ldots, k$.

Then the pdf of the random variable Y is given by

$$f_Y(y) = \sum_{i=1}^{k} f_X\left(g_i^{-1}(y)\right) \left| \frac{dg_i^{-1}(y)}{dy} \right| \mathbb{I}[y \in \mathcal{Y}_i]. \qquad (1.34)$$

For a more detailed discussion of this theorem, along with supporting examples, see Casella and Berger [2002]. For intuition, however, let us consider the case in which we do not need to partition the domain $\mathcal{X}$.

Corollary 1.1. *Let X, Y, and g be as in Theorem 1.7, and further suppose that the function g is continuously differentiable on $\mathcal{X}$ with non-zero derivative. Then the pdf of the random variable Y is given by*

$$f_Y(y) = f_X\left(g_i^{-1}(y)\right) \left| \frac{dg^{-1}(y)}{dy} \right|. \qquad (1.35)$$

1.8.2 Bivariate Transformations

Next, let us consider a transformation of two random variables X and Y. Suppose (X, Y) is a bivariate random vector which maps into (U, V) under the mapping $U = g_1(X, Y)$ and $V = g_2(X, Y)$. Given the joint distribution $f_{X,Y}(x, y)$ and the mapping functions $g : \mathbb{R}^2 \to \mathbb{R}^2$, how can we determine the joint distribution $f_{U,V}(u, v)$?

For the discrete case, the answer is quite intuitive, as seen in our next theorem.

Theorem 1.8. *Let (X, Y) be a discrete random vector with PMF $f_{X,Y}(x, y)$, and consider the vector mapping $g : \mathbb{R}^2 \to \mathbb{R}^2$, which defines a new discrete random vector $(U, V) = g(X, Y)$. Moreover, for any $u, v \in \mathbb{R}$, let us define the set*

$$g^{-1}(u, v) = \{(x, y) : g(x, y) = (u, v)\}.$$

Then the PMF of the transformed random vector (U, V) is given by

$$f_{U,V}(u, v) = \sum_{(x,y) \in g^{-1}(u,v)} f_{X,Y}(x, y). \qquad (1.36)$$

Proof. Since the random vectors are discrete, they are defined on a countable subset of $\mathbb{R}^2$. By tracing the probability back to the original space, we may express the probability mass function as

$$
\begin{aligned}
f_{U,V}(u,v) &= \mathbb{P}((U,V) = (u,v)) \\
&= \mathbb{P}((X,Y) \in g^{-1}(u,v)) \\
&= \sum_{(x,y) \in g^{-1}(u,v)} f_{X,Y}(x,y).
\end{aligned}
$$

$\square$

The generalization to more than two dimensions is straightforward.

As we saw in the one-dimensional case, transforming probability *densities* always requires additional care as compared to their discrete analogue. Let us begin by defining the following subsets of $\mathbb{R}^2$. First, let $\mathcal{A} \subset \mathbb{R}^2$ be region of the (X,Y)-space with nonzero probability

$$
\mathcal{A} = \{(x,y) : f_{X,Y}(x,y) > 0\}.
$$

Similarly, let's define

$$
\mathcal{B} = g(\mathcal{A}) = \{(u,v) : (u,v) = g(x,y), \text{ for some } (x,y) \in \mathcal{A}\},
$$

so that $f_{U,V}(u,v) > 0$ if and only if $(u,v) \in \mathcal{B}$.

Theorem 1.9. *Let (X,Y) be a continuous random vector with pdf $f_{X,Y}(x,y)$, and consider the transformation $(U,V) = g(X,Y)$. Let $\mathcal{A}, \mathcal{B} \subset \mathbb{R}^2$ be defined as above. Suppose there is a partition $\mathcal{A}_0, \mathcal{A}_1, \ldots, \mathcal{A}_k$ of $\mathcal{A}$ such that $\mathbb{P}((X,Y) \in \mathcal{A}_0) = 0$ and $f_{X,Y}$ is continuous on $\mathcal{A}_i$. Furthermore, let us define the subsets $\mathcal{B}_1, \ldots, \mathcal{B}_k \subset \mathbb{R}^2$ by the relations $\mathcal{B}_i = \{(u,v) : (u,v) = g(x,y), \text{ for some } (x,y) \in \mathcal{A}_i\}$. Finally, let us suppose that g is one-to-one and differentiable on each $\mathcal{A}_i$, i.e., suppose there exists functions $g_1(x,y), \ldots, g_k(x,y)$, defined on $\mathcal{A}_1, \ldots, \mathcal{A}_k$, respectively, such that*

1. *$g(x,y) = g_i(x,y)$, for $(x,y) \in \mathcal{A}_i$,*
2. *$g_i(x,y)$ is one-to-one on $\mathcal{A}_i$,*
3. *$g_i^{-1}(u,v)$ has a continuous derivative on $\mathcal{B}_i$, for each $i = 1, \ldots, k$.*

Then the pdf of the random vector (U,V) is given by

$$
f_{U,V}(u,v) = \sum_{i=1}^{k} f_{X,Y}\left(g_i^{-1}(u,v)\right) |J_i(u,v)| \, \mathbb{I}[(u,v) \in \mathcal{B}_i]. \tag{1.37}
$$

where the ith Jacobian is given by the determinant

$$
J_i(u,v) = \det \frac{\partial g_i^{-1}(u,v)}{\partial(u,v)} = \det \begin{bmatrix} \dfrac{\partial x}{\partial u} & \dfrac{\partial x}{\partial v} \\ \dfrac{\partial y}{\partial u} & \dfrac{\partial y}{\partial v} \end{bmatrix}.
$$

Of course, when $g : \mathcal{A} \to \mathcal{B}$ is one-to-one, there is no need to partition $\mathcal{A}$ into pieces. For an example, see the proof for Lemma 2.2.

1.8.3 Moment-generating Functions

Another useful tool that we shall encounter is sometimes useful in determining the distribution of a random variable.

Definition 1.34. *Let X be a random variable with a given* CDF *F. The moment-generating function* (MGF) *of X, denoted $M_X(t)$, is the function*

$$M_X(t) = \mathbb{E}[e^{tX}], \tag{1.38}$$

provided that such an expectation exists on some neighborhood of the origin.

The moment-generating function is a useful tool given the following.

Proposition 1.11. *Suppose the random variable X has a* MGF *$M_X(t)$, and let $M_X^{(n)}(t)$ represent the nth derivative of $M_X(t)$. Then*

$$\mathbb{E}[X^n] = M_X^{(n)}(0), \tag{1.39}$$

for $n \in \mathbb{Z}_+ = \{1, 2, \ldots\}$.

Proof. We note that the result follows as long as we can establish the slightly more general statement

$$M_X^{(n)}(t) = \mathbb{E}[X^n e^{tX}].$$

We proceed by induction. For the basis step, consider the case $n = 1$:

$$M_X'(t) = \frac{d}{dt} \int_{-\infty}^{\infty} e^{tx} f_X(x)\, dx$$
$$= \int_{-\infty}^{\infty} x e^{tx} f_X(x)\, dx$$
$$= \mathbb{E}[X e^{tX}].$$

Setting $t = 0$ yields the result for $n = 1$.

For the induction step, let us assume that $M_X^{(n)}(t) = \mathbb{E}[X^n e^{tX}]$ is true for some $n \geq 1$. This implies that

$$M_X^{(n+1)}(t) = \frac{d}{dt} \int_{-\infty}^{\infty} x^n e^{tx} f_X(x)\, dx$$
$$= \int_{-\infty}^{\infty} x^{n+1} e^{tx} f_X(x)\, dx$$
$$= \mathbb{E}[X^{n+1} e^{tX}].$$

This proves our proposition. □

Despite its ability to generate arbitrary moments $\mathbb{E}[X^n]$ for a random variable X, the MGF further possesses the following property.

Proposition 1.12. *Let X and Y be independent random variables with* MGF*s M_X and M_Y, resepctively. Then the* MGF *of the random variable $Z = X + Y$ is given by*

$$M_Z(t) = M_X(t)M_Y(t). \tag{1.40}$$

We leave the proof to the reader. We note, however, that Proposition 1.12 can sometimes lead to a simpler method for computing the PDF of the sum of two independent random variables. If we recognize the MGF of Z, we can sometimes easily draw conclusions on the distribution of the sum $X + Y$. This saves us the work of constructing a bivariate transformation of the form

$$U = X + Y$$
$$V = Y,$$

and then having to marginalize the result over V.

1.8.4 Location and Scale Families

Given any random variable with known distribution, it is natural to consider an associated family of random variables, known as the *location–scale family*, formed by horizontal translations and stretches of the underlying distribution. Specifically, we have the following.

Theorem 1.10. *Let Z be any continuous random variable with* PDF *$f_Z(z)$. Then the random variable X defined through the affine transformation*

$$X = \sigma Z + \mu,$$

for some $\mu \in \mathbb{R}$ and $\sigma \in \mathbb{R}_+$, has PDF *given by*

$$f_X(x) = \frac{1}{\sigma} f_Z\left(\frac{x - \mu}{\sigma}\right). \tag{1.41}$$

The theorem follows immediately from Corollary 1.1. We shall visit many examples in our next chapter. The transformation defined in Theorem 1.10 occurs often enough that we give special names to the parameters μ and σ, as follows.

Definition 1.35. *Given any* PDF *$f(x)$ and parameters $\mu \in \mathbb{R}$ and $\sigma > 0$, the family of distributions defined by Equation (1.41), i.e., the family of* PDF*s given by*

$$\frac{1}{\sigma} f\left(\frac{x - \mu}{\sigma}\right), \tag{1.42}$$

is called the location–scale family *with standard* PDF *$f(x)$. Moreover, we refer to μ as the* location parameter *and σ as the* scale parameter.

Note 1.10. The function defined by Equation (1.42) constitutes a PDF due to Theorem 1.10.   ▷

Note 1.11. For the special case $\sigma = 1$, we refer to the family $f(x-\mu)$ simply as a *location family.*

Similarly, for the special case $\mu = 0$, we refer to the family $(1/\sigma)f(x/\sigma)$ as a *scale family.*   ▷

Problems

1.1. Show that if the random variables X and Y are independent, then the conditional distributions reduce to

$$f(x|y) = f_X(x) \qquad \text{and} \qquad f(y|x) = f_Y(y).$$

1.2. Prove Proposition 1.5.

1.3. Find a counterexample to the assertion: uncorrelated random variables must be independent; i.e., find a pair of random variables X and Y that are *dependent,* but that nonetheless have covariance $\text{COV}(X,Y) = 0$.

1.4. Repeat Example 1.8 for the sample standard deviation S_n. In particular, for each sample size $n = 10^1, 10^2, \ldots, 10^5$, compute n_samples $= 10,000$ samples from the normal distribution $\mathcal{N}(\mu = 3, \sigma^2 = 4)$ with sample size n. Report the average sample standard deviation and the variance of the sample standard deviation across all n_samples samples. For $n = 10$ and $n = 100$, plot the histogram of sample standard deviations.

1.5. Let $X \sim \text{Unif}(a, b)$. Show that the variance of X is given by

$$\mathbb{V}(X) = \frac{(b-a)^2}{12}.$$

1.6. Prove Proposition 1.12.

1.7. The *Dvoretzky-Kiefer-Wolfowitz (DKW) inequality* states that, for any $\epsilon > 0$,

$$\mathbb{P}\left(\sup_x \left| F(x) - \hat{F}_n(x) \right| > \epsilon \right) \le 2e^{-2n\epsilon^2}.$$

Use this to derive the $1 - \alpha$ confidence band:

$$\hat{F}_n(x) \pm \sqrt{\frac{1}{2n} \ln(2/\alpha)}$$

for the empirical distribution $\hat{F}_n(x)$.

2

A Menagerie of Distributions

The goal of this chapter is to provide a lexicon of basic distributions along with their definitions and properties. We will also indicate how one may evaluate and sample these distributions using built-in Python packages.

More details, illustrative examples, and discussion are available in many introductory statistics texts, such as Casella and Berger [2002], Hogg, *et al.* [2015], or Ross [2012].

2.1 Bernouilli and Binomial

2.1.1 Bernouilli Trials

A Bernouilli trial is a mathematical generalization of a coin-flip, as shown below.

Definition 2.1. *A* Bernouilli experiment *or* Bernouilli trial *is any random experiment over a two-point sample space* $\Omega = \{S, F\}$ *with fixed probability distribution* $\mathbb{P}(S) = p$. *A random variable* X *is said to be a* Bernouilli random variable with probability of success p, *denoted* $X \sim \text{Bern}(p)$, *if* $X : \Omega \to \mathbb{R}$ *is a random variable over* Ω *defined by* $X(S) = 1$ *and* $X(F) = 0$. *The points* S *and* F *are referred to as* success *and* failure, *respectively.*

An immediate consequence of this definition is the probability distribution over the random variable X.

Proposition 2.1. *Let* $X \sim \text{Bern}(p)$ *represent a Bernouilli random variable. Then the* PMF *for* X *is given by*

$$f(x) = p^x(1-p)^{x-1},\tag{2.1}$$

for $x = 0, 1$.

Proof. Clearly, $\mathbb{P}(X = 1) = \mathbb{P}(S) = p$ and $\mathbb{P}(X = 0) = \mathbb{P}(F) = 1 - p$. Equation (2.1) is a compact way to express this fact, for if $x = 1$, then $f(1) = p$, and if $x = 0$, we have $f(0) = 1 - p$.   □

Proposition 2.2. *Let $X \sim \text{Bern}(p)$ represent a Bernouilli random variable. Then*

$$\mathbb{E}[X] = p \tag{2.2}$$
$$\mathbb{V}(X) = p(1 - p). \tag{2.3}$$

Proof. The expected value of X is given by

$$\mathbb{E}[X] = \sum_{x=0}^{1} x f(x) = 0 \cdot (1 - p) + 1 \cdot p = p.$$

The variance is given by

$$\mathbb{V}(X) = \sum_{x=0}^{1} (x - p)^2 f(x) = p^2(1 - p) + (1 - p)^2 p = p(1 - p).$$

□

Simulation in Python

The quickest way to simulate an outcome of a Bernouilli trial is to invoke `np.random.random()`, which generates a random number on the interval $[0, 1]$, and then check to see if it is less than the probability p. This is done in Code Block 2.1. If the random number generated by `np.random.random()` is less than p (which will occur with probability p), then the random variable x will equal one. Otherwise it will equal zero.

```
1  p = 0.3 # Probability of success.
2  x = int( np.random.random() < p )
```

Code Block 2.1: Simulation of a Bernouilli trial

2.1.2 Binomial Distribution

A Bernouilli trial rarely occurs in a vacuum. Typically, we are interested in a series of Bernouilli trials; e.g., we might flip a coin ten times. This gives rise to our second distribution family.

Definition 2.2. *Let* $X_1, \ldots, X_n \sim \text{Bern}(p)$ *be* IID *Bernouilli random variables. Then the random variable* $X = X_1 + \cdots + X_n$ *is a* binomial random variable, *which is denoted* $X \sim \text{Binom}(n, p)$.

Note 2.1. The binomial random variable can be interpreted as representing the total number of successes in a series of n Bernouilli trials, for which each trial has a probability of success p.   ▷

Proposition 2.3. *The* PMF *of a binomial random variable* $X \sim \text{Binom}(n, p)$ *is given by*

$$f(X = k) = \binom{n}{k} p^k (1 - p)^{n-k}, \tag{2.4}$$

where

$$\binom{n}{k} = \frac{n!}{k!(n - k)!},$$

spoken "n choose k", is the binomial coefficient.

Note 2.2. For convenience, we will sometime refer to the binomial distribution, as given by Equation (2.4), as $\text{Binom}(k; n, p)$.   ▷

Note 2.3. Normalization of the PMF given in Equation (2.4) follows directly from the binomial theorem:

$$(p + q)^n = \sum_{k=0}^{n} \binom{n}{k} p^k q^{n-k},$$

with $q = 1 - p$.   ▷

Proof. In order to determine the value of $f(k)$, the probability of observing exactly $X = k$ successes from a total of n Bernouilli trials, let us first consider the case in which those k successes occur for the first k Bernouilli trials, i.e., suppose we observe precisely:

$$\underbrace{S \cdots S}_{k \text{ times}} \cdot \underbrace{F \cdots F}_{n-k \text{ times}}.$$

The probability of this occurring is exactly the product of the individual probabilities, hence: $p^k (1 - p)^{n-k}$. However, this is only one possible way to arrive at a total of k successes. Any permutation of this result would also result in a total of k successes, simply in a different order. Since there are precisely $\binom{n}{k}$ ways of choosing k items from a set of n items, we conclude the total probability of achieving k successes is correctly given by Equation (2.4).   □

Proposition 2.4. *Let* $X \sim \text{Binom}(n, p)$ *be a binomial random variable. Then its expectation and variance are given by*

$$\mathbb{E}[X] = pn \tag{2.5}$$

$$\mathbb{V}(X) = np(1 - p). \tag{2.6}$$

Proof. This follows trivially from Propositions 1.1, 1.2, and 2.2: since

$$\mathbb{E}[X_i] = p \qquad \text{and} \qquad \mathbb{V}(X_i) = p(1-p),$$

it follows from Proposition 1.1 and 1.2 that

$$\mathbb{E}[X] = \sum_{i=1}^{n} p = np \qquad \text{and} \qquad \mathbb{V}(X) = \sum_{i=1}^{n} p(1-p) = np(1-p).$$

$\square$

Proposition 2.5. *Let* $X_1, \ldots, X_n \sim \text{Bern}(p)$ *be* IID *Bernouilli trials and consider the sample mean* $\overline{X} = \frac{1}{n} \sum_{i=1}^{n} X_i$. *Then*

$$\mathbb{E}[\overline{X}] = p \tag{2.7}$$

$$\mathbb{V}(\overline{X}) = \frac{p(1-p)}{n}. \tag{2.8}$$

Proof. From Proposition 1.1 and 1.2, it follows that

$$\mathbb{E}[\overline{X}] = \mathbb{E}\left[\frac{1}{n} \sum_{i=1}^{n} X_i\right] = \frac{1}{n} \sum_{i=1}^{n} \mathbb{E}[X_i] = p.$$

Similarly,

$$\mathbb{V}(\overline{X}) = \mathbb{V}\left(\frac{1}{n} \sum_{i=1}^{n} X_i\right) = \frac{1}{n^2} \sum_{i=1}^{n} \mathbb{V}(X_i) = \frac{1}{n^2} \cdot np(1-p) = \frac{p(1-p)}{n}.$$

$\square$

Note 2.4. Proposition 2.5 is simply an application of Theorem 1.2. $\triangleright$

Simulation in Python

We can modify Code Block 2.1 to account for multiple Bernouilli trials by setting the `size` argument of `np.random.random`, thus generating a vector of random numbers, each of which may be compared in size to the probability p. This is done in Code Block 2.2.   The statement

```python
p = 0.3 # Probability of success.
n = 10 # Number of trials.
x = np.sum( np.random.random(size=n) < p )
```

Code Block 2.2: Simulation of a series of Bernouilli trials

`np.random.random(size=n) < p` represents a Boolean vector of length n, i.e., by executing this command using $n = 5$, you might get, for example, something that looks like

array([False, True, True, False, False]).

Using the built-in `np.sum` function will then sum these values together, interpreting `True` as 1 and `False` as 0. The variable `x` therefore represents a single realization of the Binomial random variable $X \sim \text{Binom}(n, p)$.

Of course, the `numpy` library is slightly more sophisticated than this, as it offers a built-in Binomial random variable generator. The same can be achieved by using the code shown in Code Block 2.3. Using the `size`

```
p = 0.3 # Probability of success
n = 10 # Number of trials
T = 1   # Number of i.i.d. draws from Binomial distribution
x = np.random.binomial(n, p, size=T)
```

Code Block 2.3: Binomial random variable

argument, we can further control the number of samples from the binomial distribution that we'd like to simulate. For example, executing line 4 using T=5 could return a vector similar to

array([3, 1, 4, 7, 1]) .

Of course, if `T` is large, the relative proportion of any fixed value k should be approximately equal to the value given by Equation (2.4).

It turns out, the PMF given by Equation (2.4) is also a built-in function. To use it, we must import the `scipy` package. In Code Block 2.4, we pass in the entire array of possible values for the random variable $X = k$, for $k = 0, \ldots, 10$. If one is only interested in obtaining the probability of a particular value, say $k = 3$, one could simply pass $k = 3$ and obtain the value 0.266827932. This means that there will be a total of three out of ten successes approximately 26.68% of the time, if one repeats a large number of 10-trial experiments.

The plot of the PMF is shown in Figure 2.1. It was produced using `plt.bar(k, f)`.

Note 2.5. For a complete list of distributions available in `scipy`, please see the documentation available online at

https://docs.scipy.org/doc/scipy/reference/stats.html.

Each distribution has a built in CDF calculator, and a PMF or PDF calculator, depending on whether the random variable is discrete or continuous. In

```
1  p = 0.3 # Probability of success
2  n = 10 # Number of trials
3  k = np.arange(11) # Array from 0 to 10
4  f = scipy.stats.binom.pmf(k, n, p) # PMF
5  F = scipy.stats.binom.cdf(k, n, p) # CDF
```

Code Block 2.4: Distribution functions of a binomial random variable

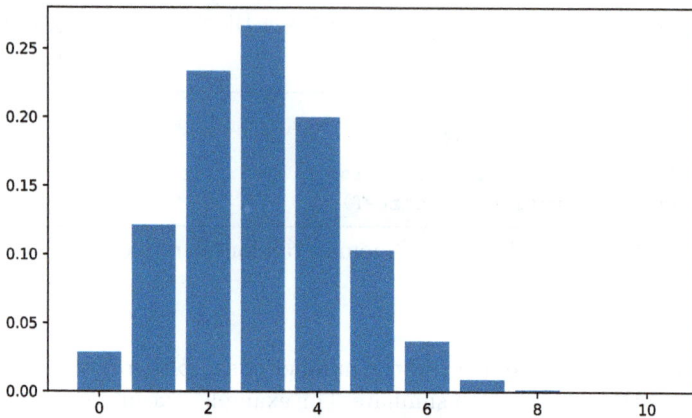

Fig. 2.1: PMF for $X \sim \text{Binom}(10, 0.30)$.

addition, each distribution has a built-in survival function (sf) and inverse survival function (isf) method, that operate similar to the above pmf and cdf methods, as well as basic statistics functions such as mean, median, var, and std. For example, executing scipy.stats.binom.var(n, p) returns a value of 2.1, which agrees with Equation (2.6).   ▷

2.2 Negative Binomial, Geometric, and Hypergeometric

In Section 2.1, we discussed Bernouilli trials and binomial random variables, which describe the probability distribution of the total number of successes out of a fixed number of repeated Bernouilli experiments. In this section, we discuss three closely related variations on that problem.

2.2.1 Negative Binomial Distribution

As we've seen, a binomial random variable represents the total number of successful outcomes from a *fixed* number of trials n. Thus, we decide *in advanced* how many experiments to make, and then set off to make that number. The negative binomial distribution is a variation on that theme.

Definition 2.3. *Let $r \in \mathbb{Z}_*$ be a fixed positive integer. A negative binomial random variable is a random variable X that represents the total number of failures that occur prior to the rth success occurs. The probability distribution for X is said to have a* negative binomial distribution, *which we write as $X \sim \text{sNBD}(r,p)$.*

A shifted negative binomial random variable is a random variable X that represents the total number of trials needed before the rth success occurs. The probability distribution for X is said to have a shifted negative binomial distribution, *which we write as $X \sim \text{NBD}(r,p)$.*

Note 2.6. The negative binomial and shifted negative binomial random variables are equivalent. If $X \sim \text{sNBD}(r,p)$, then $Y = X - r \sim \text{NBD}(r,p)$.   ▷

Note 2.7. A binomial random variable represents the total number of successes out of a fixed number of trials, whereas a negative binomial random variable represents the total number of trials (or failures) required before a fixed number of successes is observed.   ▷

Proposition 2.6 (Negative Binomial Distribution). *Let $X \sim \text{sNBD}(r,p)$ represent a shifted negative binomial random variable. Then the* PMF *for X is given by*

$$f(x) = \binom{x-1}{r-1} p^r (1-p)^{x-r}, \tag{2.9}$$

for $x = r, r+1, \ldots$. Similarly, the PMF *for $Y = X - r \sim \text{NBD}(r,p)$ is given by*

$$f(y) = \binom{y+r-1}{r-1} p^r (1-p)^y,$$

for $y = 0, 1, \ldots$.

Proof. For a fixed x, Equation (2.9) represents the probability of the rth success occurring exactly on the xth trial. This implies that on the $(x-1)$th trial, there must have been only $r - 1$ successes (which may have occurred on *any* of the preceding $x - 1$ trials). The probability of observing $r - 1$ successes (anywhere) in $x - 1$ trials is given by the binomial distribution,

$$\text{Binom}(r-1; x-1, p) = \binom{x-1}{r-1} p^{r-1} (1-p)^{x-1}.$$

The probability of observing a success on the xth trial is simply p. The probability of observing the rth success on the xth trial is therefore the product $p \cdot \text{Binom}(r-1; x-1, p)$, which is equivalent to Equation (2.9). □

Proposition 2.7. *Let* $X \sim \text{sNBD}(r, p)$ *represent a shifted negative binomial random variable. Then the mean and variance of* X *are given by*

$$\mathbb{E}[X] = \frac{r}{p} \tag{2.10}$$

$$\mathbb{V}(X) = \frac{r(1-p)}{p^2}. \tag{2.11}$$

The expected value of X given in Equation (2.10) should be exactly as we expected. For example, let us consider the case where $p = 0.10$ and $r = 10$, that is, each trial has a 10% probability of success, and we continue until we have observed exactly ten successes. It makes intuitive sense that, on average, we should require one hundred trials before observing the tenth success, which agrees exactly with Equation (2.10).

Proof. Following the elegant proof presented in Ross [2012], we can compute the expected value of the kth power of X as

$$
\begin{aligned}
\mathbb{E}[X^k] &= \sum_{n=r}^{\infty} n^k \binom{n-1}{r-1} p^r (1-p)^{n-r} \\
&= \frac{r}{p} \sum_{n=r}^{\infty} n^{k-1} \binom{n}{r} p^{r+1} (1-p)^{n-r} \\
&= \frac{r}{p} \sum_{m=r+1}^{\infty} (m-1)^{k-1} \binom{m-1}{r} p^{r+1} (1-p)^{m-(r+1)} \\
&= \frac{r}{p} \mathbb{E}\left[(Y-1)^{k-1} \right]
\end{aligned}
$$

where the second line follows since $n\binom{n-1}{r-1} = r\binom{n}{r}$, the third line follows due to the reindexing $m = n+1$, and where we recognize $Y \sim \text{sNBD}(r+1, p)$. Setting $k = 1$ yields Equation (2.10). Further, by setting $k = 2$, we have

$$\mathbb{E}[X^2] = \frac{r}{p}\mathbb{E}[Y-1] = \frac{r}{p}\left(\frac{r+1}{p} - 1 \right),$$

from which we may, upon application of Proposition 1.3[1], obtain Equation (2.11). □

Simulation in Python

Before delving into the built-in **numpy** and **scipy** methods, let us first build our own simulation from scratch. This is done in Code Block 2.5.

For this example, we threw in a couple of assertions (lines 10–12) to ensure that the input values take the appropriate form. Passing in $p = 5$ or $r = -6$ will throw an error.

---

[1] $\mathbb{V}(X) = \mathbb{E}[X^2] - \mathbb{E}[X]^2$

```
1   def nbd_single_sample(r, p):
2       # Returns a sample from NBD(r, p)
3       # Returns number of trials before observing
4       # rth success, with success probability p.
5       # Inputs:
6       #   r nonnegative int. number of success
7       #   p float in (0, 1]
8       # Output:
9       #   n_trials (int) number of trials
10      assert r >= 0
11      assert isinstance(r, int)
12      assert p > 0 and p <= 1
13
14      n_success = 0
15      n_trials = 0
16      while n_success < r:
17          n_trials += 1
18          n_success += np.random.random() < p
19
20      return n_trials
```

Code Block 2.5: Sample from sNBD

The rest of the code is straightforward: we begin our sequence of experiments with **n_success** and **n_trials** set to zero. Then we simply add Bernouilli trials, one by one, until we reach a total of r successes. Line 17 simply increments **n_trials** by 1. (x+=1 is a Pythonic shorthand for x=x+1, though both are equivalent: they increment the value stored in variable x by one.) In Line 18, the phrase np.random.random() < p is evaluated as being either **True** or **False**, depending on whether our random number (on the interval $(0, 1)$) is less than p. When adding **True** or **False** to an integer, Python instead adds the integer equivalents: 1 for **True** or 0 for **False**. Line 18 therefore simulates a single Bernouilli trial, with probability of success p, and adds the result to the running total **n_success**. Finally, notice that we use a while-loop, as opposed to a for-loop, as the total number of trials is not determined in advanced.

Recall that, for $X \sim \text{sNBD}(10, 0.10)$, $\mathbb{E}[X] = 100$. The first three times I ran this script[2], by executing **nbd_single_sample(10, 0.10)** from the command line, I received the outputs 140, 108, 140. I ran it a few more times: 115, 126, 117, 136, 95. It seemed like there were a lot of outputs greater than 100, our expected average. But was this due to random chance? Or was there a mistake in the code?

---

[2] Since it is based on random numbers, the results will be different each time. Due to the random nature of the generator, the reader should not expect to receive the same outputs as printed herein.

To answer this question, let's consider a variation that instead returns an array of samples of a given sample size, as shown in Code Block 2.6[3]. This allows us to simulate many draws from sNBD(r, p) at once. From the

```python
def nbd_sample(r, p, size=1):
    # Inputs:
    #   r nonnegative int. number of success
    #   p float in (0, 1]
    # Output:
    #   x (array) sample from nbd(r, p)


    x = np.zeros(size)
    for i in range(size):
        x[i] = nbd_single_sample(r, p)

    return x
```

Code Block 2.6: Multiple samples from sNBD

command line, I executed x = nbd_sample(10, 0.10, size=1000), and then computed x.mean() and x.var(), obtaining outputs of 99.299 for the sample mean and 815.101599 for the sample variance. These are point estimators for μ and σ^2, and they closely agree with the exact, true values of $\mu = 100$, $\sigma^2 = 900$, as given by Equations (2.10) and (2.11).

If we were to run x = nbd_sample(10, 0.10, size=1000) many times, and compute the sample mean $\overline{X}$, using x.mean(), for each one, we would have an approximation of the sample distribution for $\overline{X}$. From Theorem 1.2, we have $\mathbb{E}[\overline{X}] = \mu = 100$ and $\mathbb{V}(\overline{X}) = \sigma^2/n = 900/1000 = 0.9$. Hence the standard error is given by $\text{se}(\overline{X}) = \sigma/\sqrt{n} \approx 0.9487$. Our original sample mean of 99.299 was therefore 0.74 standard errors below the expected value.

Numpy and Scipy commands

The **numpy** and **scipy** definition for the NBDis for the standard, non-shifted definition. In order to simulate from the shifted version, one applies the adjustments as shown in Code Block 2.7.

In particular, since **np.random.negative_binomial** returns the number of *failures* before the rth success, simply adding r (which will be added to each component of the output array) will yield the total number of *trials* before the rth success. Adjusting the **scipy.stats.nbinom** is a little more tricky, but can be accomplished by passing in an additional parameter **loc=r**, which effectively shifts the x-value by this number of units. In

---

[3] Normally, Code Blocks 2.5 and 2.6 would be combined into a single function. For the purpose of illustration, however, we decided to keep them separate.

```
1  r = 10    # Number of success.
2  p = 0.10 # Probability of success.
3  T = 1000 # Sample size.
4
5  x = np.random.negative_binomial(r, p, size=T) + r
6  scipy.stats.nbinom.pmf(100, r, p, loc=r) #Probability of exactly
       100 trials.
7  scipy.stats.nbinom.cdf(100, r, p, loc=r) #Probability less than or
       equal to 100 trials.
8  scipy.stats.nbinom.sf(100, r, p, loc=r) #Probability greater than
       100 trials (survival function).
9  scipy.stats.nbinom.mean(r, p, loc=r)   #E[X] = 100
10 scipy.stats.nbinom.var(r, p, loc=r)    #Var(X) = 900
11 scipy.stats.nbinom.median(r, p, loc=r) #Median = 97
```

Code Block 2.7: Numpy and Scipy for the *shifted* negative binomial distribution.

other words, if x represents the total number of trials, adding location `loc=r` will effectively shift the input to be $x - r$, which represents the number of failures, consistent with `scipy`'s definition of the negative binomial distribution.

A plot of the PMF for the negative binomial random variable $X \sim$ sNBD$(10, 0.10)$ is shown in Figure 2.2.

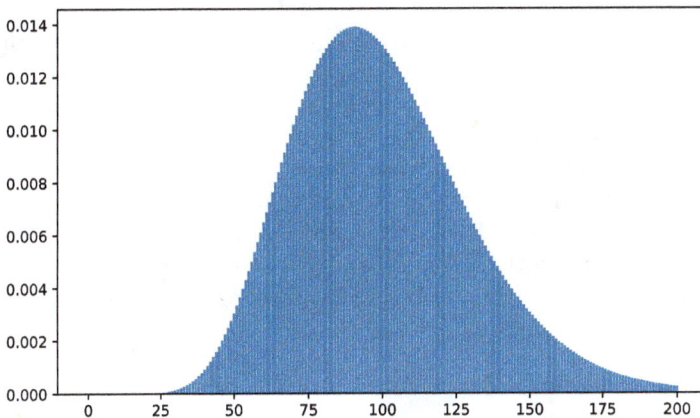

Fig. 2.2: PMF for $X \sim$ sNBD$(r = 10, p = 0.10)$.

2.2.2 Geometric Distribution

A geometric distribution is a special case of the negative binomial distribution, as defined below.

Definition 2.4. *A (shifted) geometric random variable X with probability p, denoted by either $X \sim \mathrm{Geom}(p)$ or $X \sim \mathrm{sGeom}(p)$, is a special case of the (shifted) negative binomial random variable with $r = 1$.*

Thus, a geometric random variable represents the number of trials before observing a single success. It follows from Proposition 2.7, that the first success will occur, on average, on the $1/p$th trial. The full distribution is given below.

Proposition 2.8. *The PMF for a (shifted) geometric random variable $X \sim \mathrm{sGeom}(p)$ is given by*

$$f(x) = p(1-p)^{x-1}, \tag{2.12}$$

for $x = 1, 2, 3, \ldots$. Similarly, the PMF for a geometric random variable $X \sim \mathrm{Geom}(p)$ is given by

$$f(x) = p(1-p)^x,$$

for $x = 0, 1, 2, \ldots$.

Proposition 2.9. *The geometric distribution is memoryless, since, for integers $s > t$, we have*

$$\mathbb{P}(X > s \mid X > t) = \mathbb{P}(X > s - t). \tag{2.13}$$

The memoryless property of the geometric distribution means that it "forgets" its history. This is best illustrated by means of an example: suppose we have already observed ten failures, and hence we now know that the first success will occur for some value of $X > 10$. Given what we now know, the probability of the first success occurring on trial #12 is the same as what the probability of the first success occurring on trial #2 was, before we began the experiment. Another variation of this theme: at any point in time prior to observing the first success, it doesn't matter how many subsequent failures we have already observed, observing a success x *more* units in the future is equivalent to $f(x)$.

Proof. From the definition of conditional probability, we have

$$\mathbb{P}(X > s \mid X > t) = \frac{\mathbb{P}(X > s)}{\mathbb{P}(X > t)} = (1-p)^{s-t} = \mathbb{P}(X > s - t).$$

(Note that $\mathbb{P}(X > s, X > t) = \mathbb{P}(X > s)$, since $s > t$.) □

```
1  p = 0.10 # 1\% Probability of success.
2  T = 1000 # Number of samples
3  x = np.random.geometric(p, size=T) # 1,000 samples from sGeom(0.01)
4  n = np.arange(61) # Array of Total number of trials.
5  f = scipy.stats.geom.pmf(n, p) #PMF evaluated over n
6  F = scipy.stats.geom.cdf(n, p) #CDF
```

Code Block 2.8: Geometric distribution in Python

Simulation in Python

The basic functionality to produce random samples from the geometric distribution is shown in Code Block 2.8.

The PMF for a geometric random variable with $p = 0.10$ is shown in Figure 2.3.

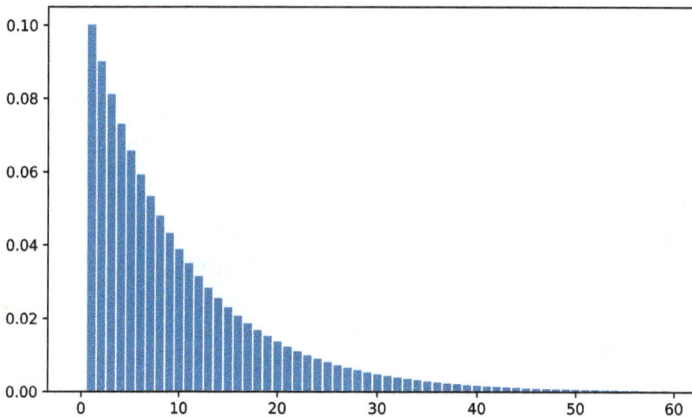

Fig. 2.3: PMF for $X \sim$ sGeom(0.10).

2.2.3 Hypergeometric Distribution

Our final variation of the binomial distribution is the *hypergeometric distribution*. Like the binomial distribution, the hypergeometric distribution also represents the number of successes given a fixed total number of trials. The difference, however, is that the trials in the hypergeometric distribution are done *without replacement*, meaning that they are not independent.

Definition 2.5. *A random variable X is said to be a* hypergeometric ran- dom variable with parameters N, M, and K, *denoted* $X \sim \text{HypGeom}(N, M, K)$, *if it represents the number of red marbles in a random selection without re- placement of K marbles from a bag initially consisting of N total marbles, M of which are red.*

We can think of the hypergeometric random variable as the outcome of drawing a fixed number K marbles from a bag containing a total of N marbles, M of which are red. The K marbles are drawn *without replacement*, meaning that the outcome of each draw changes the probability of choosing a red marble on the following draw.

Proposition 2.10. *Let $X \sim \text{HypGeom}(N, M, K)$. Then the* PMF *for the random variable X is given by*

$$f(x) = \frac{\binom{M}{x}\binom{N-M}{K-x}}{\binom{N}{K}}, \tag{2.14}$$

for $x = 0, 1, 2, \dots, K$. We say that the random variable X has a hypergeo- metric distribution.

Proof. The logic behind Equation (2.14) can be seen by counting. We start off with a bag of N marbles, M of which are red. From that bag, we draw K marbles, without replacement.

The total number of ways of selecting K marbles from a bag of N mar- bles is $\binom{N}{K}$. This represents the size of our sample space.

The probability mass $f(x)$ represents the probability that exactly $X = x$ of the K selected marbles are red. Out of the total number $\binom{N}{K}$ of possible outcomes, how many ways are there to generate an outcome with exactly x red marbles in the sample? First, we count the number of ways of selecting x red marbles out of the M red marbles in the bag; this is $\binom{M}{x}$. Second, we count the number of ways of selecting $K - x$ white marbles from the $N - M$ white marbles in the bag; this is $\binom{N-M}{K-x}$. Thus, the total number of possible ways of selecting x red marbles (and, therefore, $K - x$ white marbles) is the product $\binom{M}{x}\binom{N-M}{K-x}$. Dividing by the total number of possible outcomes (regardless of the count of red), we obtain our result. □

Proposition 2.11. *Let $X \sim \text{HypGeom}(N, M, K)$. Then the expected value and variance of X are given by*

$$\mathbb{E}[X] = \frac{KM}{N} \tag{2.15}$$

$$\mathbb{V}(X) = \frac{KM}{N}\frac{(N-M)(N-K)}{N(N-1)}. \tag{2.16}$$

We will omit the proof; though the interested reader can consult Casella and Berger [2002].

It is interesting to compare the mean and variance of the hypergeometric and binomial random variables. First, note that the ratio M/N represents the *initial* probability of drawing a red marble, i.e., the probability of drawing a red marble on the first draw. In this respect, Equation (2.15) is identical to the result in the case of a binomial random variable, i.e.,

$$\mathbb{E}[\text{HypGeom}(N, M, K)] = \mathbb{E}[\text{Binom}(n = K, p = M/N)].$$

The initial probability of drawing a white marble is of course $(N - M)/N$, which is a factor in Equation (2.16). If we were to suppose that the variance of the hypergeometric random variable was the same as the binomial random variable, we would end up with

$$np(1 - p) = \frac{KM}{N}\frac{(N - M)}{N},$$

which would be off by a factor of $(N - K)/(N - 1) < 1$. We conclude that the distribution of the hypergeometric random variable has an identical mean but smaller variance than the corresponding distribution of a binomial random variable, making the connection $p = M/N$. Therefore, the effect of the varying probability on each draw is to decrease the variance in the possible outcomes. This makes intuitive sense: each time we draw a red marble, it decreases the probability that the following marble will be red (because there is one fewer red marble in the bag). Similarly, when we draw a white marble, it increases the probability that the next marble is red. This acts as a sort of feedback mechanism that makes the total number of red marbles more likely to be closer to the expected value KM/N.

Simulation in Python

The parameters for the `numpy.random.hypergeometric` function are `ngood`, `nbad`, and `nsamples`, corresponding to M, $N - M$, and K, respectively. Usage is shown in line 5 of Code Block 2.9. The `scipy.stats.hypergeom` has the same variables presented above, i.e., $f(x) = \text{HypGeom}(x; N, M, K)$[4]. A plot of the PMF for HypGeom$(100, 30, 10)$ is shown in Figure 2.4.

Note 2.8. The output produced by the hypergeometric distribution, shown in Figure 2.4, looks very similar to the binomial distribution plotted in Figure 2.2. This makes intuitive sense, however, when we consider that we are only drawing 10 out of 100 marbles from the bag, and over the course of those ten draws, though the probability of choosing a red is changing, it is not changing greatly.

---

[4] In the `scipy` docs, however, they use different variables to name these terms

```
1  N = 100
2  M = 30
3  K = 10
4  T = 1000
5  samples = np.random.hypergeometric(M, N-M, K, size=T) # Generate T
       random samples
6  x = np.arange(11) # Array representing possible values for X
7  f = scipy.stats.hypergeom.pdf(x, N, M, K)
8  mu = scipy.stats.hypergeom.mean(N, M, K) # 3.0
9  sig2 = scipy.stats.hypergeom.var(N, M, K) # 1.9090
```

Code Block 2.9: Hypergeometric random variables in Python

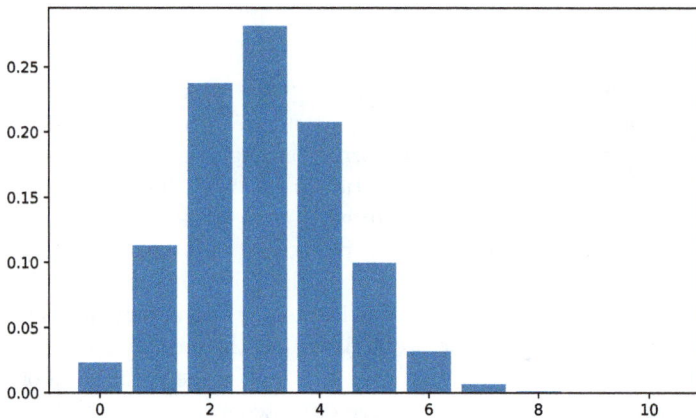

Fig. 2.4: PMF for $X \sim \mathrm{HypGeom}(N = 100, M = 30, K = 10)$.

In fact, we can take this observation a step further and, holding the sample size K and the initial probability of a red marble $p = M/N$ fixed, take the limit to obtain

$$\lim_{N \to \infty} \mathrm{HypGeom}(x; N, pN, K) = \mathrm{Binom}(x; K, p).$$

$\triangleright$

2.3 Poisson and Exponential

In this section, we build a bridge from discrete to continuous distributions, introducing both the Poisson (discrete) and exponential (continuous) distributions. These two distributions are actually closely related, through a mechanism called the Poisson process.

2.3.1 Poisson Distribution

Definition 2.6. *The random variable X is said to have a* Poisson *distribution, denoted $X \sim \text{Poiss}(\lambda)$, if there is a number $\lambda > 0$ such that the* PMF *for X is given by*

$$f(x) = \frac{e^{-\lambda}\lambda^x}{x!}, \tag{2.17}$$

for $x = 0, 1, 2, \ldots$.

The Poisson distribution is actually an interesting approximation of the binomial distribution, as shown in Proposition 2.12.

Proposition 2.12. *Let $X \sim \text{Binom}(n, p)$ be a binomial random variable. As n increases, with $\lambda = np$ held fixed, the random variable X approaches the Poisson random variable with parameter λ, i.e.,*

$$\lim_{n \to \infty} \text{Binom}(x; n, p = \lambda/n) = \text{Poiss}(x; \lambda). \tag{2.18}$$

Proof. Let $X \sim \text{Binom}(n, p)$ and $\lambda = np$. Then the PMF for X is given by

$$\text{Binom}(x; n, p = \lambda/n) = \binom{n}{x} p^x (1-p)^{n-x}$$

$$= \frac{n!}{x!(n-x)!} \left(\frac{\lambda}{n}\right)^x \left(1 - \frac{\lambda}{n}\right)^{n-x}$$

$$= \frac{n(n-1)\cdots(n-x+1)}{n^x} \frac{\lambda^x}{x!} \frac{(1-\lambda/n)^n}{(1-\lambda/n)^x}$$

Now, from calculus, we have

$$\lim_{n \to \infty} (1 - \lambda/n)^x = 1,$$

$$\lim_{n \to \infty} (1 - \lambda/n)^n = e^{-\lambda}, \text{ and}$$

$$\lim_{n \to \infty} \frac{n(n-1)\cdots(n-x+1)}{n^x} = 1.$$

The result follows.   □

Proposition 2.12 shows that for large n and moderate $\lambda = np$, the number of successes may be approximated by a Poisson random variable.

The Poisson distribution is also used to model phenomena in which we are waiting for events to occur. In this context, the number of events occurring in a fixed interval of time can be modeled using the Poisson distribution. This relies on an underlying assumption that, for small time intervals, the probability of arrival is proportional to the length of the interval. We will examine this in more detail in Section 2.3.3.

Proposition 2.13. *Let $X \sim \text{Poiss}(\lambda)$. Then the mean and variance of X are given by*

$$\mathbb{E}[X] = \mathbb{V}(X) = \lambda. \tag{2.19}$$

For a proof, see Casella and Berger [2002]. For some insights into why this may not be surprising, consider the substitution $p = \lambda/n$ into Equations (2.5) and (2.6):

$$\mathbb{E}[X] = np = \lambda, \qquad \mathbb{V}(X) = np(1 - p) = \lambda \left(1 - \frac{\lambda}{n} \right).$$

We thus obtain Equation (2.19) in the limit as $n \to \infty$.

Finally, we note that the Poisson random variable is additive, in the following sense.

Proposition 2.14. *Let $X \sim \text{Poiss}(\lambda_1)$ and $Y \sim \text{Poiss}(\lambda_2)$ be independent Poisson random variables. Then the random variable $X + Y$ is the Poisson random variable $X + Y \sim \text{Poiss}(\lambda_1 + \lambda_2)$.*

We leave the proof for Exercise 2.3.

Simulation in Python

Simulating random numbers from a Poisson distribution in **numpy** and evaluating the PMF and CDF and other statistical quantities in **scipy** are done in the obvious way, as shown in Code Block 2.10.

```
lam = 5 # lambda has a reserved meaning in Python
T = 1000
samples = np.random.poisson(lam, size=T) # T random samples from
    Poisson
x = np.arange(16)
f = scipy.stats.poisson.pmf(x, lam)
F = scipy.stats.poisson.cdf(x, lam)
```

Code Block 2.10: Simulation of a Poisson random variable

A plot of a Poisson random variable is shown in Figure 2.4.

2.3.2 Exponential Distribution

We next turn to our first example of a continuous random variable: the exponential.

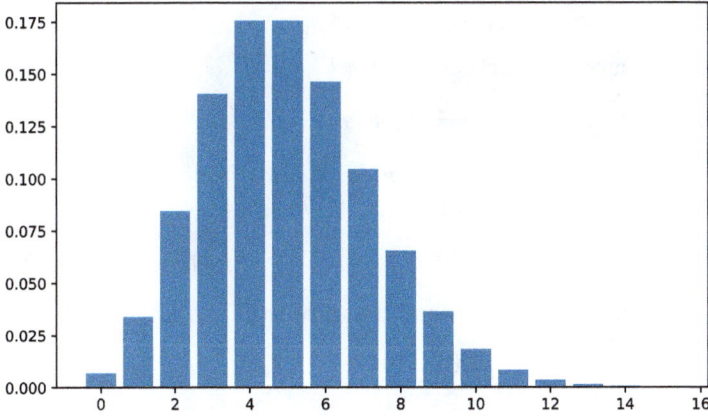

Fig. 2.5: PMF for $X \sim \text{Poiss}(5)$.

Definition 2.7. *A nonnegative, continuous random variable X is said to have an* exponential distribution *with scale parameter β, denoted $X \sim \text{Exp}(\beta)$, if its* PDF *is of the form*

$$f(x) = \frac{1}{\beta} e^{-x/\beta}, \tag{2.20}$$

where $0 < \beta < \infty$, for $x \in [0, \infty)$. Alternatively, the exponential distribution is sometimes parameterized by the rate parameter *$\lambda = 1/\beta$.*

Note 2.9. Since the exponential random variable is nonnegative, its normalization is given by

$$\int_0^\infty \frac{1}{\beta} e^{-x/\beta}\, dx = 1.$$

$\triangleright$

Proposition 2.15. *Let $X \sim \text{Exp}(\beta)$. Then its mean and variance are given by*

$$\mathbb{E}[X] = \beta, \tag{2.21}$$
$$\mathbb{V}(X) = \beta^2. \tag{2.22}$$

Proof. It is straightforward to show that

$$\mathbb{E}[X] = \int_0^\infty \frac{x}{\beta} e^{-x/\beta}\, dx = \beta \quad \text{and} \quad \mathbb{E}[X^2] = \int_0^\infty \frac{x^2}{\beta} e^{-x/\beta}\, dx = 2\beta^2.$$

These can be combined, using Proposition 1.3, to yield the variance given by Equation (2.22). $\qquad\square$

Proposition 2.16. *Let $X \sim \mathrm{Exp}(\beta)$. Then the* CDF *for X is given by*

$$F(x) = 1 - e^{-x/\beta}. \tag{2.23}$$

Similarly, its survival function *is given by*

$$S(x) = 1 - F(x) = e^{-x/\beta}. \tag{2.24}$$

Proposition 2.17. *The exponential distribution is memoryless, in the sense of Equation (2.13), i.e., for $s > t \geq 0$, we have*

$$\mathbb{P}(X > s | X > t) = \mathbb{P}(X > s - t).$$

Proof. From the definition of conditional probability, we have

$$\mathbb{P}(X > s | X > t) = \frac{\mathbb{P}(X > s)}{\mathbb{P}(X > t)} = e^{-(s-t)/\beta} = \mathbb{P}(X > s - t).$$

$\square$

Simulation in Python

Python code for generating random numbers from an exponential distribution and evaluating the PDF and CDF is shown in Code Block 2.11. A plot of the PDF is shown in Figure 2.6.

```
beta = 10
T = 1000
samples = np.random.exponential(scale=beta, size=T)
x = np.linspace(0, 30, num=100) # A linear array of values between
        0 and 30
f = scipy.stats.expon.pdf(x, scale=beta)
F = scipy.stats.expon.cdf(x, scale=beta)
```

Code Block 2.11: Simulation of the exponential random variable in Python

2.3.3 Poisson Process

The Poisson and exponential distributions are closely related to a mechanism often used in modeling a stream of events occurring at random intervals.

Definition 2.8. *For each $t \geq 0$, let N_t denote the number of occurrences of a particular event on the interval $[0, t]$. The nonnegative random variables N_t are said to constitute a* Poisson process *with parameter $\lambda > 0$ if they satisfy the following axioms*

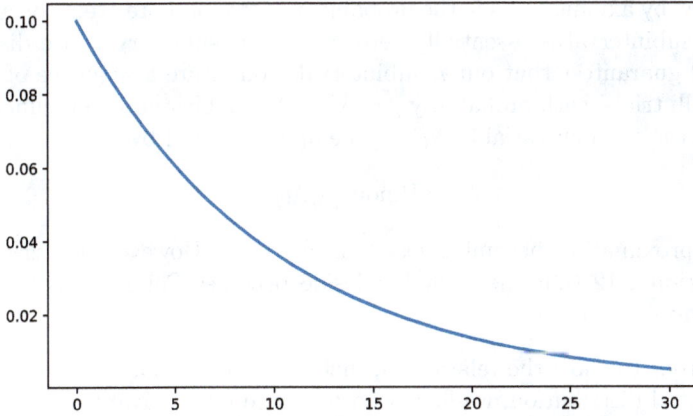

Fig. 2.6: PMF for $X \sim \text{Exp}(10)$.

1. *The process begins at $t = 0$; i.e., $N_0 = 0$;*
2. *The numbers of events in nonoverlapping intervals are independent; i.e., $N_{t_2} - N_{t_1}$ and $N_{s_2} - N_{s_1}$ are independent whenever $[t_1, t_2] \cap [s_1, s_2] = \emptyset$;*
3. *The distribution of the number of events in a given interval depends only on its length; i.e., for any $s, t > 0$, the random variables N_s and $N_{t+s} - N_t$ are identically distributed;*
4. *The probability of an event in a sufficiently short interval is proportional to its length; i.e., $\mathbb{P}(N_h = 1) = \lambda h + o(h)^5$; and*
5. *The probability of simulatneous events in a sufficiently short interval is essentially zero; i.e., $\mathbb{P}(N_h > 1) = o(h)$.*

Note 2.10. Due to axiom 2, axioms 4 and 5 apply to any sufficiently short interval, i.e., $\mathbb{P}(N_{t+h} - N_t = 1) = \lambda h + o(h)$, and $\mathbb{P}(N_{t+h} - N_t > 1) = o(h)$.
▷

Theorem 2.1. *Let N_t define a Poisson process with parameter λ. Then N_t is a Poisson random variable with parameter λt; i.e.,*

$$N_t \sim \text{Poiss}(\lambda t). \tag{2.25}$$

Proof. Fix $t > 0$. The random variable N_t represents the number of events occurring on the interval $I = [0, t]$. Let us divide the interval I into n equally spaced subintervals of length $h = t/n$, so that $I_i = [x_{i-1}, x_i]$ for $i = 1, \ldots, n$, where $x_i = ih$. If n is large enough, we may combine axioms 3 and 4 to show that the probability of an event occurring on the ith subinterval is given by

---

[5] *Little-Oh Notation: We write $f(h) = o(h)$ if $f(h)/h \to 0$ as $h \to 0$.*

$$p = \mathbb{P}(N_{x_i} - N_{x_{i-1}} = 1) \approx \lambda h.$$

Similarly, by axioms 3 and 5, the probability of two or more events occurring on any subinterval is essentially zero. Since our subintervals are disjoint, axiom 2 guarantees that our n subintervals constitute a sequence of n IID Bernouilli trials, with probability $p = \lambda h = \lambda t/n$. Therefore, for sufficiently large n, the random variable N_t may be approximated by

$$N_t \approx \text{Binom}(n, \lambda t/n),$$

that approximation becoming exact as $n \to \infty$. However, we saw from Proposition 2.12 that the right hand side becomes $\text{Poiss}(\lambda t)$ as $n \to \infty$. This proves the result. $\square$

In order to show the relationship between the Poisson process and the exponential distribution, we first need to define interarrival times.

Definition 2.9. *Let N_t represent a Poisson process with parameter λ. Then the nth interarrival time, denoted T_n, represents the time of the first event, if $n = 1$, or the time between the $(n-1)$th event and the nth event, if $n \geq 2$.*

Note 2.11. If we let

$$t_n = \inf\{t \geq 0 : N_t \geq n\}$$

represent the time of the nth event (notice that $t_0 = 0$), then the nth interarrival time T_n is given by $T_n = t_n - t_{n-1}$. $\triangleright$

Theorem 2.2. *Let N_t define a Poisson process with parameter λ, and let $\{T_n\}$ represent the sequence of interarrival times. Then the T_ns are IID exponential random variables with scale parameter $\beta = 1/\lambda$; i.e.,*

$$T_n \sim \exp(\beta = 1/\lambda). \tag{2.26}$$

Proof. Let us first consider T_1. For a given, fixed value $t > 0$, the probability that T_1 occurs after t is given by

$$\mathbb{P}(T_1 > t) = \mathbb{P}(N_t = 0) = \text{Poiss}(0; \lambda t) = e^{-\lambda t},$$

since $N_t \sim \text{Poiss}(\lambda t)$[6]. However, we recognize this from Equation (2.24) as the survival function for an exponential random variable with scale parameter $\beta = 1/\lambda$. Thus, $T_1 \sim \text{Exp}(1/\lambda)$.

Next, let us consider T_n for $n \geq 2$, and suppose that the $(n-1)$th event occurred at time $s = t_{n-1}$. Consider now any $t > s$. The probability of the nth interarrival time being greater than $t - s$ is given by

$$\mathbb{P}(T_n > t|t_{n-1} = s) = \mathbb{P}(N_t - N_s = 0) = \mathbb{P}(N_{t-s} = 0) = e^{-\lambda(t-s)}.$$

Thus, the Poisson process is memoryless, and $T_n \sim \text{Exp}(1/\lambda)$. $\square$

---
[6] Recall that $0! = 1$

Simulation in Python

We can use the theory presented in this section to build a simple simulator for a Poisson process, as shown in Code Block 2.12. Note that by appending the event time to the list of events (line 9) inside of the while-loop, we are automatically ensuring that the event times are less than the input parameter t.

In order to simulate a single sample from $N_t \sim \text{Poiss}(\lambda t)$, we can simply take the length of the output, e.g., `len(poiss_proc(10, lambda_rate=3))`.

```python
def poiss_proc(t, lambda_rate=1):
    # Inputs:
    #    t (float) -- length of time to run simulation.
    #    lambda_rate -- the rate parameter of the Poisson process.
    # Output:
    #    events -- a list of event times.
    assert lambda_rate > 0

    # Set current_time to time of first event.
    current_time = np.random.exponential(scale=1/lambda_rate)
    events = [] # List of Events

    while current_time < t:
        events.append(current_time) # Add event time to list
        delta_t = np.random.exponential(scale=1/lambda_rate)
        current_time += delta_t # Time of next event

    events = np.array(events) # Convert list to np array.

    return events
```

Code Block 2.12: Simulation of a Poisson process

The output of the `poiss_proc` can be plotted visually, as shown in Code Block 2.13. Note that line 3 leverages the `emp_dist` function, defined

```python
x = np.linspace(0, 100, num=2000)
samples = poiss_proc(100, lambda_rate=0.10)
y = emp_dist(x, samples) * len(samples)
plt.plot(x, y)
```

Code Block 2.13: Generating plot of Poisson process

previously in Code Block 1.5. The output of emp_dist is multiplied by the number of samples, in order to *un-normalize* the CDF. Using num=2000 in numpy's linspace method ensures a resolution of 0.05 for successive event times. The resulting output is shown in Figure 2.7. Four additional runs of poiss_proc(100, lambda_rate=0.10) are shown in Figure 2.8.

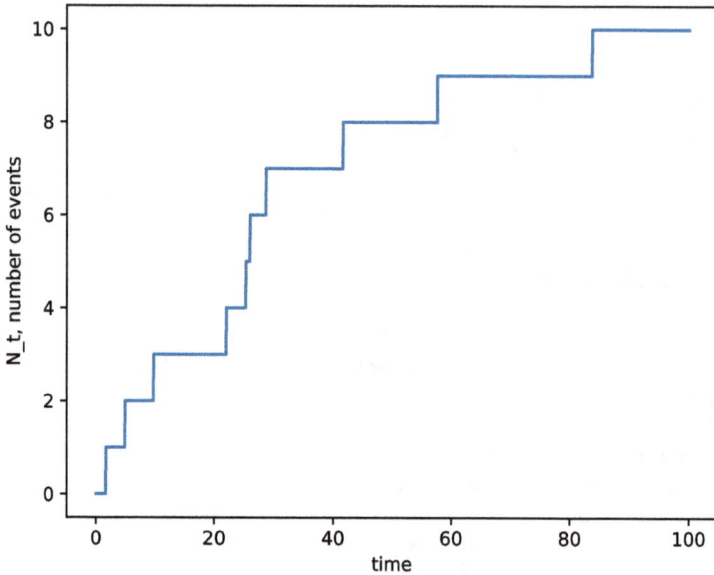

Fig. 2.7: Poisson process N_t, with rate $\lambda = 0.10$.

2.4 The Normal Distribution

In this section, we introduce the normal distribution, discuss its properties, and conclude with a discussion of one of the key results of statistics: the central limit theorem.

2.4.1 Normal Distribution

The normal distribution (or *Gaussian* distribution) is the distribution which gives rise to the familiar "bell-shaped curve" of classical statistics. It also plays a key part in sampling distributions via the central limit theorem, which we'll discuss later in the section.

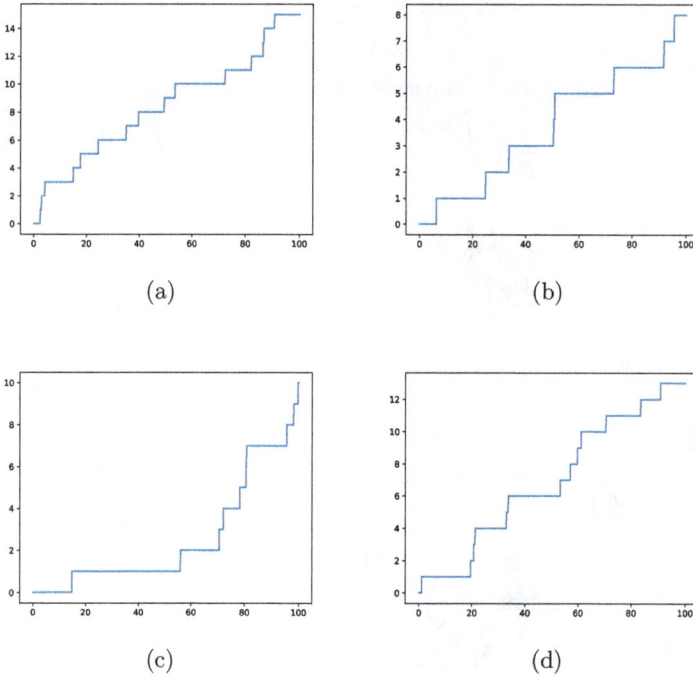

(a)

(b)

(c)

(d)

Fig. 2.8: Four simulations of a Poisson process.

Definition 2.10. *The continuous random variable X is a normal random variable with mean μ and variance σ^2, denoted $X \sim \mathrm{N}(\mu, \sigma^2)$, if its* PDF *satisfies the normal distribution, given by*

$$f(x) = \frac{1}{\sigma\sqrt{2\pi}} e^{-(x-\mu)^2/(2\sigma^2)}, \qquad (2.27)$$

for $-\infty < x < \infty$. The special case of $Z \sim \mathrm{N}(0,1)$ is referred to as a standard normal random variable.

Note 2.12. Given the normal random variable $X \sim \mathrm{N}(\mu, \sigma^2)$, the transformed random variable $Z = (X - \mu)/\sigma$ constitutes a *standard* normal random variable; i.e., $Z \sim \mathrm{N}(0,1)$. See Lemma 2.1 for a proof.   ▷

Proposition 2.18. *The normal distribution, as given by Equation (2.27), is properly normalized; i.e., the total probability under the curve is 1.*

Proof. To prove normalization, let us define the total probability as

$$I = \int_{-\infty}^{\infty} \frac{1}{\sigma\sqrt{2\pi}} e^{-(x-\mu)^2/(2\sigma^2)}\, dx.$$

By introducing the change of variable $z = (x - \mu)/\sigma$, we can rewrite I as

$$I = \int_{-\infty}^{\infty} \frac{1}{\sqrt{2\pi}} e^{-z^2/2} \, dz.$$

Next, let us consider the quantity I^2, as given by

$$I^2 = \frac{1}{2\pi} \left(\int_{-\infty}^{\infty} e^{-x^2/2} \, dx \right) \left(\int_{-\infty}^{\infty} e^{-y^2/2} \, dy \right)$$

$$= \frac{1}{2\pi} \int_{-\infty}^{\infty} \int_{-\infty}^{\infty} e^{-(x^2+y^2)/2} \, dx \, dy$$

$$= \frac{1}{2\pi} \int_{0}^{2\pi} \int_{0}^{\infty} e^{-r^2/2} r \, dr \, d\theta,$$

where the last equality follows by converting to polar coordinates. The final expression is easily evaluated to yield $I^2 = 1$. Since $I > 0$, it follows that $I = 1$ and the proposition is proved. $\square$

Proposition 2.19. *Let $X \sim N(\mu, \sigma^2)$. Then the mean and variance of X are (shockingly) given by*

$$\mathbb{E}[X] = \mu \quad \text{and} \quad \mathbb{V}(X) = \sigma^2. \tag{2.28}$$

Proof. It is a straightforward matter to show that

$$\mathbb{E}[X] = \int_{-\infty}^{\infty} \frac{x}{\sigma\sqrt{2\pi}} e^{-(x-\mu)^2/(2\sigma^2)} \, dx = \mu$$

and

$$\mathbb{E}[X^2] = \int_{-\infty}^{\infty} \frac{x^2}{\sigma\sqrt{2\pi}} e^{-(x-\mu)^2/(2\sigma^2)} \, dx = \mu^2 + \sigma^2.$$

The result follows. $\square$

Definition 2.11. *Let $Z \sim N(0,1)$ be the standard normal random variable. We shall denote the standard-normal* CDF *and survival function using the variable $\Phi(z)$ and $\Psi(z)$, respectively; i.e., we may write*

$$\Phi(z) = \mathbb{P}(Z \le z) = \int_{-\infty}^{z} \frac{1}{\sigma\sqrt{2\pi}} e^{-x^2/2} \, dx, \tag{2.29}$$

$$\Psi(z) = \mathbb{P}(Z > z) = \int_{z}^{\infty} \frac{1}{\sigma\sqrt{2\pi}} e^{-x^2/2} \, dx. \tag{2.30}$$

Further, for $\alpha \in (0,1)$, we shall use z_α to represent the point for which

$$\Psi(z_\alpha) = \mathbb{P}(Z > z_\alpha) = \alpha. \tag{2.31}$$

In other words, $z_\alpha = \Psi^{-1}(\alpha)$.

Though the CDF, survival function, and inverse survival function cannot be evaluated analytically, numerical evaluations are no challenge to the computer. Before turning to simulation, however, there is one additional feature worth pointing out.

Simulation in Python

The **numpy** and **scipy** code for handling normal random variables is shown in Code Block 2.14. First, note that **np.random.randn** represents the standard normal random variable, so samples must be amplified by the standard deviation σ and shifted by the mean μ. The result of line 4 is an array of samples from $N(\mu, \sigma^2)$.

The **scipy** library handles the mean and variance using the location and scale parameters, as shown on lines 6–11. Lines 6–8 generate the PMF, the CDF, and the survival function for the distribution $N(\mu, \sigma^2)$, evaluated at x, which may be a number or an array of numbers.

The built-in inverse survival function (line 9) generates the point x_0 at which 5% of the total probability still remains, i.e., $C(x_0) = 0.95$. This is confirmed in lines 10 and 11.

```
1  mu = 100
2  sigma = 15
3  n_samples = 1000
4  samples = np.random.randn(n_samples) * sigma + mu
5  x = np.linspace(50, 150, num=200)
6  f = scipy.stats.norm.pdf(x, loc=mu, scale=sigma) # PDF
7  F = scipy.stats.norm.cdf(x, loc=mu, scale=sigma) # CDF
8  S = scipy.stats.norm.sf(x, loc=mu, scale=sigma) # Survival Function
9  x0 = scipy.stats.norm.isf(0.05, loc=mu, scale=sigma) # Inv. Survival
10 scipy.stats.norm.sf(x0, loc=mu, scale=sigma) # 0.05
11 scipy.stats.norm.cdf(x0, loc=mu, scale=sigma) # 0.95
```

Code Block 2.14: Normal random variable

The normal random variable $X \sim N(100, 225)$, as generated by Code Block 2.14, is plotted in Figure 2.9. The area to the right of x0=124.6728 is shaded, representing a total probability of 0.05.

2.4.2 Sums of Normal Random Variables

Normal random variables have certain additive properties that are quite useful.

Lemma 2.1. *Let $X \sim N(\mu, \sigma^2)$ be a normal random variable. Then the random variable $Y = aX + b$, for $a \neq 0$, is also a random variable, and $Y \sim N(a\mu + b, a^2\sigma^2)$.*

Proof. Consider the transformation $y = g(x) = ax + b$. The mapping $g : \mathbb{R} \to \mathbb{R}$ is one-to-one, so there is no need to partition the space. From Equation (1.35) (Corollary 1.1), we have

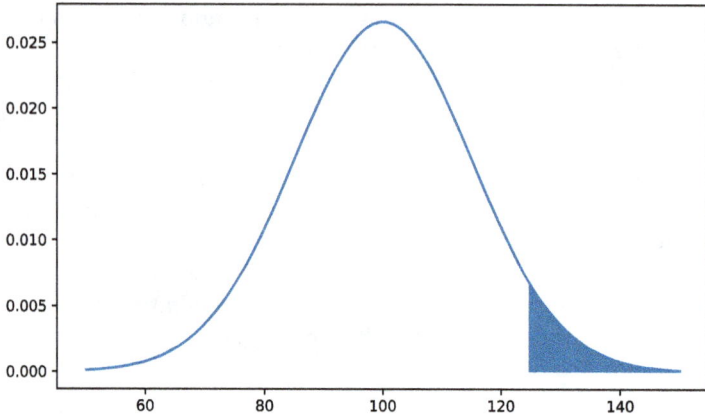

Fig. 2.9: Normal random variable $X \sim \mathrm{N}(100, 225)$.

$$f_Y(y) = f_X((y-b)/a)/a$$
$$= \frac{1}{a\sigma\sqrt{2\pi}} e^{-((y-b)/a - \mu)^2/(2\sigma^2)}$$
$$= \frac{1}{a\sigma\sqrt{2\pi}} e^{-(y-(a\mu+b))^2/(2a^2\sigma^2)},$$

which we recognize as the PDF of a random variable $Y \sim \mathrm{N}(a\mu + b, a^2\sigma^2)$.
□

This lemma immediately validates Note 2.12, which is important on its own right, so we restate the result next as a corollary to the preceding lemma.

Corollary 2.1. *Let $X \sim \mathrm{N}(\mu, \sigma^2)$ be a normal random variable. Then the transformed random variable*

$$Z = \frac{X - \mu}{\sigma} \tag{2.32}$$

is a standard normal random variable, i.e., $Z \sim \mathrm{N}(0,1)$.

Proof. This follows from Lemma 2.1 by setting $a = 1/\sigma$ and $b = -\mu/\sigma$. □

Lemma 2.2. *Let $X, Y \sim \mathrm{N}(0,1)$ be two independent and identically distributed standard normal random variables. Then the random variable $U = X + Y$ is a normal random variable, and $U \sim \mathrm{N}(0,2)$.*

Proof. Due to the independence of X and Y, the joint distribution is given by

$$f_{X,Y}(x,y) = \frac{1}{2\pi}e^{-x^2/2}e^{-y^2/2} = \frac{1}{2\pi}e^{-(x^2+y^2)/2}.$$

To proceed, we will be applying Theorem 1.9. Let us define the mapping $(u,v) = g(x,y)$ by the relations

$$u = x + y \qquad \text{and} \qquad v = x - y,$$

which has the inverse mapping

$$x = \frac{u+v}{2} \qquad \text{and} \qquad y = \frac{u-v}{2}.$$

The mapping $g : \mathbb{R}^2 \to \mathbb{R}^2$ is one-to-one, so we do not need to partition the domain. The Jacobian of this transformation is given by

$$J(u,v) = \det\frac{\partial g^{-1}(u,v)}{\partial(u,v)} = \det\begin{bmatrix} \dfrac{\partial x}{\partial u} & \dfrac{\partial x}{\partial v} \\ \dfrac{\partial y}{\partial u} & \dfrac{\partial y}{\partial v} \end{bmatrix} = \det\begin{bmatrix} 1/2 & 1/2 \\ 1/2 & -1/2 \end{bmatrix} = -1/2.$$

From Equation (1.37), we have

$$f_{U,V}(u,v) = f_{X,Y}\left(\frac{u+v}{2}, \frac{u-v}{2}\right)\left|\frac{-1}{2}\right| = \frac{1}{4\pi}e^{-(u^2+v^2)/4}.$$

This can be factored as

$$f_{U,V}(u,v) = \left[\frac{1}{2\sqrt{\pi}}e^{-u^2/4}\right]\left[\frac{1}{2\sqrt{\pi}}e^{-v^2/4}\right],$$

which we recognize as the product of two normal distributions with mean $\mu = 0$ and variance $\sigma^2 = 2$. The distribution for U can be obtained by marginalizing over the random variable V, from which we obtain

$$f_U(u) = \int_{-\infty}^{\infty} f_{U,V}(u,v)\,dv = \frac{1}{2\sqrt{\pi}}e^{-u^2/4}.$$

This proves that $U = X + Y \sim N(0,2)$. (Consequently, it also proves that $V = X - Y \sim N(0,2)$.)   □

Lemma 2.2 was a warm-up for our next lemma.

Lemma 2.3. *Let $X \sim N(0,\sigma^2)$ and $Y \sim N(0,\tau^2)$ be two independent and identically distributed standard normal random variables. Then the random variable $Z = X + Y$ is a normal random variable, and $Z \sim N(0,\sigma^2 + \tau^2)$.*

Proof. The joint distribution is given by

$$f_{X,Y}(x,u) = \frac{1}{2\pi\sigma\tau}e^{-(x^2/\sigma^2+y^2/\tau^2)/2}.$$

Consider the transformation

$$u = \frac{x+y}{\sqrt{\sigma^2 + \tau^2}} \qquad \text{and} \qquad v = \frac{\tau^2 x - \sigma^2 y}{\sigma\tau\sqrt{\sigma^2 + \tau^2}}.$$

The Jacobian of the transformation (exercise) is $|J(u,v)| = \sigma\tau$. Further, note that

$$u^2 + v^2 = \frac{x^2}{\sigma^2} + \frac{y^2}{\tau^2}.$$

It follows that the joint distribution for the random vector (U,V) is given by

$$f_{U,V}(u,v) = \frac{1}{2\pi}e^{-(u^2+v^2)/2} = \left[\frac{1}{\sqrt{2\pi}}e^{-u^2/2}\right]\left[\frac{1}{\sqrt{2\pi}}e^{-v^2/2}\right].$$

By marginalizing over V, we conclude that $U \sim N(0,1)$. From Lemma 2.1, we further conclude that the random variable $Z = X + Y \sim N(0, \sigma^2 + \tau^2)$, since $Z = \sqrt{\sigma^2 + \tau^2}U$. $\qquad\square$

Lemma 2.4. *Let $X \sim N(\mu, \sigma^2)$ and $Y \sim N(\nu, \tau^2)$ be two independent normal random variables. Then the random variable $Z = X + Y$ is normal, and $Z \sim N(\mu + \nu, \sigma^2 + \tau^2)$.*

Proof. First, consider $U = X - \mu$ and $V = Y - \nu$. By Lemma 2.1, we have $U \sim N(0, \sigma^2)$ and $V \sim N(0, \tau^2)$.

Next, consider $W = U + V$. By Lemma 2.3, it follows that $W \sim N(0, \sigma^2 + \tau^2)$. (In terms of our original variables X and Y, $W = X + Y - (\mu + \nu)$.)

Finally, consider $Z = W + (\mu + \nu)$. By Lemma 2.1, $Z \sim N(\mu + \nu, \sigma^2 + \tau^2)$. However, in terms of our original variables, $Z = X + Y$. This completes the result. $\qquad\square$

Theorem 2.3. *Suppose $X_1, \ldots, X_n$ are mutually independent normal random variables with $X_i \sim N(\mu_i, \sigma_i^2)$, for $i = 1, \ldots, n$. Then the random variable*

$$X = \sum_{i=1}^{n} c_i X_i$$

is also a normal random variable; moreover

$$X \sim N\left(\sum_{i=1}^{n} c_i \mu_i, \sum_{i=1}^{n} c_i^2 \sigma_i^2\right). \tag{2.33}$$

Proof. First, let us define $Y_1, \ldots, Y_n$ by the relations $Y_i = c_i X_i$. From Lemma 2.1, $Y_i \sim N(c_i \mu_i, c_i^2 \sigma_i^2)$. Applying Lemma 2.4 repeatedly then produces the result. $\qquad\square$

Note 2.13. There are many ways to go about proving Theorem 2.3. A very simple and elegant proof involves *moment-generating functions*, as shown in Casella and Berger [2002]. It can also be proved using convolution integrals, as done in Ross [2012]. $\qquad\triangleright$

Corollary 2.2. *Let* $X_1, \ldots, X_n \sim \mathrm{N}(\mu, \sigma^2)$ *be* IID *samples, then the sample mean* $\overline{X} = (1/n) \sum_{i=1}^{n} X_i$ *has the normal distribution* $\overline{X} \sim \mathrm{N}(\mu, \sigma^2/n)$.

Proof. This follows from Theorem 2.3 using $c_i = 1/n$, $\mu_i = \mu$, and $\sigma_i = \sigma$. □

An illustration of this Corollary can be found in Figure 1.1.

2.4.3 The Central Limit Theorem

The result from Corollary 2.2 is not only exactly true for IID samples of a normal random variable, but also approximately true for IID samples from many different distributions.

Theorem 2.4 (Central Limit Theorem). *Let* $X_1, \ldots, X_n \sim F$ *be* IID *samples from a distribution with finite mean* μ *and finite variance* σ^2. *Let* $\overline{X}_n = (1/n) \sum_{i=1}^{n} X_i$ *represent the sample mean. Then the sequence of distributions*

$$W_n = \frac{\overline{X}_n - \mu}{\sigma/\sqrt{n}} \tag{2.34}$$

approaches the standard normal distribution $\mathrm{N}(0, 1)$ *in the limit as* $n \to \infty$[7].

Note 2.14. In practice, we may approximate W using the standard normal for $n > 30$.   ▷

The central limit theorem adds to Theorem 1.2, which gave us the expected value and variance of the sample mean as $\mathbb{E}[\overline{X}_n] = \mu$ and $\mathbb{V}(\overline{X}_n) = \sigma^2/n$. Theorem 2.4 tells us that, in addition, whenever we have large samples, the sampling distribution of $\overline{X}_n$ is approximately a normal with mean μ and variance σ^2/n.

Simulation in Python

We can see Theorem 2.4 in action by constructing a simulations in Python. Let's choose the distribution that is most unlike the normal distribution: a good candidate is the geometric distribution, as shown in Figure 2.3. In particular, let's draw samples from Geom(0.10), which has $\mathbb{E}[X] = p = 10$ and $\mathbb{V}(X) = (1-p)/p^2 = 90$. The code is given in Code Block 2.15.

We collect a total of 1,000 samples, each of sample size 10, and compute the sample mean $\overline{X}_{10}$ for each one. The resulting histogram for these 1,000 sample means is shown in Figure 2.10, plotted concurrently with N(10, 9).

The result is certainly not bad. But let's see what happens if we choose a larger sample size. Let's increase the sample size to 30. The histogram for $\overline{X}_{30}$ and the PDF for the approximating normal distribution N(10, 3) are shown in Figure 2.11.

---
[7] I.e., the sequence $\{W_n\}$ converges to $Z \sim N(0, 1)$ *in distribution as* $n \to \infty$

```
1   n_samples = 1000
2   sample_size = 10
3   p = 0.10
4   mu = 1 / p
5   sigma = np.sqrt( (1 - p) / p**2 )
6   x_bar = np.zeros(n_samples) # Sample Means
7   for i in range(n_samples):
8       x = np.random.geometric(p, size=sample_size)
9       x_bar[i] = x.mean() # Sample Mean.
10
11  plt.hist(x_bar, bins=50, normed=True)
12  x = np.linspace(0, 22, num=100)
13  f = scipy.stats.norm.pdf(x, loc=1/p,
        scale=sigma/np.sqrt(sample_size))
14  plt.plot(x, f, color='b', linewidth=2)
```

Code Block 2.15: Sampling distribution for $\overline{X}_{10}$ for $X \sim \text{Geom}(0.10)$

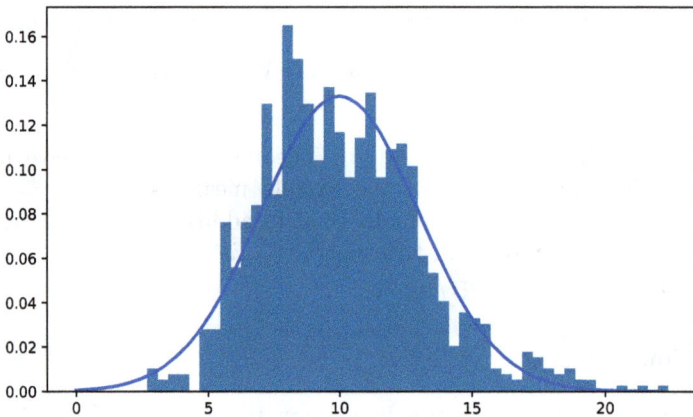

Fig. 2.10: Distribution of sample mean $\overline{X}_{10}$, with $N(10, 9)$.

As the sample size n increases, the distribution of the sample mean $\overline{X}_n$ will more closely resemble that of the normal distribution $N(\mu, \sigma^2/n)$. Note also the change of scales between Figures 2.10 and 2.11: the variance in the sample mean decreases as the sample size increases.

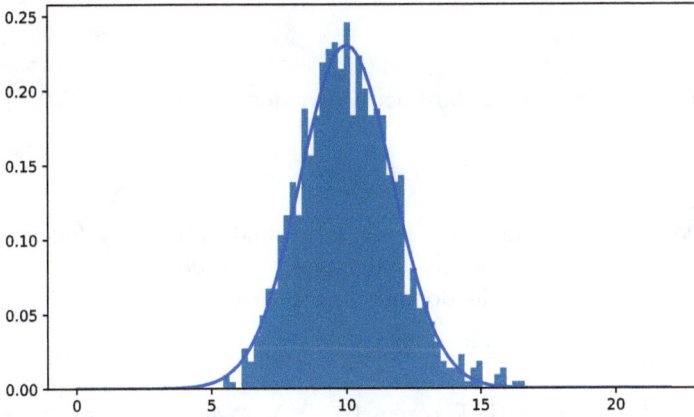

Fig. 2.11: Distribution of sample mean $\overline{X}_{30}$, with $N(10, 3)$.

2.5 Chi-Squared, T, and F

In this section, we introduce several important distributions for hypothesis testing and analysis of variance.

2.5.1 The Chi-Squared Distribution

Definition 2.12. *Let* $Z_1, \ldots, Z_p \sim N(0, 1)$ *represent* IID *standard normal random variables. Then a random variable* X *is said to be a chi-squared random variable with* p *degrees of freedom, denoted* $X \sim \chi_p^2$, *if it can be expressed as*

$$X = \sum_{i=1}^{p} Z_i^2. \tag{2.35}$$

Proposition 2.20. *If* $X \sim \chi_p^2$ *and* $Y \sim \chi_q^2$ *are two independent chi-squared distributions, with degrees of freedom* p *and* q, *respectively, then the sum is a chi-squared random variable with* $p + q$ *degrees of freedom; i.e.,* $X + Y \sim \chi_{p+q}^2$.

Proof. This follows immediately from Definition 2.12.   □

We next seek to uncover an analytical form for the PDF of a chi-squared random variable. We will proceed using mathematical induction. We state our basis step in the following lemma.

Lemma 2.5. *Let* $X \sim \chi_1^2$ *represent a chi-squared random variable with 1 degrees of freedom. Then the* PDF *for* X *is given by*

$$f(x) = \frac{1}{\sqrt{2\pi x}} e^{-x/2}, \tag{2.36}$$

for $0 < x < \infty$.

Proof. First consider a standard normal random variable $Z \sim N(0, 1)$, with PDF

$$f_Z(z) = \frac{1}{\sqrt{2\pi}} e^{-z^2/2}.$$

Let us consider the transformation $x = z^2$, and proceed by following the recipe outlined in Theorem 1.7. The mapping $x = g(z) = z^2$ is not one-to-one, so we will partition the domain into three subsets

$$\mathcal{A}_0 = \{0\}, \qquad \mathcal{A}_1 = (-\infty, 0), \qquad \text{and} \qquad \mathcal{A}_2 = (0, \infty).$$

Note that, consistent with Theorem 1.7, $\mathbb{P}(X \in \mathcal{A}_0) = 0$. Also, $\mathcal{X}_i = \{x \in \mathbb{R} : x = g(z) \text{ for some } z \in \mathcal{A}_i\} = (0, \infty)$ for $i = 1, 2$. Moreover, $g_1^{-1}(x) = -\sqrt{x}$ and $g_2^{-1}(x) = \sqrt{x}$, from which it follows

$$\left| \frac{dg_1^{-1}(x)}{dx} \right| = \left| \frac{dg_2^{-1}(x)}{dx} \right| = \frac{1}{2\sqrt{x}}.$$

Applying Equation (1.34) to this context, we have

$$f_X(x) = \sum_{i=1}^{2} f_Z(g_i^{-1}(x)) \cdot \frac{1}{2\sqrt{x}} = \frac{1}{\sqrt{2\pi x}} e^{-x/2},$$

which proves the result. (Note that the $1/2$ disappears as there are two identical terms in the sum.) □

Lemma 2.6. *Let $X \sim \chi_2^2$ represent a chi-squared random variable with 2 degrees of freedom. Then the PDF for X is given by*

$$f(x) = \frac{1}{2} e^{-x/2}, \tag{2.37}$$

for $0 < x < \infty$.

Proof. Let $X, Y \sim N(0, 1)$ be standard normal random variables, with joint distribution

$$f_{X,Y}(x, y) = \frac{1}{2\pi} e^{-(x^2+y^2)/2}.$$

Next, let us consider the transformation

$$u = x^2 + y^2 \qquad \text{and} \qquad v = x/y.$$

Note, the random variable $U = X^2 + Y^2$ is a χ_2^2 random variable, so it suffices to show that the distribution for U is given by Equation (2.37).

To proceed, we shall follow the recipe outlined in Theorem 1.9. Since the mapping $g : \mathbb{R}^2 \to (0, \infty) \times \mathbb{R}$ defined above is not one-to-one, we must partition the plane into three regions:

$$
\begin{aligned}
\mathcal{A}_0 &= \{(x, y) \in \mathbb{R}^2 : y = 0\} \\
\mathcal{A}_1 &= \{(x, y) \in \mathbb{R}^2 : y > 0\} \\
\mathcal{A}_2 &= \{(x, y) \in \mathbb{R}^2 : y < 0\}.
\end{aligned}
$$

Note that $\mathbb{P}((x, y) \in \mathcal{A}_0) = 0$. The Jacobian of the transformation can be readily be shown to equal $|J(u, v)| = 1/(2(1 + v^2))$. Applying Equation (1.37), we may write

$$
f_{U,V}(u, v) = \sum_{i=1}^{2} f_{X,Y}(g_i^{-1}(u, v)) \frac{1}{2(1 + v^2)} = \frac{1}{2\pi(1 + v^2)} e^{-u/2}.
$$

Recall from calculus that

$$
\int_{-\infty}^{\infty} \frac{dv}{1 + v^2} = \pi.
$$

Therefore, by margianlizing over the random variable V, we obtain Equation (2.37), and the result is proven.   □

Before presenting the PDF of a general chi-squared random variable, let us first define the *gamma function*, which is a useful generalization of the factorial.

Definition 2.13. *The gamma function $\Gamma(z)$ is the function defined by the integral*

$$
\Gamma(z) = \int_0^{\infty} x^{z-1} e^{-x} \, dx. \tag{2.38}
$$

By integrating by parts, it is not difficult to show that

$$
\Gamma(z + 1) = z\Gamma(z). \tag{2.39}
$$

Since $\Gamma(1) = 1$, it follows that $\Gamma(n) = (n - 1)!$. The gamma function, however, is defined for any positive real number (as well as negative non-integers). A particularly useful value to know is $\Gamma(1/2) = \sqrt{\pi}$. Using the recurrence relation given by Equation (2.39), one can easily determine values of $\Gamma(n/2)$, for any positive integer n.

Theorem 2.5. *Let $X \sim \chi_p^2$ represent a chi-squared random variable with p degrees of freedom. Then the PDF for X is given by*

$$
f(x) = \frac{1}{\Gamma(p/2)2^{p/2}} x^{p/2-1} e^{-x/2}, \tag{2.40}
$$

for $0 < x < \infty$. This distribution is known as the chi-squared distribution with p degrees of freedom.

Proof. We shall proceed by using a modified form of mathematical induction: for our induction step, we will increment the parameter p by *two* instead of by one. We will therefore require two basis steps: for $p = 0$ and for $p = 1$. The basis steps, however, have already been proved in Lemmas 2.5 and 2.6, as one can easily verify that Equations (2.36) and (2.37) are a special cases of Equation (2.40) for $p = 1, 2$, respectively.

For the *induction step*, we must prove the logical statement: if Equation (2.40) is true for a given value of p, then it must also be true for $p + 2$.

To begin, let us assume that Equation (2.40) is valid for a given value of $p \geq 1$. Let $X \sim \chi_p^2$ be a chi-squared random variable with p degrees of freedom, and let $Y \sim \chi_2^2$ be a chi-squared random variable with 2 degrees of freedom. From Equation (2.37) and our assumed distribution for X, given by Equation (2.40), we see that the joint distribution is given by

$$f_{X,Y}(x, y) = \frac{x^{p/2-1}}{\Gamma(p/2)2^{p/2+1}} e^{-(x+y)/2}.$$

Now consider the transformation $g : (0, \infty) \times (0, \infty) \to (0, \infty) \times (0, \pi/2)$ defined by

$$u = x + y \quad \text{and} \quad v = \tan^{-1}\left(\frac{y}{x}\right).$$

Note that due to Proposition 2.20, the random variable $U = X + Y$ is a chi-squared random variable with $p + 2$ degrees of freedom; i.e., $U \sim \chi_{p+2}^2$. This relation is invertible, with inverse transformation given by

$$x = \frac{u}{1 + \tan(v)} \quad \text{and} \quad y = \frac{u\tan(v)}{1 + \tan(v)}.$$

The Jacobian is therefore

$$J(u, v) = \det \begin{bmatrix} \dfrac{1}{1 + \tan(v)} & \dfrac{-u\sec^2(v)}{(1 + \tan(v))^2} \\ \dfrac{\tan(v)}{1 + \tan(v)} & \dfrac{u\sec^2(v)}{(1 + \tan(v))^2} \end{bmatrix} = \frac{u\sec^2(v)}{(1 + \tan(v))^2}.$$

The joint distribution for the random vector (U, V) is therefore given by

$$f_{U,V}(u, v) = \frac{1}{\Gamma(p/2)2^{p/2+1}} \left(\frac{u}{1 + \tan(v)}\right)^{p/2-1} e^{-u/2} \cdot \frac{u\sec^2(v)}{(1 + \tan(v))^2}$$

$$= \frac{u^{p/2}e^{-u/2}}{\Gamma(p/2)2^{p/2+1}} \frac{\sec^2(v)}{(1 + \tan(v))^{p/2+1}}$$

However, one can easily show that

$$\int_0^{\pi/2} \frac{\sec^2(v)}{(1 + \tan(v))^{p/2+1}} \, dv = \frac{2}{p}.$$

Therefore, the marginal distribution for the random variable U is given by

$$f_U(u) = \int_0^{\pi/2} f_{U,V}(u,v)\, dv = \frac{u^{p/2}e^{-u/2}}{(p/2)\cdot\Gamma(p/2)2^{p/2+1}} = \frac{u^{p/2}e^{-u/2}}{\Gamma(p/2+1)2^{p/2+1}}.$$

However, this expression is equivalent to Equation (2.40) with the replacement $p \to p+2$.

In conclusion, we have shown each of the following conditions:

- Equation (2.40) is true for $p = 1$ (Lemma 2.5);
- Equation (2.40) is true for $p = 2$ (Lemma 2.6); and
- whenever Equation (2.40) is true for p, then it must be true for $p+2$.

We conclude that Equation (2.40) holds for all positive integers p.   □

Sampling a Normal Distribution

We saw previously, in Corollary 2.2, that when sampling from a normal distribution, the sample mean is also a normal random variable. Next, we show the relation between the sample variance with the chi-squared distribution.

Theorem 2.6. Let $X_1, \ldots, X_n \sim N(\mu, \sigma^2)$ be IID samples from a normal distribution. Then the sample mean $\overline{X}_n = (1/n)\sum_{i=1}^n X_i$ and the sample variance $S_n^2 = 1/(n-1)\sum_{i=1}^n (X_i - \overline{X}_n)^2$ are independent and, moreover,

$$\frac{(n-1)S_n^2}{\sigma^2} \sim \chi_{n-1}^2 \tag{2.41}$$

is a chi-squared random variable with $n-1$ degrees of freedom.

Proof. First, to show that $\overline{X}_n$ is independent of S_n^2, we can rewrite S_n^2 as

$$S_n^2 = \frac{1}{n-1}\left(\left[\sum_{i=2}^n (X_i - \overline{X}_n)\right]^2 + \sum_{i=2}^n (X_i - \overline{X}_n)^2\right),$$

which is a function of only $(X_2 - \overline{X}_n, \ldots, X_n - \overline{X}_n)$. From there, one can show that each of these $n-1$ random variables are independent of $\overline{X}_n$, which completes the proof of the first statement. See Casella and Berger [2002] for details.

Next, in order to prove Equation (2.41), let us consider the random variable

$$W = \sum_{i=1}^n \left(\frac{X_i - \mu}{\sigma}\right)^2.$$

Now, due to Corollary 2.1 and Definition 2.12, it follows that $W \sim \chi_n^2$ is a chi-squared random variable with n degrees of freedom. Further, we have

$$W = \sum_{i=1}^{n} \left[\frac{(X_i - \overline{X}_n) + (\overline{X}_n - \mu)}{\sigma} \right]^2 = \sum_{i=1}^{n} \left(\frac{X_i - \overline{X}_n}{\sigma} \right)^2 + \frac{(\overline{X}_n - \mu)^2}{\sigma^2/n},$$

since the cross terms vanish:

$$2 \sum_{i=1}^{n} \frac{(\overline{X}_n - \mu)(X_i - \overline{X}_n)}{\sigma^2} = \frac{2(\overline{X}_n - \mu)}{\sigma^2} \sum_{i=1}^{n} (X_i - \overline{X}_n) = 0.$$

However, we already know from Corollary 2.2 that the sample mean $\overline{X}_n \sim N(\mu, \sigma^2/n)$, and therefore the random variable

$$Z = \frac{\overline{X}_n - \mu}{\sigma/\sqrt{n}} \sim N(0, 1)$$

is a standard normal random variable, independent from S_n^2.
   Combining the above, we have

$$W = \frac{(n-1)}{\sigma^2} S_n^2 + Z^2,$$

where $W \sim \chi_n^2$ and $Z^2 \sim \chi_1^2$. We conclude that $(n-1)S_n^2/\sigma^2 \sim \chi_{n-1}^2$.   □

Simulation in Python

The notation for invoking the chi-squared random variable in **numpy** and **scipy** is given in Code Block 2.16. Several chi-squared distributions are plotted in Figure 2.12.

```
1  dof = 5
2  sample_size = 100
3  samples = np.random.chisquare(dof, size=sample_size)
4  x = np.linspace(0, 20)
5  f = scipy.stats.chi2.pdf(x, dof)
```

Code Block 2.16: Chi-squared random variable in Python

2.5.2 Student's T Distribution

If we have a random sample $X_1, \ldots, X_n \sim N(\mu, \sigma^2)$, we have seen that the quantity

$$Z = \frac{\overline{X}_n - \mu}{\sigma/\sqrt{n}}$$

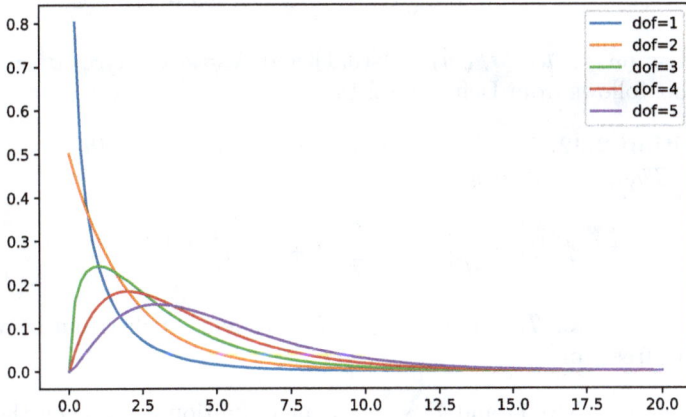

Fig. 2.12: Chi-squared distribution with $p = 1, 2, 3, 4, 5$ degrees of freedom.

is a standard normal random variable, i.e., $Z \sim \mathrm{N}(0,1)$, where $\overline{X}_n = (1/n)\sum_{i=1}^{n} X_i$ is the sample mean. But what if we do not know the true variance? Could we use the sample variance as a proxy? Let us substitute S_n for σ and then make the following clever rearrangement:

$$\frac{\overline{X}_n - \mu}{S_n/\sqrt{n}} = \frac{(\overline{X}_n - \mu)/(\sigma/\sqrt{n})}{\sqrt{S_n^2/\sigma^2}}.$$

the numerator $(\overline{X}_n - \mu)/(\sigma/\sqrt{n}) \sim \mathrm{N}(0,1)$ is a standard normal random variable, and the denominator is $\sqrt{\chi_{n-1}^2/(n-1)}$, independent of the numerator. This motivates the following definition.

Definition 2.14. *Let* $X \sim \mathrm{N}(0,1)$ *and* $Y \sim \chi_p^2$ *represent independent random variables:* X *a standard normal and* Y *a chi-squared with p degrees of freedom. Then the random variable* $T = X/\sqrt{Y/p}$ *is said to be a t random variable with p degrees of freedom, denoted* $T \sim \mathrm{t}_p$.

Proposition 2.21. *Let* $X_1, \ldots, X_n \sim \mathrm{N}(\mu, \sigma^2)$ *be* IID *normal random variables, and let* $\overline{X}_n = (1/n)\sum_{i=1}^{n} X_i$ *be the sample mean and* $S_n^2 = 1/(n-1)\sum_{i=1}^{n}(X_i - \overline{X}_n)^2$ *the sample standard deviation. Then the random variable*

$$T = \frac{\overline{X} - \mu}{S_n/\sqrt{n}} \tag{2.42}$$

is a t-random variable with $n - 1$ degrees of freedom; i.e., $T \sim \mathrm{t}_{n-1}$.

Proof. As previously shown, the random variable T can be rearranged as

$$T = \frac{(\overline{X}_n - \mu)/(\sigma/\sqrt{n})}{\sqrt{S_n^2/\sigma^2}} = \frac{Z}{\sqrt{X/(n-1)}},$$

where $Z = (\overline{X}_n - \mu)/(\sigma/\sqrt{n}) \sim N(0,1)$ and $X = (n-1)S_n^2/\sigma^2 \sim \chi_{n-1}^2$. The result follows from Definition 2.14. $\qquad\square$

Proposition 2.22. *Let $T \sim t_p$ be a t-random variable with p degrees of freedom. Then its* PDF *is given by*

$$f_T(t) = \frac{\Gamma((p+1)/2)}{\Gamma(p/2)\sqrt{p\pi}}(1 + t^2/p)^{-(p+1)/2}, \qquad (2.43)$$

for $-\infty < t < \infty$. This is known as the Student's t distribution *with p degrees of freedom.*

Proof. Let $X \sim N(0,1)$ and $Y \sim \chi_p^2$, as in Definition 2.14. Then the joint distribution is given by

$$f_{X,Y}(x,y) = \frac{1}{\sqrt{2\pi}}e^{-x^2/2}\frac{1}{\Gamma(p/2)2^{p/2}}y^{p/2-1}e^{-y/2},$$

for $-\infty < x < \infty$ and $0 < y < \infty$. Next, consider the transformation

$$t = \frac{x}{\sqrt{y/p}} \qquad \text{and} \qquad u = y.$$

It is straightforward to show that the Jacobian is given by $J(t,u) = \sqrt{u/p}$. Therefore, the joint distribution of (U, V) is given by

$$f_{U,V}(u,v) = f_{X,Y}\left(t\sqrt{u/p}, u\right)\sqrt{u/p} = \frac{u^{p/2-1/2}e^{-u/2(1+t^2/p)}}{\sqrt{2p\pi}\Gamma(p/2)2^{p/2}}$$

It can be shown that by integrating with respect to u from zero to infinity that one obtains the result. (See Casella and Berger [2002].) $\qquad\square$

Simulation in Python

The `numpy` and `scipy` calls for the Student's t distribution are shown in Code Block 2.17. Student's t distributions with several different degrees of freedom are shown in Figure 2.13. As the number of degrees of freedom increases, the Student's t distribution approaches a standard normal distribution.

2.5.3 Snedecor's F Distribution

We next introduce a distribution that is useful when comparing the variances of samples from two normal distributions.

```
1  dof = 5
2  sample_size=100
3  samples = np.random.standard_t(dof, size=sample_size)
4  x = np.linspace(-5, 5, num=100)
5  f = scipy.stats.t.pdf(x, dof)
```

Code Block 2.17: Student's t distribution in Python

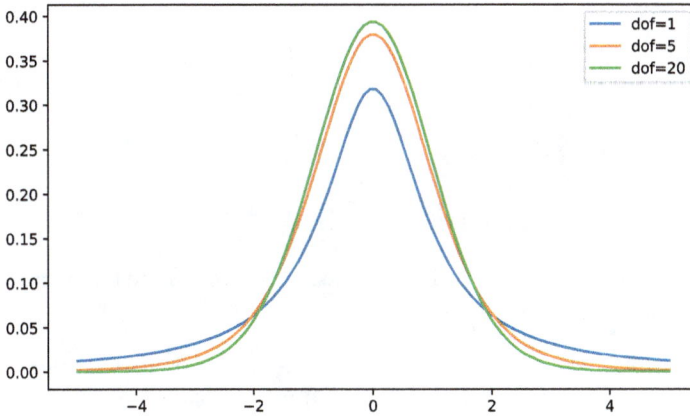

Fig. 2.13: Student's t distribution with $p = 1, 5, 20$ degrees of freedom.

Definition 2.15. *Let* $U \sim \chi_p^2$ *and* $V \sim \chi_q^2$ *be independent chi-squared distributions with* p *and* q *degrees of freedom. Then the random variable* F *defined by*

$$F = \frac{U/p}{V/q} \tag{2.44}$$

is said to be a Snedecor's F *random variable with* p *and* q *degrees of freedom, denoted* $F \sim F_{p,q}$.

Note 2.15. If $F \sim F_{p,q}$, it follows trivially that $1/F \sim F_{q,p}$.   ▷

The Snedecor's F distribution can be used to understand the relationship between the ratio of sample variances to the ratio of population (true) variances, as shown in the following proposition.

Proposition 2.23. *Let* $X_1, \ldots, X_n \sim N(\mu_X, \sigma_X^2)$ *and* $Y_1, \ldots, Y_m \sim N(\mu_Y, \sigma_Y^2)$ *constitute random samples from independent normal distributions, and let* S_X^2 *and* S_Y^2 *be the sample variances of the respective samples. Then the random variable*

$$F = \frac{S_X^2/S_Y^2}{\sigma_X^2/\sigma_Y^2} \sim F_{n-1,m-1} \tag{2.45}$$

has a Snedecor's F distribution with $n - 1$ and $m - 1$ degrees of freedom.

Proof. We may rearrange Equation (2.45) as

$$F = \frac{S_X^2/S_Y^2}{\sigma_X^2/\sigma_Y^2} = \frac{S_X^2/\sigma_X^2}{S_Y^2/\sigma_Y^2}.$$

However, from Theorem 2.6, we have that

$$U = \frac{(n-1)S_X^2}{\sigma_X^2} \sim \chi_{n-1}^2 \quad \text{and} \quad V = \frac{(m-1)S_Y^2}{\sigma_Y^2} \sim \chi_{m-1}^2,$$

thus

$$F = \frac{U/(n-1)}{V/(m-1)},$$

and the result follows from Definition 2.15.   □

The PDF for the Snedecor's F distribution is given in the following proposition, which we state without proof.

Proposition 2.24. *Let $F \sim F_{p,q}$ be a Snedecor's F distribution with p and q degrees of freedom. Then the pdf for F is called the* Snedecor's F distribution *and is given by*

$$f_F(x) = \frac{\Gamma\left(\frac{p+q}{2}\right)}{\Gamma(p/2)\Gamma(q/2)} \left(\frac{p}{q}\right)^{p/2} x^{p/2-2} \left[1 + (p/q)x\right]^{-(p+q)/2}, \tag{2.46}$$

for $0 < x < \infty$.

The F distribution is also closely related to the Student's t distribution.

Proposition 2.25. *Let $X \sim t_q$ be a Student's t random variable with q degrees of freedom. Then $X^2 \sim F_{1,q}$.*

Proof. Since $X \sim t_q$, we can write

$$X = \frac{Z}{\sqrt{Y/p}},$$

where $Z \sim N(0,1)$ and $Y \sim \chi_p^2$. Hence, it follows that

$$X^2 = \frac{Z^2}{Y/p},$$

and recalling that $Z^2 \sim \chi_1^2$, the result follows.   □

```
1  p = 7
2  q = 11
3  n_samples = 30
4  samples = np.random.f(p, q, size=n_samples)
5  x = np.linspace(0, 5, num=100)
6  f = scipy.stats.f.pdf(x, p, q)
```

Code Block 2.18: F distribution in Python

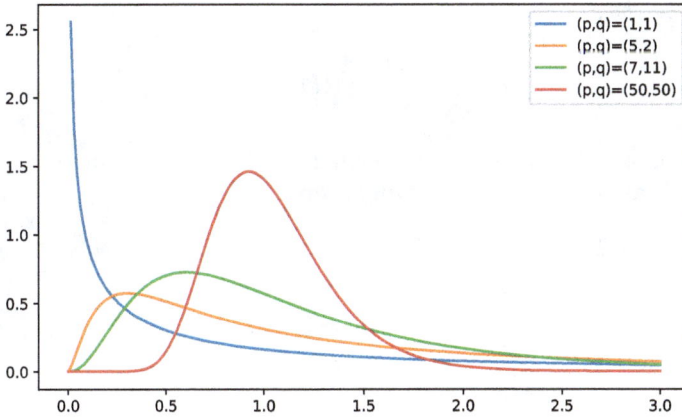

Fig. 2.14: Snedecor's F-distribution with (1,1), (2,5), (7,11), and (50,50) degrees of freedom.

Simulation in Python

The **numpy** and **scipy** commands for the Snedecor's F distribution are given in Code Block 2.18. The PDFs of the Snedecor's F-distribution for several select values of p and q are shown in Figure 2.14.

2.6 Beta and Gamma

The binomial distribution is parameterized by the probability of success p. But what if we do not know the value for p? Is it possible to specify a distribution for the parameter p? If so, what would such a distribution look like?

Questions like these are resolved using Bayesian statistics. In this context, there are two probability distributions that are quite useful: the beta and the gamma. They are used for creating *distributions for unknown parameters*. We will explore how this is done later; the purpose of this section is simply to introduce the beta and gamma distributions.

2.6.1 Beta Distribution

The beta distribution is often used a distribution for probabilities, as its domain is the open interval $(0, 1)$. Before defining the beta distribution, however, we will first define the beta function.

Definition 2.16. *The* beta function *is defined to be the following function of two variables α and β*

$$B(\alpha, \beta) = \int_0^1 x^{\alpha-1}(1 - x)^{\beta-1}\, dx. \tag{2.47}$$

The beta function is closely related to the gamma function (Definition 2.13), as shown in the following proposition.

Proposition 2.26. *The beta function can be computed using the gamma function via the identity*

$$B(\alpha, \beta) = \frac{\Gamma(\alpha)\Gamma(\beta)}{\Gamma(\alpha + \beta)}. \tag{2.48}$$

In practice, we use Equation (2.48) when evaluating the beta function, as opposed to dealing directly with Equation (2.47). Now that we have introduced the beta function, we are ready to define the beta distribution.

Definition 2.17. *The continuous random variable X defined on the interval $(0, 1)$ is said to be a* beta random variable with parameters α and β, *denoted $X \sim \text{Beta}(\alpha, \beta)$, if its* PDF *is given by the* beta distribution

$$f(x) = \frac{1}{B(\alpha, \beta)} x^{\alpha-1}(1 - x)^{\beta-1}, \tag{2.49}$$

for $0 < x < 1$, and for positive parameters $\alpha > 0$ and $\beta > 0$.

In order to determine the mean and variance, we will rely on the following lemma.

Lemma 2.7. *Let $X \sim \text{Beta}(\alpha, \beta)$. Then the expected value of X^n is given by*

$$\mathbb{E}[X^n] = \frac{\Gamma(\alpha + n)\Gamma(\alpha + \beta)}{\Gamma(\alpha)\Gamma(\alpha + \beta + n)}.$$

Proof. To compute the expected value of X^n, we take the integral

$$\mathbb{E}[X^n] = \frac{1}{B(\alpha, \beta)} \int_0^1 x^n x^{\alpha-1}(1-x)^{\beta-1}\, dx$$

$$= \frac{1}{B(\alpha, \beta)} \int_0^1 x^{\alpha+n-1}(1-x)^{\beta-1}\, dx.$$

However, from Equation (2.47), we recognize this integral as $B(\alpha + n, \beta)$, hence

$$\mathbb{E}[X^n] = \frac{B(\alpha + n, \beta)}{B(\alpha, \beta)}.$$

A simple application of Equation (2.48) produces the result.   □

Theorem 2.7. *Let* $X \sim \text{Beta}(\alpha, \beta)$ *be a beta random variable. Then the mean and variance of* X *are given by*

$$\mathbb{E}[X] = \frac{\alpha}{\alpha + \beta} \tag{2.50}$$

$$\mathbb{V}(X) = \frac{\alpha\beta}{(\alpha + \beta)^2(\alpha + \beta + 1)} \tag{2.51}$$

Proof. From Lemma 2.7, we have

$$\mathbb{E}[X] = \frac{\Gamma(\alpha + 1)\Gamma(\alpha + \beta)}{\Gamma(\alpha)\Gamma(\alpha + \beta + 1)} = \frac{\alpha}{\alpha + \beta},$$

where the last equality holds due to Equation (2.39), since $\Gamma(z+1)/\Gamma(z) = z$. A second application of this identity yields

$$\Gamma(z + 2) = (z + 1)\Gamma(z + 1) = (z + 1)z\Gamma(z).$$

Using this, we obtain

$$\mathbb{E}[X^2] = \frac{\Gamma(\alpha + 2)\Gamma(\alpha + \beta)}{\Gamma(\alpha)\Gamma(\alpha + \beta + 2)} = \frac{(\alpha + 1)\alpha}{(\alpha + \beta + 1)(\alpha + \beta)}.$$

Applying Equation (1.4) to the above two results and simplifying produces the result.   □

Simulation in Python

The `numpy` and `scipy` methods for simulating a beta distribution are given in Code Block 2.19. The PDF for several different beta random variables is shown in Figure 2.15.

```
1  alpha = 5
2  beta = 10
3  n_samples = 30
4  samples = np.random.beta(alpha, beta, size=n_samples)
5  x = np.linspace(0, 1, num=100)
6  f = scipy.stats.beta.pdf(x, alpha, beta)
```

Code Block 2.19: Beta random variable in Python

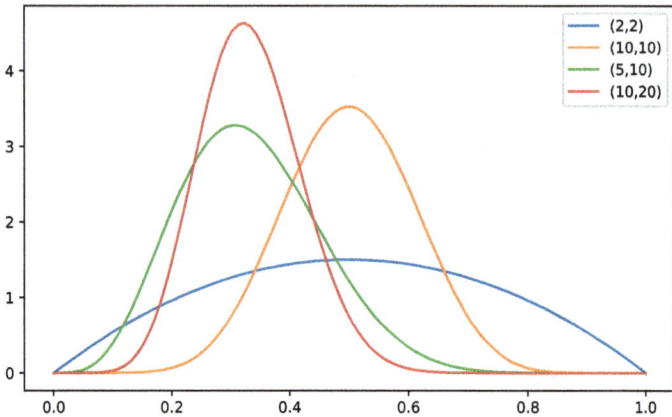

Fig. 2.15: PDF for several beta random variables.

2.6.2 Gamma Distribution

Definition 2.18. *The continuous random variable X defined on the interval $(0, \infty)$ is said to be a* gamma random variable *with parameters α and β, denoted $X \sim \mathrm{Gamma}(\alpha, \beta)$, if its PDF is given by the* gamma distribution

$$f(x) = \frac{1}{\Gamma(\alpha)\beta^{\alpha}}x^{\alpha-1}e^{-x/\beta}, \qquad (2.52)$$

for $0 < x < \infty$, and for positive parameters $\alpha > 0$ and $\beta > 0$.

Note 2.16. The parameters of a gamma random variable $X \sim \mathrm{Gamma}(\alpha, \beta)$ are sometimes referred to as the shape parameter (α) and scale parameter (β). It is important to point out that, like the exponential distribution, the gamma distribution can be parameterized either by the scale parameter or the rate parameter $\lambda = 1/\beta$. (To make matters worse, some authors use β as the rate parameter as well; sometimes k and θ are used as the shape and scale parameters.)   ▷

Note 2.17. For the special case of $\alpha = 1$, Equation (2.52) is equivalent to an exponential distribution with scale parameter β; i.e., the exponential distribution is a special case of the gamma distribution: $\text{Exp}(\beta) = \text{Gamma}(1, \beta)$.

▷

Proposition 2.27. *Let* $X \sim \text{Gamma}(\alpha, \beta)$ *represent a gamma random variable with shape* α *and scale* β. *Then the mean and variance are given by*

$$\mathbb{E}[X] = \alpha\beta \tag{2.53}$$
$$\mathbb{V}(X) = \alpha\beta^2. \tag{2.54}$$

Proof. The mean is given by

$$\mathbb{E}[X] = \frac{1}{\Gamma(\alpha)\beta^\alpha} \int_0^\infty x^\alpha e^{-x/\beta} \, dx$$
$$= \frac{1}{\Gamma(\alpha)\beta^\alpha} \cdot \Gamma(\alpha+1)\beta^{\alpha+1},$$

and the result follows by application of Equation (2.39). Note: the second line follows by recognizing $x^\alpha e^{-x/\beta}$ as the kernel of a $\Gamma(\alpha+1, \beta)$ distribution, since Equation (2.52) must integrate to unity.

The proof for Equation (2.54) follows along similar lines. □

Proposition 2.28. *The* MGF *for a Gamma random varaible* $X \in \text{Gamma}(\alpha, \beta)$ *is given by*
$$M_X(t) = (1 - \beta t)^\alpha,$$

for $t < 1/\beta$.

Proof. The MGF for a Gamma random variable can be expressed as

$$M_X(t) = \frac{1}{\Gamma(\alpha)\beta^\alpha} \int_0^\infty x^{\alpha-1} e^{-x/(\frac{\beta}{1-\beta t})} \, dx.$$

(See Exercise 2.4.) However, we recognize the integrand as the kernel of a Gamma distribution with $a = \alpha$ and $b = \frac{\beta}{1-\beta t}$. Hence

$$\int_0^\infty x^{\alpha-1} e^{-x/(\frac{\beta}{1-\beta t})} \, dx = \int_0^\infty x^{a-1} e^{-x/b} \, dx = \Gamma(a)b^a = \Gamma(\alpha) \left(\frac{\beta}{1 - \beta t} \right)^\alpha.$$

This completes the result. □

Proposition 2.29. *Given two independent Gamma random variables with the same scale parameter* $X_1 \sim \text{Gamma}(\alpha_1, \beta)$ *and* $X_2 \sim \text{Gamma}(\alpha_2, \beta)$, *the sum* $X_1 + X_2$ *is the Gamma random variable* $X_1 + X_2 \sim \text{Gamma}(\alpha_1 + \alpha_2, \beta)$.

Proof. From Proposition 1.12, the MGF of $Y = X_1 + X_2$ is

$$M_{X_1 + X_2}(t) = M_{X_1}(t)M_{X_2}(t) = (1 - \beta)^{\alpha_1 + \alpha_2},$$

where the last equality follows from Proposition 2.28. However, we recognize the result as the MGF for a $\mathrm{Gamma}(\alpha_1 + \alpha_2, \beta)$ distribution, which completes the result. □

Finally, we note a connection between the Gamma distribution and the chi-squared distribution.

Proposition 2.30. *The chi-squared distribution is a special case of the Gamma distribution:*

$$\chi_k^2 = \mathrm{Gamma}(k/2, 2).$$

Simulation in Python

Due to the varying definitions of the gamma random variable (whether it is parameterized by rate or scale), one must take an extra step of precaution to ensure that one is passing the correct arguments into the numpy and scipy methods. This is shown in Code Block 2.20. Several members of the family of gamma random variables are shown in Figure 2.16.

```
1  alpha = 5 #shape
2  beta = 2 #scale
3  n_samples = 30
4  samples = np.random.gamma(alpha, scale=beta, size=n_samples)
5  x = np.linspace(0, 30, num=200)
6  f = scipy.stats.gamma.pdf(x, alpha, scale=beta)
```

Code Block 2.20: Gamma random variable in Python

2.6.3 Gamma's Relation to the Poisson Process

We have already seen in Theorem 2.2 that the interarrival times of a Poisson process with parameter λ follow an exponential distribution with scale parameter $\beta = 1/\lambda$. Since the exponential distribution is a special case of the gamma distribution, it is natural to ask whether the gamma distribution is also related to the Poisson process. Our next theorem provides the answer.

Theorem 2.8. *Let N_t represent a Poisson process with rate parameter λ. Let X_k represent the arrival time of the kth event; i.e., $X_k = \inf\{t > 0 : N_t \geq k\}$. Then the distribution for X_k is given by*

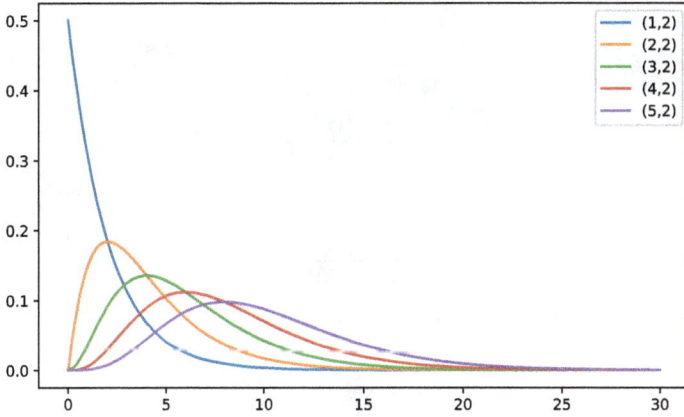

Fig. 2.16: PDF for several gamma random variables.

$$X_k \sim \text{Gamma}(k, 1/\lambda); \tag{2.55}$$

i.e., the arrival time of the kth event follows a gamma random variable with shape parameter $\alpha = k$ and scale parameter $\beta = 1/\lambda$.

Proof. We proceed by induction. The basis step is trivial, as pointed out in Note 2.17. Specifically, it follows from Equation (2.52) that a $\text{Gamma}(1, \beta)$ random variable is equivalent to a $\text{Exp}(\beta)$ random variable. Since the random variable X_1 is the first arrival time, it follows from Theorem 2.2 that $X_1 \sim \text{Exp}(1/\lambda)$, which is equivalent to $X_1 \sim \text{Gamma}(1, 1/\lambda)$.

For the induction step, let us assume that Equation (2.55) holds for some fixed $k \geq 1$. Theorem 2.2 also tells us that the interarrival time T_{k+1} for the $(k+1)$th event follows the same exponential distribution, i.e., $T_{k+1} \sim \text{Exp}(1/\lambda)$. It follows that, assuming Equation (2.55) holds for k, the joint distribution is given by

$$f_{X_k, T_{k+1}}(x, t) = \frac{\lambda^k}{\Gamma(k)} x^{k-1} e^{-\lambda x} \cdot \lambda e^{-\lambda t} = \frac{\lambda^{k+1}}{\Gamma(k)} x^{k-1} e^{-\lambda(x+t)}.$$

We will follow the transformation from the proof of Theorem 2.5, namely,

$$u = x + t \qquad \text{and} \qquad v = \tan^{-1}\left(\frac{t}{x}\right),$$

for $0 < u < \infty$ and $0 < v < \pi/2$. We saw previously that $x = u/(1 + \tan(v))$ and the Jacobian is given by

$$J(u, v) = \frac{u \sec^2(v)}{(1 + \tan(v))^2}.$$

The joint distribution for the variables $U = X_k + T_{k+1}$ and $V = \tan^{-1}(T_{k+1}/X_k)$ is therefore given by

$$
\begin{aligned}
f_{U,V}(u,v) &= \frac{\lambda^{k+1}}{\Gamma(k)} \frac{u^{k-1}}{(1+\tan(v))^{k-1}} e^{-\lambda u} \frac{u \sec^2(v)}{(1+\tan(v))^2} \\
&= \frac{\lambda^{k+1}}{\Gamma(k)} \frac{\sec^2(v)}{(1+\tan(v))^{k+1}} u^k e^{-\lambda u}.
\end{aligned}
$$

Noting that

$$
\int_0^{\pi/2} \frac{\sec^2(v)}{(1+\tan(v))^{k+1}} \, dv = \frac{1}{k},
$$

we conclude that the marginal distribution for U is

$$
f_U(u) = \frac{\lambda^{k+1}}{k\Gamma(k)} u^k e^{-\lambda u}.
$$

But $U = X_k + T_{k+1} = X_{k+1}$, the arrival time of the kth event. Since $k\Gamma(k) = \Gamma(k+1)$, we recognize the preceding equation as the gamma distribution Gamma$(k+1, 1/\lambda)$.

We have therefore proven that

1. Equation (2.55) is true for $k = 1$, and
2. if Equation (2.55) is true for any $k \geq 1$, it follows that it must also be true for $k+1$.

We conclude that the theorem holds for all positive integers k.   □

2.6.4 Beta-Binomial

Earlier in this section, we introduced the beta distribution by alluding to the fact that it can serve as a useful distribution for an unknown probability. Let's take the first step in this direction and derive our first *hierarchical* distribution: the beta-binomial distribution.

Definition 2.19. *Let $P \sim \text{Beta}(\alpha, \beta)$ be a beta random variable and let $X|(P = p) \sim \text{Binom}(n, p)$, the value of X conditioned on the realization $P = p$, be a binomial random variable. Then we say that the discrete random variable X is a beta-binomial random variable with parameters n, α, and β, which we denote as $X \sim \text{BetaBin}(n, \alpha, \beta)$.*

Note 2.18. For brevity, moving forward we shall use the shorthand $X|P \sim \text{Binom}(n, P)$ to represent the more verbose $X|(P = p) \sim \text{Binom}(n, p)$. In either case, however, the statement is interpreted as meaning that the conditional distribution of X, given the realization $P = p$, is a binomial distribution with parameters n and p.   ▷

The beta-binomial random variable is an example of a more general structure known as a hierarchical model, as defined in the following.

Definition 2.20. *A* hierarchical model *for a random variable X is any statistical model for X in which one or more of its parameters is itself a random variable; i.e., a hierarchical model is any model of the form*

$$X|Y \sim \mathcal{D}(\theta; Y)$$
$$Y \sim \mathcal{E}(\phi);$$

where $\mathcal{D}$ and $\mathcal{E}$ are given distributions with parameters $\mathcal{D}(\theta, y)$ and $\mathcal{E}(\phi)$. The parameters θ, ϕ, and y may represent scalar or vector parameters. The parameter(s) ϕ are sometimes referred to as hyperparameters *for X. The distribution for a random variable X modeled by a hierarchical model is sometimes referred to as a* mixture distribution *or* compound distribution.

Proposition 2.31. *Let $X \sim \mathrm{BetaBin}(n, \alpha, \beta)$. Then the mean and variance of X are given by*

$$\mathbb{E}[X] = \frac{n\alpha}{\alpha + \beta} \tag{2.56}$$

$$\mathbb{V}(X) = \frac{\alpha\beta n(\alpha + \beta + n)}{(\alpha + \beta)^2(\alpha + \beta + 1)}. \tag{2.57}$$

Proof. From Equation (1.19), we have

$$\mathbb{E}[X] = \mathbb{E}[\mathbb{E}[X|P]] = \mathbb{E}[nP] = \frac{n\alpha}{\alpha + \beta}.$$

We have further seen that the expected value of the square of a binomial random variable is given by

$$\mathbb{E}[X^2|P] = np(1 - p) + n^2 p^2 = np + p^2(n^2 - n)$$

and that the expected value of the square of a beta random variable is given by

$$\mathbb{E}[P^2] = \frac{\alpha(\alpha + 1)}{(\alpha + \beta + 1)(\alpha + \beta)}.$$

Thus, we can write

$$\mathbb{E}[X^2] = \mathbb{E}[\mathbb{E}[X^2|P]] = \mathbb{E}[nP + P^2(n^2 - n)] = \frac{n\alpha}{\alpha + \beta} + \frac{(n^2 - n)\alpha(1 + \alpha)}{(\alpha + \beta + 1)(\alpha + \beta)}.$$

Finally, the variance, as given by Equation (2.57), can be derived using $\mathbb{V}(X) = \mathbb{E}[X^2] - \mathbb{E}[X]^2$. □

It turns out that Equation (1.19) from Theorem 1.1 has a variance counterpart, which we state in our next theorem. For a proof, see Casella and Berger [2002].

Theorem 2.9 (Law of Total Variance). *Given random variables X, Y,*

$$\mathbb{V}(X) = \mathbb{E}[\mathbb{V}(X|Y)] + \mathbb{V}(\mathbb{E}[X|Y]), \tag{2.58}$$

provided that the expectations exist.

Proposition 2.32. *Let $X \sim \mathrm{BetaBin}(n, \alpha, \beta)$. Then the PMF for X is known as the* beta-binomial distribution *and is given by*

$$f(x) = \binom{n}{x} \frac{B(\alpha + x, \beta + n - x)}{B(\alpha, \beta)}, \tag{2.59}$$

for $x = 0, 1, 2, \ldots, n$.

Proof. First, we have $X|P \sim \mathrm{Binom}(n, p)$, and so therefore

$$f_{X|P}(x|p) = \binom{n}{p} p^x (1 - p)^{n-x}.$$

Further, $P \sim \mathrm{Beta}(\alpha, \beta)$, and thus

$$f_P(p) = \frac{1}{B(\alpha, \beta)} p^{\alpha-1} (1 - p)^{\beta-1}.$$

In order to obtain the probability mass function for f_X, we can marginalize over P, weighting each value of $P = p$ with its respective probability. Thus,

$$
\begin{aligned}
f_X(x) &= \int_0^1 f_{X|P}(x|p) f_P(p) \, dp \\
&= \binom{n}{x} \frac{1}{B(\alpha, \beta)} \int_0^1 p^{\alpha+x-1} (1 - p)^{\beta+n-x-1} \, dp.
\end{aligned}
$$

Recognizing the integrand as the kernel of the $\mathrm{Beta}(\alpha + x, \beta + n - x)$ distribution proves the result. $\qquad\square$

Simulation in Python

numpy does not contain a built-in beta-binomial random number generator; this, however, can easily be remedied, as shown in Code Block 2.21. Note that the default size of betabinomSampler is size=None, not size=1. This way, if no argument is passed in for size, the output will be a scalar, not an np.array (as would be the case if using size=1). The PMF for a BetaBin(10, 5, 10) random variable is shown in Figure 2.17.

It is of course natural to ask how the beta-binomial compares with a simple binomial distribution. Since the expected value of $P \sim \mathrm{Beta}(5, 10)$ is $1/3$, we may add the PMF for $\mathrm{Binom}(10, 0.33)$ to our plot, as shown in Figure 2.18. We observe that the beta-binomial distribution peaks at the

```
1  n = 10
2  alpha = 5
3  beta = 10
4  n_samples = 30
5  def betabinomSampler(n, alpha, beta, size=None):
6      p = np.random.beta(alpha, beta, size=size)
7      x = np.random.binomial(n, p)
8      return x
9  samples = betabinomSampler(n, alpha, beta, size=n_samples)
10  x = np.arange(11)
11  f = scipy.stats.betabinom.pmf(x, n, alpha, beta)
```

Code Block 2.21: Simulation of a beta-binomial random variable

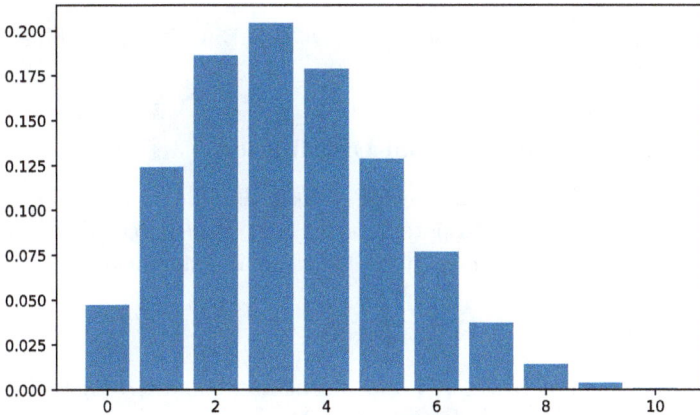

Fig. 2.17: PMF for BetaBin$(10, 5, 10)$.

same spot, but it is more spread out, as compared with the corresponding binomial distribution. This makes intuitive sense: since the probability in the beta-binomial hierarchical model is allowed to vary around $p = 0.33$, we expect to see more samples further away from the expected value, as is indeed the case.

2.7 Multivariate Distributions

We conclude the chapter with two important generalizations of distributions to higher dimensions.

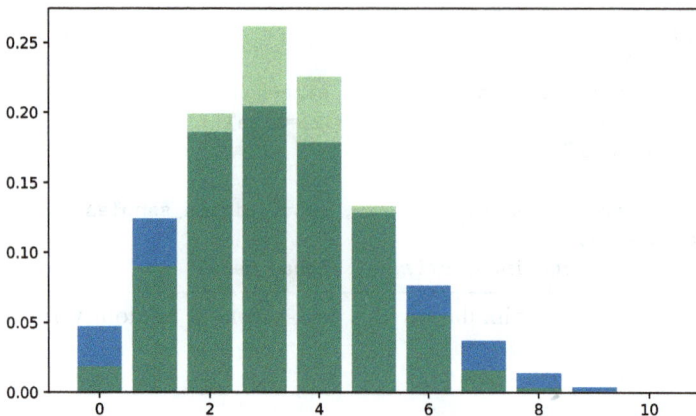

Fig. 2.18: Concurrent plot of PMF for BetaBin$(10, 5, 10)$ (blue) and Binom$(10, 0.33)$ (green).

2.7.1 The Multivariate Normal Distribution

Definition 2.21. *The random vector X is a k-dimensional multivariate normal random variable with parameters μ and Σ, denoted $X \sim N(\mu, \Sigma)$, if its* PDF *is of the form of a* multivariate normal distribution; *i.e., if*

$$f(x; \mu, \Sigma) = \frac{1}{(2\pi)^{k/2}|\Sigma|^{1/2}} e^{-(x-\mu)^T \Sigma^{-1}(x-\mu)/2}, \qquad (2.60)$$

for some constant vector $\mu \in \mathbb{R}^k$, and a constant, symmetric, positive definite matrix $\Sigma \in \mathbb{R}^{k \times k}$, with determinant $|\Sigma|$.

Proposition 2.33. *Let $X \sim N(\mu, \Sigma)$ for some $\mu \in \mathbb{R}^k$ and symmetric, positive definite $\Sigma \in \mathbb{R}^{k \times x}$. Then the mean and variance–covariance matrix of the random vector X are given by*

$$\mathbb{E}[X] = \mu, \qquad (2.61)$$
$$\mathbb{V}(X) = \Sigma. \qquad (2.62)$$

The following is a generalization of Corollary 2.1.

Proposition 2.34. *Let $X \sim N(\mu, \Sigma)$ be a k-dimensional multivariate normal random vector. Then the transformed random vector*

$$Z = \Sigma^{-1/2}(X - \mu) \qquad (2.63)$$

is a k-dimensional standard normal random vector; i.e., $Z \sim N(0_k, I_k)$, where $0_k \in \mathbb{R}^k$ is the zero vector and $I_k \in \mathbb{R}^{k \times k}$ is the identity matrix. Conversely, given $Z \sim N(0_k, I_k)$, the transformed random vector $X = \Sigma^{1/2}Z + \mu \sim N(\mu, \Sigma)$.

Proposition 2.35. *Let $X \sim N(\mu, \Sigma)$ represent a k-dimensional normal random vector. Then the components of X are independent, i.e., $X_i \perp\!\!\!\perp X_j$ for $i \neq j$, if and only if the covariance matrix Σ is diagonal, i.e., if and only if $\Sigma = \text{diag}(\sigma_1^2, \ldots, \sigma_k^2)$, for positive constants $\sigma_1^2, \ldots, \sigma_k^2$. Moreover, given independence, each component is a normal random variable with $X_i \sim N(\mu_i, \sigma_i^2)$.*

Proof. First, let us assume that the components of X are independent. This implies that they are uncorrelated, by Proposition 1.6, and thus $\text{COV}(X_i, X_j) = 0$ whenever $i \neq j$, and $\text{COV}(X_i, X_i) = \sigma_i^2$. Since $\Sigma_{ij} = \text{COV}(X_i, X_j)$, by the definition of a variance–covariance matrix, it follows that independence implies that Σ must be a diagonal matrix.

Second, let us assume that the variance–covariance matrix Σ is diagonal, which implies that the components of X are uncorrelated (but not necessarily independent). To prove independence, we must show that the PDF given by Equation (2.60) can be factored. To start, recall from linear algebra that the determinant of a diagonal matrix $\Sigma = \text{diag}(\sigma_1^2, \ldots, \sigma_k^2)$ is the product of its diagonal entries; i.e., $|\Sigma| = \sigma_1^2 \cdots \sigma_k^2$. Furthermore, we have that $\Sigma^{-1} = \text{diag}(\sigma_1^{-2}, \ldots, \sigma_k^{-2})$, and thus

$$(x - \mu)^T \Sigma^{-1} (x - \mu) = \sum_{i=1}^{k} (x_i - \mu)^2 / \sigma_i^2.$$

Combining these results, it follows that, for the case of a diagonal covariance matrix, the PDF given by Equation (2.60) can be expressed

$$f_{X_1, \ldots, X_n}(x; \mu, \Sigma) = \prod_{i=1}^{k} \frac{1}{\sigma_i \sqrt{2\pi}} e^{-(x_i - \mu_i)^2 / \sigma_i^2} = \prod_{i=1}^{k} f_{X_i}(x_i; \mu_i, \sigma_i^2).$$

Moreover, we recognize these factors as independent normal distributions, hence $X_i \sim N(\mu_i, \sigma_i^2)$. This completes the proof. ☐

To understand the multivariate normal distribution a step further, a consequence of our next result will show that the contours of the multivariate normal PDF are ellipsoids.

Theorem 2.10. *Let $X \sim N(\mu, \Sigma)$ represent a k-dimensional multivariate normal distribution, then there exists a linear transformation of the form $Y = V^T(X - \mu)$, for some matrix $V \in \mathbb{R}^{k \times k}$, such that the components of Y are independent normal random variables. Moreover, the matrix V is the orthogonal matrix whose columns are the eigenvectors of Σ.*

Proof. Since the covariance matrix Σ is symmetric and positive definite, it can be factored as

$$\Sigma = VDV^T,$$

where $D = \mathrm{diag}(\sigma_1^2, \ldots, \sigma_k^2)$ is the diagonal matrix of positive eigenvalues $\sigma_1^2, \ldots, \sigma_k^2$, and $V = [v_1, \ldots, v_k]$ is the (orthogonal) matrix of associated eigenvectors $v_1, \ldots, v_k$. Moreover, the inverse matrix is obtained simply by inverting the eigenvalues; i.e.,

$$\Sigma^{-1} = VD^{-1}V^T.$$

Now consider the tranformation $y = V^T(x - \mu)$. The exponent in Equation (2.60) is thus equivalent to

$$(x - \mu)^T \Sigma^{-1}(x - \mu) = (x - \mu)^T VD^{-1}V^T(x - \mu) = y^T D^{-1} y,$$

excluding the factor of $-1/2$ for brevity. Moreover, since the matrix V of eigenvectors is orthogonal, it follows that the determinant

$$|\Sigma| = |V|\,|D|\,|V^T| = |D|,$$

since $|V^T| = |V|^{-1}$ for orthogonal matrices. It follows that the random vector $Y \sim \mathrm{N}(0_k, D)$. Since D is diagonal, the components of Y are independent, by Proposition 2.35. Moreover, it follows that $Y_i \sim \mathrm{N}(0, \sigma_i^2)$.
□

Note 2.19. We refer to the columns of matrix V as the *principal axes* of the variance–covariance matrix Σ. Moreover, the eigenvalues σ_i^2 are referred to as the *principal variances* of Σ.   ▷

In the case of a univariate normal random variable, we saw the usefulness of the z-score as a way for determining "how far" a given value is from the mean. The generalization of the z-score to higher dimensions is given in the following definition.

Definition 2.22. *Given a symmetric, positive-definite matrix $\Sigma \in \mathbb{R}^{k \times k}$, the* Mahalanobis distance *between two vectors $x, y \in \mathbb{R}^k$ is given by*

$$d_\Sigma(x, y) = \sqrt{(x - y)^T \Sigma^{-1}(x - y)}. \tag{2.64}$$

In particular, if X is a random vector with mean $\mathbb{E}[X] = \mu \in \mathbb{R}^k$ and variance–covariance matrix $\mathbb{V}(X) = \Sigma \in \mathbb{R}^{k \times k}$, we say that the Mahalanobis distance m of an observed value $x \in \mathbb{R}^k$ is the number

$$m(x) = d_\Sigma(x, \mu). \tag{2.65}$$

Proposition 2.36. *Let $X \sim \mathrm{N}(\mu, \Sigma)$ be a k-dimensional normal random vector. The contours of the Mahalanobis distance $m(x)$ are concentric ellipsoids centered at μ.*

Proof. Consider again the transformation $Y = V^T(X - \mu)$, as given in Theorem 2.10, and let $\Sigma = VDV^T$ be the principal component decomposition of Σ. The level set $d_\Sigma(x, \mu) = m$ is equivalent to

$$m^2 = (x - \mu)^T VD^{-1}V^T(x - \mu) = y^T D^{-1}y.$$

Since D^{-1} is diagonal, this is equivalent to

$$\frac{y_1^2}{\sigma_1^2} + \cdots + \frac{y_k^2}{\sigma_k^2} = z^2, \tag{2.66}$$

the equation of an ellipsoid in $\mathbb{R}^k$. We conclude that the level sets $d_\Sigma(x, \mu) = m$ are ellipsoids, centered at μ, with principal axes $v_1, \ldots, v_k$ and principal widths $y_i = \pm m\sigma_i$. $\square$

Corollary 2.3. *Let $X \sim N(\mu, \Sigma)$ be a k-dimensional normal random vector. Then the random variable $M^2 = d_\Sigma(X, \mu)$, corresponding to the squared Mahalanobis distance of X, is a chi-squared random variable with k degrees of freedom; i.e., $M^2 \sim \chi_k^2$.*

Proof. Because the components Y_i of the random vector Y, as defined in the proofs of Theorem 2.10, are normal random variables with $Y_i \sim N(0, \sigma_i^2)$, it follows that the scaled random variables $Y_i/\sigma_i \sim N(0, 1)$ are standard normal random variables. Equation (2.66) therefore implies that M^2 is a chi-squared random variable with k degrees of freedom, following Definition 2.12. $\square$

The probability of a multivariate normal random vector X having a Mahalanobis distance $d_\Sigma(X, \mu)$ less than m is given for the case of $k = 1, 2, 3, 4$ in Table 2.1[8]. Notice, in particular, that the row $k = 1$ recovers the basic facts about a standard normal random variable: 68% probability of being within one standard deviation from the mean; 95% probability of being within two standard deviations from the mean; and 99.7% probability of being within three standard deviations from the mean. Notice that as the dimension increases, the probability of being within a fixed Mahalanobis distance to the mean diminishes.

This has dire implications for computing a 95% confidence ellipsoid for high-dimensional normal distributions. The Mahalanobis distance that contains 95% of the probability, i.e., the solution to $\mathbb{P}(d_\Sigma(X, \mu) < m) = 0.95$, is shown in Table 2.2. These data can be generated using the built-in inverse survival function, `np.sqrt(scipy.stats.chi2.isf(0.05, k))`. We see that the "two standard deviations from the mean" rule begins to fail as we move up in dimension.

Finally, we conclude with a generalization of Theorem 2.4.

---

[8] These probabilities can be computed in Python using `scipy.stats.chi2.cdf(m**2, k)`, for dimension k and Mahalanobis distance m.

	$m = 1$	$m = 2$	$m = 3$
$k = 1$	0.6827	0.9545	0.9973
$k = 2$	0.3935	0.8647	0.9889
$k = 3$	0.1987	0.7385	0.9707
$k = 4$	0.0902	0.5940	0.9389

Table 2.1: Cumulative probabilities at various Mahalanobis distances and dimensions.

$k = 1$	$k = 2$	$k = 3$	$k = 4$	$k = 5$	$k = 6$	$k = 7$	$k = 8$	$k = 9$	$k = 10$
1.9600	2.4477	2.7955	3.0802	3.3272	3.5485	3.7506	3.9379	4.1133	4.2787

Table 2.2: Mahalanobis distance containing 95% probability.

Theorem 2.11 (Multivariate Central Limit Theorem). *Suppose the k-dimensional random vectors $X_1, \ldots, X_n \sim F$ constitute an IID sample from a distribution F with finite mean $\mu \in \mathbb{R}^k$ and variance $\Sigma \in \mathbb{R}^{k \times k}$. Let $\overline{X}_n = (1/n) \sum_{i=1}^{n} X_i \in \mathbb{R}^k$ represent the sample mean. Then the sequence of random vectors*

$$Z_n = \sqrt{n} \Sigma^{-1/2} (\overline{X}_n - \mu) \tag{2.67}$$

approaches the k-dimensional standard normal random vector $Z \sim N(0_k, I_k)$ in the limit as $n \to \infty$; i.e., $Z_n \to Z$ in distribution as $n \to \infty$.

Note 2.20. An alternative way of stating Theorem 2.11 is

$$\sqrt{n}(\overline{X}_n - \mu) \to N(0_k, \Sigma), \tag{2.68}$$

as $n \to \infty$. This follows from Proposition 2.34.   ▷

Python

The Mahalanobis distance, as given by Definition 2.22, can easily be set to code, as shown in Code Block 2.22.

```python
def mahalanobis(x, y, Sigma):
    return np.sqrt( (x-y).T @ np.linalg.inv(Sigma) @ (x-y))
```

Code Block 2.22: Mahalanobis Distance

Next, we turn to the case of level sets of the Mahalanobis distance for the special case of dimension $k = 2$, which will be common in examples. Such a routine is provided in Code Block 2.23. (We will use this code later, in Example 5.6.)

```python
def ellipse(mu, Sigma, z=1, num=100):
    t = np.linspace(0, 2*np.pi, num=num)

    D, V = np.linalg.eig(Sigma)
    # D: Array of Eigenvalues
    # V: (Orthogonal) Matrix of Eigenvectors

    f1 = z * np.sqrt(D[0])
    f2 = z * np.sqrt(D[1])
    # Scale Factors along principal axes.

    Y = np.array([ f1 * np.cos(t), f2 * np.sin(t)])
    # 2 x 100 matrix

    X = V @ Y + mu.reshape((2,1)) * np.ones(num)
    # Transform back to original axes.

    return X[0, :], X[1, :]
```

Code Block 2.23: Code to generate ellipse.

Line 4 returns the diagonal matrix of eigenvalues and the matrix of eigenvectors for Σ. Lines 8–9 computes the width of the ellipse along its principal axes, as given by Equation (2.66). We can parameterize this ellipse relative to its principal axes coordinates using

$$y(t) = \begin{bmatrix} z\sigma_1 \cos(t) \\ z\sigma_2 \sin(t) \end{bmatrix},$$

for $0 \le t \le 2\pi$. This is done on line 12. Finally, we transform back into original coordinates using $x = Vy + \mu$ (line 15).

Moreover, we can take random samples and compute the PDF and CDF of a multivariate normal using the built-in numpy and scipy packages, as shown in Code Block 2.24. It should be noted that the CDF of a multivariate distribution is defined in the sense that $F([x_0, y_0]) = \mathbb{P}(X < x_0 \text{ and } Y < y_0)$.

One way to generate a contour plot of the PDF is shown in lines 18–21. The output is shown in Figure 2.19.

2.7.2 The Multinomial Distribution

We next turn to our final distribution of the chapter, completing the circle by returning to our Bernoulli roots. For additional detail, see Härdle and Simar [2019]. To motivate our definition, let us first consider a simple example.

```
1   mu = np.array([2, 3])
2   Sigma = np.array([[5/2, -3/2], [-3/2, 5/2]])
3   # D = array([4., 1.])
4   # V = array([[ 0.70710678, 0.70710678],
5   #             [-0.70710678, 0.70710678]])
6
7    n_samples=100
8   X = np.random.multivariate_normal(mu, Sigma, size=n_samples)
9   #  100 x 2
10  x0 = np.array([2,3])
11  scipy.stats.multivariate_normal.pdf([2,3], mean=mu, cov=Sigma)
12  # 0.07957747154594767
13  scipy.stats.multivariate_normal.cdf([2,3], mean=mu, cov=Sigma)
14  # P(X < 2 and Y < 3)
15  # 0.14758361765043332
16
17  ## Contour Plot of PDF
18  x, y = np.mgrid[0:5:.01, 0:5:.01]
19  pos = np.dstack((x, y))
20  f = scipy.stats.multivariate_normal.pdf(pos, mean=mu, cov=Sigma)
21  plt.contourf(x, y, f)
```

Code Block 2.24: Random samples from multivariate normal

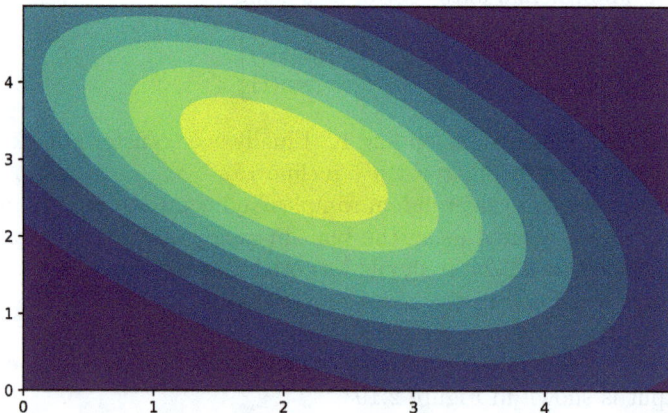

Fig. 2.19: Contour plot of multivariate normal PDF produced by Code Block 2.24.

Example 2.1. Suppose that a jar contains marbles of an assortment of k different colors, where the ith color is represented with proportion $p_i \in [0, 1]$, such that $\sum_{i=1}^{k} p_i = 1$. A total of n marbles are selected at random from the jar; i.e., we perform n independent draws *with replacement*[9]. Let $X = (X_1, \ldots, X_k)$ be the vector of counts following our n draws; i.e., let X_i represent the number of marbles of color i obtained from the n draws, such that $\sum_{i=1}^{n} X_i = n$. The vector X is said to constitute a *multinomial random variable.*   ▷

We will build up the theory to tackle such problems step-by-step. First, however, we present some useful notation.

Definition 2.23. *The standard $(k-1)$-simplex Δ^{k-1}, or probability simplex, is the subset of $\mathbb{R}_*^k$ defined by*

$$\Delta^{k-1} = \left\{ x \in \mathbb{R}_*^k : \sum_{i=1}^{k} x_i = 1 \right\},$$

where $\mathbb{R}_* = [0, \infty)$ is the set of nonnegative real numbers.

The set Δ^{k-1} is useful in describing a vector of probabilities. Specifically, a vector $p = (p_1, \ldots, p_k)$ that satisfies $p_i \geq 0$ and $\sum_{i=1}^{k} p_i = 1$ and alternatively be specified by the simple statement that $p \in \Delta^{k-1}$. Another way of viewing the set Δ^{k-1} is the *convex hull* of the set of standard basic vectors $\mathbb{U}^k \subset \mathbb{R}^k$, defined by

$$\mathbb{U}^k = \left\{ x \in \mathbb{B}^k : \sum_{i=1}^{k} x_i = 1 \right\} = \{e_1, \ldots, e_k\},$$

where e_i is the ith standard basis vector $e_i = (\delta_{i1}, \ldots, \delta_{ik})$.

Next, let us describe a single experiment that has k mutually exclusive and exhaustive outcomes $A_1, \ldots, A_k$, with probabilities $p_1, \ldots, p_k$. This is a generalization of the Bernoulli random variable, for which there were precisely two such outcomes. The random vector to describe such a scenario is therefore of the form $X = (X_1, \ldots, X_k)$.

Definition 2.24. *A random vector X is said to be a* multivariate Bernoulli *random variable* with probability $p \in \Delta^{k-1}$, *denoted $X \sim \text{MultiBern}(p)$, if X is of the form $X \in \mathbb{U}^k = \{e_1, \ldots, e_k\}$ and has* PMF *given by $f_X(e_i) = p_i$, i.e.,*

$$f_X(x_1, \ldots, x_k) = p_1^{x_1} \cdots p_k^{x_k}. \tag{2.69}$$

---

[9] *With replacement* means that the probabilities of selecting a particular color does not change as we draw more marbles; i.e., it is as though we return each marble to the jar before drawing the next.

In particular, note that a multivariate Bernoulli trial has precisely k possible outcomes. The mean and variance–covariance matrix of a multivariate Bernoulli random vector is given as follows.

Proposition 2.37. *Let $X \sim$ MultiBern(p) be a multivariate Bernoulli random vector with $p \in \Delta^{k-1}$. Then $\mathbb{E}[X] = p$ and the variance and covariance of the components of X are given by*

$$\Sigma_{ij} = \text{COV}(X_i, X_j) = \begin{cases} p_i(1 - p_i) & \text{for } i = j \\ -p_i p_j & \text{for } i \neq j \end{cases} \qquad (2.70)$$

Proof. The marginal distribution of X_i is $X_i \sim$ Bern(p_i), and hence $\mathbb{V}(X_i) = p_i(1 - p_i)$.

For the second part, consider X_i and X_j, with $i \neq j$. From Proposition 1.5, we have

$$\text{COV}(X_i, X_j) = \mathbb{E}[X_i X_j] - \mathbb{E}[X_i]\mathbb{E}[X_j].$$

Now, $\mathbb{E}[X_i] = p_i$ and $\mathbb{E}[X_j] = p_j$. Moreover, $\mathbb{E}[X_i X_j] = 0$, since $X_i X_j = 0$ for all $X \in \mathbb{U}^k$; i.e., X_i and X_j cannot *both* equal one. The result follows. $\square$

Note 2.21. The matrix Σ from Proposition 2.37 is singular due to a relation of linear dependence among its columns created by the constraint $\sum_{i=1}^{k} p_i = 1$; moreover rank$(\Sigma) = k - 1$. This is easiest to see for the case of $k = 2$, since

$$\det \begin{bmatrix} p_1(1 - p_1) & -p_1 p_2 \\ -p_1 p_2 & p_2(1 - p_2) \end{bmatrix} = p_1 p_2(1 - p_1 - p_2) = 0,$$

since $p_1 + p_2 = 1$.   ▷

In order to describe the scenario in Example 2.1, we can therefore conduct a sequence of multivariate Bernoulli trials, similar to the construction of the binomial random variable.

Definition 2.25. *Let $B_1, \ldots, B_n \sim$ MultiBern(p) represent a sequence of k-dimensional Bernoulli trials, for some $p \in \Delta^{k-1}$. The random vector $X = \sum_{i=1}^{n} B_i \in \mathbb{Z}_*^k$ of nonnegative integers, satisfying $\sum_{i=1}^{k} X_i = n$, is a multinomial random vector, denoted $X \sim$ Multi(n, p).*

Note 2.22. The sum $X = \sum_{i=1}^{n} B_i$ that defines the vector X is a vector sum, whereas the sum $\sum_{i=1}^{k} X_i = n$, which constrains the vector X, represents a sum of its components. Moreover, the latter sum is redundant and stated only for clarity: Since $\sum_{j=1}^{k} B_{ij} = 1$ for any $B_i \in$ MultiBern(p) (since B_i is a standard unit vector), it follows automatically that $\sum_{i=1}^{n} \sum_{j=1}^{n} B_{ij} = n$.
▷

In order to construct the PMF for a multinomial random vector, we will require an understanding of the multinomial coefficients.

Definition 2.26. *Given* $x \in \mathbb{Z}_*^k$, *with* $\sum_{i=1}^k x_i = n$, *the multinomial coefficient is defined by*

$$\binom{n}{x_1,\ldots,x_k} = \frac{n!}{x_1!\cdots x_k!}. \tag{2.71}$$

An important interpretation of the multinomial coefficient is given in our next proposition. This interpretation is crucial in deriving the PMF for a multinomial random vector.

Proposition 2.38. *The number of ways of distributing n objects into k distinct bins, such that x_i objects are assigned to the ith bin, is given by the multinomial coefficient defined by Equation (2.71).*

Proof. We proceed bin-by-bin. The number of ways of selecting x_1 objects for the first bin is simply

$$\binom{n}{x_1} = \frac{n!}{x_1!(n-x_1)!}.$$

Once this selection has been made, we have $(n-x_1)$ objects remaining that are eligible to be distributed into the second bin. The number of ways to select x_2 of these remaining objects is

$$\binom{n-x_1}{x_2} = \frac{(n-x_1)!}{x_2!(n-x_1-x_2)!}.$$

Therefore, the number of ways of selecting x_1 objects for the first bin *and* x_2 objects for the second bin, leaving $(n-x_1-x_2)$ remaining objects that have yet to be selected, is given by

$$\binom{n}{x_1}\binom{n-x_1}{x_2} = \frac{n!}{x_1!x_2!(n-x_1-x_2)!}.$$

Proceeding in this fashion, we find that the number of ways to assign x_1 objects to the first bin, $\ldots$, x_{k-1} objects to the $(k-1)$st bin, is given by

$$\binom{n}{x_1}\cdots\binom{n-x_1-\cdots-x_{k-2}}{x_{k-1}} = \frac{n!}{x_1!\cdots x_{k-1}!(n-x_1-\cdots-x_{k-1})!}.$$

However, $n-x_1-\cdots-x_{k-1} = x_k$, and so the above expression reduces to Equation (2.71). Finally, we note that there is precisely *one* way to select x_k objects for the final bin out of the remaining x_k objects. This completes the result. $\square$

Proposition 2.39. *The PMF of a multinomial random vector $X \sim \text{Multi}(n,p)$, for some $p \in \Delta^{k-1}$, is given by the multinomial distribution*

$$f(x_1,\ldots,x_k) = \frac{n!}{x_1!\cdots x_k!}p_1^{x_1}\cdots p_k^{x_k} = n!\prod_{i=1}^k \frac{p_i^{x_i}}{x_i!}, \tag{2.72}$$

such that $\sum_{i=1}^k X_i = n$.

Proof. Let $B_1, \ldots, B_n \sim \text{MultiBern}(p)$ represent a sequence of n IID multivariate Bernoulli trials, with $X = \sum_{i=1}^{n} B_i$, as given by Definition 2.25. We must show that the probability of outcome $X = (x_1, \ldots, x_k)$ is given by Equation (2.72).

However, stating that the random vector $X = (x_1, \ldots, x_k)$ is equivalent to requiring that

- precisely x_1 of the Bernoulli trials resulted in e_1;
- precisely x_2 of the Bernoulli trials resulted in e_2; ... and
- precisely x_k of the Bernoulli trials resulted in e_k.

Now, the probability of observing any *one* particular sequence $\{B_1, \ldots, B_n\}$ satisfying the above conditions is given by

$$p_1^{x_1} \cdots p_k^{x_k}.$$

However, from Proposition 2.38, we know that there are precisely

$$\binom{n}{x_1, \ldots, x_k} = \frac{n!}{x_1! \cdots x_k!}$$

different ways in which the n Bernoulli trials can produce a result consistent with the above requirements. The result follows.   □

The fact that the distribution Equation (2.72) is normalized is an immediate consequence of the multinomial theorem, which is stated as follows.

Theorem 2.12 (Multinomial Theorem). *For positive integers k and n, let $\mathcal{A} = \{x \in \mathbb{Z}_*^k : \sum_{i=1}^{k} x_i = n\}$. Then for any real numbers $p_1, \ldots, p_k$,*

$$(p_1 + \cdots + p_k)^n = \sum_{x \in \mathcal{A}} \frac{n!}{x_1! \cdots x_k!} p_1^{x_1} \cdots p_k^{x_k}.$$

Normalization follows since the probabilities in Definition 2.23 sum to unity; i.e., $\sum_{i=1}^{k} p_i = 1$.

Another important property of a multinomial random vector X is that the marginal distribution of any component is a binomial distribution. We state this result in the following.

Proposition 2.40. *Let $X \sim \text{Multi}(n, p)$. Then the marginal distribution of each component X_i is given by $X_i \sim \text{Binom}(n, p_i)$, for $i = 1, \ldots, k$.*

Proof. We follow the proof presented in Casella and Berger [2002]. Without loss of generality, let us compute the marginal distribution for X_k. Given a particular value $X_k = x_k \in [0, n]$, we must sum the PMF given in Equation (2.72) over all of the permissible values of $(x_1, \ldots, x_{k-1})$, which is given by the set $\mathcal{B} = \{(x_1, \ldots, x_{k-1}) : \sum_{i=1}^{k-1} x_i = m - x_k\}$. Hence

$$f_{X_k}(x_k) = \sum_{\mathcal{B}} \frac{n!}{x_1! \cdots x_k!} p_1^{x_1} \cdots p_k^{x_k} \frac{(n-x_k)!}{(n-x_k)!} \frac{(1-p_k)^{n-x_k}}{(1-p_k)^{n-x_k}}$$

$$= \frac{n!}{x_k!(n-x_k)!} p_k^{x_k} (1-p_k)^{n-x_k} \sum_{\mathcal{B}} \frac{(n-x_k)!}{x_1! \cdots x_{k-1}!} \frac{p_1^{x_1} \cdots p_{k-1}^{x_{k-1}}}{(1-p_k)^{n-x_k}}$$

$$= \text{Binom}(x_k; n, p_k) \sum_{\mathcal{B}} \frac{(n-x_k)!}{x_1! \cdots x_{k-1}!} \left(\frac{p_1}{1-p_k}\right)^{x_1} \cdots \left(\frac{p_{k-1}}{1-p_k}\right)^{x_{k-1}}$$

The last line follows since $n - x_k = x_1 + \cdots + x_{k-1}$. Now, the term in the summation is equivalent to the PMF for a Multi$(n - x_k, q)$ random vector, with reduced probability vector $q_i = p_i/(1 - p_k)$, for $i = 1, \ldots, k - 1$. This expression sums to unity, by Theorem 2.12. The result follows.   □

Finally, since we define the multinomial random vector as a sum of IID multivariate Bernoulli trials, the variance–covariance of a multinomial random vectort immediately follows from Proposition 2.37. We state the result as follows.

Proposition 2.41. Let $X \sim \text{Multi}(n, p)$, where $p \in \Delta^{k-1}$. Then $\mathbb{E}[X] = np$ and the variance-covariance matrix is given by $n\Sigma$, where Σ is the variance-covariance matrix of a Bernoulli random vector with probability $p \in \Delta^{k-1}$, as given by Proposition 2.37.

Example usage of a multinomial random variable in Python is shown in Code Block 2.25. Note that `scipy.stats.multinomial` has a limited number of methods: **pmf** and a few others we have not discussed (`logpmf`, `rvs` (random samples), **entropy**, and **cov**).

```
1   n = 10
2   p = [0.2, 0.3, 0.5]
3   n_samples = 1000
4   samples = np.random.multinomial(n, p, size=n_samples)
5   # Returns n_samples x len(p) matrix: print( samples.shape ) ->
        (1000, 3)
6   # First row: samples[0, :] --> array([3, 0, 7])
7
8   scipy.stats.multinomial.pmf([2, 3, 5], n, p) # 0.08505 # Does not
        have CDF, SF
9   scipy.stats.multinomial.cov(n, p)
10  # array([[ 1.6, -0.6, -1. ],
11  #         [-0.6,  2.1, -1.5],
12  #         [-1. , -1.5,  2.5]])
```

Code Block 2.25: Multinomial distribution in Python

Reduced Multinomial Random Vector

Let X be a k-dimensional multinomial random vector $X \in \text{Multi}(n, p)$, where $p \in \Delta^{k-1}$. We have seen that the components of X are *dependent*, since the final component can always be determined by the preceding components through the relationship

$$X_k = n - X_1 - \cdots - X_{k-1}.$$

Moreover, we've seen that this relation of linear dependence leads to a singularity in the corresponding variance–covariance matrix $\Sigma = \mathbb{V}(X)$. As a practical implication, Σ is therefore noninvertible, rendering it difficult to apply the central limit theorem to such a case. To remedy this, we define the following *reduced* multinomial random vector, as follows.

Definition 2.27. *A random vector $X \in \mathbb{R}^{k-1}$ is said to be a* reduced *multinomial random vector, denoted $X \in \text{RedMulti}(n, p)$, for a given $n \in \mathbb{Z}_+$ and $p \in \Delta^{k-1}$, if its distribution is of the form*

$$f(x_1, \ldots, x_{k-1}) = \frac{n!}{x_1! \cdots x_{k-1}! (n - x_1 - \cdots - x_{k-1})!} p_1^{x_1} \cdots p_{k-1}^{x_{k-1}} p_k^{n - x_1 - \cdots - x_{k-1}}$$

$$= \frac{n! \, p_k^n}{(n - x_1 - \cdots - x_{k-1})!} \prod_{i=1}^{k-1} \frac{1}{x_i!} \left(\frac{p_i}{p_k} \right)^{x_i},$$

on the domain $0 \le \sum_{i=1}^{k-1} X_i \le n$; as long as $p_k \ne 0$.

In particular, for the special case $n = 1$, we call the random vector X a reduced multivariate Bernoulli random vector, *denoted $X \sim \text{RedMultiBern}(p)$, with corresponding distribution*

$$f(x_1, \ldots, x_{k-1}) = p_k \prod_{i=1}^{k-1} \left(\frac{p_i}{p_k} \right)^{x_i},$$

on the domain $X \in \{0_{k-1}, e_1, \ldots, e_{k-1}\} \subset \mathbb{R}^{k-1}$.[10]

Note 2.23. The reduced multinomial distribution is obtained by substituting $X_k = n - \sum_{i=1}^{k-1} X_i$ in Equation (2.72). We are thus eliminating a redundant variable X_k. (We may continue to reference $p_k = 1 - \sum_{i=1}^{k-1} p_i$.)

▷

Note 2.24. In the construction of Definition 2.27, we require that $p_k \ne 0$. This is typically not a problem, as it is customary to have $p_i > 0$, for $i = 1, \ldots, k$. (Otherwise, what is the point of including the corresponding

---

[10] This is equivalent to $X_i \in \{0, 1\}$, such that $\sum_{i=1}^{k-1} X_i \le 1$.

outcome.) Should $p_k = 0$,[11] the situation is easily resolved by permuting the indices.
▷

Note 2.25. For the case $k = 2$, the multinomial distribution Multi(n, p), for $p \in \Delta^1$, is a function of *two* random variables: the number of successes *and* the number of failures. The *reduced* multinomial distribution is the binomial distribution, which is a function of *only* the number of successes.
▷

Proposition 2.42. *Let* $X \in$ RedMultiBern(p) *be a* $(k-1)$-*dimensional reduced multivariate Bernoulli random vector, with* $p \in \Delta^{k-1}$. *Then its mean and variance–covariance matrix* $\Sigma = \mathbb{V}(X)$ *are given by*

$$\mathbb{E}[X_i] = p_i \tag{2.73}$$
$$\Sigma_{ij} = p_i \delta_{ij} - p_i p_j, \tag{2.74}$$

respectively, where $\delta_{ij} = \mathbb{I}[i = j]$ *is the Kronecker-delta function, and where* $i, j = 1, \ldots, k-1$.

Note 2.26. The expressions given by Equations (2.73) and (2.74) are equivalent to the results of Proposition 2.37. In particular, Equation (2.74) is a restatement of Equation (2.70). The difference is that the matrix given by Equation (2.70) is the full variance–covariance matrix, which lives in $\Sigma \in \mathbb{R}^{k \times k}$ and is singular, whereas the matrix specified in Equation (2.74) is the *reduced* variance–covariance matrix, which lives in $\Sigma \in \mathbb{R}^{(k-1) \times (k-1)}$ and is nonsingular.
▷

We leave the proof to the reader; see Exercise 2.7.

Since the variance–covariance matrix of a reduced multinomial random vector is nonsingular, we can compute its inverse. This is provided in the following.

Proposition 2.43. *Let* $X \in$ RedMultiBern(p) *be a* $(k-1)$-*dimensional reduced multivariate Bernoulli random vector, with* $p \in \Delta^{k-1}$. *Then its variance–covariance matrix* $\Sigma = \mathbb{V}(X)$, *as given by Equation (2.74), is invertible. Moreover, its inverse is given by*

$$\Sigma_{ij}^{-1} = p_i^{-1} \delta_{ij} + p_k^{-1}, \tag{2.75}$$

for $i, j = 1, \ldots, k-1$.

---

[11] For instance, suppose that a small island population exists in which every member's favorite color is either red, blue, green, or yellow. Experimenters arrive and conduct a survey in which inhabitants can respond with red, blue, green, yellow, or chartreuse. There is an exactly 0% probability that any inhabitant has the favorite color of chartreuse, and therefore $p_5 = 0$. To determine the reduced multinomial distribution, the experimenters must either remove X_5 from consideration altogether, as it serves no purpose, or permute the indices (say, to chartreuse, red, blue, green, yellow) and then proceed as usual.

Proposition 2.43 is a special case of the result proved in Withers and Nadarajah [2014]. It is also given in Mood [1950].

Note 2.27. More generally, any matrix $\Sigma \in \mathbb{R}^{s \times s}$ of the form

$$\Sigma = \operatorname{diag}(p) - pp^T, \tag{2.76}$$

where pp^T is the outer product $(pp^T)_{ij} = p_i p_j$, has the matrix inverse

$$\Sigma^{-1} = \operatorname{diag}(p)^{-1} + (1-\nu)^{-1} J_s, \tag{2.77}$$

where $J_s \in \mathbb{R}^{s \times s}$ is a matrix of all ones and where $\nu = \sum_{i=1}^s p_i$, whenever $\nu \neq 1$. ▷

Note 2.28. Let us compare Proposition 2.37 and 2.42 in light of Note 2.27. Let $p = (p_1, \ldots, p_k) \in \Delta^{k-1} \subset \mathbb{R}^k$ and let $\tilde{p} = (p_1, \ldots, p_{k-1}) \in \mathbb{R}^{k-1}$ represent the first $k-1$ components of p. Note that $||p||_1 = \sum_{i=1}^k p_i = 1$, whereas $||\tilde{p}||_1 = \sum_{i=1}^{k-1} \tilde{p}_i = 1 - p_k < 1$.[12]

The variance–covariance matrix for the original multivariate Bernoulli random vector $X \sim \operatorname{MultiBern}(p)$ given by Equation (2.70) is equivalent to

$$\Sigma = \mathbb{V}(X) = \operatorname{diag}(p) - pp^T \in \mathbb{R}^{k \times k}.$$

Similarly, the variance–covariance matrix of the *reduced* random vector $\tilde{X} \sim \operatorname{RedMultiBern}(p)$ given by Equation (2.74) is equivalent to

$$\tilde{\Sigma} = \mathbb{V}(\tilde{X}) = \operatorname{diag}(\tilde{p}) - \tilde{p}\tilde{p}^T \in \mathbb{R}^{(k-1) \times (k-1)}. \tag{2.78}$$

Note that $\tilde{\Sigma}$ consists of the first $(k-1)$ rows and $(k-1)$ columns of Σ. Both of these are in the form of Equation (2.76), with one *crucial* distinction: $\nu = ||p||_1 = 1$, whereas $\tilde{\nu} = ||\tilde{p}||_1 = 1 - p_k$. Thus, the former is not invertible, whereas the inverse of the reduced variance–covariance matrix is given by Equation (2.77), which gives us

$$\tilde{\Sigma}^{-1} = \operatorname{diag}(\tilde{p})^{-1} + p_k^{-1} J_{k-1}, \tag{2.79}$$

which is equivalent to Equation (2.75).[13] ▷

---

[12] Here, the quantity $||x||_1 = \sum_{i=1}^n |x_i|$ represents the ℓ_1 *norm*, or *Manhattan norm*, of the vector $x \in \mathbb{R}^n$. In this context, the absolute values in the sum are unnecessary, since $p_i \geq 0$, for $i = 1, \ldots, k$.

[13] Recall that the inverse of a diagonal matrix is obtained by inverting each of its diagonal elements.

Problems

2.1. In the proof of Proposition 2.7, show that

$$n\binom{n-1}{r-1} = r\binom{n}{r}.$$

2.2. Prove that for any two independent continuous random variables X and Y with PDFs $f_X(x)$ and $f_Y(y)$, respectively, the PDF of the random variable $Z = X + Y$ is given by the following *convolution integral*

$$f_Z(z) = f_X * f_Y(z) \triangleq \int_{-\infty}^{\infty} f_X(w) f_Y(z - w)\, dw.$$

Hint: consider the bivariate transformation

$$Z = X + Y$$
$$W = X.$$

2.3. Prove Proposition 2.14.

2.4. Complete the first step in the proof of Proposition 2.28: Let $X \sim$ Gamma(α, β). Show that the MGF for X is given by

$$M_X(t) = \frac{1}{\Gamma(\alpha)\beta^\alpha} \int_0^{\infty} x^{\alpha-1} e^{-x/(\frac{\beta}{1-\beta t})}\, dx.$$

2.5. Let $X \sim$ Gamma(α, β) be a gamma random variable with PDF $f_X(x)$. Show that for $\sigma > 0$, the PDF $(1/\sigma)f_X(x/\sigma)$ is the PDF of a Gamma$(\alpha, \sigma\beta)$ random variable.

2.6. Variance of Beta–Binomial Random Variable
(a) Prove Theorem 2.9.
(b) Use Equation (2.58) to provide an alternate proof of Equation (2.57).

2.7. Prove Proposition 2.42.

2.8. Let $X_1, \ldots, X_n$ be independent samples, where $X_i \sim N(\mu, \sigma^2/w_i)$, for a known set of weights $w_i > 0$. Let $w = \sum_{i=1}^{n} w_i$ be the total weight, and define

$$\overline{X} = \frac{1}{w} \sum_{i=1}^{n} w_i X_i \quad \text{and} \quad S^2 = \frac{1}{n-1} \sum_{i=1}^{n} w_i \left(X_i - \overline{X}\right)^2.$$

Show that $\overline{X} \sim N(\mu, \sigma^2/w)$ and $(n-1)S^2/\sigma^2 \sim \chi^2_{n-1}$.

3

Getting Testy

The goal of this chapter is to apply the knowledge of various distributions that we built up in Chapter 2 to our first task of statistical inference: classical hypothesis testing and confidence intervals. Excellent supplemental reading, discussion, and examples can be found in Casella and Berger [2002], Hogg, *et al.* [2015], and Shao [2003].

3.1 Hypothesis Testing

The goal of this section is to introduce basic terminology surrounding classical hypothesis testing before diving into specific examples.

3.1.1 The Null Hypothesis

In general, a *hypothesis* is a statement about one or more population parameters, and a *hypothesis test* is a procedure for inferring the truth or validity of a hypothesis given a set of data.

There is a twist, however, in that hypothesis testing does not seek to prove what is true, but rather to disprove what is not true. Therefore, instead of trying to prove a hypothesis, we will be devising methods for disproving its complement, otherwise known as the *null hypothesis*. The null hypothesis earns its name as it is typically a statement about what is *not* there: in the language of experimentation, a null hypothesis is usually of the form that our experiment had no effect. Formally, we have the following.

Definition 3.1. *A hypothesis test concerning some population parameter $\theta \in \Theta$, where Θ is the parameter space, consists of a null hypothesis of the form*

$$H_0 : \theta \in \Theta_0$$

and a decision rule that determines, as a function of a sample from the population, whether we should reject the null hypothesis.

Definition 3.2. *The complement of a null hypothesis is called the* alternative hypothesis, *denoted $H_A : \theta \in \Theta_0^c$. Whenever we do not reject the null hypothesis, we say that we* accept the null hypothesis.

Definition 3.3. *A* two-sided test *is any hypothesis test with null hypothesis of the form $H_0 : \theta = \theta_0$. A* one-sided test *is any hypothesis test with null hypothesis of the form $H_0 : \theta \leq \theta_0$ or $H_0 : \theta \geq \theta_0$.*

Note 3.1. We were very purposeful in how we set up Definitions 3.1 and 3.2: a hypothesis test is a decision on whether or not to reject the null hypothesis. "Acceptance" of a null hypothesis is simply the absence of its rejection; it is an unfortunate term which should not be misinterpreted as implying the truth of the null hypothesis. For example, if the sample size is $n = 1$, i.e., our data set consists of a single datum, it should be virtually impossible to reject the null hypothesis, simply due to having an insufficient amount data to support such a claim. In such a case, we are said to *accept the null hypothesis.* But this simply means that we could not rule it out. Thus, the act of accepting the null hypothesis really means accepting the null hypothesis *for now*; i.e., we allow it to live to die another day[1].   ▷

To be slightly more careful with the meaning of a *decision rule* in Definition 3.1, we have the following.

Definition 3.4. *Let $\mathbb{X}$ be the sample space for a distribution. A* decision rule *is a statistic $r : \mathbb{X} \to \mathbb{A}$ that determines an action $a \in \mathbb{A}$ as a function of the data $\mathcal{X} \in \mathbb{X}$, where $\mathbb{A}$ is the set of permissible actions, i.e.,* action space.

In particular, the action space for any decision rule in a hypothesis test is always binary; i.e., $\mathbb{A} = \mathbb{B} \triangleq \{0, 1\}$.

Typically, however, the decision rule in a hypothesis test consists of a *test statistic $T : \mathbb{X} \to \mathbb{R}$* along with set of values $R \subset \mathbb{R}$ for which we would reject the null hypothesis. Formally, we have the following.

Definition 3.5. *Let $T : \mathbb{X} \to \mathbb{R}$ be a statistic and $R \subset \mathbb{R}$, and define and the function $\mathbb{I}_R : \mathbb{R} \to \mathbb{B}$ as the indicator function $\mathbb{I}_R(x) = \mathbb{I}[x \in R]$. Whenever a decision rule r for a hypothesis test consists of the composition $r = \mathbb{I}_R \circ T$, then the statistic T is referred to as a* test statistic *and the region R is referred to as the* rejection region *corresponding to T.*

When a rejection region is in standard form *$R_c = [c, \infty)$, the value of c is referred to as the* critical value *of the test. (In some cases, a simple transformation such as $T \to -T$ or $T \to |T|$ is required to achieve standard form.)*

Note 3.2. Some authors use *rejection region* to refer to the subset $\mathcal{R} \subset \mathbb{X}$ of the sample space defined by $\mathcal{R} = \{\mathcal{X} \in \mathbb{X} : r(\mathcal{X}) = 1\}$.   ▷

---

[1] Be weary of statisticians who accept you as their friend.

Example 3.1. Suppose we have a friend who is a magician who persistently challenges us to annoying coin tosses. Our magician friend always takes *heads*. We are beginning to suspect that he has a trick coin that is more likely to come up *heads* than it is *tails*. (The dastard!)

We devise the following hypothesis test: our null hypothesis is that the coin is fair or better, i.e., $\mathbb{P}(\text{heads}) \leq 0.50$. Our decision rule is to reject the null hypothesis if the observed rate of heads is greater than or equal to 60%. In particular, our test statistic is simply the sample mean $T(\mathcal{X}) = \overline{X}_n$, and our rejection region is the interval $R = [0.60, 1]$. Thus, we reject the null hypothesis (and call our friend a *cheat!*) whenever we observe more than 60% heads.

This, of course, is not a very *good* hypothesis test. In particular, if our magician friend makes a single coin flip and it comes up heads, we have already rejected the null hypothesis and concluded that the coin is not fair. It is, however, a very good illustration of the concepts defined thus far into the chapter.   ▷

3.1.2 Types of Errors

Next, we explore the types of errors that we might encounter when performing a hypothesis test. Indeed, any hypothesis test has four possible outcomes: our test must either accept or reject the null hypothesis, and the null hypothesis must be (actually) true or false. These are illustrated in Table 3.1.

		Truth	
		H_A	H_0
Decision	Reject H_0	Correct Decision	Type I Error
	Accept H_0	Type II Error	Correct Decision

Table 3.1: Type I and II Errors in hypothesis testing.

In particular, we have the following definition.

Definition 3.6. *We say that a hypothesis test results in a* Type I Error *whenever the decision rule falsely rejects a true null hypothesis. Similarly, we say that a hypothesis test results in a* Type II Error *whenever the decision rule accepts a false null hypothesis.*

In other words, a Type I Error occurs whenever we reject a null hypothesis that is true, and a Type II Error occur whenever we fail to reject a null hypothesis that is false. Again, Table 3.1 is a luciferous remedy should these words seem at first opaque.

Example 3.2. Our magician friend flips his coin and it lands *heads*. Using the hypothesis test from Example 3.1, we accuse him of cheating. It turns out, much to our dismay, that he in fact used an ordinary coin.

In this example, the null hypothesis was that the coin was fair (or even favored us). We falsely rejected this, concluding that his coin was weighted, making a Type I Error.

Consider now a variation: our magician friend is using a weighted coin that is more likely to turn up heads than it is tails. Despite odds being in his favor, and much to his chagrin, the coin lands *tails*. In this case, we accept the null hypothesis, that the coin is fair. However, we do so erroneously, thereby committing a Type II Error.   ▷

Note 3.3. The type of error is determined entirely upon whether or not the null hypothesis is actually true, i.e., the column of Table 3.1. If the null hypothesis is true and our conclusion is wrong, it is a Type I error. If the alternative hypothesis is true and our conclusion is wrong, it is a Type II error.   ▷

3.1.3 The Power Function

We still need a method of evaluating *how good* a particular hypothesis test is. We have an intuitive sense that the hypothesis test from Examples 3.1 and 3.2 is a poor test, but how can we quantity this? The answer lies in understanding the probability of making different types of errors. Those probabilities, however, depend on the *actual* value of the parameter θ, which is typically unknown. A useful tool for discussing these probabilities is given in our next definition.

Definition 3.7. *The* power function *of a hypothesis test with test statistic T and rejection region R is defined as the function $\beta : \Theta \to [0,1]$ given by the relation*

$$\beta(\theta) = \mathbb{P}_\theta(T(\mathcal{X}) \in R). \tag{3.1}$$

Thus, the power of a test tells us the probability of rejecting the null hypothesis as a function of the true value of parameter θ. Note that the power function is determined by the test statistic and rejection region, and is implicitly dependent on the sample size of the data but not on the data itself.

We next show how the power function is related to Type I and II Errors. First, let us assume that the null hypothesis is true; i.e., let us assume that $\theta \in \Theta_0$. In this case, the power function $\beta(\theta)$, when evaluated at the true value for θ, represents the probability of making a Type I Error.

Next, let us assume that the null hypothesis is false; i.e., let us assume that $\theta \in \Theta_0^c$. Since the power function represents the probability of rejecting the null hypothesis, the quantity $1 - \beta(\theta)$ must therefore represent the probability of accepting the null hypothesis. Thus, when $\theta \in \Theta_0^c$, one minus the power function represents the probability of making a Type II Error.

We therefore have the equality

$$\beta(\theta) = \begin{cases} \mathbb{P}(\text{Type I Error}) & \text{if } \theta \in \Theta_0 \\ 1 - \mathbb{P}(\text{Type II Error}) & \text{if } \theta \in \Theta_0^c \end{cases}.$$

Of course, the preceding equation assumes that we can actually evaluate the power function at the *true* value of the parameter θ. This is hardly the case in practice, and, as such, we must quantify the worst case scenario. Moreover, when devising tests, we do not typically prescribe the precise amount of error, but rather seek to place an upper bound on the error. This gives rise to the following definition.

Definition 3.8. *Let $\beta(\theta)$ be the power function of a hypothesis test. Then, for $0 \le \alpha \le 1$, we say that the test has* significance level α *if*

$$\sup_{\theta \in \Theta_0} \beta(\theta) \le \alpha. \tag{3.2}$$

The exact value of the supremum is referred to as the size *of the test. The number $(1 - \alpha)$ is sometimes referred to as the* confidence level *of the test.*

Note 3.4. As pointed out in Casella and Berger [2002], it is, in practice, often difficult to construct a test with an exact size. It is much simpler to require that a test is valid at a particular significance level; typical choices are $\alpha = 0.01$, 0.05, and 0.10. A hypothesis test at the 5% significance level is therefore guaranteed to have no more than a 5% probability of producing a Type I error. ▷

Note 3.5. A test with a low significance level can still have a high probability of producing a Type II Error. ▷

Example 3.3. So we've burnt through several magician friends and have decided that it's time for a modest improvement to our test. Our new test will require that we observe ten coin tosses, and if 7 or more are *heads*, we will reject the null hypothesis $H_0 : \mathbb{P}(\text{heads}) \le 0.50$. We can characterize the test statistic as $T = \sum_{i=1}^{10} X_i$, and the rejection region is $R = \{7, 8, 9, 10\}$. Of course, if the true probability of heads is p, our test statistic will have a binomial distribution; i.e., $T \sim \text{Binom}(10, p)$. Moreover, given p, the probability of observing a value within the rejection region is simply

$$\mathbb{P}_p(T \in R) = S(6) = 1 - F(6),$$

where $F(x)$ is the CDF for Binom(10, p) and $S(x) = 1 - F(x)$ is the associated survival function[2].

---

[2] Note that $\mathbb{P}_p(T \in R) = S(6)$ and not $S(7)$. Since the CDF is defined as $F(x) = \mathbb{P}(X \le x)$, its complement, the survival function, is the probability $S(x) = \mathbb{P}(X > x)$ Hence we must evaluate the survival function at 6 in order to obtain the probability that $T \in \{7, 8, 9, 10\}$.

Luckily, the plot of the survival code is easily achieved using Python, as shown in Code Block 3.1. The resulting output is shown in Figure 3.1.

```
1   p = np.linspace(0, 1, num=100)
2   beta = scipy.stats.binom.sf(6, 10, p)
3   alpha = scipy.stats.binom.sf(6, 10, 0.5)
4   plt.plot(p, beta)
5   plt.fill_between(p[:50], beta[:50], color='#1f77b4', alpha=0.5)
6   x = 0.5 * np.ones(50)
7   y = np.linspace(0, alpha)
8   plt.plot(x, y, '--', color='#1f77b4')
9   x = np.linspace(0, 0.5)
10  y = alpha * np.ones(50)
11  plt.plot(x, y, '--', color='#1f77b4')
```

Code Block 3.1: Survival code for coin-toss experiment.

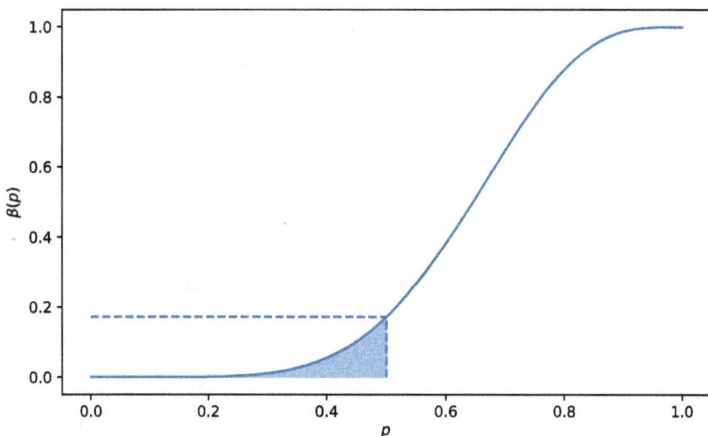

Fig. 3.1: Power function for Example 3.3.

Moreover, the size of our test is revealed by **print(alpha)**, and is (approximately) 17.2%. This point is shown with dotted lines in Figure 3.1, and the area under the power curve corresponding to our null hypothesis is shaded.

We conclude that if our null hypothesis is true, we still have up to a 17.2% probability of bearing false witness against our dear magician friend.

Moreover, it is likely that the probability of a Type I Error under the null hypothesis is exactly 17.2%, and not one of the lesser values. This is because our null hypothesis is actually a *composite hypothesis*: the null hypothesis is true if $\theta = 0.50$ (the coin is fair) *or if* $\theta < 0.50$ (the coin is weighted, but our friend is incompetent). Under the nulll hypothesis, we very well might assess that a fair coin is much more likely than a weighted coin being used the wrong way. In general, the size determines the worst-case scenario, under the null hypothesis, for the probability of committing a Type I Error. In this case, it is also the most likely value of that probability, if the null hypothesis is true.

Our magician friend proceeds to flip his coin ten times. Seven out of ten are heads. Our test says we should reject the null hypothesis, but we now know that, if the null hypothesis is true, we danger a 17.2% chance of rejecting the null hypothesis in error. *Should we reject the null hypothesis?* The answer, in this case, depends on our fondness for magicians and their scarcity.   ▷

As we might expect after examining Figure 3.1, the ideal power function will be as close as possible to zero for $\theta \in \Theta_0$ and as close as possible to one for $\theta \in \Theta_0^c$.

Essentially, the significance level is a bound we place on the probability of a Type I Error. We stress again that the power function, size, and significance level are a function of the test, not of the data. When we go to apply the test for an actual, realized set of data $\mathcal{X}_0 \in \mathbb{X}$, it is further useful to report the smallest level of significance that would result in rejection of the null hypothesis.

Definition 3.9. *Let T be a test statistic and R_c be a rejection region in standard form. For a given set of data $\mathcal{X}_0 \in \mathbb{X}$, the p-value is given by*

$$p_0 = \sup_{\theta \in \Theta_0} \mathbb{P}_\theta(T(\mathcal{X}) \in R_{c^*}), \tag{3.3}$$

where $c^* = T(\mathcal{X}_0)$.

Thus, the *p*-value is what the size of the test *would have been*, had the rejection region been chosen to be the smallest possible rejection region, consistent with the standard form, that contains the observed value of our test statistic. In other words, the *p*-value is the best-case probability under the null hypothesis of obtaining a sample at least as extreme, as measured by the test statistic T, as the data that were actually observed.

Note 3.6. Though one of the earliest and most familiar concepts the author is familiar with, we found the *p*-value difficult to define. It is often simply described as *the probability of observing data at least as extreme as your sample* and is illustrated as the tail-end probability under the distribution for the test statistic. While we admit that the definition we finally gravitated to has limits in the sense that it relies on the form of the rejection region, we feel the final result is closest to practice and therefore most intuitive.

Some authors define the p-value as any statistic $p : \mathbb{X} \to [0, 1]$ that satisfies the relation

$$\mathbb{P}_\theta(p(\mathcal{X}) \leq \alpha) \leq \alpha$$

for every $\theta \in \Theta_0$ and $\alpha \in [0, 1]$; see Casella and Berger [2002]. Thus, the p-value is immediately seen to be a test with significance level α when used in conjunction with the rejection region $R_\alpha = [0, \alpha]$. We felt that such a definition obfuscates where the p-value comes from in the first place: the original test statistic.

Other authors (Shao [2003], Wasserman [2004]) suppose that, for a given test statistic T, and for each $\alpha \in (0, 1)$, that a size α test can be constructed by choice of an appropriate rejection region R_α. They go on to define the p-value, given an observed set of data $\mathcal{X}_0 \in \mathbb{X}$, by the relation

$$p = \inf\{\alpha \in (0, 1) : T(\mathcal{X}_0) \in R_\alpha\}.$$

The p-value therefore represents the *smallest* size test that would have rejected the null hypothesis given the observed data. We were quite fond of this definition, but nevertheless felt that it suffered two shortcomings: there was no continuity in the family of rejection regions R_α and it is not always possible to construct such a region for *any* $\alpha \in (0, 1)$, in particular when the possible values of the test statistic are discrete. (Example 3.4 will constitute a counterexample to this.)

After constructing the language for Definition 3.9, our choice was later validated upon discovery that each of the above-mentioned authors ultimately came to the same formula for the special case of a standard rejection region. As this is the definition closest to practice, we ultimately decided to adopt it as our definition. We invite the reader to explore the alternative definitions and draw their own conclusions.   ▷

Example 3.4. Let us return to the hypothesis test discussed in Example 3.3. Recall that we are testing the null hypothesis $H_0 : p \leq 0.5$ by performing a sequence of ten independent coin tosses. Our test statistic is $T(\mathcal{X}) = \sum_{i=1}^{10} X_i$, the number of observed *heads*, and our rejection region is of the form $R_c = [c, \infty)$; i.e., we will reject the null hypothesis if we observe more than c heads. Furthermore, we have shown that in order to have a test at the 20% significance level, we can set $c = 7$.

Thus, our test is set: we reject the null hypothesis if we observe at least 7 out of 10 *heads*. Having a 20% significance level means that, if the coin is fair, there is (up to) a 20% probability of falsely rejecting a true null hypothesis. The significance level depends on the *test*, not the *data*.

We perform our test and observe *nine* heads! Our p-value is 1.1%. This is because, under the null hypothesis, there is only a 1.1% probability of observing a result that is so extreme, which is computed by evaluating the survival function for a binomial distribution Binom$(10, 0.5)^3$ at $x = 8$.

```
for c in range(11):
    alpha = scipy.stats.binom.sf(c-1, 10, 0.5)
    print (c, round(alpha, 3))
```

Code Block 3.2: Generate all possible p-values of coin-toss experiment.

$\sum_{i=1}^{10} X_i$	p-value
0	1.000
1	0.999
2	0.989
3	0.945
4	0.828
5	0.623
6	0.377
7	0.172
8	0.055
9	0.011
10	0.001

Table 3.2: p-values for coin-toss experiment

In fact, since there are only a handful of possible outcomes of the test statistic T, it is not difficult to compute a p-value for all of them, as shown in Code Block 3.2. The output is given in Table 3.2.   ▷

3.1.4 Confidence Intervals

Hypothesis testing is closely related to the topic of confidence intervals, which we now define.

Definition 3.10. *Given two statistics* $L, U : \mathbb{X} \to \mathbb{R}$ *satisfying* $L(\mathcal{X}) \le U(\mathcal{X})$ *for all* $\mathcal{X} \in \mathbb{X}$, *we say that the random interval* $[L(\mathcal{X}), U(\mathcal{X})]$ *is a* $(1-\alpha)$-*confidence interval for the parameter* $\theta \in \Theta$ *if it satisfies the relation*

$$\inf_{\theta \in \Theta} \mathbb{P}_\theta(\theta \in [L(\mathcal{X}), U(\mathcal{X})]) = 1 - \alpha. \tag{3.4}$$

Note 3.7. In Definition 3.10, the parameter θ is a *fixed* albeit unknown quantity; it is the interval $[L(\mathcal{X}), U(\mathcal{X})]$ that is random. Hence, the quantity

---

[3] Using a probability of $p = 0.5$ maximizes the probability of a given number of *heads* out of all $p \in \Theta_0 = [0, 0.5]$.

$$\mathbb{P}_\theta(\theta \in [L(\mathcal{X}), U(\mathcal{X})])$$

should be interpreted as the probability, given parameter θ, that the statistics L and U satisfy $L(\mathcal{X}) \leq \theta$ and $U(\mathcal{X}) \geq \theta$.   ▷

In other words, a 95% confidence interval means that, regardless the true value of the parameter θ, as least 95% of samples $\mathcal{X}$ from our distribution will produce correct upper and lower bounds on the parameter θ. Since the parameter θ is *fixed*, and not *random*, this is *not* a probability statement on θ.

Confidence intervals are oftentimes closely connected with hypothesis tests, in that each implies each other. To see this, consider a simple hypothesis of the form $H_0 : \theta = \theta_0$. Suppose a size α test is devised with test statistic T and rejection region $R_c = \{t : |t| \geq c\}$. The *acceptance region* is therefore given by $A_c = R_c^c = (-c, c)$, so that we accept the null hypothesis whenever $-c < T(\mathcal{X}) < c$. Now, since the test is size α, and since there is but a single point in Θ_0, Equation (3.2) implies that

$$\mathbb{P}_\theta(T(\mathcal{X}) \in R_c) = \alpha,$$

and hence

$$\mathbb{P}_\theta(T(\mathcal{X}) \in A_c) = 1 - \alpha.$$

It is often the case (as we shall shortly see) that the test statistic depends on the value θ_0 and, moreover, that this relation can be inverted. This inversion creates a correspondence

$$-c < T(\mathcal{X}; \theta_0) < c \qquad \text{if and only if} \qquad L(\mathcal{X}) < \theta_0 < U(\mathcal{X}),$$

for some functions L and U. Thus, under the above assumptions, a size α hypothesis test is often equivalent to a $(1 - \alpha)$-confidence interval.

3.2 Tests for Estimates

In this section we discuss various tests related to the expected value of a random variable. Additionally, we hope to further illustrate each of the concepts in Section 3.1.

3.2.1 The Mean of a Normal Random Variable

We begin with a test for the mean of a normal random variable. We will treat the cases of known and unknown variance separately.

Proposition 3.1 (z-test). *Let* $X_1, \ldots, X_n \sim N(\mu, \sigma^2)$, *with known variance* σ^2. *The hypothesis test consisting of null hypothesis*

$$H_0 : \mu = \mu_0,$$

test statistic

$$Z_n = \frac{\overline{X}_n - \mu_0}{\sigma/\sqrt{n}}, \tag{3.5}$$

and rejection region

$$R_\alpha = \{z : |z| \geq z_{\alpha/2}\},$$

where $z_{\alpha/2}$ is the solution to $\Psi(z_{\alpha/2}) = \alpha/2$, as given by Definition 2.11, constitutes a size α test.

Moreover, given an observed sample X and test statistic $Z_n = z$, the p-value of the test for the given data is given by $2\Psi(|z|)$.

Proof. Under the null hypothesis $\mu = \mu_0$, the sample mean has distribution $\overline{X}_n \sim N(\mu_0, \sigma^2/n)$, according to Corollary 2.2. Therefore, our test statistic given by Equation (3.5) is a standard normal random variable, $Z_n \sim N(0,1)$, according to Corollary 2.1.

The size of our test is therefore given by

$$\mathbb{P}_{\mu_0}(Z_n \in R_\alpha) = \alpha.$$

This follows due to symmetry of the two tails: $\mathbb{P}_{\mu_0}(Z_n \geq z_{\alpha/2}) = \alpha/2$ and $\mathbb{P}_{\mu_0}(Z_n \leq -z_{\alpha/2}) = \alpha/2$.

For a given, observed value of the test statistic $Z_n = z$, the p-value is the probability, under the null hypothesis, of observing a result as extreme. Following Definition 3.9, the p-value is given by

$$p_0 = p_{\mu_0}(|Z_n| > |z|) = \mathbb{P}(Z > |z|) + \mathbb{P}(-Z < -|z|) = 2\mathbb{P}(Z > |z|),$$

for $Z \sim N(0,1)$. The result follows since $\Psi(z) = \mathbb{P}(Z > z)$. □

Note 3.8. A straightforward variation of Proposition 3.1 is the one-sided test, with null hypothesis

$$H_0 : \mu \leq \mu_0,$$

same test statistic Equation (3.5), and rejection region

$$R_\alpha = [z_\alpha, \infty).$$

We would therefore reject the null hypothesis if $Z_n \geq z_\alpha$. Since this is a one-sided test, the full probability α of rejection must lie on one tail and not be divided among two tails. To show that this is a size α test, we must show that

$$\sup_{\mu \leq \mu_0} \mathbb{P}_\mu(Z_n \geq z_\alpha) = \alpha.$$

However, this probability is maximized by taking $\mu = \mu_0$, and so the result follows. (If the true value of μ is less than μ_0, it would shift the true probability distribution to the left, thereby decreasing the residual probability in the *right* tail $Z_n \geq z_\alpha$.)

Similarly, for the one-sided null hypothesis $H_0 : \mu \geq \mu_0$, we would use $R_\alpha = (-\infty, -z_\alpha]$ to obtain a size α test.   ▷

We stated previously that under normal circumstances there is a correspondence between hypothesis tests and confidence intervals. The confidence interval corresponding to the z-test is given below.

Proposition 3.2. *Let* $X_1, \ldots, X_n \sim N(\mu, \sigma^2)$, *with known variance* σ^2. *Then a* $(1 - \alpha)$ *confidence interval for the population mean* μ_0 *is given by the interval*

$$I_n = \left(\overline{X}_n - \frac{\sigma z_{\alpha/2}}{\sqrt{n}}, \overline{X}_n + \frac{\sigma z_{\alpha/2}}{\sqrt{n}} \right). \tag{3.6}$$

Proof. The acceptance region of the z-test is given by

$$A_\alpha = \left(-z_{\alpha/2}, z_{\alpha/2} \right),$$

so that the test statistic Z_n has a $(1 - \alpha)$ of being found within A_α. It is straightforward to show that

$$-z_{\alpha/2} < \frac{\overline{X}_n - \mu_0}{\sigma/\sqrt{n}} < z_{\alpha/2}$$

if and only if

$$\overline{X}_n - \frac{\sigma z_{\alpha/2}}{\sqrt{n}} < \mu_0 < \overline{X}_n + \frac{\sigma z_{\alpha/2}}{\sqrt{n}},$$

which completes the result.   □

If the population variance is unknown, we can use the next best thing: the sample variance. The result is given below.

Proposition 3.3 (t-test). *Let* $X_1, \ldots, X_n \sim N(\mu, \sigma^2)$, *with unknown variance* σ^2. *The hypothesis test consisting of null hypothesis*

$$H_0 : \mu = \mu_0,$$

test statistic

$$T_n = \frac{\overline{X}_n - \mu_0}{S_n/\sqrt{n}}, \tag{3.7}$$

and rejection region

$$R_\alpha = \{t : |t| \geq t_{n-1,\alpha/2}\},$$

where $t_{n-1,\alpha/2}$ *is the solution to* $S(t_{n-1,\alpha/2}) = \alpha/2$, *where S is the survival function of a* t_{n-1} *random variable, constitutes a size* α *test.*

Proof. Under the null hypothesis, the test statistic Equation (3.7) is a t_{n-1} random variable due to Proposition 2.21. The result follows.   □

Note 3.9. The one-sided variations of this test correspond to the null hypothesis $H_0 : \mu \geq \mu_0$ with rejection region $R_\alpha = (-\infty, -t_{n-1,\alpha}]$ and null hypothesis $H_0 : \mu \leq \mu_0$ with rejection region $R_\alpha = [t_{n-1,\alpha}, \infty)$.   ▷

Note 3.10. It is helpful to recall that the t_n distribution looks much like a standard normal distribution with enlarged tails, and that $t_n \to N(0, 1)$ as $n \to \infty$. Thus, for large sample sizes, say $n > 30$, the t-test given by Proposition 3.3 will not yield a different result than the z-test of Proposition 3.1. Thus, the t-test is best suited for dealing with inferences from samples from a normally distributed population with unknown variance and small sample size. ▷

Example 3.5. The difference between the z-test and t-test is relevant largely for small samples sizes. In this example, we verify the validity of the t-test over the z-test when constructing the test statistic using the sample variance. In particular, suppose our true distribution is $N(3, 1)$, and consider the null hypothesis $\mu_0 = 3$. Because we are constructing this numerical simulation, we know that the null hypothesis is in fact true. What we want to capture is, therefore, the rate that we are wrong if we were to use either test; i.e., an incorrect application of the z-test vs. a correct application of the t-test. The code is shown in Code Block 3.3.

```
mu_0 = 3
var = 1
n_samples = 10
alpha = 0.05
z_crit = scipy.stats.norm.isf(alpha/2) #1.960
t_crit = scipy.stats.t.isf(alpha/2, n_samples-1) # 2.262
significance_z_crit = 2 * scipy.stats.t.sf(z_crit, 9) # 8.165%

n_trials = 10000
rejections_z = np.zeros(n_trials)
rejections_t = np.zeros(n_trials)
p_values = np.zeros(n_trials)
for i in range(n_trials):
    samples = np.random.normal(mu, scale=var, size=n_samples)
    sample_mean = samples.mean()
    sample_var = samples.var() * n_samples / (n_samples - 1)
    T = (sample_mean - mu_0) / (np.sqrt(sample_var) /
            np.sqrt(n_samples))
    rejections_z[i] = int( abs(T) > z_crit )
    rejections_t[i] = int( abs(T) > t_crit )
    p_values[i] = 2 * scipy.stats.t.sf(abs(T), 9)

print ( 'Z_crit Rejection Rate', rejections_z.mean() ) # 0.0834
print ( 'T_crit Rejection Rate', rejections_t.mean() ) # 0.0505
plt.hist(p_values, bins=100, normed=True)
```

Code Block 3.3: t-test and z-test using sample variance

First, we define our (true) mean, variance, sample size (10), and significance level $\alpha = 0.05$ on lines 1–4. Then we compute the critical value $z_{0.025} \approx 1.960$ (incorrect) and $t_{9,0.025} \approx 2.262$ (correct), as shown in lines 5 and 6. Using the critical value for the t-distribution is correct since we will be computing our test statistic using the sample variance, as if the population variance were unknown. By incorrectly using the critical value $z_{0.025}$, we therefore expect an *actual* significance of approximately 8.165% (line 7). If our theory is correct, that our test statistic satisfies a t-distribution as opposed to a normal distribution, we should expect to see approximately 5% rejections when using the critical value $t_{9,0.025}$ and approximately 8.165% rejections when using the critical value $z_{0.025}$.

Next, we simulate 10,000 random samples, each sample having size $n = 10$, and compute the test statistic T_n (line 17) for each sample. We then record our decision on whether or not to reject the null hypothesis using either criterion (lines 18 and 19). As an added bonus, we record the observed p-value for each sample (line 20).

The results are given in lines 22 and 23. If we incorrectly apply the critical value $z_{0.025}$, we actually observe 8.34% false rejections, as opposed to the observed 5.05% false rejection rate using the correct value $t_{9,0.025}$. (These numbers will vary slightly, of course, each time we run the simulation.)

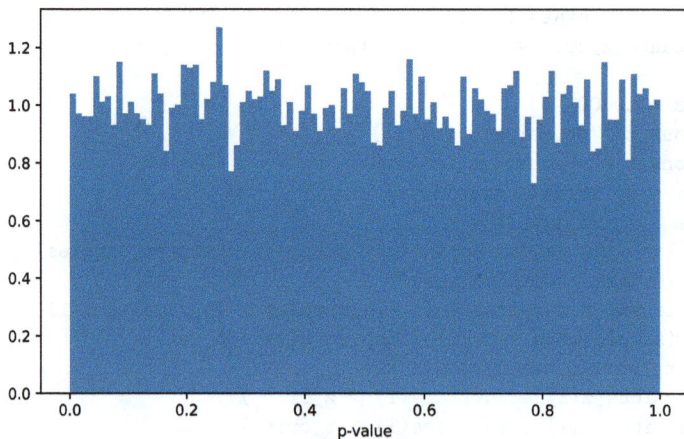

Fig. 3.2: Distribution of p-values for Example 3.5.

Finally, we can plot the distribution of our p-values, which is shown in Figure 3.2. The result is as expected: under the null hypothesis, the distribution of p-values is approximately Unif(0, 1). Our simulated data results in a distribution fairly close to this.   ▷

Note 3.11. It is true in general that whenever our test statistic has a continuous distribution, the distribution of p-values, under the null hypothesis, is Unif$(0, 1)$. Therefore, a test of the form p-value $< \alpha$ results in a size-α test. ▷

Note 3.12. Figure 3.2 further illustrates the distinction between p-value and significance level. The p-value is a measure of how unlikely a particular sample is, given the null hypothesis, and not a statement as to the Type I Error rate. For example, suppose our significance level was 20%. We would therefore reject each sample with a p-value less than 0.20. If the null hypothesis is true, we will falsely reject 20% of the samples. But, in practice, *we only get to observe one sample!* Suppose its p-value is 0.005. This does not mean that our test has a Type I Error rate of 0.5%! Our test has a Type I Error rate of 20%, because we *would have rejected* any p-value less than 0.20. It simply means that our *actual* observed sample is quite rare if the null hypothesis is true. In fact, the given sample of data *would have been rejected,* even by a test with a 1% significance level. ▷

3.2.2 The Wald Test

We have previously seen tests for comparing the sample mean from a normally distributed population against a hypothesized value for the population mean. Of course, when the population is not normally distributed, we may still use this test as an approximation, due to the central limit theorem. The result is known as the Wald test, which is stated below.

Definition 3.11 (Wald Test). *Let $\hat{\theta}_n$ be an asymptotically normal and unbiased estimator for a parameter $\theta \in \Theta$ and let $\mathrm{se}(\hat{\theta}_n)$ be the standard error. Then the size α Wald test consists of the null hypothesis*

$$H_0 : \theta = \theta_0,$$

the test statistic

$$Z_n = \frac{\hat{\theta}_n - \theta_0}{\mathrm{se}(\hat{\theta}_n)}, \tag{3.8}$$

and the rejection region

$$R_\alpha = \{Z_n : |Z_n| \geq z_{\alpha/2}\},$$

where $z_{\alpha/2}$ is the solution to $\Psi(z_{\alpha/2}) = \alpha/2$, as given by Definition 2.11. The function $W_n : \mathbb{X} \to \mathbb{R}$ given by $W_n = Z_n^2$ is known as the Wald statistic.

Theorem 3.1. *The size of a size-α Wald test converges to α, i.e.,*

$$\lim_{n \to \infty} \mathbb{P}_{\theta_0}(|Z_n| \geq z_{\alpha/2}) = \alpha.$$

Moreover, the result is still valid if we replace $\mathrm{se}(\hat{\theta}_n)$ in Z_n with any asymptotically unbiased estimate $\hat{\mathrm{se}}(\hat{\theta}_n)$.

Proof. Under the null hypothesis $\theta = \theta_0$, the statistic $Z_n = (\hat{\theta}_n - \theta_0)/\text{se}$ converges to a standard normal random variable $Z \sim N(0,1)$ as $n \to \infty$. Therefore, the probability of rejecting a true null hypothesis is given by

$$\mathbb{P}_{\theta_0}(|Z_n| > z_{\alpha/2}) = \mathbb{P}_{\theta_0}\left(\frac{|\hat{\theta}_n - \theta_0|}{\text{se}} > z_{\alpha/2}\right) \to \mathbb{P}(|Z| > z_{\alpha/2}) = \alpha.$$

Moreover, if $\hat{\text{se}}/\text{se} \to 1$ in probability, then, by Slutsky's theorem, it follows that

$$Z_n = \frac{\hat{\theta}_n - \theta_0}{\hat{\text{se}}} = \frac{\hat{\theta}_n - \theta_0}{\text{se}} \cdot \frac{\text{se}}{\hat{\text{se}}} \to N(0,1).$$

We therefore obtain the same result. □

We mentioned in the discussion following Definition 3.10 that there is usually a correspondence between hypothesis tests and confidence intervals. We next turn to our first example of such a correspondence.

Corollary 3.1. *The Wald test rejects the null hypothesis if and only if the true parameter θ_0 is not contained within the interval*

$$I_n = (\hat{\theta}_n - \hat{\text{se}}\, z_{\alpha/2}, \hat{\theta}_n + \hat{\text{se}}\, z_{\alpha/2}). \tag{3.9}$$

Moreover, the interval I_n is an asymptotic $(1 - \alpha)$ confidence interval as $n \to \infty$.

Proof. A straightforward calculation shows that Z_n is in the acceptance region

$$-z_{\alpha/2} < \frac{\hat{\theta}_n - \theta_0}{\hat{\text{se}}} < z_{\alpha/2} \qquad \text{if and only if} \qquad \hat{\theta}_n - \hat{\text{se}}\, z_{\alpha/2} < \theta_0 < \hat{\theta}_n + \hat{\text{se}}\, z_{\alpha/2}.$$

The first inequality involves the test statistic whereas the second inequality involves the true value of the parameter. Since the Wald statistic is an asymptotic size-α test, it follows that there is a $(1 - \alpha)$ probability of being within the acceptance interval as $n \to \infty$, which validates the claim. □

A straightforward generalization of the Wald statistic to higher dimensions is given in the following.

Definition 3.12 (Multivariate Wald Test). *Let $\hat{\theta}_n$ be an asymptotically normal and unbiased estimator for a k-dimensional parameter $\theta \in \Theta \subset \mathbb{R}^k$. Then the size α Wald test consists of the null hypothesis*

$$H_0 : \theta = \theta_0,$$

the test statistic

$$W_n = (\hat{\theta}_n - \theta_0)^T \mathbb{V}(\hat{\theta}_n)^{-1}(\hat{\theta}_n - \theta_0), \tag{3.10}$$

and the rejection region

$$R_\alpha = [\chi^2_{k,\alpha}, \infty),$$

where $\chi^2_{k,\alpha}$ is the solution to $S(\chi^2_{k,\alpha}) = \alpha$, where S is the survival function for a χ^2_k random variable. The function $W_n : \mathbb{X} \to \mathbb{R}$ is known as the Wald *statistic.*

Note 3.13. Definition 3.12 is consistent with Definition 3.11 since, for the special case of $k = 1$, we have

$$|Z| > z_{\alpha/2} \qquad \text{if and only if} \qquad W > \chi^2_{1,\alpha},$$

where $W = Z^2$. This follows since the square of a standard normal random variable is a chi-square random variable with one degrees of freedom. Moreover, the inequality $|Z| > z_{\alpha/2}$ captures $\alpha/2$ probability under $Z > z_{\alpha/2}$ and an additional $\alpha/2$ probability under $Z < -z_{\alpha/2}$, for a total probability of rejection equal to α. Since $W = Z^2$, we only have to worry about $W > \chi^2_{1,\alpha}$, which captures the full α amount of probability.   ▷

Theorem 3.2. *The size of a size-α multivariate Wald test converges to α, i.e.,*

$$\lim_{n\to\infty} \mathbb{P}_{\theta_0}(W_n \geq \chi^2_{k,\alpha}) = \alpha.$$

Moreover, the result is still valid if we replace $\mathbb{V}(\hat\theta_n)$ in Z_n with any asymptotically unbiased estimate $\hat{\mathbb{V}}(\hat\theta_n)$.

Proof. The result follows as long as $W_n \to \chi^2_k$ as $n \to \infty$. However, since the estimator is asymptotically normal and unbiased, it follows that, under the null hypothesis, $\hat\theta_n \to N(\theta_0, \Sigma)$, for some limiting variance Σ. The Wald statistic therefore approaches the Mahalanobis distance, $W_n \to d_\Sigma(\hat\theta_n, \theta_0)$ as $n \to \infty$, which follows a χ^2_k distribution due to Corollary 2.3. This proves the result.   □

3.2.3 Test for Proportions

Our next test is concerned with inference regarding proportions. In this context, our sample is a sample of Bernoulli random variables $X_1, \ldots, X_n \sim$ Bern(p). Of course, such a sample can simply be described using a single binomial random variable.

Proposition 3.4. *Let $X \sim$ Binom(n, p) be a binomial random variable, and consider the hypothesis test consisting of the null hypothesis*

$$H_0 : p = p_0,$$

the test statistic

$$Z_n = \frac{X/n - p_0}{\sqrt{p_0(1 - p_0)/n}}, \tag{3.11}$$

and the rejection region

$$R_\alpha = \{z : |z| \geq z_{\alpha/2}\}.$$

The size of this hypothesis test approaches α as $n \to \infty$. Moreover, for large n, the size of this test is approximately α.

Essentially, Proposition 3.4 follows from the normal approximation to the binomial distribution; i.e., from a combination of the central-limit theorem and Proposition 3.1. It further constitutes an application of the Wald test, as $\overline{X}_n = X/n$ is simply the sample mean of n Bernoulli random variables, and, under the null hypothesis, $\mathbb{V}(\overline{X}_n) = p_0(1 - p_0)/n$.

One method for constructing a $(1 - \alpha)$ confidence interval for the true probability p_0 is to approximate the standard error $\text{se}(X/n) = p_0(1 - p_0)/n$ with the estimated standard error $\hat{\text{se}}(X/n) = \hat{p}(1 - \hat{p})/n$, where we define $\hat{p} = X/n$. This yields the confidence interval for p_0:

$$p_0 \in \left(\hat{p} - z_{\alpha/2}\sqrt{\hat{p}(1 - \hat{p})/n}, \hat{p} + z_{\alpha/2}\sqrt{\hat{p}(1 - \hat{p})/n}\right).$$

3.2.4 Power and Sample Size

We have thus far addressed the measurement of Type I Errors through devising tests with a fixed significance level and by measuring p-values. We have yet to discuss how to control for Type II Errors.

In particular, we have seen that by an appropriately selected rejection region, we can control the rate of false rejections, if the null hypothesis is true. So in order to control for Type I errors, we have specified a test statistic and a rejection region. What other lever do we have to further affect the likelihood of a Type II Error? The answer lies in the ability to specify a minimum required sample size for our test: the more data we collect, the more powerful our test will be.

Before proceeding, a slight shift in perspective is required. Instead of minimizing the Type II Error rate, we will seek to maximize the power of the test under the *alternative hypothesis*. Recall that the probability of a Type II Error is the probability of accepting a false null hypothesis. This is logically equivalent to 1 minus the probability of correctly rejecting a false null hypothesis. This probability is often referred to as the *power* of a test. Thus, we seek to maximize the probability of detecting an effect when one is there.

There is, however, one complication with this approach, in that the power function varies across the parameter space θ. For bounding the probability of a Type I Error, we simply found the maximum value of the power function that was consistent with the null hypothesis. We cannot take the same approach when discussing the worst-case power under the alternative hypothesis, since the minimum power consistent with the alternative hypothesis is equivalent to the maximum power consistent with the null hypothesis, whenever the power function is continuous. This should be clear in Figure 3.1. To remedy this, we instead talk about the power of a test relative to some minimum detectable effect size. Referring back to Figure 3.1, we could argue that we have at least a 60% probability of correctly rejecting the null hypothesis, as long as the true probability is greater than 70%[4].

Power and Minimum Detectable Effect

In order to apply this to test design, we will invert the problem into the question: what is the minimum sample size needed to guarantee a certain power level for a minimum detectable effect. Formally, we proceed as follows.

Definition 3.13. *Given any metric space Θ and subspace $\Theta_0 \subset \Theta$, we define the* distance *between a point $\theta \in \Theta$ and the subspace Θ_0 as*

$$d(\theta, \Theta_0) = \inf_{\psi \in \Theta_0} ||\theta - \psi||.$$

In higher-dimensional spaces, distance is often specified relative to one of several norms.

Definition 3.14. *For $x \in \mathbb{R}^n$, common norms are the ℓ_1-norm (Manhattan norm)*

$$||x||_1 = \sum_{i=1}^{n} |x_i|,$$

the ℓ_2 norm (Euclidean norm)

$$||x||_2 = \sqrt{\sum_{i=1}^{n} x_i^2},$$

and the ℓ_∞ norm (infinity norm)

$$||x||_\infty = \max(|x_1|, \ldots, |x_n|).$$

Definition 3.15. *The* minimum detectable effect *$\delta > 0$ is a positive number that defines the* detectable effect space *$\Theta_\delta \subset \Theta$ as*

$$\Theta_\delta = \{\theta \in \Theta : d(\theta, \Theta_0) \geq \delta\}. \tag{3.12}$$

Note 3.14. The subspace Θ_δ is a subset of the alternative hypothesis, i.e., $\Theta_\delta \subset \Theta_0^c$, since $\delta > 0$ is positive and since $d(\theta, \Theta_0) = 0$ whenever $\theta \in \Theta_0$. ▷

---

[4] Note that this is well into the alternative-hypothesis territory.

Note 3.15. Unlike the size or significance level, the minimum detectable effect is not a function of the test alone, but rather a number that can be prescribed freely in conjunction with a test.   ▷

Definition 3.16. *The* power *of a hypothesis test with minimum detectable effect δ is given by*

$$\beta = \inf_{\theta \in \Theta_\delta} \beta(\theta). \tag{3.13}$$

Thus, given a minimum detectable effect, Equation (3.13) places a lower bound on the power of our test; i.e., it tells us the worst-case scenario probability of correctly rejecting a false null hypothesis at a given minimum detectable effect.

Note 3.16. In general, the power function for a particular test increases with distance to the null hypothesis space Θ_0. This makes sense, since the farther away our parameter is from the null hypothesis space, the more probable it will be that we reject the null hypothesis. Therefore, in practice, Equation (3.13) is typically evaluated for the $\theta \in \Theta_\delta$ that is closest to Θ_0.
▷

To see the relation between power, minimum detectable effect, and sample size, consider the case of a one sided null hypothesis $H_0 : \theta \le \theta_0$ and minimum detectable effect δ. Let us further consider the case where the sampling distribution for an estimator $\hat{\theta}$ is normal. The most extreme sampling distribution under the null hypothesis is therefore $N(\theta_0, \sigma^2/n)$, whereas the most extreme sampling distribution under the hypothesis of minimum effect $H_\delta : \theta \ge \theta_0 + \delta$ is $N(\theta_0 + \delta, \sigma^2/n)$. The larger the sample size n, the more separated these two distributions become, and the more likely it is to detect an effect of size δ or larger.

Note 3.17. Some typical power levels for hypothesis tests are $\beta = 0.8$, 0.9, or 0.95. For example, a test with minimum detectable effect δ and power level $\beta = 0.8$ has at least an 80% probability of rejecting the null hypothesis if the null hypothesis is false and the effect size is at least δ.   ▷

Note 3.18. The minimum detectable effect is actually a bit of a misnomer: it does *not* mean that we cannot detect smaller effects. Rather, it is the minimum required effect that would need to be present to guarantee that our test has the desired power.   ▷

Power and the Coin Toss Example

To understand this in further detail, let us return to the example of the coin toss. We will assume that our sample size, once computed, is large enough to justify the normal approximation to the binomial distribution. We state our result in the following.

Theorem 3.3. *Let $X \sim \text{Binom}(n, p)$ be a binomial random variable with unknown probability p, and consider the test consisting of the single-sided null hypothesis $H_0 : p \leq p_0$, test statistic $\hat{p} = X/n$, and approximate rejection region*

$$R_\alpha = \left[p_0 + z_\alpha \sqrt{\frac{p_0(1-p_0)}{n}}, 1 \right]. \tag{3.14}$$

Then the minimum sample size n required to achieve a power β test at minimum detectable effect δ is given by

$$n = \left(\frac{z_\alpha \sqrt{p_0(1-p_0)} - z_\beta \sqrt{p_\delta(1-p_\delta)}}{\delta} \right)^2, \tag{3.15}$$

where $p_\delta = p_0 + \delta$. The result holds as long as the sample size is large enough to justify the normal approximation at p_0 and p_δ; i.e., as long as $p_0 n, (1-p_0)n, p_\delta n, (1-p_\delta)n \geq 5$.

Proof. Under the null hypothesis, we have that $p \leq p_0$. In order to construct the rejection region for a size α test, we set $p = p_0$ and make the assumption that our sample size is large enough to justify the normal approximation $\hat{p} \sim N(p_0, p_0(1-p_0)/n)$. The solution to $\Psi(z) = \alpha$ is z_α, so that a total of α probability lives in the tail $z \geq z_\alpha$. This yields the rejection region given by Equation (3.14). In particular, define the critical value

$$p_c = p_0 + z_\alpha \sqrt{\frac{p_0(1-p_0)}{n}},$$

so that we reject the null hypothesis whenever $\hat{p} \geq p_c$.

Next, consider the hypothesis of minimum detectable effect, $p \geq p_0 + \delta$. The worst-case scenario here is the border case $p = p_0 + \delta$. Let us now assume that the null hypothesis is false, and that the effect size is δ. Hence, the normal approximation will yield $\hat{p} \sim N(p_\delta, p_\delta(1-p_\delta)/n)$, where we define $p_\delta = p_0 + \delta$. Hence, the statistic

$$Z(\hat{p}) = \frac{\hat{p} - p_\delta}{\sqrt{p_\delta(1-p_\delta)/n}}$$

is approximately a standard normal distribution.

Our next step is to ensure that under the hypothesis of minimum detectable effect, there is at least a probability β of rejecting the null hypothesis. The null hypothesis, however, is rejected whenever $Z \geq Z(p_c)$. The solution to $\mathbb{P}(Z \geq Z(p_c)) = \Psi(Z(p_c)) = \beta_0$ is simply $Z(p_c) = z_\beta$. Hence, we obtain

$$\frac{p_0 + z_\alpha \sqrt{\frac{p_0(1-p_0)}{n}} - p_\delta}{\sqrt{p_\delta(1-p_\delta)/n}} = z_\beta,$$

which is equivalent to

$$z_\alpha \sqrt{p_0(1 - p_0)} - \delta\sqrt{n} = z_\beta\sqrt{p_\delta(1 - p_\delta)}.$$

Solving for n yields the result.   □

Example 3.6. Let us return to our magician friend from Example 3.4. We will retain the null hypothesis $H_0 : p \leq 0.5$ and test statistic $\hat{p}$, representing the frequency of observed *heads*. Let us devise a test with significance level $\alpha = 0.05$ and power $\beta = 0.80$ with minimum detectable effect $\delta = 0.20$. In other words, we desire a test that will produce a false positive 5% of the time when the coin is fair, and will correctly reject the null hypothesis 80% of the time, as long as the coin is biased by at least 20%, i.e., as long as the true probability of heads is at least 70%.

From Equation (3.15), our minimum sample size is $n = 37$, and from Equation (3.14), our critical point is $p_c = 0.635$. Thus, we will require 37 coin tosses, and we will reject the null hypothesis whenever we observer greater than 24 heads out of 37.

The extreme cases of the null hypothesis and hypothesis of minimum detectable effect are shown in Figure 3.3. The blue distribution is the sampling distribution for $\hat{p}$ under the edge case for the null hypothesis, $p = 0.50$. The shaded region is our $\alpha = 0.05$ significance rejection region; i.e., we reject the null hypothesis whenever $\hat{p} \geq 0.635$. The corresponding distribution under the edge case $p = 0.7$ for the hypothesis of minimum detectable effect is shown in orange; the shaded region constitutes $\beta = 0.80$ of the total probability. The code to produce Figure 3.3 is shown in Code Block 3.4.

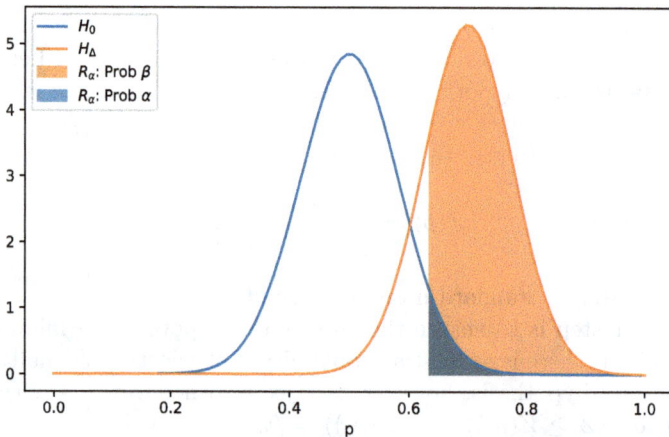

Fig. 3.3: Illustration of Theorem 3.3 for $p_0 = 0.5$, $\delta = 0.2$, $\alpha = 0.05$, and $\beta = 0.80$.

```
1  alpha = 0.05
2  beta = 0.8
3  p0 = 0.5
4  delta = 0.2
5  p1 = p0 + delta
6  za = scipy.stats.norm.isf(alpha)
7  zb = scipy.stats.norm.isf(beta)
8  n = np.ceil(((za*np.sqrt(p0*(1-p0)) - zb*np.sqrt(p1*(1-p1))) /
       delta)**2 )
9  p_crit = p0 + za*np.sqrt(p0*(1-p0)/n) # critical point
10 x = np.linspace(0, 1, num=100)
11 xr = np.linspace(p_crit, 1, num=50)
12 f_0 = scipy.stats.norm.pdf(x, p0, np.sqrt(p0*(1-p0)/n))
13 f_1 = scipy.stats.norm.pdf(x, p1, np.sqrt(p1*(1-p1)/n))
14 f_0r = scipy.stats.norm.pdf(xr, p0, np.sqrt(p0*(1-p0)/n))
15 f_1r = scipy.stats.norm.pdf(xr, p1, np.sqrt(p1*(1-p1)/n))
16 plt.plot(x, f_0, color='#1f77b4')
17 plt.plot(x, f_1, color='#ff7f0e')
18 plt.fill_between(xr, f_1r, color='#ff7f0e', alpha=0.7)
19 plt.fill_between(xr, f_0r, color='#1f77b4', alpha=0.7)
```

Code Block 3.4: Code to produce Figure 3.3.

Finally, we can verify our normal approximation by running a simulation. This is achieved by running Code Block 3.5 following Code Block 3.4. We ran the simulation twice, and observed rejection rates of 0.8000 and

```
1  n_trials = 10000
2  rejections = np.zeros(n_trials)
3  for i in range(n_trials):
4      p_hat = np.random.binomial(n, p1) / n
5      rejections[i] = int(p_hat > p_crit)
6
7  print('Rejection Rate', np.sum(rejections) / n_trials)
```

Code Block 3.5: Simulation of rejection rate under H_δ.

0.8039. This validates our theory that a sample size of 37 is required in order to correctly reject the null hypothesis 80% of the time, when the true probability of a heads is at least 70%.

This exercise was further repeated for various minimum detectable effects, holding the null hypothesis $p \leq 0.5$, significance level $\alpha = 0.05$, and power $\beta = 0.80$ fixed. The result is shown in Table 3.3.   ▷

δ	p_c	n
0.01	0.507	15,455
0.02	0.513	3,863
0.05	0.533	617
0.10	0.566	153
0.15	0.600	67
0.20	0.635	37

Table 3.3: Critical point and sample size for various minimum detectable effects; $p_0 = 0.50$, $\alpha = 0.05$, and $\beta = 0.80$.

Power and the Z-Test

Next, we turn to a power analysis of the classic z-test, as given in Proposition 3.1. This is relevant not only for the z-test, but also for similar tests, like the t-test, when there is sufficient sample size to approximate the distribution of the test statistic using the standard normal distribution.

We will consider the test statistic $Z = (\overline{X}_n - \mu_0)/(\sigma/\sqrt{n})$, as defined in Equation (3.5), under three different null hypotheses: a simple two-sided null hypothesis $H_0^2 : \mu = \mu_0$, with rejection region

$$R_\alpha^2 = \{z : |z| \geq z_{\alpha/2}\}$$

and two one-sided null hypotheses, a right-tailed and a left-tailed test. For the right-tailed test, we will consider the null hypothesis $H_0^R : \mu \leq \mu_0$, with corresponding rejection region

$$R_\alpha^R = \{z : z \geq z_\alpha\}.$$

Similarly, for the left-tailed test, will will consider the null hypothesis $H_0^L : \mu \geq \mu_0$, with corresponding rejection region

$$R_\alpha^L = \{z : z \leq -z_\alpha\}.$$

For each of these, the superscript refers to the corresponding test: superscript 2 for the two-sided test, and superscripts L and R for the left-tailed and right-tailed tests, respectively. We will follow the convention that the directionality of the tail (i.e., *left*- or *right*-tailed) follows the direction of the corresponding rejection region.

Next, let us determine the power function $\beta(\mu; n)$, as a function of the hidden true value μ and the sample size n. Recall that

$$\beta(\mu; n) = \mathbb{P}_\mu(Z \in R).$$

We must therefore obtain the distribution for Z if the true value of the parameter is actually μ. If the true parameter value is μ, then the sample mean will be distributed as

$$\overline{X}_n \sim \mathrm{N}\left(\mu, \frac{\sigma^2}{n}\right).$$

It follows that our test statistic is distributed as

$$Z \sim \mathrm{N}\left(\frac{\mu - \mu_0}{\sigma/\sqrt{n}}, 1\right).$$

For a right- and left-sided test, we therefore have

$$\beta_\alpha^R(\mu; n) = \mathbb{P}\left(Z \geq z_\alpha \,\middle|\, Z \sim \mathrm{N}\left(\frac{\mu - \mu_0}{\sigma/\sqrt{n}}, 1\right)\right)$$

$$= \Psi\left(z_\alpha - \frac{\mu - \mu_0}{\sigma/\sqrt{n}}\right) \tag{3.16}$$

$$\beta_\alpha^L(\mu; n) = \mathbb{P}\left(Z \leq z_\alpha \,\middle|\, Z \sim \mathrm{N}\left(\frac{\mu - \mu_0}{\sigma/\sqrt{n}}, 1\right)\right)$$

$$= \Phi\left(-z_\alpha - \frac{\mu - \mu_0}{\sigma/\sqrt{n}}\right), \tag{3.17}$$

respectively. Under the alternative hypothesis, we may therefore consider the test statistic Z to follow a *noncentral*, or *location-shifted* normal distribution. For the two-sided test, we must consider the quantity

$$\beta_\alpha^2(\mu; n) = \mathbb{P}_\mu\left(Z \leq -z_{\alpha/2} \text{ or } Z \geq z_{\alpha/2} \,\middle|\, Z \sim \mathrm{N}\left(\frac{\mu - \mu_0}{\sigma/\sqrt{n}}, 1\right)\right).$$

However, this can be found by combining Equations (3.16) and (3.17), while adjusting the significance level, to obtain

$$\beta_\alpha^2(\mu; n) = \Psi\left(z_{\alpha/2} - \frac{\mu - \mu_0}{\sigma/\sqrt{n}}\right) + \Phi\left(-z_{\alpha/2} - \frac{\mu - \mu_0}{\sigma/\sqrt{n}}\right). \tag{3.18}$$

Notice that for all three of these tests, we have the relationship

$$\beta_\alpha(\mu_0; n) = \alpha; \tag{3.19}$$

that is, the power function evaluated for the null hypothesis is equivalent to the significance. Furthermore, the power function of the two-sided test satisfies the following additive relation to the power functions (with reduced significance) of the two one-sided tests:

$$\beta_\alpha^2(\mu; n) = \beta_{\alpha/2}^L(\mu; n) + \beta_{\alpha/2}^R(\mu; n).$$

This relation is shown in Figure 3.4, in which Equations (3.16)–(3.18) are plotted using a significance level of $\alpha = 0.05$ for the two-sided test and $\alpha/2 = 0.025$ for each one-sided test. Note the value $z_{0.025} = 1.96$.

Next, let us determine the required sample size to achieve a two-sided test of a specific power β, given a minimum detectable effect δ. We are

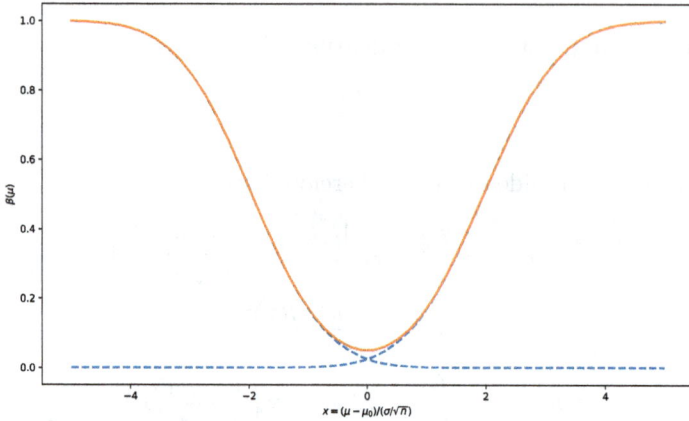

Fig. 3.4: Two-sided (β_α^2; orange solid) and both one-sided ($\beta_{\alpha/2}^L$ and $\beta_{\alpha/2}^R$; blue dashed) power functions, plotted as a function of the quantity $z = (\mu - \mu_0)/(\sigma/\sqrt{n})$; using significance level $\alpha = 0.05$.

typically interested in constructing a test with a power of at least 80%. By examining Figure 3.4, we see that

$$\beta_\alpha^2(\mu; n) \approx \begin{cases} \beta_{\alpha/2}^R(\mu; n) & \text{for } \mu \geq \mu_0 \\ \beta_{\alpha/2}^L(\mu; n) & \text{for } \mu < \mu_0 \end{cases}$$

for large values of the power, say $\beta > 0.20$ (far away from the middle $\mu = \mu_0$). We may recast the power function in terms of $\delta = |\mu - \mu_0| > 0$ to obtain the approximation

$$\beta = \beta_\alpha^2(\delta; n) \approx \Psi\left(z_{\alpha/2} - \frac{\delta}{\sigma/\sqrt{n}}\right). \tag{3.20}$$

Recalling the relation $z_\alpha = \Psi^{-1}(\alpha)$, we see that this expression is equivalent to

$$\Psi^{-1}(\alpha/2) - \Psi^{-1}(\beta) \approx \frac{\delta}{\sigma/\sqrt{n}}, \tag{3.21}$$

which we might call *the power equation*. This equation connects power, significance, sample size, and the minimum detectable effect into a single equation, which equates the survival difference between the significance and power to the normalized effect. Solving for the required sample size, we obtain

$$n = \frac{\sigma^2}{\delta^2}\left[\Psi^{-1}(\alpha/2) - \Psi^{-1}(\beta)\right]^2 \approx \frac{7.85\sigma^2}{\delta^2}. \tag{3.22}$$

The value of 7.85 was obtained using a significance of $\alpha = 0.05$ and a power of $\beta = 0.80$. To achieve the sample size required for a one-sided test, we of course simply replace $\Psi^{-1}(\alpha/2)$ with $\Psi^{-1}(\alpha)$ and 7.85 with 6.18 in the above equation.

3.3 Tests for Dual Populations

We next turn to the case of comparisons between two groups. This is a common case to consider in A/B testing, where experimenters try to measure the effect of a treatment on a particular group. A common form of hypothesis test is therefore $\theta_1 < \theta_2$; i.e., did our treatment have a positive effect on a given parameter.

3.3.1 Test for Two Means; Equal Variances

We start by comparing the means of two normal populations. As we shall see, a fantastic simplification can be made by assuming the two populations share a common variance. We begin with this case.

Theorem 3.4 (t-test). *Let $X_1, \ldots, X_n \sim N(\mu_X, \sigma_X^2)$ and $Y_1, \ldots, Y_m \sim N(\mu_Y, \sigma_Y^2)$. Let us further assume that $\sigma_X^2 = \sigma_Y^2 = \sigma^2$. Finally, let*

$$\overline{X}_n = \frac{1}{n}\sum_{i=1}^{n} X_i \qquad and \qquad \overline{Y}_m = \frac{1}{m}\sum_{i=1}^{m} Y_i$$

be the sample means and

$$S_X^2 = \frac{1}{n-1}\sum_{i=1}^{n}(X_i - \overline{X}_n)^2 \qquad and \qquad S_Y^2 = \frac{1}{m-1}\sum_{i=1}^{m}(Y_i - \overline{Y}_m)^2$$

be the sample variances, as usual. Then the test consisting of the null hypothesis

$$H_0 : \mu_X = \mu_Y,$$

test statistic

$$T = \frac{\overline{X}_n - \overline{Y}_m}{\sqrt{S_P^2 \left(\frac{1}{n} + \frac{1}{m}\right)}}, \qquad (3.23)$$

where S_P^2 is the pooled sample variance, defined by

$$S_P^2 = \frac{(n-1)S_X^2 + (m-1)S_Y^2}{n+m-2}, \qquad (3.24)$$

and rejection region

$$R_\alpha = \{t : |t| \geq t_{n+m-2,\alpha/2}\}$$

constitutes a test of size α. Moreover, under the null hypothesis, the test statistic T is a Student's t-random variable with $n+m-2$ degrees of freedom; i.e., $T \sim t_{n+m-2}$.

Proof. Normality of the population implies normality of the sample means. In particular, Corollary 2.2 implies that

$$\overline{X}_n \sim N(\mu_X, \sigma^2/n) \quad \text{and} \quad \overline{Y}_m \sim N(\mu_Y, \sigma^2/m).$$

(Notice our use of the assumption of equal variances $\sigma_X^2 = \sigma_Y^2 = \sigma^2$.) Therefore, according to Theorem 2.3, the difference $\overline{X}_n - \overline{Y}_m$ satisfies

$$\overline{X}_n - \overline{Y}_m \sim N\left(\mu_X - \mu_Y, \frac{\sigma^2}{n} + \frac{\sigma^2}{m}\right).$$

It follows that the random variable

$$Z = \frac{\overline{X}_n - \overline{Y}_m}{\sqrt{\sigma^2\left(\frac{1}{n} + \frac{1}{m}\right)}} \sim N\left(\frac{\mu_x - \mu_y}{\sqrt{\sigma^2\left(\frac{1}{n} + \frac{1}{m}\right)}}, 1\right) \tag{3.25}$$

is normally distributed and simplifies to a standard normal random variable $Z \sim N(0,1)$ under the null hypothesis $H_0 : \mu_x = \mu_y$.

Next, let us consider the sample variances. Theorem 2.6 provides that

$$\frac{(n-1)S_X^2}{\sigma^2} \sim \chi_{n-1}^2 \quad \text{and} \quad \frac{(m-1)S_Y^2}{\sigma^2} \sim \chi_{m-1}^2.$$

From Proposition 2.20, it thus follows that the random variable

$$U = \frac{(n-1)S_X^2}{\sigma^2} + \frac{(m-1)S_Y^2}{\sigma^2} \sim \chi_{n+m-2}^2$$

is a chi-squared random variable with $(n+m-2)$ degrees of freedom.

From Definition 2.14, it follows that under the null hypothesis, we have

$$T = \frac{Z}{\sqrt{U/(n+m-2)}} \sim t_{n+m-2}; \tag{3.26}$$

i.e., T is distributed a Student's t-distribution with $(n+m-2)$ degrees of freedom. It is now a matter of simple algebra to show that the distribution defined above is equivalent to the test statistic given by Equation (3.23). The result follows. $\square$

Note 3.19. Due to the central limit theorem, Theorem 3.4 applies approximately for comparing the means from any populations with common variance, as $n, m \to \infty$. In practice, a good rule of thumb is to require $n > 30$ and $m > 30$. $\triangleright$

Confidence Intervals

An immediate consequence of Theorem 3.4 is the following. We leave the proof to the reader.

Corollary 3.2. *Given the setup of Theorem 3.4, the $(1 - \alpha)$ confidence interval for the difference $\mu_X - \mu_Y$ is given by*

$$I_\alpha = (L(\mathcal{X}, \mathcal{Y}), U(\mathcal{X}, \mathcal{Y})),$$

where the lower and upper bounds are given by

$$L(\mathcal{X}, \mathcal{Y}) = \overline{X}_n - \overline{Y}_m - t_{n+m-2, \alpha/2} \sqrt{S_P^2 \left(\frac{1}{n} + \frac{1}{m} \right)}$$

$$U(\mathcal{X}, \mathcal{Y}) = \overline{X}_n - \overline{Y}_m + t_{n+m-2, \alpha/2} \sqrt{S_P^2 \left(\frac{1}{n} + \frac{1}{m} \right)},$$

respectively.

Example 3.7. In this example, we build a numerical simulation to validate the result of Theorem 3.4. In particular, we construct a simulation with true null hypothesis $\mu_X = \mu_Y = 10$ and equal variance $\sigma_X = \sigma_Y = 2$. Since the result is most interesting for small sample sizes, we select $n = 3$ and $m = 7$, with results in the degrees of freedom $(n + m - 2) = 8$. We further use a significance level of $\alpha = 0.05$, resulting in a critical value for our test $t_{8, 0.025} \approx 2.306$. The code for the simulation is given in Code Block 3.6.

We run 10,000 trials. As expected, the number of rejections is approximately 5%. (For our particular simulation, the observed rejection rate was 5.25%.) It is also interesting to note that in 48 out of 10,000 simulations, we observed a p-value less than 0.5%. The expected number of simulations that would have such a low p-value is 50, since the p-value is uniformly distributed under the null hypothesis. In fact, the smallest p-value that was observed `min(p_values)` was 0.000247, or 0.0247%. It is a stark reminder that when conducting many experiments, such small p-values are destined to arise due to mere chance.   ▷

We will discuss the power of the t-test in Section 3.3.4.

3.3.2 Test for Two Means; Unequal Variances

The case of comparing means from two normal populations with *different* variances is not as simple. In this case, the random variable Z, as defined in the proof of Theorem 3.4, would instead be given by

$$Z = \frac{\overline{X}_n - \overline{Y}_m - (\mu_X - \mu_Y)}{\sqrt{\sigma_X^2/n + \sigma_Y^2/m}}, \tag{3.27}$$

which would constitute a standard normal random variable. To proceed, we will require a certain approximation for the weighted sum of chi-squared random variables due to Satterthwaite.

```
1   mu_x = mu_y = 10
2   sigma_x = sigma_y = 2
3   n = 3
4   m = 7
5   dof = n+m-2
6   alpha = 0.05
7   t_crit = scipy.stats.t.isf(alpha/2, dof) # 2.306
8
9   n_trials = 10000
10  rejections = np.zeros(n_trials)
11  p_values = np.zeros(n_trials)
12  for i in range(n_trials):
13      samples_x = np.random.normal(mu_x, scale=sigma_x, size=n)
14      samples_y = np.random.normal(mu_y, scale=sigma_y, size=m)
15      x_mean = samples_x.mean()
16      y_mean = samples_y.mean()
17      x_var = samples_x.var() * n / (n-1)
18      y_var = samples_y.var() * m / (m-1)
19      S_P2 = ( (n-1)*x_var + (m-1)*y_var ) / dof
20      T = (x_mean - y_mean) / np.sqrt( S_P2 * (1/n + 1/m) )
21      rejections[i] = int( abs(T) > t_crit )
22      p_values[i] = 2 * scipy.stats.t.sf(abs(T), dof)
23
24  print ('Rejection Rate', rejections.sum() / n_trials) # 0.0525
25  print ('Number of P-Values < 0.005', np.sum( p_values < 0.005 )) #
        48
```

Code Block 3.6: Simulation of Two Normal Populations with Equal Variance

Satterthwaite's approximation

We describe the Satterthwaite's approximation as a method of approximating a weighted sum of random variables as a scaled random variable from the same family. If such an assumption is valid, we may match the first and second moments of the two expressions to obtain the following.

Definition 3.17 (Generalized Satterthwaite's Approximation). *Suppose $Y_1, \ldots, Y_k$ are independent random variables from the same one-parameter family, $Y_i \sim F_{\theta_i}$, for parameters $\theta_1, \ldots, \theta_k$. Then the generalized Satterthwaite's approximation for the sum $\sum_{i=1}^{k} a_i Y_i$, for arbitrary constants $a_1, \ldots, a_k$, is given by the Satterthwaite ansatz:*

$$\sum_{i=1}^{k} a_i Y_i = aY,$$

where $Y \sim F_\theta$, *such that* a *and* θ *are approximated by solving the system of equations*

$$\sum_{i=1}^{k} a_i \mathbb{E}[Y_i] = a\mathbb{E}[Y] \tag{3.28}$$

$$\sum_{i=1}^{k} a_i^2 \mathbb{V}(Y_i) = a^2 \mathbb{V}(Y). \tag{3.29}$$

Our first example is what is classically known as the *Satterthwaite's approximation*, i.e., his original use case. We have already seen that the sum of chi-squared random variables is also a chi-squared. The following provides us a way of approximated a *weighted* sum of chi-squared random variables.

Example 3.8 (Satterthwaite's Approximation). Let $Y_1, \ldots, Y_k$ be independent chi-squared random variables, with $Y_i \sim \chi^2_{r_i}$. Then the weighted sum $\sum_{i=1}^{k} a_i Y_i$ can be approximated by

$$\sum_{i=1}^{k} a_i Y_i = aY,$$

where $Y \sim \chi^2_\nu$, with parameters a and ν given by

$$a = \frac{\sum_{i=1}^{k} a_i^2 r_i}{\sum_{i=1}^{k} a_i r_i} \tag{3.30}$$

$$\nu = \frac{\left(\sum_{i=1}^{k} a_i r_i\right)^2}{\sum_{i=1}^{k} a_i^2 r_i}. \tag{3.31}$$

This follows by solving Equations (3.28) and (3.29) for a and ν, given the expected value and variance formulas for a chi-squared random variable: $\mathbb{E}[Y_i] = r_i$, $\mathbb{V}(Y_i) = 2r_i$, $\mathbb{E}[Y] = \nu$, and $\mathbb{V}(Y) = 2\nu$.   ▷

Welch's T-test

Next, we proceed with our analysis of comparing means from two populations with unequal variances $\sigma_X^2 \neq \sigma_Y^2$. When the unequal variances are unknown, Equation (3.27) is best approximated by the statistic

$$W = \frac{\overline{X}_n - \overline{Y}_m - (\mu_X - \mu_Y)}{\sqrt{S_X^2/n + S_Y^2/m}}, \tag{3.32}$$

with population variances replaced by sample variances. Under the null hypothesis $H_0 : \mu_X = \mu_Y$, the distribution of the test statistic W can be approximated as follows.

Theorem 3.5 (Welch's t-test). *Let $X_1, \ldots, X_n \sim \mathrm{N}(\mu_X, \sigma_X^2)$ and let $Y_1, \ldots, Y_m \sim \mathrm{N}(\mu_Y, \sigma_Y^2)$. Then the distribution of the test statistic W, given by Equation (3.32), is approximately a Student's t-distribution with $\hat{\nu}$ degrees of freedom, where $\hat{\nu}$ is given by*

$$\hat{\nu} = \frac{\left(\dfrac{S_X^2}{n} + \dfrac{S_Y^2}{m}\right)^2}{\dfrac{1}{n-1}\left(\dfrac{S_X^2}{n}\right)^2 + \dfrac{1}{m-1}\left(\dfrac{S_Y^2}{m}\right)^2}. \tag{3.33}$$

Moreover, the null hypothesis $\mu_X = \mu_Y$ can therefore be tested using the test statistic W and rejection region

$$R_\alpha = \{w : |w| > t_{\hat{\nu}, \alpha/2}\}.$$

The resulting test has approximate size α.

Proof. Comparing Equations (3.32) and (3.27), we seek to understand the distribution for

$$R = \frac{\dfrac{S_X^2}{n} + \dfrac{S_Y^2}{m}}{\dfrac{\sigma_X^2}{n} + \dfrac{\sigma_Y^2}{m}}.$$

Defining the random variables and constants

$$Y_1 = \frac{(n-1)S_X^2}{\sigma_X^2} \sim \chi_{n-1}^2$$

$$Y_2 = \frac{(m-1)S_Y^2}{\sigma_Y^2} \sim \chi_{m-1}^2$$

$$a_1 = \frac{\sigma_X^2}{n(n-1)}\left(\frac{\sigma_X^2}{n} + \frac{\sigma_Y^2}{m}\right)^{-1}$$

$$a_2 = \frac{\sigma_Y^2}{m(m-1)}\left(\frac{\sigma_X^2}{n} + \frac{\sigma_Y^2}{m}\right)^{-1},$$

we can express the statistic R as

$$R = a_1 Y_1 + a_2 Y_2.$$

Now let's make the Satterthwaite ansatz and assume that this sum may be approximated as $R = aY$, for $Y \sim \chi_\nu^2$, for some value of a and ν. In particular, we use Satterthwaite's approximation, which is given by Equations (3.30) and (3.31), with $r_1 = (n-1)$ and $r_2 = (m-1)$. In particular, note that

$$a_1(n-1) + a_2(m-1) = 1.$$

Thus, we see that

$$a_1 Y_1 + a_2 Y_2 \sim \frac{\chi_\nu^2}{\nu},$$

with

$$\frac{1}{\nu} = a_2^2(n-1) + a_2^2(m-1) = \left(\frac{\sigma_X^4}{n^2(n-1)} + \frac{\sigma_Y^4}{m^2(m-1)} \right) \left(\frac{\sigma_X^2}{n} + \frac{\sigma_Y^2}{m} \right)^{-2}.$$

Finally, we may approximate ν using the substitutions $\sigma_X^2 \to S_X^2$ and $\sigma_Y^2 \to S_Y^2$, thereby obtaining the desired result. □

Corollary 3.3. *Given the setup of Theorem 3.5, an approximate $(1 - \alpha)$ confidence interval for the difference $\mu_X - \mu_Y$ is therefore given by*

$$\overline{X}_n - \overline{Y}_m \pm t_{\hat{\nu},\alpha/2} \sqrt{\frac{S_X^2}{n} + \frac{S_Y^2}{m}}.$$

Note 3.20. The number of degrees of freedom given by Equation (3.33) goes to infinity like $\hat{\nu} \sim O(n)$ as $n, m \to \infty$ (commensurately). Thus, for large sample sizes, the test statistic W given by Equation (3.32) approximately follows a standard normal distribution. The Welch's t-test is therefore useful primarily in the context of small sample sizes. Moreover, for large sample sizes, the confidence interval given in Corollary 3.3 can be approximated by replacing $t_{\hat{\nu},\alpha/2}$ with $z_{\alpha/2}$.   ▷

3.3.3 Test for Two Proportions

We next turn to the case of samples from two Bernoulli populations. We again rely on the normal approximation to the binomial distribution for large sample size. We will also use the sample variance $\hat{p}(1 - \hat{p})$ in lieu of the population variance $p(1 - p)$. In this context, the test statistic is a restatement of Equation (3.32), and we have the following.

Proposition 3.5. *Consider two independent binomial random variables $X \sim \text{Binom}(n_1, p_1)$ and $Y \sim \text{Binom}(n_2, p_2)$, and define the estimates $\hat{p}_1 = X/n_1$ and $\hat{p}_2 = Y/n_2$. Then the statistic*

$$Z = \frac{\hat{p}_1 - \hat{p}_2 - (p_1 - p_2)}{\sqrt{\hat{p}_1(1 - \hat{p}_1)/n_1 + \hat{p}_2(1 - \hat{p}_2)/n_2}} \tag{3.34}$$

is approximately a standard normal distribution for large n_1, n_2. Moreover, for large n_1, n_2, the null hypothesis $H_0 : p_1 = p_2$ can be tested using Z, with $p_1 - p_2 = 0$, and with rejection region $R_\alpha = \{z : |z| \geq z_{\alpha/2}\}$. This test has approximate size α.

Finally, for large n_1, n_2, an approximate $(1 - \alpha)$ confidence interval for $p_1 - p_2$ is given by

$$\hat{p}_1 - \hat{p}_2 \pm z_{\alpha/2} \sqrt{\frac{\hat{p}_1(1 - \hat{p}_1)}{n_1} + \frac{\hat{p}_2(1 - \hat{p}_2)}{n_2}}.$$

Example 3.9. We can simulate an example of Proposition 3.5 in Code Block 3.7. We conduct simulations with a true null hypothesis $p_1 = p_2 = 0.30$, and using sample sizes $n_1 = 30$, $n_2 = 40$. The observed rejection rate is 6.39%, slightly higher than our approximate test size of 5%. To get a more accurate read on the size of this test, we repeated the simulation using 100,000 trials, obtaining a rejection rate of 5.989%.   The simula-

```
1   p1 = p2 = 0.3
2   n1 = 30
3   n2 = 40
4   alpha = 0.05
5   z_crit = scipy.stats.norm.isf(alpha/2) # 1.960
6
7   n_trials = 10000
8   rejections = np.zeros(n_trials)
9   zs = np.zeros(n_trials)
10  for i in range(n_trials):
11      p1_hat = np.random.binomial(n1, p1) / n1
12      p2_hat = np.random.binomial(n2, p2) / n2
13      z = (p1_hat - p2_hat) / np.sqrt( p1_hat * (1-p1_hat) / n1 +
            p2_hat * (1-p2_hat) / n2 )
14
15      rejections[i] = int( abs(z) > z_crit )
16      zs[i] = abs(z)
17
18  print ('Rejection Rate', rejections.sum() / n_trials ) # 0.0639
```

Code Block 3.7: Simulation of two Bernoulli trials

tion can be repeated using various sample sizes. For example, when using $n_1 = 50$ and $n_2 = 100$, our observed false rejection rate was 5.511%. Using $n_1 = 200$ and $n_2 = 300$, our observed false rejection rate was 5.267%. As $n_1, n_2 \to \infty$, the test statistic Z in Equation (3.34) will approach a standard normal distribution, and our observed rejection rate will approach α.

Further note that in this simulation, we kept track of the statistic Z for each sample. We can actually use this to construct a better critical value than the one given by $z_{\alpha/2}$. By taking the quantile `np.quantile(zs, 0.95)`, we obtain a value $z^* = 2.0687$. Repeating the simulation with `z_crit=2.0687` yields a rejection rate of 4.988%, which is much closer to our desired significance level obtained from $z_{\alpha/2}$.   ▷

3.3.4 Power of the T-test

In order to derive the power equation for the t-test, from which we can calculate the required sample size for a minimum detectable effect, we must first introduce the *noncentral t-distribution.*

Noncentral T-distribution

As we saw in Section 2.5.2, the t-distribution is formed by taking the ratio of a standard normal distribution and the square root of a chi-squared distribution (divided by its degrees of freedom). When the normal distribution in the denominator has a nonzero location parameter (i.e., it is shifted to the left or to the right), this ratio becomes the *noncentral t-distribution.*

Definition 3.18 (noncentral t-distribution). *Let $X \sim N(\lambda, 1)$ be a normal random variable centered at λ and $U \sim \chi^2_\nu$ be a chi-squared random variable with ν degrees of freedom. Then the ratio*

$$T = \frac{X}{\sqrt{U/\nu}} \tag{3.35}$$

is distributed as a noncentral t-distribution *with ν degrees of freedom and* noncentrality parameter λ, *which we denote as $T \sim \mathrm{NCT}^\lambda_\nu$.*

The form of the probability density is complex, requiring the use of certain hypergeometric functions to express; the cumulative distribution relies on incomplete beta functions. We therefore omit these forms here. To obtain values of this distribution, however, we can rely, as usual, on the `scipy.stats` package, which implements the noncentral t-distribution as `scipy.stats.nct(df, nc)`. Unfortunately, `numpy.random` does not have an implementation for the noncentral t-distribution, though this is easily remedied by combining random normal and chi-squared variables following the preceding prescription.

Despite the complexity of the distribution, there is a simple form for the mean and variance, which is given as follows.

Proposition 3.6. *Let $T \sim \mathrm{NCT}^\lambda_\nu$, then*

$$\mathbb{E}[T] = \lambda \sqrt{\frac{\nu}{2}} \frac{\Gamma((\nu-1)/2)}{\Gamma(\nu/2)}, \tag{3.36}$$

$$\mathbb{V}(T) = \frac{\nu(1+\lambda^2)}{\nu-2} - \frac{\lambda^2\nu}{2}\left[\frac{\Gamma((\nu-1)/2)}{\Gamma(\nu/2)}\right]^2, \tag{3.37}$$

such that the mean exists whenever $\nu > 1$ and the variance whenever $\nu > 2$.

Furthermore, the preceding expressions for the first two moments may be very closely approximated by

$$\sqrt{\frac{\nu}{2}}\frac{\Gamma((\nu-1)/2)}{\Gamma(\nu/2)} \approx \left(1 - \frac{3}{4\nu-1}\right)^{-1},$$

though, why would you want to?

Power Analysis for the T-test

We may easily modify our discussion leading up to Equation (3.22) to determine the required sample size for the t-test of Theorem 3.4. Under the alternative hypothesis that $\mu_x \neq \mu_y$, the test statistic T from Equation (3.26) is distributed as

$$T \sim \text{NCT}_{n+m-2}^{\lambda},$$

with noncentrality parameter λ determined from Equation (3.25):

$$\lambda = \frac{\mu_x - \mu_y}{\sqrt{\sigma^2 \left(\frac{1}{n} + \frac{1}{m}\right)}},$$

which may be estimated by replacing σ^2 with the pooled sample variance S_p^2. Thus, we may write the power equation as

$$\beta = \mathbb{P}\left(|T| \geq t_{n+m-2,\alpha/2} \,\middle|\, T \sim \text{NCT}_{n+m-2}^{\lambda}\right).$$

If we take

$$\hat{\lambda} = \frac{|\mu_x - \mu_y|}{\sqrt{S_P^2 \left(\frac{1}{n} + \frac{1}{m}\right)}},$$

we may write the approximation

$$\beta = \text{NCT}_{n+m-2}^{\hat{\lambda}}.sf(t_{n+m-2,\alpha/2}). \tag{3.38}$$

Now, if we further assume that n and m are large enough so that the noncentral t-distribution may be approximated by a normal distribution located at the noncentrality parameter, we may simplify this using the approximation

$$\beta \approx \Psi\left(t_{n+m-2,\alpha/2} - \frac{\delta}{\sqrt{S_P^2 \left(\frac{1}{n} + \frac{1}{m}\right)}}\right),$$

similar to Equation (3.20). If we further assume $n = m$, we may solve for n, obtaining

$$n \approx \frac{2S_P^2}{\delta^2}\left[\Psi^{-1}(\alpha/2) - \Psi^{-1}(\beta)\right]^2 \approx \frac{15.7 S_P^2}{\delta^2}. \tag{3.39}$$

This yields the required sample size to obtain a power β test with minimum detectable effect δ at a significance level of α, as long as $n, m > 30$ are large enough to justify the normal approximation to the t-distribution. When such approximations are unwarranted, we can fall back on Equation (3.38). Though, in practice, the right-hand side of Equation (3.39) is a good rule of thumb.

3.4 Tests for Categories

We next turn to the case of categorical data. By a *category*, we mean any one of several mutually exclusive and exhaustive outcomes for a given experiment. There are several flavors of tests that might be of interest: we shall consider the comparison of the outcome of a single experiment with a predicted set of values (Pearson's chi-squared); the comparison of the outcomes of two independent experiments (test for homogeneity); and tests concerning the independence of two or more classification attributes (test for independence). Additional details can be found in Agresti [2013, 2019].

3.4.1 Pearson's Chi-squared Test

We begin by considering an experiment that can produce c mutually exclusive and exhaustive outcomes, which we shall denote $A_1, \ldots, A_c$, and suppose that $\mathbb{P}(A_i) = p_i$, for $i = 1, \ldots, c$. We repeat our experiment n times, and count the number of times X_i that the experiment results in outcome A_i. Naturally, the random vector X is a multinomial random vector $X \sim \text{Multi}(n, p)$, with $p \in \Delta^{c-1}$. Based on an observed set of outcomes $X = x$, we would like to test the hypothesis $H_0 : p = p_0$ that the population probability vector is equal to some theoretical value p_0. Such a test is given in the following.

Definition 3.19 (Pearson's chi-squared test). *Let* $X \sim \text{Multi}(n, p)$, *where* $p \in \Delta^{c-1}$. *Pearson's chi-squared test* consists of the null hypothesis $H_0 : p = p_0$, *test statistic*

$$Q = \sum_{i=1}^{c} \frac{(X_i - np_{0i})^2}{np_{0i}}, \tag{3.40}$$

and rejection region

$$R_\alpha = [\chi^2_{c-1,\alpha}, \infty). \tag{3.41}$$

Theorem 3.6. *Under the null hypothesis* $H_0 : p = p_0$ *of Pearson's chi-squared test, the test statistic given by Equation (3.40) satisfies* $Q \to \chi^2_{c-1}$ *as* $n \to \infty$. *Therefore, for large* n, *Pearson's chi-squared test is approximately a size-α test.*

Note 3.21. A good rule of thumb is that Pearson's chi-squared test is safe to use as long as $np_{0i} \geq 5$, for $i = 1, \ldots, c$.   ▷

Before proceeding to the proof, let us first consider an easy warm-up: the case of $c = 2$. Let $X_1 \sim \text{Binom}(n, p_1)$, for $p_1 \in (0, 1)$. The normal approximation to the binomial, which follows due to the central limit theorem, implies that the statistic

$$Z = \frac{X_1 - np_1}{\sqrt{np_1(1-p_1)}}$$

is approximately a standard normal random variable for large n, particularly when $\min(np_1, n(1-p_1)) \geq 5$. It follows that the test statistic $Q = Z^2 \to \chi_1^2$ as $n \to \infty$. Now, if we define $X_2 = n - X_1$ and $p_2 = 1 - p_1$, a little algebra reveals

$$
\begin{aligned}
Q &= \frac{(X_1 - np_1)^2}{np_1(1-p_1)} \\
&= \frac{(X_1 - np_1)^2}{np_1} + \frac{(X_1 - np_1)^2}{n(1-p_1)} \\
&= \frac{(X_1 - np_1)^2}{np_1} + \frac{(X_2 - np_2)^2}{np_2}.
\end{aligned}
$$

This last line is equivalent to Equation (3.40) for the case $c = 2$. This proves that, for the case $c = 2$, $Q \to \chi_1^2$ as $n \to \infty$. Now, let us proceed to the the general proof.

Proof. Given $X \sim \text{Multi}(n, p)$ and $p \in \Delta^{c-1}$, we begin by defining the *reduced* vectors $\tilde{X} = (X_1, \ldots, X_{c-1})$ and $\tilde{p} = (p_1, \ldots, p_{c-1})$, consistent with Note 2.28. Moreover, the vector $\tilde{X}$ can be viewed as the sum of IID reduced Bernoulli random vectors $B_1, \ldots, B_n \sim \text{RedMultiBern}(p)$, such that $\tilde{X} = \sum_{i=1}^n B_i = n\bar{B}_n$.

The variance–covariance matrix $\tilde{\Sigma} = \mathbb{V}(B_i)$ is given by Equation (2.78) as

$$\tilde{\Sigma} = \text{diag}(\tilde{p}) - \tilde{p}\tilde{p}^T,$$

with inverse matrix given by Equation (2.79) as

$$\tilde{\Sigma}^{-1} = \text{diag}(\tilde{p})^{-1} + p_c^{-1} J_{c-1}.$$

Now, the central limit theorem (Theorem 2.11) tells us that

$$\sqrt{n}\tilde{\Sigma}^{-1/2}(\bar{B}_n - \tilde{p}) \to \text{N}(0_{c-1}, I_{c-1})$$

as $n \to \infty$. This in turn implies that the quantity

$$
\begin{aligned}
Q &= n(\bar{B}_n - \tilde{p})^T \tilde{\Sigma}^{-1} (\bar{B}_n - \tilde{p}) \\
&= n(\bar{B}_n - \tilde{p})^T \left[\text{diag}(\tilde{p})^{-1} + p_c^{-1} J_{c-1} \right] (\bar{B}_n - \tilde{p}) \\
&= n(\bar{B}_n - \tilde{p})^T \text{diag}(\tilde{p})^{-1}(\bar{B}_n - \tilde{p}) + np_c^{-1}(\bar{B}_n - \tilde{p})^T J_{c-1}(\bar{B}_n - \tilde{p})
\end{aligned}
$$

satisfies $Q \to \chi_{c-1}^2$ as $n \to \infty$. Let us examine each term separately.

Observing that $n(\bar{B}_n - \tilde{p}) = (\tilde{X} - n\tilde{p})$, we have, for the first term

$$n(\bar{B}_n - \tilde{p})^T \text{diag}(\tilde{p})^{-1}(\bar{B}_n - \tilde{p}) = \sum_{i=1}^{c-1} \frac{(X_i - np_i)^2}{np_i}.$$

For the second term, let us first observe that $x^T J x = \sum_{i=1}^{n} \sum_{j=1}^{n} J_{ij} x_i x_j = \left(\sum_{i=1}^{n} x_i\right)^2$, for $x \in \mathbb{R}^n$ and $J \in \mathbb{R}^{n \times n}$, with $J_{ij} = 1$. Thus, our second term is equivalent to

$$np_c^{-1}(\overline{B}_n - \tilde{p})^T J_{c-1}(\overline{B}_n - \tilde{p}) = \frac{1}{np_c}\left(\sum_{i=1}^{c-1}(X_i - np_i)\right)^2.$$

But,

$$\sum_{i=1}^{c-1}(X_i - np_i) = \sum_{i=1}^{c-1} X_i - n\sum_{i=1}^{c-1} p_i = (n - X_c) - n(1 - p_c) = -X_c + np_c.$$

Therefore,

$$np_c^{-1}(\overline{B}_n - \tilde{p})^T J_{c-1}(\overline{B}_n - \tilde{p}) = \frac{(X_c - np_c)^2}{np_c}.$$

Combining the above two results, we say that $Q = n(\overline{B}_n - \tilde{p})^T \tilde{\Sigma}^{-1}(\overline{B}_n - \tilde{p})$ is actually equivalent to Equation (3.40). The result follows. □

Example 3.10. Consider an experiment that results in one of three different outcomes with probability vector $p = (0.2, 0.3, 0.5)$. We run the experiment ten times and use Pearson's chi-squared test to test the null hypothesis $p_0 = (0.2, 0.3, 0.5)$. We simulated such an experiment in Code Block 3.8. Under this simulation, the null hypothesis is always true, and we are tracking the fraction of tests in which the chi-squared test leads to a false rejection.

The output of the code is startling, with only a 0.3% rejection rate for a test that should supposedly produce a 5% rejection rate! We also print the critical value and the bootstrap critical value: the Q-value at which 5% of our simulated tests is greater than (same approach we did in Example 3.9). The theoretical and bootstrap critical values are *way off*! This is not surprising, as such a small fraction of cases landed in our rejection region.

Since the chi-squared test is supposed to be valid for "large" sample sizes, we run our simulation again with $n = 25$ and $n = 100$, and the results get worse each time. We begin to think that, maybe, there is a mistake somewhere in our code.

We will ask the reader to find the bug in Exercise 3.4. Upon correcting the bug, and running with n=10 and n_trials=100000, we received a more reassuring output:

Observed Rejection Rate 0.05025.

Thus, the chi-squared test is valid even with such a relatively small sample size. ▷

```
1   n = 10
2   p = [0.2, 0.3, 0.5]
3   p_0 = [0.2, 0.3, 0.5]
4   alpha = 0.05
5   E_0 = n * np.array(p_0) # Expected frequencies; array REQUIRED!
6   chi_crit = scipy.stats.chi2.isf(alpha, n-1) # Critical Value
7   n_trials = 10000
8   rejections = np.zeros(n_trials)
9   Q_values = np.zeros(n_trials)
10  p_values = np.zeros(n_trials)
11
12  for i in range(n_trials):
13      x = np.random.multinomial(n, p)
14      Q = np.sum( (x - E_0)**2 / E_0 )
15      rejections[i] = int( Q > chi_crit )
16      Q_values[i] = Q
17      p_values[i] = scipy.stats.chi2.sf(Q, n-1)
18
19  print ('Observed Rejection Rate', np.sum(rejections) / n_trials)
20  print ('Critical Value', chi_crit.round(4) )
21  print ('95\% Quantile Test Statistic', np.quantile(Q_values,
            0.95).round(4))
22  # n=10, n_trials=100,000. OUTPUT ::
23  #    Observed Rejection Rate 0.00028
24  #    Critical Value 16.919
25  #    95\% Quantile Test Statistic 6.0333
```

Code Block 3.8: Chi-squared test simulation: CONTAINS A BUG!!

Note 3.22. We previously stated that Pearson's chi-squared test is valid for $\min(np_i) \geq 5$. Example 3.10, however, shows that it nevertheless produces good results for this case of $\min(np_i) = 2$. Some authors report that while $np_i \geq 5$ is a good rule-of-thumb, they have seen good results even when $np_i \geq 1$, for $i = 1, \ldots, c$; see Hogg, *et al.* [2015].   ▷

Note 3.23. The mistake in Code Block 3.8 is a very real example: this is the way I initially coded the simulation and, moreover, the discussion follows my initial thought process upon observing the results. (*Let's see what happens if we improve the sample size...*) Since it is a crucial skill in data science not only to code, but also to have a systematic approach for responding to bugs, I decided to leave the code as is and allow the reader to locate the mistake. Happy hunting!   ▷

3.4.2 Test for Homogeneity

We next turn to the case of *two* independent experiments, each capable of producing a set of c mutually exclusive and exhaustive outcomes $A_1, \ldots, A_c$. Consider two probability vectors $p_1, p_2 \in \Delta^{c-1}$ with components

$$p_{ij} = \mathbb{P}(A_j | \text{experiment } i),$$

for $i = 1, 2$ and $j = 1, \ldots, c$. Thus, we assume the outcome of experiment i may be described by $X_i \sim \text{Multi}(n_i, p_i)$, for $i = 1, 2$. We want to test whether or not these probabilities were affected by our experiment; i.e., whether or not $p_1 = p_2$.

Definition 3.20 (Two-experiment Test for Homogeneity). *Let $X_i \sim \text{Multi}(n_i, p_i)$, for $i = 1, 2$, and for $p_i \in \Delta^{c-1}$. The two-experiment test for homogeneity consists of the null hypothesis $p_1 = p_2$, the test statistic*

$$Q = \sum_{i=1}^{2} \sum_{j=1}^{c} \frac{(X_{ij} - n_i \hat{p}_j)^2}{n_i \hat{p}_j}, \tag{3.42}$$

where the quantity $\hat{p}_j$ is defined by

$$\hat{p}_j = \frac{X_{1j} + X_{2j}}{n_1 + n_2}, \tag{3.43}$$

for $j = 1, \ldots, c$, and the rejection region $R_\alpha = [\chi^2_{c-1,\alpha}, \infty)$.

Proposition 3.7. *The two-experiment test for homogeneity is approximately a size α test for large n. In particular, under the null hypothesis, the test statistic defined in Equation (3.42) satisfies $Q \to \chi^2_{c-1}$ as $n \to \infty$.*

Proof. From Theorem 3.6, we know that

$$\sum_{j=1}^{c} \frac{(X_{ij} - n_i p_{ij})^2}{n_i p_{ij}} \to \chi^2_{c-1}$$

as $n \to \infty$, for $i = 1, 2$. It follows that the quantity

$$\sum_{i=1}^{2} \sum_{j=1}^{c} \frac{(X_{ij} - n_i p_{ij})^2}{n_i p_{ij}} \to \chi^2_{2k-2}$$

as $n \to \infty$.

Now, under the null hypothesis, $p_{1j} = p_{2j}$, and we can therefore estimate this quantity using Equation (3.43). Substituting these estimates into the above results reduces the number of degrees of freedom by $c - 1$, as we are relying on c parameter estimates constrained by the single constraint $\sum_{j=1}^{c} \hat{p}_j = 1$. The result follows. $\square$

The above result can be generalized to more than two experiments, as follows.

Definition 3.21 (Many-experiment Test for Homogeneity). *Let* $X_i \sim$ *Multi*(n_i, p_i), *for* $i = 1, \ldots, m$, *and for* $p_i \in \Delta^{c-1}$. *The* m-experiment test *for homogeneity* consists of the null hypothesis $p_1 = p_2 = \cdots = p_m$, *the test statistic*

$$Q = \sum_{i=1}^{m} \sum_{j=1}^{c} \frac{(X_{ij} - n_i \hat{p}_j)^2}{n_i \hat{p}_j}, \tag{3.44}$$

where the quantity $\hat{p}_j$ *is defined by*

$$\hat{p}_j = \frac{\sum_{i=1}^{m} X_{ij}}{\sum_{i=1}^{m} n_i}, \tag{3.45}$$

for $j = 1, \ldots, c$, *and the rejection region* $R_\alpha = [\chi^2_{(m-1)(c-1),\alpha}, \infty)$.

Proposition 3.8. *The* h-experiment test for homogeneity is approximately a size α test for large n. In particular, under the null hypothesis, the test statistic defined in Equation (3.44) satisfies $Q \to \chi^2_{(m-1)(c-1)}$ as $n \to \infty$.

The proof is a straightforward generalization of Proposition 3.7.

3.4.3 Test for Independence

We return to the case of as single experiment, but add the following twist: the outcome of the experiment can be classified according to two different attributes. The first attribute assigns each outcome to one of m mutually exclusive and exhaustive events $A_1, \ldots, A_m$ whereas the second attribute assigns each outcome to one of s mutually exclusive and exhaustive events $B_1, \ldots, B_s$. We define the probability p_{ij} as

$$p_{ij} = \mathbb{P}(A_i, B_j),$$

for $i = 1, \ldots, m$ and $j = 1, \ldots, s$. Finally, let us repeat the experiment n times and define X_{ij} as the count of outcomes occurring in the bucket $A_i \cap B_j$. We may still say that $X \sim$ Multi(n, p), for $p \in \Delta^{ms-1}$, with the understanding that X and p are now *matrices* as opposed to *vectors*. We seek to test the independence of the two attributes, by testing the null hypothesis

$$H_0 : \mathbb{P}(A_i, B_j) = \mathbb{P}(A_i)\mathbb{P}(B_j),$$

for $i = 1, \ldots, m$ and $j = 1, \ldots, s$. Formally, we have the following.

Definition 3.22 (Test for Independence). *Let* $X \sim$ *Multi*(n, p), *where* $p \in \Delta^{ms-1}$, *be a multinomial random matrix arranged into* m *rows and* s *columns. Then the* test for independence *consists of the null hypothesis*

$$H_0 : p_{ij} = p_{i\cdot} p_{\cdot j},$$

where we have defined

$$p_{i\cdot} = \sum_{j=1}^{s} p_{ij} \qquad and \qquad p_{\cdot j} = \sum_{i=1}^{m} p_{ij}, \qquad (3.46)$$

the test statistic

$$Q = \sum_{i=1}^{m} \sum_{j=1}^{s} \frac{(X_{ij} - n\hat{p}_{i\cdot}\hat{p}_{\cdot j})^2}{n\hat{p}_{i\cdot}\hat{p}_{\cdot j}}, \qquad (3.47)$$

where we have defined

$$\hat{p}_{i\cdot} = \frac{1}{n}\sum_{j=1}^{s} X_{ij} \qquad and \qquad \hat{p}_{\cdot j} = \frac{1}{n}\sum_{i=1}^{m} X_{ij},$$

and the rejection region $R_\alpha = [\chi^2_{(m-1)(s-1),\alpha}, \infty)$.

Proposition 3.9. *The test for independence is approximately a size α test for large n. Moreover, the test statistic defined in Equation (3.47) satisfies $Q \to \chi^2_{(m-1)(s-1)}$ as $n \to \infty$.*

3.5 Tests for Categories II: Power Analysis

In this section, we will be exploring the power function for Pearson's chi-squared test. In particular, we will need to understand the distribution of the test statistic Q given by Equation (3.40) under the alternative hypothesis. Theorem 3.6 already tells us that Q has a χ^2_{c-1} distribution, for large n, when the null hypothesis is true. But what if $X \sim \text{Multi}(n, p_A)$, for some alternative hypothesis $p_A \in \Theta_0^c$, and we use the test assuming a particular value of $p_0 \in \Theta_0$? This immediately creates a bias in the quantity $X_i - np_{0i}$, one which is best understood with an extension to the chi-squared distribution.

3.5.1 Noncentral Chi-squared Distribution

Our ultimate goal for the remainder of the section is to understand the power function for Pearson's chi-squared test. This understanding will require some familiarity of a new distribution: the non-central chi-squared distribution.

Definition 3.23. *Let $X_1, \ldots, X_c$ be a sequence of independent unit-variance normal random variables with $X_i \sim N(\mu_i, 1)$, for $i = 1, \ldots, c$. Then the random variable*

$$X = \sum_{i=1}^{c} X_i^2,$$

consisting of the sum of squares of the X_is, is said to constitute a noncentral chi-squared random variable, *denoted $X \sim \text{NCX}_c^{\lambda}$, with c degrees of freedom* and *noncentrality parameter*

$$\lambda = \sum_{i=1}^{c} \mu_i^2.$$

In order to derive the PDF for a noncentral chi-squared random variable, we will require a few preliminary results.

Proposition 3.10. *Let $Y \sim \text{N}(\mu, \sigma^2)$ be a normal random variable. The* PDF *of the random variable $X = Y^2$ is given by*

$$f_X(x) = \frac{1}{\sigma^2} \sum_{p=0}^{\infty} \frac{e^{-\lambda/2}(\lambda/2)^p}{p!} \frac{(x/\sigma^2)^{p-1/2}e^{-x/(2\sigma^2)}}{\Gamma(p+1/2)2^{p+1/2}}, \qquad (3.48)$$

where $\lambda = \mu^2/\sigma^2$.

Proof. Consider the transformation $X = g(Y) = Y^2$, which is invertible on each of the domains $\mathcal{A}_1 = (-\infty, 0)$ and $\mathcal{A}_2 = (0, \infty)$. In fact, we may define $Y = g_1^{-1}(X) = -\sqrt{X}$ and $Y = g_2^{-1}(X) = \sqrt{X}$, with

$$\left| \frac{dg_i^{-1}(x)}{dx} \right| = \frac{1}{2\sqrt{x}},$$

for $i = 1, 2$. Additionally, the PDF for Y is given by

$$f_Y(y) = \frac{1}{\sigma} \phi \left(\frac{y - \mu}{\sigma} \right),$$

where $\phi(z) = e^{-z^2/2}/\sqrt{2\pi}$ is the PDF of a standard normal random variable. It therefore follows from Theorem 1.7 that the PDF of the random variable $X = Y^2$ is given by

$$\begin{aligned}
f_X(x) &= \frac{1}{2\sqrt{x}} \left[f_Y(\sqrt{x}) + f_Y(-\sqrt{x}) \right] \\
&= \frac{1}{2\sigma\sqrt{x}} \left[\phi \left(\frac{\sqrt{x} - \mu}{\sigma} \right) + \phi \left(\frac{-\sqrt{x} - \mu}{\sigma} \right) \right] \\
&= \frac{1}{2\sigma\sqrt{x}\sqrt{2\pi}} \left[e^{-(\sqrt{x}-\mu)^2/(2\sigma^2)} + e^{-(\sqrt{x}+\mu)^2/(2\sigma^2)} \right] \\
&= \frac{e^{-x/(2\sigma^2)}e^{-\mu^2/(2\sigma^2)}}{\sigma\sqrt{x}\sqrt{2\pi}} \cosh \left(\frac{\sqrt{x}\mu}{\sigma^2} \right),
\end{aligned}$$

where we used the definition for the hyperbolic cosine $\cosh(x) = (e^x + e^{-x})/2$. Next, let's replace the hyperbolic cosine with its Taylor series. We find

$$f_X(x) = \frac{e^{-x/(2\sigma^2)}e^{-\mu^2/(2\sigma^2)}}{\sigma\sqrt{x}\sqrt{2\pi}} \sum_{p=0}^{\infty} \frac{1}{(2p)!} \left(\frac{\sqrt{x}\mu}{\sigma^2}\right)^{2p}$$

$$= \sum_{p=0}^{\infty} \frac{e^{-x/(2\sigma^2)}e^{-\mu^2/(2\sigma^2)}(\mu^2/\sigma^2)^p}{\sigma^2\sqrt{2\pi}(2p)!} \left(\frac{x}{\sigma^2}\right)^{p-1/2}$$

Now, it might interest our readers to know that, in fact,

$$\sqrt{\pi}(2p)! = 2^{2p}\Gamma(p+1/2)p!.$$

Thus, we arrive at

$$f_X(x) = \sum_{p=0}^{\infty} \frac{e^{-x/(2\sigma^2)}e^{-\mu^2/(2\sigma^2)}(\mu^2/(2\sigma^2))^p}{\sigma^2 2^{p+1/2}\Gamma(p+1/2)p!} \left(\frac{x}{\sigma^2}\right)^{p-1/2}$$

Setting $\lambda = \mu^2/\sigma^2$ and rearranging yields our result. $\qquad\square$

Now, there's an obvious corollary to all of this.

Corollary 3.4. *The* PDF *given by Equation (3.48) represents a Poisson-weighted mixture of Gamma random variables. In particular, if $Y \sim N(\mu, \sigma^2)$ is a normal random variable, then the random variable $X = Y^2$ is given by*

$$f_X(x) = \sum_{p=0}^{\infty} \text{Poiss}(p; \lambda/2)\text{Gamma}(x; p+1/2, 2\sigma^2), \qquad (3.49)$$

where $\lambda = \mu^2/\sigma^2$.

Proof. From Equation (3.48), we have

$$f_X(x) = \frac{1}{\sigma^2} \sum_{p=0}^{\infty} \text{Poiss}(p; \lambda/2)\text{Gamma}(x/\sigma^2; p+1/2, 2).$$

The result follows from Exercise 2.4. $\qquad\square$

We may further restate Corollary 3.4 as follows.

Corollary 3.5. *Let $Y \sim N(\mu, \sigma^2)$ and define $\lambda = \mu^2/\sigma^2$. The random variable $X = Y^2$ is equivalent to the hierarchical model*

$$X|(P = p) \sim \text{Gamma}(p+1/2, 2\sigma^2)$$
$$P \sim \text{Poiss}(\lambda/2).$$

Proof. Since $P \sim \text{Poiss}(\lambda/2)$, the PMF for P is given by

$$f_P(p) = \frac{e^{-\lambda/2}(\lambda/2)^p}{p!}.$$

Similarly, the PDF of the conditional random variable $X|P$ is given by

$$f_{X|P}(x|p) = \frac{1}{\sigma^2} \frac{(x/\sigma^2)^{p-1/2} e^{-x/(2\sigma^2)}}{\Gamma(p+1/2)2^{p+1/2}}.$$

The (unconditional) distribution of X is therefore obtained by marginalizing over P,

$$f_X(x) = \sum_{p=0}^{\infty} f_{X|P}(x|p) f_P(p),$$

which yields Equation (3.48).   □

An important variation of Corollary 3.5 is the following.

Corollary 3.6. *Let $Y \sim \text{N}(\mu, \sigma^2)$ and define $\lambda = \mu^2/\sigma^2$. The random variable $X = Y^2$ is equivalent to the hierarchical model*

$$X|(P = p) \sim \sigma^2 \chi^2_{1+2p}$$
$$P \sim \text{Poiss}(\lambda/2).$$

Proof. This follows since a $\text{Gamma}(p+1/2, 2\sigma^2)$ distribution is equivalent to a scaled $\text{Gamma}(p+1/2, 2)$ distribution, which is itself equivalent to a scaled χ^2_{1+2p} distribution. Specifically, we have

$$\text{Gamma}(x; p+1/2, 2\sigma^2) = \frac{1}{\sigma^2} \text{Gamma}\left(\frac{x}{\sigma^2}; p+1/2, 2\right)$$
$$= \frac{1}{\sigma^2} f\left(\frac{x}{\sigma^2}; 1+2p\right),$$

where $f(x; 1+2p)$ is the PDF of a χ^2_{1+2p} random variable. Finally, we recall from Theorem 1.10 that the random variable that has such a distribution is the scaled chi-squared random variable $\sigma^2 \chi^2_{1+2p}$.   □

And one final variation...

Corollary 3.7. *Let $Y \sim \text{N}(\mu, \sigma^2)$ and define $\lambda = \mu^2/\sigma^2$. Then random variable $Y^2 \sim \sigma^2 \text{NCX}_1^\lambda$.*

Proof. This follows immediately from Corollary 3.6, since the scale factor σ^2 multiplies the chi-squared random variable in the breakdown.   □

Next, let us apply these results to Definition 3.23 to obtain the distribution for the noncentral chi-squared random variable.

Theorem 3.7. *Let* $X \sim \mathrm{NCX}_c^\lambda$ *be a noncentral chi-squared random variable. Then the* PDF *for* X *is given by*

$$f_X(x) = \sum_{p=0}^{\infty} \frac{e^{-\lambda/2}(\lambda/2)^p}{p!} \frac{x^{c/2+p-1}e^{-x/2}}{\Gamma(c/2+p)2^{c/2+p}}, \qquad (3.50)$$

Moreover, the random variable X *can be alternatively described using the hierarchical model*

$$X|(P = p) \sim \chi^2_{c+2p}$$
$$P \sim \mathrm{Poiss}(\lambda/2).$$

Proof. Definition 3.23 states that a noncentral chi-squared random variable X can always be expressed as the sum of squares of c normal random variables with unit variance, $X_1, \ldots, X_c$, with $X_i \sim \mathrm{N}(\mu_i, 1)$. Moreover, Corollary 3.6 tells us that the random variable X_i^2 is equivalent to the hierarchical model

$$X_i^2|(P_i = p_i) \sim \chi^2_{1+2p_i}$$
$$P_i \sim \mathrm{Poiss}(\mu_i^2/2),$$

since $\chi^2_{1+2p_i}$ is equivalent to $\mathrm{Gamma}((1 + 2p_i)/2, 2)$.

Since chi-squared random variables are additive (Proposition 2.20), it follows that our noncentral chi-squared random variable $X = \sum_{i=1}^c X_i^2$ is equivalent to

$$X|(P_1, \ldots, P_c) \sim \chi^2_{c+2(p_1+\cdots+p_c)},$$
$$P_1 \sim \mathrm{Poiss}(\mu_1^2/2),$$
$$\vdots$$
$$P_c \sim \mathrm{Poiss}(\mu_c^2/2),$$

Now, we observe that it is only the *sum* of the random variables $P_1, \ldots, P_c$ that matters, not their individual values. Given the additivity of the Poisson random variable (Proposition 2.14), this model therefore simplifies as

$$X|(P = p) \sim \chi^2_{c+2p}$$
$$P \sim \mathrm{Poiss}(\lambda/2),$$

where $\lambda = \sum_{i=1}^c \mu_i^2$. This model, however, is equivalent to Equation (3.50), which therefore completes our result. □

Corollary 3.8. *The mean and variance of* X *are given by*

$$\mathbb{E}[X] = c + \lambda \qquad and \qquad \mathbb{V}(X) = 2(c + 2\lambda). \qquad (3.51)$$

Proof. The mean of a noncentral chi-squared random variable is given by

$$\mathbb{E}[X] = \mathbb{E}[\mathbb{E}[X|P]] = \mathbb{E}[c + 2P] = c + \lambda.$$

The variance is obtained by applying Theorem 2.9. (See Exercise 3.5.) $\square$

Fortunately, we will not have occasion to revert to the analytical form of the noncentral chi-squared distribution, or its associated, unfathomably complex, CDF. Instead, basic usage in Python is given in Code Block 3.9.

```
1  nc_lambda = 2 # noncentrality parameter
2  dof = 7       # degrees of freedom
3  n_samples = 100
4  samples = np.random.noncentral_chisquare(dof, nc_lambda,
       size=n_samples)
5
6  x = np.linspace(0, 20)
7  f = scipy.stats.ncx2.pdf(x, dof, nc_lambda)
8  F = scipy.stats.ncx2.cdf(x, dof, nc_lambda)
9  alpha = 0.05
10 crit_pt = scipy.stats.ncx2.isf(alpha, dof, nc_lambda) # Critical
       point
```

Code Block 3.9: Basic operations for a noncentral chi-squared random variable

3.5.2 Classic Power Function for Chi-squared Statistic

Agresti [2013] includes without proof the following approximation for the power function of the chi-squared statistic.

Proposition 3.11. *Let Q be the chi-squared statistic (Equation (3.40)) with null hypothesis $H_0 : p = p_0$. Under the specific alternative hypothesis $H_A : p = p_A \in \Theta_0^c$, the chi-squared statistic is approximately distributed as*

$$Q \sim \mathrm{NCX}_{c-1}^{n\lambda}, \tag{3.52}$$

with noncentrality parameter

$$\lambda = \sum_{i=1}^{c} \frac{(p_{Ai} - p_{0i})^2}{p_{0i}}. \tag{3.53}$$

This approximation is appealing for several reasons. First, it follows similar form to the power function for the Z and T tests, both of which are in the form of a noncentral counterpart to the primary test statistic, as well as the noncentral F, which is the similar construction for the analysis of variance. It is also fairly easy to remember, as the noncentrality parameter is essentially the Q statistic with the observed values replaced for the true (i.e., alternate hypothesis) probability distribution.

We will, however, labor of the next several pages to develop a slightly more accurate and sophisticated estimate, culminating in Theorem 3.8.

3.5.3 Novel Power Function for Chi-squared Statistic

Our goal is to understand the distribution of the chi-squared statistic Q, as given by Equation (3.40), under the alternative hypothesis that $p = p_A$, for a particular $p_A \in \Delta^{c-1}$ with $p_A \neq p_0$. We begin with a slightly easier result.

Proposition 3.12. *Let* $X \sim \text{Multi}(n, p_A)$ *for some* $p_A \in \Delta^{c-1}$, *and let* $p_0 \in \Delta^{c-1}$ *be distinct from* p_A. *Then the statistic*

$$R = \sum_{i=1}^{c} \frac{(X_i - np_{0i})^2}{np_{Ai}} \tag{3.54}$$

tends towards $R \to \text{NCX}_{c-1}^{n\lambda}$ *as* $n \to \infty$, *where the noncentrality parameter is given by*

$$\lambda = \sum_{i=1}^{c} \frac{(p_{Ai} - p_{0i})^2}{p_{Ai}}. \tag{3.55}$$

Note 3.24. The classic power function, given by the test statistic Q and Equation (3.53), replaces the alternative probabilities p_{Ai} in the denominators of Equations (3.54) and (3.55) with the null-hypothesis probabilities p_{0i}.   ▷

Proof. We will follow the construction used in the proof of Theorem 3.6. In particular, consider a sequence of IID reduced Bernoulli random vectors $B_1, \ldots, B_n \sim \text{RedMultiBern}(p_A)$, so that $\tilde{X} = (X_1, \ldots, X_{c-1}) = n\overline{B}_n$. The central limit theorem yields

$$Z_n = \sqrt{n} \tilde{\Sigma}_A^{-1/2} (\overline{B}_n - \tilde{p}_A) \to \text{N}(0_{c-1}, I_{c-1})$$

as $n \to \infty$. Therefore, for large n, the exrpession $Z_n + \sqrt{n} \tilde{\Sigma}_A^{-1/2} (\tilde{p}_A - \tilde{p}_0)$ can be approximated by

$$\sqrt{n} \tilde{\Sigma}_A^{-1/2} (\overline{B}_n - \tilde{p}_0) \approx \text{N}(\sqrt{n} \tilde{\Sigma}_A^{-1/2} \tilde{\delta}, I_{c-1}), \tag{3.56}$$

where we have defined the vector

$$\tilde{\delta} = \tilde{p}_A - \tilde{p}_0.$$

It follows that the quantity

$$R = n(\overline{B}_n - \tilde{p}_0)\tilde{\Sigma}_A^{-1}(\overline{B}_n - \tilde{p}_0) \approx \text{NCX}_{c-1}^{\lambda_n}, \qquad (3.57)$$

where $\lambda_n = n\tilde{\delta}^T \tilde{\Sigma}_A^{-1}\tilde{\delta}$. However, by following the same steps of the proof of Theorem 3.6, one can easily show that R is equivalent to the expression given in Equation (3.54). Moreover, the noncentrality parameter can be expressed as

$$\begin{aligned}
\lambda_n &= n\tilde{\delta}^T \tilde{\Sigma}_A^{-1}\tilde{\delta} \\
&= n\tilde{\delta}^T \left(\text{diag}(\tilde{p}_A)^{-1} + p_{Ac}^{-1}J_{c-1}\right)\tilde{\delta} \\
&= n\tilde{\delta}^T \text{diag}(\tilde{p}_A)^{-1}\tilde{\delta} + np_{Ac}^{-1}\tilde{\delta}^T J_{c-1}\tilde{\delta}
\end{aligned}$$

Now, the first term is simply

$$n\tilde{\delta}^T \text{diag}(\tilde{p}_A)^{-1}\tilde{\delta} = n\sum_{i=1}^{c-1}\frac{(p_{Ai} - p_{0i})^2}{p_{Ai}}.$$

And the second term is

$$\begin{aligned}
np_{Ac}^{-1}\tilde{\delta}^T J_{c-1}\tilde{\delta} &= np_{Ac}^{-1}\left(\sum_{i=1}^{c-1}(p_{Ai} - p_{0i})\right)^2 \\
&= np_{Ac}^{-1}[(1 - p_{Ac}) - (1 - p_{0c})]^2 \\
&= \frac{n(p_{Ac} - p_{0c})^2}{p_{Ac}}.
\end{aligned}$$

It follows that $\lambda_n = n\tilde{\delta}^T \tilde{\Sigma}_A^{-1}\tilde{\delta}$ is equivalent to Equation (3.55), thereby completing the result.   □

The statistic R, defined in Proposition 3.12, is of course not equivalent to the chi-squared statistic Q, as defined by Equation (3.40), the difference being the denominator as the expected count under the null hypothesis (for the chi-squared statistic Q) or the expected count under the alternate hypothesis (for the statistic R). The remedy of such a blemish is the cause of much toil and strife. We begin by studying how Equations (3.56) and (3.57) must be modified to accommodate this change.

Lemma 3.1. *Under an alternate hypothesis $H_A : p = p_A$, with $p_A \neq p_0$, the chi-squared statistic Q, defined in Equation (3.40) is equivalent to*

$$Q = X^T X,$$

where

$$X = \sqrt{n}\tilde{\Sigma}_0^{-1/2}(\overline{B}_n - \tilde{p}_0) \approx \text{N}(\sqrt{n}\tilde{\Sigma}_0^{-1/2}\tilde{\delta}, \tilde{\Sigma}_0^{-1/2}\tilde{\Sigma}_A\tilde{\Sigma}_0^{-1/2}), \qquad (3.58)$$

following the notation of the proof of Proposition 3.12.

Proof. It is straightforward to show, following the same steps done in the proof of Proposition 3.12, that

$$Q = n(\bar{B}_n - \tilde{p}_0)\tilde{\Sigma}_0^{-1}(\bar{B}_n - \tilde{p}_0)$$

is equivalent to Equation (3.40). Equation (3.58) is obtained by premultiplying Equation (3.56) by $\tilde{\Sigma}_0^{-1/2}\tilde{\Sigma}_A^{1/2}$, using the fact that $\tilde{\Sigma}_0^{-1/2}$ is symmetric.[5]
□

Lemma 3.2. *Under an alternate hypothesis $H_A : p = p_A$, with $p_A \neq p_0$, the chi-squared statistic Q, defined in Equation (3.40) is approximately distributed like*

$$Q \sim \sum_{i=1}^{c-1} \tau_i^2 \mathrm{NCX}_1^{n\lambda_i}, \tag{3.59}$$

for large n, where $\lambda_i = \mu_i^2/\tau_i^2$, and where

$$\mu = V^T \tilde{\Sigma}_0^{-1/2}(\tilde{p}_A - \tilde{p}_0), \tag{3.60}$$

and where D and V is the eigendecomposition of $\tilde{\Sigma}_0^{-1/2}\tilde{\Sigma}_A\tilde{\Sigma}_0^{-1/2} = VDV^T$, with $D = \mathrm{diag}(\tau_1^2, \ldots, \tau_{c-1}^2)$.

Proof. Let V be the orthogonal matrix of eigenvectors for $\tilde{\Sigma}_0^{-1/2}\tilde{\Sigma}_A\tilde{\Sigma}_0^{-1/2}$ and let $D = \mathrm{diag}(\tau_1^2, \ldots, \tau_{c-1}^2)$ be the corresponding (ordered) diagonal matrix of eigenvalues. Now consider the transformation $Y = V^T X$. First, since V is orthogonal, we have

$$Y^T Y = X^T V V^T X = X^T X = Q.$$

Second, Equation (3.58) yields

$$Y \approx \mathrm{N}(\sqrt{n}V^T \tilde{\Sigma}_0^{-1/2}\delta, D).$$

If we define $\mu = V^T \tilde{\Sigma}_0^{-1/2}\delta$, then the ith component of Y is approximately

$$Y_i \approx \mathrm{N}(\sqrt{n}\mu_i, \tau_i^2),$$

since D is diagonal. The result follows from Corollary 3.7.   □

Our task is to approximate the expression for the chi-squared statistic, as given y Equation (3.59). To proceed, we shall require a number of preliminary results.

---

[5] Note that if $X \sim \mathrm{N}(\mu, \Sigma)$, then $AX \sim \mathrm{N}(A\mu, A\Sigma A^T)$.

Lemma 3.3. *The matrix* $\tilde{\Sigma}_A \tilde{\Sigma}_0^{-1}$ *may be expressed as*

$$\tilde{\Sigma}_A \tilde{\Sigma}_0^{-1} = \mathrm{diag}\left(\frac{\tilde{p}_A}{\tilde{p}_0}\right) - \tilde{p}_A \left(\frac{\tilde{p}_A}{\tilde{p}_0}\right)^T + \frac{p_{Ac}}{p_{0c}}\tilde{p}_A \tilde{1}^T, \qquad (3.61)$$

where "vector division" is understood to be component-wise, and where $\tilde{1} = \langle 1,\ldots,1\rangle \in \mathbb{R}^{c-1}$.

Proof. Using a few matrix–vector properties,

$$\mathrm{diag}(x)J = x1^T, \quad x^T J = \left(\sum x_i\right)1^T, \quad x^T \mathrm{diag}(y) = (xy)^T,$$

we have the following

$$\tilde{\Sigma}_A \tilde{\Sigma}_0^{-1} = \left(\mathrm{diag}(\tilde{p}_A) - \tilde{p}_A\tilde{p}_A^T\right)\left(\mathrm{diag}(\tilde{p}_0)^{-1} + \frac{1}{p_{0c}}J_{c-1}\right)$$

$$= \mathrm{diag}\left(\frac{\tilde{p}_A}{\tilde{p}_0}\right) + \frac{\mathrm{diag}(\tilde{p}_A)J_{c-1}}{p_{0c}} - \tilde{p}_A\tilde{p}_A^T\mathrm{diag}(\tilde{p}_0)^{-1} - \frac{\tilde{p}_A\tilde{p}_A^T J_{c-1}}{p_{0c}}$$

$$= \mathrm{diag}\left(\frac{\tilde{p}_A}{\tilde{p}_0}\right) + \frac{\tilde{p}_A\tilde{1}^T}{p_{0c}} - \tilde{p}_A\left(\frac{\tilde{p}_A}{\tilde{p}_0}\right)^T - \frac{(1-p_{Ac})\tilde{p}_A\tilde{1}^T}{p_{0c}}.$$

The result follows. □

Lemma 3.4. *The trace of the matrix* $\tilde{\Sigma}_A \tilde{\Sigma}_0^{-1}$ *is given by*

$$\mathrm{trace}\left(\tilde{\Sigma}_A \tilde{\Sigma}_0^{-1}\right) = \sum_{i=1}^{c} \frac{p_{Ai}(1-p_{Ai})}{p_{0i}}.$$

Proof. Equation (3.61) yields the diagonal elements of $\tilde{\Sigma}_A \tilde{\Sigma}_0^{-1}$, which are

$$\left[\tilde{\Sigma}_A \tilde{\Sigma}_0^{-1}\right]_{ii} = \frac{p_{Ai}}{p_{0i}} - \frac{p_{Ai}^2}{p_{0i}} + \frac{p_{Ac}}{p_{0c}}p_{Ai},$$

for $i = 1,\ldots,(c-1)$. Therefore

$$\mathrm{trace}\left(\tilde{\Sigma}_A \tilde{\Sigma}_0^{-1}\right) = \sum_{i=1}^{c-1}\left[\tilde{\Sigma}_A \tilde{\Sigma}_0^{-1}\right]_{ii}$$

$$= \sum_{i=1}^{c-1}\frac{p_{Ai}(1-p_{Ai})}{p_{0i}} + \sum_{i=1}^{c-1}\frac{p_{Ac}}{p_{0c}}p_{Ai}$$

$$= \sum_{i=1}^{c-1}\frac{p_{Ai}(1-p_{Ai})}{p_{0i}} + \frac{p_{Ac}}{p_{0c}}(1-p_{Ac}).$$

The result follows. □

Lemma 3.5. *The trace of the matrix* $(\tilde{\Sigma}_A \tilde{\Sigma}_0^{-1})^2$ *is given by*

$$\text{trace}\left((\tilde{\Sigma}_A \tilde{\Sigma}_0^{-1})^2 \right) = \sum_{i=1}^{c} \frac{p_{Ai}^2}{p_{0i}^2} (1 - 2p_{Ai}) + \left(\sum_{i=1}^{c} \frac{p_{Ai}^2}{p_{0i}} \right)^2.$$

Proof. Using the expression for $\tilde{\Sigma}_A \tilde{\Sigma}_0^{-1}$ given by Equation (3.61), we find that

$$(\tilde{\Sigma}_A \tilde{\Sigma}_0^{-1})^2 = \text{diag}\left(\frac{\tilde{p}_A^2}{\tilde{p}_0^2} \right) + S\tilde{p}_A \left(\frac{\tilde{p}_A}{\tilde{p}_0} \right)^T + \frac{p_{Ac}^2}{p_{0c}^2} (1 - p_{Ac}) \tilde{p}_A \tilde{1}^T$$
$$-2\left(\frac{\tilde{p}_A^2}{\tilde{p}_0} \right) \left(\frac{\tilde{p}_A}{\tilde{p}_0} \right)^T + 2\frac{p_{Ac}}{p_{0c}} \left(\frac{\tilde{p}_A^2}{\tilde{p}_0} \right) \tilde{1}^T - 2\frac{p_{Ac}}{p_{0c}} S\tilde{p}_A \tilde{1}^T,$$

where we define the sum S as

$$S = \left(\frac{\tilde{p}_A}{\tilde{p}_0} \right)^T \tilde{p}_A = \sum_{i=1}^{c-1} \frac{p_{Ai}^2}{p_{0i}}.$$

Summing the diagonal elements, we obtain

$$\text{trace}\left((\tilde{\Sigma}_A \tilde{\Sigma}_0^{-1})^2 \right) = \sum_{i=1}^{c-1} \left[(\tilde{\Sigma}_A \tilde{\Sigma}_0^{-1})^2 \right]_{ii}$$
$$= \sum_{i=1}^{c-1} \frac{p_{Ai}^2}{p_{0i}^2} + S^2 + \frac{p_{Ac}^2}{p_{0c}^2} (1 - p_{Ac})^2$$
$$-2\sum_{i=1}^{c-1} \frac{p_{Ai}^3}{p_{0i}^2} + 2\frac{p_{Ac}}{p_{0c}} S - 2\frac{p_{Ac}}{p_{0c}} S(1 - p_{Ac}).$$

Now, the third term may be expanded to find the cth term to complete the sum represented by the first and fourth term. Also, the fifth and sixth terms may be combined. This leaves us with

$$\text{trace}\left((\tilde{\Sigma}_A \tilde{\Sigma}_0^{-1})^2 \right) = \sum_{i=1}^{c} \frac{p_{Ai}^2}{p_{0i}^2} + S^2 + \frac{p_{Ac}^4}{p_{0c}^2} - 2\sum_{i=1}^{c} \frac{p_{Ai}^3}{p_{0i}^2} + 2\frac{p_{Ac}^2}{p_{0c}} S$$

Noting that

$$\sum_{i=1}^{c} \frac{p_{Ai}^2}{p_{0i}^2} - 2\sum_{i=1}^{c} \frac{p_{Ai}^3}{p_{0i}^2} = \sum_{i=1}^{c} \frac{p_{Ai}^2}{p_{0i}^2} (1 - 2p_{Ai})$$

and

$$\left(\sum_{i=1}^{c} \frac{p_{Ai}^2}{p_{0i}} \right)^2 = \left(S + \frac{p_{Ac}^2}{p_{0c}} \right)^2 = S^2 + 2\frac{p_{Ac}^2}{p_{0c}} S + \frac{p_{Ac}^4}{p_{0c}^2},$$

we obtain our result. $\qquad\qquad\qquad\qquad\qquad\qquad\qquad\qquad\qquad\qquad$ □

Lemma 3.6. *Let μ and D be as in Lemma 3.2. Then the quantity $\mu^T D\mu$ is given by*

$$\mu^T D\mu = \sum_{i=1}^{c} \frac{p_{Ai}(p_{Ai} - p_{0i})^2}{p_{0i}^2} - \left(\sum_{i=1}^{c} \frac{p_{Ai}(p_{Ai} - p_{0i})}{p_{0i}}\right)^2.$$

Proof. From Equation (3.60), we have

$$\mu^T D\mu = \tilde{\delta}^T \tilde{\Sigma}_0^{-1/2} V D V^T \tilde{\Sigma}_0^{-1/2} \tilde{\delta}$$
$$= \tilde{\delta}^T \tilde{\Sigma}_0^{-1} \tilde{\Sigma}_A \tilde{\Sigma}_0^{-1} \tilde{\delta}$$

Now, in general, for $x \in \mathbb{R}^{c-1}$, we have

$$x^T \tilde{\Sigma}_A x = x^T \text{diag}(\tilde{p}_A)x - x^T \tilde{p}_A \tilde{p}_A^T x$$
$$= \sum_{i=1}^{c-1} p_{Ai} x_i^2 - \left(\sum_{i=1}^{c-1} p_{Ai} x_i\right)^2.$$

Therefore, since

$$\tilde{\Sigma}_0^{-1} \tilde{\delta} = \left(\text{diag}(\tilde{p}_0)^{-1} + p_{0c}^{-1} J\right)(\tilde{p}_A - \tilde{p}_0)$$
$$= \frac{\tilde{p}_A - \tilde{p}_0}{\tilde{p}_0} + \frac{p_{0c} - p_{Ac}}{p_{0c}} \tilde{1},$$

it follows that

$$\tilde{\delta}^T \tilde{\Sigma}_0^{-1} \tilde{\Sigma}_A \tilde{\Sigma}_0^{-1} \tilde{\delta} = \sum_{i=1}^{c-1} p_{Ai} \left[\tilde{\Sigma}_0^{-1} \tilde{\delta}\right]_i^2 - \left(\sum_{i=1}^{c-1} p_{Ai} \left[\tilde{\Sigma}_0^{-1} \tilde{\delta}\right]_i\right)^2$$

$$= \sum_{i=1}^{c-1} \frac{p_{Ai}(p_{Ai} - p_{0i})^2}{p_{0i}^2} + \frac{(p_{0c} - p_{Ac})^2}{p_{0c}^2}(1 - p_{Ac})$$

$$+ \frac{2(p_{0c} - p_{Ac})}{p_{0c}} \sum_{i=1}^{c-1} \frac{p_{Ai}(p_{Ai} - p_{0i})}{p_{0i}}$$

$$- \left[\sum_{i=1}^{c-1} \frac{p_{Ai}(p_{Ai} - p_{0i})}{p_{0i}} + \frac{(p_{0c} - p_{Ac})}{p_{0c}}(1 - p_{Ac})\right]^2$$

This last term is equivalent to

$$\left[\sum_{i=1}^{c} \frac{p_{Ai}(p_{Ai} - p_{0i})}{p_{0i}} + \frac{(p_{0c} - p_{Ac})}{p_{0c}}\right]^2 = \left(\sum_{i=1}^{c} \frac{p_{Ai}(p_{Ai} - p_{0i})}{p_{0i}}\right)^2 + \frac{(p_{0c} - p_{Ac})^2}{p_{0c}^2}$$

$$+ 2\frac{(p_{0c} - p_{Ac})}{p_{0c}} \sum_{i=1}^{c} \frac{p_{Ai}(p_{Ai} - p_{0i})}{p_{0i}}$$

Combining the above, we have

$$\mu^T D\mu = \sum_{i=1}^{c-1} \frac{p_{Ai}(p_{Ai} - p_{0i})^2}{p_{0i}^2} + \frac{(p_{0c} - p_{Ac})^2}{p_{0c}^2}(1 - p_{Ac})$$

$$+ \frac{2(p_{0c} - p_{Ac})}{p_{0c}} \sum_{i=1}^{c-1} \frac{p_{Ai}(p_{Ai} - p_{0i})}{p_{0i}}$$

$$- \left(\sum_{i=1}^{c} \frac{p_{Ai}(p_{Ai} - p_{0i})}{p_{0i}}\right)^2 - \frac{(p_{0c} - p_{Ac})^2}{p_{0c}^2}$$

$$- 2\frac{(p_{0c} - p_{Ac})}{p_{0c}} \sum_{i=1}^{c} \frac{p_{Ai}(p_{Ai} - p_{0i})}{p_{0i}}$$

Combining the third and sixth terms, we have

$$2\frac{p_{Ac}(p_{0c} - p_{Ac})^2}{p_{0c}^2}.$$

Similarly, the second and forth terms combine to yield

$$-\frac{p_{Ac}(p_{0c} - p_{Ac})^2}{p_{0c}^2}.$$

Altogether, the second, third, forth, and sixth terms combine to complete the sum represented by the first term. This completes the result.   □

Before moving further, let us collect the results from Lemmas 3.4–3.6 in the following.

Lemma 3.7. *Let* $D = \text{diag}(\tau_1^2, \ldots, \tau_{c-1}^2)$ *and* μ *be defined as in Lemma 3.2. Then*

$$S_1 = \sum_{i=1}^{c-1} \tau_i^2 = \sum_{i=1}^{c} \frac{p_{Ai}(1 - p_{Ai})}{p_{0i}}$$

$$S_2 = \sum_{i=1}^{c-1} \tau_i^4 = \sum_{i=1}^{c} \frac{p_{Ai}^2}{p_{0i}^2}(1 - 2p_{Ai}) + \left(\sum_{i=1}^{c} \frac{p_{Ai}^2}{p_{0i}}\right)^2$$

$$S_3 = \sum_{i=1}^{c-1} \mu_i^2 = \sum_{i=1}^{c} \frac{(p_{Ai} - p_{0i})^2}{p_{0i}}$$

$$S_4 = \sum_{i=1}^{c-1} \mu_i^2 \tau_i^2 = \sum_{i=1}^{c} \frac{p_{Ai}(p_{Ai} - p_{0i})^2}{p_{0i}^2} - \left(\sum_{i=1}^{c} \frac{p_{Ai}(p_{Ai} - p_{0i})}{p_{0i}}\right)^2.$$

Proof. Expression S_1 is given by

$$S_1 = \sum_{i=1}^{c-1} \tau_i^2 = \text{trace}(\tilde{\Sigma}_0^{-1/2} \tilde{\Sigma}_A \tilde{\Sigma}_0^{-1/2}) = \text{trace}(\tilde{\Sigma}_A \tilde{\Sigma}_0^{-1}),$$

and is therefore given by Lemma 3.4.

Similarly, expression $S_2 = \sum_{i=1}^{c-1} \tau_i^4$ is given by Lemma 3.5. Expression S_3 is given by

$$\sum_{i=1}^{c-1} \mu_i^2 = \mu^T \mu$$

$$= (\tilde{p}_A - \tilde{p}_0)^T \tilde{\Sigma}_0^{-1/2} V V^T \tilde{\Sigma}_0^{-1/2} (\tilde{p}_A - \tilde{p}_0)$$

$$= (\tilde{p}_A - \tilde{p}_0)^T \tilde{\Sigma}_0^{-1} (\tilde{p}_A - \tilde{p}_0)$$

However, we already showed that such an expression simplifies to Equation (3.65) in the proof of Proposition 3.12.

Finally, Expression $S_4 = \mu^T D \mu$ is given by Lemma 3.6. This completes the result.   □

Theorem 3.8. *Under an alternative hypothesis $H_A : p = p_A \in \Theta_0^c$, the chi-squared statistic, for null hypothesis $H_0 : p = p_0$, is approximately distributed for large n as a location–scale shifted noncentral chi-squared distribution*

$$Q \sim a\text{NCX}_\nu^{n\lambda} + b, \tag{3.62}$$

where parameters a, ν, λ, and b are given by

$$a = \frac{S_4}{S_3} \tag{3.63}$$

$$\lambda = \frac{S_3}{a} \tag{3.64}$$

$$\nu = \frac{S_2}{a^2}, \tag{3.65}$$

$$b = S_1 - a\nu \tag{3.66}$$

where we have defined

$$S_1 = \sum_{i=1}^{c} \frac{p_{Ai}(1 - p_{Ai})}{p_{0i}} \tag{3.67}$$

$$S_2 = \sum_{i=1}^{c} \frac{p_{Ai}^2}{p_{0i}^2} (1 - 2p_{Ai}) + \left(\sum_{i=1}^{c} \frac{p_{Ai}^2}{p_{0i}} \right)^2 \tag{3.68}$$

$$S_3 = \sum_{i=1}^{c} \frac{(p_{Ai} - p_{0i})^2}{p_{0i}} \tag{3.69}$$

$$S_4 = \sum_{i=1}^{c} \frac{p_{Ai}(p_{Ai} - p_{0i})^2}{p_{0i}^2} - \left(\sum_{i=1}^{c} \frac{p_{Ai}(p_{Ai} - p_{0i})}{p_{0i}} \right)^2. \tag{3.70}$$

Note 3.25. The classic power function provided by Proposition 3.11 simply uses the parameter S_3 for the noncentrality factor λ and $c - 1$ degrees of freedom.
$\triangleright$

Proof. From Lemma 3.2, we know that under an alternative hypothesis $p = p_A$, the chi-squared statistic is approximately distributed as a weighted sum of noncentral chi-squared random variables, such that

$$Q \sim \sum_{i=1}^{c-1} \tau_i^2 \mathrm{NCX}_1^{n\lambda_i},$$

where τ_i^2 is an eigenvalue of $\tilde{\Sigma}_0^{-1/2} \tilde{\Sigma}_A \tilde{\Sigma}_0^{-1/2}$, and $\mu = V^T \tilde{\Sigma}_0^{-1/2} (\tilde{p}_A - \tilde{p}_0)$, and $\lambda_i = \mu_i^2 / \tau_i^2$. Let us make a Satterthwaite ansatz that this distribution may be approximated in the form of Equation (3.62); i.e., as the location-scale shifted noncentral chi-squared distribution

$$\sum_{i=1}^{c-1} \tau_i^2 \mathrm{NCX}_1^{n\lambda_i} \approx a \mathrm{NCX}_\nu^{n\lambda} + b,$$

for some a, ν, λ, and b. Comparing the first and second moments of both sides of this and applying Equations (3.28) and (3.29) we obtain

$$\sum_{i=1}^{c-1} \tau_i^2 (1 + n\lambda_i) = a(\nu + n\lambda) + b$$

$$\sum_{i=1}^{c-1} \tau_i^4 (1 + 2n\lambda_i) = a^2 (\nu + 2n\lambda)$$

Separating the $O(1)$ and $O(n)$ terms, we can convert this into a system of four equations:

$$\sum_{i=1}^{c-1} \tau_i^2 = a\nu + b, \qquad \sum_{i=1}^{c-1} \tau_i^2 \lambda_i = a\lambda,$$

$$\sum_{i=1}^{c-1} \tau_i^4 = a^2 \nu, \quad \text{and} \quad \sum_{i=1}^{c-1} \tau_i^4 \lambda_i = a^2 \lambda.$$

Recognizing the left-hand sides as the quantities S_1–S_4, as per Lemma 3.7, we may solve this system for our parameters, thereby obtaining our result.
$\square$

We now arrive are our key result. Now that we know the approximate distribution of the Q statistic under any given alternative hypothesis, which is the result of our previous theorem, we can compute the power function. As usual, this is tantamount to evaluating the survival function of the distribution Equation (3.62) at the critical point $\chi^2_{c-1,\alpha}$. Moreover, note that the parameters a, ν, λ, and b are each functions of the alternative probability p_A, via Equations (3.63)–(3.70). The result is given as follows.

Corollary 3.9. *The power function of Pearson's chi-squared test, as defined in Definition 3.19, with sample size* n, *is given by*

$$\beta(p_A) = S(\chi^2_{c-1,\alpha}; n, a, \nu, \lambda, b),$$

where S *is the survival function of the scaled noncentral chi-squared random variable defined in Theorem 3.8.*

Example 3.11. Suppose we are running a five-category Pearson chi-squared test with null hypothesis

$$H_0 : p = p_0 = \langle 0.2, 0.2, 0.2, 0.2, 0.2 \rangle,$$

and we want to determine the required sample size n to ensure that we achieve a power of $\beta = 0.80$ for the particular alternate hypothesis

$$p_A = \langle 0.1, 0.3, 0.2, 0.3, 0.1 \rangle.$$

We proceed as shown in Code Block 3.10.   We begin by defining our null hypothesis and our particular alternate hypothesis (lines 1–3). Next, we compute the sums S_1–S_4 (lines 5–8), which we use to compute our parameters a, λ, ν, and b (lines 10–13). We obtain values of

$$S_1 = 3.8, \qquad S_2 = 4.24, \qquad S_3 = 0.2, \qquad S_4 = 0.16,$$

and parameter values

$$a = 0.8, \qquad \lambda = 0.25, \qquad \nu = 6.625, \qquad b = -1.5.$$

Next, we compute the critical value of our test, $Q_{crit} = 9.487729$, and then solve for the value of n that will yield an 80% survival value at Q_{crit} under the alternate hypothesis (lines 15–20).

Theoretically, we are done. But let's go a step further and validate that our answer is correct. To do this, we can run 100,000 simulations of our test, using data consistent with our particular alternate hypothesis, and count the frequency at which our test (correctly) rejects the null hypothesis. The simulated power 79.90% closely matches our predicted power 79.64%, as given by the distribution Equation (3.62); this is an observed error of 0.26%. Contrast that with the classic alternative distribution Equation (3.52), which predicts a power of 77.9%; an error of 2.00%.

To further illustrate the difference between our novel power function and the classic, consider instead the alternate hypothesis

$$p_A = \langle 0, 0.4, 0.2, 0.4, 0 \rangle.$$

Our formula yields a required sample size of $n = 12$, yielding a power of $\beta = 0.7747$, whereas the classic power for a sample size of twelve would be $\beta = 0.6957$. The observed power for the resulting simulation is $\beta_{obs} =$

```
1   p_0 = np.array([0.2, 0.2, 0.2, 0.2, 0.2])
2   p = np.array([0.1, 0.3, 0.2, 0.3, 0.1])
3   dof = len(p) - 1
4
5   S1 = np.sum( p*(1 - p) / p_0 ) #3.8
6   S2 = np.sum( p**2*(1-2*p)/p_0**2 ) + (np.sum( p**2/p_0 ))**2 #4.24
7   S3 = np.sum( (p-p_0)**2/p_0 ) #0.20
8   S4 = np.sum( p*(p-p_0)**2/p_0**2 ) - (np.sum( p*(p-p_0)/p_0 ))**2
        #0.16
9
10  a = S4 / S3     #0.800
11  la = S3 / a     #0.250
12  nu = S2 / a**2 #6.625
13  b = S1 - a * nu #-1.50
14
15  alpha = 0.05
16  Q_crit = scipy.stats.chi2.isf(alpha, dof) # Critical Value 9.487729
17  beta = 0.80
18  results = scipy.optimize.root(lambda x: scipy.stats.ncx2.sf(Q_crit,
        nu, x*la, scale=a, loc=b) - beta, 10)
19  n = int(results['x']) # 57
20  beta = scipy.stats.ncx2.sf(Q_crit, nu, n*la, scale=a, loc=b) #0.7964
21  beta_classic = scipy.stats.ncx2.sf(Q_crit, dof, S3)#0.7790
22
23  n_trials = 100000
24  E_0 = n * np.array(p_0) # Expected frequencies; array REQUIRED!
25  E_A = n * np.array(p) # Expected frequencies; array REQUIRED!
26  rejections = np.zeros(n_trials)
27  for i in range(n_trials):
28      x = np.random.multinomial(n, p)
29      Q = np.sum( (x - E_0)**2 / E_0 )
30      rejections[i] = int( Q > Q_crit )
31
32  print ('Predicted Power', beta) # 0.7964
33  print ('Observed Rejection Rate', np.sum(rejections)/n_trials) #
        0.7990
```

Code Block 3.10: Power of a chi-squared test

0.7599. The error of our formula is 1.48%, whereas the classic error is 6.41%. If we were to instead follow the classic formula, we would have required a larger sample size of $n = 14$, with classic power 77.07%, novel predicted power 86.73%. The observed simulated power, for sample size of 14, actually ends up at 100%.

Finally, we note that we can further validate the theory behind our expressions for S_1–S_4, from Equations (3.67)–(3.70), as shown in Code

```
1   ps = p[:dof]
2   ps0 = p_0[:dof]
3   Sigma = np.diag(ps) - np.outer(ps, ps) # Outer product: ps @ ps.T
4   Sigma0 = np.diag(ps0) - np.outer(ps0, ps0)
5   Sigma0Rt = scipy.linalg.sqrtm(Sigma0)
6   Sigma0RtInv = np.linalg.inv(Sigma0Rt)
7   D, V = np.linalg.eig(Sigma0RtInv @ Sigma @ Sigma0RtInv)
8   mu = V.T @ Sigma0RtInv @ (ps - ps0)
9
10  S1 = np.sum( D ) # 3.8
11  S2 = np.sum( D**2 ) # 4.24
12  S3 = np.sum( mu**2 ) #0.20
13  S4 = np.sum( D * mu**2 ) # 0.16
```

Code Block 3.11: Validation of our expressions for S_1–S_4

Block 3.11.[6] Here, we compute the reduced probability vectors $\tilde{p}_A$ and $\tilde{p}_0$, and compute the matrices $\tilde{\Sigma}_A$, $\tilde{\Sigma}_0$, and $\tilde{\Sigma}_0^{-1/2}$, from which we can compute the matrices eigenvalues D and eigenvectors V, from which we can further determine the vector μ. We can then compute the four expressions given in Lemma 3.7:

$$S_1 = \sum_{i=1}^{c-1} \tau_i^2 = \texttt{np.sum(D)}, \qquad S_2 = \sum_{i=1}^{c-1} \tau_i^4 = \texttt{np.sum(D**2)}$$

$$S_3 = \sum_{i=1}^{c-1} \mu_i^2 = \texttt{np.sum(mu**2)} \qquad S_4 = \sum_{i=1}^{c-1} \tau_i^2 \mu_i^2 = \texttt{np.sum(D*mu**2)}.$$

As expected, the results produced by lines 10–13 of Code Block 3.11 produce identical results to the output of lines 5–8 of Code Block 3.10, thereby validating all of our work from Lemmas 3.3–3.6, which is further summarized in Lemma 3.7.   ▷

3.5.4 Sampling the Simplex

Let us take a moment to recap. Given a null hypothesis $H_0 : p = p_0 \in \Delta^{c-1}$ and an alternate hypothesis $p_A \in \Theta_0^c$, we have constructed an approximate power function as a function of the sample size $\beta = \beta(p, n; p_0)$ for Pearson's chi-squared test (Definition 3.19). However, this method computes the power of a test at a single point, whereas we are often interested to know the worst-case power for a given minimum detectable effect. One approach

---

[6] Note that in lines 3–4 of Code Block 3.11, we explicitly use `np.outer(ps, ps)` as opposed to `ps @ ps.T`. This is because the latter will not work correctly unless we reshape the array `ps` as a column vector.

to modeling this is to sample the simplex Δ^{c-1}, and remove those samples within a δ-ball of the point p_0; i.e., we would consider only $p \in \Delta^{c-1}$ for which $||p - p_0|| \geq \delta$, relative to a given norm. This is a form of *Monte-Carlo simulation*, a topic we shall see in numerous contexts later on. For this reason, we briefly discuss a method for sampling the standard simplex.

Recall from Definition 2.23 that the standard simplex Δ^{c-1} is equivalent to the portion of the hyperplane $\sum_{i=1}^{c} x_i = 1$ that lies within the unit cube $[0,1]^c \subset \mathbb{R}^c$. A naive approach to sample the simplex would therefore be to sample the hypercube and then normalize the result. This approach, however, does not produce a uniform sampling of the standard simplex.

One correct approach that is particularly clever relies on the concept of a *random spacing* of the unit interval, defined as follows.

Definition 3.24. *Let* $X_1, \ldots, X_{c-1} \sim \text{Unif}(0,1)$ *be a* IID *uniform sample of the unit interval* $[0,1]$, *and let* $X_{(1)} < \cdots < X_{(c-1)}$ *be the corresponding order statistics. Further, let us define the endpoints* $X_{(0)} = 0$ *and* $X_{(k)} = 1$. *Then the random variables* $S_1, \ldots, S_c$, *defined by*

$$S_i = X_{(i)} - X_{(i-1)}$$

for $i = 1, \ldots, c$, *are called a* random spacing of size c *of the unit interval.*

It is possible to show that a sample of uniform spacings constitutes a uniform sampling of the simplex, a result we present as follows.

Theorem 3.9. *The random vector* $S = (S_1, \ldots, S_c)$ *consisting of a random spacing of size c of the unit interval is distributed uniformly on the standard simplex* Δ^{c-1}.

We leave the proof to Devroye [1986]. However, note that due to the definition of a random spacing, it follows that each $S_i \geq 0$ and the sum

$$\sum_{i=1}^{c} S_i = 1,$$

which shows that the random vector $S \in \Delta^{c-1}$.

A simple function which generates a random point from the simplex is given by the first few lines of Code Block 3.12. This is then generalized to produce a sample of a given size in the subsequent lines of code. The random sample of the simplex Δ^2 is shown in Figure 3.5[7].

Sample Size for Pearson's Chi-squared Test

We can combine our sampling technique with our theory of the power function in order to approximate the power of the test as a function of the effect size

---

[7] Make sure to import: `from mpl_toolkits.mplot3d import Axes3D`.

```
1   def randomSimplex(c):
2       """ Returns a random point on the
3       simplex Delta^{c-1}
4       """
5       x = np.random.random(c-1)
6       x = np.append(x, [0, 1])
7       x.sort()
8       return np.diff(x)
9
10  def randomMultiSimplex(c, size=10):
11      """ Returns a random sample from the
12      simplex Delta^{c-1}.
13      Output is an array of shape (size, c)
14      """
15      X = np.ones((size, 2))
16      X[:, 0] = 0 # First column is 0
17      x = np.random.random((size, c-1))
18      x = np.append(x, X, axis=1)
19      x.sort(axis=1)
20      return np.diff(x, axis=1)
21
22  ## Generate and plot a random sample Delta^2.
23  fig = plt.figure(figsize=(8, 9/2))
24  ax = fig.add_subplot(111, projection='3d')
25  X = randomMultiSimplex(3, size=1000)
26  ax.scatter3D(X[:, 0], X[:, 1], X[:, 2], '.')
```

Code Block 3.12: Generate a random vector from the standard simplex

$$\delta(p) = ||p - p_0||.$$

We illustrate the approach with the following shining example.

Example 3.12. Determine the required sample size for a chi-squared test with 80% power and 5% significance level, given the null hypothesis

$$p_0 = \langle 0.2, 0.2, 0.2, 0.2, 0.2 \rangle.$$

The code is shown in Code Block 3.13.

We begin by creating a random sample of our simplex, using the result of Theorem 3.9 (line 8). We can then sort this sample so that the individual samples are ordered by the effect size (lines 9–11). For each sample, we compute the parameters a, λ, ν, and b for the approximation given by Equation (3.62) of the distribution of the chi-squared statistic under the given sample alternate hypothesis (lines 19–26). We can use the built-in `scipy.optimize.root` algorithm for root-finding in order to determine the sample size that generates a power of 80%, for the given alternate hypothesis. We then record the result and a flag for instances in which the

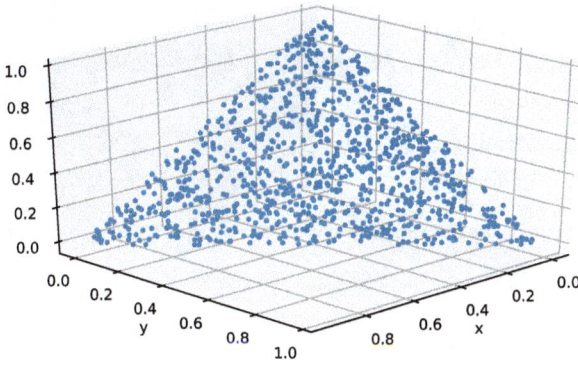

Fig. 3.5: Uniform random sample of the simplex Δ^2.

algorithm fails to converge (lines 28–30). We used an initial guess of 5, as a larger value would generate convergence issues for larger effect sizes. Finally, we can compute the required sample size for any given minimum detectable effect in lines 32–33. More interesting, however, is the plot of delta vs. ns, as shown in Figure 3.6. For instance, we see from this curve

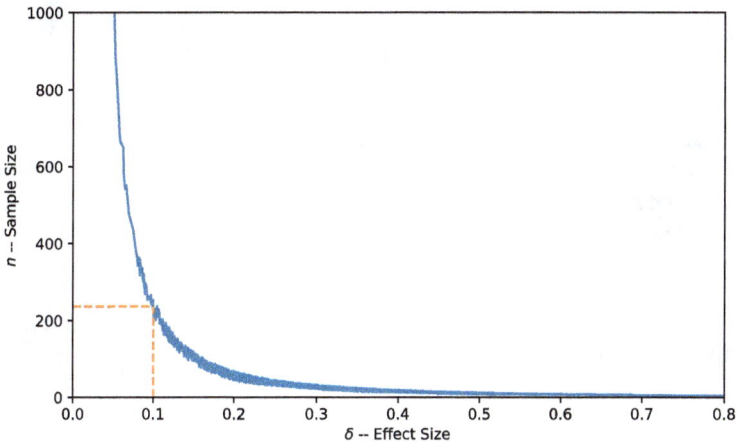

Fig. 3.6: Power function for chi-squared test.

```
1   p_0 = np.array([0.2, 0.2, 0.2, 0.2, 0.2])
2   d = len(p_0)
3   dof = d - 1
4   alpha = 0.05
5   Q_crit = scipy.stats.chi2.isf(alpha, dof)
6
7   n_samples = 10000
8   X = randomMultiSimplex(d, size=n_samples)
9   i = np.argsort( np.linalg.norm(X-p_0, axis=1) )
10  X = X[i, :]
11  delta = np.linalg.norm(X-p_0, ord=order, axis=1) #Array of deltas
12
13  beta = 0.8
14  ns = np.zeros(n_samples) # array of sample sizes
15  flags = np.zeros(n_samples)
16  for i in range(n_samples):
17      p = X[i, :]
18
19      S1 = np.sum( p*(1 - p) / p_0 )
20      S2 = np.sum( p**2*(1-2*p)/p_0**2 ) + (np.sum( p**2/p_0 ))**2
21      S3 = np.sum( (p-p_0)**2/p_0 )
22      S4 = np.sum( p*(p-p_0)**2/p_0**2 ) - (np.sum( p*(p-p_0)/p_0 ))**2
23      a = S4 / S3
24      la = S3 / a
25      nu = S2 / a**2
26      b = S1 - a * nu
27
28      results = scipy.optimize.root(lambda x: scipy.stats.ncx2.sf(
                Q_crit, nu, x*la, scale=a, loc=b ) - beta, 5)
29      flags[i] = int( not results['success'] ) # Errors
30      ns[i] = int(results['x'])
31
32  MDE = 0.1
33  sample_size = int(max(ns[delta>MDE])) # 237
```

Code Block 3.13: Power as a Function of Effect Size; Sample Size for Minimum Detectable Effect

that a minimum sample size of 237 is required to guarantee a minimum detectable effect of $\delta = 0.1$.   ▷

3.6 Tests for Distributions

Often, in practice, we have a set of data. The distribution that generated the data—or even its form—is unknown. The problem of statistical inference therefore extends beyond the simple problem of estimating a set of parameters that match an observed set of data to a parametric family of distributions: we must also infer which family of distributions the data belong to. In this section, we will discuss various tests, collectively known as goodness-of-fit tests, that seek to address this problem. These tests will each be of the following form.

Definition 3.25. *Given an* IID *sample* $X_1, \ldots, X_n \sim F$ *and a known* CDF F_0, *a goodness-of-fit test is any hypothesis test with a null hypothesis of the form*

$$H_0 : F = F_0.$$

3.6.1 Chi-squared Goodness-of-Fit Test

Pearson's chi-squared test (Definition 3.23) can readily be adopted as our first goodness-of-fit test, with the following natural modification.

Definition 3.26. *Let* $X_1, \ldots, X_n \sim F$ *be an* IID *sample from a continuous distribution* F, *and let* F_0 *be a known* CDF. *Fix* $c \in \mathbb{Z}_+$, *a positive integer (a typical value might be* $c = 5$ *or* $c = 10$). *Next, define the boundary points*

$$b_i = F_0^{-1}(i/c),$$

for $i = 1, \ldots, (c-1)$, *which partition the real line into* c *buckets*

$$A_i = (-b_{i-1}, b_i],$$

for $i = 1, \ldots, c$, *where we have further defined* $b_0 = -\infty$ *and* $b_c = \infty$. *Finally, define the counts*

$$Y_i = \sum_{j=1}^{n} \mathbb{I}[X_j \in A_i],$$

for $i = 1, \ldots, c$, *of the number of occurrences of the random numbers* X_i *being found within each bucket. Then the* chi-squared goodness-of-fit test *consists of the null hypothesis given by Definition 3.25, the test statistic*

$$Q = \sum_{i=1}^{c} \frac{(Y_i - np_0)^2}{np_0},$$

where $p_0 = 1/c$, *and rejection region*

$$R_\alpha = \left[\chi^2_{c-1,\alpha}, \infty\right).$$

Theorem 3.10. *Under the null hyptohesis, the chi-squared goodness-of-fit test is approximately a size-α test for large n. Moreover, under the null hyptohesis, the test statistic* $Q \to \chi^2_{c-1}$, *as* $n \to \infty$.

Proof. Under the null hypothesis, the sets $A_1, \ldots, A_c$ constitute equal-probability subsets of the real line, each with probability

$$\mathbb{P}(X \in A_i) = \frac{1}{c}.$$

Therefore, under the null hypothesis, $Y \sim \text{Multi}(c, p)$, where p is the equal-probability vector $p = (1/c)\langle 1, \ldots, 1 \rangle \in \mathbb{R}^c$. The result therefore follows from Theorem 3.6. □

Example 3.13. Given a sample of IID data $X_1, \ldots, X_n \sim F$, consider the null hypothesis $H_0 : F = F_0$, where $F_0 = \text{Exp}(\beta_0)$, with $\beta_0 = 7$. We will follow Definition 3.26, using $c = 10$ buckets.

We will run a simulation to see what happens when our null hypothesis is correct. The setup is shown in Code Block 3.14. We define our null-hypothesis value β_0 in line 1, and the (hidden) true value in line 2. The samples are generated in line 4 using the true value, and the boundary points are determined using the null hypothesis distribution. The `bucketCounter` function takes as inputs the samples we generated in line 4 along with the boundary points we generated in line 6, and returns a vector of counts of the number of samples that fall within each of our ten buckets. An example output is shown on line 22. Our goal is then to use Pearson's chi-squared test against the null hypothesis that each bucket has a true probability of 10%.

Next, let us repeat our simulation many times and compute the frequency at which we would (falsely) reject our (true) null hypothesis. This is shown in Code Block 3.15. Upon running the simulation 10,000 times, we see that the null hypothesis was rejected 4.96% of the time, in agreement with our value of $\alpha = 0.05$. ▷

3.7 Analysis of Variance

We conclude this chapter with an introduction to the analysis of variance (ANOVA). Though our initial theoretical groundwork is laid from the perspective of hypothesis testing, we will later find that the theory has much broader implications that span experimentation, regression theory, and generalized linear models. In the oneway ANOVA framework, observations are classified into groups, or *treatments*, along with the following modeling assumptions.

Definition 3.27. *A oneway ANOVA model identifies each datum with one of m groups, called* treatments, *such that observations satisfy the model*

$$Y_{ij} = \theta_i + \epsilon_{ij}, \qquad i = 1, \ldots, m; \qquad j = 1, \ldots, r_i, \qquad (3.71)$$

according to the following ANOVA assumptions

```
1   beta_0 = 7 # null hypothesis
2   beta = 7   # true value
3   n_samples = 1000
4   samples = np.random.exponential(scale=beta, size=n_samples) #true
        value
5   percentiles = [0.1*i for i in range(1, 10)]
6   bs = scipy.stats.expon.ppf(percentiles, scale=beta_0) # bounaries
7
8   def bucketCounter(samples, boundaries=[0]):
9       boundaries = np.sort(boundaries) #Make sure they are ordered
10      n_buckets = len(boundaries) + 1
11      n_samples = len(samples)
12      counts = np.zeros(n_buckets)
13      n_counted = 0
14
15      for i, b in enumerate(boundaries):
16          counts[i] = (samples <= b).sum() - n_counted
17          n_counted += counts[i]
18
19      counts[i+1] = n_samples - n_counted
20      return counts
21      # example output:
22      # [ 96., 91., 102., 102., 121., 100., 85., 109., 86., 108.]
```

Code Block 3.14: Simulated draws from Exp(7) and the bucket counter

1. *The* errors *have zero expected value and finite variance:* $\mathbb{E}[\epsilon_{ij}] = 0$ *and* $\mathbb{V}(\epsilon_{ij}) = \sigma_i^2 < \infty$, *for all* i, j;
2. *the errors are uncorrelated* $\text{COV}(\epsilon_{ij}, \epsilon_{kl}) = 0$ *whenever* $i \neq k$ *and* $j \neq l$;
3. *the errors* ϵ_{ij} *are independent and normally distributed;*
4. *the variance is independent of treatment, i.e.,* $\sigma_i^2 = \sigma^2$, *for all* i *(also known as* homoscedasticity*).*

Note 3.26. Sometimes the model given by Equation (3.71) is cast as

$$Y_{ij} = \mu + \tau_i + \epsilon_{ij},$$

where μ is a sort of *overall mean*, and τ_i is the *treatment effect*. This prescription, however, suffers identifiability issues, as the parameters cannot be uniquely solved for without further restraint. Nevertheless, this is the form will will predominantly use later during our discussion on experimentation.
▷

The assumption of independent and normally distributed error terms is important for defining confidence intervals. Further note that we do not require there to be an equal amount of data in each treatment. Though of

```
1   beta_0 = 7 # null hypothesis
2   beta = 7   # true value
3   n_samples = 1000
4   n_trials = 10000
5   percentiles = [0.1*i for i in range(1, 10)]
6   bs = scipy.stats.expon.ppf(percentiles, scale=beta_0) # bounaries
7   alpha = 0.05
8   dof = 9
9   Q_crit = scipy.stats.chi2.isf(alpha, dof)
10  E_0 = n_samples / (dof+1) * np.ones(dof+1)
11  rejections = np.zeros(n_trials)
12
13  for i in range(n_trials):
14      samples = np.random.exponential(scale=beta, size=n_samples)
15      x = bucketCounter(samples, boundaries=bs)
16      Q = np.sum( (x - E_0)**2 / E_0 )
17      rejections[i] = int( Q > Q_crit )
18
19  print ('Observed Rejection Rate', np.sum(rejections) / n_trials)
        #0.0496
```

Code Block 3.15: Simulation of Chi-squared goodness of fit test

limited interest in modern applications, the classic ANOVA test is given by the following.

Definition 3.28. *The* (classic) *ANOVA null hypothesis* *is the hypothesis*

$$H_0: \qquad \theta_1 = \theta_2 = \cdots = \theta_m. \qquad (3.72)$$

In this section, we shall consider the classic ANOVA null hypothesis, which states that there is no difference between any of the treatments. The alternative hypothesis is therefore that there is *some* difference between the treatments, without regard as to where that difference may lie. In practice, however, one is typically more interested in estimating the performance of various treatments to determine which is superior. We shall discuss this in our next chapter when treating the theory of experimentation.

The main idea behind analysis of variance is to partition the observed variance, known as the sum of squares, into two components: one representing the *within*-group and one representing the *between*-groups sum of squares. This is captured in the following theorem.

Theorem 3.11 (Partitioning Theorem). *For any set of numbers* y_{ij}, *where* $i = 1, \ldots, m$ *and* $j = 1, \ldots, r_i$,

$$\sum_{i=1}^{m} \sum_{j=1}^{r_i} \left(y_{ij} - \overline{\overline{y}} \right)^2 = \sum_{i=1}^{m} r_i \left(\bar{y}_{i\cdot} - \overline{\overline{y}} \right)^2 + \sum_{i=1}^{m} \sum_{j=1}^{r_i} \left(y_{ij} - \bar{y}_{i\cdot} \right)^2 ; \qquad (3.73)$$

where we have followed the dot notation *for brevity:*

$$y_{i\cdot} = \sum_{j=1}^{r_i} y_{ij}, \qquad \bar{y}_{i\cdot} = \frac{1}{r_i} \sum_{j=1}^{r_i} y_{ij}, \qquad \bar{\bar{y}} = \frac{1}{n} \sum_{i=1}^{m} \sum_{j=1}^{r_i} y_{ij}, \qquad (3.74)$$

where $n = \sum_{i=1}^{m} r_i$.

Proof. The proof is straightforward. Begin by expanding the left-hand side:

$$\sum_{i=1}^{m} \sum_{j=1}^{r_i} \left(y_{ij} - \bar{\bar{y}}\right)^2 = \sum_{i=1}^{m} \sum_{j=1}^{r_i} \left[\left(y_{ij} - \bar{y}_{i\cdot}\right) + \left(\bar{y}_{i\cdot} - \bar{\bar{y}}\right)\right]^2.$$

The result follows by expanding the right-hand side and recognizing that the cross terms vanish, which is verified in Exercise 3.6.   □

Theorem 3.11 relates the total sum of square to the between-group and within-group sums of squares, as defined in the following.

Definition 3.29. *For any set of numbers* y_{ij}, *where* $i = 1, \ldots, m$ *and* $j = 1, \ldots, r_i$, *the* between-group sum of squares *(SSB), the* within group sum of squares *(SSW), and the* total sum of squares *(SST) are defined by*

$$SSB = \sum_{i=1}^{m} r_i \left(\bar{y}_{i\cdot} - \bar{\bar{y}}\right)^2, \qquad (3.75)$$

$$SSW = \sum_{i=1}^{m} \sum_{j=1}^{r_i} \left(y_{ij} - \bar{y}_{i\cdot}\right)^2, \qquad (3.76)$$

$$SST = \sum_{i=1}^{m} \sum_{j=1}^{r_i} \left(y_{ij} - \bar{\bar{y}}\right)^2. \qquad (3.77)$$

Moreover, the between-group (MSB) *and* within-group (MSW) mean squares *are defined by the relations*

$$MSB = \frac{SSB}{m - 1} \qquad and \qquad MSW = \frac{SSW}{n - m}, \qquad (3.78)$$

respectively.

Theorem 3.11 therefore states that the total sum of squares is equal to the sum of the between-group and within-group sums of squares.

The total sum of squares describes the overall variation of the data, without regard to classification into the various treatment groups. The between-group sum of squares, SSB, captures the variation by comparing the various groups; notice that the mean of each group is compared to the overall mean, and the sum is only over the m separate groups. Finally, the within-group sum of squares yields the average variation within each group; i.e., each data point within each group is compared to that group's mean,. Thus, the within-group sum of squares measures variation with respect to random error, whereas the between-group sum of squares measures variation due to treatment differences.

We next proceed to determine a test statistic for the ANOVA null hypothesis. We begin by considering two lemmas.

Lemma 3.8. *Under the ANOVA assumptions of Definition 3.27, the following scaled within-group sum of squares is a chi-squared random variable with $(n - m)$ degrees of freedom:*

$$\frac{1}{\sigma^2} \sum_{i=1}^{m} \sum_{j=1}^{r_i} \left(Y_{ij} - \bar{Y}_{i\cdot}\right)^2 \sim \chi^2_{n-m}.$$

Proof. Following the ANOVA assumptions, we have $Y_{ij} \sim N(\theta_i, \sigma^2)$. Therefore, for each $i = 1, \ldots, m$, the quantities

$$\frac{1}{\sigma^2} \sum_{j=1}^{r_i} \left(Y_{ij} - \bar{Y}_{i\cdot}\right)^2 \sim \chi^2_{r_i-1}$$

are independent chi-square random variables. Due to their independence, their sum is also a chi-squared random variable with $\sum_{i=1}^{m}(r_i - 1) = n - m$ degrees of freedom, thus proving the result. □

Lemma 3.9. *Under the ANOVA assumptions of Definition 3.27 and the ANOVA null hypothesis of Definition 3.28, the following scaled between-group and total sums of squares are chi-squared random variables with $(m - 1)$ and $(n - 1)$ degrees of freedom, respectively.*

$$\frac{1}{\sigma^2} \sum_{i=1}^{m} r_i \left(\bar{Y}_{i\cdot} - \bar{\bar{Y}}\right)^2 \sim \chi^2_{m-1} \quad and \quad \frac{1}{\sigma^2} \sum_{i=1}^{m} \sum_{j=1}^{r_i} \left(Y_{ij} - \bar{\bar{Y}}\right)^2 \sim \chi^2_{n-1}.$$

Proof. Under the ANOVA null hypothesis, we have $\theta_i = \theta_j$, for all i, j. Thus, each observation is an independent draw from

$$Y_{ij} \sim N\left(\theta, \sigma^2\right).$$

We therefore recognize the total sum of squares as a scaled version of the sample variance. From Theorem 2.6, it therefore follows that $(1/\sigma^2)SST \sim \chi^2_{n-1}$. This proves the second relation.

For the first relation, we instead consider the sample means $\bar{Y}_{i\cdot}$. Under the null hypothesis, the sample mean is distributed according to

$$\bar{Y}_{i\cdot} \sim N\left(\theta, \frac{\sigma^2}{r_i}\right).$$

We recognize the between-group sum of squares as a scaled version of the sample variance *of the sample means*. It therefore follows (again from Theorem 2.6) that $(1/\sigma^2)SSB \sim \chi^2_{m-1}$. □

In light of Lemmas 3.8 and 3.9, the partitioning theorem, when scaled by σ^2, is actually a partitioning of a χ^2_{n-1} random variable into the sum of two independent chi-squared random variables χ^2_{m-1} and χ^2_{n-m}. Moreover, this decomposition leads us to the following theorem, which yields a test-statistic for the ANOVA null hypothesis.

Theorem 3.12. *Under the ANOVA assumptions of Definition 3.27 and the ANOVA null hypothesis of Definition 3.28, the statistic*

$$F = \frac{MSB}{MSW} = \frac{\dfrac{1}{m-1}\displaystyle\sum_{i=1}^{m} r_i\left(\bar{y}_{i\cdot} - \bar{\bar{y}}\right)^2}{\dfrac{1}{n-m}\displaystyle\sum_{i=1}^{m}\sum_{j=1}^{r_i}\left(y_{ij} - \bar{y}_{i\cdot}\right)^2}, \tag{3.79}$$

known as the F statistic, *has an F distribution with $m-1$ and $n-m$ degrees of freedom.*

Proof. This follows immediately from Lemmas 3.8 and 3.9. □

These definitions are summarized neatly in Table 3.4. In general, computation of Equation (3.79) is generally done by completion of such an ANOVA table.

	Degrees of freedom	Sum of squares	Mean square
Between groups	$m-1$	$SSB = \displaystyle\sum_{i=1}^{m} r_i(\bar{y}_{i\cdot} - \bar{\bar{y}})^2$	$MSB = \dfrac{SSB}{m-1}$
Within groups	$n-m$	$SSW = \displaystyle\sum_{i=1}^{m}\sum_{j=1}^{r_i}(y_{ij} - \bar{y}_{i\cdot})^2$	$MSW = \dfrac{SSW}{n-m}$
Total	$n-1$	$SST = \displaystyle\sum_{i=1}^{m}\sum_{j=1}^{r_i}(y_{ij} - \bar{\bar{y}})^2$	$F = MSB/MSW$

Table 3.4: ANOVA table.

Note that as a consequence of Theorem 3.12, we should reject the ANOVA null hypothesis whenever the F statistic exceeds the value

$$F > F^\alpha_{m-1,n-m},$$

where α is our significance level and $F^\alpha_{m-1,n-m}$ is the unique real number such that $\mathbb{P}(X > F^\alpha_{m-1,n-m}) = \alpha$ for $X \sim F_{m-1,n-m}$. Next, we will test our intuition versus solid statistical theory with a practical example.

Example 3.14. Consider data from five distinct treatment groups, as shown in Table 3.5. Each group consists of exactly eight data points, so that $r_i = 8$, for $i = 1, \ldots, 5$. The mean of each treatment group is given in the final column, and the overall mean of all the data is 97.175. What conclusions can we draw from these data? Is group 4 the clear and decisive victor, with groups 2 and 5 lagging sorely behind? Or is the variation in these means due to random chance?

Group	1	2	3	4	5	6	7	8	Mean
1	103	98	103	112	74	106	79	85	95
2	71	73	102	73	118	118	81	105	92.625
3	96	87	110	123	84	96	82	109	98.375
4	103	91	119	111	147	103	112	95	110.125
5	70	89	107	106	92	96	84	74	89.75

Table 3.5: Data for Example 3.14; overall mean 97.175

We begin by stating our classic ANOVA null hypothesis, that there is no difference between treatment groups:

$$H_0 : \theta_1 = \theta_2 = \theta_3 = \theta_4 = \theta_5.$$

The F statistic is computed in Table 3.6 to be $F = 1.914$. At a 5% sig-

	Degrees of freedom	Sum of squares	Mean square
Between groups	4	$SSB = 1997.65$	$MSB = 499.4125$
Within groups	35	$SSW = 9132.125$	$MSW = 260.917857$
Total	39	$SST = 11129.775$	$F = 1.914$

Table 3.6: ANOVA table for Example 3.14.

nificance level, $F_{4,35}^{0.05} = 2.64$. Therefore, we should *not* reject the null hypothesis. The data shown in Table 3.5 could have resulted from random chance. Indeed, the data *were* constructed from random chance.   Code Block 3.16 shows the Python code used to generate the data and perform the hypothesis test for this example. Lines 2–6 will load the data for the current example; to generate new examples, however, simply comment out these five lines of code.   ▷

Problems

3.1. Repeat Example 3.5 using the Z_n statistic on line 17 instead of the T_n statistic; i.e., with the sample variance replaced with the true variance. Given this modification, what is the expected rejection rate for each case (z_crit and t_crit)? Run the simulation and compare the observed rejection rates with your prediction.

```
1   X = np.random.normal(loc=100, scale=15, size=[5,8]).astype(int)
2   X = np.array([[103, 98, 103, 112, 74, 106, 79, 85],
3                 [ 71, 73, 102, 73, 118, 118, 81, 105],
4                 [ 96, 87, 110, 123, 84, 96, 82, 109],
5                 [103, 91, 119, 111, 147, 103, 112, 95],
6                 [ 70, 89, 107, 106, 92, 96, 84, 74]])
7
8   X_means = X.mean(axis=1) #mean for each group
9   X_bar = X.mean() # overall mean
10  r_i = 8
11
12  SSB = r_i * np.sum((X_means - X_bar)**2)
13  SSW = np.sum( (X - X_means.reshape((5,1)) * np.ones(8))**2 )
14  SST = np.sum( (X - X_bar)**2 )
15
16  MSB = SSB / 4
17  MSW = SSW / 35
18
19  F = MSB / MSW # 1.91406
20  alpha = 0.05
21  scipy.stats.f.isf(0.05, 4, 35) # 2.641465
```

Code Block 3.16: Simulation of an ANOVA test

3.2. Verify Equation (3.18).

3.3. Prove Corollary 3.2. *Hint*: How must Equation (3.23) be modified before proceeding?

3.4. Find the bug in Code Block 3.8. Validate your answer by reproducing the code, with the bug fixed, and showing that the rejection rate (even with $n = 10$) is close to 5%.

3.5. Use Theorem 2.9 to prove that the variance of a noncentral chi-squared random variable statisfies Equation (3.51).

3.6. Complete the proof to Theorem 3.11 by showing that

$$\sum_{i=1}^{m}\sum_{j=1}^{r_i} (y_{ij} - \bar{y}_{i.})\,(\bar{y}_{i.} - \bar{\bar{y}}) = 0.$$

3.7. Explain how the ANOVA null hypothesis is used in Lemma 3.9.

3.8. Use the code from Code Block 3.16 to build a simulation that will generate 10,000 data sets and record the fraction of instances for which the ANOVA null hypothesis was falsely rejected. *Bonus*: Update the code block to use parameters m, r_i, and $n = m*r_i$ to be specified prior to the for-loop. Repeat the simulation using other values of sample size.

4

It's Alive!

Whereas the theory of hypothesis testing focuses on statistical inference, i.e., what one may infer about a population from a random sample, the theory of experimentation focuses on elements of *design*, i.e., how one may go about constructing an experiment that will produce statistically significant results. For additional references, we refer the reader to Berger, *et al.* [2018], Kohavi, Tang, and Xu [2020], Rosenbaum [2002], and Selvamuthu and Das [2018].

4.1 Principles of the Design of Experiments

In this section, we lay out the basic principles and definitions of experimental design and present some of the mathematical framework for our subsequent discussions. We approach the subject from the perspective of *product engineering*, where various treatment groups correspond to various versions of a given product (typically a digital product, such as a website or an app). Historically, however, much of the theory was developed in the context of *clinical trials*, designed to measure the efficacy of a given drug treatment on a given disease. These two applications are practically analogous as drugs are, after all, chemical products. A slight variation is sometimes also used, when the purpose of an experiment is to study the effect of a given covariate on a product's efficacy; e.g., the effect of age range on a shampoo's ability to treat split ends. Such an application departs from the product engineering and clinical trial approach—where treatments correspond to different versionings of a product—in that treatments correspond to different properties, i.e., covariates, of the populations being tested.

Experiments

We begin by laying out the ingredients that define an experiment. We will then proceed to discuss each in turn.

Definition 4.1. *An* experiment *is a six-tuple* $(\mathcal{T}, \mathcal{S}, \mathcal{P}, x, Z, R)$ *consisting of*

- *a finite set of m possible* treatments, $\mathcal{T}$,
- *a finite set of s known* covariates *or* strata $\mathcal{X}$ *of the population of interest,*
- *a population* $\mathcal{P}$, *consisting of a set of n experimental units,*
- *a classification function* $x : \mathcal{P} \to \mathcal{X}$ *that uniquely determines the stratum for each experimental unit,*
- *an* assigment mechanicsm $Z : \mathcal{P} \to \mathcal{T}$ *that assigns a treatment to each experimental unit, and*
- *a response* $R : \mathcal{T} \times \mathcal{P} \to \mathbb{R}$, *which records the experimental results of applying a particular treatment to a particular unit.*

The purpose of defining an experiment is to study the effect of the factors, broken up into individual treatments, on a response variable, while controlling for one or more covariates. The *treatments* are the individual variations of our product. Often, the set of treatments $\mathcal{T}$ is decomposed into the Cartesian product of one or more *factors*; i.e., $\mathcal{T} = \mathcal{F}_1 \times \cdots \times \mathcal{F}_k$, where $\mathcal{F}_i$ is the set of *levels* of factor i. When such a decomposition is deployed, we call the experiment a *k-factor experiment*. For example, suppose we are studying the effect of age and gender on a particular drug's efficacy. Then the factors would be age and gender, and a reasonable set of levels would be

$$\mathcal{F}_{\text{age}} = \{18\text{--}24, 25\text{--}34, 35\text{--}44, 45\text{--}54, 55\text{-}64, 65+, \text{unknown}\},$$
$$\mathcal{F}_{\text{gender}} = \{\text{male}, \text{female}, \text{unknown}\}.$$

Here, we assume an age restriction that participants must be at least eighteen years of age, and we further allow for the possibility that some participants may choose not to disclose their age or gender. The set of possible treatment groups is therefore given by $\mathcal{T} = \mathcal{F}_{\text{age}} \times \mathcal{F}_{\text{gender}}$, which is a set of size 21.

For a more product-centric example, suppose we are testing the effect of font and background color for a website or app. The result would be a two-factor experiment, comprised of a set of fonts and a set of background colors.

The purpose of defining the treatment set as a Cartesian product, for which all possible combinations are considered, is to capture *interaction effects*. For instance, New Times Roman might be the superior choice when considering variations in fonts alone, and a blue background might be the superior color choice when considering variations of color alone, but there might be something charmingly irresistible about the enchanting *combination* of Century Gothic with a chartreuse background. Only by testing all possible combinations of font plus background together can one discover such synergies.

A *factorial design* is any multi-factor design that involves various combinations of levels from different factors. A full (or complete) factorial design involves taking all combinations of levels from each factor, as described in the Cartesian product decomposition, above.

A special case of this is that of the so-called two-level factorial design[1]. Here, each level is binary, and so the total number of possible treatments for a k-factor experiment is given by 2^k, as each factor is comprised of two levels. The special case of a one-factor design with two levels is referred to as an *A/B test*. The levels of a binary factor are customarily referred to as the *control* and the *treatment*. In clinical trials, the control typically corresponds to a placebo whereas the treatment would correspond to the drug. In product design, the control could represent the current version of the product, whereas the treatment would represent a proposed change.

With our set of treatments in hand, we next turn to *experimental units*. A *unit* is an opportunity to withhold or apply any of the various treatments. The individual experimental units are drawn from our population $\mathcal{P}$.

Experimental units may further be described by a collection of one or more *covariates*, or *nuisance factors*. The only true difference between a covariate or nuisance factor (and their corresponding strata) and the experimental factors (and their corresponding treatments) is one of intent: it is the causal relation between the experimental factors and the response we are seeking to uncover. In the context of product engineering, however, there is a more natural distinction: the experimental factors typically represent variations in the product, whereas nuisance factors represent variations in certain features of the population.

To further complicate matters, there are both *known* and *unknown* nuisance factors. To control for unknown nuisance factors, i.e., those factors which *may* exist, but over which we have no direct insight or control, we employ a design principle known as *randomization*. Randomization is achieved through the choice of the assignment mechanism. The idea here is that by randomly assigning treatments to units, we will, on average, cancel out the effect of any unknown nuisance factor, thereby making the results of the experiment more robust. Randomization essentially breaks every causal link between unknown factors and the experiment's results.

On the flip side of the coin, the known nuisance factors are the factors we expressly define as *covariates*, which are characteristics of each unit which may be measured prior to assignment or experimentation. As covariates are properties of the units themselves, they cannot be influenced by the choice of treatment. For example, in a two-factor experiment designed to test various combinations of font and background color, age and gender might be considered covariates. Here, since we have direct access of age and gender prior to our assignment of treatment, we can design an assignment mechanism to balance out the effects of age and gender across the various

---

[1] sometimes referred to as 2^k-factorial design.

treatments, so that, for example, we don't accidentally assign one particular combination predominantly to one particular stratum. Such a balancing act is known as *blocking*. Here, the various covariate strata are thought of as *blocks*, and the idea is to balance the blocks over the given set of treatments.

Another principle of design is one of *replication*. By replication, we mean that for each treatment and stratum, the experiment should be repeated across several units. In this way, we can form a measure of statistical error. If, on the other hand, each stratum contains only one unit for each treatment, assigned randomly, then the experiment is referred to as *paired randomized experiment*.

Blocking, randomization, and replication all feed into the concept of an assignment mechanism. Typically, assignment mechanisms are not functional in nature but rather stochastic, in the sense that they incorporate random numbers to properly randomize treatments across experimental units. By placing various constraints on this random process, we can further achieve a blocked-design, where, though random, the various strata are balanced over the various treatment assignments.

There are two classic assignment mechanisms that earn special consideration in our studies. The first is called a *completely randomized design*. A completely randomized experiment has no covariates, so we can say that each unit belongs to the same stratum. To carry out a completely randomized experiment, the number of replicants for each treatment group is prescribed, and the available units are assigned randomly to each treatment group. It is important to note that the *order* that the experiments are to be carried out is also randomized. The second classic assignment mechanism is called a *randomized block design*. This is similar to a completely randomized design, except for the fact that there is more than one stratum. The experimental units are divided into their respective strata, and a completely randomized design is then implemented within each stratum. In other words, each stratum is constrained to posses an equal number of replicants for each treatment. We will explore these types of design over the pursuing pages.

Finally, the experiment is executed and a response is measured. Note that the response mapping is not completely known; we only know the response for a single treatment for each experimental unit, under a particular test condition. We cannot say what the response for a particular unit *would have been* had we applied a different treatment, or even had we applied the same treatment under different operating conditions (e.g., a different time of day or day of week or with a different operator). Thus, the response is simply a record of the results of the experiment and not a full functional description; it is censored by the fact that we live in but a single universe and have no access to the results that our counterparts in other universes may have achieved using a different seed for their random assignment mechanism.

For example, suppose that experimental unit i is assigned to treatment Z_i by our random assignment mechanism. We will record the response $R_{Z_i i}$. The other *potential outcomes* of the experiment,

$$R_{ji}, \qquad \text{for } j \neq Z_i,$$

are referred to as *counterfactuals*, as they do not factually exist in reality and, as such, cannot be observed or measured. We are capable of observing only the outcome $R_{Z_i i}$ obtained by applying our chosen treatment Z_i to experimental unit i, and not the other worldly outcomes observed by our brothers and sisters in parallel universes, whose random assignment mechanisms led them to choose other treatments $j \neq Z_i$ for unit i.

Analysis

Of course, we don't just run experiments for the fun of it. The purpose of running an experiment is to learn something. While Definition 4.1 lays out the definition of an experiment, this is only half of the picture. An experiment must be coupled with a statistical analysis which supports a conclusion. Since the conclusion is itself the goal of experimentation, this analysis must also be taken into account during the design phase of the experiment. For example, if one is attempting to measure the effect of polar-magnetic vortex flares on the choice of sock color, it would be most unfortunate if an experiment was devised that measured the blood pressure of the participants and not the color of their socks.

An *experimental analysis* is an experiment coupled with a hypothesis test, such that the test statistic is a function of the experimental data. The null hypothesis we consider is typically the *hypothesis of no effect*, which states that the choice of treatment had no effect on the outcome. Given the hypothesis of no effect, we then derive an appropriate test statistic. The null hypothesis of the experiment and the test statistic must therefore be considered as part of the experimental design, to ensure that the experiment produces data that can be used to confirm or reject the hypothesis of the experiment, and to ensure that enough data is collected to achieve a particular significance and power.

Direct and Index Notation

We will encounter two types of notation during our discussion on experimentation, one based directly on the enumeration of the individual units, and one based on the index of the assignment of units to various treatments and of the classification of units into covariates. We refer to these two systems as *direct notation* and *index notation* respectively.

In direct notation, each unit of the population is given a unique index i. This serves as an ID for individual units and is independent on the eventual assignment of units into treatment groups. Direct notation goes hand-in-hand with our Definition 4.1 and our discussion thus far. Unit i receives treatment Z_i, as determined by our random assignment mechanism, resulting in observation $R_{Z_i i}$.

In index notation, we reindex all of the units based on their assigned treatment group, their covariates, and a final enumeration to distinguish replicants within a given treatment group and stratum. This notation is necessarily fluid to allow for a variable number of experimental factors and nuisance factors. To achieve this, we use the following ordering for indices:

1. *treatment group*—the innermost index is reserved for the assigned treatment group. For a one-factor experiment, we use index i; for a two-factor experiment, we use indices (i, j), and so-on.
2. *strata*—the next index (or group of indices) is reserved for the covariate classification. Typically this will constitute a single index, though it is possible to use multiple indices to study interaction effects among the nuisance factors.
3. *replicants*—the final, outermost index is reserved for enumeration of the replicants within an individual treatment group and stratum.

For example, suppose a two-factor experiment is performed using replication and a single nuisance factor. The quantity Y_{ijkl} would refer to the result obtained from the lth replicant in treatment group (i, j) in stratum k. If the experiment is conducted *without replication*, so that only a single experimental unit for each strata is given each treatment, then we can dispense with the final index l and simply write Y_{ijk} for the result.

Similarly, consider the quantity Y_{ij}. This notation would be consistent with any of the possible experimental structures:

- A one-factor completely randomized experiment (no covariates) with replication. In this case, the quantity Y_{ij} corresponds to the result obtained from the jth replicant from treatment group i.
- A two-factor completely randomized experiment with no replication. (This constitutes a poor design choice, though it is consistent with our index notation.) Here, quantity Y_{ij} represents the result of applying treatment (i, j) to the sole replicant of that treatment group.
- A one-factor experiment with one covariate and no replication. Here, quantity Y_{ij} represents the result of applying treatment i to the sole replicant of stratum j.

Though the indices will take on different meanings in different contexts, we will follow the golden rule that: treatment groups are enumerated first, then the covariate strata, then the replicants within each treatment-group and stratum.

Each set of notation has its own advantages. The direct notation is conducive to discussions on counterfactuals, treatment effect, and causality. This is especially convenient for the analysis of observational studies.

Similarly, index notation is conducive to analysis of controlled experiments, especially experiments with many-level factors, and the analysis of variance. In particular, it is not possible to have a discussion about counterfactuals using index notation, as the notation itself enumerates the units *within their respective treatment groups*; i.e., it enumerates the units *after* the units have been assigned to their respective treatments.

When we are referring to particular units, it is possible that any particular index could exceed the value of 9, especially when using the direct notation. For example, the response of applying treatment 3 to unit 171 would be R_{3171}. In order to avoid such ambiguity, we will separate the index numbers with commas whenever one of them is not a single digit; for example, $R_{3,171}$ for $i = 3$ and $j = 171$ or $R_{3,17,1}$ for $i = 3$ and $j = 17$ and $k = 1$. The only rule is that whenever commas are deployed, we shall use commas throughout: we would write $R_{1,7,19}$ for $i = 1$, $j = 7$, and $k = 19$, and not $R_{17,19}$, which could be confused with $i = 17$ and $j = 19$ or $i = 17$ and $j = 1$ and $k = 9$.

In the context of database design, direct notation directly uses the "identity key" of each unit, whereas the index notation uses a multi-column key that represents each units assigned treatment group, properties, and replication number.

4.2 A/B Tests

In this section, we go into depth to explore the design and analysis of running A/B tests. For a practical guide that explores pitfalls and nuisances of A/B testing in practice, we refer the reader to the indispensable volume Kohavi, Tang, and Xu [2020]. Recall that an A/B test is a single-factor experiment with two levels, which are referred to as a control ($i = 0$) and a treatment ($i = 1$). Our attention is thus presently restricted to single factor experiments with two levels. In the following sections, we will extend this to single and multi-factor experiments with an arbitrary number of levels. We will conclude the chapter with a higher-dimensional generalization of A/B tests to more than a single factor, during our discussion of two-level factorial design.

4.2.1 Design

We begin by discussing the design and analysis of A/B tests with no covariates; that is, we do not measure any characteristic of the units before the random assignment of the unit into a control or treatment group. We will suppose that unit i is assigned to treatment $Z_i \in \{0, 1\}$, resulting in observation $R_{Z_i i}$. We begin with a discussion of the assignment mechanism.

Assignment

We will consider two possible assignment mechanisms.

Definition 4.2. *A* Bernoulli randomized experiment *is a randomized experiment in which each unit receives the treatment with the same probability; i.e.,*

$$\mathbb{P}\left[Z_i = 1\right] = \pi,$$

for all $i = 1, \ldots, n$, where $0 < \pi < 1$.

Typically, we use $\pi = 1/2$, so that each experimental unit is equally likely to be assigned to the treatment or control. Given a population of n-units, there are 2^n possible assignments that can be achieved, each occurring with equal probability. Since each unit has the same probability of being assigned to the treatment, one shortcoming of this approach is the possibility that *all* units receive the same treatment, though this possibility becomes vanishingly small as the number of units becomes large. To remedy this, however, we can choose a different assignment mechanism.

Definition 4.3. *A* completely randomized A/B test *is an A/B test for which the total number of replicants in the treatment group has a fixed value r; i.e., an experiment in which all assignments Z_i satisfying the constraint*

$$\sum_{i=1}^{n} Z_i = r$$

are equally likely.

Typically, one requires an even number of units n and fixes the number of treatments as $r = n/2$. In a completely randomized experiment, there are $\binom{n}{r}$ possible assignments, each being equally likely to occur; i.e., we randomly select a total of r units to be assigned to the treatment group.

A completely randomized experiment therefore restricts the number of possible assignments of the Bernoulli randomized experiment. For example, in an experiment with 10 available units, a Bernoulli randomized experiment will have $2^{10} = 1024$ possible assignments, whereas a completely randomized experiment will have $\binom{10}{5} = 252$ possible assignments.

There are two standard methods for carrying out a completely randomized experiment. The first method is to simply shuffle the n units and assign the first $(n - r)$ units to the control and the final r units to the treatment. An example of this is shown in Table 4.1. In this example, units 0, 2, 3, 5, and 6 are assigned to the control group, whereas units 1, 3, 4, 7, 8, and 9 are assigned to the treatment. Once the assignment has been made, we can now introduce index notation, which refers to each unit by its treatment-group and replicant number. For example, unit 5 is the fourth replicant of the control group, so unit 5 will correspond to Y_{03}. Finally, if the experiments

Unit	0	1	2	3	4	5	6	7	8	9
Shuffled	2	6	0	5	3	1	9	4	8	7
Assignment	0	0	0	0	0	1	1	1	1	1
Index Notation	Y_{00}	Y_{01}	Y_{02}	Y_{03}	Y_{04}	Y_{10}	Y_{11}	Y_{12}	Y_{13}	Y_{14}

Table 4.1: Assignment for a completely randomized experiment

are to be performed sequentially, we would typically shuffle the final row in order to perform the experiments in a random order; i.e., we would not want to perform all the control-group experiments first followed by all the treatment-group experiments second. As long as the *original* unit keys are random (i.e., the units were not originally ordered in any way), we could alternatively perform the experiments in the original order: Y_{02}, Y_{10}, Y_{00}, Y_{04}, Y_{12}, Y_{03}, Y_{01}, Y_{14}, Y_{13}, Y_{11}.

Alternatively, one can achieve a completely randomized design by using *one-at-a-time simple random sampling*. This assignment mechanism has the same effect as shuffling, but works on each unit *sequentially*. The idea is that each unit is assigned to a treatment group with a probability that is proportional to the number of "spots" left open in that treatment group. More formally, we have the following.

Definition 4.4. One-at-a-time simple random sampling *is a method for assigning n objects into m groups, such that group z has exactly n_z assignments, with $\sum_{z=1}^{m} n_z = n$, and such that the probability of allocating object i to group z is proportional to the number of remaining spaces in the group:*

$$\mathbb{P}[Z_i = z | Z_1, \ldots, Z_{i-1}] = \frac{n_z - \sum_{j=1}^{i-1} \mathbb{I}[Z_j = z]}{n - i + 1}, \qquad (4.1)$$

for $i = 1, \ldots, n$.

It is straightforward to show that Equation (4.1) is a normalized probability mass function for each $i = 1, \ldots, n$. For the special case of a completely randomized A/B test, we have the probability of assigning unit i to the treatment group is given by

$$\mathbb{P}[Z_i = 1 | Z_0, \ldots, Z_{i-1}] = \frac{r - \sum_{j=1}^{i-1} Z_j}{n - i + 1}, \qquad (4.2)$$

for $i = 1, \ldots, n$.

Causal Effect

We are often interested in understanding the *causal effect* of a treatment, meaning, intuitively, the net effect of applying the treatment instead of the control. Before defining this, note that for the special case of an A/B test, we may write the *observed outcome* on unit i as

$$R_i = Z_i R_i^T + (1 - Z_i) R_i^C. \tag{4.3}$$

We call R_i^T and R_i^C the *potential outcomes* for unit i under the treatment $Z = 1$ and control $Z = 0$, respectively, as one of these two outcomes will come to pass. The observed outcome depends on the result of the random assignment: if unit i is assigned to the treatment $Z_i = 1$, then we will observe the outcome R_i^T. If, on the other hand, unit i is assigned to the control $Z_i = 0$, then we will observe the outcome R_i^C. The above formula accomplishes precisely this. It says that $R_i = R_i^C$ if unit i is assigned to the control $Z_i = 0$ and that $R_i = R_i^T$ if unit i is assigned to the treatment $Z_i = 1$. The potential outcome that is not observed is referred to as the *counterfactual* for unit i; i.e., we may write

$$C_i = (1 - Z_i) R_i^T + Z_i R_i^C \tag{4.4}$$

for the *counterfactual*. The counterfactual represents what the outcome of the experiment *would have been* for unit i, *had we made* the alternate assignment $1 - Z_i$ instead of Z_i. Counterfactuals exist only in our minds and in parallel universes. This handicap, however, does not prevent us from making inferences about them.

Definition 4.5. *The* causal effect *of applying the treatment over the control in an A/B test on unit i is the difference between the two potential outcomes*

$$\Delta_i = R_i^T - R_i^C. \tag{4.5}$$

Like counterfactuals, the causal effect *cannot* be observed *for any* unit. This follows since, for any unit i, we will *either* observe potential outcome R_i^C *or* potential outcome R_i^T, as we can *only* apply *either* the treatment or the control to unit i, but not both.

Though unobserved, we can estimate the average causal effect using the following.

Theorem 4.1. *In a completely randomized experiment, where each unit has the same probability $\pi = \mathbb{P}(Z = 1)$ of receiving the treatment, the estimate*

$$\hat{\Delta} = \frac{1}{r} \sum_{i=1}^{n} Z_i R_i^T - \frac{1}{n-r} \sum_{i=1}^{n} (1 - Z_i) R_i^C \tag{4.6}$$

is an unbiased estimate for the average causal effect.

Equation (4.6) represents the difference between the averages of the treatment and control groups. Notice that this estimate relies *only* on the observed outcomes, and not on any of the counterfactuals[2].

---

[2] Even though there are counterfactuals within the two sums, their coefficients are always zero.

Proof. Since the assignment probability does not vary by experimental unit, the expected treatment assignment is given by $\mathbb{E}[Z_i] = r/n$. Using this fact, we may write the expected value of our estimate as

$$
\begin{aligned}
\mathbb{E}[\hat{\Delta}] &= \mathbb{E}\left[\frac{1}{r}\sum_{i=1}^{n} Z_i R_i^T - \frac{1}{n-r}\sum_{i=1}^{n}(1-Z_i)R_i^C \right] \\
&= \frac{1}{r}\sum_{i=1}^{n}(r/n)R_i^T - \frac{1}{n-r}\sum_{i=1}^{n}(1-r/n)R_i^C \\
&= \frac{1}{n}\sum_{i=1}^{n}\left(R_i^T - R_i^C \right).
\end{aligned}
$$

However, this final quantity represents the actual average causal effect for our experiment. (Note that this quantity is cannot be directly observed, as each term in the sum consists of the difference between an actual measurement and a counterfactual.) The result follows. □

Randomization, of course, plays a crucial role in this conclusion. Only since the treatments have been properly randomized can we make this inference about *causality*. Without randomization, it is always possible that some other latent factor could have caused a certain set of units to end up in the control group and another different set of units to end up in the treatment. In such a case, the observed effect measured by Equation (4.6) could have been caused by this latent factor and not the actual experimental treatment itself. By randomizing, we are washing away all history and all unknown attributes of the experimental units, breaking their causal links to the observed outcome.

4.2.2 Analysis

Next, we discuss aspects of the analysis of A/B tests: choosing an appropriate test statistic and the roles of significance and power. As these necessarily play critical roles in the design, the analysis must be considered prior to the start of the experiment. For now, we will focus on the general case of a continuous response variable. In Section 4.2.6, however, we will discuss some additional test statistics useful specifically for the case of a binary response.

Test Statistic

Once we have randomized our units into treatment groups, we next switch to index notation. Our model for the response Y_{ij} of the jth replicant within the ith treatment group is

$$
Y_{ij} = \mu + \tau_i + \epsilon_{ij}, \tag{4.7}
$$

where $\mu = \mathbb{E}[Y_{ij}]$ is the overall average value, τ_i is the *differential effect* or *response* associated with the ith treatment, and ϵ_{ij} is the noise of the jth replicant in treatment group i.

Note 4.1. As mentioned in Note 3.26, Equation (4.7) is nonidentifiable, in the sense that the parameters cannot be uniquely determined. (In particular, the outcome, which is the true observable, is invariant under the shift $\tau_i' = \tau_i + \delta$, $\mu' = \mu - 2\delta$.) We must therefore impose an auxiliary condition so that the parameters are uniquely determined. There are two common choices: the first is to require that $\tau_0 = 0$. Under this choice, the treatment effect is identified with $\Delta = \tau_1$. This is a common choice in A/B testing, or whenever any one treatment group (the "control") is of particular interest, in that the treatment effect of the other groups is measured with respect to the former. More commonly, especially when we move beyond A/B tests, we will require

$$\sum_{i=1}^{m} \tau_i = 0,$$

so that μ is interpreted as the overall average. For the case of A/B tests, this equation reduces to $\tau_0 + \tau_1 = 0$. The average treatment effect in this case is given by $\Delta = \tau_1 - \tau_0$.   ▷

We further assume that this model satisfies the ANOVA assumptions from Definition 3.27. Since there are only two treatment groups, we soon realize that those assumptions imply the applicability of Theorem 3.4. As a result of these assumptions, we see that

$$Y_{ij} \sim \mathrm{N}(\mu + \tau_i, \sigma^2), \tag{4.8}$$

for $i = 0, 1$ and all $j = 1, \ldots, r$. Thus, the conditions of Theorem 3.4 are valid and we have the following theorem.

Theorem 4.2. *Given a completely randomized A/B test with r replicants, for which the outputs Equation (4.7) satisfy the ANOVA assumptions of Definition 3.27, then under the hypothesis of no effect $\tau_0 = \tau_1$, the test statistic*

$$T = \frac{\bar{y}_{1\cdot} - \bar{y}_{0\cdot}}{\sqrt{\lambda MSW}} \sim t_{2r-2}, \tag{4.9}$$

where $\lambda = 2/r$ and MSW is the within-group mean squared error

$$MSW = \frac{1}{2r - 2} \sum_{i=0}^{1} \sum_{j=1}^{r} (y_{ij} - \bar{y}_{i\cdot})^2, \tag{4.10}$$

is distributed as a Student's t-random variable with $2r-2$ degrees of freedom. Moreover, the rejection region

$$R_\alpha = \{t : |t| > t_{2r-2, \alpha/2}\},$$

yields a test of size α.

Proof. The result follows as the conditions of Theorem 3.4 have been satisfied.   □

If we were instead to make the inquiry as to whether the treatment was *better* than the control, our null hypothesis would be restated as $H_0 : \Delta = \tau_1 - \tau_0 \leq 0$. (The alternative hypothesis, in this case, would be $H_A : \Delta > 0$, that the treatment had a positive, non-zero effect.) In this case, we could still use the test statistic Equation (4.9), except with the alternate one-sided rejection region

$$R_\alpha = \{t : t > t_{2r-2,\alpha}\}.$$

Note 4.2. The within-group mean squared error of Equation (4.10) is analogous to the pooled sample variance of Equation (3.24). It makes intuitive sense to use the within-group error as opposed to the total error. Given the ANOVA assumption that $\sigma^2 = \sigma_0^2 = \sigma_1^2$, the within-group mean squared error is an unbiased estimate for σ^2. However, if the null hypothesis is false, and there are differences in the means, the total mean squared error (i.e., the sample variance of the entire data set, without regard to treatment group) can be biased by the between-group sum of squares.   ▷

Example 4.1. A professor, Dr. Austin, has two assistants who grade exams for her. Dr. Austin wants to determine whether her two assistants grade fairly. She therefore devises an experiment in which she gives half of the ten exams to one assistant and the other half to the other.

Let R_i^C be the score that student i would obtain *if* his or her exam were to be graded by assistant 0, and let R_i^T be the score that student i would obtain *if* his or her exam were to be graded by assistant 1. For each student, we will have the ability to observe either R_i^C or R_i^T, but not both. For the purpose of illustration, however, we can claim omniscience.

Let's suppose that both assistants grade identically; i.e., $R_i^C = R_i^T$, for all $i = 0, \ldots, 9$. Let's further suppose that the scores are not random across the students, but correlated with the student's number. For example, suppose the exams are numbered in the order in which they were turned in, and suppose that better students tend to turn in exams sooner than the students who struggle. We can generate such a set of data by taking $R_i^C \sim N(90 - 2i, 5^2)$. Random numbers pursuant to this scheme are given in Table 4.2.

At this point, let us suppose that Dr. Austin makes a faulty assumption, that the order in which she receives the exams is random. (This could even be an unconscious assumption.) She assigns the first five exams to assistant 0 and the second set of five exams to assistant 1. The results of this assignment are in the columns labeled *NR*. Upon obtaining the results, she first notices that the mean score for group 0 is 85.4, whereas the mean score for group 1 is 75.2. (Remember, she only sees R_i^C for $i = 0, \ldots, 4$ and R_i^T for $i = 5, \ldots, 9$; she does not see the counterfactuals.) She quickly computes $MSW = 40$ and $T = 2.55$. She then looks up the value $t_{8,0.025} = 2.306$, and

i	R_i^C	R_i^T	Group NR	Observed NR	Group R	Observed R
0	96	96	0	96	1	96
1	89	89	0	89	1	89
2	79	79	0	79	0	79
3	81	81	0	81	0	81
4	82	82	0	82	0	82
5	82	82	1	82	1	82
6	72	72	1	72	0	72
7	69	69	1	69	1	69
8	80	80	1	80	0	80
9	73	73	1	73	1	73

Table 4.2: Nonrandom (NR) and Random (R) assignment for exam scores; both assistants grade identically.

rejects her null hypothesis (that both assistants grade identically), reporting a statistical significance of $\alpha = 0.05$. However, her conclusion is erroneous, as the assignments were not properly randomized. It was not the *choice of assistant* the *caused* the difference in exam scores between the two groups. It was the fact that the better students tended to finish earlier that caused their exam to be assigned to the earlier group that *caused* the difference in scores between the two groups.

Since we are omniscient beings in the universe of this example, let us like angels go back in time to see how things could have played out differently. The last exam lands on Dr. Austin's desk, abruptly awakening her from her daydream in which she realized the mistake in her plan. She then quickly randomizes the order of the exam and distributes the exams to the assistants following the mechanism in the columns labeled R in Table 4.2. Upon reviewing the results, she finds that the average score of the exams graded by assistant 0 is 78.8, whereas the average score of the exams graded by assistant 1 is 81.8. She quickly calculates $MSW = 69.7$ and $T = 0.568$. She therefore fails to reject her null hypothesis and accepts that she does not have sufficient evidence to conclude that there is a difference in how the two assistants graded the exams. The assistants go happily on their ways, blissfully unaware of how close they came to mortal danger.   ▷

Example 4.2. Just down the hall, Dr. Powers has an identical group of ten students. His assistants, on the other hand, are not identical, in that assistant 1 always grades ten points higher than assistant 0. The potential outcomes for each student are given in Table 4.3.

If Dr. Powers fails to randomize the exams before assigning them to his two assistants, the average score of group 0 will be 85.4 and the average score of group 1 will be 85.2. The t-score is $T = -0.05$. Dr. Powers fails to reject the null hypothesis and believes his two assistants to be grading fairly.

i	R_i^C	R_i^T	Group NR	Observed NR	Group R	Observed R
0	96	100	0	96	1	100
1	89	99	0	89	1	99
2	79	89	0	79	0	79
3	81	91	0	81	0	81
4	82	92	0	82	0	82
5	82	92	1	92	1	92
6	72	82	1	82	0	72
7	69	79	1	79	1	79
8	80	90	1	90	0	80
9	73	83	1	83	1	83

Table 4.3: Nonrandom (NR) and Random (R) assignment for exam scores; assistant 1 always grades 10 points higher than assistant 0.

However, if Dr. Powers properly randomizes, the outcome is different. In this case, the average exam scores for assistants 0 and 1 are 78.8 and 90.6, respectively. This earns a t-score of $T = 2.587$, and Dr. Powers rejects the null hypothesis and concludes that his two assistants are not grading fairly.

In Exercise 4.1, we will simulate this and the preceding example to determine the probability of rejecting the null hypothesis in each of the four cases we discussed.   ▷

The Paradox of the One-Sided Test

The astute reader may have noticed the following oddity during our discussion on hypothesis testing: the rejection region for a standard two-sided test is

$$R_\alpha^2 = \{z < -z_{\alpha/2}\} \cup \{z > z_{\alpha/2}\},$$

and the rejection region for the corresponding one-sided (right-tailed) test is

$$R_\alpha^R = \{z > z_\alpha\}.$$

We therefore find ourselves with the peculiar outcome that if $z_\alpha < z < z_{\alpha/2}$, we would be able to reject the one-sided null hypothesis $H_0^R : \mu \leq \mu_0$, whereas we would not be able to reject the two-sided null hypothesis $H_0^2 : \mu = \mu_0$. (For $\alpha = 0.05$, this would happen if $1.65 < z < 1.96$.) From the same set of data, we would be able to conclude that the actual value of μ is greater than μ_0, since we are able to reject H_0^R, but we would not be able to conclude that μ is different than μ_0, as we are not able to reject H_0^2.

Both tests are producing the correct outcome. For the two-sided test, we will falsely reject a true null hypothesis H_0^2 exactly 5% of the time. This probability, however, is divided between the left and right tails, each of which hold 2.5% of the probability. For the one-sided test, on the other hand, we are not interested in cases for which $\mu < \mu_0$, as these cases have become incorporated into our null hypothesis. The full 5% probability is therefore entirely contained within the right tail, leading to the result that we will, again, falsely reject a true null hypothesis H_0^R exactly 5% of the time. So each test is valid for its given null hypothesis.

The conclusion, however, is still unsettling. How is it that we can conclude that μ is greater than μ_0, but yet we cannot conclude whether there is any effect at all? My view is this: the one-sided null hypothesis has done some of the lifting for us, as it has essentially removed the possibility that $\mu < \mu_0$ from our consideration. We are therefore no longer required to allocate some of our allotted $\alpha = 5\%$ significance in the pursuit of catching these cases. This results in the rejection region $\{z \geq z_\alpha = 1.65\}$, as opposed to the two-sided rejection region $\{|z| \geq z_{\alpha/2} = 1.96\}$. This is almost a Bayesian interpretation: the null hypothesis is, in some sense, our prior belief as to what the default state of the world should be. That prior belief therefore affects our posterior conclusions.

Next, let's address a practical matter: should we use one-sided tests? The answer, I think, is yes, possibly, with a caveat, but probably no. Let us suppose that we are running a test in which we are seeking to determine whether a variation in a product is beneficial, relative to some metric. (Or you could imagine we are conducting a clinical trial for a given drug against a placebo, if you like.) We are trying to improve the product, so we are only really interested in whether or not our change makes a positive impact on the product. So should we use a one-sided test? Not necessarily. It depends on our domain knowledge into the system we are studying. Consider the following two cases.

In the first case, suppose that, while we would like our product update to produce a positive result, we have no reason to rule out the possibility that it has a detrimental effect. In this case, we should run a two-sided test, and only proceed if we obtain a test statistic $z > z_{\alpha/2}$.

In the second case, suppose, on the other hand, that we have a good reason to rule out the possibility that our change can have a detrimental effect. That is, our prior belief about the system is that our change will either have no effect, or it will have a positive effect. Then we are safe to use a one-sided test with the less restrictive criterion $z > z_\alpha$.

In conclusion, our choice of whether or not to use a one- or two-sided test should not depend on our desired outcome, but on our most restrictive belief as to the possible range of outcomes. Even if we are only interested in catching positive outcomes, we would still run a two-sided test if a negative outcome is just as likely. We should not craft our null hypothesis to make our lives easier; we should craft our null hypothesis to mirror our understanding of the possible outcomes under consideration.

Example 4.3. Dr. Frankenstein is running an experiment to determine whether a corpse can be resurrected with an appropriate application of electricity. The outcome of this experiment is binary: either the corpse will remain deceased $(Y = 0)$ or it will return to the living $(Y = 1)$. Given the doctor's domain expertise, he is confident that cadavers randomly assigned to the control group are guaranteed to remain inanimate. It is therefore not possible for his treatment to produce a negative effect, as the patients in question are already deceased. He therefore constructs a one-sided test to determine whether or not his treatment has a positive effect. Furthermore, due to the vanishingly small variance, a single positive result will most likely be considered significant, especially to the test subject who finds himself stiffly awakened.   ▷

Power and Sample Size

The minimum sample size required to yield a power β test follows from Equation (3.39), which yields

$$r = \frac{2MSW}{\delta^2} \left[\Psi^{-1}(\alpha/2) - \Psi^{-1}(\beta) \right]^2 \approx \frac{15.7MSW}{\delta^2}, \qquad (4.11)$$

where the 15.7 corresponds to the choice of $\alpha = 0.05$ and $\beta = 0.80$. Here, the within-group mean-squared error has replaced the pooled sample variance, as expected.

4.2.3 Blocking

Finally, let us turn to the topic of *blocking*. Blocking is a design strategy used when different covariates are measured prior to the start of the experiment. The idea is that these covariates define certain strata (or blocks), in which units are uniform with respect to the observed set of covariates. A completely randomized design is then carried out within each stratum.

   This approach is useful when the measured covariates, though not of direct interest in the study, are known or suspected to be correlated with the outcomes of the experiment. In particular, less populous strata have a higher tendency to posses an imbalance in treatment assignments, unless the assignment mechanism is devised to explicitly guard against it. This imbalance, coupled with the correlation between the covariates and the outcomes, could skew the results. We can therefore control for the covariates by ensuring that treatment and control assignments are balanced within each strata.

Example 4.4. To illustrate the importance of blocking, let us consider the following contrived example, in which the treatment has no effect, and the population is divided into two strata, as shown in Table 4.4. Units in stratum 0 have a fixed response of 150, whereas units in stratum 1 have a fixed response of 100. Since stratum 0 is small (only three units), it is not hard to imagine the case when our random assignment places all three

i	x_i	R_i^C	R_i^T	Group NB	Observed NB	Group B	Observed B
0	0	150	150	1	150	0	150
1	0	150	150	1	150	–	150
2	0	150	150	1	150	1	150
3	1	100	100	0	100	0	100
4	1	100	100	0	100	1	100
5	1	100	100	1	100	0	100
6	1	100	100	0	100	0	100
7	1	100	100	0	100	1	100
8	1	100	100	1	100	–	100
9	1	100	100	0	100	1	100

Table 4.4: Non-block (NB) and blocked (B) outcomes in a contrived two-strata problem with no treatment effect and no noise.

units from stratum 0 into the same treatment group, as shown in the NB (nonblocked) column in the table. In this case, our observed outcomes are $\overline{Y}_{0.} = 100$ and $\overline{Y}_{1.} = 130$. In Exercise 4.2, we will show that the T-statistic is $T \approx 2.449$, so that the null hypothesis is rejected. We would therefore erroneously conclude that the treatment had an effect. However, this effect is in actuality entirely due to the fact that all units of stratum 0 have a higher response than the units of stratum 1. The treatment, by design, has no effect on the outcome.

With a block design, we would randomize within each stratum separately, enforcing the condition that we assign an equal number of units to the treatment and control groups within each stratum. (We randomly remove any excess units that would prevent us from assigning an equal number of units to each treatment group.) Such a randomized block design is shown in the final columns of the table. With this design, we see that observed outcomes between the two groups are identical: $\overline{Y}_{0.} = \overline{Y}_{1.} = 112.5$.
▷

Definition 4.6. *A block randomized A/B test is an A/B test consisting of a population with $s \geq 2$ strata (covariates) and an assignment mechanism that randomly assigns precisely half of the units of each stratum to the treatment group; i.e., if $2r_j$ units belong to stratum j, for $j = 1, \ldots, s$, then the assignment mechanism will randomly assign r_j units to the control group and r_j units to the treatment group.*

In practice, we first group the units into their respective strata *vertically*, determining the total number of units for each stratum. We then shuffle the units within each stratum in order to randomly assign half of those units to the treatment and half to the control group.

To express this using index notation, we let Y_{ijk} represent the observed result corresponding to treatment group $i = 0, 1$, stratum $j = 1, \ldots, s$, and replicant $k = 1, \ldots, r_j$. We may assume that the number $2r_j$ of units in stratum j is an even number, without loss of generality, as we can otherwise randomly omit one of the units, reducing that number to an even one. We may then model the response variable as

$$Y_{ijk} = \mu + \tau_i + \beta_j + \epsilon_{ijk}, \tag{4.12}$$

where $\epsilon_{ijk} \sim N(0, \sigma^2)$ represents noise. Here, each stratum has its own "differential effect" β_j. By balancing units across the treatment and control within each stratum, we are therefore controlling for the effect produced by the different covariates.

Note 4.3. Whether or not one uses a randomized block design does not affect the model given by Equation (4.12). Rather, it directly affects the mechanism which assigns the units to the treatments, thereby affecting the *counting* of those units; i.e., without a block design, we would enumerate the replicants for treatment i and stratum j as $k = 1, \ldots, r_{ij}$, as the total number of replicants r_{ij} for each stratum could vary treatment-to-treatment.

▷

Theorem 4.3. *In an A/B test with s strata, the average outcome $\overline{Y}_{i\cdot\cdot}$ of treatment group i under the model prescribed by Equation (4.12) is a normally distributed random variable*

$$\overline{Y}_{i\cdot\cdot} \sim N\left(\mu + \tau_i + \frac{1}{n_i}\sum_{j=1}^{s} r_{ij}\beta_j, \frac{\sigma^2}{n_i}\right), \tag{4.13}$$

where r_{ij} is the number of replicants assigned to treatment group i and stratum j and $n_i = \sum_{j=1}^{s} r_{ij}$ is the total number of units assigned to treatment group i across all strata.

Proof. According to the model given by Equation (4.12), each observation is normally distributed according to

$$Y_{ijk} \sim N\left(\mu + \tau_i + \beta_j, \sigma^2\right).$$

The sample mean of treatment group i and stratum j is therefore also normally distributed as

$$\overline{Y}_{ij\cdot} \sim N\left(\mu + \tau_i + \beta_j, \frac{\sigma^2}{r_{ij}}\right).$$

The average value of treatment group i is obtained by averaging

$$\overline{Y}_{i\cdot\cdot} = \frac{1}{n_i}\sum_{j=0}^{s-1} r_{ij}\overline{Y}_{ij\cdot}.$$

The result follows from Theorem 2.3.   □

Corollary 4.1. *In a randomized block A/B test, the expected observed effect under the model prescribed by Equation (4.12) is independent of the factors β_j determined by the covariates.*

Proof. In general, we see from Equation (4.13) that the expected treatment effect is

$$\mathbb{E}[\Delta] = \mathbb{E}\left[\overline{Y}_{1..} - \overline{Y}_{0..}\right] = (\tau_1 - \tau_0) + \sum_{j=1}^{s}\left(\frac{r_{1j}}{n_1} - \frac{r_{0j}}{n_0}\right)\beta_j. \qquad (4.14)$$

However, under a randomized block A/B test, the number of replicants in each stratum are evenly divided between the control and treatment, yielding

$$r_{0j} = r_{1j}.$$

From this, it naturally follows that $n_0 = n_1$. For a randomized block design, we therefore conclude that the expected observed effect reduces to the actual causal effect of treatment,

$$\mathbb{E}[\Delta] = (\tau_1 - \tau_0).$$

We have therefore essentially *blocked* the effect of the covariates on the final observed effect of the experiment, by ensuring that units are allocated evenly, within each stratum, to both the treatment and control.   □

Note 4.4. Without blocking, we see that the final observed effect given by Equation (4.14) is *caused* by two factors: the treatment assignment and the covariates. When there are copious quantities of data within each stratum, the effect due to the covariates is likely to balance out, all by itself. This is, after all, the principle behind randomization. When covariates are available to the experimenter, however, and are known to produce a significant effect on the outcome of the experiment, it is always prudent to control for those nuisance factors with a block design. Blocking is especially important when there is a pronounced effect β_j for one or more smaller strata, in which the imbalance of data can be large if not explicitly controlled for.   ▷

When carrying out a randomized block A/B test, we cannot apply the *t*-test of Theorem 4.2 directly. The problem is that Equation (4.10) estimates the error within each treatment group. However, since each treatment group has experimental units from multiple strata, this is not an unbiased estimate of the variance σ^2. In other words, despite having "blocked" the effect of the nuisance factor, one must nevertheless take into account both the treatment groups and strata when evaluating the t-statistic. The result is given in our next theorem.

Theorem 4.4. *Given a randomized block A/B test with s strata, following Equation (4.12), in which $r_j = r_{0j} = r_{1j}$ replicants in stratum j have been assigned to both treatment groups, define the test statistic*

$$T = \frac{\overline{y}_{1..} - \overline{y}_{0..}}{\sqrt{\lambda MSW}} \sim t_{n-2s}, \tag{4.15}$$

where $\lambda = 1/n_0 + 1/n_1$, $n = n_0 + n_1$ is the total number of experimental units, and where MSW is the within-group mean squared error

$$MSW = \frac{1}{n-2s} \sum_{i=0}^{1} \sum_{j=1}^{s} \sum_{k=1}^{r_j} \left(y_{ijk} - \overline{y}_{ij.}\right)^2. \tag{4.16}$$

Then, under the hypothesis of no effect $\tau_0 = \tau_1$, the test statistic T is distributed as a Student's t-random variable with $n - 2s$ degrees of freedom. Moreover, the rejection region

$$R_\alpha = \{t : |t| > t_{n-2s,\alpha/2}\},$$

yields a test of size α.

Proof. Under the hypothesis of no effect, $\tau_0 = \tau_1$, and for a randomized block design, $r_{0j} = r_{1j}$. Under these assumptions, Equation (4.13) yields

$$\overline{Y}_{1..} - \overline{Y}_{0..} \sim N(0, \lambda\sigma^2).$$

Thus,

$$Z = \frac{\overline{Y}_{1..} - \overline{Y}_{0..}}{\sigma\sqrt{\lambda}} \sim N(0, 1)$$

is a standard normal random variable.

Next, let

$$S_{ij}^2 = \frac{1}{r_{ij} - 1} \sum_{k=1}^{r_{ij}} \left(Y_{ijk} - \overline{Y}_{ij.}\right)^2 \tag{4.17}$$

represent the sample variance of treatment group i stratum j. Recall from Theorem 2.6 that

$$\frac{(r_{ij} - 1)S_{ij}^2}{\sigma^2} \sim \chi_{r_{ij}-1}^2.$$

It follows that the sum

$$\sum_{i=0}^{1} \sum_{j=1}^{s} \frac{(r_{ij} - 1)S_{ij}^2}{\sigma^2} = \frac{1}{\sigma^2} \sum_{i=0}^{1} \sum_{j=1}^{s} \sum_{k=1}^{r_{ij}} \left(Y_{ijk} - \overline{Y}_{ij.}\right)^2 \sim \chi_{n-2s}^2.$$

The result follows. □

Example 4.5. Let's construct an example of a randomized block A/B test with 100 experimental units and three strata with frequency distribution $\langle 0.7, 0.2, 0.1 \rangle$. Let us suppose that the hypothesis of no effect is true, so that $\delta = 0$. Further, suppose that $\mu = 100$ and $\beta = \langle 0, 100, 300 \rangle$, so that the expected value of an observation from stratum 1, 2, and 3 is 100, 200, and 400, respectively. Finally, suppose that $\sigma = 20$.

As we are following a randomized block design, the test and control group will each receive 35 units from stratum 1, 10 units from stratum 2, and 5 units from stratum 3.

Following our experiment, we make the following observations. First, the overall mean of the control and treatment groups are given by

$$\overline{Y}_{0..} = 144.3 \qquad \text{and} \qquad \overline{Y}_{1..} = 150.3,$$

respectively. Furthermore, the six individual sums of squares (i.e., the sum portion of Equation (4.17)) are given in Table 4.5. Since there are three

	Stratum 1	Stratum 2	Stratum 3
Treatment 0	8859	2714	303
Treatment 1	9870	5550	2166

Table 4.5: Sum of squares for each treatment-stratum combination.

strata and $n = 100$ experimental units, the total within-group mean-squared error is given by the sum of these values divided by 94, which works out to be

$$MSW = 313.4.$$

Since there are fifty units assigned to treatment and control, the factor $\lambda = 2/50 = 0.04$. Thus, the T-statistic is

$$T = \frac{150.3 - 144.3}{\sqrt{0.04 \cdot 313.4}} \approx 1.69.$$

The critical value for this test statistic is given by

$$T_{crit} = t_{94,0.025} \approx 1.9855.$$

Thus, we fail to reject the null hypothesis.

The data for this example were simulated using Code Block 4.1. By running this simulation many times, we find that the null hypothesis is rejected in approximately 5% of the simulations. (See Exercise 4.3.)   ▷

4.2.4 Online Randomized Block Design

We next consider a variation of randomized block design for the context of *online experiments*. An online experiment is an experiment in which information regarding the units is not known in advanced. For example, a clinical trial may be conducted over the period of several years, and units (i.e., patients) may arrive sporadically. Another example is in ad tech: an advertising campaign is run in order to test the effect of a particular creative (ad). In both of these situations, it is not known in advanced who the

```
1   n = 100
2   mu = 100
3   beta = np.array([0, 100, 300])
4   delta = 0
5   sigma = 20
6   r_j = np.array([0.35*n, 0.10*n, 0.05*n]).astype(int)
7   T_crit = scipy.stats.t.isf(0.025,n-6)
8
9   # Create simulated data set.
10  Y = {}
11  for i in range(2):
12      Y[i] = {}
13      mu_delta = mu + delta * (i == 1)
14      for j in range(3):
15          Y[i][j] = np.random.normal(mu_delta + beta[j], scale=sigma,
                  size=r_j[j])
16
17  # Create a 2 x 3 array of lists
18  Y = np.array([ list(Y[0].values()), list(Y[1].values()) ])
19
20  # All Control and Treatment Observations
21  Y0 = np.concatenate(Y[0, :])
22  Y1 = np.concatenate(Y[1, :])
23
24  MSW = 0
25  for i in range(2):
26      for j in range(3):
27          MSW += np.sum((Y[i,j] - Y[i,j].mean())**2)
28
29  MSW /= (n - 6)
30  la = 1 / len(Y0) + 1 / len(Y1)
31
32  T = ( Y1.mean() - Y0.mean() ) / np.sqrt(la*MSW)
```

Code Block 4.1: Simulated data for the randomized block A/B of Example 4.5

participants of the experiment will be. The participants show up from time to time as the experiment is underway. The covariates of each participant (unit) are only known once they arrive. Then a decision is made whether to assign each participant to the treatment or control. Though this assignment can depend on the particular covariates of each unit and the totals recorded *to date*, it cannot depend on the ultimate totals, as this number has yet to materialize. We will follow an approach presented in Efron [1971], which we call *online randomized block design*.

Definition 4.7. *An* online randomized block algorithm for A/B tests *for an online A/B test with s strata proceeds as follows:*

1. *Initialize a zero vector $b_j = 0$, for $j = 1, \ldots, s$ and choose a fixed constant $p \in (1/2, 1)$. This vector will track the* imbalance *between the treatment and control as the experiment progresses.*
2. *Upon the arrival of each subsequent unit:*
   a) *Determine the current unit's stratum j;*
   b) *Assign unit to the treatment group with probability*
      - *$1/2$, if $b_j = 0$,*
      - *p, if $b_j < 0$, and*
      - *$(1-p)$, if $b_j > 0$.*
   c) *Increment b_j by $+1$ if the assignment was to the treatment group and increment b_j by -1 if the assignment was to the control group.*

We may interpret b_j as the running total number of treatment assignments minus the running total number of control assignments. Whenever, for a given stratum, there have been more control assignments than treatments ($b_j < 0$), we assign the unit to the treatment group with probability $p > 1/2$. Conversely, if there have been more treatment assignments ($b_j > 0$), we instead favor assignment to the control group. If there have been an equal number of treatment and control assignments ($b_j = 0$), we do not favor one or the other, and instead assign the unit to the treatment group with a 50% probability.

Though the online randomized block design does not guarantee that the same number of units will be assigned to each treatment within each stratum, it creates a purposeful pressure towards balancing out units among treatment groups within each stratum. For example, suppose that $p = 0.80$. Then whenever there are more units assigned to the control, for a given stratum, subsequent units will receive an 80% probability of being assigned to the treatment, until that imbalance is remedied.

It should be cautioned that Theorem 4.4 does not apply directly, unless the condition is met that the experimental units of each stratum are equally divided among the treatment groups.

4.2.5 The Adjustment Formula

We conclude this section with a discussion on A/B tests with a stratification that is not equally split between the control and treatment groups. Such a scenario occurs frequently in *observational studies*, otherwise known as *natural experiments*, as the treatment assignment lies beyond the power of the statistician. We will return to this topic in more detail in Chapter 8.

Theorem 4.5. *Consider a population of n units, each uniquely classified into one of s strata and into one of two treatment groups (i.e., a treatment and a control). Let (i, j) represent the jth unit of the ith stratum, for $i =$*

$1, \dots, s$ and $j = 1, \dots, n_i$, so that Z_{ij} is the treatment assignment for the ijth unit, and R_{ij}^T and R_{ij}^C are the responses of the ijth unit under the treatment and control assignments, respectively. Finally, let

$$a_i = \sum_{j=1}^{n_i} Z_{ij} \tag{4.18}$$

represent the number of treatment assignments for stratum i.

If the treatment assignment probability is homogeneous throughout each stratum; i.e., if

$$\mathbb{P}(Z_{ij} = 1) = \pi_i = \frac{a_i}{n_i}, \tag{4.19}$$

then the estimate

$$\hat{\Delta} = \sum_{i=1}^{s} \frac{n_i}{n} \left[\frac{1}{a_i} \sum_{j=1}^{n_i} Z_{ij} R_{ij}^T - \frac{1}{n_i - a_i} \sum_{j=1}^{n_i} (1 - Z_{ij}) R_{ij}^C \right] \tag{4.20}$$

is an unbiased estimate for the average treatment effect.

Proof. First, we recognize that

$$\hat{\Delta}_i = \frac{1}{a_i} \sum_{j=1}^{n_i} Z_{ij} R_{ij}^T - \frac{1}{n_i - a_i} \sum_{j=1}^{n_i} (1 - Z_{ij}) R_{ij}^C$$

is an unbiased estimate of the average treatment effect for the ith stratum, due to Theorem 4.1; i.e.,

$$\mathbb{E} \left[\frac{1}{a_i} \sum_{j=1}^{n_i} Z_{ij} R_{ij}^T - \frac{1}{n_i - a_i} \sum_{j=1}^{n_i} (1 - Z_{ij}) R_{ij}^C \right] = \frac{1}{n_i} \sum_{j=1}^{n_i} \left(R_{ij}^T - R_{ij}^C \right).$$

Therefore, it follows that

$$\mathbb{E} \left[\hat{\Delta}_i \right] = \frac{1}{n} \sum_{i=1}^{s} \sum_{j=1}^{n_i} \left(R_{ij}^T - R_{ij}^C \right).$$

This proves the result.   □

4.2.6 A/B Tests with Binary Outcomes

Next, we introduce several classic test statistics that are specifically useful in for experiments with a binary response. In the case of a binary outcome, the tests we describe herein are optimal in the sense that they are the most powerful unbiased tests for the given use case.

The Original Tea Test

To motivate our first test, we begin with a quotation from Sir Ronald Fisher's *Design of Experiments*: "A lady declares that by tasting a cup of tea made with milk she can discriminate whether the milk or the tea infusion was first added to the cup." And so Fisher begun to lay out the modern theory of randomized experiments.

He goes on to describe the setup: Eight cups of tea are to be prepared, four of which shall be made by adding the milk before the tea and four of which shall be made by adding the tea to the milk. The eight cups shall then be presented to the lady in random order: "...not an order determined arbitrarily by human choice, but by the actual manipulation of the physical apparatus used in games of chance, cards, dice, roulettes, etc." The task of the lady is then to divide the eight cups, presented to her in random fashion, into two sets of four, correctly identifying the preparation order of each set.

Fisher then considered the null hypothesis, that the lady was "without any faculty of discrimination." Under such a scenario, the lady could only correctly identify all eight cups by chance or accident. Fisher then notes that there are $\binom{8}{4} = 70$ possible arrangements of the eight cups into two groups of four. Fisher concluded that under the null hypothesis, the lady has only a $1/70$ chance of correctly classifying all eight cups of tea. Thus, the significance level of the experiment is $0.014 = 1/70$.

Fisher's Exact Test

The Lady's Tea can be generalized into a form known as *Fisher's exact test*, which we state as follows.

Theorem 4.6 (Fisher's Exact Test). *Let Z_i be the treatment assignment and R_i the binary response for the ith unit in a completely randomized A/B test. Then under the null hypothesis of no effect $H_0 : \mathbb{P}(R_i|Z_i = 1) = \mathbb{P}(R_i|Z_i = 0)$, the test statistic*

$$T = \sum_{i=1}^{n} Z_i R_i \tag{4.21}$$

is distributed as a hypergeometric random variable

$$T \sim \mathrm{HypGeom}(n, R_., Z_.),$$

where $R_. = \sum_{i=1}^{n} R_i$ represents the total number of positive responses and $Z_. = \sum_{i=1}^{n} Z_i$ represents the total number of units assigned to the treatment group $Z = 1$.

The test statistic T in Equation (4.22) represents the number of positive responses within the treatment group. The proof is straightforward and left as an exercise for the reader.

For the case of the Lady's tea test, we have $n = 8$, $Z. = 4$, and $R. = 4$. Since the lady correctly identified all four of the treated tea cups, her p-value is given by $\text{HypGeom}(4; 8, 4, 4) = 1/70 \approx 1.4286\%$.

For large values of n, Fisher's exact test can be approximated using the usual chi-square statistic for a 2×2 contingency table; i.e., using Equation (3.42). These contingency tables are similar to Table 4.6, below, which is applied explicitly to a single stratum i.

The Mantel-Haenszel Test

The direct generalization of Fisher's exact test for the case of a completely randomized block design is the Mantel-Haenszel test. In this case, we instead consider the $s \times 2 \times 2$ contingency table, where s is the number of strata. For the ith stratum, the 2×2 contingency table is shown in Table 4.6.

treatment	$R = 1$	$R = 0$	totals
treatment $(T = 1)$	T_i	$a_i - T_i$	a_i
control $(T = 0)$	$R_{i\cdot} - T_i$	$n_i + T_i - a_i - R_{i\cdot}$	$n_i - a_i$
totals	$R_{i\cdot}$	$n_i - R_{i\cdot}$	n_i

Table 4.6: Contingency table for the ith stratum.

Theorem 4.7 (Mantel-Haenszel Test). *Let Z_{ij} be the treatment assignment and R_{ij} the binary response for the jth replicant in the ith strata of a block-randomized A/B test. Then under the null hypothesis of no effect $H_0 : \mathbb{P}(R_{ij}|Z_{ij} = 1) = \mathbb{P}(R_{ij}|Z_{ij} = 0)$, the test statistics*

$$T_i = \sum_{j=1}^{n_i} Z_{ij} R_{ij} \sim \text{HypGeom}(n_i, R_{i\cdot}, Z_{i\cdot}), \tag{4.22}$$

which represent the number of positive responses within the treatment group for the individual strata, are independent hypergeometric random variables, with mean and variance

$$\mathbb{E}[T_i] = \frac{a_i R_{i\cdot}}{n_i} \tag{4.23}$$

$$\mathbb{V}(T_i) = \frac{a_i R_{i\cdot}(n_i - a_i)(n_i - R_{i\cdot})}{n_i^2(n_i - 1)}, \tag{4.24}$$

where $a_i = \sum_{j=1}^{n_i} R_{ij}$ and n_i represents the number of units assigned to the treatment within the ith stratum and the total number of units within the ith stratum, respectively. Moreover, as $n \to \infty$, the test statistic

$$X_n = \frac{\left[T - \sum_{i=1}^{s} \mathbb{E}[T_i] \right]^2}{\sum_{i=1}^{s} \mathbb{V}(T_i)} \sim \chi_1^2,$$

where $T = \sum_{i=1}^{s} T_i$ is the total number of positive responses from the treatment groups, is distributed as a chi-squared distribution with one degree of freedom; i.e., $\lim_{n \to \infty} X_n \sim \chi_1^2$. In particular, this is true even if $s \to \infty$ commensurately with n.

This is an astonishing result, as asymptotic theory typically requires the number of parameters to be fixed. The conclusion here, however, remains valid, even when $s \to \infty$ commensurately with n, even if the data within each individual stratum remain sparse, a case referred to as *sparse-data asymptotics*. Notable special cases include matched pairs, for which each stratum is constrained to contain only two test instances, and case-control studies, in which multiple controls are matched to each test (e.g., in ratios 2:1, 3:1, etc.). We discuss the former forthwith.

McNemar's Test

McNemar's test is a special case of the Mantel-Haenszel test for *paired binary data*, also known as *matched pairs*. *Matching* occurs whenever one matches each treatment to a fixed number of controls. Commonly, this ratio is 1:1, resulting in each treatment unit being matched with a single control unit; i.e., *matched pairs*. We may regard each pair as an individual stratum, which fixes the parameters $n_i = 2$ and $a_i = 1$. The total number of experimental units is then given by $n = 2s$, where s is the number of matched pairs.

The test statistic

$$T = \sum_{i=1}^{s} \sum_{j=1}^{2} Z_{ij} R_{ij} \sim \mathrm{HypGeom}(2s, R_{..}, s)$$

is therefore distributed as a hypergeometric random variable. For large s, we can therefore consider the test statistic

$$X_s = \frac{\left(T - \dfrac{R_{..}}{2} \right)^2}{\dfrac{R_{..}(2s - R_{..})}{4(2s - 1)}} \sim \chi_1^2,$$

which is again asymptotically distributed as a chi-squared random variable with one degree of freedom. Note that it is sufficient to utilize a single 2×2 contingency table for McNemar's test, since the marginal totals of each stratum are fixed by design.

Note 4.5. Matched pairs is primarily useful in observational studies, for which the experimenter does not have control over the treatment assignment. For more details, see Section 8.5.   ▷

4.2.7 Sum and Sign-Score Statistics

The specific tests discussed in Section 4.2.6 fall into a broader class of test statistic called a *sum statistic*, defined as follows.

Definition 4.8. *Consider a uniform randomized experiment with s strata. A sum statistic is any statistic of the form*

$$T = \sum_{i=1}^{s} \sum_{j=1}^{n_i} Z_{ij} q_{ij}, \tag{4.25}$$

where q_{ij} is a function of the response of the ijth unit.

Similarly, a sign-score statistic is any statistic of the form

$$T = \sum_{i=1}^{s} d_i \sum_{j=1}^{n_i} c_{ij} Z_{ij}, \tag{4.26}$$

where $c_{ij} \in \mathbb{B}$ is binary and both d_i and c_{ij} are functions of the responses.

Note that any sign-score statistic is also a sum statistic, with $q_{ij} = d_i c_{ij}$, but the converse is not necessarily true.

Fisher's exact test, the Mantel-Haenszel test, and McNemar's test are all sign-score statistics, with $d_i = 1$ and $c_{ij} = R_{ij}$, since they are each tests for a binary response. (Even if the response were not binary, they would still constitute a sum statistic.) There are many more examples of test statistics that fall into these general classes.

For the case of a single stratum $s = 1$, *Wilcoxon's rank-sum test* requires one to rank the observed responses from smallest to largest, using the average rank to break any ties. The *rank-sum statistic* is then defined by summing the ranks of the treated units, so that q_i is the rank. (This is equivalent to the *Mann-Whitney test*.)

For the case of s matched pairs, so that $n_i = 2$ and $a_i = 1$, *Wilcoxon's signed rank test* involves calculating the rank d_i of the absolute differences $|R_{i1} - R_{i2}|$ of the various matched pairs, using averages to break ties. We then define the indicators

$$c_{i1} = \mathbb{I}[R_{i1} > R_{i2}] \qquad \text{and} \qquad c_{i2} = \mathbb{I}[R_{i2} > R_{i1}],$$

so that $c_{i1} = c_{i2} = 0$ if and only if $R_{i1} = R_{i2}$. With this definition, notice that the sum

$$Z_{i1}c_{i1} + Z_{i2}c_{i2}$$

is an indicator for whether or not the treated unit in pair i had a higher response than the control. The *signed-rank statistic* is then the signed-score statistic of Equation (4.26) with these given definitions. Note that pairs only contribute to the overall statistic when their responses differ.

The *stratified rank-sum statistic* is obtained by summing the individual Wilcoxon's rank-sum statistics as computed in each stratum independently. However, this method has been shown to be inefficient when s is large relative to n.

To remedy this, consider the Hodges and Lehmann's *aligned rank statistic*, which involves ranking the differences between each response and their respective stratum's average response; i.e., ranking the *aligned responses* $R_{ij} - \overline{R}_{i\cdot}$. The aligned rank statistic is then just the sum statistic obtained by summing these ranks in the treated groups.

Another example of a sign-score statistic is the *median test statistic*. Here, we let c_{ij} be the binary indicator that equals 1 whenever R_{ij} is greater than the median response of its given stratum.

The upshot is that the hypothesis of no treatment effect can easily be tested for any of these sum statistic, using the following result.

Proposition 4.1. *In a uniform randomized experiment, if the null hypothesis of no effect is true, then the expected value and variance of any sum statistic is given by the following*

$$\mathbb{E}[T] = \sum_{i=1}^{s} a_i \overline{q}_i, \tag{4.27}$$

$$\mathbb{V}(T) = \sum_{i=1}^{s} \frac{a_i(n_i - a_i)}{n_i(n_i - 1)} \sum_{j=1}^{n_i} (q_{ij} - \overline{q}_i)^2. \tag{4.28}$$

Notice that these equations reduce to Equations (4.23) and (4.24) for the special case of the Mantel-Haenszel test. Once the expected value and variance of any given sum or sign-score statistic are known, one can then use the Wald test to test the null hypothesis of no treatment effect.

4.3 Single-Factor Experiments

In this section, we extend extend the classic A/B test to the more general case of multi-level (single-factor) experiments. For example, consider a three-level clinical trial in which patients receive either a placebo, a low dose, or a high dose of a certain drug. The generalization from a two-level experiment is mostly straightforward, with the analysis of variance replacing the t-test.

4.3.1 Design

Here, we consider both completely randomized experiments and randomized block experiments for multi-level experiments of a single factor.

Completely randomized experiments

For a completely randomized experiment, we fix the number of replicants r required for each treatment group, and then randomly assign units to the treatment groups while respecting this constraint. To achieve this, we may either shuffle the data or use one-at-a-time sampling as given by Definition 4.4.

Definition 4.9. *A single-factor completely randomized experiment is a single factor experiment in which units are randomly assigned to the m treatment groups, subject to the condition that each treatment group receive the same number r of replicants.*

Index notation for a multi-level experiment is practically identical to the index notation for A/B tests, except that the index i ranges over a larger set of possible treatment groups. Following the assignment of our population into treatment groups, we may let Y_{ij}, for $i = 1, \ldots, m$ and $j = 1, \ldots, r$, represent the observed outcome of the jth replicant of the ith treatment group. This follows the model given by Equation (4.7), with the only distinction being that the index i ranges from $i = 1, \ldots, m$.

Randomized Block Design

As before, a randomized block design is achieved by implementing a completely randomized design within each of several strata. Formally, we have the following.

Definition 4.10. *A single-factor randomized block experiment is a single-factor experiment consisting of a population with $s \geq 2$ strata (covariates) and an assignment mechanism that randomly assigns an equal number of the units of each stratum to each of the m treatment groups; i.e., if mr_j units belong to stratum j, for $j = 1, \ldots, s$, then the assignment mechanism will randomly assign r_j units to each of the m treatment groups.*

Following our assignment of units into their respective treatment groups, such that the units of each stratum are equally divided among treatments, we may make the switch to index notation. As before, we let Y_{ijk}, for $i = 1, \ldots, m$, $j = 1, \ldots, s$, and $k = 1, \ldots, r_{ij}$ represent the observed outcome of the kth replicant assigned to the ith treatment group and jth stratum.

Theorem 4.3 is still valid in this general context, as nothing in the proof relied on the fact that there were only two treatment groups; the same proof applies if the index i is allowed to run over additional groups.

As before, when using a randomized block design, we have $r_{1j} = \cdots = r_{mj} = r_j$; i.e., within stratum j each treatment group receives the same number r_j of replicants, and $n_1 = \cdots = n_m = r$; i.e., each treatment group overall receives the same number of replicants. Thus, as before, we see the benefit of the randomized block design, in that the expected observed effect between any two treatment groups no longer depends on the covariates β_j.

We will table our discussion on the analysis of a single-factor experiment with randomized block design until Section 4.4.3.

Online Randomized Block Design

We next generalize the algorithm presented in Section 4.2.3 for the case of an experiment with more than two levels. To this end, we combine elements of online randomized block design for A/B tests with one-at-a-time simple random sampling (Definition 4.4) to devise an approach suitable to a multi-level experiment.

Definition 4.11. *The* online randomized block algorithm *for an online experiment with m levels and s strata proceeds as follows:*

1. *Initialize a* count matrix $C_{ij} = 0$, *for $i = 1, \ldots, m$ and $j = 1, \ldots, s$;*
2. *Upon the arrival of each subsequent unit:*
   a) Determine the current unit's stratum j;
   b) Assign unit to treatment group i with probability

$$\mathbb{P}(Z = i | C_{ij}, x = j) = \frac{||C_{\cdot j}||_\infty + 1 - C_{ij}}{m\left(||C_{\cdot j}||_\infty + 1 - \overline{C}_{\cdot j}\right)}, \qquad (4.29)$$

   where $C_{\cdot j}$ is the jth column vector of C, and $|| \cdot ||_\infty$ is the infinity norm[3] of $C_{\cdot j}$;
   c) Increment the ijth component C_{ij} by 1 based on the final treatment assignment i and stratum j for the given unit.

One can easily verify that Equation (4.29) is a normalized probability mass function over the treatment groups by summing over the index $i = 1, \ldots, m$ (See Exercise 4.4). This assignment mechanism has the following interpretation: upon the arrival of a unit, we first identify the stratum j to which that unit belongs. We then look for the treatment group that has the greatest count of assignments $||C_{\cdot j}||_\infty$ out of all the units that have arrived *thus far* belonging to stratum j. We then assign the current unit to treatment group i with a probability proportional to the difference between this greatest value (plus one) and the number of units from that stratum already assigned to that treatment group. By adding 1, we ensure that there is always a non-vanishing probability of assigning each unit to any of the treatment groups.

---

[3] The *infinity norm* of a vector $x \in \mathbb{R}^m$ is defined as $||x||_\infty = \max\{|x_1|, \ldots, |x_m|\}$.

Example 4.6. An online experiment with three levels and four strata is well underway. At a given point in time, the count matrix reads as follows:

$$C = \begin{bmatrix} 3 & 0 & 1 & 2 \\ 5 & 0 & 2 & 1 \\ 4 & 1 & 0 & 1 \end{bmatrix}.$$

What is the probability of assigning the next unit to treatment group i if the next arriving unit belongs to stratum 0?

To proceed, we first compute the infinity norm of column 0, which is given by

$$\|C_{\cdot 0}\|_\infty = 5.$$

We then construct the difference

$$\|C_{\cdot 0}\|_\infty + 1 - C_{i0} = \begin{bmatrix} 3 \\ 1 \\ 2 \end{bmatrix}.$$

By normalizing this vector, we should assign this unit to each treatment group with probability

$$\mathbb{P}(Z = i | C, x = 0) = \begin{bmatrix} 1/2 \\ 1/6 \\ 1/3 \end{bmatrix};$$

that is, a 50% probability for treatment group 0, a 16.7% probability for treatment group 1, and a 33.3% probability for treatment group 2. Suppose that we then choose a random number from this distribution, and find that our unit should be assigned to treatment group 2. We then update our count matrix to reflect this:

$$C_{20} = C_{20} + 1.$$

Finally, for purpose of illustration, we can compute the assignment probability for the next unit, as a function of the stratum. (Based on our original count matrix.) We have already computed the probability vector for stratum 0. A simple computation yields similar results for strata 1–3, which we record in the following matrix

$$\pi = \begin{bmatrix} 1/2 & 2/5 & 1/3 & 1/5 \\ 1/6 & 2/5 & 1/6 & 2/5 \\ 1/3 & 1/5 & 1/2 & 2/5 \end{bmatrix}.$$

Thus, if the next arriving unit belongs to stratum j, we use the assignment vector corresponding to column j of the probability matrix π.   ▷

Example 4.7. Write a simulation in Python that simulates an online randomized block design for three levels and four strata. Assume that the population is distributed over the strata according to the ratios $\langle 0.6, 0.2, 0.1, 0.1 \rangle$.

```
1   C = np.zeros((3,4))
2   n = 100
3
4   for u in range(n):
5       j = np.random.choice(4, p=[0.6, 0.2, 0.1, 0.1])
6       Cj_inf = max( C[:, j] )
7       p = Cj_inf + 1 - C[:, j]
8       p /= p.sum()
9       i = np.random.choice(3, p=p)
10      C[i,j] += 1
```

Code Block 4.2: Simulation of an online randomized block assignment mechanism.

We accomplish this in Code Block 4.2.

We begin by initializing our 3×4 matrix of zeros and the number of units to simulate (lines 1–2). The assignment is accomplished term-by-term within our for-loop. Upon the arrival of each subsequent unit, we first observe the stratum to which the unit belongs. This process of observation is simulated on line 5. Next, we compute $||C_{.j}||_\infty$, represented in code as Cj_inf, as the maximum value (thus far) in column j. We then form our probability mass function on lines 7–8, and carry out a random draw, according to the given probability vector, on line 9. Since unit u is assigned to treatment group i, we thus increment our count for C_{ij} by 1. This is exactly the set of code we used to simulate the count matrix in Example 4.6, except with $n = 20$.

Finally, note that this code can be modified to perform a simple random assignment, by replacing p=p in line 9 with p=[1/3,1/3,1/3].   ▷

4.3.2 Analysis

For the remainder of this section, we will restrict our attention to single-factor completely randomized experiments; i.e., single-factor experiments with no covariates. This task will be itself broken into two tasks: an analysis of variance and an analysis of pairwise comparisons. The first task uses ANOVA to determine whether or not we can reject the hypothesis of no effect. If we reject the hypothesis of no effect, we then seek to determine which specific groups show a significant effect.

We will postpone our discussion of single-factor randomized block design until Section 4.4.3, as this analysis can be constructed as a special case of a two-factor completely randomized experiment.

Analysis of Variance

When analyzing the results of a single-factor completely randomized experiment, the first question is whether or not there is any effect at all. A single-factor completely randomized experiment that satisfies the ANOVA assumptions of Definition 3.27 can be analyzed using the analysis of variance method prescribed in Theorem 3.12, following the key formulas that are summarized in Table 3.4.

Example 4.8. Run a simulation using $\alpha = 0.05$, $\mu = 100$, $\sigma = 10$, $m = 4$, and $r = 12$ under the true null hypothesis. Determine the frequency of times in which the null hypothesis is incorrectly rejected.

```
1   alpha = 0.05
2   mu = 100
3   sigma = 10
4   r = 12
5   m = 4
6   n = r * m
7   tau = np.zeros(m).reshape((m,1))
8   F_crit = scipy.stats.f.isf(alpha, m-1, n - m) #2.866
9
10  rejections = 0
11  n_sims = 10000
12  for i in range(n_sims):
13      Y = np.random.normal(loc=mu+tau, scale=sigma, size=(m, r))
14
15      SSB = np.sum( (Y.mean(axis=1) - Y.mean())**2 ) * r
16      SSW = np.sum( (Y - Y.mean(axis=1).reshape((m,1)))**2 )
17      SST = np.sum( (Y - Y.mean())**2 )
18      MSB = SSB / (m-1)
19      MSW = SSW / (n-m)
20      F = MSB / MSW
21      rejections += 1 if F > F_crit else 0
22
23  print(rejections / n_sims) #0.0503
```

Code Block 4.3: ANOVA simulation for Example 4.8.

This is achieved in Code Block 4.3. The observed rejection rate in 10,000 simulations was 5.03%, as expected. Note our use of **reshape** to convert arrays into column vectors (lines 7 and 16). This makes the computation in lines 13 and 16 possible.   ▷

Power Analysis

In Chapter 3, we saw the importance of non-central distributions in the power analysis of statistical tests. The analysis of variance is no exception. Before we begin our discussion of power analysis, we first define a generalization to the F-distribution (Definition 2.15).

Definition 4.12. *Let* $X \sim \mathrm{NCX}^\lambda_{\nu_1}$ *be a noncentral chi-squared random variable with* ν_1 *degrees of freedom and noncentrality parameter* λ, *and let* $Y \sim \chi^2_{\nu_2}$ *be a (central) chi-squared random variable with* ν_2 *degrees of freedom. Further, suppose* X *and* Y *are independent; i.e.,* $X \perp\!\!\!\perp Y$. *Then the random variable* F *defined by the ratio*

$$F = \frac{X/\nu_1}{Y/\nu_2} \tag{4.30}$$

is a noncentral F *random variable with degrees of freedom* ν_1 *and* ν_2 *and noncentrality parameter* λ. *Symbolically, we say that*

$$F \sim \mathrm{NCF}^\lambda_{\nu_1,\nu_2}.$$

Note that $\mathrm{NCF}^0_{\nu_1,\nu_2} = \mathrm{F}_{\nu_1,\nu_2}$.

Now, in order to perform a power analysis for single-factor experiments, we must determine the distribution of the F-statistic (Equation (3.79)) in the case that the null hypothesis is not true. To achieve this, we revisit Lemmas 3.8 and 3.9. However, we immediately realize that Lemma 3.8 is still valid, as its result is independent of the truth of the null hypothesis. Thus,

$$\frac{SSW}{\sigma^2} = \frac{1}{\sigma^2} \sum_{i=1}^m \sum_{j=1}^r \left(Y_{ij} - \overline{Y}_{i\cdot}\right)^2 \sim \chi^2_{n-m}. \tag{4.31}$$

However, Lemma 3.9 does rely on the truth of the null hypothesis. We therefore need to determine how the between-group sum of squares is distributed when the null hypothesis is false. (We restrict our attention to the case when $r_1 = \cdots = r_m$, which is true for completely randomized experiments.) The answer is provided in the following lemma.

Lemma 4.1. *Suppose the random variables* Y_{ij} *are distributed according to*

$$Y_{ij} \sim N(\mu + \tau_i, \sigma^2), \tag{4.32}$$

for $i = 1,\ldots,m$ *and* $j = 1,\ldots,r$. *(Let* $n = mr$ *be the total number of variables.) Without loss of generality, suppose that*

$$\sum_{i=1}^m \tau_i = 0.$$

Then the scaled between-group sum of squares is distributed as a noncentral chi-squared random variable. In particular,

$$\frac{SSB}{\sigma^2} = \frac{r}{\sigma^2} \sum_{i=1}^{m} \left(\overline{Y}_{i\cdot} - \overline{\overline{Y}} \right)^2 \sim \mathrm{NCX}_{m-1}^{\lambda}, \qquad (4.33)$$

where the noncentrality parameter is given by $\lambda = r\phi$, *where we define the noncentrality effect by the relation*

$$\phi = \frac{1}{\sigma^2} \sum_{i=1}^{m} \tau_i^2. \qquad (4.34)$$

Note 4.6. The introduction of the noncentrality effect ϕ, as opposed to simply defining λ as r times the right-hand side of Equation (4.34), may at first pass seem superfluous. However, this notation has the benefit that ϕ is independent of the sample size r, and therefore more closely resembles the concept of a *minimum detectable effect*. Since it is a scaled version of the noncentrality parameter, we name it the noncentrality effect.   ▷

Proof. From Equation (4.32), we have

$$\overline{Y}_{i\cdot} \sim \mathrm{N}\left(\mu + \tau_i, \sigma^2/r\right) \qquad \text{and} \qquad \overline{\overline{Y}} \sim \mathrm{N}\left(\mu, \sigma^2/n\right).$$

Therefore,

$$\frac{\overline{Y}_{i\cdot} - \mu}{\sigma/\sqrt{r}} \sim \mathrm{N}\left(\frac{\tau_i \sqrt{r}}{\sigma}, 1\right).$$

It follows that the quantity W, defined by

$$W = \frac{r}{\sigma^2} \sum_{i=1}^{m} \left(\overline{Y}_{i\cdot} - \mu \right)^2 \sim \mathrm{NCX}_m^{r\phi}$$

is a noncentral chi-squared random variable with m degrees of freedom and noncentrality parameter $\lambda = r\phi$, with ϕ as defined in Equation (4.34). However,

$$W = \frac{r}{\sigma^2} \sum_{i=1}^{m} \left[\left(\overline{Y}_{i\cdot} - \overline{\overline{Y}} \right) + \left(\overline{\overline{Y}} - \mu \right) \right]^2$$

$$= \frac{r}{\sigma^2} \sum_{i=1}^{m} \left(\overline{Y}_{i\cdot} - \overline{\overline{Y}} \right)^2 + \frac{n}{\sigma^2} \left(\overline{\overline{Y}} - \mu \right)^2$$

The second line follows as the cross terms sum to zero. Now, the second term on the right is distributed as a χ_1^2 random variable. But $W \sim \mathrm{NCX}_m^{r\phi}$. We conclude that the first term on the right is distributed as $\mathrm{NCX}_{m-1}^{r\phi}$. This completes the proof.   □

Theorem 4.8. *Under the model Equation (4.32), the F-statistic, defined as $F = MSB/MSW$ in Equation (3.79) is distributed as a noncentral F distribution:*

$$F \sim \mathrm{NCF}^{r\phi}_{m-1,n-m} \tag{4.35}$$

with noncentrality effect ϕ given by Equation (4.34)

Proof. This follows immediately from Lemmas 3.8 and 4.1 and Definition 4.10. □

We know that in an ANOVA test with significance level α, we reject the null hypothesis whenever

$$F > \mathrm{F}^{\alpha}_{m-1,n-m},$$

where $\mathrm{F}^{\alpha}_{m-1,n-m}$ is the unique real number that satisfies the relation

$$\mathbb{P}(X > \mathrm{F}^{\alpha}_{m-1,n-m}) = \alpha \qquad \text{whenever} \qquad X \sim \mathrm{F}_{m-1,n-m};$$

i.e., $\mathrm{F}^{\alpha}_{m-1,n-m}$ is the inverse survival function of F with residual probability α. The power of our experiment, for a specific noncentrality effect ϕ, is therefore the probability of triggering this condition if the null hypothesis is false. In other words, the power is simply the probability

$$\beta = \mathbb{P}(F > \mathrm{F}^{\alpha}_{m-1,n-m} | F \sim \mathrm{NCF}^{r\phi}_{m-1,n-m}), \tag{4.36}$$

which, in turn, implies that the power is the value of the survival function of the noncentral F distribution evaluated at the critical value for the test:

$$\beta = \mathrm{NCF}^{r\phi}_{m-1,n-m}.sf(\mathrm{F}^{\alpha}_{m-1,n-m}).$$

(We borrow Pythonic syntax here; think of the ".sf" as the survival function method for the distribution $\mathrm{NCF}^{r\phi}_{m-1,n-m}$.)

Example 4.9. Determine the power of the experiment described in Example 4.8, assuming a noncentrality effect of $\phi = 1$.

Equation (4.36) can easily be encoded as done in Code Block 4.4. We find that the power of the test is 80.2956%, for a noncentrality effect of $\phi = 1$. ▷

Example 4.10. In Example 4.8, determine a value of τ that would yield a noncentrality effect of $\phi = 1$. Run a simulation, under this alternate hypothesis, to determine the fraction of times in which the null hypothesis is correctly rejected. Does the result agree with the theoretical result obtained in Example 4.9?

We can achieve this by simply modifying line 7 of Code Block 4.3 with the expression provided in Code Block 4.5. After rerunning the simulation with this modification, we obtained an output of 0.7999, which closely agrees with our result from Example 4.9. ▷

```
1   def power(r, phi=1):
2       m = 4
3       n = r * m
4       F_crit = scipy.stats.f.isf(alpha, m-1, n-m)
5       return scipy.stats.ncf.sf(F_crit, m-1, n-m, r*phi)
6
7   r=12
8   print(f"The power is {power(r, phi=1)}") #0.802956
```

Code Block 4.4: Power calculation for the noncentral F-test

```
1   tau = np.array([0,-np.sqrt(50),+np.sqrt(50),0]).reshape((m,1))
```

Code Block 4.5: Modification of Code Block 4.3 for power simulation

4.3.3 Pairwise Comparisons

With the F-test, we are able to determine when to reject our null hypothesis, that there are no differences among the various treatment groups. In practice, however, we do not run experiments to confirm that everything is the same; rather, we run experiments to find out what is different and, in particular, what is better. The theory we developed thus far can only tell us when there is *some* difference among *some* of the treatment groups. But it fails to inform us which treatment groups are actually significantly different from each other. That is our goal for this section: to determine a test that will inform us when pairwise differences are significant.

The tools of this section should only be deployed when the null hypothesis has already been rejected. That is, it does not make sense to try to determine whether there is a significant difference between two particular treatment groups when we cannot even conclude that there is *any* significant difference in the first place.

Experimentwise Error

Consider a single-factor experiment with m treatment groups. If the ANOVA null hypothesis is rejected, we conclude that there is a significant difference between at least one pair of the m treatment groups. A *pairwise comparison analysis* seeks to determine which specific pairs of treatment groups are significantly different from each other. When performing a *pairwise* comparison, we seek to determine whether the observed effect

$$\Delta_{ii'} = \overline{Y}_{i'.} - \overline{Y}_{i.}$$

is significantly different, for each of the $c = m(m-1)$ pairs, for $i' \not\leq i$; i.e., for $i = 0, \ldots, m-1$ and $i' = i+1, \ldots, m-1$.

We know that a Type I error occurs when a true null hypothesis is falsely rejected. With multiple comparisons, however, we must carefully define whether we mean the error of a single comparison test or the overall error, a concept we now define.

Definition 4.13. *Consider an experiment that is comprised of a set of c null hypotheses $H_0^1, \ldots, H_0^c$ that are tested individually with separate test statistics. Let the event A_i be defined as accepting the ith hypothesis, and define the composite null hypothesis as $H_0 = \bigcap_{i=1}^c H_0^i$. Then the experimentwise error rate a is the probability of committing at least one Type I error among all of the tests, given that each null hypothesis is true; i.e.,*

$$a = \mathbb{P}\left(\sum_{i=1}^c \mathbb{I}[\neg A_i] \geq 1 \,\middle|\, H_0\right),$$

where $\neg A_i$ is the event that the ith null hypothesis is rejected.

If each of the c test statistics are *independent*, we have the following.

Proposition 4.2. *Suppose an experiment is comprised of a set of c individual null hypotheses $H_0^1, \ldots, H_0^c$ with independent test statistics, each with a probability α of a Type I error. Then the experimentwise error rate is*

$$a = 1 - (1 - \alpha)^c.$$

Proof. The probability that at we reject at least one null hypothesis is the complement of accepting each null hypothesis, so that

$$a = 1 - \mathbb{P}\left(\bigcap_{i=1}^c A_i \,\middle|\, H_0\right) = 1 - \prod_{i=1}^c \mathbb{P}(A_i | H_0) = 1 - (1 - \alpha)^c.$$

The second equality follows from the independence of the tests.   □

Example 4.11. Suppose an experiment consists of four independent hypothesis tests, each with significance level $\alpha = 0.05$. Then the experimentwise error rate is given by
$$a = 1 - 0.95^4 \approx 0.185.$$

▷

When we are performing pairwise comparisons for a single-factor experiment, however, the individual test hypotheses are *not* independent! The relationship between the comparisonwise and experimentwise error rates is generally not as simply stated as the expression in Proposition 4.2.

Example 4.12. Consider the null hypotheses for the pairwise comparisons

$$H_0^1 : \qquad \tau_1 = \tau_2$$
$$H_0^2 : \qquad \tau_1 = \tau_3$$
$$H_0^3 : \qquad \tau_2 = \tau_3.$$

Suppose that we fail to reject H_0^1, but we do reject H_0^2. Given this information, what do you suspect about H_0^3?

From an intuitive perspective, it is clearly now more likely that H_0^3 should be rejected than it was before we knew anything about the outcome of the first two tests. This intuition is, however, not conclusive, as pairwise comparisons are not transitive. Our intuition nevertheless illustrates that the three tests are not independent. ▷

In general, it is not straightforward to determine the experimentwise error rate a, given the individual comparison significance levels α, nor vice versa. However, if our goal is to ensure that the experimentwise error rate is below a given threshold, we can make use of the following correction.

Proposition 4.3 (Bonferroni Correction). *By setting the significance level of each comparison test at $\alpha = a^*/c$, the experimentwise error rate is bounded above by the value a^*; i.e., $a \leq a^*$.*

Proof. In the proof of Proposition 4.2, we saw that

$$a = 1 - \mathbb{P}\left(\left.\bigcap_{i=1}^{c} A_i \right| H_0\right).$$

By applying the Bonferroni inequality, we may write

$$a \leq 1 - \left[\sum_{i=1}^{c} \mathbb{P}(A_i|H_0) - (c-1)\right] = c\alpha.$$

If we select $\alpha = a^*/c$, the result follows. □

For any value of m greater than a few, this method, however, becomes inefficient, as seen in our next example.

Example 4.13. Consider the case of $m = 10$ treatment groups. In this case, there will be a total of $c = m(m-1)/2 = 45$ pairwise comparisons. In order to achieve an experimentwise error rate of $a = 0.05$, so that there is only a 5% chance of one of the pairwise comparisons of being falsely rejected, we would have to use a significance level of $\alpha = 0.05/45 \approx 0.00111$ for each comparison test. ▷

Fisher's Least Significant Difference Test

Fisher's least significant difference (LSD) test controls for the comparison-wise significance levels α, without regard for the overall experimentwise error rate a. The philosophy here is to specify the significance level α for each pairwise comparison, and let the overall experimentwise error rate be as it may. The philosophy behind Fisher's LSD test is to simply apply the standard t-test (at fixed level α) to each pairwise comparison, with the following caveat: the pooled sample variance is replaced by the overall within-group mean squared error MSW. That is, we use a single estimate for σ^2 for each comparison. This correction allows for a larger number of degrees of freedom for each individual comparison, which, in turn, increases the power of each individual test. We now state Fisher's test as follows.

Theorem 4.9 (Fisher's least significant difference test). *Given a single-factor completely randomized experiment with m treatment groups and r replicants per treatment, for which the ANOVA null hypothesis has been rejected, define the* least significant difference (LSD) *as*

$$LSD = t_{n-m,\alpha/2}\sqrt{2MSW/r}, \tag{4.37}$$

for fixed significance level α and within-group mean squared error

$$MSW = \frac{1}{n-m}\sum_{i=1}^{m}\sum_{j=1}^{r}\left(Y_{ij}-\overline{Y}_{i\cdot}\right)^2.$$

Then, by rejecting any of the $m(m-1)/2$ null hypotheses

$$H_0^{i,i'} : \tau_i = \tau_{i'},$$

for $i' \nleq i$, whenever the observed difference

$$\Delta_{ii'} = \left|\overline{Y}_{i\cdot} - \overline{Y}_{i'\cdot}\right| > LSD$$

constitutes a hypothesis test with significance level α.

In other words, once the ANOVA null hypothesis (i.e., the hypothesis of no effect) has been rejected, Fisher's LSD test entails classifying any pair $\{i, i'\}$ of treatment groups as significantly different, at level α, if the observed difference Δ is greater than the least significant difference given by Equation (4.37).

Note 4.7. Fisher's LSD test differs from the t-test of Theorem 4.2 only in that the within-group mean squared error MSW is computed using all of the data, resulting in a total of $(n-m)$ degrees of freedom. The result follows by restating the test in terms of the observed difference as opposed to the T-statistic.   ▷

Example 4.14. Consider a single-factor experiment with five-levels, and six replicants per level. The observed data are given by

$$Y = \begin{bmatrix} 52 & 65 & 89 & 60 & 93 & 56 \\ 95 & 80 & 63 & 90 & 105 & 101 \\ 72 & 107 & 103 & 117 & 99 & 71 \\ 92 & 89 & 128 & 110 & 127 & 87 \\ 113 & 91 & 138 & 104 & 160 & 101 \end{bmatrix}.$$

For these data, we reject the ANOVA null hypothesis at the $\alpha = 0.05$ significance level. (See Exercise 4.5.) For Fisher's LSD test, we first compute the treatment group means:

$$\overline{Y}_{i\cdot} = \begin{bmatrix} 69.2 & 89.0 & 94.8 & 105.5 & 117.8 \end{bmatrix}.$$

The LSD can be easily computed, and works out to be

$$LSD = t_{25,0.025}\sqrt{2MSW/r} \approx 1.708\sqrt{129.68} = 19.45.$$

Fisher's LSD test therefore generates the following statistically significant comparisons:

Group 0 and Groups:	1, 2, 3, 4
Group 1 and Groups:	4
Group 2 and Groups:	4

Notice the failure of transitivity: groups 1, 2, and 3 are not significantly different from one another. Yet groups 1 and 2 are each significantly different from group 4, although group 3 is not.   ▷

Tukey's Honestly Significant Difference Test

Tukey's honestly significant difference (HSD) test is similar to Fisher's LSD test, except that it primarily controls for the experimentwise error rate a. Its derivation makes use of the following generalization of a T-random variable.

Definition 4.14. *Suppose that m samples of size r are collected from a normal distribution $N(\mu, \sigma^2)$, and define $\overline{Y}_{\max}$ and $\overline{Y}_{\min}$ by the relations*

$$\overline{Y}_{\max} = \max_{i=1,\ldots,m} \overline{Y}_{i\cdot} \quad \text{and} \quad \overline{Y}_{\min} = \max_{i=1,\ldots,m} \overline{Y}_{i\cdot},$$

respectively. Then the random variable

$$Q = \frac{\overline{Y}_{\max} - \overline{Y}_{\min}}{\sqrt{s^2/r}}, \tag{4.38}$$

where s^2 is the pooled sample variance, is called a studentized range random variable *and its distribution the* studentized range distribution.

Note 4.8. For the case $m = 2$, the studentized range random variable is a scaled version of the T-statistic given in Equation (4.9), in that $Q = \sqrt{2}|T|$. This follows since when there are only two groups, one mean must be the maximum and the other must be the minimum.   ▷

Note 4.9. The studentized range distribution depends on two parameters: the number of groups m and the number of degrees of freedom ν of the pooled sample variance. If Q is such a random variable, we write $Q \sim Q_\nu^m$.
▷

The idea behind the studentized range distribution is as follows: the condition that there is at least one significant difference among the various pairwise comparisons is equivalent to the condition that there is a significant difference between the two most extreme group means. The logic is that if the difference between the largest and the smallest sample means is not significant, then none of the differences are significant.

Theorem 4.10 (Tukey's honestly significant difference test). *Given a single-factor completely randomized experiment with m treatment groups and r replicants per treatment, for which the ANOVA null hypothesis has been rejected, define the* honestly significant difference (HSD) *as*

$$HSD = q_a(m, n - m)\sqrt{MSW/r}, \qquad (4.39)$$

for fixed experimentwise error rate a and within-group mean squared error

$$MSW = \frac{1}{n - m} \sum_{i=1}^{m} \sum_{j=1}^{r} \left(Y_{ij} - \overline{Y}_{i\cdot} \right)^2,$$

where the quantity $q_a(m, \nu)$ is defined as the unique value such that

$$\mathbb{P}(Q > q_a(m, \nu)) = a,$$

for $Q \sim Q_\nu^m$. Then, by rejecting any of the $m(m-1)/2$ null hypotheses

$$H_0^{i,i'} : \tau_i = \tau_{i'},$$

for $i' \not\leq i$, whenever the observed difference

$$\Delta_{ii'} = \left| \overline{Y}_{i\cdot} - \overline{Y}_{i'\cdot} \right| > HSD$$

constitutes a hypothesis test with experimentwise error rate a.

Example 4.15. Let us consider once again the experiment described in Example 4.14. Instead of specifying a pairwise significance level α, let us instead require that the experimentwise error rate be set at $a = 0.05$. If all of the groups are equivalent, there is only a 5% probability of obtaining at least one false rejection. The honestly significant difference is computed as

```
1   from statsmodels.stats.libqsturng import psturng, qsturng
2   a=0.05
3   m = 5
4   r = 6
5   n = r * m
6   qsturng(1-a, m, n-m) #inverse cdf 4.15357
7   psturng(4.15357, m, n-m) # survival 0.05
```

Code Block 4.6: Inverse CDF and survival function for the studentized range distribution

$$HSD = q_{0.05}(5, 25)\sqrt{MSW/r} \approx 4.15\sqrt{64.84} = 33.4.$$

Unfortunately, the studentized range distribution is not part of the scipy.stats library. It is nevertheless available in Python. We compute the above line using the qsturng method, as shown in Code Block 4.6.

Using this value for the Tukey test, we obtain the following reduced list of significant differences:

Group 0 and Groups: 3 and 4.

That is, the following pairs are statistically significant: $\{0, 3\}$ and $\{0, 4\}$. It makes sense that we get a smaller selection than we did previously, as we are requiring the experimentwise error rate—not the individual pairwise error rate–to be capped at 5%.

It is interesting to compare this result with the Bonferroni correction established in Proposition 4.3. In order to guarantee an experimentwise error rate of $a = 0.05$, we would use a value of $\alpha = a/[m(m-1)/2] = 0.005$ for Fisher's LSD test. The value of the LSD with this more strict level of significance is $LSD = 31.74$. It is interesting that the Fisher's LSD test with Bonferroni correction is actually slightly less strict than the Tukey HSD test.   ▷

4.4 Two-Factor Experiments

4.4.1 Design

A two-factor experiment is not unlike a single-factor experiment with blocking, except instead of the blocks constituting a nuisance factor, they are an orthogonal set of treatment groups arranged for testing. The other distinction is that, unlike the case of a single-factor experiment with blocking, units can be arranged at will. This makes the design a simpler process, as units are randomly allocated across both rows and columns of the design matrix.

Suppose we have m-levels for the first factor and s-levels for the second factor. A natural question is: why not simply arrange the ms total combinations as a single factor, as opposed to burning an additional index—and additional computational complexity—on this arrangement. The answer is that a two-factor experiment allows for the study not only of the response to each combination, but also for the study of the response due to *interaction effects*.

Example 4.16. A firm manages three separate factories, each which operate at different efficiencies, and wants to test five different machines (say, machines from five different brands) for a certain production process. A completely randomized two-factor experiment is devised, which uses the factory as one factor and the machine brand as the second factor. There are thus $m = 3$ levels for the first factor, and $s = 5$ levels for the second factor, yielding 15 different treatment groups. In addition, four separate machines of each type at each location are ordered, so that the experiment can be conducted using a replication factor of $r = 4$. There are thus 60 individual outcomes that will be measured.

In this example, the factory location and type of machine represent two intrinsically different factors for the experiment, which is why all enumerations are not simply lumped together in a single row.   ▷

A completely randomized two-factor experiment is conducted in the natural way.

Definition 4.15. *A two-factor experiment with m and s levels and r replicants is said to be* completely randomized *if the total $n = msr$ experimental units are randomly allocated such that each of the ms experimental cells receive a total of r units.*

4.4.2 Analysis: Two-way ANOVA

We model the response of a completely randomized two-factor experiment using

$$Y_{ijk} = \mu + \tau_i + \beta_j + \iota_{ij} + \epsilon_{ijk}, \tag{4.40}$$

for $i = 1, \ldots, m$; $j = 1, \ldots, s$; and $k = 1, \ldots, r_{ij}$. The term ι_{ij} represents the *interaction effect* between the two factors. We usually require the added constraints to make the equation well defined:

$$\sum_{i=1}^{m} \tau_i = 0, \qquad \sum_{j=1}^{s} \beta_j = 0, \qquad \text{and} \qquad \sum_{i=1}^{m} \iota_{ij} = \sum_{j=1}^{s} \iota_{ij} = 0.$$

In the following, we will consider the general case where there are r_{ij} replicants in the ijth cell. For a completely randomized experiment, however, $r_{ij} = r$ will not vary cell to cell. We further require the ANOVA assumption $\epsilon_{ijk} \sim N(0, \sigma^2)$.

To begin, let us define the various sums of squares, analogous of Equations (3.75)–(3.77).

Definition 4.16. *For any sets of numbers Y_{ijk}, with $i = 1,\ldots,m$; $j = 1,\ldots,s$; and $k = 1,\ldots,r_{ij}$, we define the following sums of squares*

$$SST = \sum_{i=1}^{m}\sum_{j=1}^{s}\sum_{k=1}^{r_{ij}}\left(Y_{ijk} - \overline{Y}_{...}\right)^2 \qquad (4.41)$$

$$SSB_\tau = \sum_{i=1}^{m} s\hat{r}_{i\cdot}\left(\overline{Y}_{i\cdot\cdot} - \overline{Y}_{...}\right)^2 \qquad (4.42)$$

$$SSB_\beta = \sum_{j=1}^{s} m\hat{r}_{\cdot j}\left(\overline{Y}_{\cdot j\cdot} - \overline{Y}_{...}\right)^2 \qquad (4.43)$$

$$SSI = \sum_{i=1}^{m}\sum_{j=1}^{s} r_{ij}\left(\overline{Y}_{ij\cdot} - \overline{Y}_{i\cdot\cdot} - \overline{Y}_{\cdot j\cdot} + \overline{Y}_{...}\right)^2 \qquad (4.44)$$

$$SSW = \sum_{i=1}^{m}\sum_{j=1}^{s}\sum_{k=1}^{r_{ij}}\left(Y_{ijk} - \overline{Y}_{ij\cdot}\right)^2, \qquad (4.45)$$

as the total sum of squares, a between-group sum of squares for each factor, an interaction sum of squares, and, finally, a within-group sum of squares (also referred to as the error estimate, or the pooled sample variance), respectively. In the above, the "hat" operator represents the harmonic mean, so that

$$\hat{r}_{i\cdot} = \left(\frac{1}{s}\sum_{j=1}^{s} r_{ij}^{-1}\right)^{-1} \qquad and \qquad \hat{r}_{\cdot j} = \left(\frac{1}{m}\sum_{i=1}^{m} r_{ij}^{-1}\right)^{-1}.$$

For the case when we have the same number of replicants $r_{ij} = r$ in each cell, $\hat{r}_{i\cdot} = \hat{r}_{\cdot j} = r$.

Naturally, we have the following.

Theorem 4.11 (Partitioning Theorem). *For Y_{ijjk} as in Definition 4.16, we have*

$$SST = SSB_\tau + SSB_\beta + SSI + SSW. \qquad (4.46)$$

Proof. We begin by expressing the identity

$$\begin{aligned}Y_{ijk} - \overline{Y}_{...} &= \left(\overline{Y}_{i\cdot\cdot} - \overline{Y}_{...}\right)\\&+ \left(\overline{Y}_{\cdot j\cdot} - \overline{Y}_{...}\right)\\&+ \left(\overline{Y}_{ij\cdot} - \overline{Y}_{i\cdot\cdot} - \overline{Y}_{\cdot j\cdot} + \overline{Y}_{...}\right)\\&+ \left(Y_{ijk} - \overline{Y}_{ij\cdot}\right).\end{aligned}$$

Next, we square both sides and sum over all indices. As was the case in Theorem 3.11, all the cross terms cancel (see Exercise 4.6). The result follows. □

Given the sums of squares Equations (4.41)–(4.45), we next define the mean squares, analogous to Equation (3.78).

Definition 4.17. *Let Y_{ijk} be as in Definition 4.16. We define the following mean squares*

$$MSB_\tau = \frac{1}{m-1}SSB_\tau \qquad (4.47)$$

$$MSB_\beta = \frac{1}{s-1}SSB_\beta \qquad (4.48)$$

$$MSI = \frac{1}{(m-1)(s-1)}SSI \qquad (4.49)$$

$$MSW = \frac{1}{n-ms}SSW. \qquad (4.50)$$

where $n = \sum_{i=1}^{m}\sum_{j=1}^{s} r_{ij}$ is the total number of units. Note: for the case $r_{ij} = r$, the denominator $(n-ms)$ in MSW can be expressed as $ms(r-1)$.

Now, in a two-way analysis of variance (two-way ANOVA), there are three separate null hypotheses open for consideration.

Definition 4.18. *Under the model Equation (4.40), we define the three separate null hypotheses*

$$H_0^\tau : \qquad \tau_i = 0 \text{ and } \iota_{ij} = 0;$$
$$H_0^\beta : \qquad \beta_j = 0 \text{ and } \iota_{ij} = 0;$$
$$H_0^\iota : \qquad \iota_{ij} = 0.$$

For each of the ANOVA null hypotheses, we will conduct an F-test. We must therefore uncover the distributions of the various sums of squares, as we do in the following lemmas.

Lemma 4.2. *Under the ANOVA assumptions,*

$$\frac{1}{\sigma^2}SSW \sim \chi^2_{n-ms}. \qquad (4.51)$$

Proof. Under the ANOVA assumptions, we have

$$Y_{ijk} \sim N(\mu + \tau_i + \beta_j + \iota_{ij}, \sigma^2).$$

Therefore, for each $i = 1, \ldots, m$ and $j = 1, \ldots, s$, the scaled sample variance for the ijth cell satisfies

$$\frac{1}{\sigma^2}\sum_{k=1}^{r_{ij}}\left(Y_{ijk} - \overline{Y}_{ij\cdot}\right)^2 \sim \chi^2_{r_{ij}-1}.$$

Summing over indices i and j yields our result. □

Lemma 4.3. *Under the ANOVA assumptions and the null hypothesis H_0^τ from Definition 4.18, and under the additional assumption that r_{ij} is independent of i, then*

$$\frac{1}{\sigma^2} SSB_\tau \sim \chi^2_{m-1}. \tag{4.52}$$

Proof. Under the null hypothesis H_0^τ, we have $Y_{ijk} \sim N(\mu + \beta_j, \sigma^2)$. Therefore, the sample mean

$$\overline{Y}_{ij\cdot} \sim N\left(\mu + \beta_j, \frac{\sigma^2}{r_{ij}}\right).$$

Now, by averaging this equation over j, we obtain

$$\overline{Y}_{i\cdot\cdot} \sim N\left(\mu, \frac{\sigma^2}{s\hat{r}_{i\cdot}}\right),$$

where $\hat{r}_{i\cdot}$ is the harmonic mean defined in Definition 4.16. Now, if r_{ij} is independent of i, then so is $\hat{r}_{i\cdot}$, and the random variables $\overline{Y}_{i\cdot\cdot}$ are independent and identically distributed normal random variables. In this case, the quantity SSB_τ is therefore a scaled version of the sample variance of the random variable $\overline{Y}_{i\cdot\cdot}$. We therefore obtain from Theorem 2.6 that $(1/\sigma^2)SSB_\tau \sim \chi^2_{m-1}$. $\square$

Lemma 4.4. *Under the ANOVA assumptions and the null hypothesis H_0^β from Definition 4.18, and under the additional assumption that r_{ij} is independent of j, then*

$$\frac{1}{\sigma^2} SSB_\beta \sim \chi^2_{s-1}. \tag{4.53}$$

Proof. This follows from Lemma 4.3, by symmetry. $\square$

Lemma 4.5. *Under the ANOVA assumptions and the null hypothesis H_0^ι from Definition 4.18,*

$$\frac{1}{\sigma^2} SSI \sim \chi^2_{(m-1)(s-1)}. \tag{4.54}$$

We leave this final proof as an exercise for the reader. In general, however, the number of degrees of freedom for interaction terms (in two-way ANOVA as well as in higher-order ANOVA) is the product of the number of degrees of freedom in each factor. A quick way to see this is to recognize that the total sum of squares SST has $(n-1)$ degrees of freedom, and then solve for df_ι:

$$(n-1) = (m-1) + (s-1) + ms(r-1) + df_\iota.$$

We obtain $df_\iota = (m-1)(s-1)$.

Theorem 4.12. *In a completely randomized two-factor experiment, with an equal allocation of replicants $r_{ij} = r$ to each cell, the test statistics*

$$F_\tau = \frac{MSB_\tau}{MSW}, \qquad F_\beta = \frac{MSB_\beta}{MSW}, \qquad and \qquad F_\iota = \frac{MSI}{MSW} \qquad (4.55)$$

are distributed as

$$F_\tau \sim \mathrm{F}_{m-1,n-ms}$$
$$F_\beta \sim \mathrm{F}_{s-1,n-ms}$$
$$F_\iota \sim \mathrm{F}_{(m-1)(s-1),n-ms},$$

under the null hypothesis H_0^τ, H_0^β, and H_0^ι, respectively.

Proof. This result follows from the preceding three lemmas. Note that since the number of replicants per cell is independent of both i and j, both Lemmas 4.3 and 4.4 apply.   □

Example 4.17. Consider a two-factor experiment with $m = 3$ and $s = 4$ and $r = 2$ replicants per cell. Data from such an experiment has been collected and provided in Table 4.7.

	$j = 1$	$j = 2$	$j = 3$	$j = 4$
$i = 1$	72, 96	100, 101	90, 86	127, 101
$i = 2$	70, 90	82, 106	89, 100	114, 109
$i = 3$	94, 97	116, 110	139, 112	106, 132

Table 4.7: Data for Example 4.17.

The sum of squares and F-statistics are computed in Table 4.8. We observe that both F_τ and F_β exceed their respective critical values. However, the F-statistic for the interaction term does not. We may therefore reject the null hypotheses H_0^τ and H_0^β, concluding that each factor is significant. However, we fail to reject the null hypothesis H_0^ι, so we cannot conclude that there are any interaction effects.

	SSQ	df	MSE	F	F_{crit}
SSB_τ	1632.25	2	816.13	5.099	3.89
SSB_β	2430.46	3	810.15	5.062	3.49
SSI	664.42	6	110.74	0.692	3.00
SSW	1920.5	12	160.04		
SST	6647.63	23			

Table 4.8: Sum of squares and F-statistics for Example 4.17.

The code used to simulate this example and compute the test statistics is given in Code Block 4.7.   ▷

```
1   m = 3
2   s = 4
3   r = 2
4   mu = 100
5   alpha = 0.05
6   tau = np.array([-10, 0, 10]).reshape((3,1))
7   beta = np.array([-10, 0, 0, 10]).reshape((1,4))
8   sigma = 10
9
10  Y1 = np.random.normal(loc= mu+tau+beta, scale=sigma).astype(int)
11  Y1 = np.random.normal(loc= mu+tau+beta, scale=sigma).astype(int)
12
13  Y = np.array([Y1.T, Y2.T]).T
14
15  SST = np.sum( (Y - Y.mean())**2 )
16  SSB1 = s * r * np.sum( (Y.mean(axis=(1,2)) - Y.mean())**2 )
17  SSB2 = m * r * np.sum( (Y.mean(axis=(0,2)) - Y.mean())**2 )
18  SSI = r * np.sum( (Y.mean(axis=2) -
        Y.mean(axis=(1,2)).reshape((3,1)) - Y.mean(axis=(0,2)) +
        Y.mean())**2 )
19  SSW = np.sum( (Y - Y.mean(axis=2).reshape((3,4,1)))**2 )
20
21  MSB1 = SSB1 / (m-1)
22  MSB2 = SSB2 / (s-1)
23  MSI  = SSI / ((m-1)*(s-1))
24  MSW  = SSW / (m*s*(r-1))
25
26  F1 = MSB1 / MSW
27  F2 = MSB2 / MSW
28  FI = MSI / MSW
29
30  F1_crit = scipy.stats.f.isf(alpha, m-1, m*s*(r-1))
31  F2_crit = scipy.stats.f.isf(alpha, s-1, m*s*(r-1))
32  FI_crit = scipy.stats.f.isf(alpha, (m-1)*(s-1), m*s*(r-1))
```

Code Block 4.7: Simulation and calculation for Example 4.17

4.4.3 Analysis: Single Factor Experiment with Random Block Design

We took some pain to derive the results in the previous section for the general case of a varying number of replicants r_{ij} per cell. This added complexity was unnecessary for the results for two-factor experiments; however, it becomes relevant when applying those results to a single-factor randomized block design, for which r_{ij} is independent of i but not of j. Applying those results to this context, we obtain the following.

Theorem 4.13. *In a single-factor randomized block experiment, the test statistics F_τ and F_ι, as defined in Theorem 4.12, are distributed as*

$$F_\tau \sim \mathrm{F}_{m-1,n-ms}$$
$$F_\iota \sim \mathrm{F}_{(m-1)(s-1),n-ms},$$

under the null hypotheses H_0^τ and H_0^ι, respectively.

Note 4.10. The test statistic F_β is not valid for a single-factor randomized block design, as the number of replicants depends on the stratum: $r_{ij} = r_j$. Therefore, the conditions of Lemma 4.4 are not satisfied. The only exception to this rule is if there are an equal number of units per stratum. This is, however, typically not the case, as the nature of a nuisance factor is there is no control over its various counts.   ▷

Note 4.11. For a single-factor randomized block experiment, there should be no interaction effect. This is because, for each stratum, units are distributed randomly and in equal quantity to the various treatment groups. We can still perform the interaction F-test as a sanity check that our experiment is setup correctly. If we reject the null hypothesis H_0^ι, there could potentially have been an issue with our randomization methodology.   ▷

4.5 Bandits

In traditional statistical experiments, the significance and power are pre-scribed, a sample size is calculated, and the experiment is run until completion; i.e., it is not stopped prematurely to "look" at the data, as doing so could lead to inflated false positive rates (see Reinhart [2015] for a nice discussion on this). In traditional settings—predominantly clinical trials for drugs and therapies—this is absolutely necessary to maintain the rigor and validity of the results. But what if you're not to the standards of clinical trials but are instead concerned with which font looks best on your website, or if you're running a billion-dollar ad-tech startup? Should your approach be any different?

The idea is simple: experiments are expensive. Each experimental unit comes at a cost. Why not use half-baked results, once a clear leader has emerged? Can we get to the answer more quickly instead of burning money while we are waiting to achieve statistical significance? Put more gently: can we devise an algorithm that can better optimize our *true* objective of finding the optimal treatment using minimal time and expense?

To answer this, we turn next to a problem known as the *multi-armed bandit problem* (or simply *bandits*, for short), first introduced by Robbins [1952]. Essentially, a bandit algorithm seeks to maximize an objective function (total reward or payout) by allocating a fixed total amount of resources across a set of treatment groups, where the strategy (think: randomization mechanism) is allowed to adapt based on experience. The name derives from the casinos: a *single-armed bandit* is a slang term for a slot machine, whereas a multi-armed bandit represents a row of slot machines. Conceptually, the problem is stated as follows: a gambler with a finite number of coins is faced with a row of slot machines, each with an unknown reward distribution (and, hence, an unknown expected payout). Each round, the gambler must choose which machine to feed his next quarter into. The gambler has perfect memory and is allowed to adapt as he gains experience. What strategy should the gambler choose to maximize his total expected reward? Formally, we have the following.

Definition 4.19. *A multi-armed bandit problem consists of a set of real distributions* $\mathcal{B} = \{f_1, \ldots, f_m\}$, *with finite expected values* $\mu_1, \ldots, \mu_m$, *a positive integer* $n \in \mathbb{Z}^+$, *and an agent, who is tasked with selecting one of the arms* $\mathcal{A} = \{1, \ldots, m\}$ *sequentially for a total of* n *rounds.*

A strategy *or a* policy *is a method for selecting the arms based on the cumulative history; i.e., if we let* A_t *and* R_t *represent the action (choice) and reward for round* t, *respectively, so that* $R_t \sim f_{A_t}$, *we may define a policy* π *as a probability distribution over the arms* $\mathcal{A}$ *that depends on the history; i.e.,*

$$A_t \sim \pi_t = \pi_t\left(\{A_j, R_j\}_{j=1}^{t-1}\right).$$

The goal is then to maximize the total expected reward

$$R = \sum_{t=1}^{n} \mathbb{E}\left[R_t\right].$$

A binomial bandit problem is one in which the m *distributions are over the set* $\{0, 1\}$, *such that the reward is paid out with unknown probability* $p_1, \ldots, p_m$.

Bandit algorithms are related to Markov decision processes and reinforcement learning, a topic to which we will return in Part III. See Kochenderfer [2015] and Sutton and Barto [2018] for more details. Bandit algorithms get to the heart of a common dilemma in reinforcement learning known as the *exploration–exploitation tradeoff*, which balances the extent to which we *explore* or learn about our environment versus the extent to which we *exploit* what we have already learned. Traditional experiments do not address this tradeoff at all: one explores until one achieves statistical significance.

One common approach to the bandit problem is with so-called *greedy algorithms*.

Definition 4.20. *A* greedy algorithm *is one that always selects the best option, based on the available data, except for a reserved set of cases in which one uniformly selects a random action.*

The epsilon-greedy algorithm *selects the best-performing option at each point in time with a probability* $(1 - \epsilon)$, *otherwise, with probability* ϵ, *it selects an action at random. A typical value is* $\epsilon = 0.10$.

The epsilon-decreasing algorithm *is an epsilon-greedy algorithm with the modification that the value of* ϵ *decays with each choice.*

An epsilon-first algorithm *selects an action at random for the first* ϵn *rounds, and selects the best performer for the remaining* $(1 - \epsilon)n$ *rounds.*

In order to implement a greedy algorithm, we must naturally track an estimate for the value of each action as it evolves over time. For this purpose, we may define the following *action-value function*:

$$Q_t(a) = \mathbb{E}[R_t | A_t = a, \{(a_j, r_j)\}_{j=1}^{t-1}] = \frac{1}{N_t(a)} \sum_{j=1}^{t-1} R_j \mathbb{I}[A_j = a], \quad (4.56)$$

where $N_t(a)$ is the number of times action a was selected prior to round t:

$$N_t(a) = \sum_{i=j}^{t-1} \mathbb{I}[A_j = a].$$

In other words, the action-value function $Q_t(a)$ is simply, for each $a \in \mathcal{A}$, the average reward realized up to the given point in time t.

Let $\mathcal{A}_t^*$ be defined as

$$\mathcal{A}_t^* = \arg\max_{a \in \mathcal{A}} Q_t(a),$$

the set of actions that has produced the optimal reward, up to the current point in time. The ϵ-greedy policy may therefore be represented as

$$\pi_t(a) = \begin{cases} \dfrac{1 - \epsilon}{|\mathcal{A}_t^*|} & \text{for } a \in \mathcal{A}_t^*, \\ \dfrac{\epsilon}{m - |\mathcal{A}_t^*|} & \text{for } a \notin \mathcal{A}_t^* \end{cases}.$$

Typically, $\mathcal{A}_t^*$ will consist of a single action, although the above accounts for the case when multiple actions are tied in first place.

A variation of this, known as *upper-confidence bounds (UCB) algorithms*, replaces the greedy actions $\mathcal{A}_t^*$ with

$$\mathcal{U}_t^* = \arg\max_{a \in \mathcal{A}} \left[Q_t(a) + \lambda \operatorname{se}(Q_t(a)) \right],$$

where $\lambda > 0$ is a constant (e.g., $\lambda = 2$) and $\operatorname{se}(Q_t(a))$ is the standard error of our estimate $Q_t(a)$. The policy distribution for a UCB method is the same as before, except with $\epsilon = 0$. The idea behind UCB methods is instead of selecting a non-optimal action uniformly, we select each action relative to the *potential* for that action to be optimal. This resolves the exploration–exploitation tradeoff in a natural way, as we are always selecting an optimal action relative to a confidence bound for our estimate, as opposed to the raw estimate itself.

An alternative approach to using upper-confidence bounds is simply to select the action with a probability that depends on our estimation of each action's value. A common method is the *softmax* approach, which normalizes the action-value estimates into a proper probability mass function

$$\pi_t(a) = \frac{e^{Q_t(a)/\tau}}{\sum_{i=1}^{m} e^{Q_t(i)/\tau}}.$$

The parameter $\tau > 0$ is referred to as the *temperature*: a high temperate causes the probability distribution to be closer to uniform, whereas a low temperature creates a greater difference between the various actions, with the greedy action resulting in the limit as $\tau \to 0$.

We conclude with an in-depth discussion of a modern advertising problem. For this example, a *creative* is simply a particular ad, which could take the form of a text string, an image, or a video. An ad network is tasked with showing various creatives to its audience, the placement being on websites or on mobile apps within its network. An instance of serving an add to a user is called an *impression*. A *click* occurs if the user decides to click on the ad. Cost is measured two ways: the *cost-per-mille*, or CPM, is the cost per 1,000 impressions and the *cost-per-click*, or CPC, is the cost per click. The *click-through rate*, or CTR is the ratio of clicks per impression. Many ad networks operate on a CPC model, where they charge advertisers a fixed cost per click, and profit based on the arbitrage between the cost per click and the CPM. We can think of the clicks as a binary reward for each impression, as a cost is incurred for each impression served.

Example 4.18. An ad network has a set of five creatives for a particular advertiser, and wants to run an experiment to determine whether or not there is a difference among the creatives and, if so, which creative performs best. We assume that we would like to detect the following effect:

$$p = \langle 0.008, 0.009, 0.01, 0.011, 0.012 \rangle,$$

where p_i represents the expected (true) click-through rate of the ith creative. Our goal is to devise a test with a 5% significance and a 95% power; i.e., a test that will detect a difference between the click-through rates 95% of the time, when the difference is actually present.

An immediate problem arises: if we choose the impression as our experimental unit, the observed responses Y_{ij} will be Bernoulli random variables, which will invalidate our model given by Equation (4.32). Instead, let us take individual milles, or sets of 1,000 impression, as our experimental unit. If each unit represents 1,000 impressions, then the response variable Y_{ij} will constitute a binomial random variable

$$Y_{ij} \sim \text{Binom}(1000, p_i),$$

which is approximately normal with expected value $\mathbb{E}[Y_{ij}] = 1000p_i$ and variance $\mathbb{V}(Y_{ij}) = 1000p_i(1 - p_i) \approx 10$, since $p_i \approx 0.01$, for each i. We can therefore define $\mu = 10$,

$$\tau = \langle -2, -1, 0, 1, 2 \rangle,$$

and $\sigma^2 = 10$. It can be shown that the required sample size to achieve a 95% power test with the above minimum detectable effect is $r = 20$ (see Exercise 4.9). We therefore require a total of $n = rm = 100$ experimental units, which is equivalent to a total of 100,000 ad impressions. (For perspective, if impressions cost a CPM of \$10, then the experiment will cost \$1,000.)

Next, we can write a quick simulation to verify that the hypothesis of no effect is indeed rejected 95% of the time. Moreover, we can capture the probability distribution of which creative ends up in the lead at the end of the experiment. See Code Block 4.8. The simulation resulted in a rejected null hypothesis 95.6% of the time, which matches our expectation. However, the winning lot, after 100 simulated milles, had the following distribution

$$\langle 0, 0, 1.6\%, 16.2\%, 82.2\% \rangle.$$

Thus, our experiment was able to detect an effect 95% of the time, and when it did detect an effect, it correctly identified the best ad 82.2% of the time. ▷

Next, we turn the experiment described in Example 4.18 into a simulation and compare the results of the experimental approach with various bandit algorithms.

Example 4.19. Compare the results of the randomized experiment described in Example 4.18 with the epsilon-greedy, UCB, and softmax algorithms.

We begin by defining our reward function and four policies in Code Block 4.9. Based on our analysis in Example 4.18, we know that an experiment would run for 100 lots of one mille each. We therefore construct our experiment as an epsilon-first policy, where the first 100 lots are selected at randomly, and the greedy action is selected for all rounds following the initial 100. (Note that instead of using a pure epsilon-greedy policy, we added a decay parameter, so that the amount of exploration diminishes over time.)

We can then turn these policies into a simulation, as shown in Code Block 4.10. We graph the average reward as a function of time in Figure 4.1. Note that we did perform some quick parameter tuning to obtain this particular set of parameters. Though the epsilon-greedy starts with a significant (50%) amount of exploration, the decay parameter balances this out, so that after 100 rounds, there is only about 18% exploration.

```
1   n_sims = 10000
2   alpha = 0.05
3   r, m = 20, 5
4   n = r * m
5   lot_size = 1000
6   mu = 0.01
7   tau = np.array([-0.002, -0.001, 0, +0.001, +0.002]).reshape((m, 1))
8   F_crit = scipy.stats.f.isf(alpha, m-1, n - m) #2.467
9
10  count_wins = np.zeros(5)
11  rejections = 0
12  for i in range(n_sims):
13      Y = np.random.binomial(n=lot_size, p=mu+tau, size=(m, r))
14
15      SSB = np.sum( (Y.mean(axis=1) - Y.mean())**2 ) * r
16      SSW = np.sum( (Y - Y.mean(axis=1).reshape((m,1)))**2 )
17      SST = np.sum( (Y - Y.mean())**2 )
18      MSB = SSB / (m-1)
19      MSW = SSW / (n-m)
20      F = MSB / MSW
21      if F > F_crit:
22          rejections += 1
23          best = np.argmax(Y.sum(axis=1))
24          count_wins[best] += 1
25
26  print(rejections / n_sims) #95.6
27  print(count_wins / rejections) #0, 0, 0.016, 0.162, 0.822
```

Code Block 4.8: Simulation of a single-factor binomial experiment; Example 4.18

In this simulation, the upper-confidence bound algorithm is the clear winner; however, the epsilon-first strategy, which is based on a carrying out a rigorous experiment and then exploiting it results, does outperform even our tuned epsilon-greedy strategy in the long run. However, it should also be noted that the possible payouts are extremely similar in this example, which was, after all, based on our minimal detectable effect. It therefore makes sense that the bandit algorithms should do better based than the experimentation if there is a wider discrepancy between the results. For example, if some treatments are obviously bad and some are obviously winners, a good deal of cost can be saved by recognizing this early as opposed to waiting for the experiment to finish.

To visualize this, we reran the same set of simulations (with the same set of parameters) for a new "true" reward vector

$$p = \langle 0.01\%, 0.2\%, 0.5\%, 1.5\%, 3\% \rangle.$$

```
1   lot_size = 1000
2   mu = 0.01
3   tau = np.array([-0.002, -0.001, 0, +0.001, +0.002]).reshape((m, 1))
4   def get_reward(a):
5       return np.random.binomial(n=lot_size, p=mu+tau[a])[0]
6
7   def eps_first(history, size_explore=100, **kwargs):
8       if history[1, :].sum() < size_explore:
9           return np.random.randint(5)
10      else:
11          b = history[0, :] / (history[1, :]+1)
12          return np.random.choice(np.flatnonzero(b == b.max()))
13
14  def eps_greedy(history, epsilon=0.20, decay=1, **kwargs):
15      if np.random.random() < epsilon*decay**(history[1,:].sum()):
16          return np.random.randint(5)
17      else:
18          b = history[0, :] / (history[1, :]+1)
19          return np.random.choice(np.flatnonzero(b == b.max()))
20
21  def ucb(history, lambda_=3, var=10, **kwargs):
22      b = history[0, :] / (history[1, :]+1) + lambda_ * np.sqrt(var /
            (history[1, :]+1) )
23      return np.random.choice(np.flatnonzero(b == b.max()))
24
25  def softmax(history, tau=1, **kwargs):
26      b = np.exp( history[0, :] / (history[1, :]+1) / tau )
27      b /= b.sum()
28      return np.random.choice(range(5), p=b)
```

Code Block 4.9: Reward function and four policies for Example 4.19

We left all of the parameters unchanged. The results are shown in Figure 4.2. Here, the upper-confidence band algorithm wins again, followed by epsilon-greedy (with decay), and third by our epsilon-first (controlled experiment phase for exploration followed by exploitation). For this example, we tuned the parameter for the UCB, but left the other parameters the same. This leaves potential for softmax and epsilon-greedy to be further optimized. The fact that softmax levels out half way between the two best payouts (15 and 30) indicates that the temperature parameter need to be tuned to allow for more early exploration, in order to better distinguish between the top two choices.   ▷

The best approach depends largely on the problem at hand, including the particular cost and reward structure, the time horizon (e.g., can we carry out our experiment on 100 lots in five minutes or five months), as

```
1   total_rounds = 500
2   n_sims = 1000
3   kwargs = {'epsilon': 0.50,
4            'size_explore': 100,
5            'decay': 0.99,
6            'tau': 2,
7            'lambda_': 4}
8
9   i = 0
10  for get_next_action in [eps_first, eps_greedy, ucb, softmax]:
11      color = default_colors[i]
12      i += 1
13      X = np.zeros((n_sims, total_rounds))
14      simulated_rewards = np.zeros(n_sims)
15      for s in range(n_sims):
16
17          history = np.zeros((2,5))
18          for t in range(total_rounds):
19              a = get_next_action(history, **kwargs)
20              r = get_reward(a)
21              history[0, a] += r
22              history[1, a] += 1
23              X[s, t] = history[0, :].sum() / history[1, :].sum()
24          simulated_rewards[s] = history[0, :].sum()
25      print(simulated_rewards.mean()/total_rounds,
              (simulated_rewards/total_rounds).std())
26
27      plt.plot(X.mean(axis=0), color=color)
```

Code Block 4.10: Simulation for Example 4.19

well as the ability to tune the parameters for each algorithm to produce an optimal result. Variance in the final outcomes should also be taken into account: one algorithm might on average produce superior results, but if those results have high variance, a more stable algorithm might be better suited for a particular context. More advanced versions of bandit algorithms include Bayesian bandits and contextual bandits, which both are outside of our present scope. They can also be generalized to non-static problems, where the expected reward varies in time or based on context. In addition, Bandit algorithms also form a solid gateway to reinforcement learning. For further reading, see See Kochenderfer [2015] and Sutton and Barto [2018].

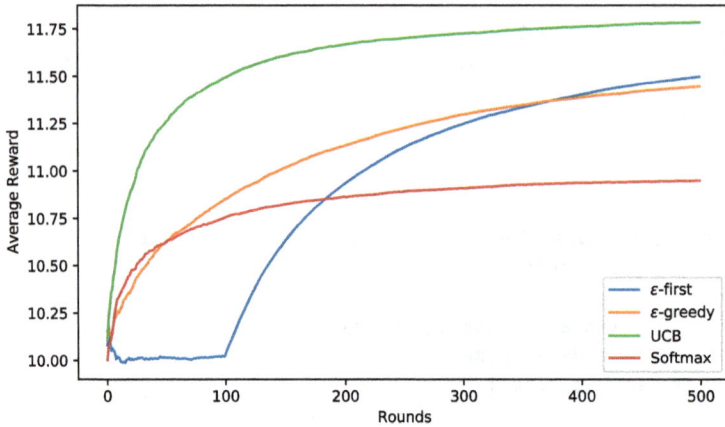

Fig. 4.1: Four bandit algorithms; Example 4.19.

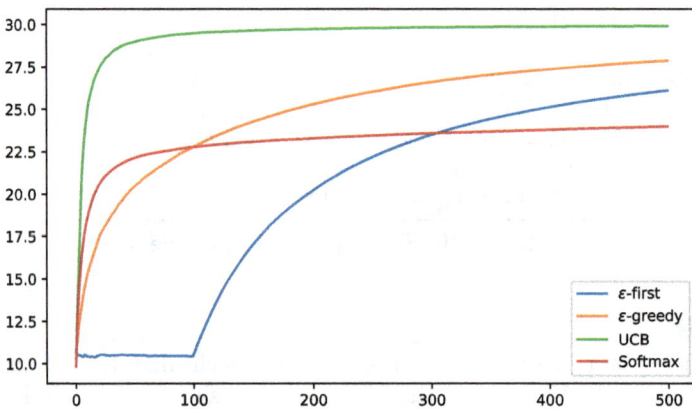

Fig. 4.2: Four bandit algorithms; Example 4.19.

Problems

4.1. Write a simulation for Examples 4.1 and 4.2 in order to numerically determine the probability of rejecting the null hypothesis for each of four cases: with and without randomization; no effect and an effect size of 10. As in the examples, assume that

$$R_{0i} \sim \mathrm{N}(90 - 2i, 25),$$

and that $R_{1i} = R_{0i} + \delta$, where δ is the effect size (either 0 or 10).

	$\delta = 0$	$\delta = 10$
no randomization		
randomization		

Discuss your findings. Based on the results, estimate the significance and power of the test?

4.2. Suppose an experiment is conducted with two treatment groups and two strata. Suppose that each treatment group contains r replicants, and that $b < r$ units are in stratum 1. Furthermore, suppose the response depends only on the stratum as follows: if a unit is in stratum 0, its response will be 1; otherwise its response will be a value $\lambda > 1$. Now suppose that all b units of stratum 1 are assigned to the treatment group 1. (Oh no!) Show that the T-statistic is given by

$$T = \sqrt{\frac{f(r-1)}{1-f}},$$

where $f = b/r$ is the ratio of stratum 1 units in the treatment group 1. This has the peculiar result that the test statistic depends on the total sample size and the relative coverage of stratum 1 within the treatment group, and not on the size of the parameter λ. Thus, we are just as likely to determine there is a statistically significant difference between the results

Control	1	1	1	1	1
Treatment	1.001	1.001	1.001	1	1

as we are between the results

Control	1	1	1	1	1
Treatment	1001	1001	1001	1	1

(For both cases, the T-statistic is $T = \sqrt{6} \approx 2.449$.)

4.3. Run a simulation of Example 4.5 in order to estimate the fraction of times the hypothesis of no effect is falsely rejected. (It should be close to 5%.)

4.4. Show that Equation (4.29) is properly normalized; i.e., show that

$$\sum_{i=0}^{m-1} \mathbb{P}(Z = i | C_{ij}, x = j) = 1.$$

Bonus: Show that for the first unit in each stratum, that

$$\mathbb{P}(Z = i | C_{ij}, x = j) = \frac{1}{m}.$$

4.5. Perform an F-test for the data in Example 4.14 using a significance level $\alpha = 0.05$. Show that the ANOVA null hypothesis is rejected. Bonus: Assuming $\mu = 100$, $\tau = \langle -20, -10, 0, 10, 20 \rangle$, and variance $\sigma^2 = 400$, determine the power of the test (using $m = 5$ and $r = 6$).

4.6. Show that the cross terms cancel in completing the proof to Theorem 4.11.

4.7. Dr. Robinson works for a lab, where she performs four single-factor randomized block experiments each day of the week, Monday through Friday. After a few months on the job, she notices that she is averaging about one significantly different interaction effect each week. Because of the randomized block design, however, there should not be any interaction effect. What is wrong with her experiment?

4.8. Run a simulation based on Code Block 4.7. Approximately what is the power, for the given values of τ and β, of the F_τ and F_β tests?

4.9. Consider a five-level, single factor experiment with $\mu = 10$, $\sigma^2 = 10$, and
$$\tau = \langle -2, -1, 0, 1, 2 \rangle.$$

Determine the sample size r, for each group, required to achieve a power of 80%, 90%, and 95%.

5

Estimator 2: Judgement Day

In this chapter, we cover a variety of commonly used methods of statistical estimation and related optimization techniques, including maximum likelihood estimation, simple linear regression, and stochastic gradient descent. We conclude with a brief introduction to regression and classification problems and associated forms of validation.

5.1 Maximum Likelihood Estimation

We begin by considering a technique for estimating the parameter of a distribution, given a sample of data arising from the distribution.

5.1.1 Likelihood and Log Likelihood

Given an IID sample $X_1, \ldots, X_n$ from a PDF (or PMF) $f(x; \theta)$, with parameter θ, recall that the joint distribution is given by

$$f_{X_1, \ldots, X_n}(x_1, \ldots, x_n; \theta) = \prod_{i=1}^{n} f(x_i; \theta).$$

For a given value of parameter θ, the joint distribution tells us how likely a particular set of data is. Conversely, if our set of data is actually observed and therefore fixed, and our parameter θ is unknown, we can make relative assertions: if the true value of the unknown parameter is θ_1 as opposed to θ_2, the data we observed would be more likely to occur. This motivates the following definition.

Definition 5.1. *Given an observed,* IID *sample* $X_1, \ldots, X_n \sim f(x; \theta)$ *from a distribution with an unknown (possibly vector) parameter θ, the* likelihood *function is the function of θ defined by*

$$\mathcal{L}(\theta) = \prod_{i=1}^{n} f(x_i; \theta). \tag{5.1}$$

It is often useful to consider the natural logarithm of this quantity, which is referred to as the log-likelihood function *and defined by the relation*

$$\ell(\theta) = \log(\mathcal{L}(\theta)) = \sum_{i=1}^{n} \log f(x_i; \theta). \tag{5.2}$$

Note 5.1. The likelihood function is equivalent to the joint distribution with the exception that it is viewed as a function of the parameter θ for a fixed set of data, as opposed to a function of the data, for a fixed parameter. ▷

Note 5.2. If $\Theta \subset \mathbb{R}^k$ is the set of possible values of the parameter θ^1, then the likelihood function is a mapping $\mathcal{L} : \Theta \to \mathbb{R}_*$ into the set of nonnegative real numbers. In particular, the likelihood function is *not* a density, i.e., it is generally not true that $\int_\Theta \mathcal{L}(\theta) \, d\theta = 1$. ▷

Definition 5.2. *Given an* IID *sample* $X_1, \ldots, X_n \sim f(x; \theta)$ *from a distribution with unknown parameter* θ, *the* maximum likelihood estimate (MLE) *for* θ, *denoted* $\hat{\theta}_n$, *is the value of* θ *that maximizes the likelihood function; i.e.,*

$$\hat{\theta}_n = \arg\max_{\theta \in \Theta} \mathcal{L}(\theta). \tag{5.3}$$

Proposition 5.1. *The* MLE *for* θ, *as given in Definition 5.2, is equivalent to*

$$\hat{\theta}_n = \arg\max_{\theta \in \Theta} \ell(\theta). \tag{5.4}$$

Proof. Since the logarithm is a monotonically increasing function, the maximum value $\hat{\theta}_n$ of $\mathcal{L}(\theta)$ is also the maximum value of $\log(\mathcal{L}(\theta)) = \ell(\theta)$. □

Example 5.1. Let $X_1, \ldots, X_n \sim \text{Bern}(p)$ be IID Bernoulli random variables with PMF $f(x; p) = p^x (1-p)^{1-x}$ for $x = 0, 1$. The likelihood function is given by

$$\mathcal{L}(p) = \prod_{i=1}^{n} f(x_i; p) = \prod_{i=1}^{n} p^{x_i} (1-p)^{1-x_i} = p^k (1-p)^{n-k},$$

where $k = \sum_{i=1}^{n} x_i$ represents the number of successes. The log-likelihood is therefore given by

$$\ell(p) = k \log(p) + (n-k) \log(1-p).$$

By setting the derivative $\ell'(p) = 0$ and solving, one finds the MLE as $\hat{p}_n = k/n$. ▷

---

[1] Here, we are considering the general case where $\theta \in \mathbb{R}^k$, though we will primarily be focused on the case of a single scalar parameter $k = 1$.

Example 5.2. Let $X_1, \ldots, X_n \sim \text{Exp}(\beta)$ be IID exponential random variables with PDF $f(x; \beta) = (1/\beta)e^{-x/\beta}$. The likelihood function is given by

$$\mathcal{L}(\beta) = \prod_{i=1}^{n} f(x_i; \beta) = \frac{1}{\beta^n} e^{-n\bar{x}/\beta},$$

where $\bar{x} = (1/n)\sum_{i=1}^{n} x_i$ is the sample mean. The log-likelihood is therefore

$$\ell(\beta) = -\frac{n\bar{x}}{\beta} - n\log(\beta).$$

The first derivative

$$\frac{d\ell(\beta)}{d\beta} = \frac{n\bar{x}}{\beta^2} - \frac{n}{\beta}$$

can be set equal to zero and solved for β to obtain the MLE $\hat{\beta} = \bar{x}$. The second derivative,

$$\frac{d^2\ell(\beta)}{d\beta^2} = -\frac{2n\bar{x}}{\beta^3} + \frac{n}{\beta^2},$$

when evaluated at the critical point $\hat{\beta} = \bar{x}$, is given by $\ell''(\hat{\beta}) = -n/\bar{x}^2 < 0$, proving that our critical point is indeed a maximum.   ▷

5.1.2 Score Statistic and Fisher Information

In studying maximum likelihood estimation and, as we shall see later, generalized linear models, it is useful to make several definitions. For additional details, see Dobson and Barnett [2018] and Dunn and Smyth [2018].

Definition 5.3. *Given an* IID *sample* $X_1, \ldots, X_n \sim f(x; \theta)$ *from a distribution with parameter* $\theta \in \mathbb{R}^p$ *and the data's associated log-likelihood* $\ell(\theta)$, *the* score function $U_n \in \mathbb{R}^p \to \mathbb{R}^{p*}$ *(see note[2]) is defined as the derivative*

$$U_n(\theta) = \frac{\partial \ell(\theta)}{\partial \theta}. \tag{5.5}$$

The score function U_n *is sometimes referred to as the* score statistic, *as it is also a function of the data.*

The equation obtained by setting the score function to zero, $U_n(\theta) = 0$, is known as the *score equation*; its solution $\hat{\theta}_n$ is the MLE estimate for the parameter θ. (We will assume that the log-likelihood functions we encounter throughout this text are unimodal and continuously differentiable, so that this is always a *maximum* solution.)

---

[2] We use the star $\mathbb{R}^{p*}$ to remind us that the quantity is a *covector*, which is represented as a row vector, not a column vector.

Theorem 5.1. *Let* $X_1, \ldots, X_n \sim f(x; \theta_0)$ *be* IID *from a distribution with true parameter* θ_0, *and let* $U_n(\theta)$ *be the data's associated score function. Then*

$$\mathbb{E}[U_n(\theta_0)] = 0; \tag{5.6}$$

i.e., when evaluated at the true parameter value θ_0, *the expected value of the score function is zero. The expectation is taken with respect to the distribution that generated the data, similar to Note 1.5.*

Proof. The score function from Equation (5.5) is equivalent to

$$U_n(\theta) = \sum_{i=1}^{n} \frac{1}{f(x_i; \theta)} \frac{\partial f(x_i; \theta)}{\partial \theta}. \tag{5.7}$$

Note that this depends on the observed data $X_1, \ldots, X_n$ and the family of distributions $f(x; \theta)$, but not the true value of $\theta = \theta_0$ that actually generated the data. The joint distribution of our sample is given by

$$f_{X_1, \ldots, X_n}(x_1, \ldots, x_n; \theta_0) = \prod_{j=1}^{n} f(x_j; \theta_0).$$

Therfore, the expected value of the score function is given, as a function of the parameter θ, by the relation

$$\mathbb{E}[U_n(\theta)] = \underbrace{\int_{-\infty}^{\infty} \cdots \int_{-\infty}^{\infty}}_{n \text{ times}} \sum_{i=1}^{n} \prod_{j=1}^{n} \frac{f(x_j; \theta_0)}{f(x_i; \theta)} \frac{\partial f(x_i; \theta)}{\partial \theta} \, dx_1 \cdots dx_n$$

$$= \sum_{i=1}^{n} \left(\int_{-\infty}^{\infty} \cdots \int_{-\infty}^{\infty} \prod_{j=1}^{n} \frac{f(x_j; \theta_0)}{f(x_i; \theta)} \frac{\partial f(x_i; \theta)}{\partial \theta} \, dx_1 \cdots dx_n \right);$$

that is, we are taking the expectation with respect to all possible realizations of the data from its true parameter θ_0. Now, if we evaluate this expression at the true value $\theta = \theta_0$, the multiple integral becomes separable, and we have

$$\mathbb{E}[U_n(\theta_0)] = \left(\int_{-\infty}^{\infty} \frac{\partial f(x_i; \theta_0)}{\partial \theta} \, dx_i \right) \cdot \prod_{\substack{j=1 \\ j \neq i}}^{n} \left(\int_{-\infty}^{\infty} f(x_j; \theta_0) \, dx_j \right).$$

However, the $(n-1)$ integrals of $f(x_j; \theta_0)$ are equal to unity, since a PDF must be normalized; similarly the lead factor,

$$\mathbb{E}[U_n(\theta_0)] = \frac{\partial}{\partial \theta} \int_{-\infty}^{\infty} f(x_i; \theta_0) \, dx_i = \frac{\partial 1}{\partial \theta} = 0,$$

vanishes. This completes the result. □

Corollary 5.1. *The variance–covariance matrix of the score function is given by*

$$\mathbb{V}(U_n(\theta_0)) = \mathbb{E}[U_n(\theta_0)^T U_n(\theta_0)], \tag{5.8}$$

when evaluated at the true value $\theta = \theta_0$.

Proof. Combining Proposition 1.7 and Theorem 5.1 yields the result. (Recall that $U_n(\theta)$ is a covector.)   □

Definition 5.4. *Consider an* IID *sample $X_1, \ldots, X_n \sim f(x; \theta_0)$ and its associated score function $U_n(\theta)$. The* observed information *is the matrix*

$$\mathcal{J}_n(\theta) = -\frac{\partial U_n(\theta)}{\partial \theta^T} = -\frac{\partial^2 \ell(\theta)}{\partial \theta^T \partial \theta}. \tag{5.9}$$

Similarly, the expected information *or* Fisher information *is the matrix*

$$\mathcal{I}_n(\theta) = \mathbb{E}[\mathcal{J}_n(\theta)]. \tag{5.10}$$

The difference between Equations (5.9) and (5.10) is that the observed information is a function of the data, i.e., a statistic, whereas the Fisher information is the expected value over all possible random samples from $f(x; \theta)$. (Both are a function of the parameter θ of the model.)

Note 5.3. The ijth component of the observed information is given by

$$\mathcal{J}_n(\theta)_{ij} = -\frac{\partial^2 \ell(\theta)}{\partial \theta_i \partial \theta_j}.$$

This is because the derivative $\partial/\partial\theta$ creates columns[3], whereas $\partial/\partial\theta^T$ creates rows.   ▷

Note 5.4. Due to the linearity of the log-likelihood and derivative operators, it follows that

$$\mathcal{I}_n(\theta) = n\mathcal{I}(\theta), \tag{5.11}$$

where we write $\mathcal{I}(\theta)$ as a short-hand for $\mathcal{I}_1(\theta)$.   ▷

Theorem 5.2. *Given an* IID *sample $X_1, \ldots, X_n \sim f(x; \theta_0)$, the variance–covariance matrix of the score function is given by*

$$\mathbb{V}(U_n(\theta_0)) = \mathcal{I}_n(\theta_0), \tag{5.12}$$

when evaluated at the true parameter $\theta = \theta_0$.

---

[3] So that $U_n(\theta) = \partial\ell/\partial\theta$ is a row vector. Also, this is consistent with the familiar case of $f : \mathbb{R}^n \to \mathbb{R}^n$, with $y = f(x)$, then $(\partial y/\partial x)_{ij} = \partial y_i/\partial x_j$.

Proof. Given Corollary 5.1, the result follows as long as we can prove that

$$\mathbb{E}\left[\frac{\partial^2 \ell}{\partial \theta^T \partial \theta}\bigg|_{\theta=\theta_0}\right] = -\mathbb{E}\left[\frac{\partial \ell}{\partial \theta^T}\frac{\partial \ell}{\partial \theta}\bigg|_{\theta=\theta_0}\right]. \tag{5.13}$$

Starting with the left-hand side, we may differentiate Equation (5.7) to find

$$\frac{\partial^2 \ell}{\partial \theta^T \partial \theta} = \sum_{i=1}^{n}\left[\frac{1}{f(x_i;\theta)}\frac{\partial^2 f(x_i;\theta)}{\partial \theta^T \partial \theta} - \frac{1}{f(x_i;\theta)^2}\frac{\partial f(x_i;\theta)}{\partial \theta^T}\frac{\partial f(x_i;\theta)}{\partial \theta}\right].$$

Now, when evaluated at the true parameter value $\theta = \theta_0$, the first term vanishes; i.e., we have

$$\mathbb{E}\left[\sum_{i=1}^{n}\frac{1}{f(x_i;\theta_0)}\frac{\partial^2 f(x_i;\theta_0)}{\partial \theta^T \partial \theta}\right] = 0;$$

for the same reason that $\mathbb{E}[U_n(\theta_0)] = 0$ in the proof of Theorem 5.1. We are thus left with

$$\mathbb{E}\left[\frac{\partial^2 \ell}{\partial \theta^T \partial \theta}\bigg|_{\theta=\theta_0}\right] = -\mathbb{E}\left[\sum_{i=1}^{n}\frac{1}{f(x_i;\theta_0)^2}\frac{\partial f(x_i;\theta_0)}{\partial \theta^T}\frac{\partial f(x_i;\theta_0)}{\partial \theta}\right],$$

which differs from the right-hand side of Equation (5.13) only by the absence of the cross terms

$$\mathbb{E}\left[\sum_{i=1}^{n}\sum_{\substack{j=1\\j\neq i}}^{n}\frac{1}{f(x_i;\theta_0)}\frac{\partial f(x_i;\theta_0)}{\partial \theta^T}\frac{1}{f(x_j;\theta_0)}\frac{\partial f(x_j;\theta_0)}{\partial \theta}\right].$$

However, the ijth term,

$$\mathbb{E}\left[\frac{1}{f(x_i;\theta_0)}\frac{\partial f(x_i;\theta)}{\partial \theta^T}\frac{1}{f(x_j;\theta_0)}\frac{\partial f(x_j;\theta)}{\partial \theta}\right],$$

is equivalent to

$$\left(\int_{-\infty}^{\infty}\frac{\partial f(x_i;\theta_0)}{\partial \theta}dx_i\right)\left(\int_{-\infty}^{\infty}\frac{\partial f(x_j;\theta_0)}{\partial \theta}dx_j\right)\prod_{\substack{k=1\\k\notin\{i,j\}}}^{n}\left(\int_{-\infty}^{\infty}f(x_k;\theta_0)dx_k\right) = 0$$

whenever $i \neq j$. (This follows similar logic as the proof to Theorem 5.1.) The result follows.   □

5.1.3 The Method of Scoring

We are concerned with finding a solution to the score equation $U_n(\theta) = 0$. One such approach uses a modification of the Newton–Raphson method, often simply referred to as Newton's method. This method is based on the following approximation. A more detailed treatment of this and related methods is discussed in Chong and Zak [2008].

Theorem 5.3. *Let $f : \mathbb{R}^n \to \mathbb{R}$ be twice continuously differentiable with invertible Hessian matrix at a given point $x_0 \in \mathbb{R}^n$. Then the global extremum of Taylor's quadratic approximation of f at x_0 is given by*

$$x^* = x_0 - D^2 f(x_0)^{-1} \cdot \nabla f(x_0), \tag{5.14}$$

where

$$\nabla f(x) = \frac{\partial f}{\partial x^T} \quad and \quad D^2 f(x) = \frac{\partial f}{\partial x^T \partial x}$$

are the gradient and Hessian of the function f, respectively. Moreover, if the Hessian is positive definite, the point x^* is the global minimum, and if the Hessian is negative definite, it is the global maximum.

Proof. The second-order Taylor approximation for f at x_0 is given by

$$f(x) \approx q(x) = f(x_0) + (x - x_0)^T \cdot \nabla f(x_0) + \frac{1}{2}(x - x_0)^T \cdot D^2 f(x_0) \cdot (x - x_0).$$

As the function $q : \mathbb{R}^n \to \mathbb{R}$ is quadratic, its extremum is straightforward to find: the gradient of q,

$$\nabla q(x) = \nabla f(x_0) + D^2 f(x_0) \cdot (x - x_0),$$

can be set equal to zero to obtain the result. $\qquad\square$

The *Newton–Raphson method*, usually referred to simply as *Newton's method*, is an iterative approach that leverages Theorem 5.3 to approximate a critical point of a multivariate function given a nearby starting point, as shown in Algorithm 5.1.

Algorithm 5.1: Newton–Raphson method.

Input : Function $f : \mathbb{R}^n \to \mathbb{R}$; starting value $x^0 \in \mathbb{R}^n$; stopping
threshold h; max iterations T.
Output: Approximation of critical value $x^* \in \mathbb{R}^n$; message.

1 Set $k = 0$
2 Set $\Delta = 1$
3 Set `message = Null`
4 **while** $k < T$ and $\Delta > h$ **do**
5 $\quad\big|\quad$ Compute $x^{k+1} = x^k - D^2 f(x^k)^{-1} \cdot \nabla f(x^k)$
6 $\quad\big|\quad$ Set $\Delta = ||\nabla f(x^k)||_1$
7 $\quad\big|\quad$ Set $k = k + 1$
8 **end**
9 Set $x^* = x^k$
10 **if** $\Delta > h$ **then**
11 $\quad\big|\quad$ Set `message = Failed to Converge`
12 **end**

We next apply this approach to the case where the function $f : \mathbb{R}^n \to \mathbb{R}$ is the log-likelihood function associated with an IID sample $X_1, \ldots, X_n \sim f(x; \theta_0)$. In this context, line 5 of Algorithm 5.1 becomes

$$\theta^{k+1} = \theta^k + \mathcal{J}_n(\theta^k)^{-1} \cdot U_n(\theta^k)^T.$$

In practice, however, the observed information $\mathcal{J}_n(\theta)$ is often difficult to compute. A modification on the preceding equation, for which the (negative) Hessian $\mathcal{J}_n(\theta^k)$ is replaced with its expected value, yields the *method of scoring* or *Fisher scoring*:

$$\theta^{k+1} = \theta^k + \mathcal{I}_n(\theta^k)^{-1} \cdot U_n(\theta^k)^T. \tag{5.15}$$

5.1.4 Asymptotic Properties of the MLE

Before we discuss the asymptotic behavior of the MLE, we first state several important properties.

Proposition 5.2 (Invariance Property). *Given an IID sample $X_1, \ldots, X_n \sim f(x; \theta_0)$, and let $\hat{\theta}_n$ be the MLE of the parameter θ. If $g : \mathbb{R} \to \mathbb{R}$ is one-to-one on the domain of the PDF f, then the MLE of the transformed parameter $\tau = g(\theta)$ is given by $\hat{\tau}_n = g(\hat{\theta}_n)$.*

Proof. Let $\mathcal{L}'(\tau) = \mathcal{L}(g^{-1}(\tau))$ represent the likelihood function expressed as a function of the new parameter τ, and define $\hat{\tau}_n = g(\hat{\theta}_n)$. Then, for any τ, we have $\mathcal{L}'(\tau) = \mathcal{L}(g^{-1}(\tau)) = \mathcal{L}(\theta) \leq \mathcal{L}(\hat{\theta}_n) = \mathcal{L}'(\hat{\tau}_n)$. This completes the proof. $\qquad\square$

Proposition 5.3 (Consistency Property). *Given an infinite sequence of* IID *random variables* $X_1, X_2, \ldots \sim f(x; \theta_0)$*, define* $\hat{\theta}_n$ *to be the* MLE *obtained by using the first n samples. Then* $\hat{\theta}_n \to \theta_0$ *in probability as* $n \to \infty$[4].

For a proof, see Wasserman [2004]. Essentially, Proposition 5.3 means that the MLE is asymptotically unbiased. We are now ready to state our main theorem.

Theorem 5.4. *Given an infinite sequence of* IID *random variables* $\{X_i\}_{i=1}^{\infty} = X_1, X_2, \ldots \sim f(x; \theta_0)$*, generated by a distribution with parameter* $\theta_0 \in \mathbb{R}^k$*, and the associated sequence* $\{\hat{\theta}_n\}_{n=1}^{\infty}$ *of estimators, where* $\hat{\theta}_n$ *is defined as the the* MLE *of the sample* $X_1, \ldots, X_n$ *consisting of the first n terms of the sequence* $\{X_i\}_{i=1}^{\infty}$*, the sequence of random vectors* $\{W_n\}_{n=1}^{\infty}$ *defined by*

$$W_n = \mathcal{I}_n(\theta_0)^{1/2}(\hat{\theta}_n - \theta_0) \tag{5.16}$$

converges in distribution to the standard normal random vector $Z \sim N(0_k, I_k)$*; i.e.,* $W_n \to Z$ *in distribution as* $n \to \infty$*.*

Proof. Consider the first-order Taylor expansion of the score statistic about $\theta = \theta_0$. We have

$$U_n(\theta) = U_n(\theta_0) - (\theta - \theta_0)^T \mathcal{J}_n(\theta_0) + O(||\theta - \theta_0||^2).$$

Now, by definition, $U_n(\hat{\theta}_n) = 0$, when evaluated at the MLE $\hat{\theta}_n$. We therefore have, to first order,

$$\sqrt{n}(\hat{\theta}_n - \theta_0) = \underbrace{n\mathcal{J}_n(\theta_0)^{-1}}_{B_n^{-1}} \underbrace{U_n(\theta_0)^T/\sqrt{n}}_{T_n}. \tag{5.17}$$

(Recall that $\mathcal{J}_n(\theta)$ is symmetric.) We have defined the statistics $B_n = \mathcal{J}_n(\theta_0)/n$ and $T_n = U_n(\theta_0)^T/\sqrt{n}$, as shown in preceding equation, for convenient usage momentarily.

Let us next define an infinite sequence of IID statistics

$$U^i(\theta) = \frac{\partial \log f(X_i; \theta)}{\partial \theta} \quad \text{and} \quad J^i(\theta) = \frac{\partial^2 f(X_i; \theta)}{\partial \theta^T \partial \theta},$$

for $i = 1, 2, \ldots$, so that

$$U_n(\theta) = \sum_{i=1}^{n} U^i(\theta) \quad \text{and} \quad \mathcal{J}_n(\theta) = \sum_{i=1}^{n} J^i(\theta).$$

The statistic B_n, defined above, is simply the sample mean of the first n IID random matrices $\{J^i\}_{i=1}^{n}$; i.e., $B_n = \bar{J}_n$. However, from Definition 5.4,

---

[4] We refer to any estimator that is asymptotically unbiased, in the sense that $\hat{\theta} \to \theta_0$ in probability, as a *consistent estimator*.

we have $\mathbb{E}[J^i(\theta)] = \mathcal{I}(\theta)$, which is, in particular, also true for $\theta = \theta_0$. By the law of large numbers (Theorem 1.3), therefore, we have $B_n(\theta_0) \to \mathcal{I}(\theta_0)$ in probability as $n \to \infty$.

Next, let us consider the statistic $T_n = U_n(\theta_0)^T / \sqrt{n} = \sqrt{n}\, \overline{U}_n(\theta_0)^T$, where the sample mean of the sequence $\{U^i\}_{i=1}^n$ is defined as $\overline{U}_n(\theta) = (1/n)U_n(\theta)$. Now, from Theorems 5.1 and 5.2, we have $\mathbb{E}[U^i(\theta_0)] = 0$ and $\mathbb{V}(U^i(\theta_0)) = \mathcal{I}(\theta)$, respectively. From the central limit theorem (Theorem 2.11), the statistic $T_n(\theta_0)$ therefore converges in distribution to $N(0_k, \mathcal{I}(\theta_0))$.

By applying Slutsky's theorem (Theorem 1.4) to Equation (5.17), we therefore have that

$$\sqrt{n}(\hat{\theta}_n - \theta_0) \to \mathcal{I}(\theta_0)^{-1} N(0_k, \mathcal{I}(\theta_0)) = N(0_k, \mathcal{I}(\theta_0)^{-1}),$$

in distribution as $n \to \infty$, where the equality holds due to a multivariate generalization of Lemma 2.1.

Applying Proposition 2.34, we may rewrite this as $\sqrt{n}\mathcal{I}^{1/2}(\theta_0)(\hat{\theta}_n - \theta_0) \to N(0_k, I_k)$. Finally, recognizing that $\mathcal{I}_n(\theta) = n\mathcal{I}(\theta)^5$, we obtain our result.   □

Corollary 5.2. *Let $X_1, \ldots, X_n \sim f(x; \theta_0)$ be an IID sample, as before, and consider the MLE $\hat{\theta}_n$. Then, for large n, we have the approximation*

$$\hat{\theta}_n \approx N(\theta_0, \mathcal{I}_n(\theta_0)^{-1}); \tag{5.18}$$

i.e., $\mathbb{V}(\hat{\theta}_n) \approx \mathcal{I}_n(\theta_0)^{-1} = \mathcal{I}(\theta_0)^{-1}/n$ and $se(\hat{\theta}_n) \approx \mathcal{I}(\theta_0)^{-1/2}/\sqrt{n}$.

In particular, note that Equation (5.16) is equivalent to the statement that

$$se(\hat{\theta}_n)^{-1}(\hat{\theta}_n - \theta_0) \to N(0_k, I_k)$$

in distribution as $n \to \infty$. We can approximate this relation one step further, as stated in the following corollary.

Corollary 5.3. *Consider $\hat{\theta}_n$ and $se(\hat{\theta}_n)$, as given in Corollary 5.2, and let the approximate standard error of the MLE be given by $\hat{se}(\hat{\theta})_n = \mathcal{I}(\hat{\theta}_n)^{-1/2}/\sqrt{n}$. Then the sequence of random vectors*

$$W_n = \mathcal{I}_n(\hat{\theta}_n)^{1/2}(\hat{\theta}_n - \theta_0) = \hat{se}(\hat{\theta}_n)^{-1}(\hat{\theta}_n - \theta_0) \tag{5.19}$$

converges to $N(0_k, I_k)$ in distribution as $n \to \infty$.

This follows from a further application of Slutsky's theorem along with the fact that, assuming $\mathcal{I}(\theta)$ is continuous, $\mathcal{I}(\hat{\theta}_n) \to \mathcal{I}(\theta_0)$ in probability as $n \to \infty$. We leave the details to Casella and Berger [2002] and Wasserman [2004].

The upshot, however, is that, for large n, the MLE can be approximated by

$$\hat{\theta}_n \approx N(\theta_0, \mathcal{I}_n(\hat{\theta}_n)^{-1}) = N(\theta_0, \mathcal{I}(\hat{\theta}_n)^{-1}/n). \tag{5.20}$$

---

[5] See Note 5.4.

Note 5.5. Recall from the definition of the variance–covariance matrix that the diagonal elements of $\mathcal{I}_n(\hat{\theta}_n)^{-1}$ approximate the variance of each individual parameter within the vector $\hat{\theta}_n = (\hat{\theta}_n^1, \ldots, \hat{\theta}_n^n)$, i.e.,

$$\mathbb{V}(\hat{\theta}_n^i) \approx \left[\mathcal{I}_n(\hat{\theta}_n)^{-1}\right]_{ii},$$

whereas the covariances are given by the off-diagonal elements

$$\mathrm{COV}(\hat{\theta}_n^i, \hat{\theta}_n^j) \approx \left[\mathcal{I}_n(\hat{\theta}_n)^{-1}\right]_{ij}.$$

In particular, if the off-diagonal elements of $\mathcal{I}(\hat{\theta}_n)$ are zero, the estimates of the individual components of $\hat{\theta}_n$ are independent.   ▷

5.1.5 Example: Gamma Random Variable

We conclude the section with an in-depth analysis of a particular example: IID gamma random variables.

Example 5.3. Let $X_1, \ldots, X_n \sim \mathrm{Gamma}(\alpha, \beta)$ be an IID gamma random variables. Determine the log-likelihood, score functions, and Fisher information.

We can use the PDF for a gamma random variable, as given in Equation (2.52), to construct the log-likelihood for the parameter $\theta = (\alpha, \beta) \in \mathbb{R}_+^2$, thereby obtaining

$$\ell(\alpha, \beta) = \sum_{i=1}^{n} \left[(\alpha - 1)\log(x_i) - \frac{x_i}{\beta}\right] - n\log(\Gamma(\alpha)) - \alpha n \log(\beta). \quad (5.21)$$

The score function is given by

$$U_n(\alpha, \beta)^T = \frac{\partial \ell(\alpha, \beta)}{\partial(\alpha, \beta)^T} = \begin{bmatrix} \sum_{i=1}^{n} \log(x_i) - n\psi(\alpha) - n\log(\beta) \\ \frac{\sum_{i=1}^{n} x_i}{\beta^2} - \frac{\alpha n}{\beta} \end{bmatrix}, \quad (5.22)$$

where we define the *digamma function*[6] $\psi(z)$ by the relation

$$\psi(z) = \frac{d\log(\Gamma(z))}{dz} = \frac{\Gamma'(z)}{\Gamma(z)}.$$

Similarly, the observed information is given by

---

[6] The digamma function is available in Python by using `scipy.special.psi(z)`. Its nth derivative is available under `scipy.special.polygamma(n, z)`.

$$\mathcal{J}_n(\alpha, \beta) = -\frac{\partial \ell^2(\theta)}{\partial \theta^T \partial \theta} = \begin{bmatrix} n\psi'(\alpha) & n/\beta \\ n/\beta & \frac{2\sum_{i=1}^{n} x_i}{\beta^3} - \frac{\alpha n}{\beta^2} \end{bmatrix}.$$

Recall from Equation (2.53) that the expected value of a gamma random variable $X_i \sim \text{Gamma}(\alpha, \beta)$ is given by $\mathbb{E}[X_i] = \alpha\beta$. Therefore, the Fisher information is given by

$$\mathcal{I}_n(\alpha, \beta) = \mathbb{E}\left[\mathcal{J}_n(\alpha, \beta)\right] = \begin{bmatrix} n\psi'(\alpha) & n/\beta \\ n/\beta & \alpha n/\beta^2 \end{bmatrix} = n\mathcal{I}(\alpha, \beta),$$

where $\mathcal{I}(\theta)$ is the Fisher information of a single observation:

$$\mathcal{I}(\alpha, \beta) = \begin{bmatrix} \psi'(\alpha) & 1/\beta \\ 1/\beta & \alpha/\beta^2 \end{bmatrix}. \tag{5.23}$$

This matrix is invertible since $\alpha\psi'(\alpha) \neq 1$, and its inverse is given by

$$\mathcal{I}(\alpha, \beta)^{-1} = \frac{\beta^2}{\alpha\psi'(\alpha) - 1} \begin{bmatrix} \alpha/\beta^2 & -1/\beta \\ -1/\beta & \psi'(\alpha) \end{bmatrix}. \tag{5.24}$$

The MLE estimates for $\theta = (\alpha, \beta)$ are obtained by solving the score equation, which is simply Equation (5.22) set to zero. Moreover, for large n, the variance of the estimates is given by the diagonal elements of Equation (5.24) divided by n, or

$$\mathbb{V}(\hat{\alpha}_n) \approx \frac{\alpha}{n(\alpha\psi'(\alpha) - 1)} \quad \text{and} \quad \mathbb{V}(\hat{\beta}_n) \approx \frac{\beta^2 \psi'(\alpha)}{n(\alpha\psi'(\alpha) - 1)}.$$

Since the off-diagonal elements are non-zero, the estimates are not independent. In fact, $\text{COV}(\hat{\alpha}_n, \hat{\beta}_n) \approx -\beta/(\alpha\psi'(\alpha) - 1)/n$. ▷

Next, we will continue this example, showing how the score function and FIsher information may be computed in Python.

Example 5.4. (Continued from Example 5.3.) Use Python to generate a random sample $X_1, \ldots, X_{100} \sim \text{Gamma}(\alpha_0, \beta_0)$, where $\alpha_0 = 2$ and $\beta_0 = 3$. Define Python functions for the score function and Fisher information. Then compute the expected variance of the MLE $\hat{\theta}_n = (\hat{\alpha}_n, \hat{\beta}_n)$.

To begin, let us generate a random sample of 100 IID gamma random variables, as shown in Code Block 5.1. This should be old hat, as we are using the `numpy.random` library to generate our random sample and the `scipy.stats` library to evaluate our PDF for an array of values. We can further plot a histogram of our sample concurrently with the PDF of Gamma(2,3), which is shown in Figure 5.1.

Next, let us define two Python functions to represent our score function, given by Equation (5.22), and the Fisher information, as given by Equation (5.23). This is done in Code Block 5.2. Notice the usage of the built-in

```
1   # Generate random sample from Gamma(2, 3)
2   alpha_0 = 2
3   beta_0 = 3
4   n_samples = 100
5   samples = np.random.gamma(alpha_0, scale=beta_0, size=n_samples)
6
7   # Compute PDF
8   x = np.linspace(0, 10)
9   y = scipy.stats.gamma.pdf(x, alpha_0, scale=beta_0)
10
11  # Plot PDF along with histogram of Sample.
12  plt.plot(x,y)
13  plt.hist(samples, bins=20, normed=True)
```

Code Block 5.1: Generate Random Sample of IID Gamma Random Variables

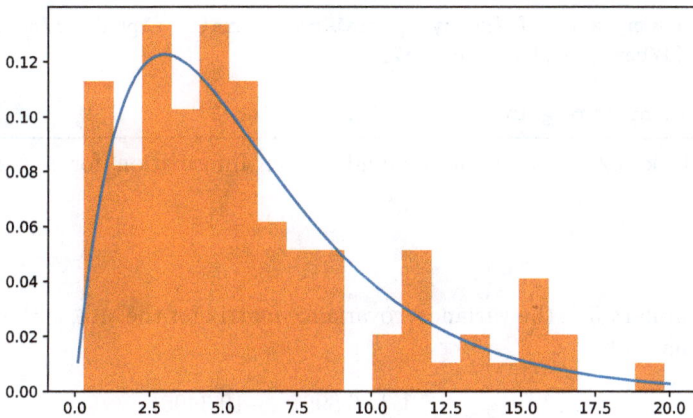

Fig. 5.1: PDF and histogram of random sample from Gamma(2,3).

functions `scipy.special.psi(z)` and `scipy.special.polygamma(n, z)` to evaluate the digamma function $\psi(z)$ and its derivatives. Furthermore, we construct the `FisherInfo` function to accept an optional argument `inverse`, which, when set to `True`, will return the inverse of the Fisher information matrix.

We can use the `FisherInfo` function to approximate the Fisher information at (α_0, β_0).

$$\mathcal{I}(2,3) \approx \begin{bmatrix} 0.6449 & 1/3 \\ 1/3 & 2/9 \end{bmatrix} \quad \text{and} \quad \mathcal{I}(2,3)^{-1} \approx \begin{bmatrix} 6.8997 & -10.3495 \\ -10.3495 & 20.0243 \end{bmatrix}.$$

```
1   def score(alpha, beta, samples):
2       """ Compute the score statistic
3       for an IID Gamma sample with parameters alpha, beta.
4       """
5       n = len(samples)
6
7       U_alpha = np.sum( np.log(samples) ) - n*scipy.special.psi(alpha)
            - n*np.log(beta)
8       U_beta = np.sum(samples) / beta**2 - alpha*n / beta
9
10      return np.array([U_alpha, U_beta]).reshape((2,1))
11
12  def FisherInfo(alpha, beta, n=1, inverse=False):
13      """ Returns the 2x2 Fisher Information for a Gamma random
            variable.
14      Set optional argument n=1 for sample size.
15      Set inverse=True to return the inverse of the Fisher Information.
16      """
17
18      I = n * np.array([ [scipy.special.polygamma(1, alpha), 1/beta],
            [1/beta, alpha / beta**2] ])
19
20      return np.linalg.inv( I ) if inverse else I
```

Code Block 5.2: Score Function and Fisher Information for a Gamma sample

From Corollary 5.2, the variance–covariance matrix for the MLE is therefore approximated by

$$\mathbb{V}((\hat{\alpha}_n, \hat{\beta}_n)) \approx \frac{1}{n} \begin{bmatrix} 6.8997 & -10.3495 \\ -10.3495 & 20.0243 \end{bmatrix}.$$

It is also interesting to note that one can use the `scipy.linalg.sqrtm` function to compute the *matrix square root* of the variance, obtaining

$$\mathbb{V}((\hat{\alpha}_n, \hat{\beta}_n))^{1/2} \approx \frac{1}{\sqrt{n}} \begin{bmatrix} 2.0214 & -1.6774 \\ -1.6774 & 4.1486 \end{bmatrix}.$$

▷

In our next example, we use Python to compute the MLE estimates for the random sample generated in Example 5.3.

Example 5.5. (Continued from Example 5.4.) Use the score and information functions from Example 5.4 to compute the MLE for the sample data generated in Example 5.3.

```
1   scipy.optimize.root(lambda x: score(x[0], x[1], samples), [1, 1])
2   ### MLE Estimate: alpha_hat = 1.73779384 beta_hat = 3.37642878
3
4   def FisherScoring(samples, score, FisherInfo, x0=[2,3],
            max_iter=1000, tol=1e-6):
5
6       k = 0
7       Delta = 1
8       theta = np.array(x0)
9       n = len(samples)
10
11      while k < max_iter and Delta > tol:
12          theta = theta + FisherInfo(theta[0], theta[1], n=n,
                inverse=True) @ score(theta[0], theta[1], samples)
13          Delta = np.linalg.norm(score(theta[0], theta[1], samples))
14          k += 1
15          print(k, theta)
16
17      if Delta > tol:
18          message = 'Failed to converge in {n_iter}
                iterations.'.format(n_iter=k)
19      else:
20          message = 'Success.'
21
22      return theta, message
23
24  beta = samples.var() / samples.mean()
25  alpha = samples.mean() / beta
26  FisherScoring(samples, score, FisherInfo, x0=[alpha,beta])
```

Code Block 5.3: Determining the MLE in Python

The code is shown in Code Block 5.3. First, we compute the MLE using the built-in root finder `scipy.optimize.root`, as shown on line 1. Note the use of a "lambda function," which is simply a way of defining a function in place[7]. For our particular sample, we obtain the estimate

$$\hat{\alpha}_{100} = 1.73779384 \qquad \text{and} \qquad \hat{\beta}_{100} = 3.37642878.$$

We then proceed to implement the method of scoring in lines 4–22. We found, however, that unless we start with values very close to the true values, this method does not converge. To facilitate this, we generate an initial "guess" for the true values of the parameters by setting Equations (2.53)

---

[7] The quantity `lambda x: score(x[0], x[1], samples)` represents a function that takes in an argument `x` and outputs the value `score(x[0], x[1], samples)`.

and (2.54) equal to the sample mean and variance and solving for α and β. This is accomplished in lines 24–26, and it produces the same estimate as line 1.

The upshot is that the built-in `scipy.optimize` routines likely do a better job at determining the MLE of a dataset than a simple implementation of the method of scoring. Note that the function `FisherScoring` takes as *inputs* the methods `score` and `FisherInfo`, defined in Code Block 5.2.

▷

Example 5.6. Using the variance–covariance matrix obtained in Example 5.4, plot the $z = 1$ and $z = 2$ Mahalanobis contours using Code Block 2.23. Then repeat Examples 5.4 and 5.5 one thousand times, using different random samples, and plot the MLE for each. What fraction of estimates are within a Mahalanobis distance of 1 or 2?

The code to accomplish this is shown in Code Block 5.4. The result in shown in Figure 5.2 (left). Moreover, we see that 39.5% of our samples had a Mahalanobis distance $z = d_\Sigma(\hat{\theta}_{100}, \theta_0)$ less than one and 83.4% had $z < 2$, agreeing nicely with the expected result from Table 2.1 for the case $k = 2$.

```
1   alpha_0 = 2
2   beta_0 = 3
3   n_samples = 100
4   n_simus = 1000
5   mu = np.array([alpha_0, beta_0])
6   Sigma = FisherInfo(alpha_0, beta_0, n=n_samples, inverse=True)
7   z_scores = np.zeros(n_simus)
8
9   for i in range(n_simus):
10      samples = np.random.gamma(alpha_0, scale=beta_0, size=n_samples)
11      output = scipy.optimize.root(lambda x: score(x[0], x[1],
            samples), [2, 2])
12      theta_hat = output['x']
13      plt.plot(theta_hat[0], theta_hat[1], '.', color='#0099ff')
14      z_scores[i] = mahalanobis(theta_hat, mu, Sigma)
15
16  x, y = ellipse(mu, Sigma, z=1)
17  plt.plot(x, y, 'r', linewidth=2)
18  x, y = ellipse(mu, Sigma, z=2)
19  plt.plot(x, y, 'r', linewidth=2)
20
21  print( 'z < 1', np.sum( z_scores < 1 ) / n_simus )
22  print( 'z < 2', np.sum( z_scores < 2 ) / n_simus )
```

Code Block 5.4: Code to run 1 000 simulations and plot the resulting maximum likelihood estimates.

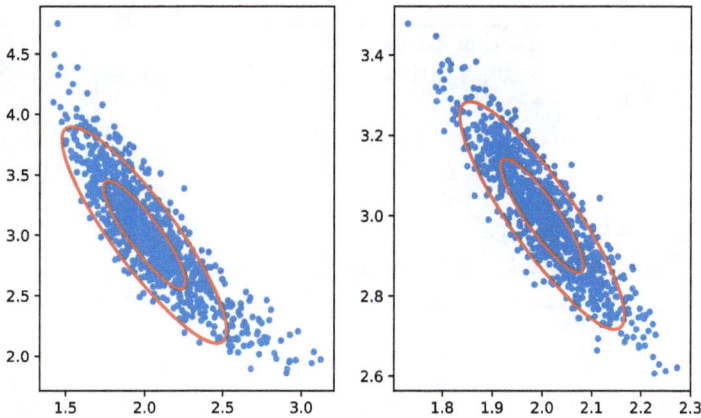

Fig. 5.2: 1,000 simulations with sample size 100 (left) and 1,000 (right).

We can repeat this game, changes **n_samples=1000** on line 3, to generate Figure 5.2 (right). (Note the different scale of Figure 5.2 (left) and (right).) The 1,000 computed MLE estimates behave much more closely to our theoretical normal distribution. However, we still observe 36.8% of simulations having $z < 1$ and 86.8% having $z < 2$. (An even closer agreement to Table 2.1, as expected.)

▷

5.2 Gradient Descent Method

The *Method of Gradient Descent* is an algorithm that locates a local minimum of an *objective function*. It is an optimization technique that can not only be used for solving maximum likelihood problems, but can also be applied to many problems in machine learning, as we will see later in this text. For more details on gradient descent and other optimization techniques, see Chong and Zak [2008].

5.2.1 Basic Gradient Descent

Classic gradient descent is an iterative technique used to solve the optimization problem

$$\min f(x),$$

for a given $f : \mathbb{R}^n \to \mathbb{R}$. Gradient descent works by taking a number of small, successive "steps" in the direction opposite of the objective function's gradient. An outline of this algorithm is provided in Algorithm 5.2.

Algorithm 5.2: Gradient Descent Algorithm. (With Line 4, Steepest Descent).

Data: A given function $f : \mathbb{R}^n \to \mathbb{R}$ and its gradient $Df : \mathbb{R}^n \to \mathbb{R}^n$; starting point $x_0 \in \mathbb{R}^n$, step size h, error tolerance `tol`.

Result: Approximate local minimum x^*.

1 $i = 0$
2 **while** $\|Df(x_i)\| > $ `tol` **do**
3 | Compute $Df(x_i)$
4 | (Optional) Let $h = \arg\min_{h \geq 0} f(x_i - hDf(x_i))$
5 | Compute $x_{i+1} = x_i - h \cdot Df(x_i)$
6 | $i = i + 1$
7 **end**
8 **return** $x_i, f(x_i)$.

The task of tuning the step size is a delicate one, as a smaller step size leads to a greater number of steps, and more computational tasks before convergence, whereas a larger step size may lead to overshooting, resulting in a more zig-zagged path. Some variations of the algorithm allow for variable step sizes. In particular, when line 4 of Algorithm 5.2 is deployed, the algorithm is referred to as *Method of Steepest Descent*. In this version of gradient descent, the task of selecting the step size at each step is tantamount to solving a one-dimensional line-search problem. Details of such an implementation can be found in Chong and Zak [2008].

It goes without saying that the method, in some cases, might fail to converge, or might converge to a local minimum that is not the global minimum. As with many optimization algorithms, success sometimes hinges on finding a suitable starting point x_0.

Example 5.7. Use gradient descent to determine the minimum of

$$f(x, y) = \frac{x^2}{5} + y^2.$$

A contour plot of this quadratic can be produced, as shown in Code Block 5.5. The contours are shown in Figure 5.3.   Gradient descent is implemented in Code Block 5.6.

Gradient descent, following Code Block 5.6, was run three times, with step sizes $h = 0.1$, 0.9, and 0.99. The results are shown in Figure 5.3. For a step size of $h = 0.1$, gradient descent follows the path of steepest descent, but it takes 311 steps to converge. (A step size of $h = 0.01$ would require 3,163 steps to converge.) A step size of $h = 0.9$ does much better, converging in only 60 steps. However, since the individual steps are larger, the algorithm results in a more zig-zagged path. Increasing the step size to $h = 0.99$, the algorithm converges in 659 steps. A step size of $h = 1$, the algorithm fails to converge. (At least, it does not converge within 10,000 steps.)

```
1   delta = 0.025
2   x = np.arange(-9.0, 9.0, delta)
3   y = np.arange(-4.0, 4.0, delta)
4   X, Y = np.meshgrid(x, y)
5   Z = X**2/5 + Y**2
6
7   fig, ax = plt.subplots(figsize=(8, 9/2))
8   CS = ax.contour(X, Y, Z)
9   ax.clabel(CS, inline=1, fontsize=10)
10  ax.set_title('Gradient Descent Example')
11  ax.set_aspect('equal')
```

Code Block 5.5: Contours for Example 5.7

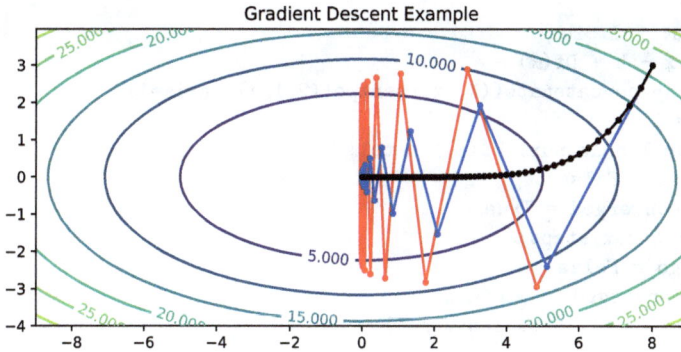

Fig. 5.3: Gradient Descent: $h = 0.1$ (black), $h = 0.9$ (blue), $h = 0.99$ (red).

This example illustrates the trade off between small and large step sizes. The number of steps required to converge is shown as a function of step size in Table 5.1. We observe that having a step size that is too small results in

step size	0.01	0.1	0.4	0.5	0.7	0.8	0.9	0.99
steps	3,163	311	73	57	39	33	60	659

Table 5.1: Converge time (number of steps) as a function of step size

an increased time to convergence, even though the path that is followed is a smooth one. Here, the converge time is a result of taking extremely small

```
1   def f(x):
2       return x[0]**2 / 5 + x[1]**2
3
4   def Df(x):
5       A = np.zeros(2)
6       A[0] = 2 * x[0] / 5
7       A[1] = 2 * x[1]
8       return A
9
10  x = np.array([8, 3]) # Initialize
11  h = 0.1
12  tol = 1e-5
13  max_steps = 10000
14  go = True
15  converged = False
16  X = x.reshape((2,1))
17
18  i = 0
19  while go: # Implement Gradient Descent
20      x = x - h * Df(x)
21      X = np.concatenate((X, x.reshape((2,1))), axis=1)
22      i += 1
23      if np.linalg.norm(Df(x)) < tol:
24          go = False
25          converged = True
26      if i > max_steps:
27          go = False
28  if not converged: # Warning
29      print ('Failed to converge')
30
31  for i in range(X.shape[1]): # Plot
32      plt.plot(X[0,i], X[1,i], 'k.')
33  plt.plot(X[0,:], X[1,:], 'k')
```

Code Block 5.6: Gradient Descent Algorithm for Example 5.7

steps. On the other hand, if the step size is too large, the path can end in a zig-zag path that bounces around before finally converging, or it might fail to converge altogether.   ▷

5.2.2 Stochastic Gradient Descent

Gradient descent is a commonly used optimization algorithm in statistical and machine learning. In this context, however, the objective function typically takes the form of a sum, as we saw with the log-likelihood function given by Equation (5.2). Thus, let us consider an objective function of the form

$$J(\theta) = \frac{1}{n} \sum_{i=1}^{n} f(\theta; x_i), \qquad (5.25)$$

where the set $\mathcal{X} = \{x_1, \dots, x_n\}$ constitutes a set of data. We will refer to the function f as the *kernel* of the objective function J, as it is used, along with the data $\mathcal{X}$, to generate the objective function J. In machine learning, the data are sometimes referred to as *training samples*.

There are three main approaches for applying gradient descent to an objective function of the form Equation (5.25), as outlined in the following definition.

Definition 5.5. *Given a set of data $\mathcal{X}$ and an optimization kernel $f : \mathbb{R}^n \times \mathbb{R} \to \mathbb{R}$, the gradient descent algorithm, given by Algorithm 5.2, may be applied to the objective function Equation (5.25) with any of the following modifications:*

1. *(Batch gradient descent) Algorithm 5.2 is implemented as written; i.e., the gradient for each step is computed using the full data set.*
2. *(Mini-batch gradient descent) A positive integer k, known as the* batch size, *is chosen (e.g., $k = 32$). For each step of gradient descent, the gradient is computed using k training examples. Once all the data have been used, the data are shuffled and the algorithm continues.*
3. *(Stochastic gradient descent) For each step of gradient descent, the gradient is computed using only a single training example. Once all the data have been used, the data are shuffled and the algorithm continues.*

A training epoch is said to occur each time the algorithm has cycled through all n training samples, regardless as to how they were grouped.

We note that batch and stochastic gradient descent are special cases of mini-batch with $k = n$ and $k = 1$, respectively. Batch gradient descent is exactly what one might expect, as the full gradient

$$DJ = \frac{1}{n} \sum_{i=1}^{n} \frac{\partial f(\theta; x_i)}{\partial \theta}$$

is computed with each step of the process. Thus, each step of gradient descent represents one training epoch. Batch gradient descent has the disadvantage that the gradient of the kernel must be calculated for each datum on each step of gradient descent. This can easily become computationally prohibitive, especially when dealing with large data sets.

On the opposite end of the spectrum is stochastic gradient descent, where the gradient is replaced with the gradient of a single datum,

$$DJ = \frac{\partial f(\theta; x_j)}{\partial \theta}.$$

The datum x_j used for the ith step is computed as $j = i \mod n$, and the ordering of the data is shuffled each time a new training epoch is reached. In this case, a training epoch corresponds to n steps of gradient descent. Due to the stochastic nature of each update, stochastic gradient descent is less likely to converge on a local minimum. However, the updates are much more noisy, as each update is determined based on a single training example.

In mini-batch gradient descent, we group the data into $n_b = \text{ceil}(n/k)$ batches[8], $K_0, \ldots, K_{n_b-1}$, where $K_i = \{ki, \ldots, k(i+1) - 1\}$, with the exception that the final batch K_{n_b-1} terminates at n instead of $(kn_b - 1)$. For each step of gradient descent, we then use the partial gradient

$$DJ = \frac{1}{|K_i|} \sum_{j \in K_i} \frac{\partial f(\theta; x_j)}{\partial \theta}, \tag{5.26}$$

where i is the step number modded by n_b, and where the data are again shuffled at the arrival of each new training epoch. In this case, a training epoch consists of precisely n_b steps of gradient descent. Mini-batch gradient descent alleviates some of the noise of stochastic gradient descent, but yet retains the advantage of the model being updated more frequently than in batch gradient descent.

In Code Blocks 5.7 and 5.8, we create an abstract class to serve as the parent for any objective function of the form of Equation (5.25). Note that this class is a subclass of ABC, which is imported from the abc package (for abstract classes). Overall, the ObjectiveFunction class looks like a typical class, except for the two methods kernel and kernelGrad, which don't seem to do anything, and which are adorned with the decorator @abstractmethod. This decorator ensures that any subclass must overwrite (and actually define) these two methods. In order to use this class, we simply create a subclass that overwrites the methods kernel and kernelGrad. (We will show how this is done in Example 5.8.)

We further utilize Python generators in order to implement all three use cases: batch, mini-batch, and stochastic. This is done within the (private) __batchGenerator method, which returns a generator object. This method is called whenever an object from this class instantiated, and the generator is saved as self.batch_gen. A generator is a function that essentially

---

[8] It would be inappropriate to use the floor function here, as the formula would not work correctly on certain edge cases. For example, consider batch size $k = 20$ and sample sizes $n = 80, 95, 100$. Naturally, the case $n = 80$ should be divided into four batches, whereas both $n = 95$ and $n = 100$ must be divided into five batches. The ceiling function of n/k yields 4, 5, and 5, whereas the floor function yields 4, 4, and 5. The point here is not that the floor function would need to be adjusted by $+1$, but rather that the floor function would group 80 and 95 into the same number of batches, which should not be the case.

```python
1   from abc import ABC, abstractmethod
2   class ObjectiveFunction(ABC):
3
4       def __init__(self, data, batch_size=0):
5           """
6               data (array or list)
7               batch_size (int):
8                   0 for full-batch gradient descent
9                   1 for stochastic gradient descent
10                  >1 for mini-batch gradient descent
11          """
12          assert batch_size >= 0, "Batch size must be non-negative"
13          assert batch_size < len(data), "Batch size too large"
14          self.data = np.array(data)
15          self.batch_size = batch_size
16          self.batch_gen = self.__batchGenerator() #generator object
17
18      def __len__(self):
19          return len(self.data)
20
21      def __batchGenerator(self):
22          cycle_index = 0
23          n_batches = int( np.ceil( len(self.data) / max(1,
                self.batch_size) ) )
24          while True:
25              if self.batch_size == 0:
26                  yield self.data
27              else:
28                  if cycle_index == n_batches:
29                      # Reset Training Epoch
30                      np.random.shuffle(self.data)
31                      cycle_index = 0
32
33                  start = self.batch_size * cycle_index
34                  stop = self.batch_size * (cycle_index + 1)
35                  cycle_index += 1
36                  yield self.data[start:stop]
```

Code Block 5.7: Abstract class for Objective Function. (Continued in Code Block 5.8.)

```python
def fEval(self, theta):
    """ Evaluates Objective Function using full data set """
    return np.sum( self.kernel(theta, self.data) ) / \
        len(self.data)

def grad(self, theta):
    """ Evaluates Gradient Function using current batch """
    data = next(self.batch_gen)
    return np.sum( self.kernelGrad(theta, data), axis=1 ) / \
        len(data)

@abstractmethod
def kernel(self, theta, x):
    pass

@abstractmethod
def kernelGrad(self, theta, x):
    pass
```

Code Block 5.8: Abstract class for Objective Function (continued).

remembers its own local variables, and executes the code until it reaches a **yield** statement. This is similar to a function that has a **return** statement, except that a generator remembers its spot in the code, as well as the values of local variables, and returns to this location the next time it is called. This way, whenever we call upon the **grad** method to obtain the gradient, it computes the gradient using only the next batch of data that is tracked by our generator.

As a final note, most generators terminate after a finite number of calls (think for-loop). However, our use of **while True**, ensures our generator will continue to run regardless of how many times it is called. For the case of (full) batch gradient descent, it simply returns the full data set each time it is called. For mini-batch (and, as a special case, stochastic) gradient descent, it defines a variable **cycle_index**, which tracks where it is within each training epoch. Each time **cycle_index** reaches the value **n_batches**, it shuffles the data set and resets the cycle index back to zero. Each time it is called, it implements Equation (5.26), returning the appropriate subset of data. When **batch_size=1**, it returns the single, next datum, and the training epoch is complete when it has returned the full data set.

The nice thing about this abstract class, is that it does all the heavy lifting for us, in terms of implementing batch, mini-batch, or stochastic gradient descent. Our actual gradient descent code can therefore be method-agnostic, as the method for tracking the batches is implemented internally within the objective function class.

Example 5.8. Next, let us continue our discussion of Example 5.5, the maximum likelihood estimation of a Gamma random variable, by showing how gradient descent can be used to estimate the parameter values. Let us start off by subclassing our `ObjectiveFunction` abstract class from Code Block 5.7, to create a class specific to the Gamma random variable. For this subclass, we only need to specify the definitions of the `kernel` and `kernelGrad` methods, which is shown in Code Block 5.9.

```python
class GammaLikelihood(ObjectiveFunction):

    def kernel(self, theta, x):
        """ theta = [alpha, beta]
            x (np.array) len n
        output:
            1 x n array
        """
        return (theta[0]-1)*np.log(x) - x / theta[1] -
                np.log(scipy.special.gamma(theta[0])) -
                theta[0]*np.log(theta[1])

    def kernelGrad(self, theta, x):
        """ theta = [alpha, beta]
            x (np.array) len n
        output:
            2 x n array
        """
        U_alpha = np.log(x) - scipy.special.psi(theta[0]) -
                np.log(theta[1])
        U_beta = x / theta[1]**2 - theta[0] / theta[1]
        return np.array([U_alpha, U_beta])
```

Code Block 5.9: Gamma Likelihood Objective Function

Note that both of these methods return matrices with as many columns as training instances. That is, `kernel` returns a $1 \times n$ array, and `kernelGrad` returns a $2 \times n$ array. The summation of these values is taken care of in the `fEval` and `grad` methods in the parent class.

Now that we have defined our subclass, we can implement gradient descent in a few lines of code, as shown in Code Block 5.10. Note that we can easily switch between (full) batch gradient descent (`batch_size = 0`), mini-batch gradient descent (e.g., `batch_size = 1000`), and stochastic gradient descent (`batch_size = 1`) by changing the input to the constructor on line 7. We also define a list L that tracks the value of the full likelihood function with each step in our gradient descent algorithm; see lines 15 and 20. We note that the true value is $\theta = (2,3)$, whereas our starting point

```
1   alpha_0 = 2
2   beta_0 = 3
3   n_samples = 1000000
4   samples = np.random.gamma(alpha_0, scale=beta_0, size=n_samples)
5
6   ## Implement Gradient Descent
7   G = GammaLikelihood(samples, batch_size=0) # set batch_size here
8   theta = np.array([3, 2])
9   h = 0.1
10  tol = 1e-3
11  max_steps = 10000
12  go = True
13  converged = False
14  X = theta.reshape((2,1))
15  L = [G.fEval(theta)]
16
17  i = 0
18  while i < max_steps and np.linalg.norm( np.sum(G.kernelGrad(theta,
        G.data), axis=1) ) / len(G) > tol:
19      theta = theta + h * G.grad(theta)
20      L.append(G.fEval(theta))
21      X = np.concatenate((X, theta.reshape((2,1))), axis=1)
22      i += 1
```

Code Block 5.10: Gradient Descent for Example 5.8

is $\theta_0 = (3, 2)$. Also, since we are trying to *maximize* the log-likelihood function, we use a *plus* instead of a *minus* on line 19, so that the algorithm allows us to drift up-hill instead of down-hill.

We ran this code three times, using full batch (batch_size = 0), mini-batch (batch_size = 32), and stochastic (batch_size = 1) gradient descent. The results are shown in Figure 5.4. We observe that the path traced by stochastic gradient descent does quite a bit more "wandering" before it finally converges. The stochastic nature of this path helps prevent the algorithm from converging on a local minimum that is not a global minimum. Using a batch size of 32 dramatically reduces the stochasticity, whereas the full batch gradient descent follows a path directly toward the maximum.

Finally, for each case we plot the log-likelihood as a function of the number of steps, in Figure 5.5. We observe that the full batch gradient descent, as expected, shows a much smoother and monotonic improvement in log-likelihood with each step, whereas the stochastic gradient descent makes numerous steps that result in a worse log-likelihood.

As a final reminder, each step of the full batch gradient descent uses all 1,000,000 pieces of data, and so there are many times more computations done with each individual step, whereas stochastic gradient descent uses

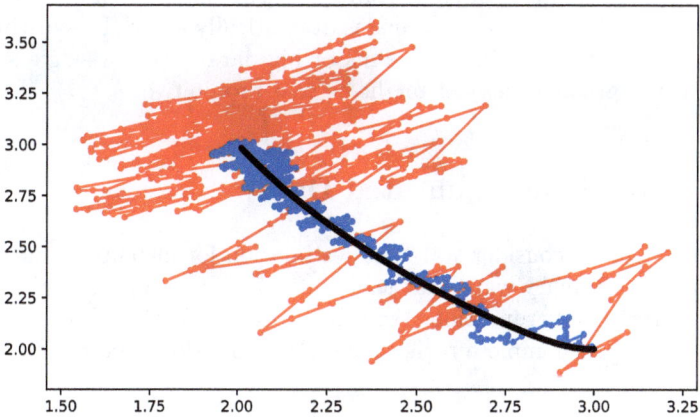

Fig. 5.4: Gradient Descent: full batch (black), mini-batch size 32 (blue), stochastic (red). Curves start at $(3, 2)$ and end at $(2, 3)$.

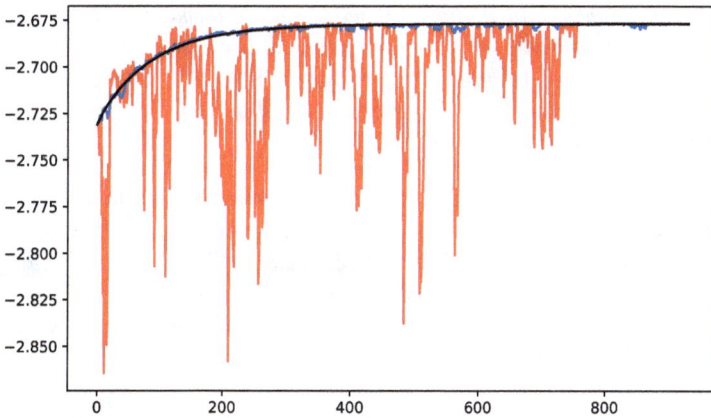

Fig. 5.5: Log-likelihood for three cases: full batch (black), mini-batch size 32 (blue), stochastic (red).

only 1 datum with each step. So even though they converge in approximately the same number of steps (full batch 931, mini-batch 900, stochastic 757), the number of *computations* is dramatically lower. This distinction is only exaggerated within the realms of big data, where a single step of the full batch gradient descent might be cost prohibitive.   ▷

5.3 Censored and Truncated Data

In this section, we consider statistical estimation for incomplete data sets. There are two primary ways that lead to (systematically) incomplete data sets: censorship and truncation. We will go over the basics of these mechanisms. For a more complete review, see Klein and Moeschberger [2003].

5.3.1 Overview

Before diving in, let us provide an overview of the distinction between censorship and truncation. As stated previously, they each deal with incomplete data. They differ, however, in the way in which data are left out.

Definition 5.6. *A set of data* $\mathcal{X} = X_1, \ldots, X_n$ *is said to be* censored *if its values are known only if they fall within an interval* $I = (x_l, x_r)$, *and if the counts of data below or above the interval are also known. Moreover, the data are considered to be* interval-censored *if both* x_l *and* x_r *are finite,* left-censored *if* $x_r = \infty$, *and* right-censored *if* $x_l = -\infty$ *(or if* $x_l = 0$, *depending on context).*

As an example of censored data, consider a small kitchen scale that can weight objects up to 3 kg. Upon weighing seven household objects, we obtain the data $\mathcal{X} = (1.27, 0.65, 3, 2.12, 3, 3, 2.86)$. Three of these data are right-censored at 3, whereas four of these data have observed values. Note that, critically, we know *how many* objects fall outside of our censorship interval $(0, 3)$. We say that these data are right-censored at $X = 3$.

Definition 5.7. *A set of data* $\mathcal{X} = X_1, \ldots, X_n$ *is said to be* truncated *if data are observed only if they fall within an interval* $I = (x_l, x_r)$. *Moreover, the data are considered to be* interval-truncated *if both* x_l *and* x_r *are finite,* left-truncated *if* $x_r = \infty$, *and* right-truncated *if* $x_l = -\infty$ *(or if* $x_l = 0$, *depending on context).*

As an example, consider the height distribution of Royal Marines. Since the Royal marines have a height restriction, only admitting persons at least 64 inches tall, the data will be left-truncated Trussell and Bloom [1979]. Note, critically, that we do not know how many individuals did not meet the height requirement and therefore did not find their way into the data set. As a result, the average height of a Royal Marine will be higher than the average height of the population. We say that the data in this case are left-truncated at $X = 64$.

Interestingly, data can be simultaneously both truncated and censored. For example, consider an auto-insurance policy that has a deductible of $500 and a maximum payout of $50,000. We would say that the set of claim values is left-truncated at 500, because individuals will not file a claim unless the total amount is above their deductible, and right-censored at 50,000, since claims greater than this amount are capped at this value. Note that we do not know how many claims there might have been under $500 (since those claims were never made), but we do know the count of claims above $50,000, as they were each capped at this maximum payout value.

5.3.2 Censored Data

In the case of censored data, we observe the values of data within our interval I, as well as the counts of data that were to the left or right of our interval. We may therefore modify our likelihood function with an appropriate number of factors of the cumulative distribution function, evaluated at the left-censorship point, and the survival function evaluated at the right censorship point. In particular, we have the following:

Proposition 5.4. *Consider a set of* IID *data* $X_1, \ldots, X_n \sim f(x; \theta)$ *that is censored by the interval* $I = (x_l, x_r)$. *Suppose that exactly* p *of the data points are left censored and* q *are right censored. Without loss of generality, we can arrange our data such that the first* $n - p - q$ *points are the values that lie within the interval* I. *Then the likelihood function of our data is given by*

$$\mathcal{L}(\theta) = [1 - S(x_l; \theta)]^p S(x_r; \theta)^q \prod_{i=1}^{n-p-q} f(x_i; \theta), \qquad (5.27)$$

where $S(x; \theta) = \int_x^\infty f(t; \theta)\, dt$ *is the survival function of the* PDF f.

The quantity $[1 - S(x_l; \theta)] = C(x_l; \theta)$ is equivalent to the CDF, which represents the total amount of probability to the left of the point $X = x_l$, whereas the survival function $S(x_r; \theta)$ represents the total probability to the right of x_r. (We express the entire formula in terms of the survival function to follow custom.) We therefore treat each censored data point by adding a factor equal to the probability that the data point would have been censored. Note that Equation (5.27) treats the case of left, right, and interval censored data using a single, compact formula.

Example 5.9. Determine the likelihood function for IID data $T_1, \ldots, T_n \sim \text{Exp}(\beta)$ that are right-censored at the point $T = \tau$. Determine an analytical formula for the MLE $\hat{\beta}$.

We begin by counting the number of data points that are censored, obtaining

$$k = \sum_{i=1}^{n} \mathbb{I}[t_i \geq \tau].$$

Since the survival function at $T = \tau$ is given by $S(\tau; \beta) = e^{-\tau/\beta}$, Equation (5.27) yields the likelihood function

$$\mathcal{L}(\beta) = \frac{e^{-k\tau/\beta}}{\beta^{n-k}} \prod_{i=1}^{n} e^{-t_i/\beta \cdot \mathbb{I}[t_i < \tau]}. \tag{5.28}$$

Note that the product in this equation is equivalent to taking the product only over the $n - k$ uncensored lifetimes. The log-likelihood is therefore given by

$$\ell(\beta) = \frac{-k\tau}{\beta} - (n - k) \log(\beta) - \frac{1}{\beta} \sum_{i=1}^{n} t_i \mathbb{I}[t_i < \tau]. \tag{5.29}$$

Differentiating with respect to the parameter β and equating the result to zero, we find the MLE

$$\hat{\beta} = \frac{1}{n - k} \left[k\tau + \sum_{i=1}^{n} t_i \mathbb{I}[t_i < \tau] \right]. \tag{5.30}$$

This represents the total sum of observed lifetimes (including the k censored lifetimes, which have a current value of k) divided by the total number of uncensored lifetimes. This expression is therefore equivalent to

$$\hat{\beta} = \frac{1}{n - k} \sum_{i=1}^{n} \min(t_i, \tau).$$

Compare this for the MLE of uncensored data, which is given by $\hat{\beta} = (1/n) \sum t_i$.

▷

Example 5.10. A cohort of 1,000 players downloads the game *Angry Penguins* on its release date. Let the random variable T represent the customer lifespan of each player, i.e., the time it takes for each customer to *churn*, and suppose that $T \sim \text{Exp}(\beta)$ is exponentially distributed. Assume that the scale parameter has a true value of $\beta = 6$, measured in months, and that nine months have passed since the game's release. Simulate the lifetime of each user, indicating which users' lifetimes are censored, and then use Equation (5.30) to estimate the value of the parameter β.

We simulate 1,000 user lifetimes from $\text{Exp}(6)$, as shown in Code Block 5.11. From this shining simulation, we estimate $\hat{\beta} = 6.146003$, which is quite close to the true value of $\beta = 6$. Without correcting for the fact that our data are

```
1  beta = 6
2  tau = 9
3  n_samples = 1000
4  lifetimes = np.random.exponential(scale=beta, size=n_samples)
5
6  k = sum( lifetimes >= tau) # 237
7  beta_hat = np.sum(np.fmin(9, lifetimes)) / (n_samples - k) #
       6.146003
```

Code Block 5.11: Simulation of user lifetime

right-censored, our estimate, obtained by `np.sum(np.fmin(9, lifetimes) / n_samples`, would equal 4.6894, which is definitely incorrect. The estimate using the full set of data, with the censorship removed, is given by `np.sum(lifetimes) / n_samples`, which results in an estimate of 6.187.
▷

5.3.3 Truncated Data

In the case of truncated data, we only observe the each datum if it lies within our interval I, and we receive no information regarding the counts of data that fell outside of our interval. In this case, we modify our likelihood function by renormalizing the probability density to only consider the probability within the interval I. In particular, we have the following:

Proposition 5.5. *Consider a set of* IID *data* $X_1, \ldots, X_n \sim f(x; \theta)$ *that are truncated by the interval* $I = (x_l, x_r)$, *so that* $x_l \leq X_i \leq x_r$, *for all* $i = 1, \ldots, n$. *Then the likelihood function of our data is given by*

$$\mathcal{L}(\theta) = \frac{1}{[S(x_l) - S(x_r)]^n} \prod_{i=1}^{n} f(x_i; \theta), \tag{5.31}$$

where $S(x; \theta) = \int_x^\infty f(t; \theta) \, dt$ *is the survival function of the* PDF f.

In Equation (5.31), we have replaced the PDF with the renormalized PDF

$$f(x; \theta) \to \frac{f(x; \theta)}{S(x_l) - S(x_r)},$$

defined on the interval I. For the special case of left-truncated data, Equation (5.31) reduces to

$$\mathcal{L}(\theta) = \frac{1}{S(x_l)^n} \prod_{i=1}^{n} f(x_i; \theta), \quad \text{(left-truncated)}$$

since $S(\infty) \to 0$. (Recall $S(x_l) = \mathbb{P}(X \ge x_l)$.) Similarly, for the case of right-truncated data, Equation (5.31) reduces to

$$\mathcal{L}(\theta) = \frac{1}{[1 - S(x_r)]^n} \prod_{i=1}^{n} f(x_i; \theta), \quad \text{(right-truncated)}$$

since $S(-\infty) \to 1$. (Recall $1 - S(x_r) = \mathbb{P}(X \le x_r)$.)

Example 5.11. Determine the likelihood function for IID data $T_1, \ldots, T_n \sim$ $\text{Exp}(\beta)$ that are right-truncated at the point $T = \tau$.

Following Equation (5.31), the likelihood function is given by

$$\mathcal{L}(\beta) = \frac{1}{[1 - e^{-\tau/\beta}]^n \beta^n} \prod_{i=1}^{n} e^{-t_i/\beta}.$$

Here, each of the t_i are observed, i.e., $t_i \le \tau$, and the number of unobserved observations, for which $T > \tau$, is unknown. The log-likelihood is given by

$$\ell(\beta) = -n \log(1 - e^{-\tau/\beta}) - n \log(\beta) - \frac{1}{\beta} \sum_{i=1}^{n} t_i.$$

By differentiating with respect to β and setting the result equal to zero, we obtain the following transcendental equation,

$$\beta - \frac{\tau}{e^{\tau/\beta} - 1} = \frac{1}{n} \sum_{i=1}^{n} t_i, \tag{5.32}$$

which may be solved for β to yield the MLE.   ▷

Example 5.12. A cohort of 10,000 players downloads the freemium game *Angry Penguins* on its release date. The game developer wants to estimate the number of players who will "convert" into paying users, i.e., the number of players who will make a purchase within the game. Suppose the random variable T represents the time until first purchase of a payer. Suppose that $T \sim \text{Exp}(\beta)$. Use the value $\beta = 6$ to simulate the first-purchase time for a total of 1,000 payers. Then truncate the data at $\tau = 9$, and estimate the value of β from the simulated data.

The code is shown in Code Block 5.12. Upon sampling 1,000 data points, corresponding to the 1,000 payers, on line 4, we then proceed to truncate the data at $\tau = 9$. That is, for the purpose of our simulation, we have two pieces of knowledge that we would not have had in real life: the true value of the parameter $\beta = 6$ and the true total number of payers 1,000.

After truncating the data (line 5), we are left with 791 observed first-purchase times. The story for the data we have simulated is as follows: 10,000 players download the game on its release date. We do not know how many will ultimately "convert" into payers. However, by day 9, we

```
1  beta = 6
2  tau = 9
3  n_samples = 1000
4  conversion_times = np.random.exponential(scale=beta, size=n_samples)
5  data = conversion_times[conversion_times < tau] #791
6  data.mean() # 3.391
7  def f(beta):
8      return beta - tau / (np.exp(tau / beta)-1) - data.mean()
9  scipy.optimize.root(f,5) # 5.8588
```

Code Block 5.12: Simulation of conversion curve

have observed 791 users who have made their first purchase. This is an example of truncated data, not censored data, as we ultimately do not know how many payers we have yet to observe. Since the data are right-truncated and exponentially distributed, we may use Equation (5.32), which is implemented in lines 7–9. Note that the mean first-purchase time of our observed data is 3.391 (line 6), whereas the MLE, when accounting for truncation, is 5.8588, an excellent approximation to the true value of $\beta = 6$.

We should mention that since these simulations are computed using random number generators, the precise values of the estimates will change each time they are run. To decrease the variance of the estimates, use a larger sample size, e.g., we ran the same code using **n_samples=1000000** and obtained an estimate $\hat{\beta} = 5.979$. Repeat simulations with a larger sample size will consistently return estimates that are much closer to the true value of $\beta = 6$.

The difference between this example and Example 5.10 is that in Example 5.10, we know the total count of users, and therefore know the total count of the users who have yet to churn. In this example, while we again know that there are a total of 10,000 players, we do not know how many of those players will ultimately make a purchase. We only know that, after nine days, a total of 791 users have already made their first purchase. However, we do not know how many of the remaining 9,209 *players* will ultimately become *payers*. Therefore, we must rely on truncation and not censorship to obtain our estimate.

With our estimate of $\hat{\beta} = 5.8588$ in hand, we can compute the survival function $S(9; 5.8588) = 0.2152$, which estimates the fraction of total payers who have yet to convert (i.e., who have yet to make their first purchase). We therefore estimate that a total of 1,008 users (792 observed / 0.7845 estimated fraction of observed) will ultimately convert into payers, which is reasonably close to the true value of 1,000, which was used to generate the data.   ▷

5.3.4 Hazard Function

The examples we have visited so far can be viewed as industry applications of a field called *survival analysis*. Survival analysis is concerned with longevity of individual items within a collective; in particular, it provides a framework for analyzing the duration until a given event happens. There are many applications in medicine, biology, and engineering, including time-to-failure of a mechanical system as well as more literal applications involving the time-to-death of patients in a clinical study[9]. There are many examples of more traditional applications in Klein and Moeschberger [2003].

One quantity often discussed in survival analysis is the *hazard function*, defined as the probability density of an item failing now, given that it has already lasted this long. Specifically, we can think of this as the following limit

$$h(x) = \lim_{\Delta x \to 0} \frac{\mathbb{P}(x \leq X \leq x + \Delta x | X \geq x)}{\Delta x}.$$

We will, however, define the hazard function in an equivalent but more practical way.

Definition 5.8. *Given a random variable* $X \sim f$, *its hazard function is defined by the ratio*

$$h(x) = \frac{f(x)}{S(x)}, \tag{5.33}$$

where $S(x)$ *is the survival function.*

Similarly, the cumulative hazard function (CHF) *is defined as the integral*

$$H(x) = \int_{-\infty}^{x} h(t)\, dt. \tag{5.34}$$

From the definition of cumulative hazard, it is straightforward to show the relation

$$H(t) = -\ln S(t). \tag{5.35}$$

Example 5.13. Show that the hazard function of an exponential random variable $X \sim \text{Exp}(\beta)$ is constant.

Recall that

$$f(x; \beta) = \frac{1}{\beta} e^{-x/\beta} \quad \text{and} \quad S(x) = e^{-x/\beta}.$$

Taking the ratio, we see that the hazard function is given by

$$h(x; \beta) = \frac{1}{\beta},$$

a constant.   ▷

---

[9] To eschew the macabre, however, we opted for applications dealing with the lifetime of players within a game.

Example 5.13 shows that an item's probability of failing eminently is always the same, regardless of its age. This is related to the fact that the exponential random variables are *memoryless*.

5.3.5 The Pareto and Weibull Distributions

We next introduce two new distributions that are commonly used in survival analysis due to the particularly simple form for their respective hazard functions.

The Pareto Distribution

Our first new distribution is a basic power-law distribution that has many applications in various areas, including the distribution of wealth. It is also the distribution that gives rise to the so-called "80-20" rule, which states that 80% of wealth of a society is held by the top 20% of its population, or that 20% of your effort typically produces 80% of your return (in work, sports, etc.). This distribution is, of course, the Pareto distribution, which is defined as follows.

Definition 5.9. *A random variable X is considered a* Pareto *random variable, denoted $X \sim \text{Pareto}(m, \alpha)$, if it satisfies the* Pareto *distribution function*

$$f(x; m, \alpha) = \frac{\alpha m^\alpha}{x^{\alpha+1}} \tag{5.36}$$

on the domain $x \in [m, \infty)$. The parameter m is called the scale parameter, and the parameter α is called the shape parameter.

The Pareto distribution has the following properties.

Proposition 5.6. *Let $X \sim \text{Pareto}(m, \alpha)$. Then*

$$S(x) = \left(\frac{m}{x}\right)^\alpha, \tag{5.37}$$

$$\mathbb{E}[X] = \frac{\alpha m}{\alpha - 1} \qquad \text{for } \alpha > 1, \tag{5.38}$$

$$\mathbb{V}(X) = \frac{m^2 \alpha}{(\alpha - 1)^2(\alpha - 2)} \qquad \text{for } \alpha > 2. \tag{5.39}$$

The expected value and variance diverges for $\alpha \leq 1$ and $\alpha \leq 2$, respectively.

As a result of Equations (5.36) and (5.37), the hazard function of the Pareto distribution is given by

$$h(x) = \frac{\alpha}{x}.$$

We can use the built-in `numpy` and `scipy` methods to sample from a Pareto distribution and draw the corresponding PDF, as shown in Code Block 5.13. Note that the built-in `numpy.random.pareto` is actually a Pareto Type II distribution, or Lomax distribution, which we correct for by adding one to the output. The resulting output is shown in Figure 5.6.

```
1   m = 1
2   alpha = 2
3   n_samples = 1000
4   samples = m * (1 + np.random.pareto(alpha, size=n_samples) )
5
6   x = np.linspace(1, 20)
7   y = scipy.stats.pareto.pdf(x, alpha, scale=m)
8   plt.hist(samples, bins=100, normed=True)
9   plt.plot(x,y)
```

Code Block 5.13: Samples from a Pareto distribution

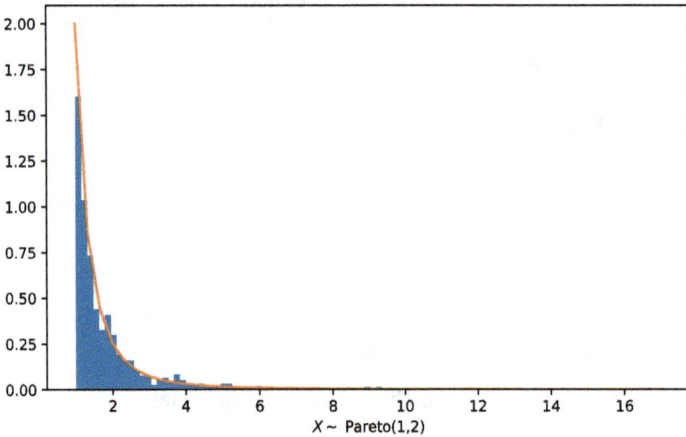

Fig. 5.6: Samples from Pareto(1,2).

The Weibull Distribution

A commonly used distribution in survival analysis is the Weibull distribution, which has a power-function hazard rate.

Definition 5.10. *A random variable X is considered a* Weibull *random variable, denoted $X \sim \text{Weibull}(\alpha, \beta)$, if it satisfies the Weibull distribution function*

$$f(x; \alpha, \beta) = \frac{\alpha}{\beta} \left(\frac{x}{\beta} \right)^{\alpha-1} e^{-(x/\beta)^\alpha}, \qquad (5.40)$$

on the domain $x \in [0, \infty)$. The parameter α is called the shape parameter, and the parameter β is called the scale parameter.

The Weibull distribution has the following properties.

Proposition 5.7. *Let $X \sim \text{Weibull}(\alpha, \beta)$. Then*

$$S(x) = e^{-(x/\beta)^{\alpha}}, \tag{5.41}$$
$$\mathbb{E}[X] = \beta\Gamma(1 + 1/\alpha), \tag{5.42}$$
$$\mathbb{V}(X) = \beta^2 \left[\Gamma(1 + 2/\alpha) + \Gamma(1 + 1/\alpha)^2 \right], \tag{5.43}$$

where Γ represents the Gamma function (Definition 2.13).

As a result of Equations (5.40) and (5.41), the hazard function of the Pareto distribution is given by

$$h(x) = \frac{\alpha}{\beta} \left(\frac{x}{\beta} \right)^{\alpha - 1}.$$

If a Weibull random variable $X \sim \text{Weibull}(\alpha, \beta)$ represents the time-to-failure or lifetime, then the behavior can be described by three separate cases.

If $\alpha < 1$, the hazard rate decays over time, indicating the longer an item survives, the less its marginal probability of demise. This is sometimes referred to as the *Lindy effect*. As the item ages, its remaining life expectancy actually increases. Often certain nonperishable goods, such as ideas, books, or technology, exhibit this kind of reverse aging: many are killed in infancy, and the longer any one survives, the more its expected additional lifetime. If this seems counterintuitive, consider the expected remaining lifespan for a recently published mass paperback novel versus the works of Shakespeare. The fact that the works of Shakespeare have already survived half a millennium is a testament to their robustness as a staple of literature, whereas the romantic tripe published just yesterday might already be off the shelves by tomorrow.

If $\alpha = 1$, the hazard rate is constant over time, and the Weibull distribution reduces to an exponential distribution as a special case, in fact careful inspection reveals $\text{Weibull}(1, \beta) = \text{Exp}(\beta)$.

Finally, if $\alpha > 1$, the hazard rate increases with time, indicative of an aging process: the more an item ages, the more likely its impending doom.

We can use the built-in `numpy` and `scipy` methods to sample from a Weibull distribution and draw the corresponding PDF, as shown in Code Block 5.14. The resulting output is shown in Figure 5.7. Finally, the Weibull distribution is plotted for various parameter values in Figure 5.8.

5.3.6 Empirical Estimates for Survival and Hazard

Next, we discuss two classic empirical estimates commonly used in survival analysis. For additional discussion, see Aalen, *et al.* [2008] or Klein and Moeschberger [2003].

```
1  alpha = 2
2  beta = 3
3  n_samples = 1000
4  samples = beta * np.random.weibull(alpha, size=n_samples)
5  x = np.linspace(0, 8)
6  y = scipy.stats.weibull_min.pdf(x, alpha, scale=beta)
```

Code Block 5.14: Samples from a Weibull distribution

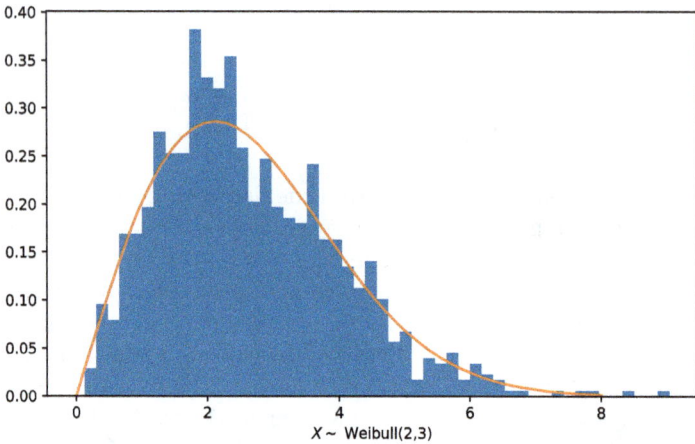

Fig. 5.7: Samples from Weibull(2,3).

The Kaplan-Meier Estimator

The Kaplan-Meier estimator, also known as the *product-limit estimator*, is an empirical estimator for the survival function. The idea is to collect all observed event times $T_1 < \cdots < T_d$ from a right-censored data set. To account for multiplicity of events, we let E_j represent the number of individuals whose observed event time is T_j, and we let Y_j represent the number of individuals at risk at time T_j. (Note that Y_j will have discontinuities due to both events *and* individuals "lost" due to censoring.) The quantity E_j/Y_j is therefore an empirical estimate of the hazard at time T_j. This quantity is critical to the following.

Theorem 5.5 (Kaplan-Meier). *Let $T_1 < \cdots < T_d$ be the distinct observed event times from a left-truncated and/or right-censored data set sampled from a distribution F_T. Let E_j be the number of observed events occurring at time T_j and let Y_j be the number of individuals at risk at time T_j. Then the Kaplan-Meier estimator*

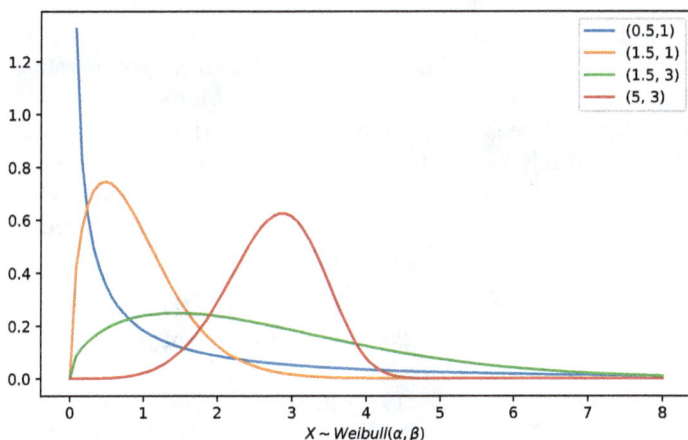

Fig. 5.8: Weibull(α, β) for various parameter values.

$$\hat{S}(t) = \prod_{T_j \leq t} \left(1 - \frac{E_j}{Y_j} \right) \tag{5.44}$$

is an unbiased estimate for the survival curve $S(t) = 1 - F(t)$. Moreover, the variance in this estimator can be estimated by

$$\mathbb{V}\left(\hat{S}(t) \right) = \hat{S}(t)^2 \sum_{T_j \leq t} \frac{E_j}{Y_j(Y_j - E_j)}, \tag{5.45}$$

known as the Greenwood formula.

Note 5.6. The number of individuals *at risk at time* T_j is the count of all individuals with an observed event time $T \geq T_j$, which therefore includes the E_j events that occurred at time T_j, and all individuals with a censoring time greater than or equal to T_j.   ▷

Note 5.7. We will take, by convention, the product of an empty set to equal one, so that
$$S(t) = 1$$
for all $t < T_1$; i.e., the survival estimate is 100% prior to the first observed event.   ▷

   When no censoring or truncation is present, the Kaplan-Meier estimator is the complement of the empirical distribution function. When the data set is right-censored, the formula still holds. The idea is that at each point in time, we are comparing the number of *observed* events to the total basis of individuals at risk.

Equation (5.44) makes intrinsic sense: since the ratio E_j/Y_j represents the hazard at time T_j, then the quantity $(1 - E_j/Y_j)$ is the fraction of the *current* population that survives through time T_j. By multiplying this latter quantity with the running estimate of the survival ratio, we therefore obtain an empirical estimate for the survival function. Moreover, the estimate is not affected by right-censorship, since it is comparing the observed deaths against all individuals known to be eligible for an event at each point in time. It is also not affected by left-truncation: we do not need to know how many individuals were at risk in the past, only the present count, which would not be possible with right-truncated data.

Corollary 5.4. *Given the Kaplan-Meier estimate $\hat{S}(t)$ of the survival function, the associated estimate of the cumulative hazard function is given by*

$$\hat{H}(t) = -\ln \hat{S}(t). \tag{5.46}$$

This follows by applying the relationship given by Equation (5.35) to the Kaplan-Meier estimate to obtain an associated estimate for the cumulative hazard.

The Nelson-Aalen Estimator

An alternate estimator that has better performance for small sample sizes is given by the following.

Theorem 5.6 (Nelson-Aalen). *Let $T_1 < \cdots < T_d$ be the distinct observed event times from a left-truncated and/or right-censored data set sampled from a distribution F_T. Let $E_j \geq 1$ be the number of observed events occurring at time T_j and let Y_j be the number of individuals at risk at time T_j. Then the Nelson-Aalen estimator*

$$\hat{H}(t) = \sum_{T_j \leq t} \frac{E_j}{Y_j} \tag{5.47}$$

is an unbiased estimate for the cumulative hazard function $H(t)$. Moreover, the variance in this estimator can be estimated by

$$\mathbb{V}\left(\hat{H}(t)\right) = \sum_{T_j \leq t} \frac{E_j}{Y_j^2}. \tag{5.48}$$

Note 5.8. Similar to Note 5.7, we take, by convention, the summation over an empty set to be zero, so that $\hat{H}(t) = 0$ for all time prior to the first observed event.   ▷

Corollary 5.5. *Given the Nelson-Aalen estimate $\hat{S}(t)$ of the cumulative hazard function, the associated estimate of the survival function is given by*

$$\hat{S}(t) = \exp\left[-\hat{H}(t)\right]. \tag{5.49}$$

5.3.7 Comparing Survival Difference Between Groups

Comparing Two Groups

The following result is a commonly used hypothesis test for comparing survival differences between two groups.

Theorem 5.7 (Mantel-Cox Test (Logrank Test)). *Let $T_1 < \cdots < T_d$ be the distinct observed pooled event times between two groups of right-censored survival data, and let Y_{0j} and Y_{1j} represent the number of individuals at risk in groups 0 and 1, respectively, at time T_j, and let E_{0j} and E_{1j} represent the number of events in groups 0 and 1, respectively, at time T_j, such that $Y_j = Y_{0j} + Y_{1j}$ and $E_j = E_{0j} + E_{1j}$ represent the total number of individuals at risk and events at time T_j, respectively. Then under the null hypothesis*

$$H_0 : h_1(t) = h_2(t)$$

that the two groups have identical hazards, the test statistic

$$Z = \frac{\displaystyle\sum_{j=1}^{d}\left(E_{ij} - \frac{Y_{ij}E_j}{Y_j}\right)}{\sqrt{\displaystyle\sum_{j=1}^{d}\frac{Y_{ij}E_j}{Y_j}\left(\frac{Y_j - E_j}{Y_j}\right)\left(\frac{Y_j - Y_{ij}}{Y_j - 1}\right)}}, \tag{5.50}$$

known as the logrank statistic, *is approximately standard-normally distributed for large samples and for $i = 0, 1$.*

Proof. Fix $i = 0$ or $i = 1$. For each event time T_j, the random variable E_{ij} is hypergeometrically distributed (Definition 2.5) under the null hypothesis:

$$E_{ij} \sim \text{HypGeom}(Y_j, E_j, Y_{ij}).$$

It follows from Proposition 2.11 that, under the null hypothesis,

$$\mathbb{E}[E_{ij}] = \frac{Y_{ij}E_j}{Y_j}$$

$$\mathbb{V}(E_{ij}) = \frac{Y_{ij}E_j}{Y_j}\left(\frac{Y_j - E_j}{Y_j}\right)\left(\frac{Y_j - Y_{ij}}{Y_j - 1}\right).$$

The result follows from the central limit theorem. □

5.4 Online Processes

We conclude this chapter with a discussion on stochastic processes and, especially, a certain type of stochastic process we call online processes, which involve censored or truncated data. We will then discuss several empirical estimates for their underlying distribution function. We will not follow the traditional treatment of stochastic processes, for which we refer the reader to, e.g., Durrett [2016] or Hajek [2015]. Rather, we will develop our theory based on a practical analysis of customer behavior for certain online products, such as websites and mobile apps. The idea is that user engagement follows a distribution over time, which can be used to generate an accompanying stochastic process. The wrinkle is that, for online products, new users can arrive each day, and the process begins anew for each user cohort. Thus, we will consider a certain type of stochastic process that comes with *two* indices, user cohort and the time that is measured within each cohort. The online nature of the process causes the observed data to be diagonally truncated or censored, depending on the specific application. The second departure from traditional stochastic processes is that we will start with the distributional nature of the process over time, using it to *generate* the ensuing process.

5.4.1 Stochastic Processes

A stochastic process is defined over a sample space and over time as follows. It is helpful to think of the probability space as consisting of the entire process itself, as opposed to a single snapshot. That is, if we flip a coin repeatedly, elements of the sample space are not $\{H, T\}$, but rather all binary strings $\{HTHTTH \cdots, HHHTHT \cdots, \ldots\}$.

Definition 5.11. *A stochastic process (or* random process*) is a collection* $\mathcal{P} = \{X_t\}_{t \in \mathcal{T}}$ *of random variables defined over a common probability space space* $(\Omega, \mathcal{E}, \mathbb{P})$ *and indexed by a set* $\mathcal{T}$*. If* $\mathcal{T} = \mathbb{N}$*, then it is called a* discrete-time random process*. If* $\mathcal{T} = \mathbb{R}$*, then it is called a* continuous-time random *process. The range of the random variables* X_t *is called the* state space *and is denoted* $\mathcal{S}$*.*

There are three equivalent ways to think of a stochastic process:

- the process X is a function over the Cartesian product $\mathcal{T} \times \Omega$, with value $X_t(\omega) = X(t, \omega)$ for any given $t \in \mathcal{T}$ and $\omega \in \Omega$;
- for each fixed $t \in \mathcal{T}$, the process $X_t(\omega)$ is a function over the sample space Ω;
- for each fixed sample outcome $\omega \in \Omega$, the process $X_t = X_t(\omega)$ is a function over the indexing set $\mathcal{T}$, known as a *realization* or *sample path* of the process.

Example 5.14. Let Ω be the sample space consisting of ten independent flips of a fair coin. Then the set of random variables $\{X_t\}_{t=0}^9$, with state space $\mathcal{S} = \mathbb{B} = \{0, 1\}$, defined so that $X_t = \mathbb{I}[i\text{th flip is heads}]$, constitutes a stochastic process. An example outcome from the sample space is $\omega =$

$HTTHTHTTHT$, which would correspond to the sample path $X_0 = 1$, $X_1 = 0$, $X_2 = 0$, $X_3 = 1$, etc. As this example shows, the sample space for a stochastic process consists of all possible read-world outcomes of the process; it is the set of all possible realizations of the process.   ▷

5.4.2 Common Types of Stochastic Processes

Before defining an online process, we first discuss some common flavors of stochastic processes.

Markov Chains

A Markov chain, or Markov process, is a certain class of stochastic process in which each random variable depends on its history only through the immediately preceding random variable.

Definition 5.12. *A discrete-time stochastic process $\{X_t\}_{t \in \mathbb{N}}$ with state space $\mathcal{S}$ satisfies the* Markov property *if*

$$\mathbb{P}(X_t \in S | X_0, X_1, \ldots, X_{t-1}) = \mathbb{P}(X_t \in S | X_{t-1}), \qquad (5.51)$$

for all measurable $S \subset \mathcal{S}$.

A Markov process, or Markov chain, is a stochastic process that satisfies the Markov property.

A Markov process is said to be homogeneous *if*

$$\mathbb{P}(X_{t+1} = x | X_t) = \mathbb{P}(X_1 = x | X_0);$$

i.e., if the probabilistic relation of one state to the next is constant over time. Moreover, the values $p_{ij}(s,t)$ defined by

$$p_{ij}(s,t) = \mathbb{P}(X_t = j | X_s = i), \qquad \text{for } t \geq s,$$

are referred to as transition probabilities, *and which are independent of time for homogeneous processes. For a finite state space, the set of transition probabilities $H(s,t) = \{p_{ij}(s,t) : s,t \in T\}$ constitutes a* state transition matrix *between times s and t. For a homogeneous process, we will typically refer to the state transition matrix as $H = H(0,1)$.*

Note 5.9. We limit our discussion to discrete-time Markov processes, as the general definition relies on generated σ-algebras, a complexity unnecessary for our present purpose. For completeness, however: A stochastic process is said to satisfy the Markov property if

$$\mathbb{P}(X_t \in S | \mathcal{F}_s) = \mathbb{P}(X_t \in S | X_s), \qquad \text{for } t \geq s,$$

where $S \in \sigma(\mathcal{S})$ is in the σ-algebra over $\mathcal{S}$ and $\mathcal{F}_s = \sigma(\{X_h | h \leq s\})$ is the σ-algebra generated by the history of the process X up until time s. In other words, for $t \geq s$, the conditional probability of X_t given the entire cumulative history of the process up until the point s is equivalent to the conditional probability of X_t given the single state X_s. In yet even other words: The probability distribution over a future state depends on its past only through its present.   ▷

Example 5.15. For fixed $p > 0$, a *random walk* is a discrete Markov process with state space $\mathcal{S} = \mathbb{Z}$ and transition probabilities

$$p_{ij} = \begin{cases} p & \text{if } j = i+1 \\ (1-p) & \text{if } j = i-1 \\ 0 & \text{otherwise} \end{cases} .$$

That is, for each point in time, the state increments by one with probability p and decrements by one with probability $(1-p)$.

```
1  n = 100
2  x = np.zeros(n)
3  x0 = 0
4  p = 0.5
5  for i in range(n):
6      if i == 0:
7          continue
8      x[i] = x[i-1] + 2 * int(np.random.random() < p) - 1
```

Code Block 5.15: Random walk.

A random walk can easily be simulated, as shown in Code Block 5.15. The results of four random walks; i.e., four realizations or sample paths of this process; are shown in Figure 5.9.

A related problem known as *the gambler's ruin* is stated as follows: Let X_t be a random walk with $X_0 = s > 0$. Determine the probability that $X_t = 0$ for some $t > 0$. We leave this as an exercise for the reader.   ▷

Most of the time, however, we will consider discrete-time Markov processes with a finite (discrete) state space. In such cases, we can formulate the state transition matrices (since the enumeration of states is finite) and rely on the following two results.

Proposition 5.8. *Given a Markov process with a finite state space, the following* Chapman–Kolmogorov *equations hold*

$$H(s,t) = H(s,\tau)H(\tau,t), \tag{5.52}$$

for any $s < \tau < t$.

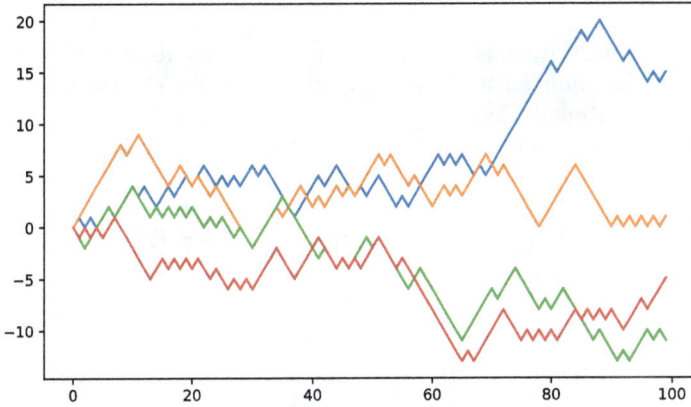

Fig. 5.9: Four random walks for Example 5.15.

Proof. This follows by considering the relation

$$\mathbb{P}(X_t = j | X_s = i) = \sum_{k \in \mathcal{S}} \mathbb{P}(X_t = j | X_\tau = k) \mathbb{P}(X_\tau = k | X_s = i),$$

which is equivalent to the result. $\qquad\qquad\qquad\qquad\qquad\qquad\qquad$ □

Proposition 5.9. *Let $\pi(t) = (\pi_i(t) : i \in \mathcal{S})$ be the PMF for the state of a homogeneous Markov process at time t, and let $H = H(0,1)$ be the single-step state transition matrix. Then*

$$\pi(t) = (H^T)^t \pi(0). \tag{5.53}$$

Proof. Let e_i be the standard basis vector for state $i \in \mathcal{S}$. If the process is in state i at time t, i.e., if $X_t = i$, then the PMF over X_{t+1} is given by the transition probabilities

$$H_{ij} = \mathbb{P}(X_{t+1} = j | X_{t=i}),$$

which comprise the ith row of the matrix H. The PMF $\pi(t+1)$ is therefore the ith column of H^T, or $\pi(t+1) = H^T \cdot e_i$.

Next, suppose that the state at time t is not known, but a probability distribution $\pi(t)$. Note that this vector is equivalent to

$$\pi(t) = \sum_{i \in \mathcal{S}} \pi_i(t) e_i.$$

It follows that the probability distribution at time $t+1$ is given by

$$\pi(t+1) = \sum_{i=\in\mathcal{S}} \pi_i(t)H^T e_i = H^T \pi(t);$$

i.e., by applying the transpose H^T to a probability distribution $\pi(t)$, we obtain the probability distribution over the states for the immediately following point in time. By successive application of this rule, we can transition a probability distribution over an initial state to the probability distribution t steps later:

$$\pi(t) = H^T \pi(t-1) = (H^T)^2 \pi(t-2) = \cdots = (H^T)^t \pi(0).$$

The result follows.   □

Counting and Arrival Processes

We have previously seen another example of a stochastic process, and that is the Poisson process. It turns out that the Poisson process is itself an example of an entire class of stochastic processes known as counting processes.

Definition 5.13. A counting process *is a stochastic process* $\{N(t)\}_{t\in\mathcal{T}}$ *with nonnegative-integer state space* $\mathcal{S} = \mathbb{N}$, *such that* $N(0) = 0$ *and such that, for any realization of the process, the sample function* N_t *is right-continuous and nondecreasing*[10].

   A Poisson process is a special type of counting process that fits under the broader umbrella of *renewal processes*.

Definition 5.14. *Given a distribution* F *over* $\mathbb{R}_+ = (0, \infty)$[11], *known as the* interarrival distribution, *an arrival process is a discrete-time Markov process* $\{T_n\}_{n\in\mathbb{N}}$ *with* $T_0 = 0$, *such that the increments* $T_{n+1} - T_n \sim F$ *are* IID *over* F, *for* $n \in \mathbb{N}$. *The value* T_n, *for* $n \geq 1$, *is referred to as the* nth *arrival time.*

Note 5.10. The initial term $T_0 = 0$ is defined purely for convenience, so that the initial increment $T_1 - T_0 = T_1$ is well defined. Otherwise, we can disregard T_0.
   Since the interarrival distribution is over $\mathbb{R}_+ = (0, \infty)$, it follows that the arrival times constitute a strictly increasing function of their index. ▷

Definition 5.15. *A continuous-time counting process* $\{N(t)\}_{t\in\mathbb{R}_*}$ *is a renewal process if there exists an arrival process* $\{T_n\}_{n\in\mathbb{N}}$ *such that*

$$N(t) = \sum_{i=1}^{\infty} \mathbb{I}[T_i \leq t]. \tag{5.54}$$

---

[10] Right-continuous: $\lim_{t\to t_0^+} N_t = N_{t_0}$; nondecreasing: $N_{t+s} \geq N_t$, for any $s > 0$.
[11] I.e., F is the CDF of a positive random variable, so that $F(0) = 0$.

Definition 5.16. *A* Poisson process *is a* renewal process with an exponentially distributed interarrival distribution.

Thus, a renewal process is a generalization of a Poisson process where the interarrival distribution is not exponentially distributed.

5.4.3 Aggregated Processes

We next turn to the case of a stochastic process that plays out independently over a collection of units, thereby generating a sample of realizations, which are then aggregated to form a more holistic picture.

Definition

The idea of an aggregate process is that it is just the sum of a sample of realizations from an underlying base process. Each sample outcome from the aggregation is therefore a set of sample outcomes from the base process. Therefore, a brief word is in order about the Cartesian product of a probability space.

Let $(\Omega, \mathcal{E}, \mathbb{P})$ be a probability space. Consider the Cartesian product of w copies of the sample space:

$$\Omega^w = \underbrace{\Omega \times \cdots \times \Omega}_{w \text{ times}}.$$

We can define the *product σ-algebra* as the σ-algebra generated[12] from the Cartesian product $\mathcal{E}^w$:

$$\sigma(\mathcal{E}^w) = \sigma\left(\{(E_1, \ldots, E_w) : E_1, \ldots, E_w \in \mathcal{E}\}\right).$$

Similarly, we can define a *product probability measure* $\mathbb{P}^w$ defined over $\sigma(\mathcal{E}^2)$. For the detailed definition of product measure, see any text on real analysis, such as DiBenedetto [2002] or Folland [1999]; or, for a definition in the context of probability theory, see Capinksi and Kopp [2005]. But essentially the product measure is a generalization of

$$\mathbb{P}^w((E_1, \ldots, E_w)) = \prod_{i=1}^{w} \mathbb{P}(E_i),$$

for $E_1, \ldots, E_w \in \mathcal{E}$, to the full σ-algebra $\sigma(\mathcal{E}^w)$.

---

[12] Let X be a set, and $\mathcal{S} \subset \mathcal{P}(X)$ a collection of subsets. Then the σ-algebra generated from $\mathcal{S}$, denoted $\sigma(\mathcal{S})$, is the smallest σ-algebra that contains the subsets $\mathcal{S}$.

Definition 5.17. *Given an underlying* base process $\{B_t\}_{t \in T}$ *over* $(\Omega, \mathcal{E}, \mathbb{P})$, *an* aggregate process with weight w *is defined as the stochastic process* $\{X_t\}_{t \in T}$ *over the product probability measure* $(\Omega^w, \mathcal{E}^w, \mathbb{P}^w)$, *such that*

$$X_t(\omega) = \sum_{i=1}^{w} B_t(\omega_i),$$

for any realization $\omega = (\omega_1, \ldots, \omega_w) \in \Omega^w$.

In other words: an aggregate process is simply the sum of a sample of realizations of an underlying base process. The sample mean of the realizations is therefore simply X_t/w.

Bernoulli Processes

Just how a binomial random variable is the sum of Bernoulli random variables, a binomial stochastic process is the sum of Bernoulli stochastic processes. It is therefore a perfect example of an aggregate process.

Definition 5.18. *A* Bernoulli process *is the stochastic process* $\{X_t\}_{t \in \mathbb{N}}$ *defined by* $X_t \sim \text{Bern}(p)$, *for some* $p \in (0, 1)$.
 A binomial process with weight w *is the stochastic process* $\{X_t\}_{t \in \mathbb{N}}$ *defined by* $X_t \sim \text{Binom}(w, p)$, *for some* $p \in (0, 1)$.
 If the parameter p is replaced by a function $p : \mathbb{N} \to (0, 1)$, *called the* generator *of the process, either process is said to be* nonhomogeneous. *Futhermore, if the series* $\sum_{t=0}^{\infty} p_t$ *converges, the process is said to be* finite.

Clearly, the binomial process is simply an aggregate process, for which the underlying base process is a Bernoulli process.

We find the finite nonhomogeneous binomial process to be quite interesting. For example, consider a mobile app in which user engagement fluctuates over time. Each user's individual behavioral pattern constitutes a Bernoulli process. However, by aggregating the patterns over a collection of users, we have a binomial process. If enough users are aggregated together, we may infer something about the underlying generator p_t for the process; i.e., what is the distribution of engagement over time for the process.

If data for a binomial process do not exist beyond some $t > \tau$, we say the process is *truncated* at $t = \tau$. The process is truncated, and not censored, as we do not know the total count of events that would ultimately transpire, if the process were allowed to continue.

Survival Processes

Definition 5.19. *A* unit survival process *is a stochastic process* $\{A_t\}_{t \in T}$ *with binary state space* $\mathcal{S} = \mathbb{B} = \{0, 1\}$ *over the probability space* $(\mathbb{R}_+, \mathcal{E}, \mathbb{P})$, *defined for sample outcome* $T \in \Omega = \mathbb{R}_+$:

$$A_t = \mathbb{I}[T > t].$$

The sample outcome T is known as the survival time. *The probability distribution F_T over $\Omega = \mathbb{R}_+$ is called the* distribution of lifetimes, *and is said to* generate the process.

A survival process with weight w *is an aggregation of w unit survival processes.*

The aggregate survival process is simply the empirical survival function of the distribution F_T that generates the process.

If data for a survival process do not exist for $t > \tau$, we say the process is *censored* at $t = \tau$. The process is censored, and not truncated, as we know a priori the ultimate tally of survival events.

5.4.4 Online Processes

Many online products (apps, websites, etc.) are constantly acquiring new users who interact with them on a regular basis. A *user cohort* is a group of users who were acquired on the same day. It is often of interest to study certain random processes over users' lifecycles (e.g., engagement, churn, monetization, etc.). It is therefore useful to think of such online processes as a function of two indices: cohort membership and cohort time. *Cohort membership* is an index counting the various acquisition cohorts chronologically, and *cohort time* is a measure of each user cohort's time since acquisition. The idea is that each user cohort undergoes the same mathematical process once they are acquired; however, the acquisition of user cohorts happens in a staggered manner over time. Formally, we have the following.

Definition 5.20. *An* online process *is a stochastic process $\{X_{ij}\}_{(i,j)\in\mathbb{N}^2}$ with a two-dimensional discrete indexing set $\mathbb{N}^2$ and a set of weights w, such that*

1. calendar time *represents the total real calendar time since the first user cohort was acquired and the overall online process commenced,*
2. *index i represents the calendar time at which cohort i was acquired;*
3. *index j represents the* cohort time, *or the time since acquisition, within each cohort;*
4. *the isochrones $t = i + j$ represent the calendar time at which each random variable X_{ij} was observed;*
5. *the process age τ represents the current (cumulative) calendar time for the overall process;*
6. *the cohort age $\tau - i$ represents the current cohort time for cohort i;*
7. *the cohort weights w constitute a set of nonnegative weights assigned to each cohort, where the weight w_i typically represents the number of users belonging to cohort i;*

8. *the random variables X_{ij} are censored or truncated, depending on application, for $t > \tau$.*

Due to future-truncation across the isochrone $t > \tau$, the random variables X_{ij} are naturally expressed within an upper-left triangular matrix known as the *process matrix*. For example, the process matrix for an online process that is currently six days old (i.e., the process age is $\tau = 6$), is represented by

$$X = \begin{bmatrix} X_{00} & X_{01} & X_{02} & X_{03} & X_{04} & X_{05} \\ X_{10} & X_{11} & X_{12} & X_{13} & X_{14} & 0 \\ X_{20} & X_{21} & X_{22} & X_{23} & 0 & 0 \\ X_{30} & X_{31} & X_{32} & 0 & 0 & 0 \\ X_{40} & X_{41} & 0 & 0 & 0 & 0 \\ X_{50} & 0 & 0 & 0 & 0 & 0 \end{bmatrix}.$$

We can visualize the calendar time $t = i + j$, which represents the real, calendar times since the first cohort was acquired, by the series of diagonals represented below:

$$\begin{bmatrix} 0 & 1 & 2 & 3 & 4 & 5 \\ 1 & 2 & 3 & 4 & 5 & * \\ 2 & 3 & 4 & 5 & * & * \\ 3 & 4 & 5 & * & * & * \\ 4 & 5 & * & * & * & * \\ 5 & * & * & * & * & * \end{bmatrix}.$$

Notice that cohort 0 was acquired at calendar time 0; cohort 1 was acquired at calendar time 1, etc. The isochrones $t = i+j$ represent the actual calendar day on which the data were collected. The most recent day is the diagonal $t = \tau$. Data below this diagonal $(t > \tau)$ are missing (censored or truncated), as they represent days yet to come.

The more recently a cohort was acquired, the fewer data we have for the given cohort. It is useful to analyze online processes to determine whether the process is changing as a function of cohort age (e.g., recent cohorts are less engaged than cohorts from a year ago) and calendar-time isochrones (e.g., weekend or holiday effects, server outages, sale promotions, etc.).

Notwithstanding cohort and calendar effects, we typically like to model the process happening within each cohort. We therefore like to endow this array of random variables with some kind of additional, overarching structure that depends on the cohort time.

Definition 5.21. *Let $\mathcal{B}(\theta) = \{B_j | \theta\}_{j \in \mathbb{N}}$ represent a stochastic process that depends on a parameter θ. This process is the* base process *for an online process $\{X_{ij}\}_{(i,j) \in \mathbb{N}^2}$ with weights $\{w_i\}_{i \in \mathbb{N}}$ if, for each $i \in \mathbb{N}$, the process $\{X_{ij}\}_{j \in \mathbb{N}}$ constitutes an aggregate process over the base process $\mathcal{B}(\theta)$ with weight w_i.*

If the parameter θ depends on the cohort i, the online process is said to have a cohort effect. *If the parameter θ depends on the calendar time $t =*

$i + j$, *the online process is said to have a* calendar effect. *An online process with no cohort or calendar effect is said to be* pure.

Note 5.11. If a calendar effect depends on $t = i + j$ only through t mod 7, it is called a *day-of-week effect*. If a day-of-week effect corresponds to weekends, it is referred to as a *weekend effect*.   ▷

Definition 5.22. *An* online Bernoulli process *is an online process whose base process is the Bernoulli process (Definition 5.18).*

An online survival process *is an online process whose base process is the unit survival process (Definition 5.19).*

Example 5.16. Simulate the following online Bernoulli process, representing the counts of a certain engagement event within a mobile app:

1. cohort weights are normally distributed with mean 100 and standard deviation 20;
2. the random process $X_{ij} \sim \text{Binom}(w_i, \alpha f_T(j))$, where f_T is the probability mass function for a shifted geometric distribution with mean 20, and alpha is a scaling parameter;
3. the scaling parameter is typically 10, but after cohort 49, a sudden shift in user acquisitions causes this parameter to be reduced by 50%.
4. there was a bug in the app on days 30–15 resulting in a 60% drop in engagement,
5. there is a 50% lift in engagement on weekends; the first weekend occurred for calendar times t=5, 6.

All of these ingredients can easily be combined in a single simulation, as shown in Code Block 5.16. The result is plotted in Figure 5.10.

The horizontal axis in Figure 5.10 represents the cohort time j, whereas the vertical axis represents the user cohort i. Note that the fraction of engagements is decreasing from left to right, as users age within their individual cohorts. This is the structured binomial distribution at work. However, we can also easily spot some additional features: there is a dark blue diagonal strip starting with cohorts 30–35 and directed diagonally (north-east). These are the isochrones corresponding to calendar times 30–35, when there was a bug in the mobile app, causing diminished engagement across all cohorts for those particular calendar days. We further see a periodic sequence of bright diagonals, corresponding to the weekend effect. Finally, we notice that there is a darker shading for all cohorts beginning with cohort 50, indicating a downward shift in the type of user being acquired (and, perhaps, a cause for alarm).   ▷

```
1   tau = 60
2   w = np.random.normal(loc=100, scale=20, size=tau).astype(int)
3   X = np.zeros((tau,tau))
4   alpha_0 = 10
5
6   for i in range(tau):
7       if i == 50: # cohort effect
8           alpha_0 *= 0.5
9       for j in range(tau-i):
10          t = i + j
11          alpha = alpha_0
12          if t % 7 in [5, 6]: # weekend effect
13              alpha *= 1.5
14          if t in range(30, 36): # calendar effect
15              alpha *= 0.4
16          p = alpha * scipy.stats.geom.pmf(j, 0.05, loc=-1)
17          X[i, j] = np.random.binomial(w[i], p)
18
19  plt.figure(figsize=(8, 9/2))
20  Y = (X / w.reshape((tau, 1))) # Divide each row by its weight
21  plt.pcolormesh(Y)
22  plt.gca().invert_yaxis()
23  pj_hat = X.sum(axis=0) / np.flip(np.cumsum(w))
24  se = np.sqrt(pj_hat * (1 - pj_hat) / np.flip(np.cumsum(w)))
```

Code Block 5.16: Simulation of an Online Process.

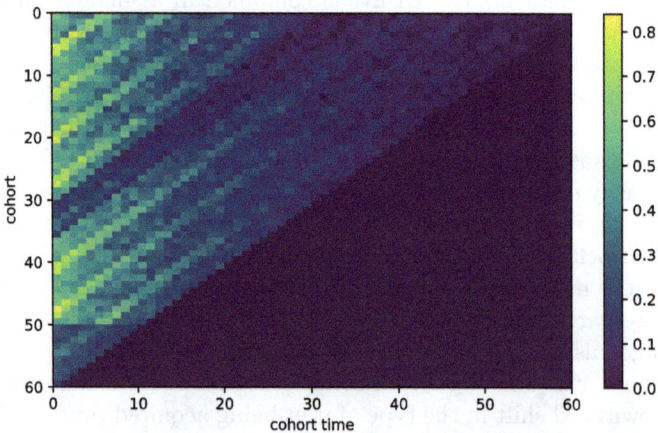

Fig. 5.10: Normalized process matrix for Example 5.16.

Empirical Estimation of Online Bernoulli Process

Naturally, we are interested in approximating the distribution of events over cohort time, as this distribution tells us how each cohort individually is expected to age. The following theorem gives us an empirical estimate.

Theorem 5.8. *Let X_{ij} be a pure online Bernoulli process of age τ. Then an unbiased empirical estimate for the truncated generator function p_j : $\{0, \ldots, \tau\} \rightarrow (0, 1)$ is given as follows:*

$$\hat{p}_j = \frac{1}{\omega_{\tau-j}} \sum_{i=0}^{\tau-j} X_{ij}, \qquad (5.55)$$

where we define the cumulative weights ω_k as

$$\omega_k = \sum_{i=0}^{k} w_i, \qquad (5.56)$$

for $k = 0, \ldots, \tau$. Moreover, the variance in the estimate is given by

$$\mathbb{V}\left(\hat{p}_j\right) = \frac{p_j(1 - p_j)}{\omega_{\tau-j}}. \qquad (5.57)$$

Proof. From the definitions, the random variables

$$X_{ij} \sim \mathrm{Binom}(w_i, p_j),$$

which has expected value $\mathbb{E}[X_{ij}] = w_i p_j$. This processed is truncated for $t = i + j > \tau$, so that, for a fixed cohort time $j = 0, \ldots, \tau$, we only have data for cohorts $i = 0, \ldots, \tau - j$. Note that the probability depends only on the calendar time, so that

$$\sum_{i=0}^{\tau-j} X_{ij} \sim \mathrm{Binom}(\omega_{\tau-j}, p_j),$$

for the cumulative weights ω_k defined in Equation (5.56). The expected value of this sum is therefore given by

$$\mathbb{E}\left[\sum_{i=0}^{\tau-j} X_{ij}\right] = \omega_{\tau-j} p_j.$$

The estimate provided in Equation (5.55) therefore constitutes an unbiased estimate for p_j. Moreover, the variance of a binomial random variable is given by

$$\mathbb{V}\left[\sum_{i=0}^{\tau-j} X_{ij}\right] = \omega_{\tau-j} p_j(1 - p_j),$$

which yields Equation (5.57). $\qquad \square$

For Example 5.16, this empirical estimate and standard error are computed in lines 23–24 of Code Block 5.16. The results are shown in Figure 5.11, plotted concurrently with the geometric base distribution used for the simulation. (Note that the estimate works quite well, despite there being some cohort and calendar effects in the data.)

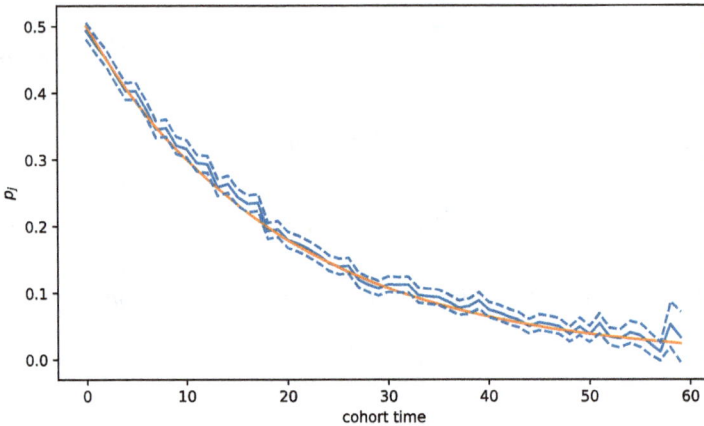

Fig. 5.11: Empirical estimate for $p_j = \alpha f_T(j)$ in Example 5.16, with 95% confidence bounds $(2\hat{\text{se}}(\hat{p}))$.

For comparison, Figure 5.12 was generated using the same simulation, except with the cohort and calendar effects removed.

Empirical Estimation of Online Survival Process

Next, we will consider online survival processes. This follows directly from the Kaplan-Meier estimate of Theorem 5.5. For an online survival process, we recognize that $X_{i0} = w_i$, for each cohort i; i.e., each cohort starts its process with full survival. For $j > 0$, we therefore have

$$E_j = \sum_{i=0}^{\tau-j} \left[X_{i(j-1)} - X_{ij} \right]$$

$$Y_j = \sum_{i=0}^{\tau-j} X_{i(j-1)}.$$

Of course, the number of units censored between $(j-1)$ and j is simply $X_{(\tau-j+1)(j-1)}$. Therefore, as a corollary to Theorem 5.5, we have the following:

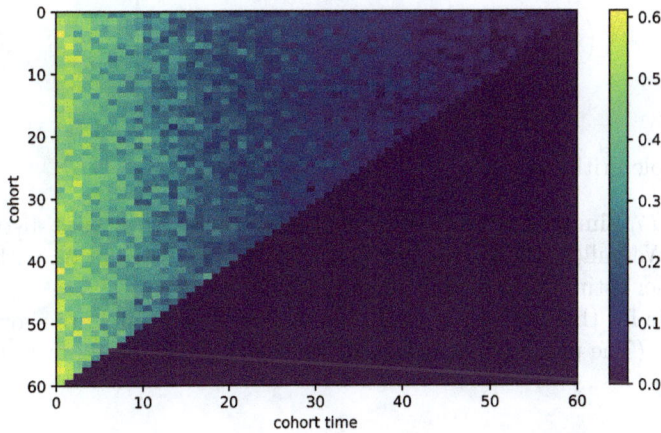

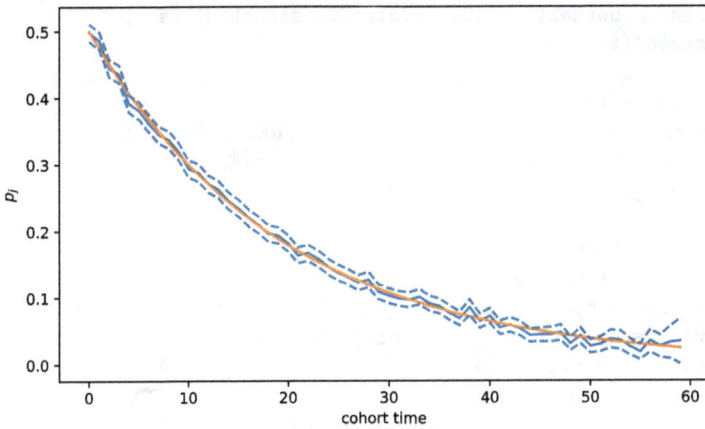

Fig. 5.12: Simulation with no calendar or cohort effects.

Theorem 5.9. *The Kaplan-Meier estimate for the survival function for an online survival process is given by*

$$\hat{S}_j = \prod_{k=1}^{j} \left(\frac{\displaystyle\sum_{i=0}^{\tau-k} X_{ik}}{\displaystyle\sum_{i=0}^{\tau-k} X_{i(k-1)}} \right). \qquad (5.58)$$

Moreover, Greenwood's formula reduces to

$$\mathbb{V}\left(\hat{S}_j\right) = \hat{S}_j^2 \sum_{k=1}^{j} \left(\frac{1}{\sum_{i=0}^{\tau-k} X_{ik}} - \frac{1}{\sum_{i=0}^{\tau-k} X_{i(k-1)}} \right). \tag{5.59}$$

Proof. Simple arithmetic.   □

Example 5.17. Simulate a pure online survival process with base distribution $F_T = \text{Weibull}(2, 40)$ for $\tau = 60$. Plot the Weibull distribution with the Kaplan-Meier estimate from the simulated data.

The code for this is given in Code Block 5.17. The results are shown in Figure 5.13. (The process matrix was plotted using lines 19–22 from Code Block 5.16.)

```python
alpha, beta, tau = 2, 40, 60
w = np.random.normal(loc=100, scale=20, size=tau).astype(int)
X = np.zeros((tau,tau))

for i in range(tau):
    samples = beta * np.random.weibull(alpha, size=w[i])
    for j in range(tau-i):
        X[i, j] = np.sum( (samples > j) )

# Kaplan Meier estimate
col_sum = X.sum(axis=0)
diag_elements = np.flipud(X).diagonal()
col_at_risk = np.roll(col_sum - diag_elements, 1)
col_at_risk[0] = col_sum[0]

S_hat = np.cumprod(col_sum / col_at_risk)
S = scipy.stats.weibull_min.sf(np.arange(tau), alpha, scale=beta)
se = S_hat * np.sqrt( np.cumsum( 1 / col_sum - 1 / col_at_risk ) )
```

Code Block 5.17: Random walk.

As expected, the Kaplan-Meier estimate provides an accurate empirical estimate for the survival function of our online survival process.   ▷

Problems

5.1. Let $X_1, \ldots, X_n \sim N(\mu, \sigma^2)$. Show that the MLE estimates for μ and σ^2 are given by

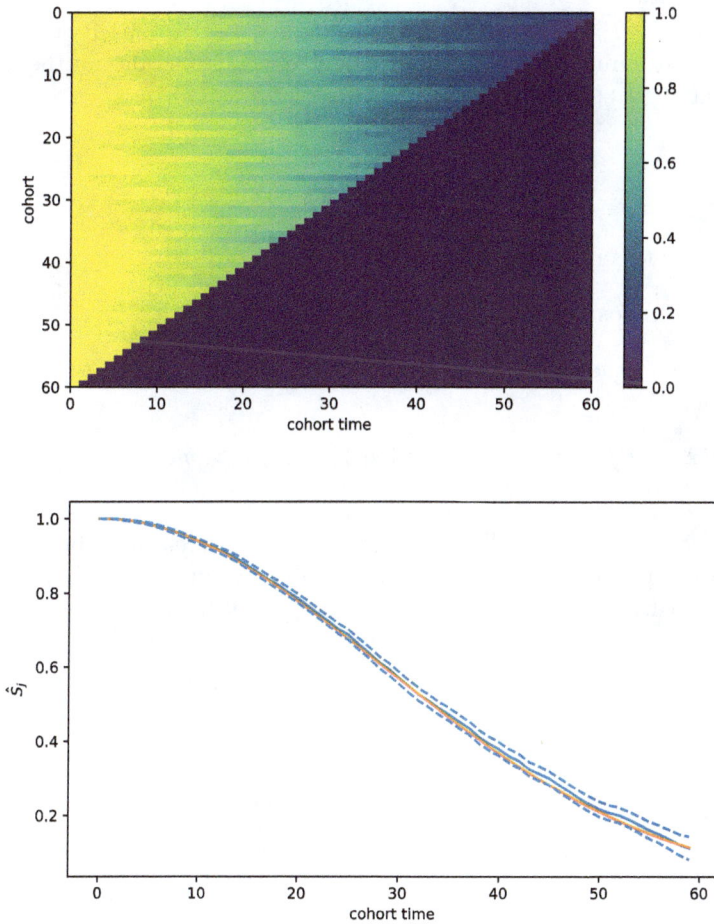

Fig. 5.13: Process matrix and Kaplan-Meier estimate for online survival process of Example 5.17.

$$\hat{\mu}_n = \frac{1}{n} \sum_{i=1}^{n} x_i,$$

$$\hat{\sigma}_n^2 = \frac{1}{n} \sum_{i=1}^{n} (x_i - \hat{\mu})^2.$$

5.2. Prove the assertion made in Note 5.4.

5.3. Show the missing steps between Equations (5.29) and (5.30).

5.4. Show that the definition of the hazard function, given by Equation (5.33), is equivalent to

$$h(x) = -d \log(S(x))/dx.$$

5.5. Let $T_1, \ldots, T_n \sim \text{Exp}(\beta)$ be IID exponential data that are left-truncated at $T = \tau$. Determine the likelihood and log-likelihood functions for the data. Show that the MLE is given by

$$\hat{\beta} = \left(\frac{1}{n} \sum_{i=0}^{n} t_i \right) - \tau.$$

Explain why this result makes sense intuitively.

5.6. Prove Proposition 5.7.

5.7. Determine the value of α for the Pareto distribution that results in the 80-20 rule.

5.8. Let $X \sim \text{Pareto}(1, \alpha)$, and define $Y = X | X > m$, for $m > 1$. Show that $Y \sim \text{Pareto}(m, \alpha)$.

5.9. Suppose that $X_1, \ldots, X_n \sim \text{Pareto}(1, \alpha)$ is IID Pareto-distributed data that represents the amounts of insurance claim payouts for a policy with a $500 deductible and $20,000 cap. Determine the likelihood function for the data.

Part II

Statistical Modeling

6

Modeling Runway

In this chapter, we lay some of the groundwork for the models we will discuss throughout the rest of this book. We begin with a discussion of basic nomenclature, and then discuss common approaches for model validation. We then discuss aspects of preprocessing that are common to most data science applications. We conclude the chapter with a discussion on engineering aspects of data science; in particular, we discuss how to take an object-oriented approach to data-science projects.

6.1 Modeling and Validation

In this section we lay out some key definitions of modeling and machine learning. We then describe various metrics and methods for model validation that will be used extensively throughout the remainder of this text.

6.1.1 Models and Model Learning

We begin by laying some terminology for modeling and machine learning.

Definition 6.1. *A* model *is a mathematical representation of a system or a process.*

A predictive model *is a model for an output (or* target*) variable as it depends on a prescribed set of inputs (or* features*).*

A statistical model *is a predictive model whose output is a probability distribution for the target variable.*

Statistical models are often expressed in the form

$$Y = f(X) + \epsilon,$$

where f is a predictive model and ϵ is an *error term*, with $\mathbb{E}[\epsilon] = 0$, so that $\mathbb{E}[Y|X] = f(x)$. Some additional assumptions regarding the noise ϵ

are typically included in the model, such as independence and assumptions about its variance or distribution. Sometimes, for brevity, we will refer to a predictive model by its deterministic component $f : X \rightarrow \hat{Y}$, where $\hat{Y} = \mathbb{E}[Y|X]$.

Definition 6.2. *A* machine learning algorithm *is any computer algorithm that is able to automatically improve its performance of a task through experience.*

Supervised learning *refers to any machine learning algorithm that learns a predictive model $f : X \rightarrow \hat{Y}$ from a set of labeled data $\mathcal{D} = \{(x_i, y_i)\}_{i=1}^{n}$. The input X is often referred to as a* feature vector, *and its components as a set of* features. *The output Y is often referred to as the* target *variable. Each datum (x_i, y_i) is referred to as an* instance.

Unsupervised learning *is any machine learning algorithm that learns patterns in unlabeled data.*

Reinforcement learning *is any machine learning algorithm in which an* agent *takes a sequence of* actions *in an* environment *in order to maximize a cumulative* reward.

In Part II of this text, we will predominantly focus on predictive models. In Part III, we will explore machine learning algorithms more closely. As a quick illustration of the distinction laid out in Definition 6.2, consider the following.

Example 6.1. A computer algorithm is trained with a set of images as inputs, each input having a label "cat" or "dog." It is tested on its ability to recognize whether or not a new image is a cat or dog. This is an example of supervised learning, as the labels ("cat" and "dog") were provided with the training set.

A different algorithm is provided the same set of input images, except without being explicitly told what is a cat and what is a dog. It is asked to seek patterns in the data. Here, the algorithm might come up with patterns that are unexpected. For example, the algorithm might classify images of the side view versus images of a front view; or it might classify partially hidden versus fully visible animals. Or it might classify head shots versus body shots. Or it might classify based on whether or not the pet is wearing a collar. Or it might come up with its own concept, something that distinguishes the images that is present in the data but would not normally be recognized by the human eye. This is an example of unsupervised learning, as the algorithm is left to its own devices to infer patterns from the data.

Finally, our third algorithm has learned how to play the game *Go*. Here, the environment is the board. The actions is the set of permissible (legal) Go moves. And the reward is issued at the end of each game as $+1$ win or 0 for loss. The computer can be trained with actual Go games, and it will continue to learn based on its experience. That is, as it plays, it becomes better at playing. In fact, two computers can be set up to play each other, with entire games occurring at lightning speed, while the algorithm learns which strategies work best.   ▷

Predictive models themselves can be grouped into two different types, based on the type of target variable they are predicting.

Definition 6.3. *A* classification algorithm *is any supervised machine learning algorithm that learns a predictive model with a categorical output variable. The enumeration of possible outputs are referred to as* labels.
A regression algorithm *is any supervised machine learning algorithm that learns a predictive model with a continuous real output variable.*

In Example 6.1, the supervised algorithm is a classification algorithm, as it learned whether each picture represented a dog or a cat. An example of a regression algorithm is an algorithm that predicts home price based on a set of features (e.g., square footage, city, number of beds / baths, school district, etc.).

In general, the feature vector of a model will consist of a mix of continuous, categorical, and ordinal features. (An ordinal feature is similar to a categorical feature in the sense that it is finite and discrete, but it differs as it constitutes an ordered collection; e.g., *small*, *medium*, and *large*.) We will explore various requirements for data preprocessing for each of these feature types in Section 6.2.

Definition 6.4. *Given a predictive model $Y = f(X) + \epsilon$, with error term $\mathbb{E}[\epsilon] = 0$ and $\mathbb{V}(\epsilon) = \sigma^2$, an instance (x, y) is said to have a weight (or prior weight) $w > 0$ if the variance of its error ϵ is reduced by a factor w; i.e., if $y_i \sim f_Y$, then $\mathbb{E}[Y] = f(x_i)$ and $\mathbb{V}(Y - f(x_i)) = \sigma^2/w$.*
A weighted data set $\mathcal{D} = \{(x_i, y_i, w_i)\}_{i=1}^n$ *is a data set with a prescribed (known) set of prior weights, so that if $y_i \sim f_{Y_i}$, then $\mathbb{E}[Y_i] = f(x_i)$ and $\mathbb{V}(Y_i - f(x_i)) = \sigma^2/w_i$. The* total weight ω *is the sum of the individual weights; i.e., $\omega = \sum_{i=1}^n w_i$.*

An instance (x, y) with positive integer weight $w \in \mathbb{Z}^+$ is equivalent to a set of w identical instances $\{(x, y)\}_{i=1}^w$, so that weights are a convenient encapsulation for multiplicity.

We conclude this section with a more detailed definition on what constitutes a machine learning model.

Definition 6.5. *A (supervised)* machine learning model *over a* feature space $\mathbb{F}$ *into a* target space $\mathbb{T}$ *is a mapping*

$$f : \mathbb{D} \to \mathscr{F}(\mathbb{F}, \mathbb{T}), \tag{6.1}$$

where $\mathbb{D} = \mathscr{P}_{\text{fin}}(\mathbb{F} \times \mathbb{T})$ is the set of all possible finite subsets (or samples) of the data, and where $\mathscr{F}(\mathbb{F}, \mathbb{T})$ is the set of functions from $\mathbb{F}$ into $\mathbb{T}$.

Given a sample of data $\mathcal{D} \in \mathbb{D}$, known as training data, *the function $f_{\mathcal{D}} : \mathbb{F} \to \mathbb{T}$ (known as the* trained model*) is a predictive model for the target variable as it depends on the feature vector.*

If the target space is a finite set $\mathbb{T} = \mathcal{C}$ of categories, we say that the model is a classification *model. Otherwise, if the target space is a subset of the reals, $\mathbb{T} \subset \mathbb{R}$, we say that the model is a* regression *model.*

An unsupervised machine learning model *is similar, except that the set of data $\mathbb{D} = \mathscr{P}_{\text{fin}}(\mathbb{F})$ consists only of the feature data.*

Another aspect of learning algorithms is that the mapping shown in Equation (6.1) need not be performed in a single step; i.e., the model can continuously *improve* as data are added. In this way, we can view a machine learning algorithm as a mapping

$$f : \mathbb{D} \times \mathscr{F}(\mathbb{F}, \mathbb{T}) \to \mathscr{F}(\mathbb{F}, \mathbb{T}),$$

which symbolizes the important aspect that the model can improve (or *learn*) with time.

6.1.2 Validation Metrics for Regression

In this section, we will consider a predictive model of the form $\mathbb{E}[Y|X] = f(X)$ that was trained using a regression algorithm. We will consider various metrics that can be used to measure the performance of the model.

Squared Errors

Consider a data set $\mathcal{D} = \{(x_i, y_i)\}$ and a model f with predictive values $\hat{y}_i = f(x_i)$. Note that the data set $\mathcal{D}$ is not necessarily the same data that was used to generate the model. We will discuss this in more detail in Section 6.1.5. We consider the following sum-of-squares statistics.

Definition 6.6. *Given a weighted data set $\mathcal{D} = \{(x_i, y_i, w_i)\}_{i=1}^n$ and model predictions $\hat{y}_i = f(x_i)$, we define the* sum of squared errors (SSE), *the re- gression sum of squares (SSR), and the* total sum of squares (SST) *as*

$$\text{SSE} = \sum_{i=1}^{n} w_i \left(y_i - \hat{y}_i\right)^2, \tag{6.2}$$

$$\text{SSR} = \sum_{i=1}^{n} w_i \left(\hat{y}_i - \overline{y}\right)^2, \tag{6.3}$$

$$\text{SST} = \sum_{i=1}^{n} w_i \left(y_i - \overline{y}\right)^2, \tag{6.4}$$

respectively.

Note 6.1. The difference $y_i - \hat{y}_i$ between the true and predicted value is referred to as a *residual*. Therefore, many authors, especially in machine learning, refer to the sum of squared errors as the *residual sum of squares*, or RSS. We will use the phrases *sum of squared errors* and *residual sum of squares* interchangeably, however we favor the notation SSE as it is consistent with other sum-of-squares notation and the *mean-squared error* MSE, defined as MSE $=$ SSE$/n$ or, for a weighted data set, as MSE $=$ SSE$/w$.   ▷

One might be tempted to assume that the residual sum of squares and the regression sum of squares sum to the total sum of squares. This assumption, however, is not true in general. In fact, it is only true if the cross terms

$$\sum_{i=1}^{n} (y_i - \hat{y}_i)(\hat{y}_i - \overline{y})$$

sum to zero. Whether or not these terms vanish clearly depends on the model. For instance, consider the model that predicts $f(x) = 0$, for all x. Then the cross terms do not cancel, but instead sum to $-n\overline{y}^2$.

A common metric used in model evaluation is the coefficient of determination, which is often denoted simply and erroneously as R^2, which we define as follows.

Definition 6.7. *The* coefficient of determination (R^2) *is defined as*

$$R^2 = 1 - \frac{\text{SSE}}{\text{SST}}, \tag{6.5}$$

where SSE *and* SST *are defined as in Definition 6.6.*

This "R-squared" statistic really isn't the square of anything, other than its own square root, which need not be real. In other words, the R^2 statistic in certain cases can be negative. It cannot, however, be larger than 1. In fact, a value of $R^2 = 1$ would be a perfect model, as this is only achieved whenever $\hat{y}_i = y_i$, for all $i = 1, \ldots, n$. Hence we have the inequality

$$R^2 \leq 1.$$

In fact, the R^2 statistic *will* be negative whenever our predictions have a greater total error than had we simply predicted the average value $f(x) = \overline{y}$ for the entire data set. (See Exercise 6.1.) So a *positive* R^2 indicates we are doing better than no model at all (where, in absence of a model, we can simply use the average value as the value of all of our predictions).

The R^2 statistic is commonly interpreted as the *fraction of variance explained* by the model. To see this, note that the total sum of squares is essentially (a multiple of) the sample variance of the data. This represents the variance observed in the data set itself. The residual sum of squares, on the other hand, is how much variance is still *unexplained* by our model, as the residual sum of squares measures the variance between our predictions and the true values of the data.

The R^2 statistic is good in that the residual sum of squares is scaled by the total sample variance. Whereas interpretation of the MSE of a model completely relies on domain expertise, the R^2 statistic is a normalized metric that has a standard meaning.

A pitfall of the R^2 statistic, however, is that there is no universally adopted notion of which values of R^2 are acceptable. This is due, in part, to the fact that R^2 depends on the variance in the data. If we hold the mse fixed, the larger the variance in the data, the higher the value of R^2. In one context, an $R^2 = 0.20$ might be acceptable, but in another, an $R^2 = 0.90$ might not. For example, if we are modeling housing prices, which show a significant variability, a value of $R^2 = 0.90$ might translate to a RMSE in the tens of thousands of dollars, which would probably not be considered acceptable to the given application (example from Kuhn and Johnson [2013]).

The R^2 statistic is, however, useful in comparing different models. But even then there is an important caveat: the R^2 statistic will increase as one adds explanatory variables in the model. This leads to the erroneous conclusion that one should keep adding features to the model: everything except the kitchen sink. One important modification that addresses this is the following.

Definition 6.8. *The* adjusted coefficient of determination $\bar{R}^2$ *is defined as*

$$\bar{R}^2 = 1 - \frac{\text{SSE}/\text{df}_e}{\text{SST}/\text{df}_t},\tag{6.6}$$

where $df_t = n - 1$ is the number of degrees of freedom of the estimate of the population variance and $df_e = n - p$ is the number of degrees of freedom of the estimate of the residual variance, where p is the total number of independent modeling parameters of the model.

Unlike regular R^2, the adjusted $\bar{R}^2$ will increase with the inclusion of additional features only if the the model improvement is more than one would expect to see by chance.

An alternate assessment metric is given as follows.

Definition 6.9. *The* mean absolute error (MAE) *of a model is given by*

$$MAE = \frac{1}{n}\sum_{i=1}^{n}|\hat{y}_i - y_i|.\tag{6.7}$$

Similarly, the mean absolute percentage error (MAPE) *of a model is given by*

$$MAPE = \frac{1}{n}\sum_{i=1}^{n}\left|\frac{\hat{y}_i - y_i}{y_i}\right|.\tag{6.8}$$

Essentially, MAPE is a measure of how close the predictions are to the true values. A MAPE of 10% indicates that the predictions are typically about ±10% of the true values. This metric also has various flaws: it cannot be used if there are any zeros in the data, and there is an asymmetry between under-predictions (which cannot exceed 100%) and over-predictions (which are unlimited).

Bias–Variance Tradeoff

We will discuss model selection in Section 6.1.5. Presently, however, we shall content ourselves with laying out some of the theoretical groundwork. We begin with two definitions.

Definition 6.10. *Given a normal predictive model $Y = f(X) + \epsilon$, for $\epsilon \sim N(0, \sigma^2)$, where $f(X)$ is an unknown function of the input X. Let $\hat{f}(X) = \hat{f}(X; \mathcal{D})$ be an estimate for the unknown function $f(X)$ that is learned on a set of data $\mathcal{D} = \{(X_i, Y_i)\}_{i=1}^n$. Then we define the bias of the model $\hat{f}$ at a point $X = x$ as*

$$\text{bias}\left[\hat{f}(x)\right] = \mathbb{E}\left[\hat{f}(x; \mathcal{D})\right] - f(x). \tag{6.9}$$

Similarly, we define the model variance at a point $X = x$ as

$$\mathbb{V}\left[\hat{f}(x)\right] = \mathbb{E}\left[\left(\hat{f}(x; \mathcal{D}) - \mathbb{E}[\hat{f}(x; \mathcal{D})]\right)^2\right]. \tag{6.10}$$

In Equations (6.9) and (6.10), all expectations are taken relative to the data $\mathcal{D} \sim f_{X,Y}^n$.

Definitions in hand, we are now ready to state our main result.

Theorem 6.1 (Bias–Variance tradeoff). *Let $Y = f(X) + \epsilon$, where $\epsilon \sim N(0, \sigma^2)$, and $\hat{f}(X) = \hat{f}(X; \mathcal{D})$ be as in Definition 6.10. Then the expected squared residual at a point $X = x$ is given by*

$$\mathbb{E}\left[\left(Y - \hat{f}(x; \mathcal{D})\right)^2\right] = \sigma^2 + \text{bias}\left[\hat{f}(x)\right]^2 + \mathbb{V}\left[\hat{f}(x)\right], \tag{6.11}$$

where the expectation on the left-hand side is taken over the joint distribution for Y and $\mathcal{D}$.

Note 6.2. The term σ^2 is often referred to as the *irreducible error*, as it is due to the white noise $\epsilon \sim N(0, \sigma^2)$ in the random variable Y; i.e., it is inherent uncertainty that cannot be remedied by modeling.   ▷

Proof. We begin by expanding the residual for fixed $X = x$ as

$$Y - \hat{f}(x; \mathcal{D}) = (Y - f(x)) + \left(f(x) - \mathbb{E}\left[\hat{f}(x; \mathcal{D}) \right] \right) + \left(\mathbb{E}\left[\hat{f}(x; \mathcal{D}) \right] - \hat{f}(x; \mathcal{D}) \right).$$

Now, the first term depends on the random variable Y, whereas the second and third terms depend on the random variable $\mathcal{D}$. Therefore, the first term is independent with the second and third terms, and we conclude that

$$\mathbb{E}\left[(Y - f(x)) \left(f(x) - \mathbb{E}\left[\hat{f}(x; \mathcal{D}) \right] \right) \right] = 0$$

and

$$\mathbb{E}\left[(Y - f(x)) \left(\mathbb{E}\left[\hat{f}(x; \mathcal{D}) \right] - \hat{f}(x; \mathcal{D}) \right) \right] = 0.$$

The second term is actually independent of $\mathcal{D}$ itself, so a similar result holds for the penultimate and final terms

$$\mathbb{E}\left[\left(f(x) - \mathbb{E}\left[\hat{f}(x; \mathcal{D}) \right] \right) \left(\mathbb{E}\left[\hat{f}(x; \mathcal{D}) \right] - \hat{f}(x; \mathcal{D}) \right) \right] = 0.$$

By squaring and taking expectation over our expression for the residual, and by recalling that $Y - f(x) = \epsilon$, we therefore arrive at the expression

$$\mathbb{E}\left[\left(Y - \hat{f}(x; \mathcal{D}) \right)^2 \right] = \mathbb{E}[\epsilon^2] + \left(\mathbb{E}\left[\hat{f}(x; \mathcal{D}) \right] - f(x) \right)^2$$

$$+ \mathbb{E}\left[\left(\hat{f}(x; \mathcal{D}) - \mathbb{E}\left[\hat{f}(x; \mathcal{D}) \right] \right)^2 \right].$$

The result follows.   □

Note 6.3. The *mean-squared error* MSE of a model can be defined by taking the expectation of Equation (6.11) over the random variable X.   ▷

We will see illustrations of this result as we progress through the chapter. Note that the bias-variance theorem is inherently concerned with regression problems, as it is a statement concerning the our expectation (literally) for the residual sum of squares.

6.1.3 Validation Metrics for Classification

In classification problems, the target variable Y constitutes a categorical response, taking one of c values in a set $\mathcal{C}$, whose elements can be labeled as $1, \ldots, c$. Oftentimes we will directly model a set of functions $f_i(X)$, for $i = 1, \ldots, c$, that are required to satisfy the constraints that $0 < f_i(X) < 1$, for all $i = 1, \ldots, c$ and $X \in \mathcal{X}$, and the normalization condition

$$\sum_{i=1}^{c} f_i(X) = 1.$$

Moreover, these functions should satisfy the property that the larger the value of $f_i(X)$, for a given class label i, the more likely it is that $Y = i$. They can be thought of as probabilities, but these outputs typically do not correspond to the actual real probability. When the raw functions $f_i'(X)$ do not satisfy the normalization conditions, we can always apply the *softmax transformation*

$$f_i(X) = \frac{e^{f_i'(X)}}{\sum_{i=1}^{c} e^{f_i'(X)}}. \tag{6.12}$$

(See Exercise 6.2.) However, our goal is not to predict a number between zero and one, but to predict class labels for a new instance X. A common method is to simply select the class with the largest value of f_i:

$$\hat{Y}(X; \mathcal{D}) - \arg\max_{i=1,\dots,c} f_i(X; \mathcal{D}). \tag{6.13}$$

However, in many cases, we will be specifically considering classification algorithms for the purpose of predicting a *binary response*, so that $Y \in \{0, 1\}$. In this case, we only need to specify a single function $f(X) = f_1(X)$. In this case, it is more common to use a threshold t to determine the final classification:

$$\hat{Y}_t(X; \mathcal{D}) = \mathbb{I}\left[f(X; \mathcal{D}) > t\right]. \tag{6.14}$$

Here, the same model will yield a different set of predictions for different thresholds. We will discuss this in more depth at the end of this subsection.

Note 6.4. We should further note that many classification algorithms *do not model an actual probability*! That is, the function $f(X)$ that is learned by the classification algorithm will satisfy the properties of a probability, but will not correspond to a probability itself. A value of $f(X) = 0.5$ means that the response $Y = 1$ is more likely than it would have been if $f(X) = 0.4$, but it does *not* mean that there is a 50% *probability* that the response is $Y = 1$. We will discuss this concern, and its remedy, in Section 6.1.4.   ▷

Standard Classification Metrics

For the remainder of this section, we shall consider the common case of binary classification problems. In this context, the response $Y = 1$ is referred to as "the event," and it typically corresponds to the target of our prediction task. Common applications include a diagnostic for a lab test and a user in a freemium mobile app making a purchase. This distinction is made without loss of any generality, as $Y = 0$ and $Y = 1$ could just as easily correspond to dogs and cats. However, for the purpose of the following, we will consider the event $Y = 1$ to be the target of our prediction problem.

Definition 6.11. *Given a classification problem and a set of true values Y_i and predictions $\hat{Y}_i$, for $i = 1, \dots, n$, where each Y_i may take one of m distinct values, the confusion matrix C is the $c \times c$ matrix of counts, where*

$$C_{ij} = \sum_{k=1}^{n} \mathbb{I}\left[\hat{Y}_k = i \ and \ Y_k = j\right].$$

Thus, in a confusion matrix, the rows correspond to the predicted classes, and the columns correspond to the actual (true) classes. For the case of binary classification, it is conventional (for some reason) to write the rows and columns of the confusion matrix in *descending* order[1]. A typ-

	$Y = 1$	$Y = 0$
$\hat{Y} = 1$	TP	FP
$\hat{Y} = 0$	FN	TN

Table 6.1: Confusion matrix for binary classification.

ical confusion matrix is shown in Table 6.1. The elements of the matrix are labeled TP, FP, FN, and TN. These correspond to true/false positive/negative, as defined below.

Definition 6.12. *In a binary classification problem, the elements of the confusion matrix are referred to as*

1. True Positive (TP)—*the number of correct positive predictions* ($\hat{Y} = 1$ *and* $Y = 1$),
2. False Positive (FP)—*the number of incorrect positive predictions* ($\hat{Y} = 1$ *and* $Y = 0$),
3. False Negative (FN)—*the number of incorrect negative predictions* ($\hat{Y} = 0$ *and* $Y = 1$), *and*
4. True Negative (TN)—*the number of correct negative predictions* ($\hat{Y} = 0$ *and* $Y = 0$).

Note 6.5. In Definition 6.12, think of the terminology in the context of a laboratory test for a medical condition: a positive result means you tested positive for the condition and a negative result means you tested negative for the condition. The True/False indicates whether or not the test result (the predicted value) was correct.   ▷

The first validation metric for a classifier is accuracy.

Definition 6.13. *The* accuracy *of a classifier is the ratio of the number of correct predictions to the number of overall predictions, i.e.,*

$$accuracy = \frac{TP + TN}{TP + FP + FN + TN}.$$

However, considering accuracy alone can be misleading.

---

[1] I suppose otherwise there wouldn't be anything confusing about it.

Example 6.2. Consider a binary classification problem for a rare event. Now consider the classifier that classifies *everything* as $\hat{Y} = 0$. The confusion matrix for 100 samples is given in Table 6.2. The accuracy of this model on

	$Y = 1$	$Y = 0$
$\hat{Y} = 1$	0	0
$\hat{Y} = 0$	1	99

Table 6.2: Confusion matrix for Example 6.2; 99% accuracy.

this data set is 99%, even though the model (by design) doesn't do anything at all.   ▷

To remedy this, statisticians typically refer to two quantities known as sensitivity and specificity, defined below.

Definition 6.14. *The* sensitivity *and* specificity *of a binary classifier are the ratios of the number of true positives or negatives, respectively, to the total number of positives or negatives, respectively; i.e.,*

$$sensitivity = \frac{true\ positives}{total\ actual\ positives} = \frac{TP}{TP + FN} = TPR,$$

$$specificity = \frac{true\ negatives}{total\ actual\ negatives} = \frac{TN}{FP + TN} = TNR = 1 - FPR.$$

The sensitivity is equivalent to the true positive rate *(TPR) and the specificity is equivalent to the* true negative rate *(TNR), which is equivalent to one minus the* false positive rate *(FPR).*

It is desirable to have high sensitivity and specificity, and a low false positive rate. However, these two metrics still do not capture the full picture, as illustrated in our next example.

Example 6.3. Consider a binary classification problem with confusion matrix given in Table 6.3. The overall accuracy ($190/210 \approx 90.47\%$) is good.

	$Y = 1$	$Y = 0$
$\hat{Y} = 1$	10	20
$\hat{Y} = 0$	0	180

Table 6.3: Confusion matrix for Example 6.3; 100% sensitivity, 90% specificity.

Similarly, the sensitivity and specificity both look healthy:

$$\text{sensitivity} = \frac{10}{10} = 100\%,$$

$$\text{specificity} = \frac{180}{200} = 90\%.$$

However, what does a positive test result $\hat{Y} = 1$ mean? Given a positive test result, we have to look at the first row of the confusion matrix. And we see that we only have a 33% probability of actually being a true positive. This discrepancy is not captured at all when looking at sensitivity and specificity alone.   ▷

As a result of Example 6.3, in the field of machine learning, it is more common to analyze the following pair of metrics.

Definition 6.15. *The* precision *and* recall *of a binary classifier are defined as*

$$precision = \frac{true\ positives}{total\ predicted\ positives} = \frac{TP}{TP + FP},$$

$$recall = \frac{true\ positives}{total\ actual\ positives} = \frac{TP}{TP + FN}.$$

Recall is the same as sensitivity.

In Example 6.3, the precision is 33% and the recall is 100%. Thus, a positive example has a 100% chance of being identified as positive; however, an example with a positive prediction only has a 33% chance of being an actual positive.

If a combined metric is required, one can use the F_1-score, which is the harmonic mean of precision and recall. More generally, we can use F_β, for $\beta > 0$, which is defined as

$$F_\beta = (1 + \beta^2) \frac{precision \cdot recall}{\beta^2 precision + recall}.$$

For the case $\beta = 1$, we recover the harmonic mean of the two metrics.

Receiver-operator Characteristic (ROC) Curves

As stated above, the decision of whether or not to classify the output $f(X)$ of a binary classifier as an event is typically determined by setting a threshold $t \in [0, 1]$ and applying Equation (6.14). All of our validation metrics thus far apply to a single predictive model, i.e., following the selection of an appropriate threshold. But how should we compare different classifiers without regard to the particular value of the threshold parameter? The answer lies in the following.

Definition 6.16. *The* receiver–operator characteristic (ROC) *curve of a predictive model* $f(X;\mathcal{D})$, *with classification function*

$$\hat{Y}_t(X;\mathcal{D}) = \mathbb{I}\left[f(X;\mathcal{D}) > t\right],$$

for $t \in [0,1]$, *is the parametric curve* $\mathbf{r} : [0,1] \to [0,1]^2$ *defined by*

$$\mathbf{r}(t) = \langle FPR(t), TPR(t) \rangle,$$

where $FPR(t)$ *and* $TPR(t)$ *are the false- and true-positive rates as a function of the threshold parameter* t.

Typically, the x-label of the ROC curve is represented as "1 - specificity," which, of course, is equivalent to the false-positive rate. All ROC curves satisfy the endpoint conditions. For an example ROC curve, see Figure 7.3.

Proposition 6.1. *Let* $\mathbf{r} : [0,1] \to [0,1]^2$ *be an ROC curve. Then*

$$\mathbf{r}(0) = \langle 1,1 \rangle \qquad and \qquad \mathbf{r}(1) = \langle 0,0 \rangle.$$

Moreover, the TPR is a nondecreasing function of the FPR.

Proof. We begin by considering the threshold $t = 0$. In this case, $\hat{Y}_0(X) = 1$, for all X, since the function $f \in (0,1)$. Therefore

$$TPR(0) = FPR(0) = 1.$$

Similarly, at the endpoint $t = 1$, we have $\hat{Y}_1(X) = 0$, for all X, so that

$$TPR(1) = FPR(1) = 0.$$

This proves the two endpoint conditions; i.e., the ROC curve connects the endpoints $\langle 0,0 \rangle$ and $\langle 1,1 \rangle$.

Next, let us examine the monotonicity of the true-positive rate, when cast as a function of the false-positive rate. Consider any two t_1, t_2, with $0 \le t_1 < t_2 \le 1$. By increasing the threshold from t_1 to t_2, any instance X with $f(X;\mathcal{D}) \in (t_1, t_2)$ will no longer be classified as a positive example. The total of each column of the confusion matrix must remain constant; however, counts will drip from row $i = 1$ into row $i = 0$. Thus, we have

$$TPR(t_2; x) \le TPR(t_1; x) \qquad and \qquad FPR(t_2; x) \le FPR(t_1; x).$$

The result follows. $\square$

There are two interesting points to note. The first is that the line $TPR = FPR$ represents a model that classifies at random. That is, if the ranking of the model outputs $f(X_1) < f(X_2) < \cdots < f(X_n)$ is random, then the probability of classifying an instance as a positive is independent of whether or not the instance actually is positive. Thus, we should expect that the actual positive instances are classified as positive (TPR) at the same rate as are the actual negative instances (FPR). This gives some guidance on interpreting the ROC curve. Curves that fall below the diagonal "$y = x$" line are fairing worse than had you just assigned probabilities at random. Curves at are entirely above the diagonal are performing better than average.

The second point is that a perfect model would have $TPR = 1$ and $FPR = 0$. Thus, the closer the curve gets to the point $\langle 1, 0 \rangle$, typically the better the model. Stated differently: the area under the curve (which is bound between 0 and 1) is a measure of predictive performance for the model. We therefore defining the following.

Definition 6.17. *The* area under the curve (AUC) *of a classifier is defined as the area bounded by the ROC curve and the lines $TPR = 0$ and $FPR = 1$.*

Clearly, it follows that $0 < AUC < 1$, for any model. Moreover, an $AUC = 1/2$ is a model that performs as good as random. And a model with an AUC close to 1 is a superior model.

6.1.4 Predicting Class Probabilites

As mentioned in Note 6.4, the function $f(X; \mathcal{D})$, produced by a binary classifier, is not a true probability, though it possesses the basic properties of probabilities. We now turn to the case where the goal of our classifier is actually to determine an accurate *probability*, as opposed to an actual label.

Calibration

A classification model is said to be *well calibrated* if its output values $f(X)$ represent the probability that $Y = 1$; i.e., a value $f(X) = 0.3$ actually means that the given instance has a 30% probability of being positive. In order to visualize how well calibrated a model is, we rely on the following.

Definition 6.18. *Given a set of data $\{X_i, Y_i\}_{i=1}^{n}$ and prediction values $f(X_i)$ of a classification algorithm. The instances are ordered relative to their prediction values, so that $f(X_1) < f(X_2) < \cdots < f(X_n)$. Then they are divided into k buckets or bins, of approximately equal size n/k. Next, let f_i represent the average prediction value of the instances in the ith bucket, and let p_i represent the fraction of positive instances within the ith bucket. Then the curve connecting the points $\langle f_i, p_i \rangle$ constitutes a* calibration curve *for the model.*

Note 6.6. The diagonal "$y = x$" line represents a perfectly calibrated model, as the prediction values are equivalent to the true probabilities.   ▷

Note 6.7. An alternative formulation of the calibration curve plots the bin number on the x-axis and concurrently plots two separate curves on the y-axis: one for the predicted values and one for the true values of each bin. This format is suitable for viewing the "calibration" of a regression model as well, and can serve as a visual diagnostic as to whether a model is well calibrated. By comparing the true and predicted curves concurrently, it is also more interpretable to business stakeholders.   ▷

Typically, a calibration curve will have a sigmoidal shape. We can therefore adjust the prediction values by fitting the calibration curve to the sigmoid function

$$p_i = \frac{1}{1 + \exp(-\beta_0 - \beta_1 f_i)}.$$

Applying the same curve to each instances prediction value $f(X)$ therefore returns a prediction value that is better calibrated as a probability. This method was introduced by Platt [2000]. This approach and an alternative approach, isotonic regression, are discussed in Niculescu-Mizil and Caruana [2005]. Isotonic regression is especially suitable when the calibration curve does not have a sigmoidal shape.

Brier Score and Cohort Probabilities

We next discuss a metric used to assess probability models, due to Brier [1950].

Definition 6.19. *The* Brier score *for a binary classifier is defined by*

$$\text{BS} = \frac{1}{n} \sum_{i=1}^{n} (y_i - p(X_i; \mathcal{D}))^2, \tag{6.15}$$

where y_i is the actual value and $p(X_i; \mathcal{D})$ is the predicted probability for of the ith instance.

The Brier score differs from the residual sum of squares, as the residual sum of squares would be defined based on the final classification—and not the predicted probability—of the each instance. The prediction probabilities are typically generated from the prediction values $f(X_i; \mathcal{D})$ of a classifier by implementing either Platt scaling or isotonic regression, as discussed in the preceding paragraph.

The Brier score has an interesting decomposition when the prediction probabilities are made for fixed *cohorts*, or groups with similar characteristics, as opposed to at the individual level. This is common when there is a finite and manageable number of permutations of the feature set.

Proposition 6.2. *Suppose a unique probability is provided for each of k cohorts. Then the Brier score is equivalent to the following two-component decomposition*

$$\text{BS} = \frac{1}{n} \sum_{i=1}^{k} n_i \left(p_i - \overline{y}_i\right)^2 + \frac{1}{n} \sum_{i=1}^{k} n_i \overline{y}_i \left(1 - \overline{y}_i\right), \tag{6.16}$$

where n_i is the count of instances in the ith cohort, and $\overline{y}_i$ is the observed event probability in the ith cohort.

The first term in Equation (6.16) is related to the calibration: how well the cohort predicted probabilities align with the observed true values. The second term is an expression of the average inherent uncertainty within each cohort. This second term represents an irreducible error, as it cannot be changed with model improvements, for a fixed set of cohorts. We leave the proof to the reader. (See Exercise 6.3.)

6.1.5 Validation Methodology

Over the preceding pages, we were a bit nonchalant regarding to the data to which each of the formulas should be applied. We now seek to remedy that coolness by discussing this issue in greater depth. For additional references, see Hastie, *et al.* [2009] and Kuhn and Johnson [2013].

Loss Functions

We begin with a discussion on *loss functions*. This will allow us to speak generally in our conversation on error, without regard to the particular type of predictive model we are addressing.

Definition 6.20. *In a predictive learning model, any function of the form* $L : \mathbb{R}^2 \to [0, \infty)$ *is called a* loss function *if it has the properties that*

$$L(y, y) = 0$$

and $L(Y, \hat{Y}(X)) \to 0$ *as* $\hat{Y}(X) \to Y$, *for any instance* $(X, Y) \sim f_{X,Y}$.
    For regression problems, common loss functions include squared error and absolute error,

$$L(y, \hat{y}) = (y - \hat{y})^2 \tag{6.17}$$
$$L(y, \hat{y}) = |y - \hat{y}|, \tag{6.18}$$

respectively.
    For a c-class classification problems, where our random target variable $Y \in \mathcal{C}$, *common loss functions include binary loss and log-loss,*

$$L(y, \hat{y}) = \mathbb{I}[y \neq \hat{y}] \tag{6.19}$$
$$L(y, p) = -\sum_{k=1}^{c} \mathbb{I}[y = k] \log p_k = -\log p_y, \tag{6.20}$$

respectively.

Unlike Equations (6.17)–(6.19), Equation (6.20) is a function of the predicted probability of a classification model, not the class predictions itself. Here, p_k is the probability that a given instance belongs to class k, for $k = 1, \ldots, c$. Suppose that the instance does indeed belong to class k, for some particular $k \in \{1, \ldots, m\}$. Then the larger the value of p_k, the better the performance of the model. Recall from basic logarithm properties, that $-\log p_k > 0$ and $-\log p_k \to 0$ as $p_k \to 1$. The smaller the predicted value p_k, the worse the model did (since $y = k$ is correct), and the larger the value of $-\log p_k$.

The loss function Equation (6.17) corresponds to the residual sum of squares from Equation (6.2), whereas the loss function Equation (6.18) corresponds to the MAE from Equation (6.7). Similarly, the loss function Equation (6.19) is related to accuracy; in fact, it is 1 minus the accuracy.

Finally, we note that, for binary classification problems, if we let p represent the predicted probability of the positive label (typically, the minority label), then the log loss of Equation (6.20) simplifies as

$$L(y, p) = -y \log p - (1 - y) \log(1 - p). \tag{6.21}$$

Training, Test, and Prediction Errors

When applying any of the above metrics (e.g., Equations (6.2) and (6.15)) to the training set $\mathcal{D}$ itself, the result is referred to as the *training error*

$$\mathrm{ERR}_{\mathcal{D}} = \sum_{i=1}^{n} L(y_i, f(x_i, \mathcal{D})). \tag{6.22}$$

This, however, is not a good metric with which to assess a model, because the model was able to train on the results that we are testing its performance on. This can lead to *overfitting* of a model, which occurs when a model has a low training error but generalizes poorly to new data. To remedy this, we next define several types of errors, that are determined based on how they are applied, but not the individual context to which they are applied. Thus, we will develop the following for a generic loss function, which will depend on the particular context.

Definition 6.21. *In a predictive learning model $f(X; \mathcal{D})$, learned from a data set $\mathcal{D}$, with loss function L, the* test error *or* generalization error *is the expected error over an independent sample*

$$\mathrm{ERR}_{\mathcal{D}} = \mathbb{E}\left[L(Y, f(X; \mathcal{D}))\right]. \tag{6.23}$$

Here, the expectation is with respect to the random variable $(X, Y) \sim f_{X,Y}$.

The test error is thus the expected loss on an *independent sample*. Note that the test error is for a particular trained model $f(\cdot; \mathcal{D})$. The predictive model, and thus the test error, might have come out differently had the algorithm learned from an alternate training set $\mathcal{D}$. Thus, measuring test error doesn't go far enough to properly assess our model: what if our model had been learned from a different training set?

Definition 6.22. *In a predictive learning model $f(X; \mathcal{D})$, which can be learned from any data set $\mathcal{D}$, with loss function L, the* prediction error *is the* expected test error

$$\text{ERR} = \mathbb{E}\left[\text{ERR}_{\mathcal{D}}\right], \tag{6.24}$$

where, the expectation is with respect to the training set $\mathcal{D}$.

Thus, the test error is the expected loss given our particular training set, and the prediction error is the expected test error of a model, without regard to the particular training set deployed. Clear as pudding[2].

Model Selection and Assessment

A predictive model does not occur in a vacuum. Commonly, we are never interested in a single model, but a family of related models that follow a similar methodology. That is to say, predictive models are typically dependent on a number of *tuning parameters* or *hyperparameters* that specify the model or how it operates. The tuning parameters may vary the complexity of the model, or otherwise determine which features should be used to train the model. Once we have determined (what we believe to be) the optimal model, we then want to determine an accurate measure of its expected performance.

We therefore find ourselves faced with two distinct tasks:

- *Model selection*: estimate the performance of the various models under consideration in order to select the best one;
- *Model assessment*: evaluate the performance of the selected model.

At first pass, the task of model assessment might seem redundant; after all, have we not already estimated each model's performance before making our selection? Such an approach, however, often leads to folly, as it fails to account for any *selection bias* incurred during the model selection phase.

Example 6.4. A set of ten models are trained on a training set $\mathcal{D}$, yielding $f_i(X)$ for $i = 1, \ldots, 10$. The models are then applied to a separate test set $\mathcal{T}$, for which the test error is estimated using the mean-squared error ($\text{MSE} = \text{SSE}/df$).

Now, suppose that each model is exactly equivalent, except that they may perform differently on different data sets. That is, suppose that the

---

[2] Perhaps I'll start a new trend of ending mathematical statements with CAP, like proofs are ended with QED.

prediction error for each of our ten models is exactly $\text{ERR} = 10$, and the model variances are all $\mathbb{V}(f_i(x)) = 9$, for each $i = 1, \ldots, 10$. We can simulate such a scenario by drawing a random sample of ten numbers from $N(10, 9)$, obtaining the values

$$8.72, \ 9.44, \ 10.13, \ 6.75, \ 10.59, \ 11.67, \ 8.61, \ 6.35, \ 11.81, \ 2.76.$$

Upon seeing these results, we have a clear winner: model 10. (Even though, in reality, the models are exactly identical.)

Now, if we stop there, and report that our model has an expected generalization error of 2.76, we have a problem. This is called selection bias, and it illustrates the importance of the model assessment task as a separate task from the model selection process.   ▷

To address the issue of selection bias in model selection and assessment, we commonly divide our data set into three subsets:

1. A *training set* $\mathcal{D}$: the data set used to train the models;
2. A *validation set* $\mathcal{V}$: the data set used to estimate the test error of the various models, used for model selection;
3. A *test set* $\mathcal{T}$: the data set used to estimate the generalization error for the final model.

A typical split may be 50–25–25, but this in reality depends on the application and the size of data. Several tools that can be used for data sets that are two small to be amenable to such a split are discussed in Hastie, *et al.* [2009].

Cross Validation

Another approach often used to estimate the prediction error of a model is k-fold cross validation. The idea is that we divide our data set randomly into k partitions (or folds). We then proceed to train our model k times, each time we leave out one of the folds as a test set. This method therefore yields k separate estimates of the test error, which can be averaged together to estimate the expected test error.

Let $\mathcal{D} = \{(X_i, Y_i)\}_{i=1}^n$ be our data set, and let $\kappa : \{1, \ldots, n\} \to \{1, \ldots, k\}$ be a function that randomly assigns each data point into one of k bins (or folds), and let $\kappa^{-1}(j) = \{i : \kappa(i) = j\}$. Now define

$$\mathcal{D}_j = \{(X_i, Y_i)\}_{\kappa(i) \neq j} \qquad \text{and} \qquad \mathcal{T}_j = \{(X_i, Y_i)\}_{\kappa(i) = j}$$

be the set with the jth fold removed. We want to train our model on $\mathcal{D}_j$ and test our model on $\mathcal{T}_j$, for $j = 1, \ldots, k$. We may estimate the test error for the model trained on the set $\mathcal{D}_j$ as

$$\hat{\text{ERR}}_{\mathcal{D}_j} = \frac{1}{|\kappa^{-1}(j)|} \sum_{i \in \kappa^{-1}(j)} L(Y_i, f(X_i; \mathcal{D}_j)), \tag{6.25}$$

for $j = 1, \ldots, k$. By averaging these results, we thus arrive at an estimate for the prediction error

$$\hat{\text{ERR}} = \frac{1}{k} \sum_{j=1}^{k} \hat{\text{ERR}}_{\mathcal{D}_j}. \tag{6.26}$$

Similarly, we can compute the sample variance of our k estimated test errors to estimate the model's variance. An illustration of five-fold cross validation is shown in Figure 6.1.

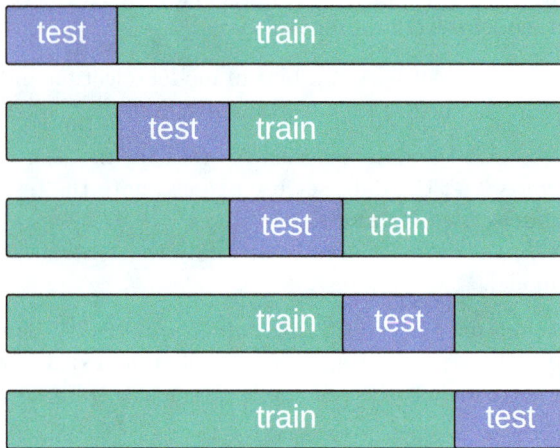

Fig. 6.1: An illustration of five-fold cross validation.

A typical choice of k may be $k = 5$ or $k = 10$. The case $k = n$ is referred to as *leave-one-out cross validation* or the *jackknife*.

6.1.6 Validation on Temporal Datasets

In Section 5.4, we discussed various kinds of stochastic processes that generate data over time. In practice, we must take extra care to perform validation on data that has an explicit time component. There are two potential problems that can arise. First, even when traditional machine-learning validation methodologies are available (e.g., cross validation), which requires that we have a well defined target variable in our training data, user behavior can still vary over time, so that a model that is valid today might not have been valid yesterday. Second, we might encounter data that is not *fully baked*, so that there is no well defined target variable. This occurs with survival processes (Section 5.3), as well as in customer lifetime value models (Section 10.3). In these models, we are learning from behavioral data we have *to date*, in order to predict future behavior. In both cases, it is therefore important to perform a historical validation over a stretch of time.

Before jumping into how we are going to validate such a model, let us first take a moment to examine how our model will be deployed in production. In order to run our model in production, we will require a well defined *training window* and a *prediction range*. In addition, we often desire to buffer the immediately trailing data from our training window. Formally, we define these as follows.

Definition 6.23. *When deploying a model over a* temporal dataset, *or a dataset with an explicit time component, the* training window *is the date range for the training set, and the* prediction range *is the date range for the prediction set.*

These are commonly specified relative to a point in time, *which defaults to the present date, and three configuration parameters*

1. delay: *the number of days between the end of the training set and the point in time;*
2. window: *the length (in days) of the training set;*
3. range: *the length (in days) of the prediction set.*

Given these definitions, our training window and prediction range can be expressed as

$$\text{training window} = [\texttt{point\_in\_time} - \texttt{delay} - \texttt{window}, \texttt{point\_in\_time} - \texttt{delay})$$
$$\text{prediction range} = [\texttt{point\_in\_time} - \texttt{range}, \texttt{point\_in\_time}).$$

This is shown schematically in Figure 6.2. Here, the training set consists of the green data, whereas the prediction set is the blue data. In the figure, the delay and range are shown as equal, though this is not a requirement. In production, the "point in time" is always taken to be the present date.

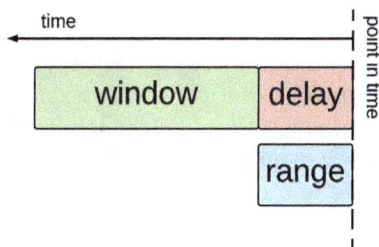

Fig. 6.2: Training (green) and prediction (blue) data for temporal datasets. In production, *point in time* is the current date.

So why don't we just say "now" instead of "point in time"? The beauty of this design, is that we are treating "now" as a parameter. This means that we can pass a "different now" into our runtime job in order to see what our model results would have been had we run our model, with the same configuration, at some point in time in the past.

We can take advantage of this configuration to perform a *backtesting validation* (or, simply, *backtest*), which consists of running our model using historical data at various points in time in the past, in order to write predictions over a period of time. Not only can we run our standard validation metrics to determine how well our model would have performed, but we can also see how our performance varied over our stretch of time. For example, we can determine how much our model predictions might fluctuate in time. A backtesting validation with four jobs is shown in Figure 6.3.

Note that each job is staggered by the range, so that we have full coverage of our predictions over a substantial period of time. In other words,

$$\mathtt{point\_in\_time}(\mathtt{job\_i}) = \mathtt{point\_in\_time} - \mathtt{range} \times i,$$

where $i \in 0, \ldots, n-1$, where n is the number of backtest jobs deployed, and $\mathtt{point\_in\_time}$ is the ending point in time.

Backtesting is especially critical for online process data, which is commonly not fully baked at the time we write our predictions (e.g., survival processes). For such a scenario, we can set the point in time for $\mathtt{job\text{-}0}$ as the latest point in time for which fully baked data is available. The model still trains using only the data it would have had available at that point in time, but we can use the fully baked data from those cohorts in order to validate our predictions. In such a scenario, cross validation is not necessary. We are training our model over a historical backtesting period in the exact same way that we would train the model during production. We can therefore see exactly how well our model would have fared over our historical backtesting period, had it been live and in production at the time.

6.2 Preprocessing

It is seldom the case that one can use the features (independent variables) to train a model in their raw form. Rather, transformations are commonly applied prior to feeding the features into the model. Now, what is good for the goose, is good for the gander: transformations applied to the training set should equally be applied to the validation and test sets prior to computing the model's predictions.

Most of this section will rely heavily on the scikit-learn package. For an introduction to scikit-learn in machine learning, see Pedregosa, *et al.* [2011]. For a discussion of the design principles of its API, see Buitinck, *et al.* [2013]. And, of course, for the latest and up-to-date changes on functionality, visit the scikit-learn user guide at

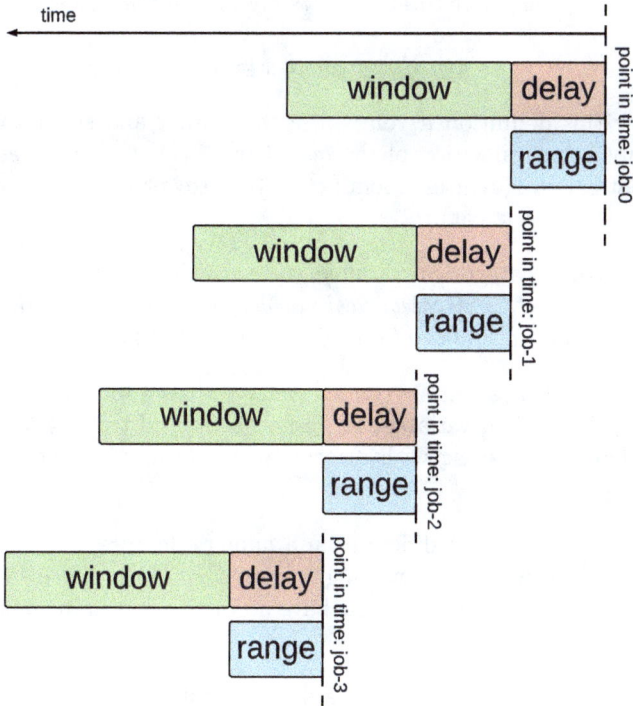

Fig. 6.3: Backtesting validation for temporal datasets.

https://scikit-learn.org/stable/user_guide.html.

6.2.1 Categorical Features

A *categorical variable* is a variable that can take on any value from a finite set $\mathcal{C}$. We will consider two types of categorical features: ordinal and nominal. To differentiate between these two, we introduce the following.

Definition 6.24. *Let $\mathcal{S}$ be a set and "$\leq$" a binary relation over $\mathcal{S}$[3]. Then the binary relation is a* total order *if it satisfies*

1. Connexity: $x \leq y$ or $y \leq x$, for all $x, y \in \mathcal{S}$;
2. Reflexivity: $x \leq x$, for all $x \in \mathcal{S}$;
3. Transitivity: $x \leq y$ and $y \leq z$ implies $x \leq z$, for all $x, y, z \in \mathcal{S}$;
4. Antisymmetry: $x \leq y$ and $y \leq x$ implies $x = y$, for all $x, y \in \mathcal{S}$.

---

[3] A binary relation of a set $\mathcal{S}$ is a subset of the Cartesian product $\mathcal{S} \times \mathcal{S}$. Connexity simply means that all pairs are in the binary relation, one way or the other.

Given a total order, a strict total order[4] *is the binary relation "$<$" defined by the condition*

$$x < y \text{ if } x \leq y \text{ and } x \neq y.$$

Gloss over this definition if you will, a total order and strict total order are simply a generalization of the *less than or equals to* and *less than* operators that act over real numbers ($\pi^2 < 10$), except now they act over elements of a set (dog $<$ cat).

Definition 6.25. *A* categorical variable *is a variable that can take on any value from a finite set $\mathcal{C}$. A categorical variable is said to be* ordinal *if its underlying set has a natural total order, otherwise it is said to be* nominal.

Note 6.8. We can always contrive an arbitrary ordering for any set, which is why we require ordinal variables to possess a *natural* total order; i.e., a total order that has meaning with respect to the meaning of the variable.
▷

Note 6.9. An ordinal feature differs from a numeric feature as it possesses no sense of scale. For example, we might know that $A < B < C$, but the nature of an ordinal variable is that we have no way to determine *how much greater* one variable is to another.
▷

Thus, an ordinal feature is simply a discrete feature that has a natural ordering. Examples include T-shirt size ($XS < S < M < L < XL$) and responses on a survey ("very dissatisfied" $<$ "dissatisfied" $<$ "neutral" $<$ "satisfied" $<$ "very satisfied"). Similarly, a nominal feature does not have such an intrinsic ordering. Examples of nominal features include color (red, blue, yellow) and country (UK, CA, JP, AU).

The presence or lack of ordering affects how we preprocess a discrete feature.

Ordinal Features

In order to make ordinal features suitable for learning algorithms, we use a technique referred to as *label encoding*.

Definition 6.26. *Let $X \in \mathcal{C}$ be an ordinal variable. Then the mapping $\iota : \mathcal{C} \to \{1, \ldots, |\mathcal{C}|\}$ is an* ordinal encoding *if it is invariant with respect to the strict total ordering on $\mathcal{C}$; i.e., if*

$$x < y \text{ implies } \iota(x) < \iota(y), \text{ for all } x, y \in \mathcal{C}.$$

---

[4] A strict total order satisfies the same set of axioms as a total order, except instead of reflexivity it is irreflexive, $x \not< x$ for all $x \in \mathcal{S}$, and instead of antisymmetry it is asymmetric, $x < y$ implies $y \not< x$ for all $x, y \in \mathcal{S}$.

Note 6.10. The comparison $x < y$ in Definition 6.26 is with respect to the ordering of the set $\mathcal{C}$, whereas the comparison $\iota(x) < \iota(y)$ is with respect to the standard Cartesian ordering of the integers.   ▷

Example 6.5. Fortunately, the scikit-learn (sklearn) package[5] has a built-in ordinal encoder, ready for use. In Code Block 6.1, we define three ordinal variables: survey, t-shirt, and color. We then instantiate an instance of the OrdinalEncoder class on line 7, declaring the required ordering of each feature as well as how to handle unknown values. The encoder fits the data frame (which is required, despite being, in this case, redundant from the object initialization) on line 17 and transforms it into an encoded matrix on line 18. Note that since we are using an ordinal encoding, we are actually saying that red $<$ yellow $<$ blue, at least for the purpose of the current illustration.

```
1   from sklearn.preprocessing import OrdinalEncoder
2
3   survey = ['very dissatisfied', 'dissatisfied', 'neutral',
            'satisfied', 'very satisfied']
4   t_shirt = ['XS', 'S', 'M', 'L', 'XL']
5   color = ['red', 'yellow', 'blue']
6
7   enc = OrdinalEncoder(
8           categories = [survey, t_shirt, color],
9           handle_unknown='use_encoded_value',
10          unknown_value=-1)
11
12  df = DataFrame({
13      'survey': np.random.choice(survey, size=10),
14      't_shirt': np.random.choice(t_shirt, size=10),
15      'color': np.random.choice(color, size=10)})
16
17  enc.fit(df)
18  X = enc.transform(df)
19  enc.categories_
```

Code Block 6.1: Syntax for sklearn's ordinal encoder

For this example, we generated our data frame by randomly choosing (with replacement) from our possible classes for each feature. A randomized data frame and its encoding is given in Table 6.4.   ▷

---

[5] For an up-to-date guide on using the scikit-learn package, see https://scikit-learn.org/stable/user_guide.html.

row	survey	t-shirt	color	X[:, 0]	X[:, 1]	X[:, 2]
0	satisfied	L	red	3	3	0
1	very dissatisfied	XS	red	0	0	0
2	very dissatisfied	M	red	0	2	0
3	satisfied	L	yellow	3	3	1
4	satisfied	S	blue	3	1	2
5	satisfied	L	blue	3	3	2
6	neutral	L	red	2	3	0
7	very satisfied	S	yellow	4	1	1
8	very dissatisfied	XL	yellow	0	4	1
9	very satisfied	M	yellow	4	2	1

Table 6.4: Example of an ordinal encoding.

Note 6.11. In a multi-class classification problem, one must further en-code the target variable. This is achieved with a *label encoder*, available in sklearn's `LabelEncoder` class. The label encoder functions almost like the ordinal encoder, with a few exceptions:

- The label encoder does not accept any arguments into its constructor (except for, optionally, specifying the number of classes),
- The fit method may only be applied to a one-dimensional array or a pandas Series (equivalently, a single column of a data frame).

The label encoder is specifically used for encoding a *dependent* (i.e., target) variable. ▷

Nominal Features

Similar to how we encoded ordinal features into integers, we encode nominal categorical features into binary vectors.

Definition 6.27. *Let $X \in C$ be a nominal categorical variable. Then the mapping $\iota : C \to \mathbb{B}^{|C|}$ is a* one-hot encoding *if it is an bijection (i.e., if it is one-to-one and onto) and if it satisfies the normalization condition*

$$||\iota(x)||_1 = 1, \text{ for all } x \in C,$$

where $||\cdot||_1$ is the ℓ_1 norm[6] and the set $\mathbb{B} = \{0, 1\}$ is the set of binary digits (bits).

Note 6.12. The normalization condition in Definition 6.27 implies that the encoding $\iota(x)$ is a binary vector consisting of all zeros except for a solitary one. ▷

---

[6] Here, the quantity $||x||_1 = \sum_{i=1}^{n} |x_i|$ represents the ℓ_1 norm, or *Manhattan norm*, of the vector $x \in \mathbb{R}^n$.

Example 6.6. Consider the set of colors $\mathcal{C} = \{R, B, Y\}$. The mapping defined by

$$\iota(R) = \begin{bmatrix} 1 \\ 0 \\ 0 \end{bmatrix}, \qquad \iota(B) = \begin{bmatrix} 0 \\ 1 \\ 0 \end{bmatrix}, \qquad \iota(Y) = \begin{bmatrix} 0 \\ 0 \\ 1 \end{bmatrix},$$

constitutes a one-hot encoding.   ▷

Example 6.7. Let's return to Example 6.5, but treat the variables as nominal, and implement a one-hot encoding. This is achieved in Code Block 6.2 (using the dataframe definition from Code Block 6.1). Here, we squash the fit and transform methods into a single command **fit_transform**, which is available to both encoders. Note, we pass in **sparse=False** if we actually want to visually see or print the output array X. Otherwise, the transform method will return a sparse array object, which is computationally more efficient for large data sets.

```
1  from sklearn.preprocessing import OneHotEncoder
2
3  enc = OneHotEncoder(sparse=False)
4  X = enc.fit_transform(df)
5  enc.categories_
```

Code Block 6.2: Syntax for sklearn's one-hot encoder

neut.	sat.	vy. dissat.	vy. sat.	L	M	S	XL	XS	blue	red	yellow
0	1	0	0	1	0	0	0	0	0	1	0
0	0	1	0	0	0	0	0	1	0	1	0
0	0	1	0	0	1	0	0	0	0	1	0
0	1	0	0	1	0	0	0	0	0	0	1
0	1	0	0	0	0	1	0	0	1	0	0
0	1	0	0	1	0	0	0	0	1	0	0
1	0	0	0	1	0	0	0	0	0	1	0
0	0	0	1	0	0	1	0	0	0	0	1
0	0	1	0	0	0	0	1	0	0	0	1
0	0	0	1	0	1	0	0	0	0	0	1

Table 6.5: Example of a one-hot encoding.

The resultant matrix is shown in Table 6.5. Now, the actual output matrix X is just a two-dimensional numpy array. In order to determine the column headings, we have to look at the attribute **enc.categories_**.

Notice that the columns do not follow the same ordering that we specified in Example 6.5. We could have passed in a `categories` argument into the constructor to have specified the ordering of the encoding. However, this is tantamount to shuffling the columns and does not typically make a difference in practice, as it does with ordinal encoding, where the relative size matters.

Notice, also, that there is not a column for *dissatisfied*, as that category does not occur in our example data frame, from Table 6.4. The `fit` method, therefore, does not create space for this feature. To include *dissatisfied*, the categories must be passed into the constructor, as in Code Block 6.1.   ▷

Even though a one-hot encoding returns a much larger matrix, which requires a separate column for each category within each nominal feature, it is necessary when encoding nominal data, i.e., data without any intrinsic ordering. Use of an ordinal encoder on nominal data, as so often with youth, leads to folly, as a learning algorithm will treat the encoded data as ordered.

6.2.2 Continuous Features

Continuous features are features which can take any real value. Unlike categorical features, they are already numeric. However, they often require their own form of manipulation prior to feeding into a learning algorithm. As was the case with categorical features, many of those transformations are built in to scikit-learn.

Discretization

Our first method for continuous features is a method that can be used to convert a continuous feature into a categorical feature. The quintessential example is age range: the raw age of a customer is converted into an ordinal age-range feature: 18–14, 25–34, 35–44, 45–54, 55–64, 65+. Instead of working with a customer's raw age, which, as an integer, is equivalent to dealing with an ordinal feature with $O(100)$ values, we work with the age-range, which is a more manageable set of seven buckets. In addition, there is usually an eighth bucket for *unknown* values.

Definition 6.28. *A mapping $\iota : \mathbb{R} \to C$ shall constitute a* discretization *or* binning *of the real numbers if C is a finite, totally ordered set and if the mapping ι is invariant with respect to the total ordering, i.e., if*

$$x \leq y \text{ if and only if } \iota(x) \leq \iota(y).$$

A discretization *is an* ordinal discretization *if the set C is ordinally encoded, and it is a* one-hot discretization *if the set C is one-hot encoded.*

Typically, when we speak about discretization, we will implicitly refer to ordinal discretization, as the real numbers have a natural ordering.

The `sklearn.preprocessing` package has a built-in class, `KBinsDiscretizer`, that can handle most of our discretization needs for us. The constructor of this class has the following keyword arguments:

1. *n_bins* (int, default=5): The number of bins to use, must be an integer greater than 1.
2. *encode* ('onehot' (default), 'onehot-dense', 'ordinal'): Determines whether to return an ordinal encoding, or a one-hot encoding, which can be either dense or sparse.
3. *strategy* ('uniform', 'quantile' (default), 'kmeans'): Determines how the widths of the bins are calculated: uniform strategy will produce bins with equal widths; a quantile strategy will bin the data points (approximately) equally into bins; and a kmeans strategy will implement a clustering technique known as *k*-means, which we will not discuss at the present.

As with the OrdinalEncoder, LabelEncoder, and OneHotEncoder classes, the KBinsDiscretizer has a `fit`, `transform`, and a `fit_transform` method. The KBinsDiscretizer can discretize multiple numeric features at once; in this case, an object from this class should be instantiated with an array of integers passed in for the `n_bins` argument. Attributes of this class include `n_bins_`, an array with the number of bins for each feature, and `bin_edges_`, an array of arrays that determines the edges of the bins for each feature.

Example 6.8. Consider student data generated from a large university course: the student's age, exam score, and attendance percentage. The data are generated in Code Block 6.3. Age is then encoded into seven bins, exam score into twelve bins, and attendance into five bins. A quantile strategy is used, so that approximately an equivalent number of data points should appear in each bucket.

We can view the bin edges and the first few rows of our data frame and output matrix, as shown in lines 14–15.   ▷

Scaling Numeric Features

Many machine learning algorithms are highly sensitive to the scaling of numeric features and perform much better when the feature set is appropriately scaled. For example, if an algorithm were to use the Euclidean distance between two numeric features for the purpose of predicting whether or not a home was in a good school district, we should expect things to break down if one feature is the number of bedrooms and another is the home price: these two features live on radically different scales.

The `sklearn.preprocessing` package has three main types of built-in scalers that represent the three most commonly used forms of scaling.

```
1   from sklearn.preprocessing import KBinsDiscretizer
2
3   n_rows = 100
4   df = DataFrame({
5           'age': 18 + np.random.exponential(10,
                    size=n_rows).astype(int),
6           'score': np.fmin(100, np.random.normal(loc=70, scale=20,
                    size=n_rows).astype(int)),
7           'attendance': np.random.beta(5, 1, size=n_rows)
8           })
9
10  K = KBinsDiscretizer(n_bins=[7, 12, 5], encode='ordinal',
            strategy='quantile')
11  X = K.fit_transform(df)
12  # K.fit(df)
13  # X = K.transform(df)
14  print(K.bin_edges_)
15  print(df.head(10), X[:10, :])
```

Code Block 6.3: Syntax for sklearn's discretizer

Definition 6.29. *Let* $X_1, \ldots, X_n \sim f_X$ *be a sample of* IID *data. Then the transformed data set*

$$Z_i = \frac{X_i - \overline{X}_n}{S_n},$$

where $\overline{X}_n$ *is the sample mean and* S_n^2 *is the sample variance, is said to be a* z-score normalized *or* standardized *data set. Similarly, the transformed data set*

$$Y_i = \frac{X_i - \min(X_i)}{\max(X_i) - \min(X_i)}$$

is said to be a min-max normalized *data set. Finally, the transformed data set*

$$W_i = \frac{X_i - Q_{0.5}}{Q_{1-\alpha} - Q_\alpha},$$

where $\alpha \in (0, 1/2)$, *typically* $\alpha = 0.2$ *or* $\alpha = 0.25$, *and* Q_α *is the* αth *quantile of the sample, is said to be a* quantile normalized *data set. The value* $Q_{0.5}$ *is the median of the original data set. For the case* $\alpha = 0.25$, *the denominator* $Q_{0.75} - Q_{0.25}$ *is referred to as the* interquartile range.

Common methods available in the `sklearn.preprocessing` module are

- `StandardScaler`: standardization produces scaled data with zero mean and unit variance;
- `MinMaxScaler`: min-max normalization produces scaled data on range $[0, 1]$;

- `MaxAbsScaler`: divides by the largest absolute value of the data, without shifting; scaled data on range $[-1, 1]$. Can be used with sparse data.
- `RobustScaler`: quantile normalization does not limit range.

The StandardScaler takes arguments `with_mean` and `with_std` (default True). If used with sparse data, `with_mean` should be set to False. The MinMaxScalar takes arguments `feature_range` (default $(0, 1)$) and `clip` (default False). If `clip` is set to True, test data (not used during the original fit) is clipped to the same range $[0, 1]$ as the input data. Finally, the RobustScaler has arguments `with_centering` (default True), `with_scaling` (default True), and `quantile_range` (default $(0.25, 0.75)$).

Once an object of any of these classes is constructed, the standard `fit`, `transform`, and `fit_transform` methods are available to transform a data set.

In addition to the aforementioned scalers, scikit-learn has two nonlinear transformation methods, `QuantileTransformer` and `PowerTransformer`, which transform the data set into a uniform distribution or a Gaussian distribution, respectively.

6.2.3 Data Pipelines

In building a model with a large feature set, one typically encounters multiple transformations prior to training a model. This process will be simplified with a structure known as a pipeline. But first, we will consider two additional preprocessing tasks: handling missing values and randomizing the data into training and test sets.

Imputing missing values

Data sets often contain missing values. Instead of deleting an entire datum, we can define a strategy for handling missing data. To achieve this, we will use the `SimpleImputer` from the `sklearn.impute` package. The constructor has an argument `strategy`, which can take a value from 'mean,' 'median,' 'most_frequent,' and 'constant.' If 'constant' is selected, the argument `fill_value` should also be specified.

In Code Block 6.4, we define an outer join of two data frames, which results in two columns (b and c) with missing values. We then demonstrate imputing with the median and most frequent strategies. Note that the *mean* and *median* strategy are not available for categorical features.

Often, in practice, it is useful to use the constant strategy, in order to give a unique and separate value to the missing data.

Building a Pipeline

So far, we have reviewed many types of transformers: encoders, discretizers, scalers, and imputers. Each of these classes has a `fit` and a `transform`

```
1  from sklearn.impute import SimpleImputer
2
3  df_1 = DataFrame({'a':[1,2,5,6], 'b':['x','y','z','z']})
4  df_2 = DataFrame({'a':[0, 2, 3, 4, 6], 'c':[4, 8, 15, 16, 23]})
5  df = pd.merge(df_1, df_2, how='outer')
6
7  imp = SimpleImputer(strategy='median')
8  X_num = imp.fit_transform(df[['a', 'c']])
9  imp = SimpleImputer(strategy='most_frequent')
10 X_cat = imp.fit_transform(df[['b']])
11 imp = SimpleImputer(strategy='constant', fill_value='MISSING')
12 X_cat = imp.fit_transform(df[['b']])
```

Code Block 6.4: Imputing missing values

method, along with, for convenience a `fit_transform` method, which is equivalent to applying both methods to the same data set in order. The actual machine learning machinery lives in classes as well, a collection of classes known as *estimators*. Estimators typically have both a *fit* and a *predict* method, which operate separately: `fit` is used with the training data, and `predict` is used with the test data.

A *pipeline* is a series of transformers with an optional estimator in its last position. The `fit` method only need be called once on a pipeline object: The dataset is passed through the `fit_transform` methods of each transformer in series. The pipeline object will have all the functionality as its last estimator or transformer.

A pipeline is constructed by passing a list of key-value pairs: the key being an arbitrary (but useful) name for each step of the pipeline, and the value being the transformer or estimator object that is to be called at that step.

A simple pipeline is constructed in Code Block 6.5. The `mask` method on line 6 randomly (with 10% probability) *masks* or removes each value of the data frame. The pipeline consists of two transformers: impute missing values with the median, and then perform standardization.

A similar pipeline for categorical data is constructed in Code Block 6.6. Here, the pipeline consists of three steps: impute missing values, apply an ordinal encoder, apply a one-hot encoding. The application of a one-hot encoding following an ordinal encoding is completely redundant, and could have (and should have) been accomplished with the one-hot encoding alone. The redundancy was only for the purpose of illustration.

In practice, the pipelines in Code Blocks 6.5 and 6.6 would contain an estimator at the end of the pipeline, so that a model could have been built. We will see examples of this soon enough.

```
1  from sklearn.pipeline import Pipeline
2  from sklearn.preprocessing import StandardScaler
3
4  X = np.random.randint(0, 20, size=(10, 4))
5  df = DataFrame(X, columns=['a','b','c','d'])
6  df = df.mask(np.random.random(df.shape) < .1)
7
8  transformers = [
9          ('impute', SimpleImputer(strategy='median')),
10         ('standardize', StandardScaler())]
11
12 pipe = Pipeline(transformers)
13 pipe.fit_transform(df)
```

Code Block 6.5: A simple pipeline for continuous features

```
1  s = """Our revels now are ended. These our actors,
2  As I foretold you, were all spirits and
3  Are melted into air, into thin air"""
4
5  s = s.replace(",", "").replace(".","").replace("\n"," ").split(' ')
6  df = DataFrame(np.random.choice(s, size=(10,4)), columns=['a', 'b',
       'c', 'd'])
7  df = df.mask(np.random.random(df.shape) < .1)
8
9  transformers = [
10         ('impute', SimpleImputer(strategy='constant',
              fill_value='UNKNOWN')),
11         ('ordinal',
              OrdinalEncoder(handle_unknown='use_encoded_value',
              unknown_value=-1)),
12         ('ohe', OneHotEncoder(sparse=False,
              handle_unknown='ignore'))]
13
14 pipe = Pipeline(transformers)
15 pipe.fit_transform(df)
```

Code Block 6.6: A simple pipeline for categorical features

The astute reader may be wondering how to handle a feature set that contains mixed data types. This, too, is easily accomplished with the construction of a `ColumnTransformer` object, which can then be inserted as a pipe in the pipeline. The column transformer simply specifies which transformations to apply to which columns. Finally, a keyword argument `remainder` tells the transformer what to do with columns not specified: 'drop' (default) or 'passthrough'. An example pipeline using a column transformer is constructed in Code Block 6.7.

```
1   from sklearn.compose import ColumnTransformer
2
3   class_levels = ['freshman', 'sophomore', 'junior', 'senior']
4   majors = ['science', 'engineering', 'math', 'computer science']
5
6   n_rows = 100
7   df = DataFrame({
8       'level': np.random.choice(class_levels, size=n_rows),
9       'major': np.random.choice(majors, size=n_rows),
10      'age': 18 + np.random.exponential(10,
            size=n_rows).astype(int),
11      'score': np.fmin(100, np.random.normal(loc=70, scale=20,
            size=n_rows).astype(int)),
12      'attendance': np.random.beta(5, 1, size=n_rows)})
13
14  column_transformer = ColumnTransformer(
15      [('ordinal', OrdinalEncoder(categories=[class_levels],
            handle_unknown='use_encoded_value', unknown_value=-1),
            ['level']),
16      ('ohe', OneHotEncoder(sparse=False,
            handle_unknown='ignore'), ['major']),
17      ('bins', KBinsDiscretizer(n_bins=8, encode='ordinal',
            strategy='quantile'), ['age', 'score'])],
18      remainder='passthrough')
19
20  column_transformer.fit(df)
```

Code Block 6.7: A column transformer can handle mixed data types

Now, ideally, you get some ideas. For example, following the illustration in Code Block 6.7, we might think to ourselves that we should construct two column transformers—one to impute missing values and one to apply the various encodings—and connect them together in a pipeline. A problem with this is that the `transform` method outputs an *array*, not a pandas data frame.

To remedy this, I devised a clever workaround. The idea is we can define our own subclass of `ColumnTransformer` that returns a pandas data frame from its `transform` and `fit_transform` methods, instead of an array. The code is given in Code Block 6.8. A final caveat: the `OrdinalEncoder` does not seem to actually handle unknowns as advertised, which requires us to redefine the `class_levels` list to include our unknown value. (It is important to actually redefine line 22 with the updated list of `class_levels`.) Finally, this method only works if *every* column is explicitly handled in the column imputer; hence the use of `remainder='drop'`.

```
class ColumnImputer(ColumnTransformer):
    def transform(self, X):
        X = ColumnTransformer.transform(self, X)
        cols = []
        for t in self.transformers_:
            cols += t[2]
        return DataFrame(X, columns=cols)

    def fit_transform(self, X, y=None):
        X = ColumnTransformer.fit_transform(self, X, y)
        cols = []
        for t in self.transformers_:
            cols += t[2]
        return DataFrame(X, columns=cols)

column_imputer = ColumnImputer([
        ('impute_cat', SimpleImputer(strategy='constant',
            fill_value='UNKNOWN'), ['level', 'major']),
        ('impute_num', SimpleImputer(strategy='median'), ['age',
            'score', 'attendance']),
        remainder='drop'])

class_levels = ['UNKNOWN', 'freshman', 'sophomore', 'junior',
    'senior']
column_transformer = ColumnTransformer( # ... Same as before.

pipe = Pipeline([
        ('impute', column_imputer),
        ('encode', column_transformer)])

pipe.fit_transform(df)
```

Code Block 6.8: Multiple column transformers in a pipeline; continuation of Code Block 6.7.

Training and Test Sets

Naturally, scikit-learn also has a built-in tool for handling the randomization of a data set (features and labels) into a training and test set. The syntax is shown in Code Block 6.9.

```
from sklearn.model_selection import train_test_split
from sklearn.metrics import r2_score

X = np.random.randint(0, 20, size=(100, 4))
y = np.random.randint(0, 2, size=100)
X_train, X_test, y_train, y_test = train_test_split(X, y,
        test_size=0.20, random_state=42)

pipe = Pipeline() #.....
pipe.fit(X_train, y_train)
y_pred = pipe.predict(X_test)
print(f"R2 score: {r2_score(y_test, y_pred)}")
```

Code Block 6.9: Train-test split

Note 6.13. In Code Block 6.9, notice that `train_test_split` is used directly without being instantiated. This is because it is a function, not a class. ▷

Here, we suppose that we have built a pipeline with the various pre-processing operations, topped with an actual model, or estimator, with a `predict` method. The code shows how we can automatically create a train-test split, train the model on the training data, and use the holdout data for our validation step.

6.3 Object-Oriented Data Science

Object-oriented programming (OOP) is a software-engineering paradigm centered around the use of modular, reusable code. OOP has allowed engineers to develop sophisticated software architectures and platforms that would otherwise be intractable. Though much of this text is focused around the statistical aspects of data science, we devote this section to some of the engineering aspects. We advocate for object-oriented design in the development of data-science projects in order to both better leverage the reusability aspects and to better manage sophisticated projects that are developed across a team. We conclude this section with a discussion of *agile*, which is a process for managing engineering projects that was developed around a set of principles known as the *agile manifesto*.

For more in depth introduction to data structures in Python, see Lambert [2014] and Lee and Hubbard [2015]. For more on data structures and algorithms, see Lafore [2003] (JAVA) or Cormon *et al.* [2009].

6.3.1 Classes and Objects

Object-oriented programming is focused on the use of *classes* and their specific realizations, *objects*. The class encapsulates a set of instructions for creating (or instantiating) individual objects. The class is the cookie cutter and the objects are the cookies.

Definition 6.30. *In object-oriented programming, a* class *is a code block that consists of a number of functions, called* methods, *and variables, called* attributes, *that provides instructions for how to construct, or instantiate any number of* objects. *An* object *(or, instance of a class) is a variable that is created from a class, that stores all of the class's methods and attributes. Two instances (objects) from the same class may have different values for their attributes, but maintain the same structure, as defined by the underlying class.*

In Python, the instructions for instantiating an object from a class are given by the __init__ method, which is reserved for such use. This method is not called directly, rather we use the name of the class like a function that takes in the prescribed input parameters and returns an object of that class. In addition, all methods of a class must take at least one argument, called `self`, which represents the object itself. It is through this argument that any method from a class can access the values stored as attributes and any of the other methods for any particular object. To illustrate, consider the following.

Example 6.9. Consider the class defined in Code Block 6.10, which defines a concept of a *car*.

Each "car object" that is constructed from this class has six attributes: name, color, max speed, position, velocity, and time. The values of those attributes will differ from car to car, though the structure itself remains constant. In addition, each object will possess four methods: speed up, time lapse, stop, and repaint.

Individual cars may be instantiated using the __init__ method, though this method is not called directly. Two cars are instantiated in Code Block 6.11. In addition, we provide a method called __str__ which returns a string representation of an individual object[7]. This is used with the `print` command, as shown in Code Block 6.11.

In Code Block 6.11, the objects are represented by the variables car_1 and car_2. In addition to printing the object, as defined by the internal

---

[7] It is not common to define a __str__ method for data-science applications, but it seemed helpful for this example

```python
class Car:
    def __init__(self, name, color, max_speed=100):
        self.name = name
        self.color = color
        self.max_speed = max_speed

        self.position = 0
        self.velocity = 0
        self.time = 0

    def __str__(self):
        s  = f"name: {self.name}\n"
        s += f"color: {self.color}\n"
        s += f"max speed: {self.max_speed}\n"
        s += f"position: {self.position}\n"
        s += f"velocity: {self.velocity}\n"
        s += f"time: {self.time}\n"
        return s

    def speedUp(self, delta_v=0):
        self.velocity = min(self.max_speed, self.velocity + delta_v)

    def timeLapse(self, t):
        self.time += t
        self.position += t * self.velocity

    def stop(self):
        self.velocity = 0

    def repaint(self, new_color):
        self.color = new_color
```

Code Block 6.10: Car class for Example 6.9

__str__ method defined in the class, we may also print specific attributes using, for example, print(car_1.speed). Finally, we may modify any of the attributes directly, using, for example, car_1.speed += 200.   ▷

6.3.2 Principles of OOP

Object-oriented programming is built around the following four principles:

1. *Inheritance*: We may define new classes ("children") from old ("parents") by updating only the new aspects, whereas all functionality not specifically modified is *inherited* from the parent class;
2. *Encapsulation*: An object only exposes certain methods and attributes to the outside world, while keeping unnecessary details hidden;

```
1   # Instantiate two car objects
2   car_1 = Car('1GAT123', 'blue', max_speed=140)
3   car_2 = Car('EIPI+1', 'red', max_speed=160)
4
5   car_1.speedUp(100)
6   car_1.timeLapse(3)
7   car_2.speedUp(80)
8   car_2.timeLapse(3)
9   car_1.stop()
10  car_1.repaint('black')
11  print(car_1) # invokes __string__ method
12  print(car_2) # invokes __string__ method
13  print(car_1.speed) # access particular attribute
14  car_1.speed += 200 # modify attribute directly
```

Code Block 6.11: Creating objects from the Car class

3. *Abstraction*: The operations of an object can be defined without reference to its internal implementation;
4. *Polymorphism*: Code can be agnostic with respect to which class it is operating on, as long as it is operating from a given inheritance hierarchy; similarly, two classes defined within an inheritance may have different internal definitions for the same method.

While encapsulation and abstraction both hide data, they differ in a crucial manner: encapsulation hides data at an implementation level, whereas abstraction hides data at a design level. We discuss each of these four principles in turn, focusing on how they are implemented in Python.

Inheritance

Inheritance allows us to define new classes from old, changing only what is needed. In this context, the new class is called a *child* of the old class, which is called the *parent*. For example, our **Car** classof Example 6.9 might be a subclass from a **Vehicle** class. The **Vehicle** class would define functionality and attributes common to all kinds of vehicles: bicycles, trains, boats, cars, planes, and so forth. The **Car** class could then overwrite—or freshly define—those elements that are unique to automobiles. This process can continue: the **Car** class might then be a parent to a number of subclasses representing different makes of cars (BMW, Subaru, Ford, etc.). Perhaps the **Car** class has a **engine** method, which could be uniquely defined for each of its subclasses.

In addition, we can allow for *multiple inheritance*, in which a new class is defined from an ordered sequence of parents. This can lead to ambiguity if one is not careful, as demonstrated by the *diamond problem* of Figure 6.4.

In this figure, class D inherits from both classes B and C, which are each,

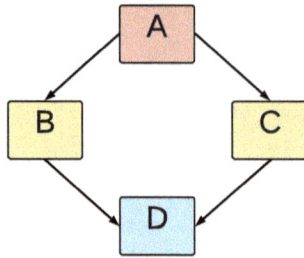

Fig. 6.4: Diamond problem: Class D inherits from both B and C.

in turn, children of the class A. In Python 3, the method resolution order is $D \to B \to C \to A$. (This differs from Python 2, which used a DLR—depth-first left-to-right—method, which would have favored A over C.)

In other words, suppose that class A has a method, which is overwritten by class C, but not by class B or by class D. Objects of type D would then follow the overwritten method as defined by class C.

Example 6.10. We can construct a subclass from the **Car** class (Code Block 6.10) that provides for a more realistic **speedUp** method, in the sense that it takes the vehicle's acceleration into account, so that the velocity jump is not instantaneous. The result is shown in Code Block 6.12.

```
1   class RealCar(Car):
2       def __init__(self, name, color, max_speed=100, acc=5):
3
4           self.acc = acc
5           super().__init__(name, color, max_speed=max_speed)
6
7       def speedUp(self, delta_v=0):
8           delta_v = min(delta_v, self.max_speed - self.velocity)
9           delta_v = max(-self.velocity, delta_v)
10          delta_t = abs(delta_v) / self.acc
11          self.velocity += delta_v
12          self.timeLapse(delta_t)
```

Code Block 6.12: Subclass of **Car**

Notice that we indicate the parent class in line 1. All methods not explicitly redefined within this code are inherited from the parent class, so that we may still access the **timeLapse**, **stop**, and **repaint** methods.

In our `RealCar` class, we update the `__init__` method, to allow for an additional input parameter, representing the vehicle's acceleration. We access the parent class using Python's built-in `super()` method. Finally, we update the `speedUp` method of the parent class with the code shown on lines 7–12.   ▷

Encapsulation

The concept of *encapsulation* refers to the art of only providing access to certain functionality to the outside world. Many programming languages have a concept of *private* and *public* attributes and methods: public attributes and methods are exposed to the outside world, whereas private attributes and methods are only available from within the class itself. For example, when building a car, there are many operations that are required for it to run. Only a handful of these methods are exposed to the driver: gas pedal, brakes, steering wheel. Even though a car requires many other operations to run—piston and cylinders, spark ignition, diverting energy to battery, and so forth—those functions are not directly exposed to the driver.

Python does not offer private attributes and methods. Instead, we may follow the convention that attributes and methods not intended to be exposed to the outside world be preceded by an underscore. For instance, we could signify our intent of keeping `timeLapse` as an internal, private method by renaming it as `_timeLapse`.

For stronger protection of private attributes and methods, we may instead use a double underscore; for example, `__timeLapse`. Attributes and methods with names beginning with a double underscore are not *directly* available from the outside, but need to be preceded by an additional underscore and the class name. For instance, to access `__timeLapse` from *outside* of the class, we first instantiate an object `my_car = Car(..)`, and then we call `my_car._Car__timeLapse(5)`. This process is known as *name mangling*, and is a layer of protection Python offers for variables intended to be kept private.

When defining classes within an inheritance hierarchy, we take the following approach: for private attributes and methods meant to be accessible to (or overwritten by) the various children and descendants, we use the single underscore, whereas private attributes and methods that are meant to be attached to a single class, we use the double underscore. The double underscore mangles the variable name when referred to from any subclass, making it more difficult to access (or overwrite) the original functionality.

Abstraction

Whereas encapsulation hides specific chunks of code or data from the outside world, at the implementation level, abstraction hides the details of an implementation at the design level. In other words, abstraction is used to define the inputs and outputs without any reference to any specific implementation. In this way, abstraction defines the interface that any user of a class must follow. Such an interface is often referred to as an *application user interface*, or *API*.

In practice, abstraction is used whenever we want to define a family of classes that follow a common API. We achieve this by constructing an *abstract class*, which contains one or more *abstract methods*, which are simply methods that have not been implemented. This defines a set of methods, along with their inputs and outputs, that must be defined for any subclass of the abstract class. Abstract classes can be written in Python using the abc (abstract base class) package, as shown in Code Block 6.13.

```python
from abc import ABC, abstractmethod

class AbstractVehicle(ABC):

    @abstractmethod
    def speedUp(self, delta_v: int) -> None:
        pass

    @abstractmethod
    def timeLapse(self, t: float) -> None:
        pass

    @abstractmethod
    def stop(self) -> None:
        pass
```

Code Block 6.13: Abstract classes and methods in Python.

Thus, to create an abstract base class in Python, we simply subclass the ABC class from the abc package. To define abstract methods, we *decorate* the method by adding the text @abstractmethod, which is referred to as a *decorator*, to the line prior to the method definition. Decorators are also known as *wrappers*, as they can be used to wrap our functions in a blanket of code. We will discuss wrapper functions and decorators shortly, but for now, think of the extra line as a magic spell that prevents the class, or any of its subclasses, from instantiating objects unless they have overridden the abstract methods with specific implementations.

Code Block 6.13 defines three methods that must be common to all vehicles: speedUp, timeLapse, and stop. Since they are abstract methods, Python will throw an error if we try to instantiate an object from the AbstractVehicle class. However, when we defined Car, in Code Block 6.10, we could replace the first line with

```
class Car(AbstractVehicle):
```

to ensure that our `Car` class satisfies the API defined by our concept of an abstract vehicle. We have also established a lengthy hierarchy of subclasses:

$$\text{ABC} \to \text{AbstractVehicle} \to \text{Car} \to \text{RealCar}.$$

It's not hard to imagine a hierarchical tree structure: we might have many types of vehicles (cars, boats, airplanes, submarines), which each might have many types of manifestations (car alone might branch to BMW, Mercedes, Subaru, Ford, Ferrari, Toyota, Trabant, etc.).

Finally, we used *type hinting* to specify the types of inputs and outputs we should expect from our abstract methods. Type hinting is not limited to abstract classes; it can be used when defining methods for any class. We typically omit type hinting from our examples to save space, though it is typically beneficial to use in practice.

Polymorphism

Our final principle of OOP is polymorphism, which simply means that objects from different classes can be designed to share behaviors. For example, all vehicles descendant from the `AbstractVehicle` method defined in Code Block 6.13 must have an implementation for `speedUp`, `timeLapse`, and `stop`. Their internal workings might vary dramatically: how an airplane *speeds up* is quite different than how a skateboarder speeds up. Polymorphism allows us to instantiate a fleet of vehicles from our `AbstractVehicle` family—cars, bicycles, trains, planes, ships, submarines, and skateboarders—and use them agnostically in regards to their internal workings. We will return to this concept in our discussion of factory methods, and again when we discuss how the object-oriented paradigm can be used in a data-science context.

6.3.3 Generators, Wrappers, and Factories

Special Methods

Certain methods of a class have names that are both preceded and followed by double underscore, such as `__init__` and `__str__`. These are referred to as *special methods*, and each have specific meaning related to how objects can be used in a Python environment. Commonly used special methods are shown in Tables 6.6 and 6.7. For a more comprehensive list, see `docs.python.org`[8]. Though many of these special methods are not used in practice, it is useful to be aware of them.

---

[8] https://docs.python.org/3/reference/datamodel.html#special-method-names

method	usage	explanation
__init__(self,...)	x=X(...)	instantiates an object from class X
__str__(self)	print(x)	returns a string representation of object
__len__(self)	len(x)	returns the length of object
__eq__(self, other)	x==y	compares two objects for equality
__getitem__(self, key)	x[key]	gets the value associated with a key
__setitem__(self, key, value)	x[key] = value	sets value of a key
__contains__(self, item)	item in x	determine if item in object
__iter__(self)	iter(x)	returns an iterator object
__next__(self)	next(x)	returns next item from container

Table 6.6: Special methods in Python for class X with objects x and y.

method	usage	explanation
__add__(self, other)	x + y	add
__sub__(self, other)	x - y	subtract
__mul__(self, other)	x * y	multiply
__truediv__(self, other)	x / y	divide
__pow__(self, other)	x ** y	power
__lt__(self, other)	x < y	less than
__le__(self, other)	x <= y	less than or equal to
__eq__(self, other)	x == y	equals to
__ne__(self, other)	x != y	not equals to
__gt__(self, other)	x > y	greater than
__ge__(self, other)	x >= y	greater than or equal to

Table 6.7: Math-type special methods in Python for class X with objects x and y.

Iterators and Generator Functions

In python, an *iterable object* is simply an object that is an object that is capable of returning its data, one element at a time. Iterables are often used in the context of a **for** loop. An iterable object must implement two special methods, __iter__ and __next__, collectively known as the *iterator protocol*. Lists, tuples, and dictionaries are all *iterable* objects.

An *iterator* is an object that represents a specific stream of data. The __iter__ method of an iterable object must return an iterator. Once an iterator has been instantiated, repeated calls to its __next__ method return successive items from the stream, until a **StopIteration** exception is raised.

For example, if we define a list my_list = [1,2,3,4,5], calling next(my_list) throws an error, as my_list is not itself an iterator. Instead, if we call my_iter = iter(my_list), we can then call next(my_iter) multiple times, receiving, in turn, the values 1–5, and the **StopIteration** exception thereafter.

Example 6.11. An iterable that represents the Fibonacci sequence is shown in Code Block 6.14. Once an object has been instantiated, we can iterate

```
1   class Fibonacci:
2       def __init__(self, max_length=10):
3           self.max_length = max_length
4       def __iter__(self):
5           self.position = 0
6           self.lag1 = 0
7           self.lag2 = 0
8           return self
9       def __next__(self):
10          self.position += 1
11          if self.position > self.max_length:
12              raise StopIteration
13
14          next_item = 1 if self.position == 1 else self.lag1 +
                  self.lag2
15          self.lag1, self.lag2 = next_item, self.lag1
16          return next_item
17
18  fib = Fibonacci()
19  for x in fib:
20      print(x) # prints 1, 1, 2, 3, 5, 8, 13, 21, 34, 55
```

Code Block 6.14: Defining an iterable in Python.

over it using a for-loop, as shown on lines 18–20.   ▷

The code required to define an iterator is, however, a bit verbose. A commonly used simplification is that of a *generator function*, which creates an iterator without the need to explicitly invoke the iterator protocol. Generator functions are similar to regular Python functions, except they have one or more `yield` statements, instead of the typical **return** statement.

Example 6.12. The Fibonacci numbers can be generated using the generator function of Code Block 6.15, achieving the same result as Code Block 6.14.

```python
def fibonacci(max_length=10):
    position = 0
    lag1 = 0
    lag2 = 0
    while position < max_length:
        position += 1
        next_item = 1 if position == 1 else lag1 + lag2
        lag1, lag2 = next_item, lag1
        yield next_item

for x in fibonacci():
    print(x) # prints 1, 1, 1, 3, 5, 8, 13, 21, 34, 55

f = fibonacci()
next(f) # 1
next(f) # 1
next(f) # 2
# .... 3, 5, and so forth...

sum(fibonacci()) # 143
```

Code Block 6.15: Fibonacci numbers using generators

The generator function `fibonacci` defined in Code Block 6.15 can be used in the context of a for loop, as shown on lines 11–12. The function itself returns a fresh iterator object, which can be used to get successive values through the **next** method, as shown on lines 14–17. Additionally, we can simply sum the iterator object returned by a call to `fibonacci()`, as shown on line 19.   ▷

Wrapper Functions and Decorators

In Python, a *wrapper function* is a special type of function that takes a function as input[9] and returns a new function as output. They can be

---

[9] In Python, functions are objects, and therefore they may be passed into other functions as inputs.

used to "wrap" around other functions by adding a decorator—a line of code with the @ symbol followed by the name of the wrapper function— to the line *preceding* the function definition, in the same way we used the @abstractmethod command to decorate various functions in Code Block 6.13. An example of a simple wrapper is shown in Code Block 6.16.

```python
def greetings(func):

    def wrapper(*args, **kwargs):
        print("Hello, there!")
        result = func(*args, **kwargs)
        print("Googdbye!")
        return result

    return wrapper

@greetings
def pow2(x, power=2):
    return x**power

t = pow2(5, power=3) # Prints Hello / Goodbye; sets t = 125
```

Code Block 6.16: A simple wrapper.

Wrappers are often used when defining hierarchies. For example, there might be a number of assertions, or data checks, that many methods of various subclasses share. Instead of writing these assertions into each individual method, we can use a single wrapper function in the parent class that is then inherited to each subclass. For example, we can define a wrapper in the AbstractVehicle abstract class defined in Code Block 6.13. The code is shown for defining the wrapper function is shown in lines 5–13 of Code Block 6.17.

The _timeWrap decorator can be used to wrap any of the methods from the Car class, which is subclassed from AbstractVehicle. Proper usage of the decorate is shown on line 19. Notice that in order to access the decorator, we must make reference to the class in which the method is defined. (Technically, the method _timeWrap is implemented as a static method, which we discuss next.)

Static Methods and Class Methods

Sometimes it is useful to define methods, within a class, that can be called *without reference to any particular object*. This scenario occurs with two distinct flavors: static methods and class methods. *Static methods* are simply

```
1   class AbstractVehicle(ABC):
2
3       # .... Methods from AbstractVehicle
4
5       def _timeWrap(func):
6
7           def wrapper(*args, **kwargs):
8               t = time.time()
9               results = func(*args, **kwargs)
10              print(f"Time to execute {func.__name__} is: {time.time()
                    - t}")
11              return results
12
13          return wrapper
14
15  class Car(AbstractVehicle):
16
17      # ....
18
19      @AbstractVehicle._timeWrap
20      def speedUp(self, delta_v=0):
21          # ....
```

Code Block 6.17: Adding a wrapper to the **AbstractVehicle** class.

methods attached to a class that can be used without instantiating an object from that class. *Class methods* are similar, except that a class method is allowed to modify any of the *class attributes*, or variables attached to the class itself, as opposed to individual objects. Class attributes are uniform and accessible across all objects from the class; this is a particularly handy way to store constant that do not vary object-to-object. Class methods must take the variable `cls`, instead of `self`, as their first argument, which refers to the class itself, and not the object. Static methods should have neither `cls` nor `self` as arguments, as they are pure functions attached to the class, and can modify neither object nor class.

Class methods should be preceded by the `@classmethod` decorator. Static methods should be preceded by the `@staticmethod` operator. The exception to the latter rule is when defining static decorators, as in lines 5–13 of Code Block 6.17. Though _timeWrap is a static method, the absence of the `@staticmethod` decorator simply means it is not available when attached to any object of the class.

Factories

Static methods are commonly used to define *factories*, which are certain methods used to generate new objects from a particular family of classes.

Example 6.13. Suppose that we have defined a variety of subclasses from the `AbstractVehicle` class, including `Bmw`, `Submarine`, and `Airplane`. A simple factory method and its usage is shown in Code Block 6.18.

```python
class VehicleFactory:

    @staticmethod
    def get(name, *args, **kwargs):
        lookup = {'bmw': Bmw,
                  'sub': Submarine,
                  'airplane': Airplane}

        return lookup[name](*args, **kwargs)

c = VehicleFactory.get('bmw', 'my sports car', 'red')
```

Code Block 6.18: Factory method to generate various objects from the `AbstractVehicle` family.

Note that, unlike other classes we've studied, the `VehicleFactory` never gets instantiated. We *directly* invoke the `get` method of the `VehicleFactory` class, without reference to any specific instance of the class. This method takes as argument a name, used as the key to a lookup map, and returns an instantiated object of the appropriate type.

We could, alternatively, define the lookup map *outside* of the `get` method, making it a class attribute. If we did so, we would instead define `get` as a class method, and reference `cls.lookup`.

Another variation is to use a factory method to *dynamically* generate new Python classes on the fly, using the built-in `type` function. Suppose, in addition to our `AbstractVehicle` class, we also had an `AbstractEngine` class, with various subclasses, such as `Diesel`, `Turbine`, `Combustion`, and `Rocket`. Further, suppose that any engine can be used for any type of vehicle, as they each follow a consistent API. By following the convention `vehicle_engine`, we can dynamically generate an object that inherits the vehicle components from the `vehicle` class and the engine components from the `engine` class, as shown in Code Block 6.19.

This factory returns an instantiated object from the new class, inheriting properties from both the vehicle as well as the engine. This is a convenient way to handle the case in which a class has multiple components,

```
1  class VehicleFactory:
2
3      # class variables
4      vehicles = {'bmw': BMW,
5                  'sub': Submarine,
6                  'airplane': Airplane}
7
8      engines = {'diesel': Diesel,
9                 'turbine': Turbine,
10                 'combustion': Combustion,
11                 'rocket': Rocket}
12
13     @classmethod
14     def get(cls, name, *args, **kwargs):
15         assert '_' in name
16         vehicle, engine = name.split('_')
17
18         return type(name, (cls.vehicles[vehicle],
19             cls.engines[engine]), {})(*args, **kwargs)
20  c = VehicleFactory.get('bmw_rocket', 'my rocket car', 'red')
21  type(c) # prints: __main__.bmw_rocket
```

Code Block 6.19: Dynamic class creation in a factory method

and each component can have multiple variations. Instead of defining each permutation as an individual subclass, we can define an abstract class for each component, and dynamically cast one subclass from each component together to generate a new mix-and-match object.   ▷

6.3.4 OOP in Data Science

We next consider how we can deploy the principles of object-oriented programming to data science practice. In doing so, we seek to benefit from the various benefits offered by an object-oriented mindset, including use of modular, reusable code as well as a structure for building data-science products in teams.

By leveraging the four principles of object-oriented design, we can deploy greatly simplified runtime scripts, as shown in Code Blocks 6.20 and 6.21. Py3912 shows the overall structure of the file, whereas Code Block 6.20 defines the main function. Notice that all of the complexity is abstracted within three classes: our database class D, which can read and write to our database, our schema class S, which defines the table structure (e.g., column types, partitions, etc.) for each table we wish to write to, and our model class M, which encapsulates all of the implementation details of our

```
 1  def main(project, model, mode, db_src, exec_id, **config):
 2
 3      D = DataBaseFactory.get(db_src)
 4      S = SchemaFactory.get(project)
 5
 6      S.set('run_config')
 7      df_config = DataFrame({'project': project,
 8                             'model': model,
 9                             'mode': mode,
10                             'exec_id': exec_id,
11                             'config': str(config)}, index=[0])
12      D.put(S, df_config)
13
14      input_table = config.pop('input_table')
15      target_table = config.pop('target_table')
16      cols = config.pop('cols')
17      target_cols = config.pop('target_cols')
18      time_col = config.pop('time_col')
19      point_in_time = config.pop('point_in_time',
20          datetime.datetime.utcnow().strftime('%Y-%m-%d'))
20      delay = config.pop('delay', 30)
21      window = config.pop('window': 30)
22      range_ = config.pop('range_': 90)
23
24      df_train = D.get(table=input_table,
25                       cols=cols,
26                       time_col=time_col,
27                       time=[point_in_time, delay, window])
28
29      df_labels = D.get(table=target_table,
30                        cols=cols,
31                        time_col=time_col,
32                        time=[point_in_time, delay, window])
33
34      df_predict = D.get(table=input_table,
35                         cols=target_cols,
36                         time_col=time_col,
37                         time=[point_in_time, 0, range_])
38
39      M = ModelFactory.get(project, model, **config)
40      M.train(df_train, df_labels)
41      df_out = M.predict(df_predict)
42      df_out['exec_id'] = exec_id
43
44      S.set('run_output')
45      D.put(S, df_out)
```

Code Block 6.20: Example of a runtime script.

```
1   import datetime, argparse
2   import SchemaFactory, DataBaseFactory, ModelFactory
3
4   def main(project, model, mode, db_src, exec_id, **config):
5       # ...
6       # main script here....
7
8   if __name__ == '__main__':
9       parser = argparse.ArgumentParser(description='Inputs for
            Project.')
10      parser.add_argument('--project', type=str, required=True)
11      parser.add_argument('--model', type=str, required=True)
12      parser.add_argument('--mode', type=str, default='TEST')
13      parser.add_argument('--db_src', type=str, default='hive')
14      parser.add_argument('--exec_id', type=int, required=True)
15      # .... additional model configuration parameters
16
17      args = parser.parse_args()
18      main(**args)
```

Code Block 6.21: Example of a runtime script (continued)

model. Additionally, by defining an abstract Model class, as shown in Code Block 6.22, we can ensure that whatever model we deploy will be compatible with our main script, as long as it is subclassed from Model with an implementation of the train and predict methods.

Moreover, by leveraging the principle of polymorphism, we construct our three classes using factory methods, so that our code is agnostic as to which particular subclass of each we are using, as long as that subclass follows the API laid out by its abstract base class. In particular, note that a single script can be leveraged across multiple projects, and many models of the same kind within a project. The model, itself, is simply passed as a parameter into the script, and the factory method delivers the appropriate subclass corresponding to that model. Moreover, we parameterized the point_in_time, so that the same runtime script can be reused in the backtesting framework, as discussed in Section 6.1.6. Another unexpected advantage occurred for one data science team, when their organization migrated to Hive, as we see in our next example.

Example 6.14. A large organization is planning on migrating all its data from Vertica to Hive. One data science team in the organization consists of a number of data scientists who operate mostly independently, writing their own scripts and methods for accessing the database. Moreover, many projects that have been finished have the database queries and connections hard-coded into dusty scripts that are scattered about. The team spends

```python
from abc import ABC, abstractmethod
class Model:

    def __init__(self, **kwargs):
        # override to provide for default parameter values
        self.params = kwargs


    @abstractmethod
    def train(self,
              df: DataFrame,
              y: np.array,
              weights: np.array=None):
        return


    @abstractmethod
    def predict(self,
                df: DataFrame):
        return
```

Code Block 6.22: Abstract Model class

several months determining all of their production dependencies and over-writing the code in order to point to the new data source. They also have to determine how, exactly, they are going to switch over to Hive, given that the rollout occurs over time and not all of the data is available yet, though they are still constrained to hit the deprecation timeline. The team loses an entire quarter of productivity handling the migration.

Another team, one which embraced the principles of object-oriented design from the ground up, has a much simpler task ahead of it. First, it must write a new subclass from following their **AbstractDataBase** API, replacing the functionality of **DataBase_Vertica** with **DataBase_Hive**. In particular, they must connect to a new data source and update their queries, as there are slight syntactical differences between the two languages. Next, they add one line to their **DataBaseFactory** lookup map:

$$\text{'hive'} : \text{DataBaseHive}.$$

Finally, they update a single config file that stores **db_src** for all production jobs, changing it to point from **vertica** to **hive**. And Voila! With a few simple changes, the team has migrated all jobs across all projects to the new database. Moreover, they only required help from a single member of their team to write the new database subclass. By using the principles of inheritance, encapsulation, abstraction, and polymorphism, the actual runtime scripts don't need to be changed at all, as they are agnostic with regard to the actual database engine that delivers its data.   ▷

In addition to the migration problem of Example 6.14, there are two other benefits of generating a database class from a factory method. The first is that it might not be possible to connect to Hive from local laptops. Instead, the team might connect to Presto. The factory method allows them to continue to run their jobs locally, by simply passing db_src='presto' instead of db_src='hive'. The second advantage is when writing unit tests for the code. Sample data can be stored in the form of .csv files in a data folder within the project. A new database subclass, DataBase_Csv, can then be created to read from the sample datasets in the local folder, requiring no connection to any database. The testing scripts can then pass db_src='csv' into the runtime script, so that unit tests will always have access to sample data, without wasting time creating database connections or managing the different connections required based on whether the tests are run locally or on a virtual workstation.

The final advantage of an object-oriented design is that it lubricates collaboration within teams. Since everything follows the same API, different members can easily spin up different models, or try different approaches, in a seamless fashion. One can easily spin up a new model subclass, add the new subclass to the model factory, and then deploy a backtest to see how well the new model would have fared compared to the current production model. Moreover, the modular, a la carte nature of the object-oriented design allows anyone on the team to run any model from anyone else. Such a system is therefore more manageable and maintainable as individual contributors on the team migrate switch roles or migrate to other teams.

Problems

6.1. Construct an counterexample to the claim that $R^2 \geq 0$; i.e., construct a data set and a predictive model with a negative coefficient of determination.

6.2. Show that the softmax transformation defined in Equation (6.12) has the properties $f_i(X) \in (0,1)$ and $\sum_{i=1}^{m} f_i(X) = 1$.

6.3. Prove Proposition 6.2.

6.4. Determine which of the following discrete variables should be ordinal versus categorical:

1. Age range: {18–24, 25–34, 35–44, 45–54, 55-64, 65+};
2. Gender: {male, female};
3. Year in college: {freshman, sophomore, junior, senior};
4. Brand: {Nike, Adidas, Asics};
5. Compass heading {north, south, east, west};
6. Blood type {A, B, AB, O}.

6.5. Prove that every label encoder is invertible.

6.6. Write your own transformer class in Python that handles one-hot encoding of a categorical variable.

6.7. Write an abstract class for each of the three classes required for the runtime script Code Block 6.20: `AbstractSchema`, `AbstractDataBase`, `AbstractModel`. *Hint*: The SQL syntax for `SELECT`, `CREATE TABLE`, and `INSERT` might vary from database to database, so that `AbstractDataBase` should house the abstract methods to generate the given SQL strings.

7

To Regression and Beyond

7.1 Linear Regression

And now for some mathematics.

7.1.1 The Model

Linear regression is the cornerstone of statistical analysis. We present the basic analysis here; for more details see Dunn and Smyth [2018], Hastie, *et al.* [2009], or James, *et al.* [2013]. For details of many of the related proofs, see Olive [2017] and Seber and Lee [2003].

Regression on Random Variables

Definition 7.1. *A* linear regression *of a random variable Y over a set of random variables $\{X_1, \ldots, X_p\}$, all defined over a common probability space, is the random variable*

$$\hat{Y} = \beta_0 + \sum_{j=1}^{p} X_j \beta_j, \tag{7.1}$$

where the regression coefficients $\beta_0, \ldots, \beta_p$ *are obtained by minimizing the* expected squared error

$$\text{MSE} = \mathbb{E}\left[\left(Y - \beta_0 - \sum_{j=1}^{p} X_j \beta_j\right)^2\right]. \tag{7.2}$$

In a linear regression, the random variables

$$\epsilon = Y - \hat{Y}$$

are referred to as the residuals *or* residual errors.

A *linear regression model is* simple *if if consists of a single explanatory variable (p = 1). Otherwise, it is said to be a* multiple linear regression model.

Note 7.1. In a linear regression, the regressors $X_1, \ldots, X_p$ are also referred to as *explanatory variables* or *features*. A linear regression model with p explanatory variables has a total of $p+1$ unknown model parameters. The parameter β_0 is referred to as the *intercept*, whereas the parameters $\beta_1, \ldots, \beta_p$ are referred to as *slopes*.   ▷

Proposition 7.1. *The regression coefficients $\beta_0, \ldots, \beta_p$ to a linear regression of Y on $\{X_1, \ldots, X_p\}$ are the solutions to the system of equations*

$$\mathbb{E}\left[\left(Y - \beta_0 - \sum_{j=1}^{p} \beta_j X_j\right) X_k\right] = 0, \tag{7.3}$$

for $k = 0, \ldots, p$, where we take $X_0 = 1$.

Proof. Equation (7.3) is obtained by differentiating Equation (7.2) with respect to the parameters β and setting the result equal to zero.   □

Note that the system of equations Equation (7.3) is equivalent to

$$\beta_0 + \sum_{j=1}^{p} \beta_j \mathbb{E}[X_j] = \mathbb{E}[Y] \tag{7.4}$$

$$\beta_0 \mathbb{E}[X_k] + \sum_{j=1}^{p} \beta_j \mathbb{E}[X_j X_k] = \mathbb{E}[Y X_k], \tag{7.5}$$

for $k = 1, \ldots, p$. This can be expressed in matrix form as

$$\begin{bmatrix} 1 & \mathbb{E}[X_1] & \cdots & \mathbb{E}[X_p] \\ \mathbb{E}[X_1] & \mathbb{E}[X_1^2] & \cdots & \mathbb{E}[X_1 X_p] \\ \vdots & & \ddots & \vdots \\ \mathbb{E}[X_p] & \mathbb{E}[X_p X_1] & \cdots & \mathbb{E}[X_p^2] \end{bmatrix} \cdot \begin{bmatrix} \beta_0 \\ \beta_1 \\ \vdots \\ \beta_p \end{bmatrix} = \begin{bmatrix} \mathbb{E}[Y] \\ \mathbb{E}[Y X_1] \\ \vdots \\ \mathbb{E}[Y X_p] \end{bmatrix}. \tag{7.6}$$

The matrix on the left can be inverted to solve for the coefficients $\beta_0, \ldots, \beta_p$.

Note that by applying the law of total expectation (Equation (1.19)) to the first equation in our system Equation (7.3), we can alternatively this equation as

$$\mathbb{E}[\mathbb{E}[Y|X] - \hat{Y}] = 0. \tag{7.7}$$

When the values of X are regarded as fixed, as opposed to random, we can drop the outer expectation and express this as the class *regression equation*

$$\hat{Y} = \mathbb{E}[Y|X]. \tag{7.8}$$

However, when X is regarded as random, the regression equation is only true *on average* (i.e., Equation (7.7)) or in special cases (e.g., the joint distribution is Gaussian).

Conditional Covariance and Regression

We next discuss a useful result that links regression with conditional covariance. First, we define conditional covariance as a straightforward generalization of Definition 1.17.

Definition 7.2. *The* conditional covariance *of random variables X and Y, conditioned on the random variable Z, is defined as*

$$\mathrm{COV}(X, Y|Z) = \mathbb{E}\left[(X - \mathbb{E}[X|Z])(Y - \mathbb{E}[Y|Z])|\, Z\right]. \tag{7.9}$$

It is easy to show that the conditional covariance is equivalent to

$$\mathrm{COV}(X, Y|Z) = \mathbb{E}[XY|Z] - \mathbb{E}[X|Z]\mathbb{E}[Y|Z], \tag{7.10}$$

in analogy to Equation (1.13).

Proposition 7.2. *Given random variables X, Y, and Z. Then the expected conditional covariance is given by*

$$\mathbb{E}\left[\mathrm{COV}(X, Y|Z)\right] = \mathrm{COV}(X - \mathbb{E}[X|Z], Y - \mathbb{E}[Y|Z]). \tag{7.11}$$

Note that the expectation on the left-hand side of Equation (7.11) is with respect to the random variable Z.

Proof. First, note that

$$\mathbb{E}[X - \mathbb{E}[X|Z]] = 0 \qquad \text{and} \qquad \mathbb{E}[Y - \mathbb{E}[Y|Z]] = 0,$$

from the law of total expectation. Therefore, the covariance of $X - \mathbb{E}[X|Z]$ and $Y - \mathbb{E}[Y|Z]$ is given by the expectation of the product

$$\mathrm{COV}(X - \mathbb{E}[X|Z], Y - \mathbb{E}[Y|Z]) = \mathbb{E}[(X - \mathbb{E}[X|Z])(Y - \mathbb{E}[Y|Z])]$$
$$= \mathbb{E}[XY - X\mathbb{E}[Y|Z] - Y\mathbb{E}[X|Z] + \mathbb{E}[X|Z]\mathbb{E}[Y|Z]].$$

Now, by applying the law of total expectation, and noting that both $\mathbb{E}[X|Z]$ and $\mathbb{E}[Y|Z]$ are functions of Z, we obtain

$$\mathrm{COV}(X - \mathbb{E}[X|Z], Y - \mathbb{E}[Y|Z]) = \mathbb{E}\left[\mathbb{E}[XY|Z] - \mathbb{E}[X|Z]\mathbb{E}[Y|Z]\right].$$

Finally, we recognize the right-hand side as the expected conditional covariance expressed in the form of Equation (7.10), thereby completing the result. □

Regression on Data

Oftentimes we are interested in estimating the regression coefficients from a set of data. We can do this by replacing the expected squared error Equation (7.2) with the sum of squared errors, and then solving for the coefficients using elementary calculus. The result of such a procedure is defined as follows.

Definition 7.3. *Given a weighted data set* $\{(x_{i1}, \ldots, x_{ip}, y_i, w_i)\}_{i=1}^n$, *the least-squares solution is an estimate for the linear regression coefficients that consists of the unique set of values* $\hat{\beta}_0, \ldots, \hat{\beta}_p$ *that minimizes the sum of squared errors*

$$\text{SSE}(\beta) = \sum_{i=1}^n w_i \left(y_i - \beta_0 - \sum_{j=1}^p x_{ij}\beta_j \right)^2. \tag{7.12}$$

Note that we can recast Equation (7.12) in matrix form as

$$\text{SSE}(\beta) = (\mathbf{y} - \mathbf{X}\beta)^T \mathbf{W}(\mathbf{y} - \mathbf{X}\beta), \tag{7.13}$$

where we define the matrix $\mathbf{X} \in \mathbb{R}^{n \times (p+1)}$ by $\mathbf{X}_{i0} = 1$ and $\mathbf{X}_{ij} = x_{ij}$, for $j \neq 0$, known as the *design matrix*, and the $n \times n$ diagonal matrix $W = \text{diag}(w_1, \ldots, w_n)$. Naturally, $\mathbf{y} \in \mathbb{R}^n$ and $\beta \in \mathbb{R}^{p+1}$.

It is a straightforward exercise to show the following.

Proposition 7.3. *The estimate* $\hat{\beta}$ *is a global minimum of Equation* (7.13) *if and only if*

$$(\mathbf{X}^T\mathbf{W}\mathbf{X})\,\hat{\beta} = \mathbf{X}^T\mathbf{W}\mathbf{y}. \tag{7.14}$$

This equation is known as the normal equation

We will leave the proof as an exercise for the reader (Exercise 7.1). It follows from the normal equation that the least-squares solution to the linear regression problem is given by

$$\hat{\beta} = (\mathbf{X}^T\mathbf{W}\mathbf{X})^{-1} \mathbf{X}^T\mathbf{W}\mathbf{y}. \tag{7.15}$$

Note 7.2. It is more computationally efficient to solve the system of equations given by Equation (7.14) than it is to solve Equation (7.15) directly. Python built-in solvers are, however, more efficient still. The scikit-learn method `sklearn.linear_model.LinearRegression` is actually a wrapper around the least-squares solver `scipy.linalg.lstsq`, which is based on an efficient form of the singular-value decomposition. ▷

Corollary 7.1. *For a simple linear regression problem, the least-squares coefficients are given by*

$$\hat{\beta}_0 = \bar{y} - \hat{\beta}_1 \bar{x},$$

$$\hat{\beta}_1 = \frac{SS_{xy}}{SS_x} = \frac{\sum_{i=1}^{n} w_i \left(x_i - \bar{x}_i \right) y_i}{\sum_{i=1}^{n} w_i \left(x_i - \bar{x}_i \right)^2},$$

where $\bar{x}$ and $\bar{y}$ are the weighted sample means

$$\bar{x} = \frac{1}{\omega} \sum_{i=1}^{n} w_i x_i \qquad and \qquad \bar{y} = \frac{1}{\omega} \sum_{i=1}^{n} w_i y_i,$$

where $\omega = \sum_{i=1}^{n} w_i$ is the total weight, as before.

We leave the proof to the reader (Exercise 7.2).

7.1.2 Analysis of the Estimates

A multilinear regression model has $p + 1$ unknown coefficients, $\beta_0, \ldots, \beta_p$, which are estimated by Equation (7.15), and an additional parameter σ^2 that represents the unknown variance. In this section, we discuss basic results regarding estimates for these parameters.

Standard Errors

We begin with a basic result about our estimates Equation (7.15) for the unknown coefficients $\beta_0, \ldots, \beta_p$.

Theorem 7.1. *Let $\hat{\beta} \in \mathbb{R}^{p+1}$ be the least-squares solution of a linear regression problem. Then*

$$\mathbb{E}[\hat{\beta}] = \beta \tag{7.16}$$

$$\mathbb{V}(\hat{\beta}) = \sigma^2 (\mathbf{X}^T \mathbf{W} \mathbf{X})^{-1}. \tag{7.17}$$

Proof. We know $\mathbb{E}[\mathbf{y}] = \mathbf{X}\beta$; hence Equation (7.16) follows immediately from Equation (7.15).

To derive Equation (7.17), first recall that $\mathbb{V}(\mathbf{y}) = \mathbf{W}^{-1}\sigma^2$. In addition, it is a general property of the variance–covariance matrix that $\mathbb{V}(\mathbf{A}\mathbf{y}) = \mathbf{A}\mathbb{V}(\mathbf{y})\mathbf{A}^T$. The result follows by setting $\mathbf{A} = \left(\mathbf{X}^T \mathbf{W} \mathbf{X}\right)^{-1} \mathbf{X}^T \mathbf{W}$.   □

Our next result is to show that, for linear regression, the total sum of squares equals the sum of the sum of squared errors and the regression sum of squares. This result, however, will rely on the following lemma.

Lemma 7.1. *Let $\mathbf{H} = \mathbf{X}(\mathbf{X}^T \mathbf{W} \mathbf{X})^{-1} \mathbf{X}^T \in \mathbb{R}^{n \times n}$, so that the least squares estimates are given by*

$$\hat{\mathbf{y}} = \mathbf{H}\mathbf{W}\mathbf{y}.$$

Then the following results hold:

(a) $\mathbf{H}$ *is symmetric and* $\mathbf{HWH} = \mathbf{H}$,
(b) $\mathbf{X}^T\mathbf{W}(\mathbf{y} - \hat{\mathbf{y}}) = \mathbf{0}$,
(c) $\sum_{i=1}^n w_i(y_i - \hat{y}_i) = 0$.
(d) Assuming $n > p + 1$, *the sum* $\text{trace}(\mathbf{HW}) = p + 1$.

Proof. Symmetry of $\mathbf{H}$ follows from its definition. Similarly, the result $\mathbf{HWH} = \mathbf{H}$ is straightforward to verify.

For (b), we use $\hat{\mathbf{y}} = \mathbf{X}\hat{\boldsymbol{\beta}}$ with Equation (7.15) to obtain

$$\mathbf{X}^T\mathbf{W}(\mathbf{y} - \hat{\mathbf{y}}) = \mathbf{X}^T\mathbf{W}\left(\mathbf{y} - \mathbf{X}(\mathbf{X}^T\mathbf{WX})^{-1}\mathbf{X}^T\mathbf{Wy}\right)$$
$$= \mathbf{X}^T\mathbf{Wy} - \mathbf{X}^T\mathbf{Wy} = \mathbf{0}.$$

Part (c) follows directly from part(b). (This is the first component of the vector equation given in part (b), since the first row of $\mathbf{X}^T$ is a row of 1s.)

For part (d), let λ be an eigenvalue of $\mathbf{HW}$, so that there exists a nonzero $\mathbf{v}$ with $\mathbf{HWv} = \lambda\mathbf{v}$. Next, consider the quantity

$$\lambda\mathbf{v}^T\mathbf{v} = \mathbf{v}^T\mathbf{WHv}$$
$$= \mathbf{v}^T\mathbf{WHWHv}$$
$$= (\mathbf{HWv})^T\mathbf{WHv}$$
$$= \lambda\mathbf{v}^T\mathbf{WHv}$$
$$= \lambda^2\mathbf{v}^T\mathbf{v}.$$

It follows that $\lambda(1 - \lambda) = 0$. Since $\text{rank}(\mathbf{HW}) = p + 1$, it follows that there are exactly $p + 1$ unity eigenvalues and $n - p - 1$ zero eigenvalues of the matrix $\mathbf{HW}$. Since the trace is the sum of the eigenvalues, the result follows. □

Theorem 7.2. *In a linear regression problem, the total sum of squares is the sum of the residual and regression sums of squares; i.e.,*

$$\text{SST} = \text{SSE} + \text{SSR}.$$

Proof. This follows as long as the cross terms sum to zero; i.e., as long as

$$\sum_{i=1}^n w_i(y_i - \hat{y}_i)(\hat{y}_i - \bar{y}) = 0.$$

Separating this into two sums, we have

$$\sum_{i=1}^n w_i(y_i - \hat{y}_i)(\hat{y}_i - \bar{y}) = \sum_{i=1}^n w_i(y_i - \hat{y}_i)\hat{y}_i - \bar{y}\sum_{i=1}^n w_i(y_i - \hat{y}_i).$$

The second term on the right-hand side vanishes due to part (c) of Lemma 7.1. The first term on the right-hand side vanishes as well: since $\hat{\mathbf{y}} = \mathbf{HWy}$, it follows that

$$\sum_{i=1}^{n} w_i(y_i - \hat{y}_i)\hat{y}_i = (\mathbf{HWy})^T \mathbf{W}(\mathbf{I} - \mathbf{HW})\mathbf{y}$$

$$= \mathbf{y}^T \mathbf{WHW}(\mathbf{I} - \mathbf{HW})\mathbf{y}$$
$$= \mathbf{y}^T \mathbf{W}(\mathbf{HW} - \mathbf{HWHW})\mathbf{y}$$
$$= \mathbf{y}^T \mathbf{W}(\mathbf{HW} - \mathbf{HW})\mathbf{y},$$

where we used part (a) of Lemma 7.1. The result follows.   □

Theorem 7.3. *The quantity*

$$\hat{\sigma}^2 = \text{MSE} = \frac{\text{SSE}}{n - p - 1} \qquad (7.18)$$

is an unbiased estimator for σ^2.

Proof. This result is equivalent to the statement

$$\mathbb{E}[\text{SSE}] = \sigma^2(n - p - 1).$$

We start by writing the sum of squared errors as

$$\text{SSE} = (\mathbf{y} - \mathbf{X}\hat{\beta})^T \mathbf{W}(\mathbf{y} - \mathbf{X}\hat{\beta})$$
$$= \left[(\mathbf{y} - \mathbf{X}\beta) + (\mathbf{X}\beta - \mathbf{X}\hat{\beta})\right]^T \mathbf{W} \left[(\mathbf{y} - \mathbf{X}\beta) + (\mathbf{X}\beta - \mathbf{X}\hat{\beta})\right]$$
$$= (\mathbf{y} - \mathbf{X}\beta)^T \mathbf{W}(\mathbf{y} - \mathbf{X}\beta) + (\mathbf{X}\beta - \mathbf{X}\hat{\beta})^T \mathbf{W}(\mathbf{X}\beta - \mathbf{X}\hat{\beta})$$
$$+ 2(\mathbf{y} - \mathbf{X}\beta)^T \mathbf{W}(\mathbf{X}\beta - \mathbf{X}\hat{\beta}).$$

Now, $\mathbb{E}[\mathbf{y}] = \mathbf{X}\beta$ and $\mathbb{E}[\hat{\beta}] = \beta$, so that

$$\mathbb{E}[\text{SSE}] = \mathbb{E}[\mathbf{y}^T \mathbf{W}(\mathbf{y} - \mathbf{X}\beta)] - \mathbb{E}[\hat{\beta}^T \mathbf{X}^T \mathbf{W}(\mathbf{X}\beta - \mathbf{X}\hat{\beta})]$$
$$+ 2\mathbb{E}[\mathbf{y}^T \mathbf{W}(\mathbf{X}\beta - \mathbf{X}\hat{\beta})]$$
$$= \mathbb{E}[\mathbf{y}^T \mathbf{Wy}] - \beta^T \mathbf{X}^T \mathbf{WX}\beta$$
$$- \beta^T \mathbf{X}^T \mathbf{WX}\beta + \mathbb{E}\left[\hat{\beta}^T \mathbf{X}^T \mathbf{WX}\hat{\beta}\right]$$
$$+ 2\beta^T \mathbf{X}^T \mathbf{WX}\beta - 2\mathbb{E}\left[\mathbf{y}^T \mathbf{WX}\hat{\beta}\right].$$

However, recalling the fact that $\mathbf{X}\hat{\beta} = \mathbf{HWy}$, we find that both

$$\hat{\beta}^T \mathbf{X}^T \mathbf{WX}\hat{\beta} = \mathbf{y}^T \mathbf{WHWy} \text{ and } \mathbf{y}^T \mathbf{WX}\hat{\beta} = \mathbf{y}^T \mathbf{WHWy},$$

which therefore reduces our expression to

$$\mathbb{E}[\text{SSE}] = \mathbb{E}[\mathbf{y}^T \mathbf{Wy}] - \mathbb{E}\left[\mathbf{y}^T \mathbf{WHWy}\right]. \qquad (7.19)$$

To complete the proof, we rely on a result from linear algebra:

$$\mathbb{E}[\mathbf{y}^T \mathbf{A} \mathbf{y}] = \text{trace}(\mathbf{A} \mathbb{V}(\mathbf{y})) + \mathbb{E}[\mathbf{y}]^T \mathbf{A} \mathbb{E}[\mathbf{y}]. \qquad (7.20)$$

(See Seber and Lee [2003] for a proof.) Now, $\mathbb{E}[\mathbf{y}] = \mathbf{X}\boldsymbol{\beta}$ and $\mathbb{V}(\mathbf{y}) = \mathbf{W}^{-1}\sigma^2$. We therefore have

$$\mathbb{E}[\mathbf{y}^T \mathbf{W} \mathbf{y}] = \text{trace}(\mathbf{W} \mathbf{W}^{-1}\sigma^2) + \boldsymbol{\beta}^T \mathbf{X}^T \mathbf{W} \mathbf{X} \boldsymbol{\beta}$$

and

$$\mathbb{E}\left[\mathbf{y}^T \mathbf{W} \mathbf{H} \mathbf{W} \mathbf{y}\right] = \text{trace}(\mathbf{W} \mathbf{H} \mathbf{W} \mathbf{W}^{-1}\sigma^2) + \boldsymbol{\beta}^T \mathbf{X}^T \mathbf{W} \mathbf{H} \mathbf{W} \mathbf{X} \boldsymbol{\beta}.$$

But

$$\boldsymbol{\beta}^T \mathbf{X}^T \mathbf{W} \mathbf{H} \mathbf{W} \mathbf{X} \boldsymbol{\beta} = \boldsymbol{\beta}^T \mathbf{X}^T \mathbf{W} \mathbf{X} (\mathbf{X}^T \mathbf{W} \mathbf{X})^{-1} \mathbf{X}^T \mathbf{W} \mathbf{X} \boldsymbol{\beta} = \boldsymbol{\beta}^T \mathbf{X}^T \mathbf{W} \mathbf{X} \boldsymbol{\beta},$$

so that Equation (7.19) is equivalent to

$$\mathbb{E}[\text{SSE}] = \text{trace}(\mathbf{I}\sigma^2) - \text{trace}(\mathbf{W} \mathbf{H} \sigma^2) = \sigma^2 (n - (p+1)),$$

following part (d) of Lemma 7.1. This completes the result. $\qquad \square$

By replacing σ^2 with $\hat{\sigma}^2$ in Equation (7.17), we can estimate the variance of our least-squares solution $\hat{\boldsymbol{\beta}}$. The square roots of the diagonal elements are therefore estimates of our standard errors:

$$\hat{\text{se}}(\hat{\beta}_j) = \sqrt{\hat{\sigma}^2 (\mathbf{X}^T \mathbf{W} \mathbf{X})_{jj}^{-1}}. \qquad (7.21)$$

Distributional Results

So far, the only assumptions we've made on the random variables Y_i are independence and that $\mathbb{E}[\mathbf{y}|\mathbf{X}] = \mathbf{X}\boldsymbol{\beta}$ and $\mathbb{V}(\mathbf{y}) = \mathbf{W}^{-1}\sigma^2$. In order to derive results regarding the distribution of the least-squares estimates $\hat{\boldsymbol{\beta}}$, we must further make an assumption regarding the distribution of the error; i.e., we will next consider the case of *normal* linear regression problems. For normal error, we have

$$Y_i = \beta_0 + \sum_{j=1}^{p} X_{ij} \beta_j + \epsilon_i,$$

$$\epsilon_i \sim N(0, \sigma^2/w_i).$$

Alternatively, this is sometimes expressed as

$$Y_i \sim N\left(\beta_0 + \sum_{j=1}^{p} X_{ij} \beta_j, \frac{\sigma^2}{w_i}\right).$$

Theorem 7.4. *The least squares estimates $\hat{\beta}$ in a normal linear regression model are normally distributed as*

$$\hat{\beta} \sim N\left(\beta, \sigma^2(\mathbf{X}^T\mathbf{W}\mathbf{X})^{-1}\right). \tag{7.22}$$

Proof. From Equation (7.15), we know that the coefficients $\hat{\beta}$ are linear combinations of the training target variables $Y_1, \ldots, Y_n$, which are normally distributed. Therefore the coefficient vector $\hat{\beta}$ must have a multivariate normal distribution.

From Equations (7.16) and (7.17) we further know the expected value and variance matrix for $\hat{\beta}$. The form given in Equation (7.22) follows. □

Theorem 7.2 tells us that the total sum of squares in a linear regression model is equal to the sum of the residual sum of squares (sum of squared errors) and the regression sum of squares. Under the normality conditions, each of these components is further distributed as a chi-squared distribution, as shown in our next theorem.

Theorem 7.5. *In a normal linear regression model,*

$$\frac{\text{SSR}}{\sigma^2} \sim \chi_p^2 \tag{7.23}$$

$$\frac{\text{SSE}}{\sigma^2} \sim \chi_{n-p-1}^2 \tag{7.24}$$

$$\frac{\text{SST}}{\sigma^2} \sim \chi_{n-1}^2. \tag{7.25}$$

For a proof, we direct the reader to Seber and Lee [2003]. Given the number of degrees of freedom of each of the sum of squares, we define their respective mean-squared errors as

$$\text{MSR} = \frac{\text{SSR}}{p}, \qquad \text{MSE} = \frac{\text{SSE}}{n-p-1}, \qquad \text{and} \qquad \text{MST} = \frac{\text{SST}}{n-1}.$$

Note 7.3. The adjusted R^2 metric given by Definition 6.8 for linear regression should be

$$\bar{R}^2 = 1 - \frac{\text{MSE}}{\text{MST}} = 1 - \frac{\text{SSE}/(n-p-1)}{\text{SST}/(n-1)},$$

as there are a total of $p+1$ (not p) independent parameters. ▷

It can further be shown that $\hat{\sigma}^2$ and $\hat{\beta}$ are statistically independent.

Hypothesis Tests and Confidence Intervals

The sums of squares and mean squares are summarized nicely in an analysis of variance (ANOVA) table, as shown in Table 7.1. Moreover, due to Theorem 7.5, we know that under our normality assumptions, the F-statistic is distributed as $F \sim F_{p,n-p-1}$. We can therefore use this statistic to test whether or not the result of our regression should be accepted, as opposed to modeling the output with the average value.

Source	Degrees of freedom	Sum of squares	Mean square
Regression	p	$\text{SSR} = \sum_{i=1}^{n} w_i \left(\hat{y}_i - \bar{y}\right)^2$	$\text{MSR} = \dfrac{SSB}{p}$
Error	$n - p - 1$	$\text{SSE} = \sum_{i=1}^{n} w_i \left(y_i - \hat{y}_i\right)^2$	$\text{MSE} = \dfrac{SSW}{n - p - 1}$
Total	$n - 1$	$\text{SST} = \sum_{i=1}^{n} w_i \left(y_i - \bar{y}\right)^2$	$F = \text{MSR}/\text{MSE}$

Table 7.1: ANOVA table.

Theorem 7.6. *For a normal linear regression problem, the F-statistic,*

$$F = \frac{\text{MSR}}{\text{MSE}} = \frac{\text{SSR}/p}{\text{SSE}/(n - p - 1)} \tag{7.26}$$

is distributed as $F \sim F_{p,n-p-1}$.

Intuitively, a large value of F means that our predictions are farther off from the overall mean than they are to their true values. The F statistic can therefore be used to test whether or not our model performs better than had we simply used a constant prediction $\hat{y} = \bar{y}$; i.e., had we chosen $\hat{\beta}_1 = \cdots = \hat{\beta}_p = 0$. For a significance level α, we should therefore reject this null hypothesis whenever $F > F_{p,n-p-1}^{\alpha}$.

To test a null hypothesis about a particular coefficient, $H_0 : \beta_j = 0$, the distribution given by Equation (7.22) implies that, under the null hypothesis, the statistic

$$Z_j = \frac{\hat{\beta}_j}{\text{se}(\hat{\beta}_j)}$$

has a standard normal distribution. The standard error, however, depends on the variance σ^2, which is typically unknown. When we replace our estimate Equation (7.18) for the variance, we obtain the statistic

$$T_j = \frac{\hat{\beta}_j}{\hat{\text{se}}(\hat{\beta}_j)}, \tag{7.27}$$

which uses the estimated standard error given in Equation (7.21). This statistic has a Student's T distribution with $(n - p - 1)$ degrees of freedom: $T_j \sim t_{n-p-1}$. Naturally the null hypothesis is to be rejected whenever $|T_j| > t_{n-p-1}^{\alpha/2}$. A $(1 - \alpha)$ confidence interval is therefore given by

$$\hat{\beta}_j \pm \hat{\text{se}}(\hat{\beta}_j) t_{n-p-1}^{\alpha/2}.$$

Often, we may be interested in testing not only if a single parameter is significant, but if a subset of parameters are significant. For example, suppose a categorical feature is one-hot encoded into a set of k binary features. We may want to test whether that set of k features, as a group, is beneficial to the model. To proceed, we consider a *reduced model* with those features eliminated.

In general, consider a reduced model consisting of the constant term and the first q coefficients: $\{\beta_0, \ldots, \beta_q\}$, with $q < p$. Our null hypothesis is

$$H_0 : \beta_{q+1} = \cdots = \beta_p = 0,$$

against the alternative hypothesis H_A that at least one $\beta_j \neq 0$, for some $j > q$. In short, our null hypothesis is that the reduced model is better than the full model. Rejecting the null hypothesis implies that it is beneficial to include (at least some of) the additional features $\beta_{q+1}, \ldots, \beta_p$.

To test this null hypothesis, we train both models, and compute the sum of squared errors for both cases: SSE_0 is the sum of squared errors when using the reduced model and SSE_1 is the sum of squared errors when using the full model. We then express our F-statistic as

$$F = \frac{(\text{SSE}_0 - \text{SSE}_1)/(p - q)}{\text{SSE}_1/(n - p - 1)}. \tag{7.28}$$

Under the null hypothesis, $F \sim F_{p-q, n-p-1}$. We call this the *partial F test*. For the special case $q = 0$, $\text{SSE}_0 = \text{SST}$ and Equation (7.28) reverts back to Equation (7.26).

The the selection of a submodel for the partial F test must be carried out *before* the regression fit. If one performs a regression and then formulates a test with the worst performers removed, the resulting F statistic will be too high and the submodel will be prone to selection bias.

Example 7.1. Load the Boston housing price dataset from `sklearn.datasets` and fit a linear regression model using a variable number of features. Plot the test error and the generalization error (expected test error) as a function of the number of features. Use `print(data.DESCR)` for a description of the dataset. Use both the given features as well as their squared values.

We can accomplish this in Code Block 7.1. We build a pipeline (lines 11–20) that imputes missing values with the feature's median value, passes through the `X[:, 0]` column (of all 1s), and filters out the columns in excess of `n_features`. We then conduct `n_sims` simulations, where we use a test size of 80%, so that each model is trained on approximately 100 randomly selected instances. Note that fitting the intercept β_0 is done automatically by the `LinearRegression` class, so we do not need to append a column of 1s in front of X.

The code to produce the plots is given in Code Block 7.2. The results are shown in Figure 7.1.

In Fig3702, we plot the test (blue) and training (red) error for our first fifty simulations. We then plot the average test error (dark blue) and average

```
1   data = load_boston()
2   X, y = data.data, data.target
3   X = np.concatenate([X, X**2], axis=1)
4   n_sims = 500
5   n, p = X.shape
6   err_test = np.zeros((p+1, n_sims))
7   err_train = np.zeros((p+1, n_sims))
8
9   for n_features in range(1, p+1):
10
11      select_columns = ColumnTransformer(
12              [('standardize', RobustScaler(quantile_range=(0.2,0.8)) ,
                    np.arange(n_features))],
13              remainder='drop')
14
15      transformers = [
16              ('impute', SimpleImputer(strategy='median')),
17              ('scale_select', select_columns),
18              ('model', LinearRegression(normalize=False))]
19
20      pipe = Pipeline(transformers)
21
22      for i in range(n_sims):
23          X_train, X_test, y_train, y_test = train_test_split(X, y,
                  test_size=0.80)
24          pipe.fit(X_train, y_train)
25          y_hat = pipe.predict(X_test)
26          y_hat_train = pipe.predict(X_train)
27          err_test[n_features, i] = mean_squared_error(y_test, y_hat)
28          err_train[n_features, i] = mean_squared_error(y_train,
                  y_hat_train)
```

Code Block 7.1: Linear Regression and Bias–Variance Tradeoff

training error (dark red) over all 500 of our simulations. These latter two plots are unbiased estimates for the expected test error and the expected training error.

Note that the training error continues to decrease as we add additional dimensions to the feature set. However, after about 14 dimensions, the variance of our model increases dramatically, resulting in an increased generalization error. This is why it is important to select a model that minimizes its generalization error.   ▷

```
1  for i in range(50):
2      plt.plot(range(1,p+1), err_test[1:, i], color='#1f77b4',
           alpha=0.3)
3      plt.plot(range(1,p+1), err_train[1:, i], color='#d62728',
           alpha=0.3)
4
5  plt.plot(range(1,p+1), err_test.mean(axis=1)[1:], color='b',
       linewidth=2)
6  plt.plot(range(1,p+1), err_train.mean(axis=1)[1:], color='r',
       linewidth=2)
7  plt.axis([0, 26, 0, 100])
```

Code Block 7.2: Code to produce Figure 7.1.

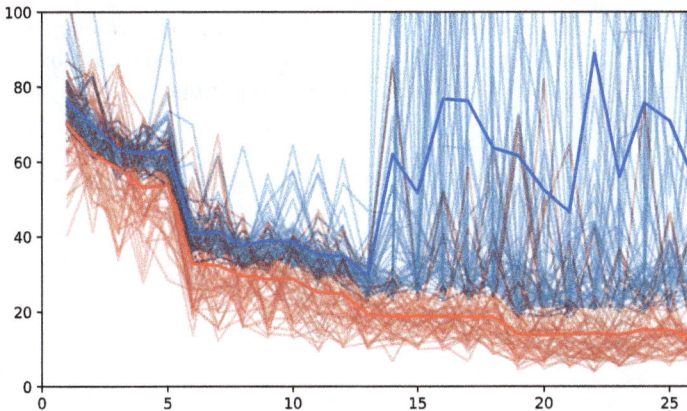

Fig. 7.1: Bias–variance tradeoff. Test error (blue) and training error (red) as a function of number of features. On the left: high bias, low variance models. On the right: low bias, high variance models.

7.1.3 Regularization

Linear regression suffers from two primary afflictions: it is a high-variance model and suffers from interpretability when many features are used. The presence of correlated explanatory variables leads to a high model variance: it is all too-often the case that two correlated variables receive large coefficients with opposite sign, the effect being one of cancellation. Slight perturbations in the training data set are therefore likely to throw the estimates off. It is therefore often the case that we are interested in finding a *subset* of features for our model that lead to a better generalization error than our full feature set.

Lasso and Ridge Regression

Definition 7.4. *Given a loss function L for a parametric predictive model f and the projection $\pi : \mathbb{R}^{k+1} \to \mathbb{R}^k$ defined by $\pi(\beta_0, \beta_1, \ldots, \beta_k) = (\beta_1, \ldots, \beta_k)$, the ℓ_p-regularized loss for a given fixed $\lambda > 0$ is the augmented loss*

$$L_p(Y, f(X; \beta)) = L(Y, f(X; \beta)) + \lambda ||\pi(\beta)||_p, \tag{7.29}$$

where $|| \cdot ||_p$ is the ℓ_p norm, defined, for $x \in \mathbb{R}^k$, by

$$||x||_p = \left(\sum_{i=1}^{k} |x_i|^p \right)^{1/p}.$$

The ℓ_1 regularization of a linear regression problem is referred to as lasso *regression, whereas the ℓ_2 regularization is referred to as* ridge *regression.*

Conceptually, it is useful to keep the following equivalency in mind: minimization of the regularized loss is equivalent to the minimization of the training loss subject to an inequality constraint, as laid out in the following proposition.

Proposition 7.4. *There is a one-to-one correspondence between the optimization problems*

$$\min_{\beta \in B} L_p(Y, f(X; \beta))$$

and

$$\min_{\beta \in B} L(Y, f(X; \beta)) \text{ subject to } ||\pi(\beta)||_p \leq t,$$

such that every $\lambda > 0$ of the first corresponds to a $t > 0$ of the second, and such that

$$\lim_{t \to \infty} \lambda(t) = 0 \quad \text{and} \quad \lim_{\lambda \to \infty} t(\lambda) = 0.$$

Each of these problems is known as the other problem's dual.

Outside of the strict confines of Definition 7.4 and Proposition 7.4, we will refer to only the ℓ_1 and ℓ_2 regularized loss, and resume our use of the variable p to represent the number of explanatory variables or features. For the squared-error loss used in linear regression models, we will consider the regularized training error functions

$$L_1(\beta; \mathcal{D}) = \sum_{i=1}^{n} w_i (y_i - f(x_i; \beta))^2 + \lambda \sum_{j=1}^{p} |\beta_j|, \tag{7.30}$$

$$L_2(\beta; \mathcal{D}) = \sum_{i=1}^{n} w_i (y_i - f(x_i; \beta))^2 + \lambda \sum_{j=1}^{p} \beta_j^2. \tag{7.31}$$

Note 7.4. The intercept β_0 is *not* penalized in the regularized loss function. The intercept helps control the overall balance and does not correspond to any of the explanatory variables. In fact, under $\lambda = \infty$ (or $t = 0$), all of the slopes $\beta_1 = \cdots = \beta_p = 0$ would be turned off, and $\beta_0 = \overline{y}$ would represent the average value of the target variable over the training set.   ▷

Note 7.5. The standard least squares solution to an ordinary linear regression problem is found by minimizing the training error Equation (7.12) over the parameter set β. Regularization can be therefore viewed as a method that penalizes large coefficients. Both ridge and lasso regression are tunable through the parameter λ, which controls the overall complexity of the model. The case $\lambda = 0$ corresponds to the ordinary least squares solution, whereas increased values of λ result in diminished coefficient scales and thus reduced levels of complexity.   ▷

The lasso method has the tendency of pushing coefficients to zero, whereas ridge regression tends to shrink the coefficients more uniformly. This is visualized in Figure 7.2 for the case $p = 2$. The darkly shaded diamond corresponds to the region $||\pi(\beta)||_1 \leq 1$, and the lightly shaded circle corresponds to the region $||\pi(\beta)||_2 \leq 2$. The ordinary least squares loss is marked with an x at $(3, 1)$, whereas the solutions to the constrained optimization dual problems (Proposition 7.4) are represented by dots. For

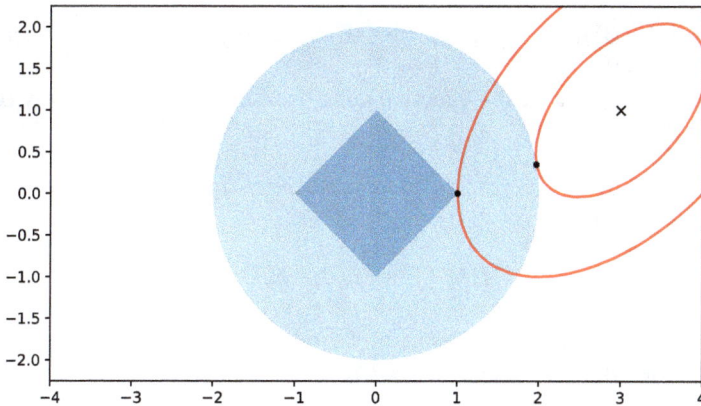

Fig. 7.2: Least squares solution with lasso and ridge regression.

$p > 2$, the region $||\pi(\beta)||_1 \leq t$ represents a family of *rhomboids* in $\mathbb{R}^p$. The solution of the lasso regression occurs when the contours of the average training loss function first touch this rhomboidal surface, providing ample opportunity for a selection of the βs to be set to zero. It is for this reason, lasso regression serves as a form of continuous subset selection, with the benefit of automatically selecting the most prevalent subset of features.

Effective Degrees of Freedom

Now, as anyone can see, the solution to the ridge regression problem is

$$\hat{\beta}^{\text{ridge}} = \left(\mathbf{X}^T \mathbf{W} \mathbf{X} + \lambda \boldsymbol{\Pi}\right)^{-1} \mathbf{X}^T \mathbf{W} \mathbf{y}, \tag{7.32}$$

where

$$\boldsymbol{\Pi} = \frac{\partial \pi(\beta)}{\partial \beta} = \mathbf{I}_{p+1} - \mathbf{e}_0 \mathbf{e}_0^T$$

is the $(p+1) \times (p+1)$ identity with the (00) component removed[1]. Inspired by Lemma 7.1, we therefore define the *effective degrees of freedom* of the model as

$$df(\lambda) = \operatorname{trace}(\mathbf{X} \left(\mathbf{X}^T \mathbf{W} \mathbf{X} + \lambda \boldsymbol{\Pi}\right)^{-1} \mathbf{X}^T \mathbf{W}). \tag{7.33}$$

From the lemma, we have $df(0) = p + 1$. However, this number shrinks as we increase λ.

The lasso regression problem, on the other hand, has no closed-form solution. Instead, we use a numeric solver, typically involving the gradient

$$\frac{\partial L_1}{\partial \beta} = -2 \mathbf{X}^T \mathbf{W} (\mathbf{y} - \mathbf{X}\beta) + \lambda \operatorname{sign}(\pi(\beta)),$$

where $\operatorname{sign}(\pi(\beta))_j = \mathbb{I}[\beta_j > 0] - \mathbb{I}[\beta_j < 0]$, for $j = 1, \ldots, p$, and $\operatorname{sign}(\pi(\beta))_0 = 0$.

In order to quantify the effective degrees of freedom, we will rely on the following definitions.

Definition 7.5. *For a given coefficient vector β in a linear regression problem, define the* active set $\mathcal{B}$ *as*

$$\mathcal{B} = \{j : \operatorname{sign}(\beta_j) \neq 0\}. \tag{7.34}$$

If $\hat{\beta}(\lambda)$ is the solution to a regularized linear regression problem, as it depends on the parameter λ, we similarly define the active set in the obvious way: $\mathcal{B}_\lambda = \{j : \operatorname{sign}(\beta_j(\lambda)) \neq 0\}$.

For a lasso regression problem, there is a discrete set of points

$$\lambda_0 > \lambda_1 > \cdots > \lambda_k = 0,$$

called the transition points, such that

---

[1] The quantity $\mathbf{e}_0 \mathbf{e}_0^T$ is dyadic notation, where $\mathbf{e}_0 = \langle 1, 0, \ldots, 0 \rangle^T \in \mathbb{R}^{p+1}$ is a standard (column) unit vector.

- For $\lambda > \lambda_0$, $\hat{\beta}_j(\lambda) = 0$ for $j > 0$,
- The active sets $\mathcal{B}_\lambda$ are constant on the interior intervals $(\lambda_m, \lambda_{m+1})$,
- As λ decreases through λ_m, new features are added to the active set.

The transition points are ordered in terms of model complexity.

These definition were proposed in Zou *et al.* [2007], along with the following proposition regarding the degrees of freedom for the lasso problem.

Proposition 7.5. *For a lasso regression problem, the number of effective degrees of freedom is*

$$df(\lambda) = \mathbb{E}\left[|\mathcal{B}_\lambda|\right],$$

where the expectation is relative to $\mathbf{y} \sim N(\boldsymbol{\beta}, \mathbf{W}^{-1}\sigma^2)$, *which determines the* $\hat{\boldsymbol{\beta}}(\lambda)$, *and hence* $\mathcal{B}_\lambda$.

Corollary 7.2. *For a lasso regression problem, the estimate*

$$\hat{df}(\lambda) = |\mathcal{B}_\lambda| \tag{7.35}$$

is an unbiased estimate of the effective degrees of freedom.

7.2 Logistic Regression

We next consider a model for classification. We will primarily consider the case of binary classification, but will show how the model can be extended to a general classification problem at the end of the section.

7.2.1 The Model

Whereas linear regression is the cornerstone of regression problems, logistic regression is the cornerstone of classification problems. Logistic regression is also considered a *linear model*, as it is linear in the model parameters. It often results in a set of linear *decision boundaries* (more on this later), though this is not the case when applying the model on nonlinear features.

Definition 7.6. *A* (binary) logistic regression model *is a predictive model, for a binary target variable* $Y \in \mathbb{B}$ *over a p-dimensional feature vector* X, *of the form*

$$Y \sim \text{Bern}(\mu) \tag{7.36}$$

$$\text{logit}(\mu) = \beta_0 + \sum_{j=1}^{p} X_j \beta_j, \tag{7.37}$$

where $\beta_0, \ldots, \beta_p$ *are unknown coefficients and* $\text{logit}(\mu)$ *is the* logit function, *defined as the* log-odds

$$\text{logit}(\mu) = \ln\left(\frac{\mu}{1-\mu}\right). \tag{7.38}$$

(Recall that the ratio $\mu/(1-\mu)$ is referred to as the odds ratio.*)*

Note that the inverse of the logit function is the *logistic function*; i.e., if $\text{logit}(\mu) = \eta$, then

$$\mu = \text{logit}^{-1}(\eta) = \frac{1}{1+e^{-\eta}} = \frac{e^{\eta}}{1+e^{\eta}}. \tag{7.39}$$

Note that the logit function is an invertible mapping $\text{logit} : (0,1) \to \mathbb{R}$.

Note 7.6. The form of the logistic regression problem, as defined in Definition 7.6, is the Bernoulli form. For the general case in dealing with proportions, please see Section 7.5, where we will discuss the binomial formulation of the problem.   ▷

It can be shown that, under the modeling assumptions of Definition 7.1, the least-squares solution (Definition 7.3) for the linear regression model is equivalent to the maximum-likelihood solution (see Exercise 7.4). It is therefore natural to proceed by determining the value of the coefficients β that maximizes the likelihood function for a given set of data. We do, however, require a slightly updated concept of likelihood for the given context.

Definition 7.7. *Let $(X_1, Y_1), \ldots, (X_n, Y_n)$ constitute a set of pairs of random variables, such that the random variables $Y_1, \ldots, Y_n$ are conditionally independent given the Xs and $Y_i \sim F_{Y|X}(Y|X_i; \theta)$, for some parameter θ. Then the* conditional likelihood *is defined as*

$$\mathcal{L}(\theta) = \prod_{i=1}^{n} f_{Y|X}(Y_i|X_i; \theta).$$

In the context of logistic regression problems, we will exclusively consider the conditional likelihood and, therefore, typically refer to it simply as the *likelihood*. Note that we do not require any independence assumptions for the explanatory (feature) variables.

Proposition 7.6. *Given a weighted data set $\{(x_{i1}, \ldots, x_{ip}, y_i, w_i)\}_{i=1}^{n}$, with binary target variable Y, the conditional likelihood function for the coefficients $\beta = \langle \beta_0, \ldots, \beta_p \rangle$ of a logistic regression model is given by*

$$\mathcal{L}(\beta) = \prod_{i=1}^{n} \mu^{w_i y_i} (1-\mu)^{w_i(1-y_i)}. \tag{7.40}$$

The log-likelihood may therefore be expressed as

$$\ell(\beta) = \sum_{i=1}^{n} w_i y_i \ln\left(\mu(\mathbf{x}_i; \beta)\right) + \sum_{i=1}^{n} w_i(1-y_i) \ln\left(1 - \mu(\mathbf{x}_i; \beta)\right), \tag{7.41}$$

or, upon further simplification, as

$$\ell(\boldsymbol{\beta}) = \sum_{i=1}^{n} w_i \left[y_i \eta_i - \ln(1 + e^{\eta_i})\right], \qquad (7.42)$$

where $\eta_i = \beta_0 + \sum_{j=1}^{p} x_{ij}\beta_j = \mathbf{x}_i^T\boldsymbol{\beta} = \hat{\mathbf{e}}_i^\mathbf{T} \cdot \mathbf{X} \cdot \boldsymbol{\beta}$, with matrix $\mathbf{X}$ and vector $\boldsymbol{\beta}$ defined as before (see text around Equation (7.13)).

We will leave the proof as an exercise for the reader (see Exercise 7.5). Note, however, that for a positive-integer weight $w \in \mathbb{Z}_+$, the likelihood function is equivalent to simply having a repeated training instance (x, y) with multiplicity w. Moreover, Equation (7.41) is equivalent to the negative log-loss.

7.2.2 Solution

As we saw in Chapter 5, the Newton–Raphson method is typically employed to solve maximum-likelihood problems. However, this method requires the score statistic and information matrix for a given set of data. Luckily, these are easily derived from Equation (7.42).

Proposition 7.7. *Given a weighted data set, as in Proposition 7.6, the score statistic and observed information are*

$$U_n(\boldsymbol{\beta}) = \sum_{i=1}^{n} w_i \left[y_i - \mu(\mathbf{x}_i; \boldsymbol{\beta})\right] \mathbf{x}_i^T, \qquad (7.43)$$

$$\mathcal{J}_n(\boldsymbol{\beta}) = \sum_{i=1}^{n} w_i \mu(\mathbf{x}_i; \boldsymbol{\beta}) \left[1 - \mu(\mathbf{x}_i; \boldsymbol{\beta})\right] \mathbf{x}_i \mathbf{x}_i^T, \qquad (7.44)$$

where $\mu(\mathbf{x}_i; \boldsymbol{\beta}) = \mathrm{logit}^{-1}(\mathbf{x}_i^T\boldsymbol{\beta})$. Moreover, the Fisher information is equivalent to the observed information: $\mathcal{I}_n(\boldsymbol{\beta}) = \mathcal{J}_n(\boldsymbol{\beta})$.

Proof. If we let $\eta_i = \mathbf{x}_i^T\boldsymbol{\beta}$, first observe that

$$\frac{\partial \eta_i}{\partial \boldsymbol{\beta}} = \mathbf{x}_i^T.$$

The results given in Equations (7.43) and (7.44) follow from simple calculus and the definitions

$$U_n(\boldsymbol{\beta}) = \frac{\partial \ell(\boldsymbol{\beta})}{\partial \boldsymbol{\beta}} \qquad \text{and} \qquad \mathcal{J}_n(\boldsymbol{\beta}) = -\frac{\partial^2 \ell(\boldsymbol{\beta})}{\partial \boldsymbol{\beta}^T \partial \boldsymbol{\beta}}.$$

Since the observed information does not depend on Y_i, it is its own expected value, so that $\mathcal{I}_n(\boldsymbol{\beta}) = \mathcal{J}_n(\boldsymbol{\beta})$. $\qquad \square$

The logistic regression model is, however, built into the scikit-learn package. Our next example shows how to train a logistic regression model using k-fold cross validation for a sample data set.

Example 7.2. Import the data set `sklearn.datasets.load_breast_cancer` that predicts a binary target variable (diagnosis for breast cancer) based on a 30-dimensional numeric feature set.

We can use the scikit-learn library to train this model, using 5-fold cross validation, as shown in Code Block 7.3. The average accuracy of the five test sets is 95.96%, with a standard deviation of 1.62%. The ROC curve is shown in Figure 7.3, with AUC of 98.68%.

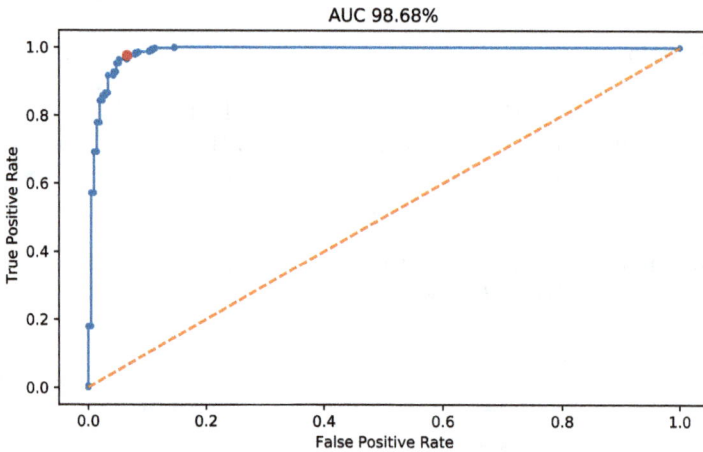

Fig. 7.3: ROC curve for logistic regression model for breast-cancer diagnostic. Red dot corresponds to the final cutoff threshold used in the `sklearn` model.

The calibration curve is plotted in Figure 7.4. Since the probability model is directly incorporated into the logistic regression model, the final probability predictions are well calibrated. (This is not necessarily the case with other classification algorithms, for which the calibration curve will diverge more dramatically from the diagonal.)

In addition, the precision and recall for the model can easily be computed to be 96.13% and 97.48%, as shown in lines 32 and 33.   ▷

Now that we've seen the overarching picture of training a logistic regression model–from model pipelining to AUC to cross validation—let's take a look under the hood and try to build our own logistic regression model from scratch, using the techniques of maximum likelihood.

```
1  from sklearn.datasets import load_breast_cancer
2  from sklearn.linear_model import LogisticRegression
3  from sklearn.model_selection import KFold
4  from sklearn.calibration import calibration_curve
5  from sklearn.metrics import roc_curve, auc
6
7  data = load_breast_cancer()
8  X, y = data.data, data.target
9  transformers = [
10         ('standardize', RobustScaler(quantile_range=(0.2,0.8))),
11         ('model', LogisticRegression(max_iter=500))]
12
13 pipe = Pipeline(transformers)
14 kf = KFold(n_splits=5)
15 results, accuracy = [], []
16
17 for train_index, test_index in kf.split(X):
18     X_train, X_test = X[train_index], X[test_index]
19     y_train, y_test = y[train_index], y[test_index]
20     pipe.fit(X_train, y_train)
21     y_hat = pipe.predict(X_test)
22     p_hat = pipe.predict_proba(X_test)
23     results.append({'y_test': y_test, 'y_hat': y_hat, 'p_hat':
               p_hat[:, 1]})
24     accuracy.append(pipe.score(X_test, y_test))
25
26 y_total = np.concatenate([result['y_test'] for result in results])
27 y_hat   = np.concatenate([result['y_hat'] for result in results])
28 p_hat   = np.concatenate([result['p_hat'] for result in results])
29 p_true, p_pred = calibration_curve(y_total, p_hat, n_bins=5)
30 fpr, tpr, thresholds = roc_curve(y_total, p_hat)
31 print(auc(fpr, tpr), np.mean(accuracy)) # 98.68%, 95.96%
32 precis=np.sum((y_total==1)&(y_hat==1))/np.sum((y_hat==1)) # 96.13%
33 recall=np.sum((y_total==1)&(y_hat==1))/np.sum((y_total==1)) # 97.48%
```

Code Block 7.3: Logistic Regression Problem

Example 7.3. Solve the logistic regression problem from Example 7.2 directly using Newton's method and Equations (7.43) and (7.44).

We can create our own class to handle logistic regression problems using the method of scoring; see Code Block 7.4.

With the `fit`, `predict`, and `predict_proba` methods, this class is designed to fit into the same scikit-learn pipeline shown in Code Block 7.3. Precision and recall were 95.2% and 95.0%, respectively, with an overall accuracy of 93.8%. This is slightly worse than scikit-learn's built-in methods, though our method converges in fewer iterations.   ▷

```python
class NewtonLogisticRegression:
    def __init__(self, max_iter=20, tol=1e-6, intercept=True):
        self.max_iter = max_iter
        self.tol = tol
        self.intercept = intercept

    def logistic(self, eta):
        return 1 / (1 + np.exp(-eta))

    def train(self, X, y):
        n, m = X.shape
        if self.intercept:
            X = np.concatenate([np.ones(n).reshape((n,1)), X], axis=1)
            m += 1
        y = y.reshape((n, 1))
        beta = np.zeros(m).reshape((m, 1))
        beta[0] = 1
        i = 0
        err = 1
        while (i < self.max_iter) and (err > self.tol):
            UT = X.T @ (y - self.logistic(X@beta))
            p_hat = self.logistic(X@beta)
            FI = X.T @ np.diag( (p_hat * (1 - p_hat)).reshape(n)) @ X
            err = np.linalg.norm(UT)
            i += 1
            beta += np.linalg.inv(FI) @ UT

        self.converged = err < self.tol
        self.beta = beta

    def predict(self, X, threshold=0.5):
        return (self.predict_proba(X) > threshold).astype(int)

    def predict_proba(self, X):
        if self.intercept:
            X = np.concatenate([np.ones(len(X)).reshape((len(X),1)),
                X], axis=1)
        return self.logistic(X @ self.beta)
```

Code Block 7.4: Newton's Method Solution to Logistic Regression Problem

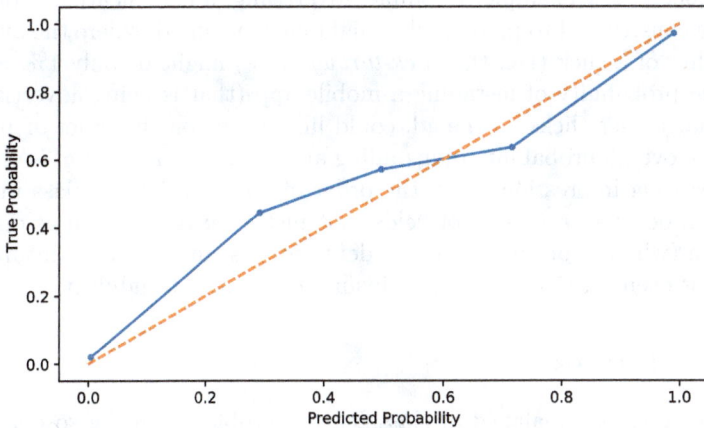

Fig. 7.4: Calibration plot for logistic regression model for breast-cancer diagnostic; 5 calibration bins and $n = 569$ data points. Orange dashed line represents perfect calibration.

7.2.3 Imbalanced Data Sets

The relative frequency of classes in the training instances of a classification problem can have a significant impact on the efficacy of the model, especially if one of the target classes is under or overrepresented. In the case of binary classification, we will, without loss of generality, think of the minority class as being the positive instances, as is common in most applications. (One typically tries to detect when an instance is going to be a *rare* event, not when it is going to be a typical event.) We will present some basic definitions and approaches for handling class imbalance here; for additional detail, see Kuhn and Johnson [2013].

Definition 7.8. *In a classification model, a* class imbalance *is said to exist whenever there is an unequal distribution among the various classes comprising the target variable. The severity of such an imbalance can vary from mild to extreme.*

In a binary classification problem, we define the class ratio *as the ratio of majority to minority classes. A* moderate class imbalance *is said to exist if the class ratio exceeds 3:1 (25% positive instances). A* severe class imbalance *is said to exist if the class ratio exceeds 19:1 (5% positive instances). An* extreme class imbalance *is said to exist if the class ratio exceeds 99:1 (1% positive instances).*

It is not uncommon in practice to encounter cases where it is necessary to model extreme class imbalances, exceeding 100:1. In the era of big data, one can even encounter extreme class imbalances on the order of 1,000:1 or even 10,000:1. For example, in online advertising, a classification problem could be constructed to predict who might click on an ad, where the overall probability of a click (i.e., the *click-through rate*) might be only 1%. Similarly, the probability of installing a mobile app (that is being advertised), given that a user clicks on the ad, could likewise be on the order of 100:1. Thus, the overall probability of installing an app, given that a user is served an advertising impression, is on the order of 10,000:1. Severe class imbalances can occur in a variety of fields, not just advertising, such as outlier detection (where it occurs almost by definition), spam detection, insurance claims, or even predicting when an insurance claim is fraudulent.

Accuracy Paradox

To see why class imbalance in classification problems can be so delicate, consider the case of a severe class imbalance of 200:1. Your current production model provides 99.5% accuracy, for which your business stakeholders are immeasurably pleased. Is it worth precious bandwidth to look into and improve this model? It depends. Quite possibly it is an absolute imperative.

  If the model was selected for its accuracy, one needn't have looked far to find a highly accurate model: the model

$$\hat{Y} = 0,$$

which predicts *every* instance to be in the negative class, has an accuracy of 99.5%. The model is going to be right 99.5% of the time, because it is trying to detect something that is so rare that it only occurs in about 0.5% of all cases. Note, the precision of this model is undefined, as there are no positive predictions, and the recall is 0%; i.e., 0% of the actual positives were correctly classified. Similarly, the model has 0% sensitivity (which is the true-positive rate, or recall) and 100% specificity (which is the true-negative rate, or one minus the false positive rate). On an ROC curve, this model would be at the origin $(0,0)$, quite far from the ideal model at $(0,1)$.

  A simple approach to this problem might be to tune the model to optimize for recall; i.e., tune the model to maximize the accuracy of the minority class. This approach, however, leads to the opposite problem: always predicting $\hat{Y} = 1$ leads to 100% recall, because it correctly predicts *all* positive instances. (The problem is that it all negative instances are also classified as positives.) In the remainder of this section, we will explore some other (better) approaches for handling class imbalance.

ROC Curve and Threshold Tuning

As discussed in the preceding paragraph, problems can easily arise when tuning classification models with imbalanced data sets to optimize for a single metric, such as accuracy or recall. This is why classification metrics often come in pairs: *precision and recall* or *sensitivity and specificity*. One common way to optimize for both simultaneously is to optimize for the area under the ROC curve; i.e., the AUC. Not only does optimizing for AUC incorporate both sensitivity and specificity into the optimization problem, but also it eliminates the particular classification threshold from model comparison. Recall from Equation (6.14) that a binary classification model outputs a probability-like function $f(X; \mathcal{D})$, that is then used in conjunction with a classification threshold t, so that an instance is classified according to

$$\hat{Y}_t(X; \mathcal{D}) = \mathbb{I}\,[f(X; \mathcal{D}) > t]\,.$$

A single predictive model might perform dramatically different based on the classification threshold that is ultimately used in classification. The AUC method therefore compares different models *without regard for* their particular classification thresholds. A model with a superior AUC will be able to generate better precision and recall with an appropriate choice of threshold cutoff. A higher AUC means there is more area under the curve, which means the curve must pass more closely to the ideal model $(0, 1)$. This approach therefore entails two steps in model selection:

1. Optimize for AUC.
2. Once the optimal model is chosen, choose the classification threshold corresponding to the point on the ROC curve that minimizes the distance to the ideal model $(0, 1)$.

Sampling Methods

Another approach is simply to adjust for the class imbalance in the training dataset. This typically entails either oversampling the underrepresented class or undersampling the overrepresented class[2]. Oversampling can also be achieved by adjusting the training weights in the algorithm, which is typically equivalent to duplicating training instances.

A more recent approach introduced by Chawla *et al.* [2002] is known as SMOTE, or *synthetic minority over-sampling technique*, which incorporates both upsampling and downsampling. Essentially, overrepresented classes are downsampled, and underrepresented classes are upsampled *synthetically*. The new synthetic instances are generated by using a random combination of features from a randomly selected data point from the minority class and its "nearest neighbors."

Regardless of the particular sampling method deployed, adjusting for class imbalance must be limited to the training set. It is crucial that the imbalance is preserved in the test and validation sets in order to accurately estimate the generalization error. Furthermore, the functional com-

---

[2] Dunn was under Oveur and I was under Dunn.

ponent $f(X; \mathcal{D})$ of the model should no longer be interpreted as a probability, as the training set was artificially altered to control for the relative frequency between the classes.

Cost-sensitive Training

Instead of optimizing for a performance metric, such as AUC, we could instead optimize for a given loss function. *Cost-sensitive training* is a method in which different types of errors are given different weights in the loss function used for training a classification model. The decision on how to weight various types of errors depends on the particular business use case. For example, for a model used to classify whether images of a tumor are cancerous, the cost of a false negative is much greater than that of a false positive; whereas a false positive could lead to some undue emotional distress for the patient, a subsequent test can easily catch the error, whereas missing the detection of an actual positive could lead to dire consequences.

7.2.4 Multinomial Logistic Regression

Next, we show how the multinomial Bernoulli distribution can be used to extend logistic regression to the case of multiclass classification.

Definition 7.9. *A multiclass (or multinomial) logistic regression model is a predictive model, for a categorical target variable $Y \in \{0, \ldots, k-1\}$ over a p-dimensional feature vector $\mathbf{x}$, of the form*

$$Y \sim \text{MultiBern}(\boldsymbol{\pi}) \tag{7.45}$$

$$\text{alr}(\boldsymbol{\pi}) = \boldsymbol{\eta} = \mathbf{x}^T \mathbf{B}, \tag{7.46}$$

where $\mathbf{B} \in \mathbb{R}^{(p+1) \times (k-1)}$ is a matrix of unknown coefficients, $\boldsymbol{\eta} \in \mathbb{R}^{k-1}$, and alr$(\boldsymbol{\pi})$ is the additive logratio transform alr $: \Delta^{k-1} \to \mathbb{R}^{k-1}$ defined by

$$\text{alr}(\boldsymbol{\pi}) = \left\langle \ln\left(\frac{\pi_1}{\pi_0}\right), \ldots, \ln\left(\frac{\pi_{k-1}}{\pi_0}\right) \right\rangle. \tag{7.47}$$

As usual, $\Delta^{k-1} = \{\boldsymbol{\pi} \in \mathbb{R}_*^k : \sum_{i=0}^{k-1} \pi_i = 1\}$ is the probability simplex in $\mathbb{R}^k$.

Note 7.7. The choice of feature index used in the denominators for Equation (7.47) is an arbitrary choice, typically chosen as some reference point. For this reason, this transform is referred to as the *baseline-category logits* by Agresti [2019]. ▷

Proposition 7.8. *The inverse of the alr transform is given by*

$$\text{alr}^{-1}(\boldsymbol{\eta}) = \left(1 + \sum_{l=1}^{k-1} e^{\eta_l}\right)^{-1} \langle 1, e^{\eta_1}, \ldots, e^{\eta_{k-1}} \rangle. \tag{7.48}$$

Proof. Let $\text{alr}(\boldsymbol{\pi}) = \boldsymbol{\eta}$, for $\boldsymbol{\pi} \in \Delta^{k-1}$ and $\boldsymbol{\eta} \in \mathbb{R}^{k-1}$. From Equation (7.47), this implies that

$$\ln\left(\frac{\pi_i}{\pi_0}\right) = \eta_i,$$

for $i = 1, \ldots, k-1$. It follows that $\pi_i = \pi_0 e^{\eta_i}$, again for $i = 1, \ldots, k-1$. Moreover, from our normalization condition, we have

$$\sum_{i=0}^{k-1} \pi_i = \pi_0 + \sum_{i=1}^{k-1} \pi_i = \pi_0 \left(1 + \sum_{i=1}^{k-1} e^{\eta_i}\right) = 1,$$

which yields $\pi_0 = \left(1 + \sum_{i=1}^{k-1} e^{\eta_i}\right)^{-1}$. The result follows.   □

Note 7.8. The *central logratio transform* is defined as the mapping clr : $\Delta^{k-1} \to U \subset \mathbb{R}^k$ defined by

$$\text{clr}(\boldsymbol{\pi}) = \left\langle \ln\left(\frac{\pi_0}{g(\boldsymbol{\pi})}\right), \ldots, \ln\left(\frac{\pi_{k-1}}{g(\boldsymbol{\pi})}\right) \right\rangle, \tag{7.49}$$

where $g(\boldsymbol{\pi})$ is the geometric mean:

$$g(\boldsymbol{\pi}) = \left(\prod_{i=0}^{k-1} \pi_i\right)^{1/k}.$$

It can be shown (see Exercise 7.6) that the inverse transform is the softmax:

$$\text{clr}^{-1}(\boldsymbol{\eta}) = \left\langle \frac{e^{\eta_0}}{\sum_{i=0}^{k-1} e^{\eta_i}}, \ldots, \frac{e^{\eta_{k-1}}}{\sum_{i=0}^{k-1} e^{\eta_i}} \right\rangle.$$

The reason the additive logratio transform is preferred, is that the model resulting from the central logratio transform is not identifiable, meaning the resulting parameter matrix $\mathbf{B} \in \mathbb{R}^{(p+1) \times k}$ would have a linearly dependent column. (Notice it would require k columns, not $k-1$, as we had before.) To see this, note that the softmax transform is invariant under scalar addition; i.e., $\text{clr}^{-1}(\boldsymbol{\eta} + \alpha\mathbf{1}) = \text{clr}^{-1}(\boldsymbol{\eta})$. Therefore the probability vector corresponding to the coefficient matrix $\mathbf{B} = [\boldsymbol{\beta}_0, \ldots, \boldsymbol{\beta}_{k-1}]$ is invariant relative to vector translation of each of its columns; i.e., $\mathbf{B} = [\boldsymbol{\beta}_0 + \mathbf{c}, \ldots, \boldsymbol{\beta}_{k-1} + \mathbf{c}]$. By selecting $\mathbf{c} = -\boldsymbol{\beta}_0$, we recover the same inverse transform given by Equation (7.48). (Of course with a redefined set of $\boldsymbol{\beta}$s.) The additive logratio transform is therefore a special case of the central logratio transform, and its inverse softmax transform, with the redundancy removed.

This has the effect of shifting $\eta_i \to \eta_i - \eta_0$, which further reduces Equation (7.49) to Equation (7.47) (with an extra zero component).   ▷

Proposition 7.9. *Given a weighted data set $\{(\mathbf{x}_i, y_i, w_i)\}_{i=1}^n$, where $\mathbf{x}_i \in \mathbb{R}^p$, with categorical target variable Y, the conditional likelihood function for the coefficients $\mathbf{B} \in \mathbb{R}^{(p+1)\times(k-1)}$ of a logistic regression model is given by*

$$\mathcal{L}(\mathbf{B}) = \prod_{i=1}^{n}\prod_{j=0}^{k-1} \pi_j(\mathbf{x}_i; \mathbf{B})^{w_i y_{ij}}, \tag{7.50}$$

where $y_{ij} = \mathbb{I}[y_i = j]$, for $i = 1,\dots,n$ and $j = 0,\dots,k-1$, and where

$$\boldsymbol{\pi}(\mathbf{x}_i, \mathbf{B}) = \mathrm{alr}^{-1}(\mathbf{x}_i^T \mathbf{B}).$$

The log-likelihood may therefore be expressed as

$$\ell(\boldsymbol{\beta}) = \sum_{i=1}^{n}\sum_{j=0}^{k-1} w_i y_{ij} \ln \pi_j(\mathbf{x}_i; \mathbf{B}). \tag{7.51}$$

or, upon further simplification, as

$$\ell(\boldsymbol{\beta}) = \sum_{i=1}^{n} w_i \left[y_{ij}\eta_{ij} - \ln\left(1 + \sum_{j=1}^{k-1} e^{\eta_{ij}}\right) \right], \tag{7.52}$$

where $\eta_{ij} = \mathbf{x}_i^T \boldsymbol{\beta}_j$, such that $\boldsymbol{\beta}_j$ is the jth column of $\mathbf{B}$, for $j = 1,\dots,k-1$, and where we define $\boldsymbol{\beta}_0 = \mathbf{0}$ for convenience.

Compare Equation (7.52) with Equation (7.42). Note again that Equation (7.51) is just the negative of the log-loss, defined in Equation (6.20); minimizing log-loss is equivalent to maximizing the likelihood of the multinomial distribution.

As it turns out, scikit-learn's `LogisticRegression` class also handles multinomial logistic regression, following the same interface we saw in Example 7.4.

Example 7.4. The NIST handwritten digit data set consists of a set of 1,797 instances of handwritten digits (i.e., 0, 1, 2, 3, 4, 5, 6, 7, 8, 9) along with labels representing their true values. Each handwritten digit is represented as a 64-dimensional feature vector, with each component being an integer between 0 and 16. If the 64-dimensional feature vector is rearranged into an 8×8 square, the value represents the shading of the handwritten sample for that pixel. Some sample feature vectors, represented visually, from the data set are shown in Figure 7.5.

We see that the multinomial logistic regression problem can be trained with minor modification as compared with how we trained a binary logistic regression problem; see Code Block 7.5. Scikit-learn's logistic regression class achieves a 91.6% accuracy on the holdout sets from 5-fold cross validation.

```
1   from sklearn.datasets import load_digits
2   from sklearn.metrics import confusion_matrix
3   data = load_digits()
4   X, y = data.data, data.target
5
6   plt.figure(figsize=(8, 9/2)) # Plot sample feature vectors.
7   for i in range(30):
8       ax = plt.subplot(5,6,i+1, xticklabels=[], yticklabels=[],
            xticks=[], yticks=[])
9       ax.matshow(X[i,:].reshape(8,8))
10
11  pipe = LogisticRegression(multi_class='multinomial', max_iter=500)
12  kf = KFold(n_splits=5)
13  accuracy = []
14  results = []
15
16  for train_index, test_index in kf.split(X):
17      X_train, X_test = X[train_index], X[test_index]
18      y_train, y_test = y[train_index], y[test_index]
19      pipe.fit(X_train, y_train)
20      y_hat = pipe.predict(X_test)
21      results.append({'y_test': y_test, 'y_hat': y_hat})
22      accuracy.append(pipe.score(X_test, y_test))
23  np.mean(accuracy) # 91.598
24  y_total = np.concatenate([result['y_test'] for result in results])
25  y_hat  = np.concatenate([result['y_hat'] for result in results])
26  C = confusion_matrix(y_hat, y_total) # Predictions rows, actual
            columns
```

Code Block 7.5: Multinomial Logistic Regression Problem

		\multicolumn{10}{c}{Actual Digit}
		0	1	2	3	4	5	6	7	8	9
	0	174	0	0	0	0	0	0	3	0	1
	1	0	160	7	0	2	2	2	1	11	4
Predicted Digit	2	1	0	170	1	0	1	0	0	2	0
	3	0	1	0	160	0	2	0	1	1	3
	4	1	1	0	0	169	1	1	3	0	0
	5	1	0	0	4	0	165	3	0	5	1
	6	1	5	0	0	6	1	173	0	0	0
	7	0	0	0	2	1	1	0	161	0	2
	8	0	7	0	12	0	1	2	1	153	8
	9	0	8	0	4	3	8	0	9	2	161

Table 7.2: Confusion matrix for Example 7.4.

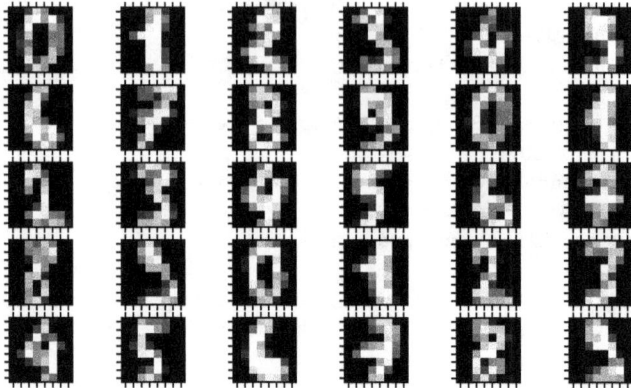

Fig. 7.5: Examples of NIST handwritten digit data set.

Our code also combines the result from each holdout set and computes the resulting confusion matrix, as shown in Table 7.2. Note that in order to get the predicted values on the rows, in agreement with Definition 6.11, one reverses the order of the inputs as per scikit-learn's documentation (which suggests entering `confusion_matrix(y_true, y_pred)`.)   ▷

Previously, we saw how the linear regression model can be broken up into two primary components:

$$Y_i \sim \mathrm{N}(\mu_i, \sigma^2/w_i) \qquad \text{random component}$$
$$g(\mu_i) = \mu_i = \mathbf{x}_i^T \boldsymbol{\beta} \qquad \text{systematic component.}$$

The logistic regression model falls neatly into a similar structure:

$$Y_i \sim \mathrm{Bern}(p_i) \qquad \text{random component}$$
$$g(p_i) = \mathrm{logit}(p_i) = \mathbf{x}_i^T \boldsymbol{\beta} \qquad \text{systematic component.}$$

It is natural to ask whether or not this structure can be generalized to a broader class of problems. And, if so, what types of distributions are conducive to such a generalization? The answer, of course, is that the structure is indeed generalizable for a specific class of distributions. We will begin by discussion the exponential family of distributions. We will go on to lay out the general theory of *generalized linear models*. Finally, we will discuss specific examples of our general theory that are suited to specific types of data: count data, proportions, positive-continuous data, and so forth. For additional references on generalized linear models, we recommend Agresti [2015], Dobson and Barnett [2018], Dunn and Smyth [2018], or Gill and Torres [2020], with Dunn and Smyth [2018] being our favorite.

7.3 Exponential Dispersion Models

Exponential dispersion models are simply a generalization of the exponential family of distributions, in which a parameter for the dispersion is called out explicitly, rather than being treated implicitly as a nuisance parameter.

7.3.1 Canonical Form

All distributions within the family of exponential dispersion models can be converted into a certain form known as the *canonical form*. This conversion entails a reparameterization of the model using what is known as the *canonical parameter*. We will see examples of this conversion later on; to begin, however, let us examine what these models look like in their canonical form.

Definition 7.10. *A distribution is said to belong to the* exponential dispersion model (EDM) *family of distributions if its* PDF *or* PMF *can be written (via a reparameterization) in the form*

$$f(y; \theta, \phi) = a(y, \phi) \exp \left[\frac{y\theta - \kappa(\theta)}{\phi} \right], \tag{7.53}$$

for a random variable $Y \sim$ EDM, *where*

- *the parameter* θ *is called the* canonical parameter,
- *the parameter* ϕ *is called the* dispersion parameter, *and*
- *the function* $\kappa(\theta)$ *is called the* cumulant function.

When an EDM *is written in the form of Equation* (7.53), *it is said to be in* canonical form.

Cumulants of an EDM

Given an EDM in canonical form, it is easy to derive its cumulants; i.e., its mean, variance, and so forth. The result is given in the following.

Theorem 7.7. *The* rth *cumulant* κ_r *of a random variable in the* EDM *family is determined by*

$$\kappa_r = \phi^{r-1} \frac{d^r \kappa(\theta)}{d\theta^r}. \tag{7.54}$$

In particular, the mean and variance (i.e., the first and second cumulants) are given by

$$\mathbb{E}[Y] = \mu = \kappa'(\theta), \tag{7.55}$$

$$\mathbb{V}(Y) = \sigma^2 = \phi\kappa''(\theta), \tag{7.56}$$

respectively. The function $\theta = (\kappa')^{-1}(\mu)$ *is referred to as the* canonical link.

Proof. From the definition of the moment generating function(MGF), we have

$$M(t) = \mathbb{E}[e^{tY}] = \int a(y, \phi) \exp \left[\frac{y\theta - \kappa(\theta)}{\phi} + ty \right] dy.$$

Next, define $\psi = \theta + t\phi$, so that

$$M(t) = \int a(y, \phi) \exp \left[\frac{y\psi - \kappa(\theta) + \kappa(\psi) - \kappa(\psi)}{\phi} \right] dy$$

$$= \exp \left[\frac{\kappa(\psi) - \kappa(\theta)}{\phi} \right] \int a(y, \phi) \exp \left[\frac{y\psi - \kappa(\psi)}{\phi} \right] dy$$

$$= \exp \left[\frac{\kappa(\psi) - \kappa(\theta)}{\phi} \right]. \tag{7.57}$$

The last equality follows because the integral in an EDM in its own right (and therefore normalized).

Now, the *cumulant-generating function* $K(t)$ is defined by the relation

$$K(t) = \ln M(t),$$

and has the property that the rth cumulant is determined by its rth derivative:

$$\kappa_r = K^{(r)}(0).$$

It follows that, for an EDM, the cumulant-generating function is given by

$$K(t) = \frac{\kappa(\theta + t\phi) - \kappa(\theta)}{\phi}.$$

The result given in Equation (7.54) follows. $\square$

Corollary 7.3. *The mean μ of an EDM is a monotonically increasing (and therefore one-to-one) function of the canonical parameter θ.*

Proof. Due to Equation (7.56), it follows that $\kappa''(\theta) > 0$, which means that $\mu = \kappa'(\theta)$ is an increasing function of θ. The result follows. $\square$

Corollary 7.4. *A GLM is uniquely specified by its dispersion parameter ϕ and cumulant function $\kappa(\theta)$.*

Proof. A distribution is uniquely determined by knowing its cumulants, which are given in Equation (7.54), expressed in terms of ϕ and $\kappa(\theta)$. Therefore, ϕ and $\kappa(\theta)$ uniquely determine each cumulant of the distribution, hence the distribution itself. This proves the result. $\square$

Note 7.9. As a result of Corollary 7.4, the coefficient function $a(y, \phi)$ must be uniquely specified based on the cumulant function and dispersion. For this reason, it is sometimes referred to as a *normalizing function*. This term is not consistent in the literature, however, as some authors refer to the cumulant function as the normalizing function. $\triangleright$

Our next result provides an interpretation for the *weight* of an instance from a GLM, as being synonymous of the average value of an IID sample from a fixed GLM.

Theorem 7.8. *Let* $Y_1, \ldots, Y_w \sim$ EDM(μ, ϕ) *represent an* IID *sample from a single* EDM. *Then the sample mean*

$$\overline{Y} = \frac{1}{w} \sum_{i=1}^{w} Y_i$$

is also distributed as an EDM *from the same family. Moreover,*

$$\overline{Y} \sim \text{EDM}(\mu, \phi/w). \qquad (7.58)$$

Proof. Since the data $Y_1, \ldots, Y_w$ are IID, the MGF of their sample mean is given by

$$M_{\overline{Y}}(t) = \mathbb{E}\left[e^{t\overline{Y}}\right] = \mathbb{E}\left[e^{t(Y_1 + \cdots + Y_w)/w}\right] = [M_Y(t/w)]^w .$$

Combining this expression with Equation (7.57), we find that

$$M_{\overline{Y}}(t) = \exp\left[\frac{\kappa(\theta + t\phi/w) - \kappa(\theta)}{\phi/w}\right].$$

However, this is identical to the MGF for Y, except with ϕ replaced with ϕ/w. We conclude that $\overline{Y} \sim$ GLM$(\mu, \phi/w)$, which shows our result.   □

Definition 7.11. *Given a cumulant function* $\kappa(\theta)$ *for an* EDM, *the* variance function *is defined as*

$$V(\mu) = \kappa''\left((\kappa')^{-1}(\mu)\right). \qquad (7.59)$$

Equation (7.59) is simply the derivative $d\mu/d\theta$ expressed as a function of μ. To see this, recall that $\mu = \kappa'(\theta)$, so that $\theta = (\kappa')^{-1}(\mu)$ (the canonical link). Moreover, $d\mu/d\theta = \kappa''(\theta)$. Combining these results yields Equation (7.59). The variance of a random variable $Y \sim$ EDM(μ, ϕ) may therefore be expressed in terms of its mean by the relation

$$\mathbb{V}(Y) = \phi V(\mu), \qquad (7.60)$$

which is an equivalent expression of Equation (7.56). Moreover, it follows that when θ is expressed as a function of μ, we have

$$\frac{d\theta}{d\mu} = \frac{1}{V(\mu)}.$$

This follows since $V(\mu)$ is simply $d\mu/d\theta$ expressed as a function of μ.

7.3.2 Dispersion Form

Corollary 7.3 guarantees that the mean μ of an EDM has a one-to-one relationship to its canonical parameter θ. We will often be interested in parameterizing an EDM with respect to its mean, as opposed to its canonical parameter. In order to facilitate this formalism, we first introduce an important function.

Definition 7.12. *The* unit deviance *of an* EDM *is the function*

$$d(y, \mu) = 2 \left[t(y, y) - t(y, \mu) \right], \tag{7.61}$$

where $t(y, \mu)$ is defined by the relation

$$t(y, \mu) = y\theta(\mu) - \kappa(\theta(\mu)), \tag{7.62}$$

where $\theta(\mu) = (\kappa')^{-1}(\mu)$, as before.

Our next proposition reveals that the unit deviance can be thought of as a measure of distance between y and μ.

Proposition 7.10. *The unit deviance satisfies the inequality*

$$d(y, \mu) \geq 0,$$

where there equality holds if and only if $\mu = y$.

Proof. From the definition of $t(y, \mu)$, observe that, for fixed y,

$$\frac{\partial t(y, \mu)}{\partial \mu} = [y - \kappa'(\theta(\mu))] \frac{d\theta}{d\mu}.$$

However, since $\theta = (\kappa')^{-1}(\mu)$, and since $d\theta/d\mu = 1/V(\mu)$, we have

$$\frac{\partial t(y, \mu)}{\partial \mu} = \frac{y - \mu}{V(\mu)},$$

which vanishes only if $\mu = y$. Moreover, the second derivative is given by

$$\frac{\partial^2 t(y, \mu)}{\partial \mu^2} = -\frac{V(\mu) + (y - \mu)V'(\mu)}{V(\mu)^2}.$$

Now, since

$$\frac{\partial^2 t(y, y)}{\partial \mu^2} = -V(\mu)^{-1},$$

it follows that $t(y, \mu)$ must, for fixed y, have its global maximum at the point $\mu = y$. It therefore follows that the unit deviance, defined in Equation (7.61), is positive for $\mu \neq y$ and zero for $\mu = y$. $\square$

Proposition 7.11. *An* EDM *may be expressed in terms of μ as*

$$f(y; \mu, \phi) = b(y, \phi) \exp \left[\frac{-d(y, \mu)}{2\phi} \right], \tag{7.63}$$

where $b(y, \phi) = a(y, \phi) \exp[t(y, y)/\phi]$. This is called the dispersion form *of the* EDM.

7.3.3 A Catalogue of EDMs

We next consider a number of common EDMs. A summary is provided in Table 7.3.

Normal Distribution

From the definition of the normal distribution, we have

$$
\begin{aligned}
f(y; \mu, \sigma^2) &= \frac{1}{\sqrt{2\pi\sigma^2}} \exp\left[\frac{-(y-\mu)^2}{2\sigma^2}\right] \\
&= \frac{\exp[-y^2/(2\sigma^2)]}{\sqrt{2\pi\sigma^2}} \exp\left[\frac{y\mu - \mu^2/2}{\sigma^2}\right].
\end{aligned}
$$

We thus identify the canonical parameter $\theta = \mu$, the cumulant function $\kappa(\theta) = \theta^2/2$, and the dispersion parameter $\phi = \sigma^2$. Since $\kappa''(\theta) = 1$, it follows that $V(\mu) = 1$. Moreover, the function $a(y, \theta)$ is given by the coefficient

$$
a(y, \phi) = \frac{\exp[-y^2/(2\phi)]}{\sqrt{2\pi\phi}}.
$$

Finally, the unit deviance is obtained by computing

$$
d(y, \mu) = 2\left[(y^2 - y^2/2) - (y\mu - \mu^2/2)\right] = (y - \mu)^2.
$$

Bernoulli Distribution

From the definition of the Bernoulli distribution, we have

$$
f(y; \mu) = \mu^y (1 - \mu)^{1-y} = \exp\left[y \ln\left(\frac{\mu}{1 - \mu}\right) + \ln(1 - \mu)\right].
$$

From this, it is apparent that $a(y, \phi) = 1$ and $\phi = 1$. Furthermore, the canonical parameter is given by

$$
\theta = \ln\left(\frac{\mu}{1 - \mu}\right),
$$

which we recognize as the logit function, which we may invert to obtain

$$
\mu = \frac{e^\theta}{1 + e^\theta}.
$$

The cumulant function is therefore obtained by replacing this expression for μ in $-\ln(1 - \mu)$:

$$
\kappa(\theta) = \ln(1 + e^\theta).
$$

Since $d\mu/d\theta = \mu(1 - \mu)$ (since $\mu(\theta)$ is the logistic function), it follows that the variance function is given by

EDM	$d(y,\mu)$	$\kappa(\theta)$	ϕ	$\theta=\theta(\mu)$	$V(\mu)$	$\mathcal{Y}$	$\mathrm{dom}(\mu)$	$\mathrm{dom}(\theta)$
Normal	$(y-\mu)^2$	$\theta^2/2$	σ^2	μ	1	$\mathbb{R}$	$\mathbb{R}$	$\mathbb{R}$
Bernoulli	$2\left[y\ln\left(\frac{y}{\mu}\right)+(1-y)\ln\left(\frac{1-y}{1-\mu}\right)\right]$	$\ln(1+e^\theta)$	1	$\ln\left(\frac{\mu}{1-\mu}\right)$	$\mu(1-\mu)$	$\mathbb{B}$	$(0,1)$	$\mathbb{R}$
Poisson	$2\left[y\ln\left(\frac{y}{\mu}\right)-(y-\mu)\right]$	$\exp(\theta)$	1	$\ln(\mu)$	μ	$\mathbb{N}$	$\mathbb{R}_+$	$\mathbb{R}$
Negative Binomial	$2\left[y\ln\left(\frac{y}{\mu}\right)-(y+k)\ln\left(\frac{y+k}{\mu+k}\right)\right]$	$-k\ln(1-e^\theta)$	1	$\ln\left(\frac{\mu}{\mu+k}\right)$	$\mu+\mu^2/k$	$\mathbb{N}$	$\mathbb{R}_+$	$\mathbb{R}_-$
Gamma	$2\left[-\ln\left(\frac{y}{\mu}\right)+\frac{y-\mu}{\mu}\right]$	$-\ln(-\theta)$	ϕ	$-1/\mu$	μ^2	$\mathbb{R}_+$	$\mathbb{R}_+$	$\mathbb{R}_-$
Inverse Gaussian	$\dfrac{(y-\mu)^2}{\mu^2 y}$	$-\sqrt{-2\theta}$	ϕ	$-1/(2\mu^2)$	μ^3	$\mathbb{R}_+$	$\mathbb{R}_+$	$\mathbb{R}_-$

Table 7.3: Common EDMs.

$$V(\mu) = \mu(1 - \mu).$$

Finally, the unit deviance works out to be

$$d(y, \mu) = 2\left[y \ln \frac{y}{\mu} + (1 - y) \ln \frac{1 - y}{1 - \mu}\right].$$

Poisson Distribution

We may express the Poisson distribution as

$$f(y; \mu) = \frac{e^{-\mu}\mu^y}{y!} = \frac{1}{y!} \exp\left[y \ln \mu - \mu\right].$$

We thus identify

$$\theta = \ln \mu$$
$$\phi = 1$$
$$a(y, \phi) = \frac{1}{y!}$$
$$\kappa(\theta) = e^\theta$$
$$V(\mu) = \mu.$$

Finally, the unit deviance is easily shown to yield

$$d(y, \mu) = 2\left[y \ln \frac{y}{\mu} + (\mu - y)\right].$$

Gamma Distribution

The gamma distribution is given by

$$f(y; \alpha, \beta) = \frac{1}{\Gamma(\alpha)\beta^\alpha} y^{\alpha-1} e^{-y/\beta},$$

with $\mathbb{E}[Y] = \alpha\beta$ and $\mathbb{V}(Y) = \alpha\beta^2$. By substituting $\alpha = 1/\phi$ and $\beta = \mu\phi$, we can reparametrize this distribution as

$$f(y; \mu, \phi) = \frac{\phi^{-1/\phi}y^{1/\phi-1}}{\Gamma(1/\phi)} \exp\left[\frac{-y/\mu - \ln(\mu)}{\phi}\right].$$

From this, we find

$$\theta = -1/\mu$$
$$\kappa(\theta) = \ln(-1/\theta)$$
$$V(\mu) = \mu^2$$
$$d(y, \mu) = 2\left[\frac{y}{\mu} - 1 + \ln\frac{\mu}{y}\right].$$

7.4 Generalized Linear Models

Now that we have defined and explored the class of distributions known as exponential dispersion models, we are ready for our main course. In this section, we introduce the main definition of generalized linear models and then proceed to lay the theory on how such models can be analyzed and applied.

7.4.1 Generalized Linear Models

We would like to develop a model in which the target variable Y is distributed relative to a member of the EDM family, such that each observation is allowed to be taken from a distribution with a different mean and dispersion:

$$Y_i \sim \text{EDM}(\mu_i, \phi/w_i).$$

As stated, if there are n observations, there would be n unknown parameters μ_i to solve for, which would make the inference task ill posed and prone to overfitting. To remedy this, we require some additional structure on *how* the parameter μ is allowed to vary, based on a lower-dimensional feature vector $\mathbf{x} \in \mathbb{R}^p$, in analogy to the structure we established for linear and logistic regression models. This is achieved with the following.

Definition 7.13. *A generalized linear model* (GLM) *is a predictive model for a random variable Y consisting of three components:*

1. A random component: *each observed target variable is conditionally distributed relative to the same member of the EDM family:*

$$Y_i|\mathbf{x}_i \sim \text{EDM}(\mu_i, \phi/w_i), \tag{7.64}$$

   where w_i is the (known) weight of the ith observation, and μ_i varies per observation, based on the feature vector $\mathbf{x}_i$;
2. A linear predictor: *a scalar parameter η_i is constructed using the ith feature vector $\mathbf{x}_i$ and an (unknown) vector of coefficients $\boldsymbol{\beta}$:*

$$\eta_i = \mathbf{x}_i^T \boldsymbol{\beta}; \tag{7.65}$$

   moreover, when $x_{i0} = 1$, then β_0 is called the intercept *of the model; and*
3. A link function: *a function g that links the expected mean of the EDM to the linear predictor:*

$$g(\mu_i) = \eta_i. \tag{7.66}$$

The particular link function $g = (\kappa')^{-1}$, where κ is the cumulant function of the EDM, is called the canonical link function.

Note 7.10. When using the canonical link function, the linear predictor coincides η with the canonical parameter θ:

$$\eta = g(\mu) = (\kappa')^{-1}(\mu) = \theta,$$

which follows from $\mu = \kappa'(\theta)$, as stated in Equation (7.55).   ▷

Example 7.5. The linear regression model is a GLM using the normal distribution and the identity link $\mu = \eta$; i.e.,

$$Y_i \sim \mathrm{N}(\mu_i, \sigma^2/w_i)$$
$$\mu_i = \mathbf{x}_i^T \boldsymbol{\beta}.$$

Note that the link function is the identity: $g(\mu) = \mu$. This is also the canonical link function for the normal distribution, so that μ and η each coincide with the canonical parameter: $\mu = \eta = \theta$.   ▷

Example 7.6. The logistic regression model is a GLM using the Bernoulli distribution and the logit link function; i.e.,

$$Y_i \sim \mathrm{Bern}(\mu_i)$$
$$\mathrm{logit}(\mu_i) = \mathbf{x}_i^T \boldsymbol{\beta}.$$

Note that the link function is the logit function: $g(\mu) = \mathrm{logit}(\mu)$. Moreover, this represents the canonical link function, so that η is the canonical parameter of the Bernoulli GLM.   ▷

Saturated Models and Total Deviance

A useful benchmark for comparison is obtained by breaking the link function and overfitting a model to the data.

Definition 7.14. *Given a set of data* $(\{\mathbf{x}_i, y_i, w_i\})_{i=1}^n$ *and a dispersion parameter* ϕ, *we define the* saturated model, *relative to a particular member of the* EDM *family, as the model*

$$\hat{Y}_i \sim \mathrm{EDM}(y_i, \phi/w_i);$$

i.e., the saturated model is the model obtained by overfitting the parameters μ_i *with* $\mu_i = y_i$.

Now, for a given set of data $(\{\mathbf{x}_i, y_i, w_i\})_{i=1}^n$, the likehood of a GLM relative to the canonical parameter θ is given by

$$L_\theta(\boldsymbol{\theta}; \mathbf{y}) = \prod_{i=1}^n f(y_i; \theta_i, \phi/w_i)$$

$$= \exp\left[\frac{1}{\phi}\left(\sum_{i=1}^n w_i y_i \theta_i - \sum_{i=1}^n w_i \kappa(\theta_i)\right)\right] \prod_{i=1}^n a(y_i, \phi/w_i).$$

The subscript L_θ reminds us that we are expressing the likelihood relative to the canonical parameters θ. The log likelihood is therefore given by

$$\ell_\theta(\boldsymbol{\theta}; \mathbf{y}) = \frac{1}{\phi} \sum_{i=1}^{n} w_i \left[y_i \theta_i - \kappa(\theta_i) \right] + \sum_{i=1}^{n} \ln a(y_i, \phi/w_i). \tag{7.67}$$

In order to write the log likehood in terms of the parameters μ_i, we can use Equation (7.62), thereby obtaining

$$\ell_\mu(\boldsymbol{\mu}; \mathbf{y}) = \frac{1}{\phi} \sum_{i=1}^{n} w_i t(y_i, \mu_i) + \sum_{i=1}^{n} \ln a(y_i, \phi/w_i). \tag{7.68}$$

This motivates the following.

Definition 7.15. *Given a set of data* $(\{\mathbf{x}_i, y_i, w_i\})_{i=1}^{n}$ *and a member of the* EDM *family, the* total deviance *is the function*

$$D(\mathbf{y}, \boldsymbol{\mu}) = \sum_{i=1}^{n} w_i d(y_i, \mu_i), \tag{7.69}$$

where the function d is the unit deviance, defined in Equation (7.61).

Proposition 7.12. *The total deviance is equivalent to the log likelihood ratio between a given model and its saturated counterpart:*

$$\frac{1}{\phi} D(\mathbf{y}, \boldsymbol{\mu}) = -2 \ln \frac{L_\mu(\boldsymbol{\mu}, \mathbf{y})}{L_\mu(\mathbf{y}, \mathbf{y})} = -2 \left[\ell_\mu(\boldsymbol{\mu}, \mathbf{y}) - \ell_\mu(\mathbf{y}, \mathbf{y}) \right].$$

The term on the left is called the scaled (total) deviance.

   This result follows directly from Equations (7.68) and (7.69). As a result, the total scaled deviance is equivalent to the likelihood ratio test statistic of a given model as compared with its saturated counterpart. As we discussed earlier, $D(\mathbf{y}, \boldsymbol{\mu}) \geq 0$, with larger values representing poorer fits. In some cases, the scaled deviance has an approximate chi-squared distribution, with degrees of freedom equal to the difference of the number of parameters used to fit the two models; i.e., $df = n - p - 1$.

7.4.2 Maximum Likelihood Redux

We next consider the special form of some key results from maximum-likelihood theory as applied to generalized linear models. We begin with the score statistics.

Theorem 7.9. *Given a set of data* $(\{\mathbf{x}_i, y_i, w_i\})_{i=1}^{n}$ *and a* GLM, *the score statistics (Definition 5.3) are given by*

$$U_j = \frac{\partial \ell}{\partial \beta_j} = \frac{1}{\phi} \sum_{i=1}^{n} \frac{w_i(y_i - \mu_i)x_{ij}}{V(\mu_i)g'(\mu_i)}, \tag{7.70}$$

for $j = 0, \ldots, p$. *Moreover, when the canonical link function is used, these equations reduce to*

$$U_j = \frac{\partial \ell}{\partial \beta_j} = \frac{1}{\phi} \sum_{i=1}^{n} w_i(y_i - \mu_i)x_{ij}. \tag{7.71}$$

Here, $\ell = \ell(\boldsymbol{\beta})$ *is the log-likelihood function with respect to the unknown modeling parameters* $\boldsymbol{\beta}$.

Proof. The log-likelihood function is given by expressing Equation (7.67) in terms of the unknown modeling parameters $\boldsymbol{\beta}$ via the chain

$$\ell(\boldsymbol{\beta}) = \ell_\theta(\theta(\mu(\eta(\boldsymbol{\beta})))).$$

Note that the second term in Equation (7.67) is a function of $\mathbf{y}$ and ϕ only, and therefore vanishes when differentiating with respect to the parameters $\boldsymbol{\beta}$.

To proceed, we invoke the chain rule:

$$\frac{\partial \ell}{\partial \beta_j} = \sum_{i=1}^{n} \frac{\partial \ell}{\partial \theta_i} \frac{\partial \theta_i}{\partial \mu_i} \frac{\partial \mu_i}{\partial \eta_i} \frac{\partial \eta_i}{\partial \beta_j}.$$

Now, from Equation (7.67), the first factor is given by

$$\frac{\partial \ell}{\partial \theta_i} = \frac{w_i(y_i - \kappa'(\theta_i))}{\phi} = \frac{w_i(y_i - \mu_i)}{\phi}.$$

For the second factor, recall from the definition of the variance function that

$$\frac{d\theta}{d\mu} = \frac{1}{V(\mu)}.$$

For the penultimate factor, recall from the definition of the link function that $\mu = g^{-1}(\eta)$, so that

$$\frac{d\mu}{d\eta} = (g^{-1})'(\eta) = \frac{1}{g'(g^{-1}(\eta))} = \frac{1}{g'(\mu)},$$

with the second equality following from elementary calculus. Finally, due to the definition of the linear predictor (Equation (7.65)), it follows that

$$\frac{\partial \eta_i}{\partial \beta_j} = x_{ij},$$

for $i = 1, \ldots, n$ and $j = 0, \ldots, p$. Combining the preceding equations yields the first result.

For the second result, note that when the canonical link function is used to define the GLM, we have $\theta = g(\mu)$, which implies that

$$\frac{d\theta}{d\mu} = g'(\mu).$$

It therefore follows that, for the case of the canonical link function, $g'(\mu)V(\mu) = 1$, thereby yielding the simplification of Equation (7.71).   □

Theorem 7.10. *Given a set of data* $(\{\mathbf{x}_i, y_i, w_i\})_{i=1}^n$ *and a GLM, the Fisher information (Definition 5.4) is given by*

$$\mathcal{I}_{jk} = \mathbb{E}\left[-\frac{\partial^2 \ell}{\partial \beta_k \partial \beta_j} \right] = \frac{1}{\phi} \sum_{i=1}^n \frac{w_i x_{ij} x_{ik}}{V(\mu_i) g'(\mu_i)^2}. \tag{7.72}$$

When $g(\mu)$ *is the canonical link, this expression reduces to*

$$\mathcal{I}_{jk} = \frac{1}{\phi} \sum_{i=1}^n \frac{w_i x_{ij} x_{ik}}{g'(\mu_i)}. \tag{7.73}$$

Proof. From Equation (7.70), we have

$$\frac{\partial^2 \ell}{\partial \beta_k \partial \beta_j} = \frac{1}{\phi} \sum_{i=1}^n (y_i - \mu_i) \frac{\partial}{\partial \beta_k} \left[\frac{w_i x_{ij}}{V(\mu_i) g'(\mu_i)} \right] - \frac{1}{\phi} \sum_{i=1}^n \frac{w_i x_{ij}}{V(\mu_i) g'(\mu_i)} \frac{\partial \mu_i}{\partial \eta_i} \frac{\partial \eta_i}{\partial \beta_k}.$$

The expectation in Equation (7.72) is the conditional expectation of Y_i given $\mathbf{x}_i$, which is simply $\mathbb{E}[Y_i | \mathbf{x}_i] = \mu_i$. We therefore find that

$$\mathbb{E}\left[-\frac{\partial^2 \ell}{\partial \beta_k \partial \beta_j} \right] = \frac{1}{\phi} \sum_{i=1}^n \frac{w_i x_{ij}}{V(\mu_i) g'(\mu_i)} \frac{\partial \mu_i}{\partial \eta_i} \frac{\partial \eta_i}{\partial \beta_k},$$

which simplifies to Equation (7.72) following the same steps from Theorem 7.9. Equation (7.73) again follows from the relation $V(\mu_i) g'(\mu_i) = 1$, when $g(\mu)$ is the canonical link function. This completes the result.   □

Matrix Formalism

We may economically represent the score statistic, which is really a gradient vector, and Fisher information using matrix notation as follows.

Proposition 7.13. *Let* $\mathbf{V}$ *and* $\mathbf{L}$ *represent the* $n \times n$ *diagonal matrices with diagonal elements*

$$V_{ii} = \frac{V(\mu_i)}{w_i} \qquad and \qquad L_{ii} = g'(\mu_i),$$

respectively. Then the score statistic and Fisher information are equivalent to the following expressions

$$\mathbf{u} = \frac{1}{\phi}\mathbf{X}^T\mathbf{V}^{-1}\mathbf{L}^{-1}(\mathbf{y} - \boldsymbol{\mu}) \tag{7.74}$$

$$\mathcal{I} = \frac{1}{\phi}\mathbf{X}^T\mathbf{V}^{-1}\mathbf{L}^{-2}\mathbf{X}, \tag{7.75}$$

respectively, where $\mathbf{X}$ is the design matrix. For the canonical link, we further obtain the simplification $\mathbf{VL} = \mathbf{I}$.

The diagonal matrix $\mathbf{W} = \mathbf{V}^{-1}\mathbf{L}^{-2}$ is sometimes referred to as the matrix of working weights.

Proof. It's obvious. □

Method of Scoring

As we saw in Section 5.1, the maximum likelihood solution is the solution to the score equation

$$U_j = 0.$$

Moreover, the Newton–Raphson method, which requires a calculation of the *observed information*, is typically replaced with the method of scoring, which uses the Fisher information. As we saw in Equation (5.15), the method of scoring is an iterative algorithm that updates the current "guess" for the parameters $\boldsymbol{\beta}$ with

$$\boldsymbol{\beta}^{k+1} = \boldsymbol{\beta}^k + \mathcal{I}(\boldsymbol{\beta}^k)^{-1} \cdot \mathbf{u}(\boldsymbol{\beta}^k). \tag{7.76}$$

Of course, the score statistic and Fisher information depend on the unknown parameters explicitly through the parameters $\boldsymbol{\mu}$ and the link function.

Estimating Dispersion

Oftentimes the dispersion parameter ϕ is not known, but must be estimated. As we saw in the case of linear regression, the MLE for ϕ tends to be a biased estimate, hence it is seldom used in practice. Another approach is to estimate ϕ using the *mean deviance estimator*

$$\hat{\phi} = \frac{D(\mathbf{y}, \hat{\boldsymbol{\mu}})}{n - (p + 1)},$$

where $p+1$ is the total number of parameters. This estimate, however, only behaves well under restricted circumstances: the GLM must be normal or inverse Gaussian, or the dispersion must be small. Details of the pros and cons of these estimates are discussed in Dunn and Smyth [2018].

In practice, the go-to estimate for ϕ, due to its ease-of-use and robustness across use cases, is based on the following.

Definition 7.16. *Given a set of data* $(\{\mathbf{x}_i, y_i, w_i\})_{i=1}^n$, *let* $\hat{\boldsymbol{\beta}}$ *represent the* MLE *for a* GLM, *and let* $\hat{\eta}_i = \mathbf{x}_i^T \hat{\boldsymbol{\beta}}$ *and* $\hat{\mu}_i = g(\hat{\eta}_i)$. *Then the* Pearson *statistic is defined as the quantity*

$$X^2 = \sum_{i=1}^n \frac{w_i(y_i - \hat{\mu}_i)^2}{V(\hat{\mu}_i)}. \tag{7.77}$$

Moreover, the Pearson *estimator for* ϕ *is defined by*

$$\hat{\phi} = \frac{X^2}{n - (p+1)}, \tag{7.78}$$

where $p+1$ *is the total number of unknown parameters; i.e.,* $\boldsymbol{\beta} \in \mathbb{R}^{p+1}$.

Note 7.11. The idea behind the Pearson statistic is to construct a quantity similar to the SSE, defined in Equation (6.2), using the working weights from the matrix $\mathbf{W}$ and the differential response in the η-space, to obtain

$$\text{SSE} = \sum_{i=1}^n W_i(z_i - \hat{\eta}_i)^2,$$

where

$$W_i = \frac{w_i}{V(\mu_i)g'(\mu_i)^2}$$

is the working weight and $z_i = g(y_i)$. Next, when $z_i \approx \hat{\eta}_i$, we can use the approximation

$$g'(\mu_i) \approx \frac{\Delta \eta_i}{\Delta \mu_i} = \frac{z_i - \hat{\eta}_i}{y_i - \hat{\mu}_i}$$

to obtain the result. ▷

7.4.3 Inference

We next consider the sampling distribution for our estimates. From this we will construct approximate confidence intervals.

Asymptotic Distribution of Maximum-likelihood Estimates

As we saw in Chapter 5, the maximum-likelihood estimates of our parameter set are approximately normally distributed for large n. In particular, we have the following.

Proposition 7.14. *Given a set of data* $(\{\mathbf{x}_i, y_i, w_i\})_{i=1}^n$, *let* $\hat{\boldsymbol{\beta}}$ *represent the* MLE *for a* GLM. *For large* n, *the* MLE *is approximately normally distributed*

$$\hat{\boldsymbol{\beta}} \approx \text{N}(\boldsymbol{\beta}_0, \boldsymbol{\mathcal{I}}(\boldsymbol{\beta}_0)^{-1}), \tag{7.79}$$

where $\boldsymbol{\beta}_0$ *is the true value of the parameter* $\boldsymbol{\beta}$.

Proof. This result is a generalization of Corollary 5.2.   □

Corollary 7.5. *Given a set of data* $(\{\mathbf{x}_i, y_i, w_i\})_{i=1}^n$, *let* $\hat{\boldsymbol{\beta}}$ *represent the* MLE *for a* GLM. *For large* n, *the variance of this estimate is approximately given by*

$$\mathbb{V}(\hat{\boldsymbol{\beta}}) = \boldsymbol{\mathcal{I}}^{-1}. \tag{7.80}$$

Moreover, the standard error of any individual $\hat{\beta}_j$ *is given by*

$$\text{se}(\hat{\beta}_j) = \sqrt{\phi\,(\mathbf{X}^T\mathbf{W}\mathbf{X})_{jj}^{-1}}, \tag{7.81}$$

where $\mathbf{W}$ *is the working weight matrix defined in Proposition 7.13.*

Proof. This follows directly form Proposition 7.14.   □

Corollary 7.6. *Given* $\mathbf{x} \in \mathbb{R}^{p+1}$ *and the* MLE $\hat{\boldsymbol{\beta}}$, *construct* $\eta = \mathbf{x}^T\hat{\boldsymbol{\beta}}$, *as usual. The variance of this estimate is given by*

$$\mathbb{V}(\hat{\eta}) = \mathbf{x}^T\boldsymbol{\mathcal{I}}^{-1}\mathbf{x}. \tag{7.82}$$

Proof. This follows from Equation (7.80) and properties of variance:

$$\mathbb{V}(\mathbf{x}^T\hat{\boldsymbol{\beta}}) = \mathbf{x}^T\mathbb{V}(\boldsymbol{\beta})\mathbf{x}.$$

The result follows.   □

Tests for Individual Estimates

Given that the MLE $\hat{\boldsymbol{\beta}}$ is an asymptotically normal and unbiased estimate of the true parameter values, we may use the Wald test (Definition 3.11) to check the null hypothesis $H_0 : \beta_j = \beta_j^0$. Often, one selects β_j^0, in order to test whether or not the jth coefficient is nonzero, in which case the model can be simplified. When the dispersion parameter ϕ is known, the Wald statistic is given by Equation (3.8) as

$$Z = \frac{\hat{\beta}_j - \beta_j^0}{\text{se}(\hat{\beta}_j)}. \tag{7.83}$$

Theorem 3.1 guarantees that the Wald statistic Z is approximately a standard normal distribution for large n. We may therefore construct a $(1 - \alpha)$ confidence interval for $\hat{\beta}_j$, or for $\hat{\eta}$, by the relations

$$\hat{\beta}_j \pm z_{\alpha/2}\text{se}(\hat{\beta}_j)$$
$$\hat{\eta} \pm z_{\alpha/2}\text{se}(\hat{\eta}), \tag{7.84}$$

respectively, where the standard errors are given in Equations (7.81) and (7.82). A confidence interval for $\hat{\mu}$ is then obtained by applying the mapping

g^{-1} to the upper and lower bounds of the confidence interval for $\hat{\eta}$. Note that the confidence interval for $\hat{\eta}$ is symmetric, whereas the confidence interval for $\hat{\mu}$ is typically asymmetric, due to the nonlinearity of the link function.

When the dispersion parameter ϕ is not known, we must estimate the standard error using a suitable estimate $\hat{\phi}$ of the dispersion; typically the Pearson estimator. For a consistent estimator and a very large sample size, $\hat{\phi} \approx \phi$, and the normal approximation is still often suitable. For small or moderate sample sizes, however, we instead formulate the T-statistic (Proposition 3.3) as

$$T = \frac{\hat{\beta}_j - \beta_j^0}{\hat{se}(\hat{\beta}_j)}, \tag{7.85}$$

which is distributed as a t_{n-p-1} distribution. The corresponding $(1 - \alpha)$ confidence intervals are therefore obtained from the relations

$$\hat{\beta}_j \pm t_{n-p-1,\alpha/2}\hat{se}(\hat{\beta}_j)$$
$$\hat{\eta} \pm t_{n-p-1,\alpha/2}\hat{se}(\hat{\eta}). \tag{7.86}$$

The corresponding confidence region for $\hat{\mu}$ can again be constructed by applying the function g^{-1} to the upper and lower bounds of the confidence region for $\hat{\eta}$.

7.4.4 Implementation

We can easily write an abstract class for a generic GLM, with the score function, Fisher information, and method of scoring built in. This is achieved in Code Blocks 7.6 and 7.7. Note that GLM is a subclass of our abstract Model class, defined in Code Block 6.22. Compare with the implementation of logistic regression in Code Block 7.4.

7.5 Models for Proportions

In Section 7.2, we formulated the logistic regression problem in terms of a Bernoulli random variable:

$$Y_i \sim \text{Bern}(\mu_i)$$
$$\text{logit}(\mu_i) = \mathbf{x}_i^T \boldsymbol{\beta}.$$

In Equation (7.40), we formulated the likelihood and log-likelihood functions for a weighted set of data. In this formulation, the target variable Y_i still takes on a binary value $Y_i \in \mathbb{B}$, and a specific weight w_i is applied to each observation. This formulation, however, is more commonly used for purely binary data, for which the weights are all $w_i = 1$. Purely binary data are likely to turn up when the feature vector has at least one continuous

```python
class GLM(Model):
    def __init__(self, max_iter=20, tol=1e-6, intercept=True,
            canonical_link=False, classifier=False):
        self.max_iter = max_iter
        self.tol = tol
        self.intercept = intercept
        self.beta = []
        self.canonical_link = canonical_link # Override default in
            subclass
        self.classifier = classifier # Override default in subclass

    @abstractmethod
    def variance(self, mu):
        pass

    @abstractmethod
    def link(self, mu):
        pass

    @abstractmethod
    def linkinv(self, eta):
        pass

    @abstractmethod
    def dlink(self, mu):

        if self.canonical_link:
            return 1 / self.variance(mu)

    def score(self, X, y, mu, w=None):
        if not w:
            w = np.ones(len(y))

        return X.T @ (w*(y - mu) / self.variance(mu) /
            self.dlink(mu))

    def information(self, X, y, mu, w=None):
        if not w:
            w = np.ones(len(y))

        D = np.diag(w / self.variance(mu) / self.dlink(mu)**2 )
        return X.T @ D @ X
## Continued...
```

Code Block 7.6: Abstract parent class for GLM implementation. (Continued in Code Block 7.7.)

```
1    def train(self, X, y, weights=None):
2        n, p = X.shape
3        w = weights if weights else np.ones(n)
4        y = y.reshape((n, 1))
5        w = w.reshape((n, 1))
6
7        if self.intercept:
8            X = np.concatenate([np.ones(n).reshape((n,1)), X], axis=1)
9            p += 1
10       beta = np.zeros(p).reshape((p, 1))
11       beta[0] = 1
12       i = 0
13       err = 1
14       while (i < self.max_iter) and (err > self.tol):
15           i += 1
16           mu = self.linkinv(X@beta)
17           score = self.score(X, y, mu, w=w)
18           info = self.information(X, y, mu, w=w)
19           err = np.linalg.norm(score)
20           beta += np.linalg.inv(info) @ score
21
22       self.converged = err < self.tol
23       self.beta = beta
24
25   def predict(self, X, threshold):
26       if self.intercept:
27           X = np.concatenate([np.ones(len(X)).reshape((len(X),1)),
                   X], axis=1)
28       assert self.converged, "Method of Scoring failed to converge"
29
30       mu = self.linkinv(X@self.beta)
31       if self.classifier:
32           return (mu > threshold).astype(int)
33       return mu
34
35   def predict_proba(self, X):
36       if self.intercept:
37           X = np.concatenate([np.ones(len(X)).reshape((len(X),1)),
                   X], axis=1)
38       assert self.converged, "Method of Scoring failed to converge"
39       return self.linkinv(X@self.beta)
```

Code Block 7.7: Abstract parent class for GLM implementation. (Continued from Code Block 7.6.)

variable. When the feature vector consists purely of categorical features, it is more likely to observe data with weights and proportions. When a weight is ascribed to each datum, we prefer a slight reformulation of the problem as a binomial random variable. As it turns out, the expressions derived for a Bernoulli random variable are still valid, except now the target variable is allowed to vary on the unit interval $[0, 1]$.

7.5.1 Binomial GLM

We begin by showing how a binomial random variable can be cast as a GLM. The result is closely related to the Bernoulli random variable formulation.

Proposition 7.15. *Let $wY \sim \text{Binom}(w, \mu)$ represent a binomial random variable, for $w \in \mathbb{Z}_+$, so that the random variable Y represents the proportion of successes. Then Y is distributed as a weighted Bernoulli GLM, with weight w; i.e., the distribution of Y is given by*

$$f_Y(y) = a(y, 1/w) \exp\left[yw\theta + w\ln(1 + e^\theta)\right],$$

where $\theta = \text{logit}(\mu)$ is the canonical parameter for the Bernoulli GLM.

Note 7.12. Thus, when Y represents the proportion of successes out of an IID sequence of w Bernoulli trials, the Bernoulli GLM, as represented in Table 7.3, can still be used by applying weight w and allowing the target variable Y to range over the unit interval $Y \in [0, 1]$, as opposed to over the set $Y \in \mathbb{B}$.

Technically, for a fixed observation (i.e., fixed μ) with weight w, the target variable Y must take on a discrete set of values $Y \in \{0, 1/w, 2/w, \ldots, 1\}$. However, as w typically varies for each observation, it is simpler to represent the acceptable values of Y as the unit interval $[0, 1]$, with it being understood that the value of Y, for each observation, may only take on an appropriate discrete subset of values.   ▷

Proof. This result follows directly from Theorem 7.8, as the target variable Y represents the sample mean of a sequence of w IID Bernoulli trials. □

Though the preceding result follows from Theorem 7.8, it is nonetheless to see how it may be obtained directly. Since $wY \sim \text{Binom}(w, \mu)$, it follows that the PMF for Y is given by

$$
\begin{aligned}
f_Y(y) &= \binom{w}{wy} \mu^{wy}(1 - \mu)^{w(1-y)} \\
&= \binom{w}{wy} \exp\left[wy\ln(\mu) + w(1 - y)\ln(1 - \mu)\right] \\
&= \binom{w}{wy} \exp\left[wy\ln\left(\frac{\mu}{1 - \mu}\right) + w\ln(1 - \mu)\right].
\end{aligned}
$$

Here, we identify $a(y, 1/w) = \binom{w}{wy}$, along with the canonical variable $\theta = \text{logit}(\mu)$ and cumulant function $\kappa(\theta) = \ln(1 + e^{\theta})$. The only difference between this GLM and the Bernoulli GLM is the weight w and the fact that the variable y is now allowed to take on the set of values $Y \in \{0, 1/w, 2/w, \ldots, 1\}$.

The other simplification gained from reformulating this as a binomial GLM is found in the representation of the data. In the weighted Bernoulli formulation of Equation (7.40), the results for a given feature vector $\mathbf{x}_0$ would be broken into two individual records: one for successes and one for failures. For example, we might have:

$$w_0 = 70, \qquad y_0 = 0, \qquad \mathbf{x}_0 = \mathbf{x}^*$$
$$w_1 = 30, \qquad y_1 = 1, \qquad \mathbf{x}_1 = \mathbf{x}^*$$

in the formulation of Equation (7.40), for which Y is only allowed to take on the values $Y \in \mathbb{B} = \{0, 1\}$. In this new formulation, we would represent the same datum as

$$w_0 = 100, \qquad y_0 = 0.3, \qquad \mathbf{x}_0 = \mathbf{x}^*.$$

Thus, instead of requiring two records for each unique feature vector $\mathbf{x}$, one representing the positive outcomes $Y = 1$ and one representing the negative outcomes $Y = 0$, we now only require a single record, in which Y represents the proportion of positive outcomes.

7.5.2 Link Functions for Logistic Regression

In Section 7.2, we used the logit function as our link function. We now recognize that the logit link function for binary or binomial (proportion) data represents the canonical link function, so that the linear predictor coincides with the natural parameter of the Bernoulli or binomial GLM. However, there are several alternatives that are worth being aware of. We enumerate a few main link functions that are sometimes used with binary data as follows.

1. The *logit link function*, as we saw previously, represents the canonical link function and is defined by the log-odds

$$g(\mu) = \text{logit}(\mu) = \ln\left(\frac{\mu}{1 - \mu}\right).$$

2. The *probit link function* is defined by

$$g(\mu) = \Phi^{-1}(\mu),$$

where $\Phi(\cdot)$ is the CDF of the standard normal distribution.

3. The *complementary log-log link function* is defined by

$$g(\mu) = \log\left[-\log(1-\mu)\right],$$

with inverse function $g^{-1}(\eta) = 1 - \exp[-\exp(\eta)]$.

4. The *log-log link function* is defined by

$$g(\mu) = -\log\left[-\log\left(\mu\right)\right],$$

with inverse function $g^{-1}(\eta) = \exp[-\exp(-\eta)]$.

Each of the preceding link functions have an important quality in common, and that is they each represent a mapping $g : (0,1) \to \mathbb{R}$, so that the expected probability may be expressed without constraint in terms of the linear predictor. Each of these link functions is plotted concurrently in Figure 7.6. Aside from this important quality, there are some similarities and some differences among the various link functions.

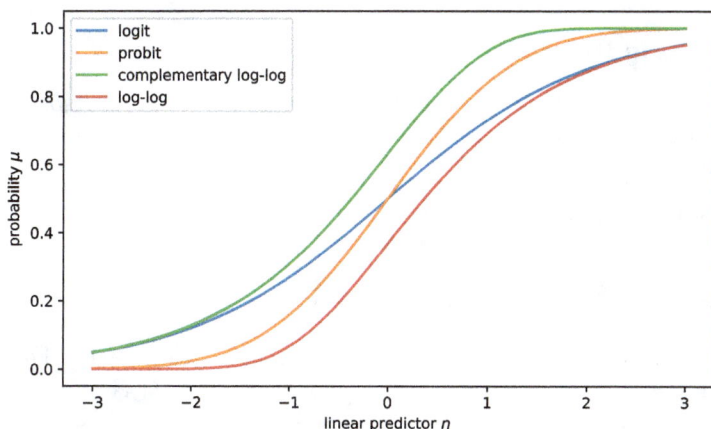

Fig. 7.6: Different (inverse) link functions for logistic regression; $\mu = g^{-1}(\eta)$.

To begin, both the probit and logit link functions are antisymmetric about the probability $\mu = 0.50$; i.e.,

$$\operatorname{logit}(\mu) = -\operatorname{logit}(1-\mu) \qquad \text{and} \qquad \operatorname{probit}(\mu) = -\operatorname{probit}(1-\mu).$$

In particular, this implies that $\operatorname{logit}(1/2) = \operatorname{probit}(1/2) = 0$. While both the probit and logit functions follow the same similar shape, the probit function is slightly more sensitive to changes in the underlying linear predictor.

This symmetry is broken for the log-log and complementary log-log link functions. These final two link functions are closely related, as the complementary log-log link function is the log-log link function applied to the complementary probability $1 - \mu$. Note in Figure 7.6 that the complementary log-log link function closely approximates the logit function for small values of $\mu \approx 0$. Similarly, the log-log link function closely approximates the logit function for large values of $\mu \approx 1$.

7.6 Models for Counts

In this section, we consider generalized linear models for which the target variable represents *count* data. People in erudite circles count things all the time; we love to count! And when we're not counting, we're modeling what our counts might total. Such is the topic of the present section.

Example of count data include: the number of potholes per mile of street or highway; the number of typos in a book; the number of signups for a website; the number of text messages; the number of sales in a particular region. Count data also frequently appear in contingency tables, for which observations are cross-classified relative to a number of factors. This arises, for instance, in analyzing the results of a multi-level experiment (Section 4.4).

7.6.1 Poisson Regression

The simplest model for count data is obtained using the Poisson distribution, which is parameterized by a single parameter μ, equivalent to the distribution's mean.

Definition 7.17. *The* Poisson GLM *is the* GLM *defined by*

$$Y_i = \text{Poiss}(\mu_i), \tag{7.87}$$
$$g(\mu_i) = \mathbf{x}_i^T \boldsymbol{\beta}, \tag{7.88}$$

for $Y_i \in \mathbb{N}$, $\mu_i \in \mathbb{R}_+$, and for a given link function $g : \mathbb{R}_+ \to \mathbb{R}$. When the link function is the canonical link $g(\mu_i) = \ln(\mu_i)$, the GLM *is referred to as the* (Poisson) loglinear model.

Since the canonical link function, which connects the mean to the linear predictor, is the log function, the model, under the canonical link, is sometimes referred to as a *loglinear model*. As this is the only link function we will consider for the Poisson regression problem, we will often simply refer to the model as Poisson, with *loglinear* being understood.

Proposition 7.16. *The score function and Fisher information for the Poisson loglinear model are given by*

$$U_j = \sum_{i=1}^{n} (y_i - \mu_i) x_{ij}, \tag{7.89}$$

$$\mathcal{I}_{jk} = \sum_{i=1}^{n} \mu_i x_{ij} x_{ik}. \tag{7.90}$$

respectively.

Proof. Since the canonical link function is used, these relations follows from Equations (7.71) and (7.73), since $\phi = 1$ and $V(\mu) = 1/g'(\mu) = \mu$. □

Note that for a Poisson GLM, the Pearson statistic, defined in Equation (7.77), simplifies to

$$X^2 = \sum_{i=1}^{n} \frac{(y_i - \hat{\mu}_i)^2}{\hat{\mu}_i},$$

which is equivalent to the Pearson chi-squared test statistic, defined in Definition 3.19.

The relationship between Poisson regression and logistic and multinomial models is discussed in Agresti [2015].

7.6.2 Contingency Tables

Count data often arise in the context of contingency tables. Interestingly, the method of data collection affects the choice of model, depending on the constraints used during the data collection process. We will focus on two-factor contingency tables, though the methods presented herein can easily be generalized to higher-order tables. A typical two-factor contingency table is shown in Table 7.4.

	$j=1$	$j=2$	$\cdots$	$j=s$	row totals
$i=1$	Y_{11}	Y_{12}	$\cdots$	Y_{1s}	$Y_{1\cdot}$
$i=2$	Y_{21}	Y_{22}		Y_{2s}	$Y_{2\cdot}$
$\vdots$	$\vdots$		$\ddots$	$\vdots$	$\vdots$
$i=m$	Y_{m1}	Y_{m2}	$\cdots$	Y_{ms}	$Y_{m\cdot}$
column totals	$Y_{\cdot1}$	$Y_{\cdot2}$	$\cdots$	$Y_{\cdot s}$	$Y_{\cdot\cdot}$

Table 7.4: Layout of a two-factor contingency table.

The random variables Y_{ij}, for $i = 1, \ldots, m$ and $j = 1, \ldots, s$ represent the total observed counts for the ith level of the first factor and the jth level of the second factor. As usual, the quantities

$$Y_{i\cdot} = \sum_{j=1}^{s} Y_{ij} \quad \text{and} \quad Y_{\cdot j} = \sum_{i=1}^{m} Y_{ij}$$

represent the marginal counts for row i and column j, respectively, and the quantity $Y_{\cdot\cdot} = \sum_{i=1}^{m} \sum_{j=1}^{s} Y_{ij}$ represents the total count of all the observations.

Contingency Tables with No Constraints

We begin by analyzing an *unconstrained* contingency table. This is common, for example, when data are collected for a specified period of time, and the results are random and tabulated at the end of the data. (We will discuss constrained variations, e.g., when the total number of counts is specified a priori, momentarily.)

In modeling a two-factor contingency table, the responses are modeled as independent Poisson random variables by the relation

$$Y_{ij} = \text{Poiss}(\mu_{ij}). \tag{7.91}$$

Further, we assume independence in the two factors, such that the Poisson parameters μ_{ij} satisfy the relation

$$\mu_{ij} = \mu \phi_i \psi_j, \tag{7.92}$$

where the parameters ϕ_i and ψ_j are normalized by the conditions

$$\sum_{i=1}^{m} \phi_i = 1 \quad \text{and} \quad \sum_{j=1}^{s} \psi_j = 1, \tag{7.93}$$

so that $\mu = \sum_{i=1}^{m} \sum_{j=1}^{s} \mu_{ij}$. In the unconstrained case, note that the total count $Y_{..}$ is distributed as a Poisson random varaible, according to

$$Y_{..} \sim \text{Poiss}(\mu). \tag{7.94}$$

When the log link function is used, the independence assumption in Equation (7.92) results in

$$\ln \mu_{ij} = \beta_0 + \beta_i^A + \beta_j^B, \tag{7.95}$$

where we use the superscript A to denote the first factor (the rows) and B to denote the second factor (the columns). Moreover, the constraints defined in Equation (7.93) place a constraint on β_i^A and β_j^B. Without loss of generality, we may take $\beta_0^A = \beta_0^B = 0$ to eliminate the dependency among the parameters. Note that the feature vector consists of the multi-index iteself; i.e., it is the rows and columns of the contingency table that constitute an encoding of the two categorical features. To see this, we may write the following one-hot encoding for the features as

$$\mathbf{x}_{ij} = \begin{bmatrix} 1 \\ \mathbb{I}[i=2] \\ \vdots \\ \mathbb{I}[i=m] \\ \mathbb{I}[j=2] \\ \vdots \\ \mathbb{I}[j=s] \end{bmatrix} \quad \text{and} \quad \boldsymbol{\beta} = \begin{bmatrix} \beta_0 \\ \beta_2^A \\ \vdots \\ \beta_m^A \\ \beta_2^B \\ \vdots \\ \beta_s^B \end{bmatrix},$$

so that Equation (7.95) is equivalent to the more familiar

$$\ln \mu_{ij} = \mathbf{x}_{ij}^T \boldsymbol{\beta}.$$

Note that $\boldsymbol{\beta} \in \mathbb{R}^{m+s-1}$, as there is one (1) parameter due to the intercept, $m - 1$ parameters due to the first factor, and $s - 1$ parameters due to the second factor.

The likelihood function for the data is given by

$$L = \prod_{i=1}^{m} \prod_{j=1}^{s} \frac{e^{-\mu_{ij}} \mu_{ij}^{y_{ij}}}{y_{ij}!} = \exp\left[\sum_{i=1}^{m} \sum_{j=1}^{s} (y_{ij} \ln \mu_{ij} - \mu_{ij})\right] \prod_{i=1}^{m} \prod_{j=1}^{s} \frac{1}{y_{ij}!}.$$

Next, we can substitute in the link function Equation (7.95) to express the log-likelihood in terms of $\boldsymbol{\beta}$ as

$$\ell(\boldsymbol{\beta}) = y_{..}\beta_0 + \sum_{i=1}^{m} y_{i.}\beta_i^A + \sum_{j=1}^{s} y_{.j}\beta_j^B - \sum_{i=1}^{m} \sum_{j=1}^{s} \exp(\beta_0 + \beta_i^A + \beta_j^B) + \ln a(y_{ij}),$$

where $a(y_{ij}) = \prod_{i=1}^{m} \prod_{j=1}^{s} (y_{ij}!)^{-1}$. The score equations follow:

$$\frac{\partial \ell}{\partial \beta_0} = y_{..} - \sum_{i=1}^{m} \sum_{j=1}^{s} \exp(\beta_0 + \beta_i^A + \beta_j^B)$$

$$\frac{\partial \ell}{\partial \beta_i^A} = y_{i.} - \sum_{j=1}^{s} \exp(\beta_0 + \beta_i^A + \beta_j^B)$$

$$\frac{\partial \ell}{\partial \beta_j^B} = y_{.j} - \sum_{i=1}^{m} \exp(\beta_0 + \beta_i^A + \beta_j^B).$$

Recognizing that $\exp(\beta_0 + \beta_i^A + \beta_j^B) = \mu_{ij} = \mu \phi_i \psi_j$, and upon setting the preceding equations to zero, we have the following MLE estimates:

$$\hat{\mu} = y_{..}, \qquad \hat{\phi}_i = \frac{y_{i.}}{y_{..}}, \qquad \hat{\psi}_j = \frac{y_{.j}}{y_{..}}.$$

From this, it follows that

$$\hat{\mu}_{ij} = \frac{y_{i.} y_{.j}}{y_{..}}, \tag{7.96}$$

as one might well expect.

Contingency Tables with Constrained Overall Total

Next, let us reformulate the problem supposing that the total count $Y_{..}$ is not random, but fixed $y_{..} = n$ in advanced. For example, it is determined *in advanced* that the data shall stop being collected once the first $n = 1000$

samples have arrived. How those samples sort among the rows and columns is random, but the overall total count is held fixed.

The conditional probability of Y_{ij} given the total $Y_.. = n$ is given by

$$f(y_{ij}|y_.. = n) = \frac{f(y_{ij})}{\mathbb{P}(y_.. = n)}.$$

However, recalling Equation (7.94), we have

$$f(y_{ij}|y_.. = n) = \frac{n!}{e^{-\mu}\mu^n} \prod_{i=1}^{m} \prod_{j=1}^{s} \frac{e^{-\mu_{ij}}\mu_{ij}^{y_{ij}}}{y_{ij}!}$$

$$= \frac{n!}{\prod_{i=1}^{m}\prod_{j=1}^{s} y_{ij}!} \frac{1}{e^{-\mu}\mu^n} \prod_{i=1}^{m} \prod_{j=1}^{s} e^{-\mu_{ij}}\mu_{ij}^{y_{ij}}$$

$$= \frac{n!}{\prod_{i=1}^{m}\prod_{j=1}^{s} y_{ij}!} \prod_{i=1}^{m} \prod_{j=1}^{s} \pi_{ij}^{y_{ij}},$$

$$= \frac{n!}{\prod_{i=1}^{m}\prod_{j=1}^{s} y_{ij}!} \exp\left[\sum_{i=1}^{m} \sum_{j=1}^{s} y_{ij} \ln \pi_{ij}\right],$$

where $\pi_{ij} = \mu_{ij}/\mu$ represents the probability of the ijth cell. We immediately recognize this as the multinomial distribution:

$$f(Y_{ij}|Y_.. = n) = \text{Multi}(\pi_{ij}).$$

We may therefore express the log-likelihood kernel as

$$\ell(\phi, \psi) = \sum_{i=1}^{m} y_{i\cdot} \ln(\phi_i) + \sum_{j=1}^{s} y_{\cdot j} \ln(\psi_j),$$

which is subject to the constraints $\sum_{i=1}^{m} y_{i\cdot} = \sum_{j=1}^{s} y_{\cdot j} = n$. Solving the constrained MLE problem yields the estimates

$$\hat{\phi}_i = \frac{y_{i\cdot}}{n} \qquad \text{and} \qquad \hat{\psi}_j = \frac{y_{\cdot j}}{n},$$

which results in the final probability estimate $\pi_{ij} = y_{i\cdot} y_{\cdot j}/n^2$.

7.6.3 Negative-binomial GLMs

In the Poisson model, the variance of the response is equal to the mean; i.e., $\mathbb{V}(Y) = \mu$. In practice, however, the observed variance often exceeds this amount, resulting in a phenomenon known as *overdispersion*. A common approach to dealing with overdispersion is by using a *mixture model*. The idea is that the Poisson parameter μ_i, even for a fixed feature vector, is not homogeneous across the group, but rather exhibits a certain level of heterogeneity, which can be modeled by a secondary distribution. Mixing the Poisson distribution with a gamma distribution, in order to model the population heterogeneity, yields the negative binomial distribution. For details of this derivation, see Agresti [2015] or Dunn and Smyth [2018].

Definition 7.18. *The* negative-binomial distribution (NBD) *is the distribution for a discrete random variable* $Y \in \mathbb{N}$ *with probability mass function*

$$f(y; \mu, k) = \frac{\Gamma(y + k)}{\Gamma(k)\Gamma(y + 1)} \left(\frac{\mu}{\mu + k}\right)^y \left(\frac{k}{\mu + k}\right)^k, \tag{7.97}$$

for parameters $\mu > 0$ *and* $k > 0$.

Proposition 7.17. *For fixed* $k > 0$, *the negative-binomial distribution is an* EDM *with canonical parameter, cumulant function, and variance function given by*

$$\theta = \ln\left(\frac{\mu}{\mu + k}\right) \tag{7.98}$$

$$\kappa(\theta) = -k \ln\left(1 - e^\theta\right) \tag{7.99}$$

$$V(\mu) = \mu + \mu^2/k, \tag{7.100}$$

respectively.

Proof. The proof is straight-forward. We begin by rewriting Equation (7.97) as

$$f(y; \mu, k) = \frac{\Gamma(y + k)}{\Gamma(k)\Gamma(y + 1)} \exp\left[y \ln\left(\frac{\mu}{\mu + k}\right) + k \ln\left(\frac{k}{\mu + k}\right)\right].$$

Immediately, we recognize the canonical link, as given in Equation (7.98). Moreover, since

$$\frac{k}{\mu + k} = 1 - \frac{\mu}{\mu + k} = 1 - e^\theta,$$

the cumulant function is clearly given by Equation (7.99). Finally, the variance function given in Equation (7.100) follows from the relation $V(\mu) = 1/g'(\mu)$, which holds for the canonical link $g(\mu) = \ln[\mu/(\mu + k)]$. This completes the result. $\square$

Note 7.13. Notice, by the nature of Equation (7.98), that the canonical parameter takes values in $\mathbb{R}_-$. For this reason, the canonical link is seldom, if ever, used as the link function when the negative-binomial distribution is used as part of a GLM. Instead, the log link function $g(\mu) = \ln(\mu)$ is typically used, ensuring that the parameter η, representing the linear predictor, can take on all values in $\mathbb{R}$. $\triangleright$

Proposition 7.18. *For the* GLM *consisting of the negative-binomial distribution and the log link function* $g(\mu) = \ln(\mu)$, *the score and Fisher information are given by*

$$U_j = \sum_{i=1}^{n} \frac{(y_i - \mu_i)x_{ij}}{1 + \mu_i/k}, \qquad (7.101)$$

$$\mathcal{I}_{jk} = \sum_{i=1}^{n} \frac{\mu x_{ij} x_{ik}}{1 + \mu_i/k}, \qquad (7.102)$$

respectively.

Proof. This follows from Equations (7.70) and (7.72) and the fact that $V(\mu) = \mu + \mu^2/k$ and $g'(\mu) = 1/\mu$. □

7.7 Models for Positive Continuous Data

In this section, we discuss models for positive continuous data. Such models are useful when the target variable measures a physical quantity.

7.7.1 Gamma GLM

We have already seen how the gamma distribution can be represented as an EDM in Section 7.3. However, since the canonical parameter takes on strictly negative values, another link function is in order. Typically, one uses the log link function.

Proposition 7.19. *For the* GLM *consisting of the gamma distribution and log link function* $g(\mu) = \ln(\mu)$, *the score and information are given by*

$$U_j = \frac{1}{\phi} \sum_{i=1}^{n} \frac{w_i(y_i - \mu_i)x_{ij}}{\mu} \qquad (7.103)$$

$$\mathcal{I}_{jk} = \frac{1}{\phi} \sum_{i=1}^{n} w_i(y_i - \mu_i)x_{ij}, \qquad (7.104)$$

respectively.

Proof. This result follows from Equations (7.70) and (7.72) and the fact that $V(\mu) = \mu^2$ and $g'(\mu) = 1/\mu$. □

7.7.2 Inverse Gaussian GLM

When the target variable is even more right-skewed than the gamma distribution, a popular alternative is the inverse Gaussian distribution, which is defined as follows.

Definition 7.19. *The* inverse Gaussian distribution *is the probability distribution defined over* $(0, \infty)$ *by the* PDF

$$f(y; \mu, \phi) = \frac{1}{\sqrt{2\pi y^3 \phi}} \exp \left[\frac{-(y - \mu)^2}{2\phi y \mu^2} \right], \tag{7.105}$$

for $y > 0$ *and positive parameters* $\mu, \phi > 0$.

We next verify the result that is tabulated in Table 7.3.

Proposition 7.20. *The inverse Gaussian distribution is an* EDM *with dispersion parameter* ϕ *and canonical parameter, cumulant function, and variance function given by*

$$\theta - \frac{-1}{2\mu^2}, \tag{7.106}$$

$$\kappa(\theta) = -\sqrt{-2\theta}, \tag{7.107}$$

$$V(\mu) = \mu^3, \tag{7.108}$$

respectively, where $\theta < 0$.

Proof. The PDF given in Equation (7.105) can be rewritten as

$$\frac{\exp[-1/(2\phi y)]}{\sqrt{2\pi y^3 \phi}} \exp \left[\frac{-y}{2\phi \mu^2} + \frac{1}{\phi \mu} \right].$$

Immediately, we recognize the canonical parameter θ, as given in Equation (7.106). This relationship can be inverted to find $\mu = 1/\sqrt{-2\theta}$. The cumulant function Equation (7.107) follows. Finally, we note that the derivative of the canonical link function given by Equation (7.106) is $g'(\mu) = 1/\mu^3$, from which Equation (7.108) follows. This completes the result. $\square$

Since the canonical parameter θ only takes negative values, the canonical link function is not preferred when constructing a GLM. Instead, we can use the log-link function, as we did with the gamma distribution.

Proposition 7.21. *For the* GLM *consisting of the inverse Gaussian distribution and log link function* $g(\mu) = \ln(\mu)$, *the score and information are given by*

$$U_j = \frac{1}{\phi} \sum_{i=1}^{n} \frac{w_i (y_i - \mu_i) x_{ij}}{\mu^2} \tag{7.109}$$

$$\mathcal{I}_{jk} = \frac{1}{\phi} \sum_{i=1}^{n} \frac{w_i (y_i - \mu_i) x_{ij}}{\mu}, \tag{7.110}$$

respectively.

Proof. This result follows from Equations (7.70) and (7.72) and the fact that $V(\mu) = \mu^3$ and $g'(\mu) = 1/\mu$. $\square$

Problems

7.1. Compute the gradient of Equation (7.13), set it equal to zero, and show that the only critical point is the one given by Equation (7.14). Why must this be the global minimum?

7.2. Prove Corollary 7.1.

7.3. Prove Equation (7.32).

7.4. Derive the likelihood function and maximum-likelihood solution for the linear regression problem, given by Definition 7.1. Show that your result is equivalent to the least-squares solution.

7.5. Show that the likelihood function for a logistic regression model is given by Equation (7.40).

7.6. Show that the inverse of the central logratio transform, defined in Equation (7.49), is the softmax transform.

7.7. Prove that both the probit and logit link functions are antisymmetric about $\mu = 0.50$; i.e., that

$$g(\mu) = -g(1 - \mu).$$

7.8. Prove that the MLE estimator for the natural parameter θ of an EDM (eq4001) is related to the sample mean by

$$\frac{1}{n} \sum_{i=1}^{n} y_i = \kappa'(\hat{\theta}).$$

It's not my Fault

Correlation does not imply causation is the mantra heard 'round the world—at least in statistics courses. Two random variables can be correlated without one begin a *cause* of the other. Ice cream sales might be correlated with people going to the beach, but you can't boost beach attendance in the middle of winter by selling ice cream. In other words, it is not the ice cream vendors who *cause* beachgoers to don their trunks and lather up in SPF 42; instead, the correlation is explained by the fact that sunbathing and ice cream sales have a common cause: hot weather.

If correlation is not causation, then what is? As it turns out, the framework for causality lies outside the scope of traditional statistics, instead relying on graph theory and graphical models. Moreover, causality can be used to answer important questions related to organic data sets, i.e., data sets that are not the result of a properly controlled experiment. We begin by introducing these *observational studies* and the problem they present to classical statistics. We then go on to define graphs, introduce causal models, and show how causal implications can be inferred from graph structures and data.

Our go-to reference is Pearl, *et al.* [2016], is it serves as a concise introduction to the field. From their, we recommend both Rosenbaum [2002] for a light introduction to observational studies and Pearl [2009] for a more in-depth test on causality. For introductory texts into causality that feature social science and biomedical applications, see Imbens and Rubin [2015] and Morgan and Winship [2015].

8.1 Simpson's Paradox

We begin with a brief introduction to the definition of an observational study. We then discuss the shortcoming of the ability for traditional statistical techniques to draw inferences from such data. This will establish the need for a proper framework around the notion of causality, the topic of this chapter.

8.1.1 Observational Studies

We discussed experimentation in depth in Chapter 4. In particular, we discussed the importance of randomization in constructing certain *controlled* experiments. But what if it is not possible to control an experiment, for either ethical or logistical reasons? For instance, how would you prove that smoking causes lung cancer? It is not ethical to take a large sample of a population and randomly select half of them to smoke three packs of cigarettes a day for the next twenty years. It is readily apparent, for such cases, that a controlled experiment is not possible. We instead have entered the realm of observational studies.

Definition 8.1. *An* observational study *is a statistical analysis which draws a conclusion from a set of data about the effect of a treatment, for which the assignment of treatment to individuals is not under control of the experimenter.*

Instead of choosing our subject's fate by toss of a coin, an observational study would seek to infer the effect of smoking by looking at large amounts of data, consisting of smokers and nonsmokers alike, and seeking to draw an appropriate conclusion about the effect of smoking. In this context, the idea of *causality*—i.e., the notion that one event causes another—becomes critically important. For example, let us suppose that the data show that smokers have a dramatically increased rate of lung cancer as compared with their nonsmoking piers. Are we able to draw a conclusion from this fact alone? *Correlation does not imply causation.* What if the tendency to smoke is itself genetic? What if we have a gene that determines whether or not we ultimately become smokers later in life? And what if this gene is also responsible for lung cancer? We would then be back in the realm of our failed business enterprise to drive sunbathers to the beach by opening ice cream stands during the winter, except, this time, ice cream is smoking and the beach is lung cancer. In short, how can we draw conclusions from data when there is potentially an unknown common cause that explains the data? How do we know that there is not an undiscovered gene that both *causes* smoking and *causes* lung cancer? We will answer that question momentarily. But first, we will explore another substantial obstacle to drawing conclusions from data, and that is the certain reversal of trends at different levels of aggregation known as Simpson's paradox.

8.1.2 Simpson's Paradox

Simpson's paradox represents a conundrum for the field of traditional statistics, that must be accounted for if we are to formulate a proper theory of causality. In short, we may state it as follows.

Definition 8.2. Simpson's paradox *is the phenomenon that occurs when the aggregation of data at one level results in a different apparent conclusion than would be obtained by an aggregation at a different level.*

That is, Simpson's paradox occurs whenever different slices of the data seem to support different, often opposing, conclusions. Simpson's paradox is one danger of observational studies, and is best understood through an illustration.

Example 8.1. Consider the data shown in Table 8.1. We seek to draw a conclusion about the effect of a treatment for a given medical condition. The data are aggregated by gender, and we observe that treatment A has a higher recovery rate than treatment B for both men and women. Women recover 70% of the time with treatment A, as opposed to 65% of the time with treatment B, whereas men recover 50% of the the time with treatment A, as opposed to only 45% with treatment B. Thus, it seems that regardless of whether a patient is male or female, the optimal prescription for that patient should be treatment A.

Treatment	Women	Men
A	70/100 (70%)	200/400 (50%)
B	260/400 (65%)	45/100 (45%)

Table 8.1: Recovery rates under two treatment options, broken down by gender.

Seems straightforward enough, but let's aggregate all of our data together, as shown in Table 8.1. Here, we are surprised to observe the exact *opposite* conclusion: the overall recover rate of treatment B is 61%, whereas the overall recovery rate of treatment A is only 54%.

Treatment	Overall
A	270/500 (54%)
B	305/500 (61%)

Table 8.2: Overall recovery rates.

How can this be? Our overall recovery rate of treatment B is clearly better than our overall recovery rate of treatment A. But yet, when we divide the data by gender, we observe that treatment A in fact does better, for *both* men *and* women!   ▷

The conundrum presented by Example 8.1 is an instance of Simpson's paradox: the apparent reversal of a conclusion by aggregating at different levels.

Now, it is tempting to say: *It's obvious! We just go with the conclusion of Table 8.1. Treatment A is better.* It is, after all, clear what is happening. Women have a higher recovery rate than men, but women are more likely to have been prescribed treatment B than men. This leads to the fact that treatment B in the overall aggregate has a better outcome: because it oversamples women, with their higher recovery rate, than men.

The reason, however, that this is a paradox is that such a conclusion is not supported by traditional statistics, but by our intuition. And not only is our intuition responsible for this conclusion, it is, in particular, our intuition on causality. That is, we have intuitively constructed a *causal* story about the data.

Nonsense! you say. *Let's just go to the most granular slice that has statistical significance and go with that.* Such an approach seems logical, except—

Example 8.2. Let us consider the data of Table 8.3, aggregated based on whether a patient exhibits side effects following the treatment.

Treatment	No side effect	Side effect
A	70/100 (70%)	200/400 (50%)
B	260/400 (65%)	45/100 (45%)

Table 8.3: Recovery rates under two treatment options, broken down by side effect.

Note that the data are identical to Table 8.1, only we have a different interpretation of the aggregation groups. Can we draw the same conclusion? An important difference is that we are now aggregating by an effect of the treatment, the appearance of side effect. On one hand, any given patient will experience side effects or they will not, and in either case it looks like treatment A does better. But on the other hand, the side effects appear *after* the treatment is administered, and so they could be caused in part by the ultimate outcome. It therefore seems logical to go with the overall winner, treatment B.   ▷

So it seems like our intuition in Example 8.2 is the exact opposite of what it was in Example 8.1, despite the underlying data being identical. The difference is the causal story that underlies our intuition. In one case, our intuition tells us that gender is a causal factor in outcome. In the other case, our intuition leads us to at least suspect that the outcome might be a causal factor, along with the choice of treatment, in whether or not a patient experiences side effect.

Since different interpretations of the data lead to different conclusions, this truly is a paradox for classical statistics. Instead, our conclusion heavily relies on the causal story we prescribe for our data. We therefore need a new set of tools for weaving these causal stories, which we will develop shortly. But first, we conclude our introduction with a final twist on Example 8.1.

Example 8.3. Let us continue Example 8.1, with our original gender interpretation of Table 8.1. Our conclusion was that treatment A was the better option, as treatment A exhibited higher recovery rates for both men and women. However, Simpson's paradox can sometimes reoccur upon further subdivision of the data. To illustrate this, let us suppose that our data are further subdivided by the severity of illness, as shown in Table 8.4.

Treatment	Women		Men	
	Mild	Severe	Mild	Severe
A	64/80 (80%)	6/20 (30%)	128/160 (80%)	72/240 (30%)
B	204/240 (85%)	56/160 (35%)	17/20 (85%)	28/80 (35%)

Table 8.4: Recovery rates under two treatment options, broken down by gender and severity of illness.

It appears that treatment B is the better choice after all. When we consider both gender and severity of illness, treatment B wins across the board. This effect was reversed at the gender level (Table 8.1), since men are more likely to present with a severe case of the ailment as compared to women. Moreover, if we look at the treatment–severity matrix, we notice that it is in fact the same for both men and women! Thus, gender actually has no effect on outcome. Patients with a mild case who take treatment A are 80% likely to recover, and so forth. The four recovery rates—80%, 30%, 85%, and 35%—do not depend on gender at all. Instead, the likelihood of a serious illness does depend on gender: 64% of men have a serious version of the illness, as opposed to only 36% of women. This imbalance is what leads to the faulty conclusion of Table 8.1. But when properly considering severity, it is clear, according to our causal intuition, that treatment B is the better choice.   ▷

Example 8.4. As a final example of Simpson's paradox, consider data from an observational study broken down by age and gender, as shown in Table 8.5.

Again, we see that for each age-gender combination, the average treatment effect is about 5%, with treatment B favored over treatment A. By aggregating all of the data together, we reach the opposite conclusion: the overall average recovery rates for treatments A and B are 74.6% and 65.9%, respectively; an 8.7% advantage for treatment A.

Instead of viewing the overall average, we can also look at the breakdown when based on age or gender alone, as shown in Table 8.6. We observe that when we look at the data based on age, treatment A has an advantage over treatment B. Moreover, that advantage is present, but rather thin, for

Treatment	Women	
	Young	Old
A	7,295/8,106 (90.0%)	12,952/16,090 (80.5%)
B	1,798/1,894 (94.9%)	3,322/3,910 (85.0%)

Treatment	Men	
	Young	Old
A	3,213/4,036 (79.6%)	5,021/9,928 (50.6%)
B	13,583/15,964 (85.1%)	22,043/40,072 (55.0%)

Table 8.5: Recovery rates under two treatment options, broken down by age and gender.

younger folks, at 0.4%, but a whopping 11.4% for older folks. Either way, it looks as though treatment A is better.

Treatment	Young	Old
A	10,508 / 12,142 (86.5%)	17,973 / 26,018 (69.1%)
B	15,381 / 17,858 (86.1%)	25,365 / 43,982 (57.7%)

Treatment	Women	Men
A	20,247/24,196 (83.7%)	8,234 / 13,964 (59.0%)
B	5,120/5,804 (88.2%)	35,626 / 56,036 (63.6%)

Table 8.6: Recovery rates under two treatment options, broken down by age (top) and gender (bottom) separately.

This conclusion is reversed when looking at the data by gender, instead of age. Here, it appears that treatment B has a 4.5% advantage for both women and men. This is consistent with the most granular level, in which treatment B also has a similar advantage. This example shows that depending on which dimension we "collapse" the data on, we get conflicting results. Does this mean we should always look at the data at the most granular level? We shall return to this example in Section 8.5.

▷

8.2 Graphs and Causal Models

In this section, we lay out several key definitions of graph theory.

8.2.1 Graphs Defined

Definition 8.3. *A graph G is a collection of* nodes $\mathcal{N}$ *and edges* $\mathcal{E}$. *A* node *is a point, and an edge $e \in \mathcal{E}$ connects two points; i.e., an edge e is uniquely specified by two distinct points $a, b \in \mathcal{N}$, such that $e = (a, b)$.*

An edge is said to be directed *if the order matters, in which case we use the notation $|a, b\rangle$, so that $|a, b\rangle$ is distinct from $|b, a\rangle$. A graph with directed edges is called a* directed graph. *Two nodes are* adjacent *if they are connected by an edge.*

Given a directed edge $|a, b\rangle$, the node a is referred to as the parent *of node b, whereas node b is referred to as the* child *of node a. We write $a = \mathrm{par}(b)$ and $b \in \mathrm{child}(a)$.*

A directed edge $e = (a, b)$ is represented visually with an arrow traveling from a to b, whereas an undirected edge is represented by a line segment. In studying graphs, one often wishes to traverse a graph from one point to another. The following definitions are therefore in order.

Definition 8.4. *Given a graph $G = \{\mathcal{N}, \mathcal{E}\}$, a* path *between two nodes $a, b \in \mathcal{N}$ is a sequence of nodes, beginning with a and ending with b, such that each node is connected to the next by an edge; the* length *of a path is the number of edges comprising the path.*

A path is simple *if no node is repeated.*

A path in a directed graph is called a directed path *if, in addition, its connecting edges are directed edges of the graph; i.e., if $|x_{i-1}, x_i\rangle \in \mathcal{E}$ for $i = 1, \ldots, l$.*

A cycle *is a directed path that starts and returns to the same node: $x_0 \to x_1 \to x_{l-1} \to x_0$. A directed graph with no cycles is a* directed acyclic graph *or* DAG.

Given two points a and b in a directed graph, we say that a is an ancestor *of b, denoted $a \in \mathrm{anc}(b)$, and that b is a* descendant *of a, denoted $b \in \mathrm{desc}(a)$, if and only if there is a directed path starting at a and ending at b.*

Examples of a directed cyclic and a directed acyclic graph are shown in Figure 8.1.

In other words, a path of length l between nodes a and b is an ordered sequence $x_0, x_1, \ldots, x_l$, such that $x_0 = a$, $x_l = b$, and each pair $(x_{i-1}, x_i) \in \mathcal{E}$, for $i = 1, \ldots, l$. With paths defined, we next define the important concept of connectedness.

Definition 8.5. *A graph is* connected *if there is a path between any two nodes; i.e., the graph does not have any isolated islands. A graph is* singly connected *if there is only one simple path between any two nodes.*

Singly connected graphs, commonly called *trees* are an important concept that are used extensively in statistical modeling and machine learning. They also constitute an important type of data structure in computer science. However, we typically require some additional structure.

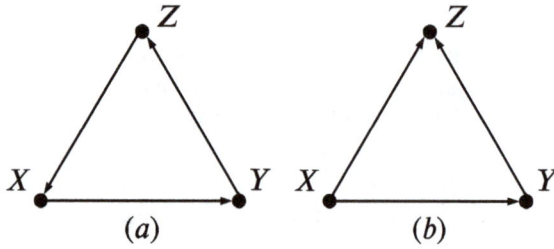

Fig. 8.1: Example of a directed cyclic graph (a) and a directed acyclic graph (b).

Definition 8.6. *A* (rooted) tree *is a directed, singly connected graph that contains a single node without parents, known as the* root, *which determines the direction of the edges, such that all edges are directed away from the root.*

A branch *of a tree is a node that has children; a* leaf *of a tree is a node without children.*

The height *of a node is the length of the longest directed path from that node to a leaf. The* depth *of a node is the length of the path connecting the root node to that node. The* height *of the tree is the height of its root.*

A binary tree *is a tree in which each node has at most two children.*

An example of a tree is shown in Figure 8.2. Here, the root node is shaded red, and the leaves are shaded green. The height of the tree is three.

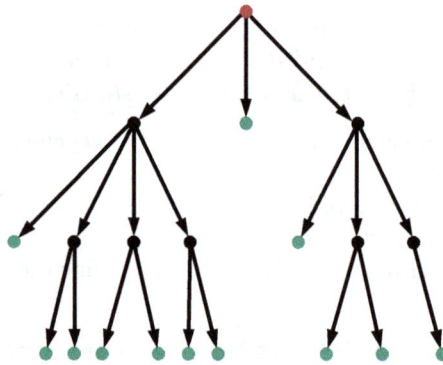

Fig. 8.2: Example of tree of height 3. Root node is red and the twelve leaves are green.

8.2.2 Causality, Structure, and Graphs

We next show how graphs can be applied to weave the *causal story* for a data set.

Causal Models

We begin by defining a mathematical object used to model causal structure in data.

Definition 8.7. *A* causal structure *over a set of variables* $\mathcal{V}$ *is a directed graph* $G = \{\mathcal{V}, \mathcal{E}\}$, *in which each node corresponds to one of the variables and each directed edge represents a direct functional relationship among the corresponding variables.*

A causal feedback structure *is a causal structure that contains cycles. A* feedback loop *is any cycle in such a structure. A* causal acyclic structure *is a causal structure that does not contain cycles.*

Feedback loops occur frequently in engineering control systems. For example, consider a thermostat: the room temperature determines whether or not the thermostat will turn on or off a heater, but the heater, in turn, affects the room temperature. This is an example of a closed feedback loop.

In general, we will assume a causal structure to be acyclic unless otherwise specified.

Causal relationships can now be defined within the context of a given causal structure.

Definition 8.8. *Given a causal structure* G *of a set of variables, variable* X *is a* direct cause *of variable* Y *if the graph* G *contains a directed edge from* X *to* Y.

Two variables X *and* Y *are said to be* causally connected *if there is a directed path connecting them. If* X *and* Y *are causally connected, we say that the ancestor is a* potential cause *of its descendant. We denote the condition that* X *is an ancestor (hence a potential cause) of* Y *with the notation* $X \prec Y$, *which defines a natural partial ordering on the causal structure*[1].

We futhermore say that a potential cause X *of a variable* Y *is, in fact, a* cause *of* Y *if a variation in* X *results in a variation in* Y, *such that variations are only allowed to propagate directionally through the graph; i.e., the variation in* X *does not change the values for any of* X's *parents. If a potential cause is not a cause, we say that the situation is* intransitive.

---

[1] The ordering is only partial as two variables are comparable relative to the ordering if and only if they are causally connected. In general, a total ordering can be defined on any directed acyclic graph; such total ordering is, however, not uniquely defined. In particular, Kahn's algorithm can be used to define a total ordering on any DAG; see Kahn [1962].

Thus, a causal structure is a mathematical object used to encode our causal picture of the world. Given a causal structure, we can now introduce the notion of a causal model.

Definition 8.9. *A structural causal model (SCM) is a triple $(\mathcal{U}, \mathcal{V}, \mathcal{F})$ that consists of*

1. *a set of n variables $\mathcal{U}$, referred to as* exogenous variables, *that represent factors outside of the model;*
2. *a set of n variables $\mathcal{V}$, referred to as* endogenous variables, *that represent variables internal to the model;*
3. *and a set of functions $\mathcal{F}$, in one-to-one correspondence with the variables in $\mathcal{V}$, that yield the* structural equations

$$V_i = f_{V_i}(P(V_i), U_i), \tag{8.1}$$

for $i = 1, \ldots, n$; where $P(V_i) \subset \mathcal{V} \setminus \{V_i\}$ are the endogenous parents of the variable V_i;

such that any specification of the variables $\mathcal{U}$ uniquely determines the values of the variables $\mathcal{V}$, so that, as a whole, $V = V(U)$.

Any causal model $M = (\mathcal{U}, \mathcal{V}, \mathcal{F})$ determines a unique causal structure $G(M) = (\mathcal{U} \cup \mathcal{V}, \mathcal{E})$, for which the edges $\mathcal{E}$ are comprised of the functional dependencies of $\mathcal{F}$; i.e.,

$$\mathcal{E} = \bigcup_{i=1}^{n} \bigcup_{P \in \mathrm{par}(V_i)} |P, V_i\rangle,$$

where $\mathrm{par}(V_i) = P(V_i) \cup \{U_i\}$ represents the complete set of dependent variables of the function f_{V_i}.

The condition in Definition 8.9 that the variables $\mathcal{U}$ uniquely determine the values of the variables $\mathcal{V}$ is equivalent to the requirement on the graph that each variable (node) in $\mathcal{V}$ have an ancestor in $\mathcal{U}$.

In general, however, the values of the exogenous variables are not known, but rather are characterized by a probability distribution. This is expressed as follows.

Definition 8.10. *A probabilistic causal model (or causal model) is a structural causal model $M = (\mathcal{U}, \mathcal{V}, \mathcal{F})$ and a joint probability function $f : \mathcal{U} \to \mathbb{R}_*$ defined over the sample space $\mathcal{U}$.*

Recalling that the endogenous variables in a causal model are uniquely determined by the exogenous variables, the probability distribution $f(u)$ over $\mathcal{U}$ determines a probability distribution over $\mathcal{V}$ by the relation

$$\mathbb{P}(V = v) = \int_{\{u : V(u) = v\}} f(u) \, du,$$

where the domain of integration consists of all possible combinations of u that yield the value $V = v$, for any $V \subset \mathcal{V}$.

Example 8.5. Let us again consider the Simpson's paradox example from Example 8.3. The causal model for the data is shown in Figure 8.3. The nodes are labeled G, S, T, and R for gender, severity, treatment, and recovery, respectively.

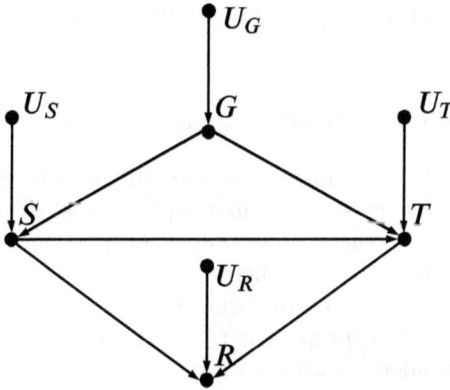

Fig. 8.3: The causal structure from Example 8.3. G represents gender; S represents severity; T represents treatment; R represents recovery.

Recall that gender influenced both treatment and severity, and that severity was another factor influencing treatment (52% of severe patients were assigned treatment A, whereas only 48% of mild patients were assigned the same treatment, reflecting a bias to assign severe cases to treatment A). Moreover, recovery rate is determined by severity and treatment, but not gender. Moreover, each of these factors could potentially be influenced by other (unknown) exogenous variables, represented by the Us. We leave the causal graphs for Examples 8.1 and 8.2 as an exercise for the reader.   ▷

Spurious Association and Confounding Variables

We have previously examined the danger of making inferences from observational data during our discussion on Simpson's theory. Now that we have established a language for causality, we will add a few definitions to help describe what we observed.

Definition 8.11. *A* spurious association *is said to exist between two variables X and Y in a causal model if those variables are dependent, yet not causally connected.*

Any common ancestor (whether or not observed) of two spuriously associated variables is said to be a confounding factor, *or* confounder.

A classic example of a spurious association is ice cream sales and beach attendance. We discussed in the chapter introduction that both are caused by a confounding factor: the weather. In order to understand the causal effect that ice cream has on beach attendance, we therefore need to *control* for the sun. But, alas, we're getting ahead of ourselves. We will return to this issue shortly, but first we will examine several common structures in graphs and detail a few powerful results regarding their (conditional) independence.

8.2.3 Causal Building Blocks and Independence

Now that we have defined causal structures and models, we need a calculus for analyzing dependency relations in complex causal structures. There are three main classes of structures we will identify in our graphical models: chains, forks, colliders. Each of these has its own rules governing dependencies, which we will discuss in turn. We will conclude the section with a powerful result called d-separation. The goal is to be able to quickly identify likely dependencies within a given causal structure. Astonishingly, these building blocks will be valid regardless of the specific set of structural equations attributed to our model.

Chains

We begin with a definition of the most basic structure one can spot in a graph: the chain.

Definition 8.12. *A chain of length $n \geq 2$ is a connected, directed acyclic graph comprised of $n + 1$ nodes and n directed edges.*

Because the DAG is connected, and because there is one fewer edge than nodes, all chains are in the form of a *unidirectional path*. The simplest example consists of three nodes (X, Y, Z) that are directional, in the sense that $X \rightarrow Y \rightarrow Z$. Such a chain is shown in Figure 8.4.

Fig. 8.4: A chain of length 2.

Of course, in the context of causal models, chains do not occur in a vacuum, but rather as part of a larger causal model. We can illustrate the dependency on the exogenous variables by adding them to our graph, as shown in Figure 8.5. As should be intuitively clear, the variables X and Y are likely dependent. Similarly, the variables Y and Z are likely dependent, as well as the variables X and Z. We cannot conclude any of the variables are dependent, as there are certain intransitive cases. For example, suppose that the variable X can only take on negative values, and the function $f_Y(X) = \text{sign}(X)$ is the sign function. In this case, the variable Y will take the value $Y = -1$, regardless of the value of the variable X. We remind the reader of such degenerate cases by saying that X and Y are *likely* dependent.

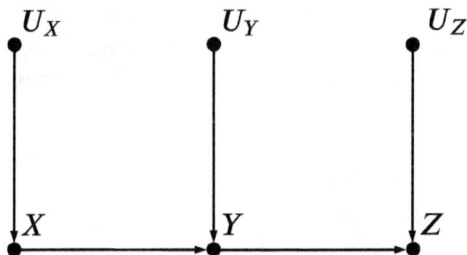

Fig. 8.5: A chain in a graphical causal model.

The other observation that should follow intuitively from Figure 8.5 is that the variables X and Z are *conditionally independent*, given the variable Y. In other words, variations of X and Z consistent with a single fixed Y will be independent of each other. This is illustrated with the following example.

Example 8.6. Consider the causal model shown in Figure 8.5, and suppose the functional relationships are given by

$$f_X(U_X) = U_X,$$
$$F_Y(X, U_Y) = X + U_Y$$
$$F_Z(Y, U_Z) = Y + U_Z.$$

If we fix the value $Y = c$, then changes in X will be commensurate with the opposite change in U_Y, but will not affect the variable Y. Since the variable Y is held fixed, and hence not affected by changes in X, the variable Z will also not be affected by changes in X, as Z depends on X only through its effect on Y. Thus, X and Z will vary independently of one another, when the value for Y is fixed.   ▷

We can state this principle more formally as follows.

Proposition 8.1 (Conditional Independence in Chains). *Let X and Y represent two variables in a causal model that are connected by a single path, and let Z represent any set of variables that intercepts that path. If the unique path connecting X and Y is unidirectional (i.e., a chain), then X and Y are conditionally independent given Z.*

Forks

We now move on to another commonly found structure.

Definition 8.13. *A fork in a directed graph is a triple of nodes (X, Y, Z) such that one of the nodes is a common parent to the other two. In a causal model, we call that common parent a* common cause *to the other two variables. If X is the common cause of Y and Z, we may represent this fork as $Y \leftarrow X \rightarrow Z$.*

The nodes (X, Y, Z) in Figure 8.6 constitute a fork in a graphical causal model. As before, the variables X, Y, and Z are all likely dependent, meaning the value of any one of the variables likely depends on the value of any other.

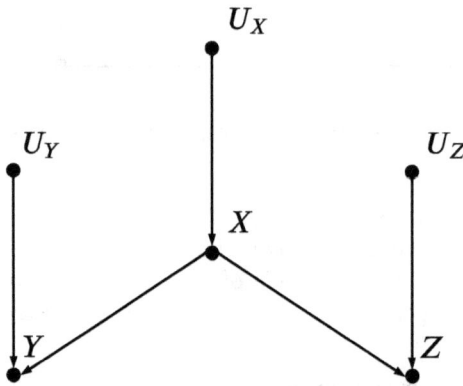

Fig. 8.6: A fork in a graphical causal model.

In addition, similar to the case of forks, the variables Y and Z are *conditionally independent*, given X. When X is held fixed, variations in Y and Z must be due to U_Y and U_Z, respectively, and, therefore, a variation in one will not influence a variation in the other. This is stated more formally as follows.

Proposition 8.2 (Conditional Independence in Forks). *If variable X is a common cause to variables Y and Z in a causal model, and if there is only one path connecting the variables Y and Z, then Y and Z are conditionally independent given X.*

Colliders

Our third structure is slightly more sophisticated, though it is simply the causal reversal of a fork.

Definition 8.14. *A collider in a directed graph is a triple of nodes* (X, Y, Z) *such that one node, referred to as the* collision node, *is a simultaneous child of the other two nodes. If Z is a collision node, we may represent this collider as* $X \to Z \leftarrow Y$.

An example of a collider in a graphical model is shown in Figure 8.7. Note that by reversing the arrows $|X, Z\rangle$ and $|Y, Z\rangle$, we would instead have a fork. The analysis, however, is slightly more subtle.

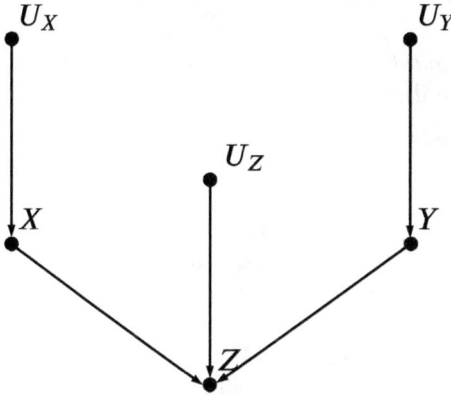

Fig. 8.7: A collider in a graphical causal model.

For the collider shown in Figure 8.7, the variables X and Z are likely dependent. Similarly, the variables Y and Z are likely dependent. (This follows because X and Y are both direct causes of the variable Z.) The variables X and Y themselves, however, are independent. The only factor that influences the variable X is the variable U_X; therefore, a variation in Y cannot manifest an influence in the variable X.

Now, in a fork, the two children are conditionally independent, given a fixed value for the common parent. Since a collider is a causally reversed fork, we might expect, therefore, that the two common parents in a collider are conditionally *dependent*, given a fixed value for their child. This turns out to be precisely the case: when we fix the value of the variable Z, we create a dependency among the variables X and Y.

Despite following formally from reversing the logic of a fork, this result might seem at first counterintuitive. To understand it on a more intuitive level, consider the following example.

Example 8.7. Consider the collider shown in Figure 8.7. Suppose the variables have the following real-world meanings:

1. X represents whether or not it is raining today;

2. Y represents whether or not my sprinkler is on;

3. Z represents whether or not my sidewalk is wet.

Clearly, X and Y are independent: the weather is not moved by my lawn-care habits. However, if we know that the sidewalk is wet, this creates a dependency in the random variables X and Y! If my sidewalk is wet, but it is not raining, it is much more likely I am running my sprinkler, and vice versa! ▷

Proposition 8.3 (Independence and Conditional Dependence in Colliders). *If a variable Z in a causal model is the collision node between two variables X and Y, and there is only one path between X and Y, then X and Y are unconditionally independent, but dependent conditionally on Z or any descendant of Z.*

d-separation

So far, we have considered three basic causal structures: chains, forks, and colliders. Rarely, however, are causal models so simple. More commonly, one is likely to find multiple paths between nodes, each containing a combination of chains, forks, and colliders. Our goal then should be to determine a process or rule that can be applied to any graphical causal model, regardless its complexity, that reveals likely dependencies among the variables for any data set that is generated by the graph. The tool we will use to discover such dependencies is called d-separation, which we define as follows.

Definition 8.15. *A path p in a directed graph is* blocked *by a set of nodes $\mathcal{Z}$ if and only if*

1. *the path p contains a chain $A \rightarrow B \rightarrow C$ or a fork $A \leftarrow B \rightarrow C$, such that the middle node $B \in \mathcal{Z}$; or*
2. *the path p contains a collider $A \rightarrow B \leftarrow C$ such that neither the collision node B, nor any of its descendants, are in $\mathcal{Z}$.*

If the set $\mathcal{Z}$ blocks every path between X and Y, we say that X and Y are directionally separated, or d-separated, *conditional on $\mathcal{Z}$.*

Two points X and Y in a directed graph that are not d-separated are said to be d-connected.

The idea behind this definition is that if two points are d-separated, they must be independent. On the other hand, if they are d-connected, they are likely, but not necessarily, dependent.

Theorem 8.1. *If two nodes in a causal graph are d-separated by a set of nodes $\mathcal{Z}$, then those two nodes are independent conditionally on the set $\mathcal{Z}$.*

Proof. Consider a single path between two nodes X and Y. Since X and Y are d-separated conditional on $\mathcal{Z}$, we know that X and Y must be conditionally independent, given $\mathcal{Z}$, as the conclusion of at least one of the results from Proposition 8.1–8.3 must hold. Therefore, X and Y would be conditionally independent, if the graph consisted of only this path. But since $\mathcal{Z}$ blocks *every* path between X and Y, therefore X and Y must be conditionally independent, given $\mathcal{Z}$, considering the full set of paths between them. Since causal affect travels only through paths, the result holds.   □

Example 8.8. Consider the nodes X and Y in the causal graph shown in Figure 8.8. There are two paths (green and blue) connecting these two nodes. (Note that these paths need not be unidirectional.) In order to determine whether X and Y are d-separated, let us consider each path in turn.

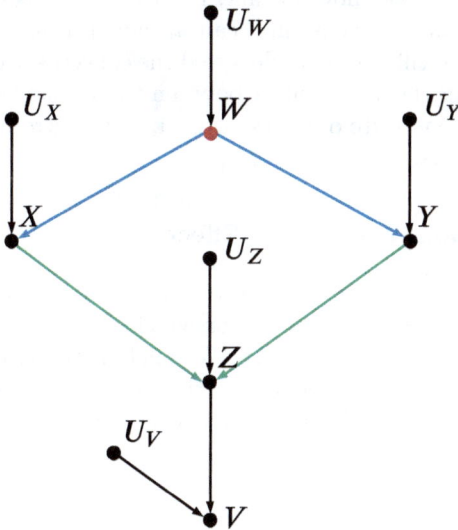

Fig. 8.8: d-separation in a causal graph; Example 8.8.

First, let us consider the *blue path*, consisting of the fork $X \leftarrow W \rightarrow Y$. Since this path contains (in fact, is) a fork, we know that it is blocked if and only if the branching point W is in our set $\mathcal{Z}$. Therefore, from this path alone, we know that the condition $W \in \mathcal{Z}$ is a necessary condition for conditional independence.

Next, let us consider the *green path*, consisting of the collider $X \rightarrow Z \leftarrow Y$. Moreover, we note that Z has a descendant, V. Our second necessary condition for conditional independence is, therefore, that neither Z nor V be included in our blocking set $\mathcal{Z}$.

Since these are our only two paths, we can conclude that X and Y are d-separated, conditional on a set $\mathcal{Z}$, if and only if $W \in \mathcal{Z}$ and $Z, V \notin \mathcal{Z}$. Now, since we have exhausted the entire graph, except for exogenous variables, we can state our conclusion, in its simplest form, that X and Y are conditionally independent given the set $\mathcal{Z} = \{W\}$, which is highlighted in red.   ▷

8.3 A Calculus of Interventions

So far, we have discussed observational studies and the danger of drawing conclusions from their data without a proper understanding of the underlying causal structure. We then went on to formalize how one can encode such a structure in mathematics, using graphical representations of the causal network. We then showed how one may make conclusions about conditional independence of various events in a causal model, using the concept of d-separation. But we still have not addressed the concerns of our main inquiry: how might we infer the causal effect between two events from observational data? This shall constitute our present goal, as we introduce a calculus for inferring such effects.

8.3.1 Interventions and Causal Effect

We begin by introducing the concept of interventions and the do-calculus. We then consider a simple example for which we are able to quantify the causal effect from observational data. We conclude the section with a generalization of the result. In subsequent sections of the chapter, we will further generalize the result to make it as broadly applicable as possible.

Graph Surgeries

Before formally defining interventions, we first require a few useful definitions regarding certain manipulations of graphs.

Definition 8.16 (Backdoor/Frontdoor). *Given a directed graph $G = \{\mathcal{V}, \mathcal{E}\}$, a* backdoor *of a node $X \in \mathcal{V}$ is any directed edge leading into X. Similarly, a* frontdoor *of a node X is any directed edge emanating from X.*
Moreover, we define the sets

$$\mathcal{E}^X = \{|P, X\rangle : P \in \mathrm{par}(X)\},$$
$$\mathcal{E}_X = \{|X, C\rangle : C \in \mathrm{child}(X)\}$$

as the sets of backdoors and front doors to X, respectively.

Definition 8.17 (Graph Surgeries). *Given a directed graph* $G = \{\mathcal{V}, \mathcal{E}\}$ *and a node* $X \in \mathcal{V}$*, we define the* manipulated graphs

$$G_{\overline{X}} = \{\mathcal{V}, \mathcal{E} \setminus \mathcal{E}^X\},$$
$$G_{\underline{X}} = \{\mathcal{V}, \mathcal{E} \setminus \mathcal{E}_X\}$$

as the graphs obtained by (surgically) removing all backdoors or front doors of the node X*, respectively.*

Graph surgeries are additive; for example, given the nodes X *and* Y*, we can define the manipulated graph*

$$G_{\overline{X}\underline{Y}} = \{\mathcal{V}, \mathcal{E} \setminus \mathcal{E}^X \setminus \mathcal{E}_Y\}$$

in the obvious way: by removing all backdoors of X *and front doors of* Y*.*

Each of the graph surgeries from this definition are illustrated in Figure 8.9.

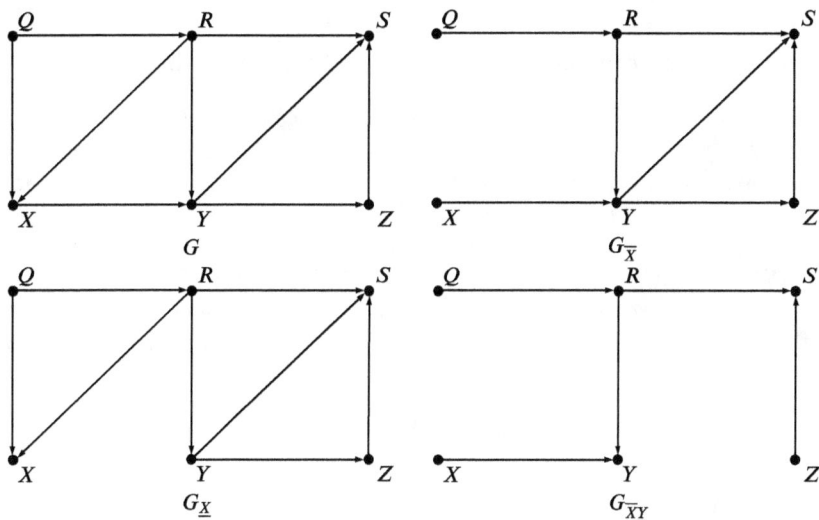

Fig. 8.9: Illustration of the graph surgeries defined in Definition 8.17.

Interventions

In observational studies, scientists are not able to properly randomize any aspect of the data. We therefore seek an alternative approach to properly *control* for potential spurious effects in order to draw the correct conclusion. In particular, we seek to draw conclusions about what effect a choice on the value of a variable would have made on a particular outcome, if we were able to make that choice. Such a choice is called en *intervention.*

In the context Example 8.1, we wish to know what the effect our choice of treatment has on recovery. The problem is that the data are purely observational: we had no *choice* on the treatment. Thus, an intervention is a form of counterfactual: what would the effect have been, had we been able to intervene on a variable X, removing all influences on it and forcing it to take a specific value? Alternatively: what would the effect have been, had we been able to go in and surgically remove all influences on a variable (i.e., the treatment) and instead *intervene* by prescribing a treatment?

There is an important distinction between intervention and conditioning:

1. When we intervene on a variable X, we change the system by fixing the value $X = x$ of a variable, thereby causing a change in the values of the other variables as a result;
2. When we condition on a variable X, we do not change the system, but instead narrow our focus on the subset of data for which $X = x$ organically.

To make this concept more formal, we have the following.

Definition 8.18. *Let X be an endogenous variable in a causal model $M = (\mathcal{U}, \mathcal{V}, \mathcal{F}, \mathbb{P})$. Then an* intervention *on X with value x, denoted* do$(X = x)$*, represents a counterfactual reality described by a specific manipulated model $M_{X=x}$ obtained by replacing G with $G_{\overline{X}}$ (see Definition 8.17) and the structural equation for X, i.e., $X = f_X(\mathrm{par}(X), U_X)$, with the constant equation $X = x$.*

Given an intervention do$(X = x)$*, we define its* causal effect *(or, causal influence) on the variable Y (alternatively, the* postintervention distribution *of outcome Y) as*

$$\mathbb{P}(Y = y | \mathrm{do}(X = x)) = \mathbb{P}_{M_{X=x}}(Y = y),$$

where $\mathbb{P}_{M_{X=x}}$ represents the probability under the manipulated model $M_{X=x}$.

Thus, an intervention on X does two things: removes all parents (i.e., direct causes) of X from the causal structure, and replaces the structural equation that determines X with a constant, as prescribed by the intervener.

Note 8.1. Interventions are additive: the intervention do$(X = x, Y = y)$ is obtained by the manipulated model $M_{X=x, Y=y}$, which is defined by replacing the causal structure

$$G \to G_{\overline{XY}}$$

and the *two* structural equations

$$\left\{ \begin{array}{l} X = f_X(\text{par}(X), U_X) \\ Y = f_Y(\text{par}(Y), U_Y) \end{array} \right\} \quad \rightarrow \quad \left\{ \begin{array}{l} X = x \\ Y = y \end{array} \right\}.$$

In other words, we remove all backdoors for X *and* Y, and replace the corresponding structural equations with their respective constants. ▷

Note 8.2. Conditional probabilities under interventions are defined in the obvious way; i.e.,

$$\mathbb{P}(Y = y | \text{do}(X = x), Z = z) = \mathbb{P}_{M_{X=x}}(Y = y | Z = z);$$

i.e., by the conditional probability in the manipulated model. ▷

Note 8.3. An alternate definition of an intervention is as follows: an intervention on X is a node I_X that can take on a null value $\emptyset$ or any value that the variable X can take on. The structural equation for X is then modified to account for the possibility of an intervention:

$$X = \hat{f}_X(\text{par}(X), U_X, I_X) = \left\{ \begin{array}{ll} f_X(\text{par}(X), U_X) & \text{if } I_X = \emptyset \\ I_X & \text{otherwise} \end{array} \right..$$

In this case, the node I_X acts as a switch that, when turned on, ignores all natural effects on the variable X, replacing them instead with a constant prescribed by the intervener. ▷

Example 8.9. A classic example of an intervention involves three random variables: R represents whether or not it rained today; S represents whether or not the sprinklers rand; and W represents whether or not the driveway pavement is wet. However, suppose that we have a smart sprinkler that is supposed to detect whether or not it has rained. It is not perfect, but the probability that the sprinkler runs does depend on whether or not it has rained. Naturally, whether or not the pavement is wet depends on both R and S. A simple causal structure for this model is shown in Figure 8.10.

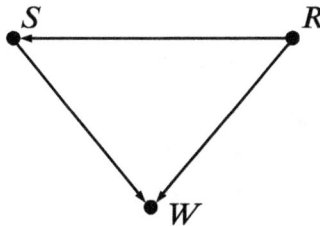

Fig. 8.10: DAG for Example 8.9: R represents rain, S represents sprinkler, and W represents wet pavement.

Next, let us assign some specific probabilities. Let us suppose that there is a 30% chance of rain. If it does rain, the sprinkler will be activated with only a 10% probability. If it does not rain, the sprinkler will be activated with a 30% probability. (The sprinkler is programmed to run about twice a week.)

Finally, the probability that the pavement is wet depends on both the sprinkler and weather conditions. On days in which it does not rain, the pavement will be wet with 0% probability if the sprinkler is not on and a 10% probability if the sprinkler is on. On days in which it does rain, the pavement will be wet with 85% probability if the sprinkler is not on and a 90% probability if the sprinkler is on. These probabilities are illustrated graphically in the decision diagram of Figure 8.11.

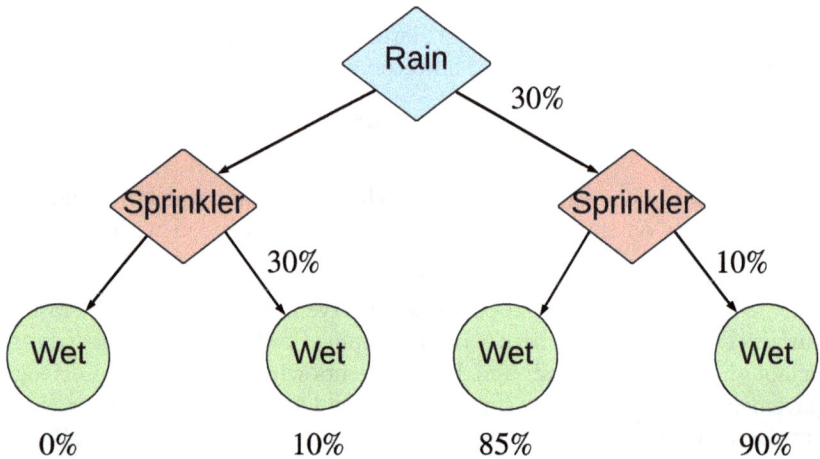

Fig. 8.11: Conditional probabilities in a decision diagram for Example 8.9. Right arrows coming out of decision nodes (diamonds) represent true; left arrows represent false.

In this model, we wish to illustrate the difference between the conditional probability $\mathbb{P}(W = 1|S = 1)$ and the causal effect $\mathbb{P}(W = 1|\text{do}(S = 1))$. Since we haven't yet discussed the formulas to determine the causal effect, we will construct a data set that conforms to our prescribed probability structure. Such a data set, comprised of 1,000 total records, is given in Table 8.7.

First, let us consider the conditional probability $\mathbb{P}(W = 1|S = 1)$; i.e., what is the probability that the pavement is wet given that the sprinkler has run? To compute this, we must condition on the outcomes in which the sprinkler was running, which occurred 240 times. Out of those 240 times in which the sprinkler was running, the pavement was wet a total of 48 times. Hence, the conditional probability is given by

	No Rain 700 (70%)		Rain 300 (30%)	
	No Sprinkler 490 (70%)	Sprinkler 210 (30%)	No Sprinkler 270 (90%)	Sprinkler 30 (10%)
Wet Pavement	0/490 (0%)	21/210 (10%)	229.5/270 (85%)	27/30 (90%)

Table 8.7: Data following our probability model for Example 8.9.

$$\mathbb{P}(W = 1|S = 1) = \frac{21 + 27}{210 + 30} = 20\%.$$

In particular, notice that we by conditioning we have narrowed our focus on a subset of the overall data.

Next, let us compute the probability of wet pavement *given an intervention* do($S = 1$). In other words, what is the probability that the pavement is wet, had we stepped in and ensured that the sprinkler was running?

Under the intervention do($S = 1$), it will still rain 30% of the time, as the weather does not care whether or not we are running our sprinkler. What changes, however, is whether or not our sprinkler is running. Under an intervention, our sprinkler *is* running, 100% of the time. Moreover, our probability of wet pavement *given* both the weather and sprinkler condition is unchanged. (We are changing whether or not the sprinkler is running, not whether or not the pavement will get wet *given* that the sprinkler is running.) Thus, in our altered reality, we would have obtained the data set shown in Table 8.8, if we had intervened to make sure that the sprinkler was running every single day.

	No Rain 700 (70%)		Rain 300 (30%)	
	No Sprinkler 0 (0%)	Sprinkler 700 (100%)	No Sprinkler 0 (0%)	Sprinkler 300 (100%)
Wet Pavement	0/0 (0%)	70/700 (10%)	0/0 (85%)	270/300 (90%)

Table 8.8: Data for Example 8.9 *under the intervention* do($S = 1$); i.e., what our data would have looked like in the altered world in which we had ensured our sprinkler had run every day.

In our altered world of Table 8.8, it is clear that we now have wet pavement on 340 days out of 1,000, or 34% of the time. Since this altered world is the world we obtained by conditioning on our intervention, the causal effect of running our sprinkler on the pavement is therefore

$$\mathbb{P}(W = 1|\text{do}(S = 1)) = 34\%.$$

This simple example illustrates the divergence between the conditional probability $\mathbb{P}(W = 1|S = 1) = 20\%$ and the intervention effect $\mathbb{P}(W = 1|\mathrm{do}(S = 1))$.

▷

8.3.2 The Adjustment Formula

Now that we have defined what an intervention is, we will develop a preliminary formula that can be used for computation. In order to account for spurious associations, we must devise a method to control for, or block, confounding factors. The most simplest approach is to control for a variable's direct causes: this basic result is known as the adjustment formula. We will first develop it for a simple case, and then more generally. Subsequent sections in the chapter will continue to build on this basic adjustment formula, extending it to more general contexts.

Adjustment for Direct Causes

The adjustment formula will be our first, simple tool that will allow us to control for, or adjust for, confounding factors. With it, we will be able to describe the influence of an intervention based solely on the observed probabilities in our actual universe (as opposed to the probabilities within the manipulated model). This, however, is but a first step; we will extend this to more general settings in upcoming sections.

The adjustment formula, however, will rely on the following definition.

Definition 8.19 (Direct Sum). *Let* $\mathcal{X} = \{X_1, \ldots, X_n\}$ *be a finite set of* n *random variables over a common probability space* $(\Omega, \mathcal{E}, \mathbb{P})$. *Then the direct sum over* $\mathcal{X}$, *denoted* $\oplus \mathcal{X}$, *is the random vector*

$$\oplus \mathcal{X} = \bigoplus_{i=1}^{n} X_i = \langle X_1, \ldots, X_n \rangle,$$

with the natural range

$$\mathrm{range}(\oplus \mathcal{X}) = \{\langle x_1, \ldots, x_n \rangle : x_i \in \mathrm{range}(X_i) \text{ for } i = 1, \ldots, n\};$$

i.e., the set of all possible combinations of the individual random variable.

For example, if random variables X and Y are binary, the direct sum $X \oplus Y$ is a two-dimensional random binary vector that takes on four possible values in $\mathbb{B}^2$. More generally, if $\mathcal{X} = \{X_1, \ldots, X_n\}$ is a set of n random binary variables, the direct sum $\oplus \mathcal{X}$ will be an n-dimensional random binary vector, that takes on 2^n possible values in $\mathbb{B}^n$.

We next devise a method method for determining causal effect of a variable X on a variable Y by explicitly controlling for each of the parents of variable X. We begin with two lemmas.

Lemma 8.1. *Let X and Z be two variables in an acyclic causal model M, such that Z is a direct cause of X; i.e., $Z \in \mathrm{par}(X)$. Let $m = M_{X=x}$ represent the manipulated model representing the intervention $\mathrm{do}(X = x)$. Then*

$$\mathbb{P}_m(Z = z | X = x) = \mathbb{P}_m(Z = z) = \mathbb{P}(Z = z); \tag{8.2}$$

i.e., X and Z are unconditionally independent in the modified model and, moreover, the probability distribution of Z is invariant under the intervention $\mathrm{do}(X = x)$.

Proof. First, let us consider the unconditional independence of X and Z in the modified model. All paths between X and Z have been removed, so the only possible way they are dependent is through a path connecting X and Z indirectly. However, since the model is acyclic, we know there is no directed path between X and Z, as such a path, when added to the missing link $|X, Z\rangle$, would constitute a cycle. Any path connecting X and Z in the modified model must therefore necessarily contain a collider, making X and Z unconditionally independent. See Figure 8.12 for an illustration. Thus, in the modified model, $\mathbb{P}_m(Z = z | X = x) = \mathbb{P}_m(Z = z)$.

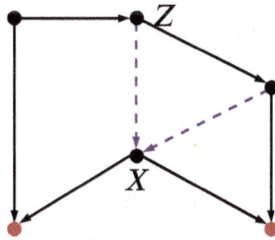

Fig. 8.12: Try as we might, with the backdoors of X (shown in purple, dashed lines) removed, we cannot construct a path between X and Z that does not contain a collider (red nodes). Also, no directed path from Z to X exists, as all edges leading into X have been removed.

Now, let us prove that the probability distribution of Z is invariant under the distribution. The value of Z is obtained from the structural equation

$$Z = f_Z(\mathrm{par}(Z), U_Z).$$

Since each parent variable Z is determined by a similar structural equation, we may more generally state that the value of Z is uniquely determined given the values of all ancestors of Z and their respective exogenous counterparts. Since the graph is acyclic, $\mathrm{anc}(Z)$ does not contain any descendants of X, as X is a child of Z. (If one of Z's ancestors were a descendant of X, this would create a cycle.)

Now, the intervention removes the causal links between X and Z, fixing the value of X. This potentially affects the probability distribution of the descendants of X, but cannot affect the value of any of the ancestors or parents of Z. Thus, the value of Z is already determined, and remains invariant under the intervention. □

Lemma 8.2. *Let X and Y be two variables in an acyclic causal model M, such that Y is not a direct cause of X; i.e., $Y \notin \mathrm{par}(X)$, and let $Z = \oplus\mathrm{par}(X)$ be the random vector consisting of all direct causes (parents) of X. Let $m = M_{X=x}$ represent the manipulated model representing the intervention $(X = x)$. Then*

$$\mathbb{P}_m(Y = y | X = x, Z = z) = \mathbb{P}(Y = y | X = x, Z = z); \qquad (8.3)$$

i.e., the conditional probability of Y given the value of X and each of its parents in invariant under the intervention $\mathrm{do}(X = x)$.

Proof. The manipulated model changes the functional relationship between X and each of its parents. Such an intervention therefore cannot affect the conditional probability distribution over Y, as long as we condition on the values of both X and its parents. Essentially, we have sealed off the effect of the intervention—the removal of the causal connections between X and its parents—from the rest of the graph, by conditioning on the full set of variables that contain the intervention. We illustrate this in Figure 8.13. □

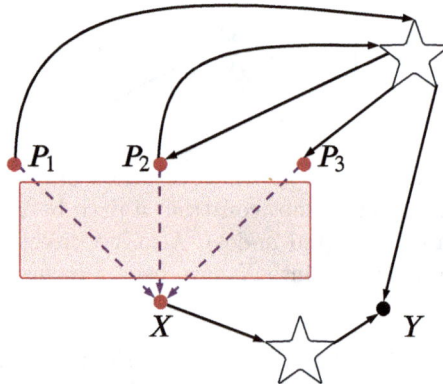

Fig. 8.13: Illustration of Lemma 8.2. By conditioning on both X and its parents (P_1, P_2, and P_3), the interactions within the control box are irrelevant, from the perspective of Y. The star shapes represent subgraphs of arbitrary complexity.

Together, these lemmas allow us to capture a powerful result, which allows us to quantify the causal effect of an intervention with reference solely to the original (non-manipulated) model that generated our observational data.

Theorem 8.2 (The Adjustment Formula). *Let X be a variable in an acyclic causal model M and let $Z = \oplus\mathrm{par}(X)$ be the random vector consisting of all direct causes (parents) of X. Let Y be a variable (or a set of variables) disjoint from $\{X\} \cup \mathrm{par}(X)$. The causal effect of X on Y is then given by the following adjustment formula:*

$$\mathbb{P}(Y = y|\mathrm{do}(X = x)) = \sum_{z \in \mathcal{R}_Z} \mathbb{P}(Y = y|X = x, Z = z)\mathbb{P}(Z = z), \quad (8.4)$$

where $\mathcal{R}_Z = \mathrm{range}(Z)$ is the set of all possible values of the random vector Z.

When we apply the adjustment formula, we say we are *adjusting for*, or *controlling for, the direct causes* of X.

We may alternatively express Equation (8.4) as

$$\mathbb{P}(Y = y|\mathrm{do}(X = x)) = \sum_{z \in \mathcal{R}_Z} \frac{\mathbb{P}(X = x, Y = y, Z = z)}{\mathbb{P}(X = x|Z = z)}, \quad (8.5)$$

the denominators of which are known as the propensity scores.

Proof. By definition,

$$\mathbb{P}(Y = y|\mathrm{do}(X = x)) = \mathbb{P}_m(Y = y|X = x).$$

By applying the law of total probability, we can now write

$$\mathbb{P}_m(Y = y|X = x) = \sum_{z \in \mathcal{R}_Z} \mathbb{P}_m(Y = y|X = x, Z = z)\mathbb{P}_m(Z = z|X = x).$$

However, in the manipulated model, the causal connection between X and Z is broken, so that

$$\mathbb{P}_m(Z = z|X = x) = \mathbb{P}_m(Z = z).$$

Our result follows by applying Lemmas 8.1 and 8.2. In summary:

$$\mathbb{P}(Y = y|\mathrm{do}(X = x)) = \mathbb{P}_m(Y = y|X = x)$$
$$= \sum_{z \in \mathcal{R}_Z} \mathbb{P}_m(Y = y|X = x, Z = z)\mathbb{P}_m(Z = z|X = x)$$
$$= \sum_{z \in \mathcal{R}_Z} \mathbb{P}_m(Y = y|X = x, Z = z)\mathbb{P}_m(Z = z)$$
$$= \sum_{z \in \mathcal{R}_Z} \mathbb{P}(Y = y|X = x, Z = z)\mathbb{P}(Z = z).$$

Equation (8.5) is obtained by multiplying and dividing the summand in Equation (8.4) by the propensity score $\mathbb{P}(X = x|Z = z)$, and then simplifying, using the definition of conditional probability. This completes the proof. $\square$

Notice the difference between the adjustment formula Equation (8.4), which conditions on an intervention, versus regular conditioning,

$$\mathbb{P}(Y = y|X = x) = \sum_{z \in \mathcal{R}_R} \mathbb{P}(Y = y|X = x, Z = z)\mathbb{P}(Z = z|X = x),$$

which clearly yields a different result.

Example 8.10. Let us consider again our rain–sprinkler–pavement Example 8.9. Let us recompute both the conditional probability $\mathbb{P}(W = 1|S = 1)$ and the causal effect $\mathbb{P}(W = 1|do(S = 1))$ again, this time using the appropriate formulas as opposed to a made-up data set.

The conditional probability is given by

$$\begin{aligned} \mathbb{P}(W = 1|S = 1) &= \mathbb{P}(W = 1|S = 1, R = 0)\mathbb{P}(R = 0|S = 1) \\ &+ \mathbb{P}(W = 1|S = 1, R = 1)\mathbb{P}(R = 1|S = 1) \\ &= (0.10)(7/8) + (0.90)(1/8) = 0.20, \end{aligned}$$

which agrees with our previous result. (We will discuss how we obtained the probabilities $\mathbb{P}(R = 0|S = 1)$ and $\mathbb{P}(R = 1|S = 1)$ momentarily.)

The causal effect, on the other hand, is obtained by

$$\begin{aligned} \mathbb{P}(W = 1|do(S = 1)) &= \mathbb{P}(W = 1|S = 1, R = 0)\mathbb{P}(R = 0) \\ &+ \mathbb{P}(W = 1|S = 1, R = 1)\mathbb{P}(R = 1) \\ &= (0.10)(0.70) + (0.90)(0.30) = 0.34, \end{aligned}$$

which again agrees with our previous result.

The conditional probability $\mathbb{P}(R = 1|S = 1)$ used in the first calculation can be obtained using an application of Bayes' law, which we will discuss in Chapter 10. For now, however, we will simply rely on the result, which states

$$\mathbb{P}(R = 1|S = 1) = \frac{\mathbb{P}(S = 1|R = 1)\mathbb{P}(R = 1)}{\mathbb{P}(S = 1)}.$$

Now,

$$\begin{aligned} \mathbb{P}(S = 1) &= \mathbb{P}(S = 1|R = 1)\mathbb{P}(R = 1) + \mathbb{P}(S = 1|R = 0)\mathbb{P}(R = 0) \\ &= (0.10)(0.30) + (0.30)(0.70) = 0.24. \end{aligned}$$

Therefore,

$$\mathbb{P}(R = 1|S = 1) = \frac{\mathbb{P}(S = 1|R = 1)\mathbb{P}(R = 1)}{\mathbb{P}(S = 1)} = \frac{(0.10)(0.30)}{(0.24)} = 0.125,$$

or 1/8. Naturally, $\mathbb{P}(R = 0|S = 1) = 1 - \mathbb{P}(R = 1|S = 1) = 7/8$.

It is interesting to note that, with Bayes' rule, we see that our formula for the regular conditional probability is equivalent to

$$\mathbb{P}(W = 1|S = 1) = \mathbb{P}(W = 1|S = 1, R = 0)\mathbb{P}(R = 0)\frac{\mathbb{P}(S = 1|R = 0)}{\mathbb{P}(S = 1)}$$
$$+\mathbb{P}(W = 1|S = 1, R = 1)\mathbb{P}(R = 1)\frac{\mathbb{P}(S = 1|R = 1)}{\mathbb{P}(S = 1)},$$

which differs from the formula for the causal effect by the factors

$$\frac{\mathbb{P}(S = 1|R = 0)}{\mathbb{P}(S = 1)} \quad \text{and} \quad \frac{\mathbb{P}(S = 1|R = 1)}{\mathbb{P}(S = 1)}.$$

This provides some additional insight, as it shows that the causal effect $\mathbb{P}(W = 1|do(S = 1))$ is equivalent to the regular conditional probability $\mathbb{P}(W = 1|S = 1)$ precisely when R and S are independent. This makes a good deal of sense: if R and S are independent, then R is not a parent of S, and removing the link between R and S does absolutely nothing.   ▷

Example 8.11. Let us return to our single-layer Simpson's paradox Example 8.1, which considers only gender, treatment, and recovery. The DAG for this model is shown in Figure 8.14 (a).

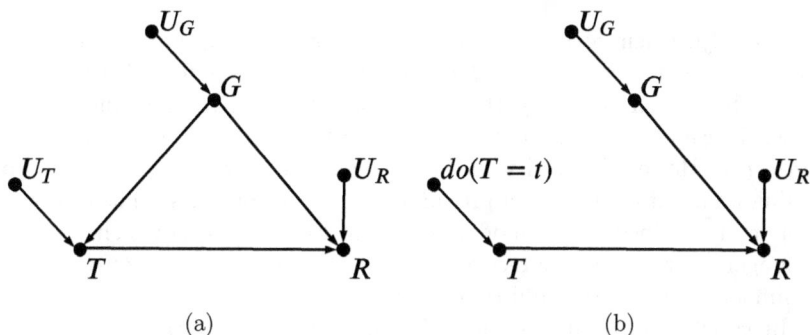

Fig. 8.14: Simpson's paradox example, considering gender G, treatment T, and recovery R. (a) Original model. (b) Manipulated model, showing an intervention on T.

The aggregated probabilities in Table 8.2 can be found by conditioning on gender and applying the law of total probability:

$$\mathbb{P}(R = 1|T = A) = \mathbb{P}(R = 1|T = A, G = F)\mathbb{P}(G = F|T = A)$$
$$+\mathbb{P}(R = 1|T = A, G = M)\mathbb{P}(G = M|T = A)$$
$$= (0.70)(0.20) + (0.50)(0.80) = 0.54,$$

where the individual probabilities are given in Table 8.1. Similarly,

$$\mathbb{P}(R = 1|T = B) = \mathbb{P}(R = 1|T = B, G = F)\mathbb{P}(G = F|T = B)$$
$$+\mathbb{P}(R = 1|T = B, G = M)\mathbb{P}(G = M|T = B)$$
$$= (0.65)(0.80) + (0.45)(0.20) = 0.61,$$

which agrees with our aggregated result from Table 8.2.

However, these results are contrary to the more granular results when we look at the data by gender. To resolve this paradox, we can *control for gender* by using the adjustment formula:

$$\mathbb{P}(R = 1|do(T = A)) = \mathbb{P}(R = 1|T = A, G = F)\mathbb{P}(G = F)$$
$$+\mathbb{P}(R = 1|T = A, G = M)\mathbb{P}(G = M)$$
$$= (0.70)(0.50) + (0.50)(0.50) = 0.60,$$
$$\mathbb{P}(R = 1|do(T = B)) = \mathbb{P}(R = 1|T = B, G = F)\mathbb{P}(G = F)$$
$$+\mathbb{P}(R = 1|T = B, G = M)\mathbb{P}(G = M)$$
$$= (0.65)(0.50) + (0.45)(0.50) = 0.55.$$

These probabilities reflect the recovery rates when we control for gender by intervening on the treatment, as shown in Figure 8.14 (b). These results show that treatment A is in fact superior to treatment B when we control for gender, a confounding factor.   ▷

The adjustment formula can control for confounding variables; however, *the adjustment formula cannot control for our ignorance!* Example 8.11 shows that when we apply the adjustment formula to the causal model shown in Figure 8.14, we obtain the corrected result, *as far as gender is concerned.* However, recall that we changed the story in Example 8.3 by adding an additional layer of granularity: severity of illness. The data now support a modified causal model (shown in Figure 8.3), in which gender is no longer a direct cause of recovery; instead it affects recovery only through its influence on severity and treatment.

In conclusion, applying the adjustment formula at the layer of gender did resolved Simpson's paradox *in the context of the paradoxical result between Tables 8.1 and 8.2.* The adjustment formula did not, however, extend its magic hand into further subdivisions of the data based on severity, which, when taken into account, led us to a different causal structure and conclusion.

8.3.3 The Backdoor Criterion

So far, we have seen how the adjustment formula can control for direct causes of a given variable whose causal effect we are trying to measure. As it turns out, there is a more general (and more powerful) alternative than controlling for each of the a variable's parents. The idea is as follows: if we are trying to measure the causal effect that X has on Y, instead of controlling for each of the parents of X, we can instead control for an alternate set of variables that

1. blocks all spurious paths between X and Y,
2. does not affect any directed path between X and Y,
3. does not create any new spurious relationships between X and Y.

In other words, we seek to control for a minimal set of confounding variables in order to remove any spurious relationship between X and Y. To achieve this, we must first identify a certain set of path between the two variables.

Definition 8.20 (Backdoor/Front-door Path). *Given a node X in a directed graph G, a backdoor path of X is any path that contains X as an endpoint and is furthermore connected to X through a backdoor of X; i.e., the path connects to X with an arrow that is directed into X.*

Similarly, a front-door path of X is any path that is connected to X, as an endpoint, through a front door of X.

When a backdoor/front-door path of X connects to a second endpoint Y, we call it a backdoor/front-door path from X to Y.

Warning! The language "from X to Y" does not imply that a backdoor/front-door path is unidirectional; rather the directional implication "from X" merely implies that it is a backdoor/front-door path of X, but it is free to connect to its secondary endpoint Y as it will.

An example backdoor path from X to Y is

$$X \leftarrow A \rightarrow B \rightarrow C \leftarrow D \rightarrow E \leftarrow Y;$$

but if we change the direction of the first arrow, $X \rightarrow A \cdots$, we would instead have a front-door path.

Since spurious associations between X and Y must operate through the backdoor paths of X, we next make the following useful definition.

Definition 8.21 (Backdoor Criterion). *Given an ordered[2] pair of variables (X, Y) in a directed acyclic graph G, a set of variables Z satisfies the* backdoor criterion *(BDC) relative to (X, Y) if*

1. *Z does not contain any descendants of X; i.e., $Z \cap \mathrm{desc}(X) = \emptyset$; and*
2. *Z blocks all backdoor paths from X to Y.*

Naturally, we should not need to control for each parent of X, as long as we can control for a set of variables that satisfies the backdoor criterion, thereby blocking all possible spurious paths between X and Y. This is the result of the following theorem.

---

[2] relative to the partial ordering of the set; i.e., $Y \in \mathrm{desc}(X)$.

Theorem 8.3. *Let (X, Y) be an ordered pair of variables in a causal model $M = (\mathcal{U}, \mathcal{V}, \mathcal{F}, \mathbb{P})$, and let $\mathcal{Z}$ be a set of variables that satisfies the backdoor criterion relative to (X, Y). Then the causal effect of X on Y is given by the* backdoor adjustment formula

$$\mathbb{P}(Y = y | do(X = x)) = \sum_{z \in \mathcal{R}_Z} \mathbb{P}(Y = y | X = x, Z = z)\mathbb{P}(Z = z); \quad (8.6)$$

where $Z = \oplus \mathcal{Z}$ and $\mathcal{R}_Z = \text{range}(Z)$, as usual. When $\mathcal{Z} = \emptyset$ is the empty set, this degenerates to

$$\mathbb{P}(Y = y | do(X = x)) = \mathbb{P}(Y = y | X = x), \quad \text{when } \mathcal{Z} = \emptyset. \quad (8.7)$$

Note 8.4. Equations (8.4) and (8.6) are identical in form; the only distinction is what constitutes the set $\mathcal{Z}$. ▷

Note 8.5. Theorem 8.2 is a special case of Theorem 8.3, as the full set of parents of a variable X automatically satisfy the backdoor criterion relative to (X, Y); i.e., the parents of X must necessarily block all backdoor paths from X to Y, as X can only communicate through backdoor channels via its parents. ▷

Example 8.12. To illustrate the usefulness of this result, consider Figure 8.15. Using the original adjustment formula, we would have to control

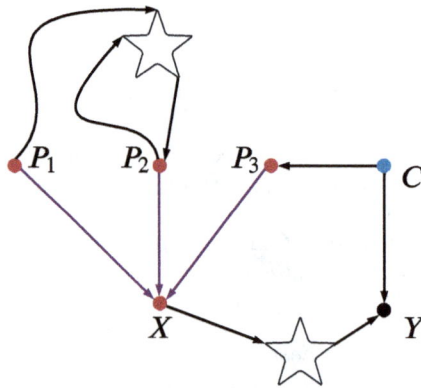

Fig. 8.15: Illustration of Theorem 8.3. Instead of controlling for parents (P_1, P_2, P_3), we can instead control for the single confounding variable C.

for each of the three parent variables P_1, P_2, and P_3. However, we notice that there are no backdoor paths from X to Y that travel through P_1 or P_2; thus, we have cut out unnecessary work for ourselves. It suffices to control for P_3 alone, which blocks the backdoor path

$$X \leftarrow P_3 \leftarrow C \rightarrow Y,$$

as it blocks the chain $X \leftarrow P_3 \leftarrow C$. Let's take this a step further, however, and notice that we could, alternatively, control for C, as the variable C blocks the fork $P_3 \leftarrow C \rightarrow Y$. It is always attractive to have alternatives. Suppose, for sake of argument, that the parent variable P_3 possesses a high cardinality, taking on, say, ten or twenty possible values. If variable C, on the other hand, were to be low cardinality, perhaps even binary, we immediately see the savings in terms of the number of enumerations—and hence subdivisions of our data—when carrying out the summation.   ▷

Example 8.13. Let us again return to our original Simpson's paradox example, but the double-paradox example illustrated in Example 8.3 that took into account both gender and illness severity. We illustrated the causal structure of these data in Example 8.5; in particular, by the DAG represented in Figure 8.3.

By careful examination of the DAG in Figure 8.3, we immediately see our folly in trying to control for gender alone. In seeking to understand the causal effect of treatment on recovery, i.e., $\mathbb{P}(R = 1|\text{do}(T = t))$, it is now immediately clear that gender G blocks the backdoor path

$$p_1 : T \leftarrow G \rightarrow S \rightarrow R,$$

but it leaves the backdoor path

$$p_2 : T \leftarrow S \rightarrow R$$

unblocked! Controlling for severity alone, on the other hand, blocks both backdoor paths. In fact, we see that either set $\mathcal{Z} = \{S\}$ or set $\mathcal{Z} = \{S, G\}$ accomplish the same goal in blocking both of our backdoor paths. We can therefore obtain our desired result by aggregating by illness severity, and then controlling for severity.   ▷

8.3.4 The Front-door Criterion

We saw in Example 8.11 that the backdoor adjustment formula cannot control for our ignorance. For example, if we *only* consider gender in the gender–severity–illness–recovery problem, we will not catch the true culprit: severity. But what if you are in a legal battle with a foe using ignorance as a defense—not negligible ignorance, but the ignorance of certain *unknown unknowns* that we could never possibly control for? Such was the case in the years prior to 1970, when the tobacco industry made such a defense in order to successfully lobby *against* antismoking legislation.

Example 8.14. Consider the causal model depicted by the DAG in Figure 8.16. In trying to assess the causal effect that smoking has on lung cancer, the devil's advocates argued successfully (at least, for some time)

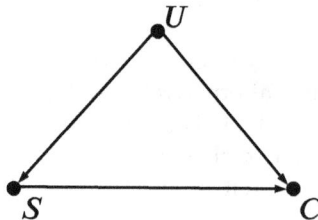

Fig. 8.16: Causal relation between smoking S and lung cancer C, with an unobserved confounder U (an *unknown unknown*).

that it was possible that an unobserved confounding variable could be leading to a spurious association between smoking and cancer. If such an idea seems laughable and transparent from our modern perspective, try proving, for example, that gender does *not cause* recovery in our Simpson's paradox example, by using Table 8.1 alone; i.e., without reference to the further subdivision of illness severity (Table 8.4). Come to think of it, how can we prove that any of our results are valid? Isn't there always the possibility of an unseen hand wreaking havoc over our data?

In the case of the tobacco industry, lobbyists argued that it is possible that an as-of-yet undiscovered genotype (U in the diagram) could be causing *both* the propensity of an individual to have a penchant towards smoking *and* the likelihood for that individual to end up with lung cancer later in life. In other words, the same gene that makes a person likely to smoke is also responsible for that individual's later ailments. And since it would be unethical to performed a controlled experiment, it seemed like hope was lost in proving the effect that smoking has on lung cancer.   ▷

The resolution should be obvious: if the backdoor criterion doesn't work, maybe we could try a front-door criterion? Consider the modified causal diagram of Figure 8.17, which shows the addition of a new variable Z, representing the amount of tar deposits in an individual's lungs. The idea is to use this intermediate variable with two applications of the backdoor criterion in order to infer the causal effect that smoking has on lung cancer. To achieve this, we next define the front-door analog to Definition 8.21.

Definition 8.22 (Front-door Criterion). *Given an ordered pair of variables (X, Y) in a directed acyclic graph G, a set of variables $\mathcal{Z}$ satisfies the* front-door criterion (BDC) *relative to (X, Y) if*

1. *$\mathcal{Z}$ intercepts all directed paths from X to Y;*
2. *there are no unblocked backdoor paths from X to any variable in $\mathcal{Z}$;*
3. *all backdoor paths from the variables in $\mathcal{Z}$ to Y are blocked by X.*

For example, the single variable Z satisfies the front-door criterion relative to (S, C) in Figure 8.16.

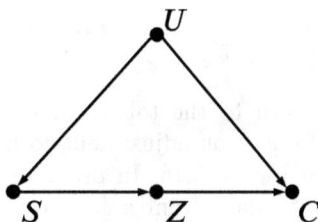

Fig. 8.17: Causal relation between smoking S, tar deposits Z, and lung cancer C, with an unobserved confounder U (an *unknown unknown*).

Theorem 8.4. *Let (X, Y) be an ordered pair of variables in a causal model $M = (\mathcal{U}, \mathcal{V}, \mathcal{F}, \mathbb{P})$, and let $\mathcal{Z}$ be a set of variables that satisfies the front-door criterion relative to (X, Y). Then the causal effect of X on Y is given by the* front-door adjustment formula

$$\mathbb{P}(Y = y | \mathrm{do}(X = x)) = \tag{8.8}$$
$$\sum_{z \in \mathcal{R}_Z} \sum_{s \in \mathcal{R}_X} \mathbb{P}(Y = y | X = s, Z = z)\mathbb{P}(X = s)\mathbb{P}(Z = z | X = x).$$

where $Z = \oplus \mathcal{Z}$ and $\mathcal{R}_Z = \mathrm{range}(Z)$; $\mathcal{R}_X = \mathrm{range}(X)$, as usual.

Proof. The proof follows in two steps. First, since there are no unblocked backdoor paths from X to $\mathcal{Z}$, the degenerate form of Theorem 8.3 implies that

$$\mathbb{P}(Z = z | \mathrm{do}(X = x)) = \mathbb{P}(Z = z | X = x).$$

Second, since all backdoor paths from $\mathcal{Z}$ to Y are blocked by conditioning on X, Theorem 8.3 implies that

$$\mathbb{P}(Y = y | \mathrm{do}(Z = z)) = \sum_{s \in \mathcal{R}_X} \mathbb{P}(Y = y | Z = z, X = s)\mathbb{P}(X = s),$$

where we have used a dummy variable s for purpose of summation.

We have thus determined both the causal effect of X on Z, as well as the causal effect of Z on Y. We can combine these results using the law of total probability in the form

$$\mathbb{P}(Y = y | \mathrm{do}(X = x)) = \sum_{z \in \mathcal{R}_Z} \mathbb{P}(Y = y | \mathrm{do}(Z = z))\mathbb{P}(Z = z | \mathrm{do}(X = x)).$$

Replacing the terms on the right-hand side with our earlier two formulas yields the result. □

Note that the term $\mathbb{P}(Z = z | X = x)$ can be factored out of the inner summation for convenience, so that the right-hand side of Equation (8.8) may be equivalently expressed as

$$\sum_{z \in \mathcal{R}_Z} \left[\mathbb{P}(Z = z | X = x) \sum_{s \in \mathcal{R}_X} \mathbb{P}(Y = y | X = s, Z = z) \mathbb{P}(X = s) \right].$$

Example 8.15. Let us return to the tobacco Example 8.14, and suppose we wish to apply the front-door adjustment formula to account for tar deposits Z, as shown in Figure 8.16. In order to provide some numbers to this story, consider the data from a fictitious observational study of 1,000,000 individuals as shown in Table 8.9. We can apply the front-door criterion and our causal story Figure 8.16 in order to determine the causal effect of smoking on cancer. (The logic is that patients who smoke are much more likely to develop tar in their lungs, and the development of tar similarly makes a patient more likely to have lung cancer.)

	No tar $(Z = 0)$	Tar $(Z = 1)$
Nonsmokers $(S = 0)$	14/665 (2.1%)	2/35 (5.7%)
Smokers $(S = 1)$	1/15 (6.7%)	40/285 (14%)

Table 8.9: Observational data on the relation between smoking and tar deposits on lung cancer. Numbers in thousands.

In order to compute the probability of cancer given the intervention of smoking or non-smoking, we apply the front-door adjustment formula to obtain

$$\begin{aligned}
\mathbb{P}(C = 1 | \mathrm{do}(S = 0)) = {}& \mathbb{P}(Z = 0 | S = 0) \left[\mathbb{P}(C = 1 | S = 0, Z = 0) \mathbb{P}(S = 0) \right. \\
& \left. + \mathbb{P}(C = 1 | S = 1, Z = 0) \mathbb{P}(S = 1) \right] \\
{}+ {}& \mathbb{P}(Z = 1 | S = 0) \left[\mathbb{P}(C = 1 | S = 0, Z = 1) \mathbb{P}(S = 0) \right. \\
& \left. + \mathbb{P}(C = 1 | S = 1, Z = 1) \mathbb{P}(S = 1) \right] \\
\mathbb{P}(C = 1 | \mathrm{do}(S = 1)) = {}& \mathbb{P}(Z = 0 | S = 1) \left[\mathbb{P}(C = 1 | S = 0, Z = 0) \mathbb{P}(S = 0) \right. \\
& \left. + \mathbb{P}(C = 1 | S = 1, Z = 0) \mathbb{P}(S = 1) \right] \\
{}+ {}& \mathbb{P}(Z = 1 | S = 1) \left[\mathbb{P}(C = 1 | S = 0, Z = 1) \mathbb{P}(S = 0) \right. \\
& \left. + \mathbb{P}(C = 1 | S = 1, Z = 1) \mathbb{P}(S = 1) \right]
\end{aligned}$$

Now, from Table 8.9, we have

$$\mathbb{P}(S = 0) = 0.7 \quad \text{and} \quad \mathbb{P}(S = 1) = 0.3;$$

i.e., the data were collected from a population comprised of 30% smokers. Furthermore, we have

$$\mathbb{P}(Z = 1 | S = 0) = 5\%$$
$$\mathbb{P}(Z = 1 | S = 1) = 95\%,$$

from which we can easily obtain $\mathbb{P}(Z = 0|S = s)$ for $s = 0, 1$. Combining the above, we find that

$$\mathbb{P}(C = 1|\mathrm{do}(S = 0)) \approx 0.0371$$
$$\mathbb{P}(C = 1|\mathrm{do}(S = 1)) \approx 0.0797,$$

which shows that smokers have double the risk[3] than nonsmokers.

The upshot is we were able to prove the causal link between smoking and cancer using the mediating variable of tar deposit, ruling out the possibility of a spurious association between the two.   ▷

8.4 Counterfactuals

The English word *counterfactual* is an adjective that means *contrary to fact* or *expressing what has not happened or what isn't the case*. We used the word in this way in our definition of an intervention (Definition 8.18), which makes reference to a "counterfactual reality;" i.e., what would have the overall results looked like had we, contrary to fact, swooped in and intervened to enforce a particular value of a variable. Thus, unlike prediction (i.e., conditioning), we are not limiting the focus of the data, but rather enforcing our intervention across all the data and reporting an overall result. We are, however, missing part of the story.

A *counterfactual* (as a noun) is a particular instance in our counterfactual reality. Whereas intervention is a population-level report (i.e., *what is the probability the pavement would have been wet, had we intervened to make sure the sprinkler was turned on?*), a counterfactual is an individual-level statement limited to a single instance (i.e., *What is the probability that the pavement would have been wet, had we intervened to make sure the sprinkler was turned on, when, in fact, the sprinkler was turned off and and it had rained and the pavement was not wet?*).

A quick glance at Tables 8.7 and 8.8 in Example 8.9 confirms that, in the context of interventions, we moved the entire population of rain and no sprinkler (85% chance of wet pavement) to rain and sprinkler (90% chance of wet pavement). We did not, however, answer the question for particular individuals: what is the probability that the pavement would have been wet under our intervention of sprinkler for the instances of rain and no sprinkler and wet vs. dry pavement?

To answer this, we first craft a precise definition for counterfactuals, and then we discuss the three-step process for evaluating them.

8.4.1 Counterfactuals: An Illustration

We begin with an example to illustrate the meaning of a counterfactual statement.

---

[3] using our made-up numbers; in reality, the increased risk is much greater.

Example 8.16. Let us revisit the rain–sprinkler–pavement problem of Example 8.9. We have recaptured the data of Tables 8.7 and 8.8 in Table 8.10, separating the factual and counterfactual worlds. In this version, however, we track individuals in the counterfactual world according to who they were in reality. Additionally, the individuals are labeled by their actual real-world values, so we can understand how their conditions have changed under the intervention $\mathrm{do}(S = 1)$. In the counterfactual world, we have an added row that shows the overall effect of the intervention. For example, in the counterfactual world, no rain and no sprinkler has a total 49/490 (10%) that end up being wet, as we have ensured that the sprinkler is running in the counterfactual world. (The 10% is from the observed rate of wet pavement for no rain and sprinkler.) This row (49/490, 21/210, 243/270, 27/30) is the same as Table 8.8, except that we are accounting for individuals based on whether or not the sprinkler was running in the real world, even though the sprinkler for all of them was running in our alternate world.

		No Rain 700 (70%)			Rain 300 (30%)		
Real World	No Sprinkler 490 (70%)		Sprinkler 210 (30%)		No Sprinkler 270 (90%)		Sprinkler 30 (10%)	
	Dry	Wet	Dry	Wet	Dry	Wet	Dry	Wet
	490	0	189	21	40.5	229.5	3	27
	(100%)	(0%)	(90%)	(10%)	(15%)	(85%)	(10%)	(90%)
Counterfactual World		No Rain 700 (70%)			Rain 300 (30%)		
	No Sprinkler 490 (70%)		Sprinkler 210 (30%)		No Sprinkler 270 (90%)		Sprinkler 30 (10%)	
	Frac. Wet 49/490 (10%)		Frac. Wet 21/210 (10%)		Frac. Wet 243/270 (90%)		Frac. Wet 27/30 (90%)	
	Dry	Wet	Dry	Wet	Dry	Wet	Dry	Wet
	49/490	0/0	0/189	21/21	13.5/40.5	229.5/229.5	0/3	27/27
	(10%)	(100%)	(0%)	(100%)	(33%)	(100%)	(0%)	(100%)

Table 8.10: Counterfactuals in the Rain–Sprinkler–Pavement Example. The counterfactual world is defined by the intervention $\mathrm{do}(S = 1)$; i.e., the sprinkler is running in our counterfactual world for all cases, though we are organizing individuals according to their position in the real world, highlighting the labels in red when their is risk of conflict.

We can capture the full counterfactuals by examining our final row in the table. These numbers are obtained as follows. We first make the reasonable assumption that if the pavement was wet in the real world, it will also be wet in our counterfactual world. (After all, we are turning the sprinkler *on.*) Further, if the sprinkler was on in the real world, the results should be unchanged—for those individuals—in our counterfactual world. Finally, since we are only concerned with the no sprinkler cases, and since we know that our intervention will leave wet the pavement that was actually wet, and since we also know the total count of wet pavement excepted under our intervention, we can simply attribute the remainder of this count to the cases that corresponded to dry pavement in the real world.

For example, consider the 270 individuals that fall under rain and no sprinkler. Since we have intervened to ensure the sprinkler was running in our counterfactual world, we know that 243 of these individuals (90%) will have wet pavement in our counterfactual world. Moreover, the 229.5 individuals who had wet pavement in the real world, due to the rain, will still have wet pavement in our counterfactual world. The difference is 13.5, which must constitute the number of individuals who have wet pavement in our alternate world, out of the 40.5 individuals who had dry pavement, despite it being raining and their sprinkler being off. We can state this counterfactual as follows: there is a 33% probability that the pavement will be wet, if we intervene to ensure the sprinkler is turned on, for individuals who, in fact, had the sprinkler turned off and had dry pavement on a day in which it had rained.   ▷

8.4.2 A Crisis of Notation

In Example 8.16, we would like capture our counterfactual statement—*the probability of wet pavement under the sprinkler intervention, if, in fact, the pavement was dry, the sprinkler was off, and it was raining*—in mathematical notation, but we are immediately confronted with a problem. For instance, we would like to state

$$\mathbb{P}(W|\text{do}(S), \neg S, R, \neg W)$$

for this precise case: is the pavement wet if we make sure the sprinkler is on, given that the sprinkler is off, it had rained, and the pavement is not wet. Despite the immediate contradiction (what is the probability that the pavement is wet, given that the pavement is not wet), we also find such notation to fall short under our simple calculus of intervention. For example,

$$\mathbb{P}(W|\text{do}(S), R, \neg S) = \mathbb{P}(W|S, R),$$

that is, our intervention do(S) *eats* the reality $\neg S$ when we are searching for the correct probabilities to apply.

These conflicts arise as we are referring to events occurring in separate worlds, the real world (from which we gathered observational data) and the alternate counterfactual world (in which we are making certain surgical alterations, such as making sure that our sprinkler was running when, in fact, it was not).

Pearl [2009] resolves this with his *counterfactual notation*, instead writing

$$\mathbb{P}(W_{S=1}|R=1, S=0, W_{S=0}=0),$$

where W_S represents the event of wet pavement, as dependent on the sprinkler's condition. In part, this negates the good work we have done in establishing the *do* operator, as it slides the *do* portion into the subscript. We will advocate for and introduce an alternate notation, which we construct with heavy reference to Pearl's logical construction and definition of counterfactuals.

Note 8.6. Pearl treats interventions and counterfactuals as distinct. The logic is that a counterfactual, to Pearl, involves probability statements from conflicting worlds. We take a different view. The intervention is, in fact, a statement about a counterfactual reality, which is composed of many individual counterfactuals. Thus, an intervention is a sort of aggregation over counterfactuals. An intervention is, therefore, a counterfactual statement, as it is a statement about a world that is contrary to our reality. A counterfactual (noun), on the other hand, is simply an atomic instance from that world.   ▷

8.4.3 Defining Counterfactuals

Pearl [2009] shows how counterfactuals should be defined and computed using a three step process: abduction, intervention, and prediction. The logic is that since a counterfactual is a statement about an individual, we should first trace back the values of that individual's exogenous variables (abduction), then perform our minimal graph surgery (intervention), and finally determine the modified probability of a variable of interest, using that individual's value for $U = u$ and the modified graph (prediction). Analogous to Pearl's *do* operator, we therefore define an *if* operator, to capture the meaning of an abduction. The *if* operator traces a statement regarding the endogenous variables back to the exogenous ones, thereby defining a new probability distribution over $\mathcal{U}$. We state this definition formally as follows.

Definition 8.23. *Let $\mathcal{E}$ be a set of endogenous variables ("the evidence") in a causal model $M = (\mathcal{U}, \mathcal{V}, \mathcal{F}, \mathbb{P})$. Then an* abduction *on $E = \oplus \mathcal{E}$ with value e, denoted* if$(E = e)$, *represents the conditional probability distribution*

$$\mathbb{P}_{E=e}(U = u) = \mathbb{P}(U = u|E = e)$$

over $\mathcal{U}$, such that the probability statement

$$\mathbb{P}(V = v|\text{if}(E = e)) = \mathbb{P}_{E=e}(V = v) \qquad (8.9)$$

on an endogenous variable $V \in \mathcal{V}$ represents the the resultant probability distribution over V arising from the modified distribution $\mathbb{P}(U = u|E = e)$ over $\mathcal{U}$; i.e., it represents a modification of the probability distribution over $\mathcal{U}$ that is consistent with the evidence $E = e$.

Thus, our *if* operator is a shorthand for the statement *if, in fact, we observed* ____. The *if* operator then takes a statement about our observation and translates it back to an updated probability distribution over $\mathcal{U}$. Recall that each possible value of $U = \oplus \mathcal{U}$ uniquely determines all values in $\mathcal{V}$. Having captured our broadest understanding of the variable U, for an individual with evidence $E = e$, we are now free to alter the graph and make predictions about what would have happened for this given individual had we intervened to do X.

Given this definition, we may now define a counterfactual as follows.

Definition 8.24. *Let X and Y be endogenous variables in a causal model $M = (\mathcal{U}, \mathcal{V}, \mathcal{F}, \mathbb{P})$, and let $\mathcal{E} \subset \mathcal{V}$ be a set of endogenous variables, possibly and most likely containing X and Y. Then a* counterfactual *is a triple* $(\text{if}(E = e), \text{do}(X = x), Y = y)$ *consisting of*

1. *an abduction if$(E = e)$,*
2. *an intervention do$(X = x)$,*
3. *and a prediction $Y = y$,*

that defines the quantity

$$\mathbb{P}(Y = y | \text{do}(X = x), \text{if}(E = e)) = \mathbb{P}_{\substack{E=e \\ M_{X=x}}} (Y = y) \qquad (8.10)$$

as the probability distribution over Y obtained by propagating the modified probability distribution $\mathbb{P}(U = u | E = e)$ over $\mathcal{U}$ through the modified model $M_{X=x}$ (Definition 8.18), which represents the counterfactual statement "the probability that $Y = y$, if we intervene to do $X = x$, if, in reality, we observed $E = e$".

Example 8.17. So far, we have yet to prescribe the actual structural equations for the rain–sprinkler–pavement problem from Example 8.9, referencing instead the conditional probabilities. These, however, are easy enough to write down:

$$R = f_R(U_R) = \mathbb{I}[U_R < 0.30],$$

$$S = f_S(R, U_S) = \begin{cases} \mathbb{I}[U_S < 0.3] & \text{if } R = 0 \\ \mathbb{I}[U_S < 0.1] & \text{if } R = 1 \end{cases},$$

$$W = f_W(R, W, U_W) = \begin{cases} 0 & \text{if } R = 0 \text{ and } S = 0 \\ \mathbb{I}[U_W < 0.10] & \text{if } R = 0 \text{ and } S = 1 \\ \mathbb{I}[U_W < 0.85] & \text{if } R = 1 \text{ and } S = 0 \\ \mathbb{I}[U_W < 0.90] & \text{if } R = 1 \text{ and } S = 1 \end{cases},$$

where each exogenous variable is an independent and identically distributed uniform random variable $U_R, U_S, U_W \sim \text{Unif}(0, 1)$.

Consider now the counterfactual statement, *the probability that the pavement is wet, if we had intervened to ensure the sprinkler was on, if, in fact, the sprinkler was off, it was raining, and the pavement was dry*, as represented using our counterfactual notation as

$$\mathbb{P}(W|\mathrm{do}(S), \mathrm{if}(R, \neg S, \neg W)).$$

In order to compute this, we first perform our abduction. It is simple to show that our abduction yields

$$U_R|R \sim \mathrm{Unif}(0, 0.30),$$
$$U_S|R, \neg S \sim \mathrm{Unif}(0.10, 1.0),$$
$$U_W|R, \neg S, \neg W \sim \mathrm{Unif}(0.85, 1.0).$$

Moreover, our intervention $\mathrm{do}(S)$ replaces the structural equation $S = f_S(R, U_S)$ with $S = 1$.

Finally, to evaluate our counterfactual, we propagate the reduced probability distribution over $\mathcal{U}$ through the modified model. First, since $U_R|R \sim \mathrm{Unif}(0, 0.30)$, we see that $R = 1$. Second, our intervention replaces the structural equation for S with $S = 1$. Thus, we see from our third structural equation, that

$$W = \mathbb{I}[U_W < 0.90],$$

since $R = 1$ (from abduction) and $S = 1$ (from intervention). Now, our abducted distribution for U_W was $U_W|R, \neg S, \neg W \sim \mathrm{Unif}(0.85, 1.0)$, so we see that

$$\mathbb{P}(W|\mathrm{do}(S), \mathrm{if}(R, \neg S, \neg W)) = \frac{0.90 - 0.85}{1.00 - 0.85} = \frac{1}{3},$$

agreeing with our previous result of Example 8.16.   ▷

8.5 Observational Studies

Earlier in the chapter, we briefly introduced the concept of *observational studies* or *natural experiments*, in which the statistician does not have direct control over the treatment assignment. In this section, we flesh out this idea in more depth. We begin by revisiting the experimental design principle of randomization, which we previously discussed in Chapter 4, from the perspective of causal graph theory. Next, we discuss the two types of biases that can present in natural experiments: overt biases, which can be measured by observed covariates, and hidden biases, which are not observed. We relate both types of bias back to Simpson's paradox, introducing a new concept of *propensity score*. We then discuss how overt biases may be controlled for by using the adjustment formula. Finally, we discuss the more difficult problem of drawing inferences regarding hidden biases.

8.5.1 Randomized and Natural Experiments

The objective of experimentation is, of course, to draw inferences about the effect of a treatment T on a response R for a given population. As we saw earlier in the chapter, one must take care to guard against the possibility of any confounding variables X, that might have a causal implication for both the treatment and the response.

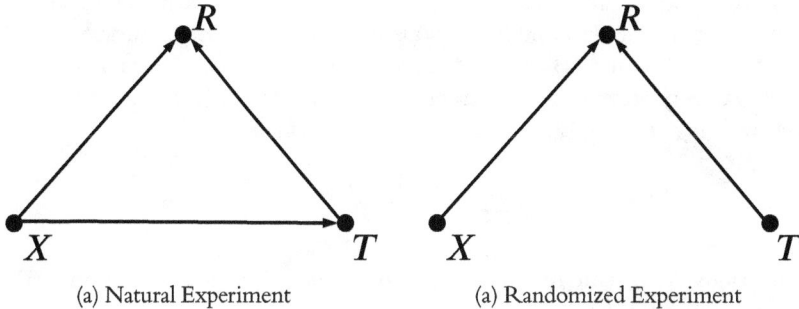

(a) Natural Experiment   (a) Randomized Experiment

Fig. 8.18: Causal graph for (left) a natural experiment and (right) a randomized experiment; R is the response, T is the treatment, and X is a confounder.

The situation is depicted in Figure 8.18 (left). The presence of a confounding variable in a natural experiment can create a spurious correlation between the treatment groups and the observed response, thereby biasing any inferences regarding the causal effect of T on R. In Chapter 4, we discussed the principle of randomization, which essentially erases all causal connections into the choice of treatment. This results in the modified graph shown in Figure 8.18 (right). Since treatment is determined solely on a properly randomized assignment mechanism, nothing else can effect whether or not any particular test subject lands in any of the treatment groups. This effectively controls for any potential confounder, as the experimenter has intervened and selected the treatment group, relying on chance alone to make each assignment.

Moreover, because treatments have been properly randomized, the composition of any confounding covariates should be similar within each treatment group. One may further implement a randomized block design, to further guard against the possibility of any known covariate being imbalanced among the treatment groups due to chance. This is especially important for small sample sized experiments with known rare covariates that exhibit on average an abnormal response.

8.5.2 The Adjustment Formula: Controlling for Overt Bias

The Adjustment Formula

We have previously seen two separate incarnations of the adjustment formula (Equations (4.20) and (8.4)). These are, of course, equivalent. It is worthwhile, however, discussing their relation and, in particular, the assumptions behind them.

In particular, let us consider the adjustment formula of Equation (8.4) in the context of measuring the causal effect of a treatment T on a response R, while controlling for various covariates X. Let us further restrict our attention to an A/B test, in which there are only two treatment groups, which we may refer to as treatment ($T = 1$) and control ($T = 0$). In this case, we may rewrite Equation (8.4) in the form

$$\mathbb{E}[R|\mathrm{do}(T = t)] = \sum_{x \in \mathcal{R}_X} \mathbb{E}[R|T = t, X = x]\mathbb{P}(X = x). \tag{8.11}$$

In particular, we can recast the *average causal effect* (Definition 4.5) in terms of our *do*-calculus as

$$\Delta = \mathbb{E}[R|\mathrm{do}(T = 1)] - \mathbb{E}[R|\mathrm{do}(T = 0)]. \tag{8.12}$$

Combining the previous two equations then yields

$$\Delta = \sum_{x \in \mathcal{R}_X} \Big(\mathbb{E}[R|T = 1, X = x] - \mathbb{E}[R|T = 0, X = x]\Big)\mathbb{P}(X = x). \tag{8.13}$$

Consider now the case of s strata, enumerated by $i = 1, \ldots, s$, which represent a unique enumeration of the covariates $\mathcal{R}_X$. Let $j = 1, \ldots, n_i$ enumerate the n_i experimental units within each strata. Further let Z_{ij} represent the assignment of the ijth unit into the treatment ($Z_{ij} = 1$) or control ($Z_{ij} = 0$) groups and let a_i represent the total number of units within the ith stratum assigned to the treatment, as given by Equation (4.18). Finally, let R_{ij}^T and R_{ij}^C represent the response of the ijth experimental unit under the treatment and control, respectively, such that the actual response is given by

$$R_{ij} = Z_{ij}R_{ij}^T + (1 - Z_{ij})R_{ij}^C.$$

Now, unbiased estimators for the expected response of units within the ith stratum under the treatment and control are given by

$$\mathbb{E}[R|T = 1, X = x_i] = \frac{1}{a_i} \sum_{j=1}^{n_i} Z_{ij}R_{ij}^T$$

$$\mathbb{E}[R|T = 0, X = x_i] = \frac{1}{n_i - a_i} \sum_{j=1}^{n_i} (1 - Z_{ij})R_{ij}^C,$$

respectively. Moreover, the ratio n_i/n represents an unbiased estimator for $\mathbb{P}(X = x_i)$. Combining these into Equation (8.13) yields the adjustment formula of Equation (4.20), repeated here for convenience:

$$\hat{\Delta} = \sum_{i=1}^{s} \frac{n_i}{n} \left[\frac{1}{a_i} \sum_{j=1}^{n_i} Z_{ij} R_{ij}^T - \frac{1}{n_i - a_i} \sum_{j=1}^{n_i} (1 - Z_{ij}) R_{ij}^C \right] \qquad (8.14)$$

This shows the equivalence between the two equations, at least in principle. However, the set of underlying assumptions of Theorems 4.5 and 8.2 are quite different, and it will prove instructive to understand these differences more precisely, which we venture to do forthwith.

Example 8.18. Applying the adjustment formula to the observed recovery rates of Table 8.5, we can estimate the average treatment effect of treatment B over treatment A as

$$\hat{\Delta} = (10\%)(4.9\%) + (20\%)(4.5\%) + (20\%)(5.5\%) + (50\%)(4.4\%) = 4.69\%.$$

This is close to the actual true treatment effect of 5.00%, which we only know as the data shown in the table were generated using random numbers. Notice that without the adjustment formula, the law of total expectation would have led us to the opposite conclusion, that treatment A has an 11.4% advantage over treatment B, with probabilities of recovery being 74.6% and 65.9%, respectively.   ▷

The Propensity Score

The underlying assumption of Theorem 4.5, which yields Equation (8.14), is that the probability of assignment to the treatment group is constant within each stratum. Contrast this with the underlying assumption of Theorem 8.2, which states that the strata should consist of all causal parents of the treatment assignment. These conditions are then generalized in Theorem 8.3, which states that it is sufficient to collect strata X that satisfy the backdoor criterion relative to the treatment T and response R.

In order to better understand this distinction, we define the following.

Definition 8.25. *The* propensity score *is the function* $\lambda : \mathcal{X} \to [0,1]$ *over the set of covariates that is defined by the conditional probability*

$$\lambda(x) = \mathbb{P}(T = 1 | X = x). \qquad (8.15)$$

Notice this is consistent with the denominator in Equation (8.5).

Notice that an unbiased estimate for the propensity score of a particular stratum $X = x_i$ (Equation (8.15)) is given by $\hat{\lambda}_i = \hat{\lambda}(x_i) = a_i/n_i$. This is true regardless of whether or not each individual within the stratum has the same probability of treatment. More on this in a moment.

In a randomized block experiment, the propensity score is known as it is determined in advanced by design. It is further known that the probability of treatment assignment for the jth unit within stratum i is given precisely by $\pi_{ij} = \lambda_i$, as the treatment assignment is specified directly by the controller. In an observational study, however, the propensity score is not known in advanced, nor is it known whether or not the probability of treatment assignment is homogeneous within each stratum. This distinction gives rise to the following fundamental definition.

Definition 8.26. Hidden bias *is said to be present within an observational study whenever*

$$\pi_u \neq \lambda(x(u)) \qquad (8.16)$$

for any experimental unit $u \in \mathcal{P}$; i.e., whenever the probability of treatment for an individual is not a function of the observed covariates.

The adjustment formula controls for *overt bias*, which occurs whenever the propensity scores vary across the observed covariates, but does not account for the possibility of hidden bias. For example, based on the gender covariates of Table 8.1 alone, one might conclude that treatment A is superior to treatment B. However, this is a hidden bias, which is apparent when examining Table 8.4, which further stratifies based on severity of illness.

Stratification on the Basis of Propensity Score

We are now well posed to answer our question of when it is appropriate to "collapse" along particular dimensions or covariates. Underlying the calculation of expectation in Equation (8.12) is the probability distribution

$$\mathbb{P}(R = r \mid T = t, X = x)\mathbb{P}(X = x) = \frac{\mathbb{P}(R = r, T = t, X = x)}{\mathbb{P}(T = t \mid X = x)},$$

which also appears in Equation (8.5). We see that the covariates affect the conditional expectation only to the extent that they affect the propensity score $\lambda(x) = \mathbb{P}(T = 1 \mid X = x)$. To see this, consider the transformation from X to a new categorical covariate $Y = \lambda(X)$, whose values are the enumerated values of propensity of all the various permutations of X. Under such a transformation, it is easy to show that

$$\sum_{y \in \mathcal{R}_Y} \frac{\mathbb{P}(R = r, T = t, Y = y)}{\mathbb{P}(T = t \mid Y = y)} = \sum_{x \in \mathcal{R}_X} \frac{\mathbb{P}(R = r, T = t, X = x)}{\mathbb{P}(T = t \mid X = x)},$$

since, whenever two distinct x_1 and x_2 both map to the same value y, the denominators are, by definition, the same, and the joint densities in the numerators are additive, i.e.,

$$\mathbb{P}(R = r, T = t, X = x_1) + \mathbb{P}(R = r, T = t, X = x_2) = \mathbb{P}(R = r, T = t, Y = y),$$

since $Y = y$ is simply the event that X equals x_1 or x_2.

This conclusion also makes sense from a causal perspective. The whole point of the adjustment formula, or the more general backdoor criterion, is to control for variables that have a causal effect on treatment. If two ancestor variables X_1 and X_2 have the same causal effect on the treatment, as revealed by the equality of their respective propensities, they are, in fact, indistinguishable from one another. They may therefore be combined and handled as a single causal agent.

Example 8.19. Let us return again to Example 8.4. From Table 8.5 we can easily calculate the propensity scores for each stratum:

$$\mathbb{P}(T = B|\text{Female, Young}) = 20\%$$
$$\mathbb{P}(T = B|\text{Female, Old}) = 20\%$$
$$\mathbb{P}(T = B|\text{Male, Young}) = 80\%$$
$$\mathbb{P}(T = B|\text{Male, Old}) = 80\%.$$

It follows that the propensity score is constant across age groups; i.e., gender has a causal effect on treatment assignment, but age does not. This is why Table 8.6 (bottom), in which we collapse the data based on gender, yields the correct conclusion, whereas collapsing on age (Table 8.6 (top)) does not. An example in which propensity varies by age, and not gender, is given in Rosenbaum [2019], along with a very in depth and intuitive discussion on propensity scores and observational studies in general.   ▷

Balancing Property

One important property of stratification based on propensity is that it balances out the distribution of observed covariates over the control and treatment groups. Whatever imbalance there is among observed covariates is due to chance and not due to any systematic imbalance. The same conclusion, however, does not generally apply to hidden variables. We state this result precisely as follows.

Proposition 8.4 (Balancing Property). *Let $\Lambda \in (0,1)$ be one of the propensity scores, realized by at least one covariate stratum. Then*

$$\mathbb{P}(X = x|\lambda(X) = \Lambda, Z = z) = \mathbb{P}(X = x|\lambda(X) = \Lambda); \qquad (8.17)$$

thus, observed covariates are conditionally independent of treatment assignment, given the propensity score.

Stated differently, the probability distribution over the covariates, given the propensity score, is identical for treatment and control. Naturally, for any given propensity score Λ, this distribution is supported by the subset $\{i_1, \ldots, i_p\} \subset \{1, \ldots, m\}$ of covariates for which $\lambda(x_{i_j}) = \lambda(x_{i_k}) = \Lambda$, for $j, k = 1, \ldots, p$.

Proof. From Bayes' law (Proposition 10.1), we have

$$\mathbb{P}\left(X = x | \lambda(X) = \Lambda, Z = z\right) = \frac{\mathbb{P}\left(Z = z | \lambda(X) = \Lambda, X = x\right) \mathbb{P}\left(X = x | \lambda(X) = \Lambda\right)}{\mathbb{P}(Z = z | \lambda(X) = \Lambda)}.$$

However, it is obvious that

$$\mathbb{P}\left(Z = 1 | \lambda(X) = \Lambda, X = x\right) = \mathbb{P}\left(Z = 1 | \lambda(X) = \Lambda\right) = \Lambda.$$

A similar statement can be made for $z = 0$. The result follows.   □

Distribution of Treatment Assignments

In this paragraph, we provide a brief justification of the equivalence between a properly randomized experiment and an observational study stratified on a set of covariates, such that there is no hidden bias. Let π_{ij} represent the probability of treatment for the jth unit in the ith stratum, such that the probability distribution over the possible treatment assignments is given by

$$\mathbb{P}(Z_{ij} = z_{ij}) = \prod_{i=1}^{s} \prod_{j=1}^{n_i} \pi_{ij}^{z_{ij}} (1 - \pi_{ij})^{1-z_{ij}}. \tag{8.18}$$

Now, when the study is free of hidden bias, we have $\pi_{ij} = \lambda_i = \lambda(x_i)$. However, the preceding probability distribution is still unknown, since the true propensity scores λ_i are unknown and since $A_i = \sum_{j=1}^{n_i} Z_{ij}$ are random variables. However, if we condition on the number of units assigned to the treatment groups, $A_1, \dots, A_s$, we find that each treatment assignment has the same probability

$$\mathbb{P}(Z_{ij} = z_{ij} | A_i = a_i) = \prod_{i=1}^{s} \lambda_i^{a_i} (1 - \lambda_i)^{n_i - a_i} = \left[\prod_{i=1}^{s} \binom{n_i}{a_i} \right]^{-1}. \tag{8.19}$$

The final equality follows by counting the number of ways of arranging $A_i = a_i$ treatments out of a collection of n_i units, for each individual stratum. (Since the conditional probabilities are constant, they must equal the inverse of the total number of permissible permutations.) This is, however, equivalent to the probability distribution obtained from a uniform block-randomized experiment. Though this is a very nice result, a word of caution is in order: this conclusion heavily relies on the assumption that our observational study is free of hidden bias; i.e., our observed covariates cover the causal parents of the treatment assignment. Such an assumption, in practice, must be met with intense scrutiny, as there is no reason to believe, *a priori*, in the absence of such hidden bias.

Matched Pairs

In the age of big data, it is not uncommon to have hundreds, thousands, or millions of features to describe a user base, that is often the target of an observational study or experiment. If one considers covariates described using f distinct factors, each with l levels, one quickly realizes the exponential nature in the number of required strata, as shown in Table 8.11. If individual users of an online or web-based product are to constitute units in an experiment or observational study, one finds that strata quickly become sparse or empty as the number of relevant features increases. (Consider that country alone is the equivalent of eight binary features.) This presents us with a problem: do we ignore most of the features we have collected? Or can we use them in an interesting way?

Number of factors f	$l = 2$ levels	$l = 3$ levels
1	2	3
2	4	9
10	1,024	59,049
20	$\approx$ 1 million	$\approx$ 3.5 billion
30	$\approx$ 1 billion	$\approx$ 206 trillion
50	$\approx 1.1 \times 10^{15}$	$\approx 7.2 \times 10^{23}$
100	$\approx 1.3 \times 10^{30}$	$\approx 5.2 \times 10^{47}$

Table 8.11: Number ($s = l^f$) of individual covariate strata for f factors, each with l levels.

One elegant approach is a technique referred to as *matched pairs*. The idea is that we can first build a probability model for the propensity score, which can be trained on a large set of features. Classically, this was done using logistic regression, but more recently, with the advent of machine learning, one can use tree-based methods (e.g., random forest or XGBoost) or deep neural networks, two topics we will discuss in Part III. Once we have a model for the propensity score for all units in our observational study, we can then "match" units into pairs on the basis of propensity score, such that each pair consists of one control and one treatment unit.

8.5.3 Sensitivity Analysis: Implications of Hidden Bias

So far, we have discussed multiple methods for controlling for overt bias in observational studies, including the adjustment formula and matching. But how can we understand *hidden bias*? This will be our focus for the remainder of the section. In particular, we will break this discussion into two parts. The first part is sensitivity analysis, which seeks to quantify the effects of hidden bias as a function of the extent to which it might be present. The second part is to discuss ways in which we can infer whether or hidden bias is actually present.

A *sensitivity analysis* is any analysis that seeks to quantify the extent to which the conclusions of an observation study would differ as a function of the degree of hidden bias. For example, a study that is *sensitive* is one in which even a small amount of hidden bias would dramatically throw off the conclusions. On the other hand, a study that is insensitive or robust is one in which a small (or even medium) amount of hidden bias would not change the ultimate conclusions. In short, we seek to quantify how much hidden bias would have to be present before our conclusion would no longer be valid.

We will make use of a parameter $\gamma \geq 1$ to parameterize our sensitivity analyses for observational studies, such that

$$\frac{1}{\gamma} \leq \frac{\pi_{ij}(1 - \lambda_i)}{\lambda_i(1 - \pi_{ij})} \leq \gamma, \tag{8.20}$$

where $\lambda_i = \lambda(x_{ij})$ is the propensity score for the jth unit in the ith stratum; i.e., γ is an upper bound on the odds ratio of each unit's true treatment probability π_i compared with its propensity score λ_i, which is a function of the observed covariates. If there is no hidden bias, we would obtain $\gamma = 1$, which implies that $\pi_i = \lambda_i$; i.e., the probability of treatment assignment is determined solely based on the observed covariates.

One typically repeats the analysis using various values of γ, in each case, quantifying how the conclusion would differ from the nominal case of no hidden bias. Before proceeding, we first note the following, alternative formulation of hidden bias.

Proposition 8.5. *Equation* (8.20) *is equivalent to the relation*

$$\ln\left(\frac{\pi_{ij}}{1 - \pi_{ij}}\right) = \ln\left(\frac{\lambda_i}{1 - \lambda_i}\right) + h_{ij} \ln \gamma, \tag{8.21}$$

for hidden covariates $h_{ij} \in [-1, 1]$.

Note that Equation (8.21) compares the logits[4] of the true treatment probability π_i and the propensity score λ_i. Thus, the proposition states that under the model Equation (8.20), the log-odds of the treatment probability, for each individual unit, is equal to the log-odds of the propensity score plus an additive constant, that may vary unit-to-unit, but is bounded above by the log-sensitivity $\ln \gamma$. We leave the proof as an exercise for the reader.

The degree of hidden bias is traditionally defined a bit differently, by comparing the odds ratios between any two arbitrary units from any given cohort. To understand this distinction in more detail, consider the following.

---

[4] The logit function is defined in Equation (7.38).

Proposition 8.6. *Let π_{ij} and π_{ik} be the probability of treatment assignment for the jth and kth unit in the ith cohort, respectively. Assuming the upper and lower bounds provided by Equation (8.20) are actually realized within each stratum, these bounds are equivalent to the requirement*

$$\frac{1}{\gamma^2} \leq \frac{\pi_{ij}(1 - \pi_{ik})}{\pi_{ik}(1 - \pi_{ij})} \leq \gamma^2, \qquad (8.22)$$

such that the propensity score λ_i of the ith stratum satisfies

$$\left(\frac{\lambda_i}{1 - \lambda_i}\right)^2 = \frac{\pi_i^+}{(1 - \pi_i^+)} \frac{\pi_i^-}{(1 - \pi_i^-)}, \qquad (8.23)$$

where π_i^+ and π_i^- are the maximum and minimum true treatment assignment probabilities

$$\pi_i^+ = \max_{j=1,\ldots,n_i} \pi_{ij} \quad and \quad \pi_i^- = \min_{j=1,\ldots,n_i} \pi_{ij} \qquad (8.24)$$

for the ith cohort, respectively. We call Equation (8.23) the geometric odds-balancing property.

Note 8.7. The sensitivity parameter is traditionally (and outside the scope of this text) defined instead by $\Gamma = \gamma^2$ and Equation (8.22). We use the lower-case γ to disambiguate the two[5]. We prefer our formulation as we feel it is more natural to compare each unit's treatment probability with its propensity score. Moreover, Equation (8.21) is directly amenable for numerical simulations (e.g., bootstrap estimates).   ▷

Note that Equation (8.20) is stronger than Equation (8.22) by itself: the latter simply places a bound on the odds ratio between any two units within any stratum, but is silent regarding the distributional balance within the stratum. For example, if the treatment probabilities of all but one unit are identical, but there is a single outlier with a dramatically different treatment probability, the sensitivity parameter in Equation (8.22) would be quite large, despite there being virtually hidden bias. Conversely, consider Equation (8.20). Since everything is anchored relative to the propensity score, it is a stronger statement, as it states that the average treatment probability within each stratum (i.e., the propensity score) satisfy Equation (8.23), which states that *the odds ratio of the propensity score is equal to the geometric mean of the odds ratios of the maximum and minimum treatment probabilities within each stratum.* Momentarily, we will address whether balancing the hidden biases within each stratum in such a way is optimal, but for now, let us see why our definition of the sensitivity parameter implies such a balance.

---

[5] So if you read a study with $\Gamma = 100$, it should be equated with $\gamma = 10$.

Proof. Let us assume that the upper and lower bounds of Equation (8.20) are realized for each stratum. This is a reasonable and typical assumption for a sensitivity analysis, as we are explicitly trying to determine what would happen if we have the most extreme hidden bias allowed by our sensitivity parameter. Applying Equation (8.20) to the units with maximum π_i^+ and minimum π_i^- treatment probabilities yields

$$\frac{\pi_i^+(1-\lambda_i)}{\lambda_i(1-\pi_i^+)} = \gamma \quad \text{and} \quad \frac{1}{\gamma} = \frac{\pi_i^-(1-\lambda_i)}{\lambda_i(1-\pi_i^-)}, \tag{8.25}$$

respectively, where we have replaced the inequalities with equalities, as these bounds are actually realized. Taking the reciprocal of the latter equation and multiplying, we obtain

$$\frac{\pi_i^+(1-\pi_i^-)}{\pi_i^-(1-\pi_i^+)} = \gamma^2.$$

But, for any two units of a given stratum, we have

$$\frac{1}{\gamma^2} = \frac{\pi_i^-(1-\pi_i^+)}{\pi_i^+(1-\pi_i^-)} \leq \frac{\pi_{ij}(1-\pi_{ik})}{\pi_{ik}(1-\pi_{ij})} \leq \frac{\pi_i^+(1-\pi_i^-)}{\pi_i^-(1-\pi_i^+)} = \gamma^2.$$

Thus Equation (8.20) implies Equation (8.22).

Now, Equation (8.22) does not, by itself, imply Equation (8.20). For this, we require the geometric balancing property of Equation (8.23). But, alas, this relation can be obtained by multiplying together Equation (8.24) and rearranging. This yields the result. □

So, we are proposing to usurp or replace the traditional classic sensitivity parameter defined by Equation (8.22) with the one anchored on the propensity score (Equation (8.20)), which yields the geometric odds-balancing property given by Equation (8.23). Naturally, it is reasonable ask whether or not such a balancing is optimal. We answer that question in the affirmative as follows.

Proposition 8.7. *Suppose that the distribution of treatment probabilities within each stratum is extreme, in the sense that either $\pi_{ij} = \pi_i^-$ or $\pi_{ij} = \pi_i^+$, for all $j = 1,\ldots,n_i$, for a given stratum i. Let us parameterize this mixture using a parameter $\alpha_i \in (0,1)$, such that*

$$\lambda_i = \alpha_i\pi_i^+ + (1-\alpha_i)\pi_i^-; \tag{8.26}$$

i.e., α_i represents the fraction of units with the high probability π_i^+. Then the value of α_i that maximizes the amount of hidden bias within the ith stratum also results in a distribution for which the propensity score λ_i satisfies the geometric odds-balancing property of Equation (8.23).

Proof. For simplicity, and without loss of generality, let us drop the subscript i for the following. We may instead consider the argument existing within a single stratum, with overall propensity score λ, or for an unstratified population.

In order to formulate the amount of hidden bias, let us consider the hidden variable H, such that $\pi = \pi^+$ whenever $H = 1$ and $\pi = \pi^-$ otherwise. Our assumptions yield the 2×2 contingency table shown in Table 8.12.

	$T = 1$	$T = 0$	$\mathbb{P}(H = h)$
$H = 1$	$\alpha\pi^+$	$\alpha(1 - \pi^+)$	α
$H = 0$	$(1 - \alpha)\pi^-$	$(1 - \alpha)(1 - \pi^-)$	$(1 - \alpha)$
$\mathbb{P}(T = t)$	λ	$(1 - \lambda)$	1

Table 8.12: Contingency table for treatment T and hidden variable H, showing the joint distribution probabilities $\mathbb{P}(T, H)$.

Moreover, let us assume our response is linear in treatment and hidden covariate H, yielding

$$R = \rho + \Delta \cdot H + \delta \cdot T.$$

Now, since the variable H is hidden, we cannot control for it. Therefore, our estimated treatment effect is given by considering

$$\mathbb{E}[R|T = 1] = \sum_{h \in \mathcal{R}_H} \mathbb{E}[R|T = 1, H = h]\mathbb{P}(H = h|T = 1)$$

$$= (R + \Delta + \delta)\frac{\alpha\pi^+}{\lambda} + (R + \delta)\frac{(1 - \alpha)\pi^-}{\lambda}$$

$$\mathbb{E}[R|T = 0] = \sum_{h \in \mathcal{R}_H} \mathbb{E}[R|T = 0, H = h]\mathbb{P}(H = h|T = 0)$$

$$= (R + \Delta)\frac{\alpha(1 - \pi^+)}{1 - \lambda} + R\frac{(1 - \alpha)(1 - \pi^-)}{1 - \lambda}.$$

It follows that failing to control for the hidden factor H would result in the biased estimate for the treatment effect given by

$$\hat{\delta} = \mathbb{E}[R|T = 1] - \mathbb{E}[R|T = 0] = \delta + \Delta\left[\frac{\alpha\pi^+}{\lambda} - \frac{\alpha(1 - \pi^+)}{1 - \lambda}\right].$$

We may therefore define the contents of the square brackets as the *specific bias* β[6], given by

$$\beta = \frac{\alpha\pi^+}{\lambda} - \frac{\alpha(1 - \pi^+)}{1 - \lambda}.$$

Taking the derivative with respect to α, we obtain

---

[6] *specific* in the sense that we are normalizing by the size Δ of the bias.

$$\frac{\partial \beta}{\partial \alpha} = \left[\frac{\pi^+}{\lambda} - \frac{(1-\pi^+)}{(1-\lambda)}\right] - \left[\frac{\pi^+}{\lambda^2} + \frac{(1-\pi^+)}{(1-\lambda)^2}\right]\alpha\frac{\partial\lambda}{\partial\alpha}.$$

However, one can easily verify the identity

$$\alpha\frac{\partial\lambda}{\partial\alpha} = \lambda - \pi^-.$$

Setting the derivative $\beta'(\alpha)$ to zero and rearranging, we therefore obtain

$$\frac{\lambda\pi^+ - \pi^+(\lambda - \pi^-)}{\lambda^2} = \frac{(1-\lambda)(1-\pi^+) + (1-\pi^+)(\lambda - \pi^-)}{(1-\lambda)^2}$$

$$\frac{\pi^-\pi^+}{\lambda^2} = \frac{(1-\pi^-)(1-\pi^+)}{(1-\lambda)^2},$$

which is equivalent to Equation (8.23). To show that this critical point is the maximum, note that the bias may be written as

$$\beta = \frac{\alpha(\pi^+ - \lambda)}{\lambda(1-\lambda)},$$

which satisfies $\beta(\alpha) > 0$ for $\alpha \in (0,1)$ and $\beta(0) = 0$. This completes the proof. $\square$

For a given sensitivity analysis with parameter $\gamma > 1$, we begin by measuring the observed propensities λ_i. Then Equation (8.25) yields the maximum and minimum values π_i^+ and π_i^-. Finally, Equation (8.26) yields the mixture parameter α_i. Since we are only interested in reporting whether our observational conclusion would differ for the most extreme variation due to hidden bias consistent with a given sensitivity size γ, we need consider no other mixture. Proposition 8.7 guarantees that a mixture so constructed produces the maximal impact.

Distribution of Treatment Assignments

Under Equation (8.21), the distribution of treatment assignments given by Equation (8.18) is equivalent to

$$\ln\mathbb{P}(Z_{ij} = z_{ij}) = \sum_{i=1}^{s}\sum_{j=1}^{n_i}\left[z_{ij}\ln\left(\frac{\pi_{ij}}{1-\pi_{ij}}\right) + \ln(1-\pi_{ij})\right]$$

$$= \sum_{i=1}^{s}\sum_{j=1}^{n_i}\left[z_{ij}\ln\left(\frac{\lambda_i}{1-\lambda_i}\right) + z_{ij}h_{ij}\ln\gamma + \ln(1-\pi_{ij})\right]$$

Now, the final term in square brackets is a constant, as it does not depend on z_{ij}. Moreover, given the marginal totals $A_i = a_i$ of treated units within each stratum, the first term is also a constant, following similar logic as

Equation (8.19). We conclude that the marginal distribution of treatment assignments given the marginal total count of treated units is given by

$$\mathbb{P}(Z_{ij} = z_{ij}|A_i = a_i) \propto \prod_{i=1}^{s} \prod_{j=1}^{n_i} \exp\left(z_{ij}h_{ij} \ln \gamma\right).$$

To normalize, consider the set Ω_i of the $\binom{n_i}{a_i}$ possible ways to arrange a_i 1s among the n_i units in the ith stratum. This yields

$$\mathbb{P}(Z_{ij} = z_{ij}|A_i = a_i) = \prod_{i=1}^{s} \frac{\displaystyle\prod_{j=1}^{n_i} \exp\left(z_{ij}h_{ij} \ln \gamma\right)}{\displaystyle\sum_{w_{ij} \in \Omega_i} \prod_{j=1}^{n_i} \exp\left(w_{ij}h_{ij} \ln \gamma\right)}. \tag{8.27}$$

Thus, within each stratum, we weight our product of exponentials, reflecting the actual treatment assignments, against its sum over all possible treatment assignments that result in a_i treated units from the n_i total units.

For $\gamma > 1$, this distribution is no longer a constant, confirming that stratification on the observed covariates is not equivalent to a randomized experiment whenever hidden bias lurks. Stratification, however, did eliminate part of the uncertainty of the treatment assignment probabilities π_{ij} due to the propensity scores λ_i. Naturally, since γ and h_{ij} are unknown, our sensitivity analysis will assume the most extreme values of hidden covariate $h_{ij} \in [-1, 1]$ for several fixed values of γ.

Sensitivity Analysis for Matched Pairs

For the special case of matched pairs, we have $n_i = 2$ and $a_i = 1$, for all $i = 1, \ldots, s$. Each stratum (i.e., pair) therefore has only two possible treatment arrangements, parameterized by $z_{i1} = 0$ or $z_{i1} = 1$[7]. Equation (8.27) therefore simplifies as

$$\mathbb{P}(Z_{ij} = z_{ij}|A_i = 1) = \prod_{i=1}^{s} \frac{\exp(z_{i1}h_{i1} \ln \gamma)\exp(z_{i2}h_{i2} \ln \gamma)}{\exp(h_{i1} \ln \gamma) + \exp(h_{i2} \ln \gamma)}. \tag{8.28}$$

In particular, this implies that in the presence of hidden bias the matched pairs are no longer equally likely to receive treatment. Rather, we have

$$\pi_{i1} = \frac{\gamma^{h_{i1}}}{\gamma^{h_{i1}} + \gamma^{h_{i2}}} \quad \text{and} \quad \pi_{i2} = \frac{\gamma^{h_{i2}}}{\gamma^{h_{i1}} + \gamma^{h_{i2}}}. \tag{8.29}$$

Note, in particular, that whenever $h_{i1} = h_{i2}$, the probability of treatment assignment is again 50-50. Now, consider any sign-score statistic (Equation (4.26)), which, for matched pairs, takes the form

---

[7] We may obtain $z_{i2} = 1 - z_{i1}$.

$$T = \sum_{i=1}^{s} d_i \left(c_{i1} Z_{i1} + c_{i2} Z_{i2} \right),$$

where c_{ij} is binary. For example, in McNemar's test, $d_i = 1$ and c_{ij} is the binary response, so that T represents the number of $+1$ responses among the treated units. In Wilcoxon's signed-rank test, d_i represents the rank of the absolute difference $|R_{i1} - R_{i2}|$ and c_{ij} is the indicator that the jth response in the ith pair is greater than its paired response. The Wilcoxon signed-rank test statistic therefore represents the sum of the ranked absolute differences over all treatments that had a greater response than their matched control.

Now, the ith term of the sign-score statistic T will equal d_i with probability

$$p_i = c_{i1} \pi_{i1} + c_{i2} \pi_{i2},$$

and 0 otherwise, where π_{i1} and π_{i2} are given by Equation (8.29). The ith matched pair is said to be *concordant* if $c_{i1} = c_{i2}$ and *discordant* otherwise. Notice that concordant pairs contribute a fixed quantity (either d_i, if $p_i = 1$, or 0 otherwise) to the test statistic T. We therefore only need to understand the extent to which hidden bias affects T through the various discordant pairs.

Since the actual treatment probabilities (Equation (8.29)) are unknown, then so too is the distribution of T under the null hypothesis of no treatment effect. Nevertheless, for a fixed sensitivity level, one can place upper and lower bounds on this probability distribution. In particular, define the probabilities p_i^+ and p_i^- by setting

$$p_i^+ = \frac{\gamma^2}{1 + \gamma^2} \qquad \text{and} \qquad p_i^- = \frac{1}{1 + \gamma^2}$$

for discordant pairs ($c_{i1} \neq c_{i2}$) and by setting both p_i^+ and p_i^- to c_{i1} for concordant pairs. Since $-1 \leq h_{ij} \leq 1$, it follows that $p_i^- \leq p_i \leq p_i^+$, for each stratum. This bounds our unknown distribution for T by two known distributions T^- and T^+, obtained by summing d_i for the ith stratum with probability p_i^+ and p_i^-, respectively, and zero otherwise. Under the hypothesis of no treatment effect and fixed sensitivity parameter γ, we therefore obtain the bounds on the significance level

$$\mathbb{P}(T^+ \geq \alpha) \geq \mathbb{P}(T \geq \alpha) \geq \mathbb{P}(T^- \geq \alpha).$$

For additional details and examples, as well as generalizations beyond matched pairs and more general discussion, see Rosenbaum [2002]. For a detailed discussion on how one might *design* observational studies to maximize the possibility of detecting hidden effects, see Rosenbaum [2020].

8.5.4 Methods for Uncovering Hidden Bias

Sensitivity analysis is a mathematical way to quantify how the conclusions of an observational study might change as a function of the degree of hidden bias: literally, it answers how sensitive our conclusions are to hidden bias. An observational study is sensitive if even a small amount of hidden bias would cast the conclusions into doubt. On the other hand, a study is robust if only a large degree of hidden bias would change the results of the study. A sensitivity analysis does not, however, address whether or not there actually is any hidden bias. We will refer the reader to Rosenbaum [2002] for a discussion of the three primary methods here, namely: known effects, case-referent studies, and multiple control groups. See also Rosenbaum [2019] for a more intuitive, less mathematical discussion of the concepts.

Problems

8.1. In Example 8.9, determine the causal effect of *not running the sprinkler* on the wetness of the pavement; i.e., determine

$$\mathbb{P}(W = 1 | \text{do}(S = 0)).$$

What is the difference $\mathbb{P}(W = 1 | \text{do}(S = 1)) - \mathbb{P}(W = 1 | \text{do}(S = 0))$?

8.2. How should one define $\mathbb{E}[Y | \text{do}(X = x)]$?

8.3. Show that Equation (8.4) implies that

$$\mathbb{E}[Y | \text{do}(X = x)] = \sum_{z \in \mathcal{R}_Z} \mathbb{E}[Y | X = x, z = z] \mathbb{P}(Z = z),$$

assuming you arrived at the correct definition in Exercise 8.2.

8.4. Prove Proposition 8.5.

8.5. In the proof of Proposition 8.7, show that by controlling for the hidden variable H, one would instead obtain the unbiased treatment effect δ.

Time to Get Serious

Thus far, we have discussed many topics of statistics and statistical modeling. The element of time, however, has yet to play a role in our discussion, aside from our brief introduction to stochastic processes in Section 5.4. When we consider IID draws from a random variable, we imagine an infinite urn with an infinite number of marbles that somehow remains unchanged no matter how many (or for how long) we draw. In reality, however, many statistical processes play our over time. We devote the following pages to an introduction to such processes. Our favorite references are Box, *et al.* [2016], Shumway and Stoffer [1999], and Wei [2019]. An additional detailed and comprehensive classic introduction is Hamilton [1994]. Finally, for a data-science introduction to time series, see Nielson [2019].

9.1 Basic Concepts

In this section, we first lay out our basic definitions (time series, basic statistics, and stationarity), and then introduce white-noise time series and random walks (Section 9.1.4). The anxious reader, averse to abstraction, may at times glance ahead to Section 9.2 to gander examples.

9.1.1 Time Series

We previously introduced a stochastic process as an index collection of random variables defined over a common probability space (see Definition 5.11). A time series is a special case of such processes, which has grown into a field in its own right. We formally define time series as follows.

Definition 9.1. *A* time series *is a discrete stochastic process indexed by the integers $\mathbb{Z}$, or a subset thereof; i.e., a time series is a collection $\{X_t\}_{t \in \mathbb{Z}}$ of random variables defined over a common probability space $(\Omega, \mathcal{E}, \mathbb{P})$.*

Moreover, we shall normally consider the indexing set $\mathbb{Z}$ to represent a discretization of time.

Recall that the observed values of any stochastic process are referred to as a *realization* or *sample path* of that process. The phrase *time series* is often used interchangeably to refer to either the process or its realization, as the meaning is typically clear in context.

9.1.2 Common Time Series Statistics

We next introduce three basic time series statistics: mean, autocovariance, and autocorrelation.

Mean

The most basic statistic of a time series $\{X_t\}$ is its mean,

$$\mu_t = \mathbb{E}[X_t].$$

Since each X_t is a separate random variable, the mean of a time series will, in general, vary over time.

Autocovariance

Since a time series is a sequence of random variables, we are naturally interested in how those random variables are related to each other. The most basic measure of this is autocovariance, which describes the variance of a time series with itself.

Definition 9.2. *Given a time series $\{X_t\}$, its autocovariance function is defined by the second moment*

$$\gamma_X(s,t) = \text{COV}(X_s, X_t) = \mathbb{E}\left[(X_s - \mu_s)(X_t - \mu_t)\right], \qquad (9.1)$$

where $\mu_t = \mathbb{E}[X_t]$ is the mean at time t. When there is no danger of ambiguity, we will drop the subscript and refer to the autocovariance simply as $\gamma(s,t)$.

Note that for the special case $s = t$, we have the variance of the time series

$$\gamma(t,t) = \mathbb{V}(X_t). \qquad (9.2)$$

Autocorrelation

We are often interested in normalizing the autocovariance to the interval $[-1, 1]$, similar to the normalization of the covariance of two random variables into their correlation. We thus arrive at the following definition.

Definition 9.3. *Given a time series* $\{X_t\}$, *its* autocorrelation function (ACF) *is defined by*

$$\rho_X(s, t) = \frac{\gamma(s, t)}{\sqrt{\gamma(s, s)\gamma(t, t)}}, \tag{9.3}$$

where $\gamma(s, t)$ *is the autocovariance of the time series. When there is no danger of ambiguity, we will drop the subscript and refer to the* ACF *simply as* $\rho(s, t)$.

Now, the Cauchy–Schwarz inequality can be used to show that

$$\gamma(s, t)^2 \leq \gamma(s, s)\gamma(t, t),$$

which, in turn, implies that the autocorrelation of a time series is bounded between $-1 \leq \rho(s, t) \leq 1$.

Note that the autocorrelation measures the *linear* predictability of the time series' value X_t based on X_s. When there is a perfect linear relation, $X_t = \beta_0 + \beta_1 X_s$, the correlation will be perfect $\rho(s, t) = \pm 1$, with the sign depending on the sign of β_1. Recall that zero correlation $\gamma(s, t) = 0$ implies only that there is no *linear* dependence between X_s and X_t, which does not rule out the possibility of a *nonlinear* dependence between the two random variables.

9.1.3 Stationarity

An important concept is that of stationarity. Stationarity intuitively means that the process is time invariant, or stationary. There are two flavors of stationarity: strong and weak. Strong stationarity can be thought of as more theoretical, whereas weak stationarity is more practical.

Definition 9.4. *A time series* $\{X_t\}$ *is said to be* strongly stationary (or strictly stationary) *if the joint distribution of any subset*

$$\{X_{t_1}, \ldots, X_{t_k}\}$$

is invariant under any temporal shift $t_i \to t_i + h$, *for* $h \in \mathbb{Z}$; *i.e., if the joint cumulative distribution functions are invariant:*

$$\mathbb{P}(X_{t_1} \leq c_1, \ldots, X_{t_k} \leq c_k) = \mathbb{P}(X_{t_1+h} \leq c_1, \ldots, X_{t_k+h} \leq c_k),$$

for any $h \in \mathbb{Z}$.

See... impractical.

Definition 9.5. *A time series $\{X_t\}$ is said to be* weakly stationary *(or* stationary*) if*

1. *its mean is invariant with time; i.e., $\mu_t = \mu_s$, for any $t, s \in \mathbb{Z}$;*
2. *its variance is finite; i.e., $\mathbb{V}(X_t) < \infty$ for all $t \in \mathbb{Z}$; and*
3. *its autocovariance function $\gamma(s, t)$ depends on s and t only through their absolute difference $|s - t|$.*

See... practical. We will therefore refer to weakly stationary time series simply as *stationary time series*, as weak stationarity is the flavor commonly used in practice. Whenever we require strict stationarity, we will explicitly call it as such.

We can simplify our notation for autocovariance and autocorrelation for stationary time series as follows.

Definition 9.6. *The* autocovariance function *for a stationary time series is defined as*

$$\gamma(h) = \gamma(0, h), \tag{9.4}$$

where $\gamma(s, t)$ is the autocovariance function defined in Definition 9.2. Similarly, the autocorrelation function ACF *of a stationary time series is defined as*

$$\rho(h) = \frac{\gamma(h)}{\gamma(0)}, \tag{9.5}$$

where $\gamma(h)$ is the autocovariance function defined by Equation (9.4).

These definitions are well defined since, for a stationary time series,

$$\gamma(t, t + h) = \gamma(0, h),$$

since the autocovariance depends on its independent variables only through their difference. Similarly, the autocorrelation for a stationary time series simplifies as

$$\rho(t, t + h) = \frac{\gamma(t, t + h)}{\sqrt{\gamma(t, t)\gamma(t + h, t + h)}} = \frac{\gamma(0, h)}{\gamma(0, 0)}.$$

Equation (9.5) follows.

9.1.4 Seedling Examples

We next introduce two cornerstone examples of time series: white noise and the random walk. We will start building more sophisticated examples in Section 9.2, when we introduce moving averages and autoregressive processes.

White Noise

White noise is a basic building block of more complex time series, and it derives its name from white light, which is a superposition of oscillations over the entire frequency spectrum.

Definition 9.7. *A stationary time series $\{W_t\}$ is referred to as* white noise *if its mean is zero*

$$\mathbb{E}[W_t] = 0$$

and if it is uncorrelated; i.e., if its autocovariance is given by

$$\gamma(t) = \begin{cases} \sigma_w^2 & \textit{if } t = 0 \\ 0 & \textit{otherwise} \end{cases},$$

for some constant variance σ_w^2. We denote such a process by $W_t \sim$ WN$(0, \sigma_w^2)$.

If, in addition, the random variables $\{W_t\}$ are independent and identically distributed, we will refer to the time series as white independent noise *or* IID noise.

When the random variables $\{W_t\}$ are independent and identically distributed normal random variables, $W_t \sim$ N$(0, \sigma^2)$, we refer to the time series as Gaussian white noise.

Example 9.1. In this example, we use a generator function that constructs iterators that output values from a Gaussian white noise sequence. Though this is unnecessary for such a simple example, the simplicity allows us to highlight usage of such function prior to the more complex generators we will construct in the next section.

In Code Block 9.1, we define two functions, `whiteNoise` and `tsPlot`, that return a white-noise generator and that, given a time-series generator as input, produce a set of time-series graphs. We rely on the built-in `statsmodels.tsa.stattools` package to provide a numerical estimator for both ACF and PACF, the latter of which we discuss in Section 9.1.6. The output graphs are shown in Figure 9.1.

We note that the sample ACF (and PACF) both have errors for $h > 0$, as the true values of both of these functions is zero for $h > 0$. The time series itself is comprised of independent and identically distributed draws from a Gaussian distribution. ▷

Random Walk

We next present an alternate version of the random walk from Example 5.15.

```
1   from statsmodels.tsa.stattools import acf, pacf
2
3   # Generator function
4   def whiteNoise(sigma_w=1):
5       while True:
6           yield np.random.normal(scale=sigma_w)
7
8   # time-series plot function
9   def tsPlot(Z, n=500, lag=40):
10      z = [next(Z) for i in range(n)]
11      rho = acf(z, nlags=lag, fft=True)
12      phihh = pacf(z, nlags=lag)
13
14      plt.figure(figsize=(8, 9/2))
15      plt.subplot(2,1,1)
16      plt.plot(z)
17
18      plt.subplot(2,2,3)
19      plt.bar(np.arange(lag+1), rho)
20
21      plt.subplot(2,2,4)
22      plt.bar(np.arange(lag+1), phihh)
23
24  # code
25  W = whiteNoise()
26  tsPlot(W)
```

Code Block 9.1: Generator function for Gaussian white noise

Definition 9.8. *Let $\{W_t\}$ represent a white-noise process. Then a* random walk with drift *is a time series $\{X_t\}$ defined recursively by the relation*

$$X_t = X_{t-1} + \delta + W_t, \tag{9.6}$$

for a parameter $\delta \in \mathbb{R}$, referred to as the drift.
When $\delta = 0$, we refer to the time series $\{X_t\}$ simply as a random walk.

Note that the drift δ represents a secular term that builds up over time. The random walk can be integrated as follows.

Proposition 9.1. *Given the initial condition $X_0 = 0$, the random walk defined by Equation (9.6) is equivalent to*

$$X_t = \delta t + \sum_{s=1}^{t} W_s. \tag{9.7}$$

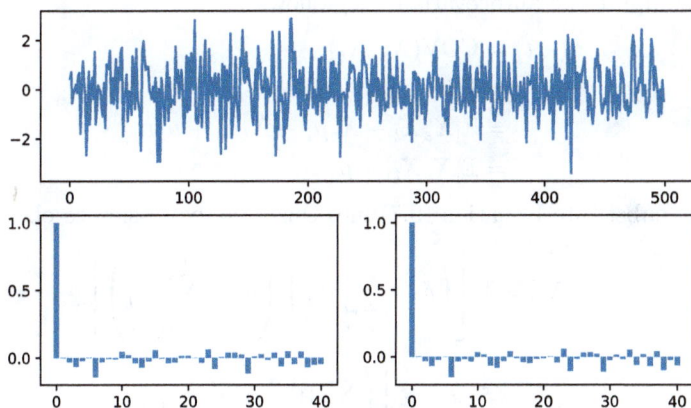

Fig. 9.1: Gaussian white noise (top); plotted with sample ACF (bottom left) and PACF (bottom right).

Proof. This follows from mathematical induction: For the basis step, note that our formula yields $X_0 = 0$. For the induction step, assume that Equation (9.7) is true for some $t \geq 0$. Then by applying Equation (9.6) to obtain X_{t+1}, we obtain

$$X_{t+1} = X_t + \delta + W_{t+1}$$
$$= \left(\delta t + \sum_{s=1}^{t} W_s \right) + \delta + W_{t+1}$$
$$= \delta(t+1) + \sum_{s=1}^{t+1} W_s,$$

which is equivalent to Equation (9.7) for $t + 1$. The result follows by induction. □

Proposition 9.2. *The mean and autocovariance of a random walk with drift δ and variance σ^2 are given by*

$$\mu_t = \delta t \tag{9.8}$$
$$\gamma(s, t) = \min(s, t)\sigma^2, \tag{9.9}$$

respectively.

Proof. Equation (9.8) follows directly from Equation (9.7), since

$$\mathbb{E}\left[\delta t + \sum_{s=1}^{t} W_s \right] = \delta t,$$

which follows since the mean of any white noise process is zero, by definition.
We compute the autocovariance as follows

$$
\begin{aligned}
\gamma(s,t) &= \mathrm{COV}(X_s, X_t) \\
&= \mathbb{E}[(X_s - \mu_s)(X_t - \mu_t)] \\
&= \mathbb{E}[X_s X_t - X_s \mu_t - X_t \mu_s + \mu_s \mu_t] \\
&= \mathbb{E}[X_s X_t] - \mu_s \mu_t.
\end{aligned}
$$

Next, we substitute Equation (9.7) into this final expression to obtain

$$
\begin{aligned}
\mathbb{E}[X_s X_t] &= \mathbb{E}\left[\left(\delta s + \sum_{i=1}^{s} W_i \right) \left(\delta t + \sum_{j=1}^{t} W_j \right) \right] \\
&= \delta^2 st + \mathbb{E}\left[\left(\sum_{i=1}^{s} W_i \right) \left(\sum_{j=1}^{t} W_j \right) \right] \\
&= \delta^2 st + \mathbb{E}\left[\sum_{i=1}^{s} \sum_{j=1}^{t} W_i W_j \right].
\end{aligned}
$$

But $\mathbb{E}[W_i W_j] = \sigma^2 \mathbb{I}[i = j]$. And since $\mu_s \mu_t = \delta^2 st$, the result follows.   □

Not only is the random walk nonstationary (as the mean clearly depends on time), but its variance grows linearly with time as well, since

$$
\mathbb{V}(X_t) = \sigma^2 t, \tag{9.10}
$$

from Equation (9.9). Note that this expression is independent of the drift.

Example 9.2. A simple generator function for the random walk is given in Code Block 9.2, with an initial condition $X_0 = 0$. An example output is shown in Figure 9.2.

```
1  def randomWalk(delta=0, sigma_w=1):
2      X = 0
3      while True:
4          X = X + delta + np.random.normal(scale=sigma_w)
5          yield X
```

Code Block 9.2: Generator function for the random walk

Despite the driftless nature of the random walk of Figure 9.2, the time series has wandered quite a bit away from its expected value (the t-axis). This is a powerful visual reminder of the linear growth in the walk's variance, as given by Equation (9.10). We also note that the autocorrelation only gradually diminishes with time, whereas the partial autocorrelation, which we will discuss shortly, drops off after lag 1.   ▷

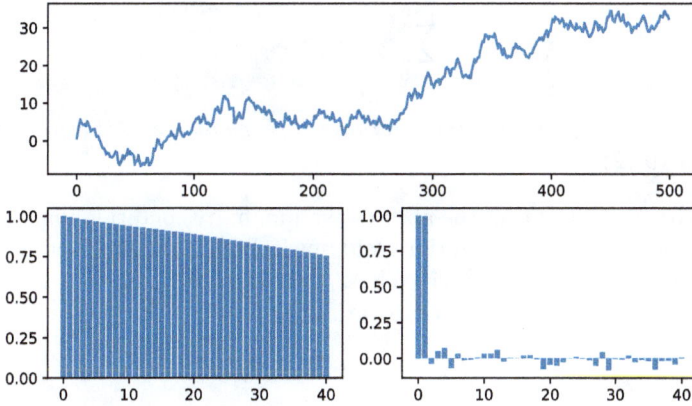

Fig. 9.2: Random walk with zero drift ($\delta = 0$) (top); plotted with sample ACF (bottom left) and PACF (bottom right).

Linear Processes and Invertibility

We may define time series as arbitrary linear combinations of white noise processes as follows.

Definition 9.9. *A mean-zero linear process is a time series defined as a linear combination of white noise; i.e., it is a time series that can be expressed by*

$$Z_t = \sum_{j=-\infty}^{\infty} \psi_j W_{t-j}, \qquad where \qquad \sum_{j=-\infty}^{\infty} |\psi_j| < \infty, \qquad (9.11)$$

for a white noise process $W_t \sim \mathrm{WN}(0, \sigma_w^2)$ *and coefficients* $\{\psi_j\}$. *A linear process with nonzero mean can be established by setting* $X_t = Z_t + \mu$, *for* $\mu \neq 0$.

A causal linear process *is a linear process with* $\psi_j = 0$ *for* $j < 0$; *i.e.*,

$$Z_t = \sum_{j=0}^{\infty} \psi_j W_{t-j}, \qquad (9.12)$$

where we may take $\psi_0 = 1$.

Note that the values for a general linear process are determined by the past ($j > 0$), the present ($j = 0$), and the future ($j < 0$). For this reason, we call such a process *causal* if its values are *not* influenced by the future; i.e., if $\psi_j = 0$ for $j < 0$. We will only consider causal linear processes in this text, as they are real-world rooted.

Proposition 9.3. *The absolute summability condition*

$$\sum_{j=0}^{\infty} |\psi_j| < \infty, \tag{9.13}$$

is a sufficient condition to prove stationarity for any (causal) linear process Equation (9.12).

Proof. We must show that the absolute summability condition implies each of the three stationarity conditions laid out in Definition 9.5. First, stationarity of the mean is obvious, since $\mathbb{E}[W_t] = 0$, for all $t \in \mathbb{Z}$.

Next, let us consider the autocovariance,

$$\gamma(t, t - h) = \mathbb{E}[Z_t Z_{t-h}].$$

We first note that

$$Z_{t-h} = \sum_{i=0}^{\infty} \psi_i W_{t-h-i} = \sum_{j=h}^{\infty} \psi_{j-h} W_{t-j},$$

which follows by reindexing the summation using $j = i + h$. It follows that

$$Z_t Z_{t-h} = \left(\sum_{i=0}^{\infty} \psi_i W_{t-i} \right) \left(\sum_{j=h}^{\infty} \psi_{j-h} W_{t-j} \right)$$

$$= \sum_{i=0}^{\infty} \sum_{j=h}^{\infty} \psi_i \psi_{j-h} W_{t-i} W_{t-j}$$

Noting that $\mathbb{E}[W_t W_s] = 0$ whenever $t \neq s$, we next take the expectation to uncover

$$\gamma(t, t - h) = \mathbb{E} \left[\sum_{j=h}^{\infty} \psi_j \psi_{j-h} W_{t-j}^2 \right] = \sigma_w^2 \sum_{j=h}^{\infty} \psi_j \psi_{j-h}, \tag{9.14}$$

which, of course is independent of t. This proves the third condition of stationarity, that the autocovariance is time independent.

Finally, we may express the variance from Equation (9.14) as

$$\mathbb{V}(Z_t) = \gamma(0, 0) = \sigma_w^2 \sum_{j=0}^{\infty} = \psi_j^2,$$

which converges since the coefficients ψ_j are absolute summable. This completes the proof. □

Corollary 9.1. *The convergence of the complex power series*

$$\psi(z) = \sum_{j=0}^{\infty} \psi_j z^j$$

on the unit disk $|z| \leq 1$ is a sufficient condition for the stationarity of the linear process Equation (9.12).

Proof. This follows since the complex power series converges on the unit disk $|z| \leq 1$ if and only if the series of coefficients $\psi(1)$ is absolutely convergent. For details regarding the theory of complex variables, see, for example, Brown and Churchill [2013] or Conway [1978]. $\square$

We also make a distinction whenever the white noise process can be written as a linear combination of the time series, as follows.

Definition 9.10. *A mean-zero linear process $\{Z_t\}$ is said to be* invertible *if it can be expressed as*

$$W_t = \sum_{j=0}^{\infty} \pi_j Z_{t-j}, \tag{9.15}$$

where $\sum_{j=0}^{\infty} |\pi_j| < \infty$, for white noise $W_t \sim \mathrm{WN}(0, \sigma_w^2)$.

Corollary 9.1 has a direct analog for invertible time series: convergence of the complex power series

$$\pi(z) = \sum_{j=0}^{\infty} \pi_j z^j$$

on the unit disk $|z| \leq 1$ implies absolute convergence of the coefficients $\sum_{j=0}^{\infty} |\pi_j| < \infty$.

9.1.5 Autoregression

We are often interested in the best linear estimate for a particular value in a time series based on its history.

Definition 9.11. *Given a stationary time series $\{X_t\}$, we define the* backward *and* forward autoregression *of length h and lag k as the linear regressions*

$$\hat{X}_t^{\langle h]k} = \alpha_0 + \alpha_1 X_{t-k} + \cdots + \alpha_h X_{t-k-h+1} \tag{9.16}$$

$$\hat{X}_t^{k[h\rangle} = \alpha_0 + \alpha_1 X_{t+k} + \cdots + \alpha_h X_{t+k+h-1}, \tag{9.17}$$

respectively, for $k \geq 1$. When no lag is specified, the lag is taken as one, and we use an alternate double-subscript notation for the regression coefficients; i.e.,

$$\hat{X}_t^{\langle h]} = \phi_{h0} + \phi_{h1} X_{t-1} + \cdots + \phi_{hh} X_{t-h} \tag{9.18}$$

$$\hat{X}_t^{[h\rangle} = \phi_{h0} + \phi_{h1} X_{t+1} + \cdots + \phi_{hh} X_{t+h}. \tag{9.19}$$

The regression coefficients of Equations (9.16) and (9.17) are equal due to stationarity; similarly for Equations (9.18) and (9.19). Here, ϕ_{hi} is the ith regression coefficient when regressing on the h lagging (or leading) values.

Note 9.1. The subscript in Equations (9.16)–(9.19) represents the time at which the prediction is made; e.g., $\hat{X}_t^{\langle h]k}$ is a prediction for the target variable X_t. The ordering of the superscript notation visually represents the timeline, as anchored at time t:

$$\underbrace{\langle \quad h \quad]}_{\text{backward window}} \underbrace{\quad k \quad}_{\text{backward lag}} \underbrace{X_t}_{\text{anchor}} \underbrace{\quad k \quad}_{\text{forward lead}} \underbrace{[\quad h \quad \rangle}_{\text{forward window}} .$$

The backward or forward window lengths are encapsulated within $\langle \cdot]$ or $[\cdot \rangle$, respectively, with the lag sitting on the square-bracket side of the window, closest to the anchor point.   ▷

   Without loss of generality, we will consider mean-zero stationary time series. For a non-mean-zero time series $\{X_t\}$, we can always obtain a mean-zero time series by subtracting out the mean

$$Z_t = X_t - \mu.$$

Proposition 9.4. *Let $\{Z_t\}$ be a mean-zero stationary time series. Then the regression coefficients for $\hat{Z}_t^{\langle h]}$ and $\hat{Z}_t^{[h\rangle}$ are the solutions to the linear system*

$$\begin{bmatrix} 1 & \rho_1 & \rho_2 & \cdots & \rho_{h-2} & \rho_{h-1} \\ \rho_1 & 1 & \rho_1 & \cdots & \rho_{h-3} & \rho_{h-2} \\ \vdots & \vdots & \vdots & \ddots & \vdots & \vdots \\ \rho_{h-1} & \rho_{h-2} & \rho_{h-3} & \cdots & \rho_1 & 1 \end{bmatrix} \cdot \begin{bmatrix} \phi_{h1} \\ \phi_{h2} \\ \vdots \\ \phi_{hh} \end{bmatrix} = \begin{bmatrix} \rho_1 \\ \rho_2 \\ \vdots \\ \rho_h \end{bmatrix}, \tag{9.20}$$

where $\rho_1, \ldots, \rho_h$ are the values of the autocorrelation function.

Proof. We will prove the result for Equation (9.19). A similar calculation yields the identical result for Equation (9.18).

   From Equation (7.5), we have

$$\phi_{h1}\mathbb{E}[Z_{t+1}Z_{t+k}] + \phi_{h2}\mathbb{E}[Z_{t+2}Z_{t+k}] + \cdots + \phi_{hh}\mathbb{E}[Z_{t+h}Z_{t+k}] = \mathbb{E}[Z_t Z_{t+k}], \tag{9.21}$$

for $k = 1, \ldots, h$. However, since the time series has zero mean, $\mathbb{E}[Z_t] = 0$, the expected products are equivalent to their covariances:

$$\mathbb{E}[Z_t Z_s] = \mathrm{COV}(Z_t, Z_s) = \gamma_{|t-s|},$$

for any $t, s \in \mathbb{Z}$, where the last equality holds due to the stationarity of the time series. Therefore, the regression coefficients satisfy the equations

$$\phi_{h1}\gamma_{k-1} + \phi_{h2}\gamma_{k-2} + \cdots + \phi_{hh}\gamma_{h-k} = \gamma_k,$$

for $k = 1, \ldots, h$. Dividing these equations by γ_0 yields the system Equation (9.20).

A similar calculation shows that the regression coefficients for $\hat{Z}_t^{\langle h]}$ satisfy the exact same set of equations. □

As one might expect, the autocorrelations together with the general form of Equation (9.20) allow us to generate the regression coefficients recursively. The following linear algebra definitions, though slightly off the beaten path, will prove useful.

Definition 9.12. *The n-dimensional identity matrix I_n and exchange matrix J_n are defined componentwise by the relations*

$$[I_n]_{ij} = \mathbb{I}[i = j]$$
$$[J_n]_{ij} = \mathbb{I}[i + j = n + 1],$$

for $1 \le i, j \le n$; i.e., the matrix J_n is the identity matrix with columns in reversed order. For a vector $x \in \mathbb{R}^n$, the quantity $J_n x$ is obtained by reversing the order of the elements of x; $[J_n x]_i = x_{n+1-i}$.

Given an $n \times n$ matrix A, its counterdiagonal transpose *is the matrix A_T defined by the relation*

$$A_T = J_n A^T J_n,$$

which is formed by reflecting A about its counterdiagonal *(the northeast to southwest diagonal). (A^T is the regular matrix transpose.) The matrix A is* symmetric *if $A^T = A$ and* persymmetric *if $A_T = A$. A matrix that is both symmetric and persymmetric is* bisymmetric.

Note that the coefficient matrix in Equation (9.20) is bisymmetric. Also, note that since

$$\det(J_n) = (-1)^{n(n-1)/2},$$

it follows from properties of the determinant that $\det(A_T) = \det(A)$; i.e., the matrix determinant is invariant under the operation of counterdiagonal transpose. We will also make use of the following row and column substitution operations.

Definition 9.13. *Let $A \in \mathbb{R}^{n \times n}$ and let $x \in \mathbb{R}^n$. We define the* row-substitution (row-sub) *and* column-substitution (col-sub) *operations as*

$$[\mathscr{S}_s(A,x)]_{ij} = \begin{cases} x_j & \text{if } i = s \\ A_{ij} & \text{otherwise} \end{cases}$$

$$[\mathscr{S}^s(A,x)]_{ij} = \begin{cases} x_i & \text{if } j = s \\ A_{ij} & \text{otherwise} \end{cases}$$

respectively; i.e., $\mathscr{S}_i(A,x)$ is obtained by replacing the ith row of matrix A with vector x and $\mathscr{S}^j(A,x)$ is obtained by replacing the jth column of A with vector x. Note we use a subscript or superscript to represent the row or column operation, respectively.

Notice that for any bisymmetric matrix A, we have

$$\mathscr{S}^j(A,x) = \mathscr{S}_{n+1-j}(A, J_n \cdot x)_T,$$

and therefore

$$\det\left[\mathscr{S}^j(A,x)\right] = \det\left[\mathscr{S}_{n+1-j}(A, J_n \cdot x)\right]. \tag{9.22}$$

We next derive a recursion relation that allows us to generate the autoregression coefficients for a stationary time series. The following result is due to Durbin [1960].

Theorem 9.1. *Let $\{Z_t\}$ be a mean-zero stationary time series. The autoregression coefficients ϕ_{hi}, for $h \in \mathbb{Z}^+$, $i = 1, \ldots, h$, can be generated using the initial condition*

$$\phi_{11} = \rho_1 \tag{9.23}$$

along with the following recursion relations

$$\phi_{h+1,h+1} = \frac{\rho_{h+1} - \displaystyle\sum_{j=1}^{h} \phi_{hj}\rho_{h+1-j}}{1 - \displaystyle\sum_{j=1}^{h} \phi_{hj}\rho_j} \tag{9.24}$$

$$\phi_{h+1,j} = \phi_{hj} - \phi_{h+1,h+1}\phi_{h,h+1-j}, \tag{9.25}$$

for $j = 1, \ldots, h$.

Proof. Equation (9.23) follows immediately from Equation (9.20) for the case $h = 1$.

To proceed, we define the $h \times h$ bisymmetric matrix R_h and the h-dimensional vectors ϕ_h and r_h as

$$[R_h]_{ij} = \rho_{i-j}, \qquad [\phi_h]_i = \phi_{hi}, \qquad \text{and} \qquad [r_h]_i = \rho_i,$$

respectively, so that Equation (9.20) is equivalent to the matrix equation

$$R_h \cdot \phi_h = r_h.$$

By Cramer's rule, from linear algebra, the coefficient ϕ_{hj} is given by the ratio of determinants

$$\phi_{hj} = \frac{\det(\mathscr{S}^j(R_h, r_h))}{\det(R_h)}.$$

Next, note that R_h is the $n \times n$ submatrix of R_{h+1} comprised of the latter matrix's first n columns and n rows:

$$R_{h+1} = \begin{bmatrix} 1 & \rho_1 & \rho_2 & \cdots & \rho_{h-2} & \rho_{h-1} & \rho_h \\ \rho_1 & 1 & \rho_1 & \cdots & \rho_{h-3} & \rho_{h-2} & \rho_{h-1} \\ \vdots & \vdots & \vdots & \ddots & \vdots & \vdots & \vdots \\ \rho_{h-1} & \rho_{h-2} & \rho_{h-3} & \cdots & \rho_1 & 1 & \rho_1 \\ \rho_h & \rho_{h-1} & \rho_{h-2} & \cdots & \rho_2 & \rho_1 & 1 \end{bmatrix}$$

To compute the determinant of this matrix, we perform cofactor expansion along the $(h+1)th$ column, starting at the bottom, thereby obtaining

$$\det(R_{h+1}) = \det(R_h) - \sum_{j=1}^{h} \rho_j \det(\mathscr{S}_{h+1-j}(R_h, J_h \cdot r_h)).$$

The index j in the summation coincides with the index of ρ_j in the final column of matrix R_{h+1}. Note that the first h columns of the $(h+1)$th row of R_{h+1} can be expressed as $J_h \cdot r_h$, and when eliminating the $(h+1-j)$th row of R_{h+1}, we can move $J_h \cdot r_h$ to replace it with precisely $(j-1)$ row swaps, such that the terms in the summation no longer alternate signs. However, from Equation (9.22), we see this is equivalent to

$$\det(R_{h+1}) = \det(R_h) - \sum_{j=1}^{h} \rho_j \det(\mathscr{S}^j(R_h, r_h)).$$

Similarly, by replacing the coefficients (elements of the $(h+1)$th column) with r_{h+1}, we obtain

$$\det(\mathscr{S}^{h+1}(R_{h+1}, r_{h+1})) = \rho_{h+1} \det(R_h) - \sum_{j=1}^{h} \rho_{h+1-j} \det(\mathscr{S}^j(R_h, r_h)).$$

But, by Cramer's rule, we can replace

$$\det(\mathscr{S}^j(R_h, r_h)) = \phi_{hj} \det(R_h),$$

so that the factor $\det(R_h)$ cancels out when computing the ratio

$$\phi_{h+1,h+1} = \frac{\det(\mathscr{S}^{h+1}(R_{h+1}, r_{h+1}))}{\det(R_{h+1})}.$$

Equation (9.24) follows.

We leave the proof for Equation (9.25) to the motivated reader.   □

9.1.6 Partial Autocorrelation Function

Another useful statistic for stationary time series is defined as follows.

Definition 9.14. *Given a stationary time series* $\{X_t\}$*, the* partial auto-correlation function (PACF) *is defined as*

$$P_h = \text{CORR}\left(X_t - \hat{X}_t^{[h-1\rangle}, X_{t+h} - \hat{X}_{t+h}^{\langle h-1]}\right), \qquad (9.26)$$

for $h \in \mathbb{Z}^+$*, where* $\hat{X}_t^{[h-1\rangle}$ *and* $\hat{X}_{t+h}^{\langle h-1]}$ *are the regressions of* X_t *and* X_{t+h} *on the intermediary variables* $X_{t+1}, \ldots, X_{t+h-1}$*, respectively, as defined by Definition 9.11; i.e.,*

$$\hat{Z}_{t+h}^{\langle h-1]} = \beta_1 Z_{t+h-1} + \beta_2 Z_{t+h-2} + \cdots + \beta_{h-1} Z_{t+1} \qquad (9.27)$$

$$\hat{Z}_t^{[h-1\rangle} = \beta_1 Z_{t+1} + \beta_2 Z_{t+2} + \cdots + \beta_{h-1} Z_{t+h-1}, \qquad (9.28)$$

where we used the coefficients $\beta_i = \phi_{h-1,i}$*, for brevity.*

Note 9.2. If $\{X_t\}$ is a Gaussian process, the PACF is equivalent to

$$P_h = \mathbb{E}\left[\text{CORR}(X_t, X_{t+h}|X_{t+1}, \ldots, X_{t+h-1})\right],$$

where the expectation is taken with respect to the intermediary variables $X_{t+1}, \ldots, X_{t+h-1}$. This follows form Proposition 7.2 and the fact that

$$\mathbb{E}[X|Z] = \hat{X},$$

where $\hat{X}$ is the liner regression of X on Z, whenever the joint distribution of X and Z is Gaussian. ▷

In order to derive a formula for the partial autocorrelation function, we first require two lemmas.

Lemma 9.1. *Let* $\{Z_t\}$ *be a mean-zero stationary time series. Then the variance of the difference* $Z_t - \hat{Z}_t^{[h-1\rangle}$ *is given by*

$$\mathbb{V}\left(Z_t - \hat{Z}_t^{[h-1\rangle}\right) = \gamma_0 - \beta_1 \gamma_1 - \cdots - \beta_{h-1}\gamma_{h-1}, \qquad (9.29)$$

where $\beta_i = \phi_{h-1,i}$*. Moreover, due to the stationarity, it follows that*

$$\mathbb{V}\left(Z_t - \hat{Z}_t^{[h-1\rangle}\right) = \mathbb{V}\left(Z_{t+h} - \hat{Z}_{t+h}^{\langle h-1]}\right).$$

Proof. Since $\mathbb{E}\left[Z_t - \hat{Z}_t^{[h-1\rangle}\right] = 0$, we have

$$\mathbb{V}\left(Z_t - \hat{Z}_t^{[h-1\rangle}\right) = \mathbb{E}\left[(Z_t - \beta_1 Z_{t+1} - \cdots - \beta_{h-1} Z_{t+h-1})^2\right]$$
$$= \mathbb{E}[Z_t(Z_t - \beta_1 Z_{t+1} - \cdots - \beta_{h-1} Z_{t+h-1})]$$
$$- \beta_1 \mathbb{E}[Z_{t+1}(Z_t - \beta_1 Z_{t+1} - \cdots - \beta_{h-1} Z_{t+h-1})]$$
$$- \cdots - \beta_{h-1} \mathbb{E}[Z_{t+h-1}(Z_t - \beta_1 Z_{t+1} - \cdots - \beta_{h-1} Z_{t+h-1})].$$

However, all but the first term vanishes, due to Equation (9.21), which can be rewritten for $h - 1$, replacing $\phi_{h-1,i}$ with β_i, as

$$\beta_1 \mathbb{E}[Z_{t+1} Z_{t+k}] + \beta_2 \mathbb{E}[Z_{t+2} Z_{t+k}] + \cdots + \beta_{h-1} \mathbb{E}[Z_{t+h-1} Z_{t+k}] = \mathbb{E}[Z_t Z_{t+k}], \tag{9.30}$$

for $k = 1, \ldots, h - 1$. Thus,

$$\mathbb{V}\left(Z_t - \hat{Z}_t^{[h-1\rangle}\right) = \mathbb{E}[Z_t(Z_t - \beta_1 Z_{t+1} - \cdots - \beta_{h-1} Z_{t+h-1})].$$

The result follows from the definition of the autocovariance function. □

Lemma 9.2. *Let $\{Z_t\}$ be a mean-zero stationary time series, and let $\hat{Z}_{t+h}^{\langle h-1]}$ and $\hat{Z}_t^{[h-1\rangle}$ be the regressions defined by Equations (9.27) and (9.28), respectively. Then the covariance of the difference $Z_t - \hat{Z}_t^{[h-1\rangle}$ and $Z_{t+h} - \hat{Z}_{t+h}^{\langle h-1]}$ is given by*

$$\mathrm{COV}\left(Z_t - \hat{Z}_t^{[h-1\rangle}, Z_{t+h} - \hat{Z}_{t+h}^{\langle h-1]}\right) = \gamma_h - \beta_1 \gamma_{h-1} - \cdots - \beta_{h-1} \gamma_1. \tag{9.31}$$

Proof. The covariance in Equation (9.31) is equivalent to the expression

$$\mathbb{E}\left[(Z_t - \beta_1 Z_{t+1} - \cdots - \beta_{h-1} Z_{t+h-1})(Z_{t+h} - \beta_1 Z_{t+h-1} - \cdots - \beta_{h-1} Z_{t+1})\right].$$

However, recalling again Equation (9.30), this simplifies as

$$\mathbb{E}\left[(Z_t - \beta_1 Z_{t+1} - \cdots - \beta_{h-1} Z_{t+h-1}) Z_{t+h}\right].$$

The result follows. □

Theorem 9.2. *The partial autocorrelation function for a mean-zero stationary time series $\{Z_t\}$ at lag h is given by the ratio of determinants*

$$P_h = \frac{\det \begin{bmatrix} 1 & \rho_1 & \rho_2 & \cdots & \rho_{h-2} & \rho_1 \\ \rho_1 & 1 & \rho_1 & \cdots & \rho_{h-3} & \rho_2 \\ \vdots & \vdots & \vdots & \ddots & \vdots & \vdots \\ \rho_{h-1} & \rho_{h-2} & \rho_{h-3} & \cdots & \rho_1 & \rho_h \end{bmatrix}}{\det \begin{bmatrix} 1 & \rho_1 & \rho_2 & \cdots & \rho_{h-2} & \rho_{h-1} \\ \rho_1 & 1 & \rho_1 & \cdots & \rho_{h-3} & \rho_{h-2} \\ \vdots & \vdots & \vdots & \ddots & \vdots & \vdots \\ \rho_{h-1} & \rho_{h-2} & \rho_{h-3} & \cdots & \rho_1 & 1 \end{bmatrix}}. \tag{9.32}$$

Proof. From the definition of the partial autocorrelation function,

$$P_h = \text{CORR}\left(Z_t - \hat{Z}_t^{[h-1\rangle}, Z_{t+h} - \hat{Z}_{t+h}^{\langle h-1]}\right)$$

$$= \frac{\text{COV}\left(Z_t - \hat{Z}_t^{[h-1\rangle}, Z_{t+h} - \hat{Z}_{t+h}^{\langle h-1]}\right)}{\sqrt{\mathbb{V}\left(Z_t - \hat{Z}_t^{[h-1\rangle}\right)\mathbb{V}\left(Z_{t+h} - \hat{Z}_{t+h}^{\langle h-1]}\right)}}$$

$$= \frac{\gamma_h - \beta_1\gamma_{h-1} - \cdots - \beta_{h-1}\gamma_1}{\gamma_0 - \beta_1\gamma_1 - \cdots - \beta_{h-1}\gamma_{h-1}},$$

where the last equality follows from Lemmas 9.1 and 9.2. Now, by dividing the numerator and denominator by γ_0, we obtain

$$P_h = \frac{\rho_h - \beta_1\rho_{h-1} - \cdots - \beta_{h-1}\rho_1}{1 - \beta_1\rho_1 - \cdots - \beta_{h-1}\rho_{h-1}}.$$

However, noting that $\beta_j = \phi_{h-1,j}$, we see that this expression is equivalent to the expression for ϕ_{hh}, by replacing $h+1$ with h in Equation (9.24).

We conclude that the partial autocorrelation with lag h, P_h, is equivalent to the final regression coefficient of the hth order autoregression, so that

$$P_h = \phi_{hh}.$$

Since ϕ_{hh} is standard notation for the partial autocorrelation function, we will use it in place of P_h.   □

9.1.7 Sample Estimates

Definition 9.15. *Let $(x_1, \ldots, x_n)$ be a finite sample from a stationary time series $\{X_t\}$. Then the sample mean is defined by*

$$\bar{x} = \frac{1}{n}\sum_{t=1}^{n} x_t.$$

The sample autocovariance function is defined by

$$\hat{\gamma}(h) = \frac{1}{n}\sum_{t=1}^{n-h}(x_{t+h} - \bar{x})(x_t - \bar{x}), \tag{9.33}$$

with $\hat{\gamma}(-h) = \hat{\gamma}(h)$ for $h = 0, \ldots, n-1$.

Similarly, the sample autocorrelation function is defined by

$$\hat{\rho}(h) = \frac{\hat{\gamma}(h)}{\hat{\gamma}(0)}. \tag{9.34}$$

9.2 Stationary Time Series

In this section, we introduce an important class of stationary time series, known as the autoregressive moving average process. We will begin by discussing each of its components—autoregressive processes and moving average processes—in turn.

Backshift and Finite-Difference Operators

To facilitate our discussion, we introduce the following operator.

Definition 9.16. *Let $\{X_t\}$ be a time series. We define the* backshift *operator B by the relation*

$$BX_t = X_{t-1}, \tag{9.35}$$

where we define the integer powers of B in the natural way:

$$B^h X_t = X_{t-h},$$

for $h \in \mathbb{Z}$. In particular, we will refer to B^{-1} as the forward-shift *operator.*

In addition, we will be interested in discussing *finite differences*; the first-order compatriot of which being the difference $(1-B)X_t = X_t - X_{t-1}$. Higher order differences are easily generalized with the following operator.

Definition 9.17 (Finite Differences). *The difference of order d is the operator*

$$\nabla^d = (1 - B)^d, \tag{9.36}$$

where the superscript is dropped for $d = 1$.

9.2.1 Moving Average Processes

Definition

The simplest way of constructing a stationary time series is by computing a moving average over the trailing values from a white noise process $\{W_t\}$. We define such a process as follows.

Definition 9.18. *A* moving-average *(MA) process of order q, denoted* MA(q), *is a time series of the form*

$$Z_t = W_t + \theta_1 W_{t-1} + \cdots + \theta_q W_{t-q}, \tag{9.37}$$

for constant coefficients $\theta_1, \ldots, \theta_q$, where $\theta_q \neq 0$, and white noise process $W_t \sim \mathrm{WN}(0, \sigma_w^t)$.

The associated moving-average *polynomial is defined as*

$$\theta(z) = 1 + \theta_1 z + \cdots + \theta_q z^q, \tag{9.38}$$

for $z \in \mathbb{C}$, and the moving-average operator *is defined as*

$$\theta(B) = 1 + \theta_1 B + \cdots + \theta_q B^q, \tag{9.39}$$

where B is the backshift operator (Definition 9.16).

Given our operator notation Equation (9.39), we may equivalently express a MA(q) process in the form

$$Z_t = \theta(B)W_t. \tag{9.40}$$

Autocorrelation and Partial Autocorrelation

By comparing with our definition of a causal linear process Equation (9.12), we immediately have the following.

Proposition 9.5. *The* MA(q) *process given by Equation (9.40) is always stationary, but invertible if and only if*

$$\theta(z) \neq 0 \qquad for \qquad |z| \leq 1;$$

i.e., a moving-average process is invertible if and only if the roots of the moving-average polynomial defined by Equation (9.38) lie outside of the unit circle on the complex plane.

Proof. Since a MA process is a finite linear process, stationarity follows from Proposition 9.3, as there are only a finite number of coefficients; i.e.,

$$\sum_{j=0}^{\infty} |\psi_j| = \sum_{j=0}^{q} |\theta_j| < \infty.$$

By inverting the polynomial $\theta(z)$, we may express Equation (9.40) as

$$\theta(B)^{-1} Z_t = W_t.$$

From the theory of complex variables (e.g., see Corollary 9.1), the power series $\pi(z) = \theta(z)^{-1}$ must converge for $|z| \leq 1$ for the coefficients to be absolutely summable; i.e., for $\sum_{j=0}^{\infty} |\pi_j| < \infty$. This is only true if $\theta(z)$ does not have any zeros on the unit disk, thus completing the result. $\square$

Proposition 9.6. *The* ACF *of a* MA(q) *process is given by*

$$\rho_h = \begin{cases} \dfrac{-\theta_h + \theta_1 \theta_{h+1} + \cdots + \theta_{q-h}\theta_q}{1 + \theta_1^2 + \cdots + \theta_q^2} & for\ h = 1, \ldots, q \\ 0 & for\ h > q \end{cases} \tag{9.41}$$

The PACF *of a* MA(q) *process tails off as a mixture of exponential decay and/or damped sine waves, depending on the roots of the moving-average polynomial; damped oscillations occurring if and only if the polynomial possesses complex roots.*

Proof. We begin by recalling that a MA(q) process is a linear combination of the trailing q terms of a white-noise process. The autocovariance between Z_t and Z_{t+h} can therefore be expressed as

$$
\begin{aligned}
\gamma_h &= \text{COV}(Z_t, Z_{t-h}) \\
&= \text{COV}(W_t + \theta_1 W_{t-1} + \cdots + \theta_q W_{t-q}, \\
&\qquad\qquad W_{t-h} + \theta_1 W_{t-h-1} + \cdots + \theta_q W_{t-h-q}).
\end{aligned}
$$

Now, by definition, $\text{COV}(W_t, W_s) = 0$ whenever $t \neq s$. It follows that, when $h > q$, there is absolutely no overlap between Z_t and Z_{t-h}, so that $\text{CORR}(Z_t, Z_{t-h}) = 0$ for $h > q$.

For the case $h \leq q$, on the other hand, there are precisely $q - h + 1$ overlapping terms, so that (if we define $\theta_0 = 1$),

$$
\begin{aligned}
\gamma_h &= \text{COV}\left(\sum_{i=0}^{q-h} \theta_{h+i} W_{t-h-i}, \sum_{j=0}^{q-h} \theta_i W_{t-h-j} \right) \\
&= \sum_{i=0}^{q-h} \theta_{h+i} \theta_i \text{COV}(W_{t-h-i}, W_{t-h-i}) \\
&= \sum_{i=0}^{q-h} \theta_{h+i} \theta_i \sigma_w^2.
\end{aligned}
$$

Now, it is clear that the variance is given by

$$
\gamma_0 = \sigma_w^2 \sum_{i=0}^{q-h} \theta_i^2,
$$

and Equation (9.41) follows.

$\square$

Examples

Example 9.3. We can easily construct a simple *generator function* that represents a MA(q) process, as shown in Code Block 9.3. We use the `queue.Queue`[1] class to construct a queue that remembers the trailing q value of the white-noise process W_t.

Think of a queue object as a list with $O(1)$ get and put operations, and a maximum length. The put method adds an element to the end of the queue, whereas the get method pops the first element at the beginning of the queue. For this reason, we reverse the list of θ coefficients, as the queue, at any point, represents the values

---

[1] See https://docs.python.org/3/library/queue.html for more details.

```
1  from queue import Queue
2  def ma(theta=[], sigma_w=1, burn_length=100, max_iter=1e6):
3      q, t = len(theta), 0
4      assert q > 0
5      W = Queue(maxsize=q)
6      theta = list(reversed(theta))
7
8      while not W.full():
9          W.put(np.random.normal(scale=sigma_w))
10
11     while t < max_iter:
12         t += 1
13         assert q==0 or W.full()
14         W_t = np.random.normal(scale=sigma_w)
15         Z_t = np.dot(theta, W.queue) + W_t
16         W.get()
17         W.put(W_t)
18         if t > burn_length:
19             yield Z_t
```

Code Block 9.3: Generator function for $\text{MA}(q)$ process

$$[W_{t-q}, \ldots, W_{t-1}],$$

whereas our input parameter is

$$\theta = [\theta_1, \ldots, \theta_q].$$

```
1  Z = ma(theta=[0.9])
2  z = [next(Z) for i in range(100)] # generate a list of values
3  tsPlot(Z) # Input is the iterator, not the values
```

Code Block 9.4: Using the generator to construct values from a time series

The generator function itself returns an iterator object, that can be used to construct an arbitrary sequence from our $\text{MA}(q)$ process, as shown in Code Block 9.4. (We reused **tsPlot** function from Code Block 9.1.) The **burn_length** parameter discards the first so-many terms, which is relevant for the $\text{AR}(p)$ process, which we will discuss in the next section. The **max_iter** parameter prevents the sort of infinite loops that would otherwise occur when summing the iterator object.   ▷

Example 9.4. The $\text{MA}(1)$ process is given by

$$Z_t = W_t + \theta W_{t-1}.$$

We use the generator function in Code Block 9.3 and the `tsPlot` function we defined in Code Block 9.1 to plot the MA(1) process for $\theta = \pm 0.9$, as shown in Figures 9.3 and 9.4. Compare with the white noise shown in Figure 9.1.

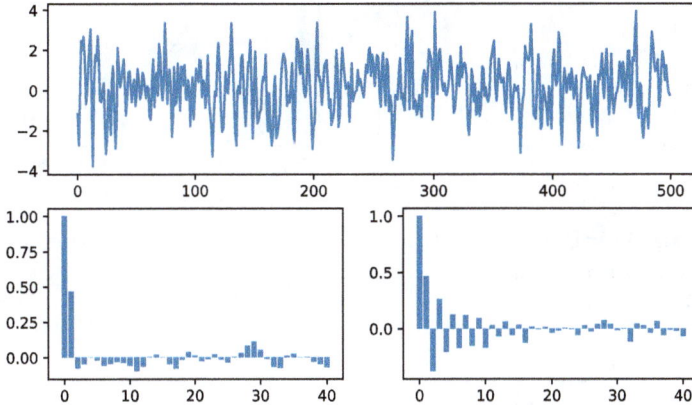

Fig. 9.3: MA(1) with $\theta = 0.9$ (top); plotted with sample ACF (bottom left) and PACF (bottom right).

For each of these, the ACF vanishes after lag 1, except for error from the numerical estimator. The PACF functions trail off   ▷

Example 9.5. Consider the MA(2) process

$$Z_t = W_t - W_{t-1} + 0.5W_{t-2},$$

corresponding to the moving-average polynomial $\theta(z) = 1 - z + 0.5z^2$ with complex roots $z = 1 \pm i$. Notice the damped oscillations in the PACF shown in Figure 9.5.   ▷

9.2.2 Autoregressive Processes

Definition

In Section 9.1.5, we discussed the linear autoregression of time series, as a regression Equation (9.18) of the variable X_t on its h lagging values $X_{t-1}, \ldots, X_{t-h}$. In this section, we show how this regression equation can be used to construct a special class of time series known as *autoregressive processes*, which we define as follows.

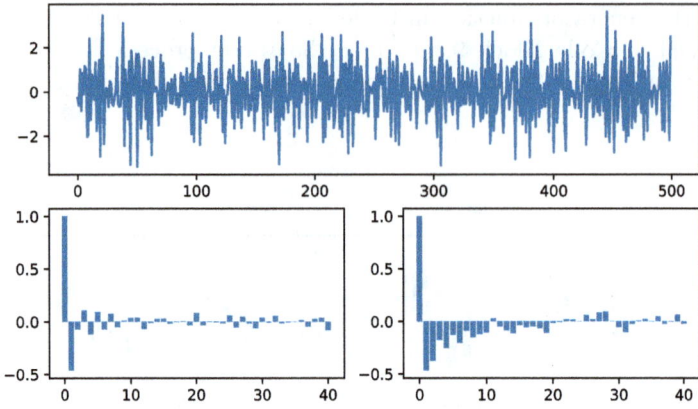

Fig. 9.4: MA(1) with $\theta = -0.9$ (top); plotted with sample ACF (bottom left) and PACF (bottom right).

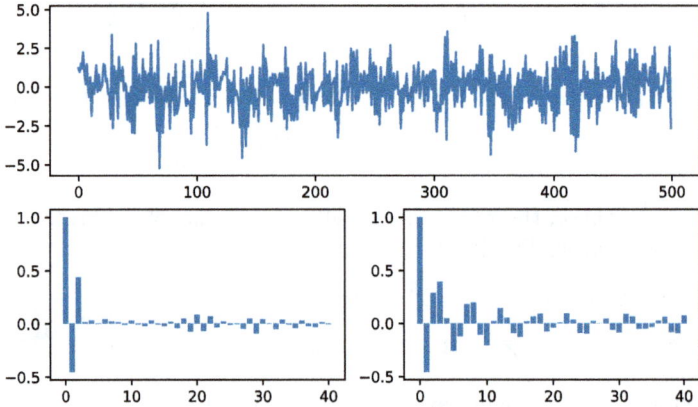

Fig. 9.5: MA(2) process $(1 - B + 0.5B^2)Z_t = W_t$ (top); plotted with sample ACF (bottom left) and PACF (bottom right).

Definition 9.19. *An* autoregressive (AR) *process of order* p, *denoted* AR(p), *is a time series of the form*

$$Z_t = \phi_1 Z_{t-1} + \phi_2 Z_{t-2} + \cdots + \phi_p Z_{t-p} + W_t, \tag{9.42}$$

for constant coefficients $\phi_1, \ldots, \phi_p$, *where* $\phi_p \neq 0$, *and white noise process* $W_t \sim$ WN$(0, \sigma_w^t)$.

The associated autoregressive polynomial *is defined as*

$$\phi(z) = 1 - \phi_1 z - \cdots - \phi_q z^q, \tag{9.43}$$

for $z \in \mathbb{C}$, *and the* autoregressive operator *is defined as*

$$\phi(B) = 1 - \phi_1 B - \cdots - \phi_p B^p, \tag{9.44}$$

where B *is the backshift operator (Definition 9.16).*

Given our operator notation Equation (9.44), we may equivalently express an AR(p) process in the form

$$\phi(B) Z_t = W_t. \tag{9.45}$$

Autocorrelation and Partial Autocorrelation

By comparing with our definition of an invertible process Equation (9.15), we immediately have the following.

Proposition 9.7. *The* AR(p) *process given by Equation (9.45) is always invertible, but stationary if and only if*

$$\phi(z) \neq 0 \qquad for \qquad |z| \leq 1;$$

i.e., an autoregressive process is stationary if and only if the roots of the autoregressive polynomial defined by Equation (9.43) lie outside of the unit circle on the complex plane.

Proof. Invertibility follows from the fact that an AR(p) process, as defined by Equation (9.42), is already in the form of Equation (9.15), with a finite set of coefficients.

The stationarity condition follows again from Corollary 9.1. Note that Equation (9.45) is equivalent to the causal linear process

$$Z_t = \phi(B)^{-1} W_t.$$

However, we have seen that this process is invertible only if $\psi(z) = \phi(z)^{-1}$ converges on the unit disk $|z| \leq 1$, which can only be true if $\phi(z) \neq 0$ on the unit disk, thus completing the result. $\square$

A simple counterexample, showing that not all autoregressive processes are stationary, is the random walk of Equation (9.6), which has a single unity root and time dependent mean and autocovariance (Proposition 9.2).

Proposition 9.8. *The* ACF *of an* AR(p) *process tails off as a mixture of exponential decay and/or damped sine waves, depending on the roots of the moving-average polynomial; damped oscillations occurring if and only if the polynomial possesses complex roots.*

The PACF *of an* AR(p) *process vanishes after lag* p; *i.e.,*

$$\phi_{hh} = 0 \qquad for \qquad h > p.$$

Proof. To show the second part, we begin by considering the autoregression Equation (9.18) as applied to the AR(p) process Equation (9.42). It is immediately clear that the regression coefficients coincide with the coefficients of Equation (9.42), when defined:

$$\phi_{hi} = \phi_i, \qquad for \qquad i \leq p,$$

and $\phi_{hi} = 0$ otherwise. In particular, when $h > p$, it follows that

$$Z_t - \hat{Z}_t^{\langle h-1]} = Z_t - \hat{Z}_t^{\langle p]} = W_t,$$

so that

$$\begin{aligned} \phi_{hh} &= \text{CORR}\left(Z_t - \hat{Z}_t^{\langle h-1]}, Z_{t+h} - \hat{Z}_{t+h}^{\langle h-1]}\right) \\ &= \text{CORR}\left(Z_t - \hat{Z}_t^{\langle p]}, Z_{t+h} - \hat{Z}_{t+h}^{\langle p]}\right) \\ &= \text{CORR}(W_t, W_{t+h}) = 0, \end{aligned}$$

for $h > p$. We conclude that the PACF of an AR(p) process vanishes for $h > p$.

The ACF follows a certain recurrence relation. To see this, let us multiply Equation (9.42) by Z_{t-h}, for $h > 0$, to obtain

$$Z_t Z_{t-h} = \phi_1 Z_{t-1} Z_{t-h} + \cdots + \phi_p Z_{t-p} Z_{t-h} + W_t Z_{t-h}.$$

Taking expectation, and noting that $\mathbb{E}[W_t Z_{t-h}] = 0$ for $h > 0$, we obtain

$$\gamma_h = \phi_1 \gamma_{h-1} + \cdots + \phi_p \gamma_{h-p}.$$

Dividing by γ_0 yields the relation

$$\rho_h = \phi_1 \rho_{h-1} + \cdots + \phi_p \rho_{h-p}, \tag{9.46}$$

for $h > 0$. □

Examples

We leave it as an exercise for the reader to write a generator function for an AR(p) process that is similar to Code Block 9.3.

Example 9.6. Consider the AR(1) process

$$Z_t = \phi Z + W_t,$$

for $\phi = \pm 0.9$. These are shown in Figures 9.6 and 9.7

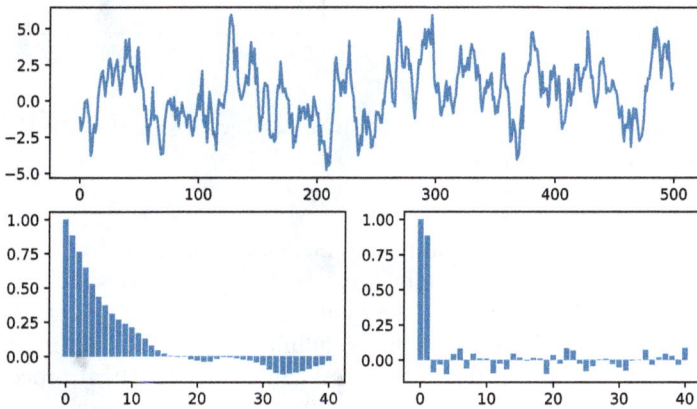

Fig. 9.6: AR(1) process $\phi = 0.9$ (top); plotted with sample ACF (bottom left) and PACF (bottom right).

$\triangleright$

9.2.3 ARMA Processes

The Dual Nature of Autoregressive and Moving-Average Processes

Autoregressive and moving-average processes are flip sides of the same coin. A MA(q) process, as defined in Equation (9.37), is a finite version of a causal stationary linear process Equation (9.12). Similarly, an AR(p) process is a finite version of the general invertible process of Equation (9.15). It follows that an invertible MA(q) process, one for which $\theta(z)$ has no roots in the unit disk of the complex plane, can be expressed as an infinite autoregressive process. Similarly, a stationary AR(p) process, one for which $\phi(z)$ has no

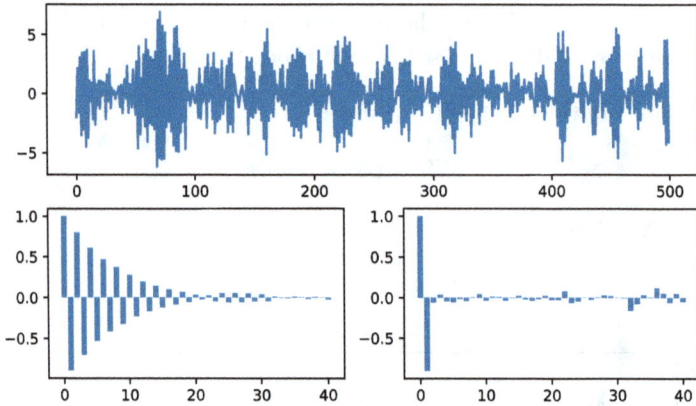

Fig. 9.7: AR(1) process $\phi = -0.9$ (top); plotted with sample ACF (bottom left) and PACF (bottom right).

roots in the unit disk of the complex plane, can be expressed as an infinite moving-average process (shown for the case AR(1) in Exercise 9.2).

This duality also reflects in the dual nature of the ACF and PACF. A MA(q) process has a finite ACF, but an infinite PACF, reflecting its equivalence to an infinite autoregressive process. Similarly, an AR(p) process has a finite PACF, but an infinite ACF, also reflecting its dual nature.

It follows that, thus far, we are leaving a class of stationary time series off the table, those that exhibit both elements of autoregressive and elements of moving-average processes. We turn to this topic next.

Autoregressive Moving-Average Processes

Definition 9.20. *An* autoregressive moving-average (ARMA) *process of orders* p, q, *denoted* ARMA(p, q), *is a time series of the form*

$$Z_t = \phi_1 Z_{t-1} + \cdots + \phi_p Z_{t-p} + W_t + \theta_1 W_{t-1} + \cdots + \theta_q W_{t-q}, \quad (9.47)$$

for constant coefficients $\phi_1, \ldots, \phi_p; \theta_1, \ldots, \theta_q$, *where* $\phi_p \neq 0$ *and* $\theta_q \neq 0$, *and white noise process* $W_t \sim \text{WN}(0, \sigma_w^t)$.

Given the operators Equations (9.39) and (9.44), we may equivalently express an ARMA(p, q) process in the form

$$\phi(B)Z_t = \theta(B)W_t. \quad (9.48)$$

By combining Proposition 9.5 and 9.7, we arrive at the follwoing.

Proposition 9.9. *The* ARMA*(p,q) process given by Equation* (9.47) *is*

1. *stationary if and only if $\phi(z) \neq 0$ for $|z| \leq 1$, and*
2. *invertible if and only if $\theta(z) \neq 0$ for $|z| \leq 1$.*

Proof. Stationarity follows as long as we can express Equation (9.48) as

$$Z_t = \phi(B)^{-1}\theta(B)W_t,$$

so that $\psi(B) = \phi(B)^{-1}\theta(B)$, with $||\psi||_2 < \infty$. Invertibility follows if we can express it as

$$\theta(B)^{-1}\phi(B)Z_t = W_t,$$

so that $\pi(B) = \theta(B)^{-1}\phi(B)$. The condition on the roots of the associated polynomials follows from complex-variable theory. □

Autocorrelation and Partial Autocorrelation

Proposition 9.10. *The* ACF *of an* ARMA*(p,q) depends on both sets of coefficients, θ and ϕ, for lag $h \leq q$, and decays like an* AR*(p) process for lag $h > q$.*

The PACF *of an* ARMA*(p,q) decays like a* MA*(q) process for lag $h > p$.*

Proof. Multiplying both sides of Equation (9.47) by Z_{t-h}, we obtain

$$Z_t Z_{t-h} = \phi_1 Z_{t-1} Z_{t-h} + \cdots + \phi_p Z_{t-p} Z_{t-h}$$
$$+ W_t Z_{t-h} + \theta_1 W_{t-1} Z_{t-h} + \cdots + \theta_q W_{t-q} Z_{t-h},$$

so that

$$\gamma_h = \phi_1 \gamma_{h-1} + \cdots + \phi_p \gamma_{h-p}$$
$$+ \mathbb{E}[Z_{t-h} W_t] + \theta_1 \mathbb{E}[Z_{t-h} W_{t-1}] + \cdots + \theta_q \mathbb{E}[Z_{t-h} W_{t-q}].$$

Recall that an invertible ARMA model $\phi(B)Z_t = \theta(B)W_t$ can always be written in the form $Z_t = \psi(B)W_t$, where $\psi(B) = \phi(B)^{-1}\theta(B)$ and where $\phi(B)^{-1}$ an infinite series. Using this relation, we can show that

$$\mathbb{E}[Z_{t-h}W_{t-k}] = \begin{cases} 0 & \text{for } k < h \\ \psi_{k-h}\sigma_w^2 & \text{for } k \geq h \end{cases}$$

Note that the condition $\mathbb{E}[Z_{t-h}W_{t-k}] = 0$ for $k < h$ follows from causality: only past values of the shocks can affect the time series; i.e., when $k < h$, $t - k$ represents a time after $t - h$, so that the effect of W_{t-k} cannot be felt by Z_{t-h}. Putting this together, we have

$$\gamma_h = \phi_1 \gamma_{h-1} + \cdots + \phi_p \gamma_{h-p} + \sigma_w^2 \left(\theta_h \psi_0 + \theta_{h+1} \psi_1 + \cdots + \theta_q \psi_{q-h} \right).$$

Note that there are $q - h + 1$ terms in the final term. In particular, note that This implies that

$$\gamma_h = \phi_1 \gamma_{h-1} + \cdots + \phi_p \gamma_{h-p},$$

for $h > q$. Dividing by γ_0, we obtain the recurrence relation

$$\rho_h = \phi_1 \rho_{h-1} + \cdots + \phi_p \rho_{h-p} \qquad (9.49)$$

for the ACF. Next, let $z_1, \ldots, z_r$ represent the roots of the polynomial $\phi(z) = 1 - \phi_1 z - \cdots - \phi_p z^p$, with multiplicities $m_1, \ldots, m_r$, such that $m_1 + \cdots + m_r = p$. It can be shown that the general solution to Equation (9.49) is given by

$$\rho_h = z_1^{-h} P_1(h) + z_2^{-h} P_2(h) + \cdots + z_r^{-h} P_r(h), \qquad (9.50)$$

for $h \geq q$, where P_j is a polynomial of order $m_j - 1$. In particular, if each root is a single root, than the functions $P_1(h), \ldots, P_r(h)$ are simply constant coefficients. For details of the proof, see, for example, Shumway and Stoffer [1999] or Wei [2019].

For a stationary process, each root of $\phi(z)$ must lie outside of the unit circle, so that $|z_j| \geq 1$. If the roots are real, the solution Equation (9.50) will decay exponentially with h. If the roots are complex-conjugate pairs, the solution will exhibit damped oscillations; i.e., sinusoidal waves with an exponentially decaying envelope. We conclude that the ACF for an ARMA(p, q) process decays similarly to an AR(p) process for $h > q$. $\qquad \square$

A summary of results for AR(p), MA(q), and ARMA(p, q) models provided in Table 9.1[2].

Examples

Example 9.7. We can easily extend Code Block 9.3 to accommodate a general ARMA(p, q) process, as shown in Code Block 9.5. We note that a `Queue` object with a `maxsize=0` has infinite length, so we must take care to handle the edge cases $p = 0$ or $q = 0$ whenever we interact with our queues. Other than that, it functions in the same way as our previous code.

The ARMA$(1, 1)$ process defined by

$$Z_t = 0.25 Z_{t-1} + W_t + 0.5 W_{t-1}$$

is shown in Figure 9.8. Similarly, the ARMA$(2, 6)$ process defined by

$$Z_t = 1.5 Z_{t-1} - 0.75 Z_{t-2}$$
$$+ W_t - 0.6 W_{t-1} + 0.05 W_{t-2} + 0.15 W_{t-3} + 0.025 W_{t-4} - 0.025 W_{t-6}$$

---

[2] Note that both Box, *et al.* [2016] and Wei [2019] claim that the ACF and PACF for the ARMA process tail off after $q - p$ and $p - q$ lags, respectively. This claim is related to the solution Equation (9.50) to the linear difference equation Equation (9.49), and a similar equation for the PACF, though neither author, in our opinion, provides a convincing justification for this claim; we invite the interested reader to explore further.

	AR(p)	MA(q)	ARMA(p,q)
model	$\phi(B)Z_t = W_t$	$Z_t = \theta(B)W_t$	$\phi(B)Z_t = \theta(B)W_t$
stationarity condition	$\phi(z) \neq 0$ for $|z| \leq 1$	always	$\phi(z) \neq 0$ for $|z| \leq 1$
invertibility condition	always	$\theta(z) \neq 0$ for $|z| \leq 1$	$\theta(z) \neq 0$ for $|z| \leq 1$
ACF	tails off	cuts off after lag q	tails off after lag $q - p$
PACF	cuts off after lag p	tails off	tails off after lag $p - q$

Table 9.1: Summary of stationary models. *Tails off* means that the series tails off as exponential decay or dampened sinusoidal waves.

is depicted in Figure 9.9. Note that it is no longer as straightforward to infer the values of p and q from the ACF and PACF plots. Note that we create this plot using our generator Code Block 9.5, instantiated using the command

Z = arma(phi $= [1.5, -0.75]$, theta $= [-0.6, 0.05, 0.15, 0.025, 0, -0.025])$.

As a final example, consider the ARMA($3,2$) process

$$Z_t = 2Z_{t-1} - 1.5Z_{t-2} + 0.375Z_{t-3} + W_t - W_{t-1} + 0.5W_{t-2},$$

which is shown in Figure 9.10. We will use this model to illustrate an integrated process in the next section.

$\triangleright$

9.2.4 Inferring Random Shocks

Many authors liken the white-noise process W_t to a sequence of *random shocks* that ultimately build up and determine the values of the time series Z_t. These shocks are, however, unobserved, though they may be inferred from the primary time series data, given a sufficient sample size.

To proceed, note that, for a given ARMA(p,q) process, the terms of Equation (9.47) may be rearranged to express the current shock W_t as a function of the current time-series value Z_t and the trailing history $Z_{t-1}, \ldots, Z_{t-p}, W_{t-1}, \ldots, W_{t-q}$, via the relation

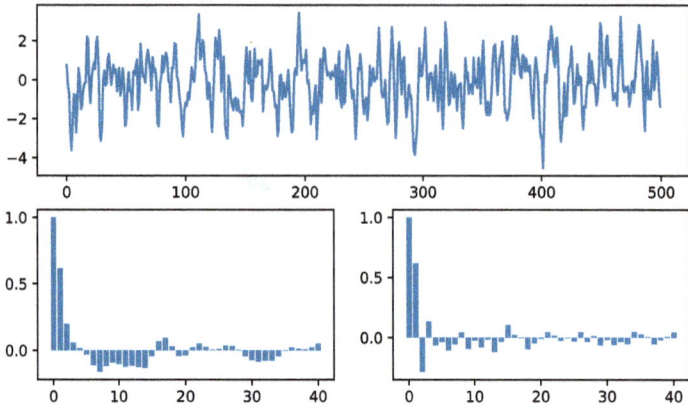

Fig. 9.8: ARMA(1, 1) process (top); plotted with sample ACF (bottom left) and PACF (bottom right).

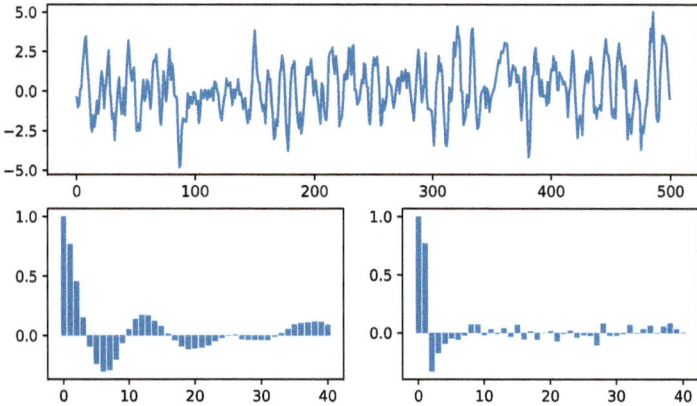

Fig. 9.9: ARMA(2, 6) process (top); plotted with sample ACF (bottom left) and PACF (bottom right).

```
 1  def arma(phi=[], theta=[], sigma_w=1, burn_length=100,
            max_iter=1e6):
 2      p, q, t = len(phi), len(theta), 0
 3      Z = Queue(maxsize=p)
 4      W = Queue(maxsize=q)
 5      theta, phi = list(reversed(theta)), list(reversed(phi))
 6      while (p > 0) and (not Z.full()):
 7          Z.put(np.random.normal(scale=sigma_w))
 8      while (q > 0) and (not W.full()):
 9          W.put(np.random.normal(scale=sigma_w))
10
11      while t < max_iter:
12          t += 1
13          assert p==0 or Z.full()
14          assert q==0 or W.full()
15          W_t = np.random.normal(scale=sigma_w)
16          Z_t = np.dot(phi, Z.queue) + np.dot(theta, W.queue) + W_t
17
18          if p > 0:
19              Z.get()
20              Z.put(Z_t)
21          if q > 0:
22              W.get()
23              W.put(W_t)
24          if t > burn_length:
25              yield Z_t
```

Code Block 9.5: Generator function for an ARMA(p, q) process

$$W_t = \phi(B)Z_t + (1 - \theta(B))W_t$$
$$= Z_t - \phi_1 Z_{t-1} - \cdots - \phi_p Z_{t-p}$$
$$-\theta_1 W_{t-1} - \cdots - \theta_q W_{t-q}.$$

By knowing the previous values of the shocks $W_{t-1}, \ldots, W_{t-q}$ and the previous and current value of the time series $Z_t, Z_{t-1}, \ldots, Z_{t-p}$, we can therefore determine the current shock W_t from the time-series equations.

Since we do not know any of the values of the series W_t, we can initialize our estimates by setting the initial q values of W_t to their expected value: $\hat{W}_t = 0$ for $t = p - q + 1, \ldots, p$. We may then infer $\hat{W}_{p+1}$ using the p values $Z_1, \ldots, Z_p$, and proceed according to Algorithm 9.1.

Note that, in general, stationary time series may be expressed solely in terms of their current and previous shocks, according to

$$Z_t = W_t + \sum_{t=1}^{\infty} \psi_j W_{t-j},$$

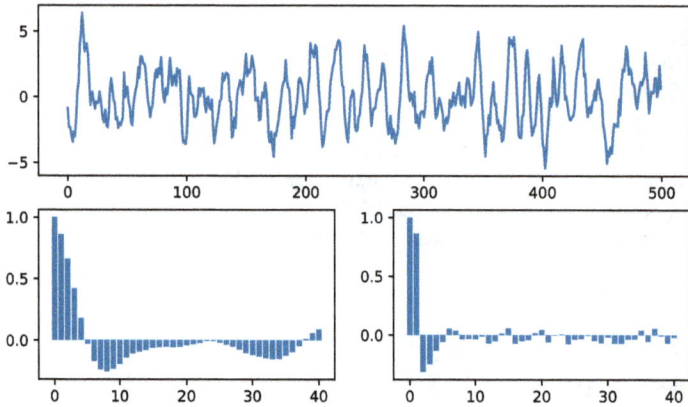

Fig. 9.10: ARMA$(3,2)$ process (top); plotted with sample ACF (bottom left) and PACF (bottom right).

where the series $\sum \psi_j$ is absolutely convergent. The absolute convergence property therefore implies that impact of the shocks must decay to zero as time goes by. Our errors induced by our incorrect estimates $\hat{W}_t = 0$ for $t \leq p$ will therefore decay to zero as well. We applied this algorithm to an ARMA(p,q) process and compared our estimated shocks $\hat{W}_t$ with the actual, true shocks W_t. The result is shown in Figure 9.11. (See Exercise 9.4 for additional details.) Note that our estimates converge fairly rapidly to the underlying true values.

Algorithm 9.1: Estimating the white-noise variables for a stationary process.

Input: time-series data Z_t, for $t = 1, \ldots, n$;
        an ARMA(p,q) model with known parameters θ, ϕ, σ_w^2
Output: Estimates $\hat{W}_t$ of the underlying random shocks.

1   Set $\hat{W}_t = 0$, for $t = p - q + 1, \ldots, p$
2   **for** $t = p + 1, \ldots, n$ **do**
3 | Set $\hat{W}_t = \phi(B)(Z_t) + (1 - \theta(B))\hat{W}_t$
4   **end**
5   **return** $\{\hat{W}_t\}$

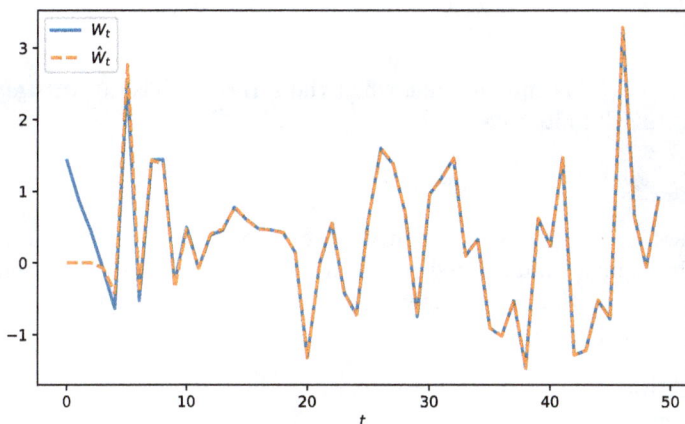

Fig. 9.11: Random shocks W_t and their estimates $\hat{W}_t$ for an ARMA$(3,2)$ process.

9.3 Nonstationary Time Series

In this section, we discuss two important nonstationary generalizations of the classic ARMA model. The first is an *integrated* version of an ARMA process, where integration refers to the finite-difference counterpart to the classic integration of a function from calculus. The second is a seasonal version of the model, that can be used to model seasonal effects; i.e., those effects that reoccur with a certain frequency.

9.3.1 ARIMA Process

One of our earliest examples of a time series was that of a nonstationary time series: the random walk. However, by considering the definition of this time series (Equation (9.6)), it is clear that the differenced time series

$$D_t = X_t - X_{t-1}$$

is stationary. (In fact, this differenced series is simply a white-noise process.) This observation leads us to an obvious question: can we construct a class of nonstationary time series that can be "differentiated" into a stationary time series? This motivates the following definition.

Definition 9.21. *An* autoregressive integrated moving-average process, *denoted* ARIMA(p, d, q), *is a time series* $\{Z_t\}$ *that satisfies the difference equation*

$$\phi(B)\nabla^d Z_t = \theta(B)W_t, \tag{9.51}$$

where the functions θ and ϕ are defined in Equations (9.39) and (9.44), respectively, and where the finite-difference operator ∇ *is defined by* $\nabla = 1 - B$.

Equation (9.51) implies that when the series $\{Z_t\}$ is an ARIMA(p, d, q) process, its dth finite difference

$$D_t = \nabla^d Z_t = (1 - B)^d Z_t$$

is an ARMA(p, q) process. Now, instead of expanding out Equation (9.51) to obtain a single complicated recurrence relation for the ARIMA series Z_t, our next proposition shows how we can start with an ARMA process and integrate to achieve the same result.

Proposition 9.11. *Let $\{D_t\}$ be an ARMA(p, q) process. Then the process $\{Z_t\}$, defined by*

$$Z_t = D_t - \sum_{j=1}^{d} (-1)^j \binom{d}{j} Z_{t-j}. \tag{9.52}$$

for $d \in \mathbb{Z}_+$, is an ARIMA(p, d, q) process and is referred to as the dth integral of the process D_t.

Proof. We see from Equation (9.51) that if Z_t is an ARIMA(p, d, q) process, then $D_t = \nabla^d Z_t$ is an ARMA(p, q) process. The binomial expansion of ∇^d, however, yields

$$D_t = \nabla^d Z_t = (1 - B)^d Z_t = \sum_{j=0}^{d} (-1)^j \binom{d}{j} B^j Z_t.$$

By recognizing $B^j Z_t = Z_{t-j}$ and subtracting all but the $j = 0$th term from both sides, we obtain our result. $\quad\square$

In order to provide some insight into Equation (9.52), let us consider the cases $d = 1$ and $d = 2$. The first integral X_t of an ARMA(p, q) process D_t is defined by

$$X_t = D_t + X_{t-1},$$

which itself is equivalent to a summation of all historic values of D_t: $X_t = \sum_{j=0}^{\infty} D_{t-j}$. Now, if we take the first integral of the series X_t, we obtain

$$Y_t = X_t + Y_{t-1}.$$

From this it must follow, also, that $Y_{t-1} = X_{t-1} + Y_{t-2}$. By substituting these last two equations into the equation for X_t, we obtain

$$\underbrace{(Y_t - Y_{t-1})}_{X_t} = D_t + \underbrace{(Y_{t-1} - Y_{t-2})}_{X_{t-1}}.$$

By rearranging, we find

$$Y_t - 2Y_{t-1} + Y_{t-2} = D_t,$$

equivalent to Equation (9.52) for the case $d = 2$. Higher-order integrals follow in kind.

Equation (9.52) shows that to determine the next term of an ARIMA process, we need only the current term of the base ARMA process and the trailing d terms of the ARIMA process. This is the key observation for modifying Code Block 9.5 for ARIMA models. A generator function for a general ARIMA(p, d, q) process is shown in Code Block 9.6. We use `scipy.special.comb` in order to compute the binomial coefficients. The coefficients for the queue of Z values is calculated on line 8; this is equivalent to removing the first value (`range(1,d+1)` instead of `range(d)`), reversing (which alternates each of the signs), and then multiplying by -1 (which alternates the signs again). We also simplified some of the queue operations with the `put` function.

The once integrated process derived from the ARMA$(3, 2)$ stationary process of Figure 9.10 is shown in Figure 9.12. This is constructed with the line

$$Z = \texttt{arima}(d = 1, \texttt{phi} = [2, -1.5, 0.375], \texttt{theta} = [-1, 0.5]),$$

as usual. Notice the spike in ϕ_{11} in the PACF, and the slow decay of the ACF. These are characteristic features of integrated / nonstationary models which hint at their want for differencing.

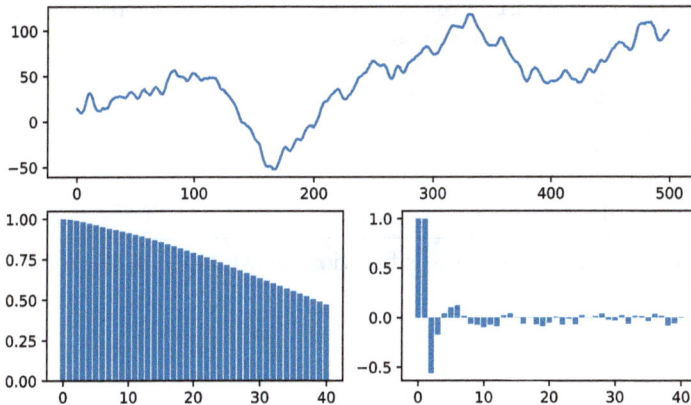

Fig. 9.12: ARIMA$(3, 1, 2)$ process (top); plotted with sample ACF (bottom left) and PACF (bottom right).

```
1   from scipy.special import comb
2   def put(Q, value=0, sigma_w = 1, fill_random=False):
3       if Q.maxsize == 0:
4           return
5       if fill_random:
6           while not Q.full():
7               Q.put(np.random.normal(scale=sigma_w))
8       if value:
9           Q.get()
10          Q.put(value)
11
12  def arima(d=0, phi=[], theta=[], sigma_w=1, burn_length=100,
        max_iter=1e6):
13      p, q, t = len(phi), len(theta), 0
14      D = Queue(maxsize=p)
15      W = Queue(maxsize=q)
16      Z = Queue(maxsize=d)
17      theta, phi = list(reversed(theta)), list(reversed(phi))
18      z_coeff = [(-1)**i * comb(d, i, exact=True) for i in range(d)]
19      put(D, sigma_w=sigma_w, fill_random=True)
20      put(W, sigma_w=sigma_w, fill_random=True)
21      put(Z, sigma_w=sigma_w, fill_random=True)
22
23      while t < max_iter:
24          t += 1
25          assert p==0 or D.full()
26          assert q==0 or W.full()
27          assert d==0 or Z.full()
28          W_t = np.random.normal(scale=sigma_w)
29          D_t = np.dot(phi, D.queue) + np.dot(theta, W.queue) + W_t
30          Z_t = D_t + np.dot(z_coeff, Z.queue)
31
32          put(D, D_t)
33          put(W, W_t)
34          put(Z, Z_t)
35          if t >= burn_length:
36              yield Z_t
```

Code Block 9.6: Generator function for ARIMA(p, d, q) process

9.3.2 Seasonality

Another situation commonly encountered in nonstationary time series is that of *seasonality*, or periodic deviations from the set behavior. Seasonal effects could exist at a variety of frequencies: weekly (e.g., users of a mobile app might have higher engagement during the weekend), monthly (e.g., pension accounts might deposit funds into the stock market at the end/beginning of each month), quarterly (e.g., company quarterly reports), annually (e.g., holiday effects), or at other frequencies.

A simple approach of including seasonality is to decompose a time series into three components: a baseline trend component B_t, a season component S_t, and a random component W_t, so that

$$Z_t = B_t + S_t + W_t.$$

The baseline trend could be linear or polynomial in time, and seasonal, for example, could be a Fourier series representation. Coefficients may then be inferred by regression. This approach, however, often yields poor results, as it is too rigid to accurately model the time-series data. A more modern approach, which is the topic of this section, is to incorporate seasonality directly into the time-series model.

Inspirations: A Purely Seasonal Model

To get a handle on seasonal effects, we begin with a model that is purely seasonal. Fundamentally, seasonal effects occur when data that differ by an integer multiple of a given period s behave in a similar fashion. If a given time series exhibits periodicity at lag s, it is therefore natural to assume that the operator B^s, defined by $B^s Z_t = Z_{t-s}$, plays an important role in capturing the seasonal effect. In particular, let us consider the following *purely seasonal autoregressive moving-average process*

$$\Phi(B^s)S_t = \Theta(B^s)W_t, \tag{9.53}$$

where $W_t \sim \mathrm{WN}(0, \sigma_w^2)$ is a white-noise process and the polynomials Φ and Θ are order P and Q polynomials defined analogously to the polynomials ϕ and θ defined in Equations (9.39) and (9.44), so that the operators

$$\Phi(B^s) = 1 - B^s - B^{2s} - \cdots - B^{Ps}, \tag{9.54}$$
$$\Theta(B^s) = 1 + B^s + B^{2s} + \cdots + B^{Qs} \tag{9.55}$$

constitute a *seasonal autoregressive operator* and a *seasonal moving average operator*, respectively.

To illustrate this, we consider the purely seasonal moving average process

$$S_t = 0.95 S_{t-12} + W_t,$$

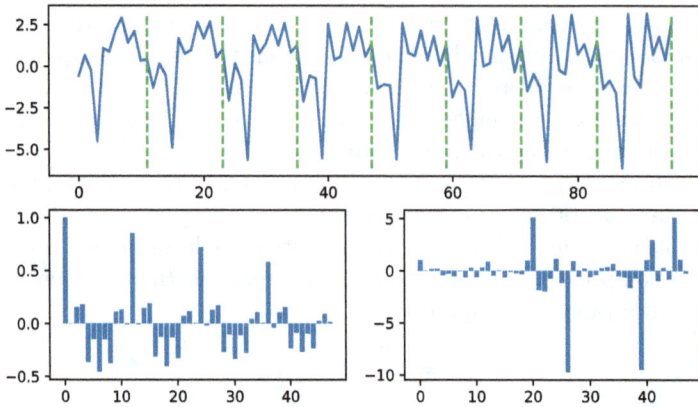

Fig. 9.13: A purely seasonal moving average process with lag 12 and coefficient $\Phi_1 = 0.9$. Increments of period $s = 12$ are illustrated with the vertical
dashed green lines.

with $W_t \sim \text{WN}(0, 0.25)$, as shown in Figure 9.13. Notice the slowly decaying
peaks in the ACF at lags $12k$, for $k = 1, 2, 3, \ldots$. Theoretically, there is a
spike in the PACF at lag 12; however, our numerical simulations failed to
clearly detect this peak. We conclude that the periodic peaks in the ACF
are the clearest signal in identifying the periodicity of a seasonal effect.

SARIMA Models

In order to define our general seasonal autoregressive integrated moving-
average process, we next require two additions to the purely seasonal model
discussed in the previous paragraph: integrating the seasonal effect and
incorporating nonseasonal trends. To achieve the former, we consider the
seasonal finite-difference operator $\nabla_s = 1 - B^s$, so that we may express a
purely seasonal ARIMA model by the equation

$$\Phi(B^s)\nabla_s^D S_t = \Theta(B^s)W_t.$$

This is completely analogous to the basic ARIMA model, except that the
backshift operator B is replaced by its seasonal counterpart B^s. To achieve
the latter (i.e., to integrate nonseasonal effects into our model), we have
the following.

Definition 9.22. *The* seasonal autoregressive moving-average (SARIMA)
process *of* trend order (p, d, q) *and* seasonal order (P, D, Q) *and* seasonal

period (periodicity) s, *denoted* SARIMA $(p,d,q) \times (P,D,Q)_s$, *is the time series* Z_t *defined by equation*

$$\Phi(B^s)\phi(B)\nabla_s^D \nabla^d Z_t = \Theta(B^s)\theta(B)W_t, \tag{9.56}$$

where $W_t \sim \mathrm{WN}(0, \sigma_w^2)$.

We can add an autoregressive seasonal component to the ARIMA$(3,1,2)$ model of Figure 9.12, to obtain

$$(1 - 0.9B^{12})(1 - 2B + 1.5B^2 - 0.375B^3)(1 - B)Z_t = (1 - B + 0.5B^2)W_t.$$

Expanding, we find this model is equivalent to

$$Z_t = 3Z_{t-1} - 3.5Z_{t-2} + 1.875Z_{t-3} - 0.375Z_{t-4}$$
$$+ 0.9Z_{t-12} - 2.7Z_{t-13} + 3.15Z_{t-14} - 1.6875Z_{t-15} + 0.3375Z_{t-16}$$
$$+ W_t - W_{t-1} + 0.5W_{t-2}.$$

The autoregressive component has complex roots $2, 1 \pm i/\sqrt{3}$, and the moving-average component has complex roots $1 \pm i$. We can use our existing `arima` class to simulate this time series by using the command

```
Z = arima(d = 1,
    phi = [2, −1.5, 0.375, 0, 0, 0, 0, 0, 0, 0, 0, 0.9, −1.8, 1.35, −0.3375],
    theta = [−1, 0.5], burn_length = 0, sigma_w = 1).
```

(The ϕ coefficients are obtained by expanding $(1 - 0.9B^{12})(1 - 2B + 1.5B^2 - 0.375B^3)$.) A simulation from this model is shown in Figure 9.14.

Note the general trends of an integrated process: a slowly decaying ACF and a peak in the PACF at ϕ_{11} that immediately cuts off.

9.4 Model Identification, Estimation, and Forecasting

So far, we have explored both stationary and nonstationary time series, showing how we can build and simulate common time series from their definitions. We next discuss the problem from the other direction: given a time series, how do we form an appropriate model and, having decided upon such a model, how do we estimate the coefficients? We will conclude this section with a discussion on forecasting.

9.4.1 Model Identification

So far, we have discussed many types of time series. In this section, we will discuss how to identify which particular model represents a good fit to a given problem; i.e., given an actual real-life time series, how can we go about determining the parameters p, d, and q to fit a general ARIMA(p,d,q) model to the data? Naturally, we will make use of the time series along with its sample ACF and sample PACF in order to determine an appropriate model for our data.

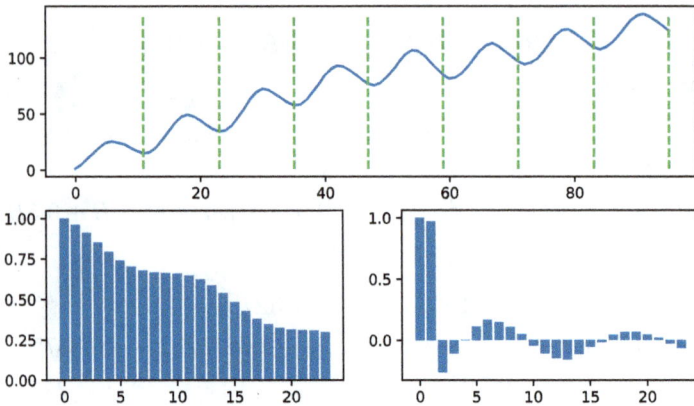

Fig. 9.14: A SARIMA$(3, 1, 2) \times (1, 0, 0)_{12}$ model.

Overview

Our task is to determine the hyperparameters p, d, and q in order to fit a reasonable ARIMA model to a given set of time-series data. A good rule of thumb is that we require at least $n = 50$ observations to fit a good model, and that we consider the sample ACF and PACF only up to a lag of $n/4$. Assuming we have a sufficient set of observations, we may proceed to the model identification stage, which we outline in this section.

Model identification is a process and not an exact science; it requires us to roll up our sleeves and don our detectives' caps in order to uncover ideal candidates for appropriate models based on our data set and its properties. At a high level, the process proceeds according to the following steps:

1. Exploratory analysis
2. Transformation
3. Differencing
4. Fitting a stationary model.

We next discuss each of these topics in turn.

Exploratory Analysis

We begin by plotting our data. By visual inspection of the time series data, we often gain a good feel for which ingredients will be needed to produce an appropriate model. In particular, we examine the data to get a handle on several key questions:

1. Does the time-series plot exhibit any seasonality?

2. Does the series contain a trend?
3. Does the series contain any outliers?
4. Is the variance constant, or does it vary over time?
5. Does the series exhibit any other unusual nonstationary phenomena?

Variance-stabilizing Transforms

A common approach to remedy nonconstant variance is the *Box-Cox power transform*, which typically takes the form of a logarithmic transformation on the dependent variable Z_t. When such transformations are necessary, they should, as a rule of thumb, be performed prior to differencing, as they sometimes require positive values of the series.

In particular, it is common for the variance in a nonstationary process to change based on its expected value; i.e.,

$$\mathbb{V}(Z_t) = cf(\mu_t),$$

for some function f. A common remedy is the Box-Cox power transform, which we define as follows.

Definition 9.23. *The* Box-Cox power transform *is the transformation T defined by*

$$T(Z_t) = \frac{Z_t^\lambda - 1}{\lambda}. \tag{9.57}$$

Common values of the parameter λ are shown in Table 9.2 along with their corresponding transformation. Note that the logarithmic transformation for the case $\lambda = 0$ is defined by taking the limit of Equation (9.57) as $\lambda \to 0$, which, of course, as everybody knows, yields the natural logarithm. For details of how Equation (9.57) may stabilize the variance, see, for example, Box, *et al.* [2016] or Wei [2019].

λ	$T(Z_t)$
-1	$\dfrac{1}{Z_t}$
-0.5	$\dfrac{1}{\sqrt{Z_t}}$
0	$\ln(Z_t)$
0.5	$\sqrt{Z_t}$
1	Z_t

Table 9.2: Common forms of the Box-Cox power transformation.

If such a transformation is necessary, it is typically performed before further analysis. It is therefore common for some authors to refer to the transformed series as the original series.

Finite Differencing

Our next task is to deduce a minimum necessary degree d of differencing, such that our transformed series $X_t = \nabla^d Z_t$ is a stationary ARMA(p, q) series, for some value of p and q. Typically, the degree of differencing d will be either 0, 1, or 2. In order to determine whether or not differencing (or an additional degree of differencing) is required, we examine the sample ACF and sample PACF, and invoke the following rule of thumb.

Proposition 9.12 (Rule of Thumb). *A strong indicator of a nonstationary time series, which therefore requires at least one degree of differencing* $(1 - B)Z_t$, *is the concurrence of a slowly decaying sample* ACF *and a corresponding* PACF *that cuts off sharply after lag 1.*

For example, see Figures 9.12 and 9.14. In practice, it is usually sufficient to inspect the first 20 or so sample ACF coefficients, to determine whether or not they rapidly decay. To see why this rule of thumb is true, consider again Equation (9.50). For a stationary time series, each root of the function $\phi(z)$ lies outside the unit circle, so that the ACF decays exponentially. For a nonstationary ARIMA model, on the other hand, the polynomial coefficient $\varphi(z) = \phi(z)(1 - B)^d$ of Z_t in the ARIMA time series $\varphi(B)Z_t = \theta(B)W_t$ has precisely d roots of unity. If we consider a root $z_0 = 1 + \delta$, where δ is a small positive coefficient, we see that

$$\rho_h \approx a(1 - h\delta),$$

for large h as $\delta \to 0^+$, for some coefficient a, illustrating the slow decay of the ACF.

To understand the statement regarding PACF, consider the following. First, note that the PACF and ACF of lag 1 are equivalent, since

$$\phi_{11} = \mathrm{COV}(X_t, X_{t+1}) = \rho_1,$$

so that ϕ_{11} should be close to 1, since the ACF slowly decays. For lags $h > 1$, the PACF conditions on the intermediary values. Again assuming a root $z_0 = 1 - \delta$, so that $\rho_h \approx 1 - h\delta$, we have

$$\rho_1^2 \approx 1 - 2\delta + \delta^2$$
$$\rho_2 = 1 - 2\delta.$$

Now, from the recurrence relation Equation (9.24) for the PACF, we have

$$\phi_{22} = \frac{\rho_2 - \rho_1^2}{1 - \rho_1^2} \approx \frac{\delta}{2 - \delta},$$

which goes to zero as $\delta \to 0^+$. This shows that the correlation between values of the time series is broken when conditioning on at least one intermediary value.

By considering the ACF and PACF of the differenced series $(1 - B)^d Z_t$ for $d = 0, 1, 2, \ldots$, we can then choose the minimum degree of differencing d that breaks these patterns.

The Peril of Overdifferencing

It is important not to difference our time series more times than required; though some authors have argued these perils are much less than the perils of *underdifferencing*. Nonetheless, overdifferencing may lead to a tendency to overfit our model. Take, for example, the random walk, our first encounter with a nonstationary time series. The differenced equation is, of course, simple white noise. However, if we apply finite differencing to a white-noise process, we actually obtain a MA(1) process! (See Exercise 9.5.) Thus, by applying two rounds of differencing, we would incorrectly identify the random walk as an ARIMA$(0, 2, 1)$ process, as opposed to its simpler and correct representation as an ARIMA$(0, 1, 0)$ process.

Fitting a Stationary Model

Now that we have applied a sufficient degree of differencing, our next task is to fit a stationary ARMA model to the resultant time series. In order to fit the parameters p and q, we examine the ACF and PACF of the differenced time series $X_t = (1 - B)^d Z_t$, and compare them with our theoretical results from Table 9.1. In practice, we typically find that we can fit a reasonable model with orders of p and q less than or equal to three.

9.4.2 Estimation

Given a stationary ARMA(p, q) time series with known (or suspected) values of the hyperparameters p and q, we next turn to the problem of estimating the model's $p + q + 1$ parameters $\xi = (\phi_1, \ldots, \phi_p, \theta_1, \ldots, \theta_q, \sigma_w^2)$. We will discuss two methods for achieving such a result: method of moments and maximum likelihood. The latter takes several forms, as it is more common to solve for an approximation of the likelihood function than the true likelihood itself.

Method of Moments for Autoregressive Processes

Our first approach involves the method of moments, which we previously discussed in the form of Satterthwaite's approximation (Definition 3.17). Formally, we define the approach as follows.

Definition 9.24. *The* method of moments *is a method for estimating a model's parameters that involves computing the sample moments (mean, variance, etc.) and equating them to their theoretical counterparts, expressed as a function of the model's parameters.*

In particular, for time-series data, we consider the sample mean $\overline{Z}$, sample variance $\hat{\gamma}_0$, and the sample autocorrelations $\hat{\rho}_j$. Note that we may use the estimate $\hat{\mu} = \overline{Z}$, and assume, without loss of generality, that our time series has zero mean.

In particular, the method of moments is commonly used with an AR(p) process, as follows. The method of moments estimates for such a process are based on the following result.

Proposition 9.13. *An* AR*(p) process satisfies the following system of equations*

$$\rho_1 = \phi_1 + \phi_2\rho_1 + \cdots + \phi_p\rho_{p-1}$$
$$\rho_2 = \phi_1\rho_1 + \phi_2 + \cdots + \phi_p\rho_{p-2}$$
$$\vdots \qquad \vdots$$
$$\rho_p = \phi_1\rho_{p-1} + \phi_2\rho_{p-2} + \cdots + \phi_p, \qquad (9.58)$$

which are known as the Yule–Walker equations.

Proof. The p Yule–Walker equations are equivalent to Equation (9.46), explicitly expressed for $h = 1, \ldots, p$. □

If we let

$$\rho = \begin{bmatrix} \rho_1 \\ \rho_2 \\ \vdots \\ \rho_p \end{bmatrix}, \qquad \phi = \begin{bmatrix} \phi_1 \\ \phi_2 \\ \vdots \\ \phi_p \end{bmatrix}, \qquad R = \begin{bmatrix} 1 & \rho_1 & \rho_2 & \cdots & \rho_{p-1} \\ \rho_1 & 1 & \rho_1 & \cdots & \rho_{p-2} \\ \rho_2 & \rho_2 & 1 & \cdots & \rho_{p-3} \\ \vdots & \vdots & \vdots & \ddots & \vdots \\ \rho_{p-1} & \rho_{p-2} & \rho_{p-3} & \cdots & 1 \end{bmatrix},$$

we see that the Yule–Walker equations may be expressed in matrix form as

$$\rho = R\phi.$$

The *Yule–Walker estimates* $\hat{\phi}_1, \ldots, \hat{\phi}_p$ for the autoregression parameters $\phi_1, \ldots, \phi_p$ are given by inverting this relation and substituting the sample autocorrelations for $\rho_1, \ldots, \rho_p$:

$$\hat{\phi} = \hat{R}^{-1}\hat{\rho}. \qquad (9.59)$$

In order to estimate the variance σ_w^2 of the white noise process, we consider

$$\gamma_0 = \mathbb{E}[Z_t Z_t]$$
$$= \mathbb{E}[Z_t(\phi_1 Z_{t-1} + \cdots + \phi_p Z_{t-p} + W_t)]$$
$$= \phi_1\gamma_1 + \cdots + \phi_p\gamma_p + \sigma_w^2,$$

which may be rearranged to obtain

$$\sigma_w^2 = \gamma_0 \left(1 - \sum_{j=1}^{p} \phi_j \rho_j \right).$$
(9.60)

By substituting the sample variance $\hat{\gamma}_0$ for γ_0 and the sample autocorrelations $\hat{\rho}_j$ for their theoretical counterparts ρ_j, we obtain an estimate $\hat{\sigma}_w^2$ for the variance of the white noise, completing our estimate for the AR(p) parameters.

Asymptotically, we have

$$\sqrt{n} \left(\hat{\phi} - \phi \right) \rightarrow N \left(0, \sigma_w^2 R \right)$$

in distribution and $\hat{\sigma}_w^2 \rightarrow \sigma_w^2$ in probability as $n \rightarrow \infty$.

Maximum Likelihood Estimation

For a stationary ARMA(p, q) process, we wish to express the likelihood function $L(\xi; Z)$ for the parameters

$$\xi = (\phi_1, \ldots, \phi_p, \theta_1, \ldots, \theta_q, \sigma_w^2),$$

so that we may then apply standard techniques, such as the Newton–Raphson method or the method of scoring (see Section 5.1), in order to determine the value $\hat{\xi}$ that maximizes the likelihood function. The exact likelihood function for a general ARMA process, however, is quite complicated; see Newbold [1974] for details. We will instead consider two approximations, known as conditional and unconditional likelihoods, where conditioning refers to the option of conditioning the likelihood on an assumed set of initial conditions.

Conditional Maximum Likelihood

In order to derive the likelihood equation, we rewrite Equation (9.47) in terms of W_t as

$$W_t = Z_t - \phi_1 Z_{t-1} - \cdots - \phi_p Z_{t-p} - \theta_1 W_{t-1} - \cdots - \theta_q W_{t-q}.$$
(9.61)

Assuming the random variables W_t constitute an IID Gaussian white-noise process, we may express the joint density of $w = (w_1, \ldots, w_n)$ as

$$p(w|\theta, \phi, \sigma_w^2) = (2\pi\sigma_w^2)^{-n/2} \exp \left[\frac{-1}{2\sigma_w^2} \sum_{t=1}^{n} w_t^2 \right].$$
(9.62)

The problem with this expression is that the values of the variables $w_1, \ldots, w_n$ cannot be solved using Equation (9.61) due to the lack of initial conditions. However, if we *assume* prior values

$$Z_* = \langle Z_{1-p}, \ldots, Z_{-1}, Z_0 \rangle \tag{9.63}$$
$$W_* = \langle W_{1-q}, \ldots, W_{-1}, W_0 \rangle, \tag{9.64}$$

we may then successively calculate $w_1, \ldots, w_n$, using Equation (9.61) with the initial conditions Equations (9.63) and (9.64), to obtain

$$w_t(\phi, \theta | w_*, z_*, z),$$

where $z = (z_1, \ldots, z_n)$ represents the observed values of the time series. If we define the conditional sum of squares

$$S_*(\phi, \theta) = \sum_{t=1}^{n} w_t^2(\phi, \theta | w_*, z_*, z), \tag{9.65}$$

we may then express the conditional log-likelihood of the density Equation (9.62) as

$$l_*(\phi, \theta, \sigma_w^2) = -\frac{n}{2} \ln(2\pi\sigma_w^2) - \frac{S_*(\phi, \theta)}{2\sigma_w^2}. \tag{9.66}$$

Note, in particular, that the dependence on $l_*(\phi, \theta, \sigma_w^2)$ on ϕ and θ is only through the sum-of-squares term $S_*(\phi, \theta)$. It is therefore sufficient to minimize $S_*(\phi, \theta)$ to determine the maximum-likelihood estimates of ϕ and θ.

One approach to assuming w_* and z_* is to replace these initial conditions with their expected values; i.e., $z_* = \bar{z}$ and $w_* = 0$.

Alternatively, if we have a sufficient amount of data, we may assume that $w_p = w_{p-1} = \cdots = w_{p+1-q} = 0$ and then calculate w_t for $t \geq (p+1)$ using the difference formula of Equation (9.61) (see Algorithm 9.1). We then replace the sum of squares with

$$S_*(\phi, \theta) = \sum_{t=p+1}^{n} w_t^2(\phi, \theta | z). \tag{9.67}$$

After estimating the parameters, we then estimate the variance of the white-noise process using

$$\hat{\sigma}_w^2 = \frac{S_*(\hat{\phi}, \hat{\theta})}{df},$$

where df is the number of degrees of freedom, given by the difference $df = (n-p) - (p+q+1)$ between the number of data points used to estimate S_* and the number of parameters estimated.

Unconditional Maximum Likelihood

Box, *et al.* [2016] proposed an interesting alternative approach to the problem of initial conditions, which has become known as the *unconditional maximum likelihood estimate.* For an additional description, see also Wei [2019].

The idea, in short, is to replace the conditional sum of squares, given by Equation (9.65), in the log-likelihood equation Equation (9.66), with an unconditional sum of squares

$$S(\phi, \theta) = \sum_{t=-\infty}^{n} \mathbb{E}\left[W_t | \phi, \theta, z\right]^2 \approx \sum_{t=-m}^{n} \mathbb{E}\left[W_t | \phi, \theta, z\right]^2,$$

for some sufficiently large $m > 0$. Here, $\mathbb{E}[W_t|\phi,\theta,z]$ represents the conditional expected values for W_t, given the parameters ϕ and θ and the observed time-series values z. For $t \leq 0$, we rewrite the forward form of the ARMA(p,q) model (Equation (9.47)) in its equivalent *backward form*:

$$\phi(F)Z_t = \theta(F)W_t,$$

where F is the *forwardshift operator*, defined analogously to Definition 9.16:

$$FZ_t = Z_{t+1}.$$

This allows us to *backcast* our time series to a sequence of historic values $Z_0, Z_{-1}, Z_{-2}, \ldots$.

9.4.3 Forecasting

The point of time series analysis is, of course, not simply to build a model that describes variables we have already observed, but also, and most importantly, to predict, or *forecast*, future values from the given series of study. We take on this topic presently; for additional references, see the usual suspects Shumway and Stoffer [1999] and Wei [2019], as well as Brockwell and Davis [2016]. Our goal is to predict both the expected value as well as the variance—from which we may obtain error bars—of future unseen values of a given time series model.

Bootstrap Forecasts

The simplest way to approximate a time-series forecast and its error bounds is using the bootstrap. From our observed history $Z_1, \ldots, Z_n$, we may compute d finite differences to obtain the differenced series D_t, from whence we may infer the values of the white-noise process W_t using Algorithm 9.1. We may then use the trailing d values of Z, p values of D, and q values of W as initial conditions for our time-series model, from which we may run n_{boot} simulations using a random variable generator for the future values of W_t. The steps for this process are given in Algorithm 9.2.

In order to implement this approach in Python, we require the ability to simulate the future of a time series with a *prescribed* set of initial conditions.

Algorithm 9.2: Generating bootstrap forecasts.

Input: time-series data Z_t, for $t = 1, \ldots, n$;
an ARIMA(p, d, q) model with known parameters d, θ, ϕ, σ_w^2;
number n_{boot} of bootstrap samples.

Output: Forecasts $\hat{Z}_t$ for $t = n + 1, \ldots, n + m$.

1 Compute the differenced time series $X_t = (1 - B)^d Z_t$

2 Use Algorithm 9.1 to estimate the shocks $\hat{W}_t$ for $t = n - q + 1, \ldots, n$

3 **for** $b = 1, \ldots, n_{\text{boot}}$ **do**

4 $\quad$ Set $\tilde{Z}_t^b = Z_t$ for $t = 1, \ldots, n$

5 $\quad$ Set $V_t = \hat{W}_t$ for $t = 1, \ldots, n$

6 $\quad$ Generate a sequence of random shocks $V_t \sim \text{WN}(0, \sigma_w^2)$ for $t = n + 1, \ldots, n + m$

7 $\quad$ Calculate the bootstrap sample $\{\tilde{Z}_t^b\}$ for $t = n + 1, \ldots, n + m$ using the ARIMA equation $\phi(B)(1 - B)^d \tilde{Z}_t = \theta(B) V_t$.

8 **end**

9 **return** $\{\{\tilde{Z}_t^b\}_{t=n+1}^{n+m}\}_{b=1}^{n_{\text{boot}}}$

```
def arima(d=0, phi=[], theta=[], sigma_w=1, burn_length=100,
    max_iter=1e6, D0=[], W0=[], Z0=[]):
    ### Modify lines 19-21
    if len(D0)>0 or len(W0)>0:
        burn_length = 0
        assert len(D0) == p
        assert len(Z0) == d
        assert len(W0) == q
        for d0 in D0:
            D.put(d0)
        for w0 in W0:
            W.put(w0)
        for z0 in Z0:
            Z.put(z0)
    else:
        put(D, sigma_w=sigma_w, fill_random=True)
        put(W, sigma_w=sigma_w, fill_random=True)
        put(Z, sigma_w=sigma_w, fill_random=True)
    ###
```

Code Block 9.7: Modified section of generator function for ARIMA(p, d, q) process; modified from Code Block 9.6.

This requires a small modification to our ARIMA generator function (Code Block 9.6), which is shown in Code Block 9.7.

We implement Algorithms 9.1 and 9.2 in Code Block 9.8. Note that we set the first column of the matrix z_boots to equal the final value of the observed time series, as to connect our forecasts and their error bars with the anchoring point when we produce our graphs.

```
1   phi, theta = [2, -1.5, 0.375], [-1, 0.5]
2   n, m, n_boot = 100, 40, 1000
3   Z = arima(phi=phi, theta=theta)
4   z = [next(Z) for i in range(n)]
5   w_hat = np.zeros(n)
6   for i in range(3, n):
7       w_hat[i] = z[i] - 2*z[i-1] + 1.5*z[i-2] - 0.375*z[i-3] +
            w_hat[i-1] - 0.5*w_hat[i-2]
8   Z0, D0, W0 = [], z[-3:], w_hat[-2:]
9   z = z + [next(Z) for i in range(m)]  # actual future-values
10  z_boots = np.zeros((n_boot, m+1))
11  z_boots[:, 0] = z[n-1]
12  for i in range(n_boot):
13      Z_b = arima(phi=phi, theta=theta, Z0=Z0, D0=D0, W0=W0)
14      z_boots[i, 1:] = [next(Z_b) for i in range(m)]
```

Code Block 9.8: Bootstrap forecasts with error bounds for an ARMA(3, 2) process; results shown in Figure 9.15.

The result is shown in Figure 9.15 (top). The solid blue line represents the actual time series, and the red line represents the average bootstrap (solid red) and its 95% error bounds (dashed red), for the bootstrap forecasts based on time $n = 100$. The similar result for an ARIMA(3, 1, 2) process is also shown in Figure 9.15 (bottom). We leave it as an exercise for the reader to modify Code Block 9.8 to produce a bootstrap forecast for the integrated model (see Exercise 9.6).

From these bootstrap simulations, we note that the confidence intervals for a stationary ARMA process level out at a constant, whereas the confidence intervals for the integrated ARIMA process grow linearly over time. Moreover, the forecasts for the stationary ARMA process quickly decay back to the process mean. We will establish a theoretical basis for these observations over the coming pages.

In generating our bootstrap forecasts and, in general, in the other forecasting techniques we will soon discuss, we have assumed that the model parameters $\phi_1, \ldots, \phi_p$, $\theta_1, \ldots, \theta_q$, and σ_w^2 are known. One advantage to the bootstrap forecast is that when we include error bounds in our modeling parameters, we can still produce bootstrap forecasts by simply sampling the

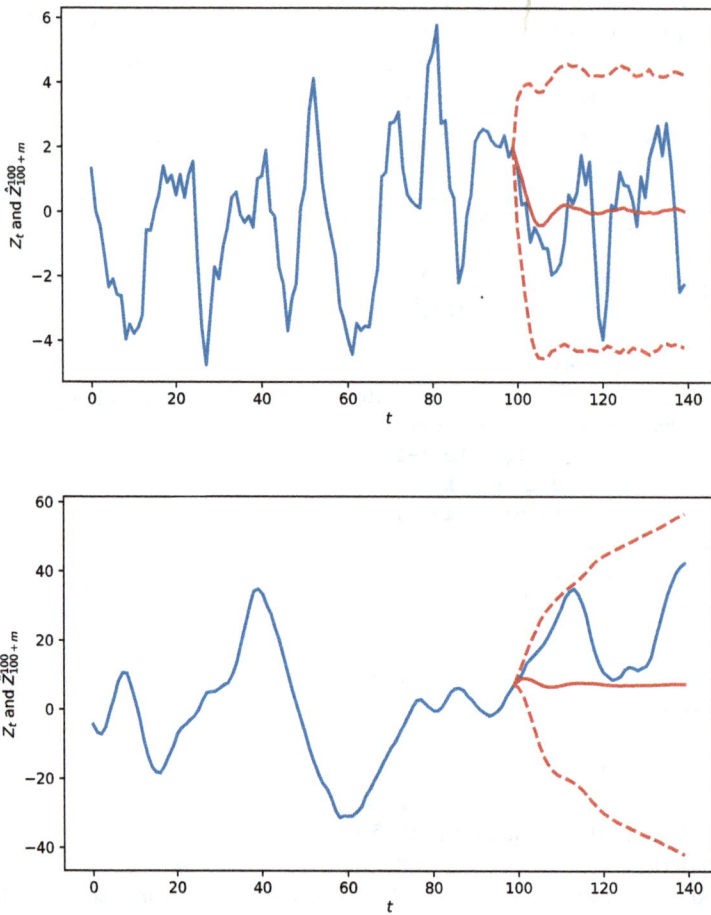

Fig. 9.15: An ARMA(3, 2) process (top) and an ARIMA(3, 1, 2) process (bottom), observed up to time $n = 140$ (blue), with forecasts (red) and 95% error bounds (red dashed) anchored at time $n = 100$.

joint distribution from our parameter space for each individual trace. Thus, the bootstrap forecast, with minor modification, constitutes a straightforward and elegant way to incorporate the errors of our parameter estimates into our forecasts.

Best Linear Predictors

In the context of forecasting, we wish to use a time series' entire observed history, $x_1, \ldots, x_n$, to predict future values. Our prediction window, in the context of Equation (9.16), is therefore coincident with our forecast origin $t = n$. We are therefore interested in determining the *m-step ahead predictor based at forecast origin n*

$$\hat{X}^n_{n+m} = \mathbb{E}\left[X_{n+m}|X_1,\ldots,X_n\right] \tag{9.68}$$

for $m \geq 1$. When this quantity is expressed as a linear regression on the values $X_1, \ldots, X_n$, it is also referred to as the *best linear predictor* and may be expressed as

$$\hat{X}^n_{n+m} = \phi^m_{n0} + \phi^m_{n1} X_n + \cdots + \phi^m_{nn} X_1, \tag{9.69}$$

where $\phi^m_{n0} = 0$ for a mean-zero process. For the case of $m = 1$ (i.e., the *best one-step ahead predictor*), we drop the superscript so that $\phi_{ni} = \phi^1_{ni}$, consistent with Equation (9.18).

For theoretical reasons, it is also useful to define the m-step ahead predictor at forecast origin n *based on the infinite past* as the quantity

$$\tilde{X}^n_{n+m} = \mathbb{E}\left[X_{n+m}|X_n, X_{n-1},\ldots\right]. \tag{9.70}$$

In general, $\hat{X}^n_{n+m} \neq \tilde{X}^n_{n+m}$; for large sample sizes, however, we may regard the two predictors as approximately indistinguishable.

Note that Equations (9.68) and (9.70) may be expressed formally in terms of the backward autoregression (Equation (9.16)) as

$$\hat{X}^n_{n+m} = \hat{X}^{\langle n]m}_{n+m} \quad \text{and} \quad \tilde{X}^n_{n+m} = \hat{X}^{\langle \infty]m}_{n+m},$$

respectively.

Long-range Results for Stationary Processes

For a mean-zero stationary process, we may write

$$Z_{n+m} = \sum_{j=0}^{\infty} \psi_j W_{n+m-j}, \tag{9.71}$$

where we have defined $\psi_0 = 1$. Taking the conditional expectation based on the infinite past anchored at point $t = n$, we have

$$\tilde{Z}^n_{n+m} = \sum_{j=m}^{\infty} \psi_j w_{n+m-j}, \tag{9.72}$$

where $w_n, w_{n-1}, \ldots$ represent the observed (inferred) values of the white-noise process up to time $t = n$. Since the coefficients ψ_j constitute an absolutely convergent series, we conclude that

$$\lim_{m \to \infty} \tilde{Z}^n_{n+m} = 0.$$

Moreover, by combining Equations (9.71) and (9.72), we see that the prediction errors P^n_{n+m} must satisfy

$$P^n_{n+m} = \mathbb{E}\left[(Z_{n+m} - \tilde{Z}^n_{n+m})^2\right] = \sigma^2_w \sum_{j=0}^{m-1} \psi^2_j,$$

from whence it follows that

$$\lim_{m \to \infty} P^n_{n+m} = \gamma_0.$$

These theoretical results are illustrated in the Figure 9.15 (top), where our forecasts decay to zero and the error bounds decay to a constant.

   This makes a great deal of intuitive sense: since the autocorrelations of a stationary process decay exponentially fast, the ability to predict the future is limited to a finite horizon, after which our predictions relax to the process mean and variance.

   It can further be shown that the error bounds for a nonstationary ARIMA(p, d, q) process grow as $P^n_{n+m} \sim O(m^d)$ as $m \to \infty$, similar to the behavior of the $d = 1$ model we examined in Figure 9.15 (bottom).

Recursive Relation for Invertible Processes

For an invertible process, we may write

$$W_{n+m} = \sum_{j=0}^{\infty} \pi_j Z_{n+m-j},$$

where we take $\pi_0 = 1$. Now, by recalling that

$$\mathbb{E}[W_{n+m}|X_n, X_{n-1}, \ldots] = 0,$$

we see that by taking the conditional expected value of both sides based on the infinite past anchored at time $t = n$ and rearranging, we obtain

$$\tilde{Z}^n_{n+m} = -\sum_{j=1}^{m-1} \tilde{Z}^n_{n+m-j} - \sum_{j=m}^{\infty} \pi_j Z_{n+m-j}. \qquad (9.73)$$

This equation may be applied recursively for $m = 1, 2, \ldots$ to obtain estimates $\tilde{Z}^n_{n+1}, \tilde{Z}^n_{n+2}, \ldots$. Invertibility implies that the weights π_j form a convergent series. In practice, they decay sufficiently rapidly, such that the preceding equation may be approximating using only a finite lookback window.

The Prediction Equations

We next discuss how one may calculate the coefficients of Equation (9.69) by minimizing the squared error. We will then show how these resulting prediction equations can be expressed in matrix format.

Proposition 9.14. *Given time-series data $x_1, \ldots, x_n$, up to the forecast origin n, the coefficients α_j of the best linear predictor Equation (9.68) that minimize the expected square error*

$$P_{n+m}^n = \mathbb{E}\left[\left(X_{n+m} - \hat{X}_{n+m}^n\right)^2\right] \tag{9.74}$$

are obtained by solving the following prediction equations

$$\mathbb{E}\left[X_{n+m} - \hat{X}_{n+m}^n\right] = 0 \tag{9.75}$$

$$\mathbb{E}\left[\left(X_{n+m} - \hat{X}_{n+m}^n\right)X_k\right] = 0, \tag{9.76}$$

where $k = 1, \ldots, n$, for the coefficients $\alpha_0, \alpha_1, \ldots, \alpha_n$.

Proof. Equations (9.75) and (9.76) are obtained by differentiating the expected squared-error loss Equation (9.74) with respect to the coefficients $\alpha_0, \ldots, \alpha_n$. (See Exercise 9.7.) $\square$

For a stationary process, by substituting Equation (9.69), we can show that Equations (9.75) and (9.76) are equivalent to

$$\phi_{n0}^m = \mu\left(1 - \sum_{i=1}^n \phi_{ni}^m\right),$$

$$\gamma_{n+m-k} = \phi_{n1}^m \gamma_{n-k} + \cdots + \phi_{nn}^m \gamma_{k-1},$$

for $k = 1, \ldots, n$. In matrix notation, we may express this second equation as

$$\Gamma_n \phi_n^m = \gamma_n^m, \tag{9.77}$$

where we have defined the $n \times n$ matrix Γ_n and the $n \times 1$ vectors ϕ_n^m and γ_n^m as

$$[\Gamma_n]_{ij} = \gamma_{i-j}, \text{ for } i, j = 1, \ldots, n,$$
$$\phi_n^m = \langle \phi_{n1}^m, \ldots, \phi_{nn}^m \rangle,$$
$$\gamma_n^m = \langle \gamma_m, \ldots, \gamma_{m+n-1} \rangle,$$

respectively. Moreover, one can show that the prediction error is given by

$$P_{n+m}^n = \gamma_0 - (\gamma_n^m)^T \Gamma_n^{-1} \gamma_n^m. \tag{9.78}$$

The Durbin–Levinson Algorithm

For large n, the system of equations Equation (9.77) may be difficult and time consuming to solve. We will conclude this chapter with an introduction to two recursive approaches, that allows one to use a sequence of one-step predictors to generate forecasts recursively. We refer the reader to Brockwell and Davis [2016] for a discussion of these methods based on the theory of projection operators.

Proposition 9.15 (Durbin–Levinson algorithm). *The one-step ahead prediction coefficients ϕ_{nj} for a mean-zero stationary process can be computed recursively using the equations*

$$\phi_{11} = \rho_1 \tag{9.79}$$

$$\phi_{nn} = \frac{\rho_n - \sum_{j=1}^{n-1} \phi_{n-1,j}\rho_{n-j}}{1 - \sum_{j=1}^{n-1} \phi_{n-1,j}\rho_j} \ for \ n \geq 2 \tag{9.80}$$

$$\phi_{nj} = \phi_{n-1,j} - \phi_{nn}\phi_{n-1,n-j} \ for \ n \geq 2, j = 1, \ldots, n-1, \tag{9.81}$$

$$P_1^0 = \gamma_0 \tag{9.82}$$

$$P_{n+1}^n = P_n^{n-1}(1 - \phi_{nn}^2) \ for \ n \geq 1. \tag{9.83}$$

Note that Equations (9.79)–(9.81) constitute a restatement of Theorem 9.1.

The Innovations Algorithm

Finally, we state a result due to Brockwell and Davis [1992]. An *innovation* is the one-step prediction error, defined by

$$U_n = X_n - \hat{X}_n^{n-1}.$$

Now, it can be shown that the innovations are related to the one-step predictions via a lower-triangular matrix Θ with null diagonals:

$$
\begin{bmatrix} \hat{X}_1^0 \\ \hat{X}_2^1 \\ \vdots \\ \hat{X}_n^{n-1} \end{bmatrix}
=
\begin{bmatrix}
0 & 0 & 0 & \cdots & 0 \\
\theta_{11} & 0 & 0 & \cdots & 0 \\
\theta_{22} & \theta_{21} & 0 & \cdots & 0 \\
\vdots & \vdots & \vdots & \ddots & \vdots \\
\theta_{n-1,n-1} & \theta_{n-1,n-2} & \theta_{n-1,n-3} & \cdots & 0
\end{bmatrix}
\cdot
\begin{bmatrix} U_1 \\ U_2 \\ U_3 \\ \vdots \\ U_n \end{bmatrix},
$$

such that

$$\hat{X}_{n+1}^n = \sum_{j=1}^{n} \theta_{nj} U_{n+1-j},$$

for $n = 1, 2, \ldots$. For additional discussion, see Brockwell and Davis [2016] or Shumway and Stoffer [1999].

Proposition 9.16 (Innovations algorithm). *Given a time series with finite second moments, the coefficients θ_{nj} may be computed recursively via the relations*

$$\theta_{n,n-j} = (P_{j+1}^j)^{-1} \left[\gamma(n+1, j+1) - \sum_{i=0}^{j-1} \theta_{j,j-1} \theta_{n,n-i} P_{i+1}^i \right], \quad (9.84)$$

$$P_1^0 = \gamma(1,1), \quad (9.85)$$

$$P_{n+1}^n = \gamma(n+1, n+1) - \sum_{j=0}^{n-1} \theta_{n,n-j}^2 P_{j+1}^j. \quad (9.86)$$

Moreover, for an m-step predictor, the innovations algorithm can be used to show

$$\hat{X}_{n+m}^n = \sum_{j=m}^{n+m-1} \theta_{n+m-1,j} U_{n+m-j}, \quad (9.87)$$

$$P_{n+m}^n = \gamma(n+m, n+m) - \sum_{j=h}^{n+m-1} \theta_{n+m-1,j}^2 P_{n+m-j}^{n+m-j-1}. \quad (9.88)$$

For details, see references.

Finally, we note that while the Durbin–Levinson algorithm generates the coefficients to express the one-step predictors in the form

$$\hat{X}_{n+1}^n = \sum_{j=1}^n \phi_{nj} X_{n+1-j},$$

the innovations algorithm represents the analogous formula,

$$\hat{X}_{n+1}^n = \sum_{j=1}^n \theta_{nj} \left(X_{n+1-j} - \hat{X}_{n+1-j}^{n-j} \right),$$

expressed in terms of linear combinations of the innovations. This is useful as the innovations, unlike the individual predictors, are uncorrelated. Moreover, the innovations algorithm has a tidy representation for the stationary ARMA(p, q) process.

Problems

9.1. Prove Equation (9.25).

9.2. By exploiting the recursive nature of the AR(1) process

$$Z_t = \phi Z_{t-1} + W_t,$$

show that it is equivalent to

$$Z_t = \phi^k Z_{t-k} + \sum_{j=0}^{k-1} \phi^j W_{t-j},$$

for any $k \in \mathbb{Z}_+$. Conclude that the AR(1) process is a stationary linear process (Equation (9.12)) if and only if $|\phi| < 1$, and show that this is equivalent to the condition $\phi(z) = 1 - \phi z \neq 0$ for $|z| \leq 1$.

9.3. Write a generator function for an AR(p) process, following a similar structure as in Code Block 9.3.

9.4. Reproduce Figure 9.11. By simulating the ARMA(3, 2) process

$$Z_t = 2Z_{t-1} - 1.5Z_{t-2} + 0.375Z_{t-3} + W_t - W_{t-1} + 0.5W_{t-2};$$

i.e., using

$$\phi(B) = 1 - 2B + 1.5B^2 - 0.375B^3$$
$$\theta(B) = 1 - B + 0.5B^2.$$

Repeat for an ARIMA(3, d, 2) process using the same polynomials $\phi(B)$ and $\theta(B)$ and $d = 1, 2$. What do you observe?

9.5. Overdifferencing In this exercise, we will illustrate the danger of *overdifferencing* when fitting an ARIMA model.
(a) Show that the differenced series for a random walk is a white noise process. Explain why this implies that the random walk is equivalent to an ARIMA(0, 1, 0) process.
(b) Show the the differenced series for white noise is a MA(1) process. Conclude that by overdifferencing a random walk, one might incorrectly identify a random walk as an ARIMA(0, 2, 1) process, thereby overfitting the model.

9.6. Modify Code Block 9.8 to obtain bootstrap forecasts for a simulated ARIMA(3, 1, 2) model.

9.7. Show that by differentiating the expected squared-error loss Equation (9.74), one obtains the prediction equations Equations (9.75) and (9.76).

10

Night at the Casino

Bayesian inference constitutes a radical departure from the frequentist point of view, which we have exclusively dealt with up until now.

Our main go-to reference for Bayesian inference is Gelman, *et al.* [2014]. For a discussion of Bayesian inference from a graph-theory point of view, see Barber [2012]. Finally, an important reference on probabilistic programming, i.e., how we should go about performing Bayesian inference using simulation, is Davidson-Pilon [2016]. Another excellent source that covers many classic topics in statistics from a Bayesian viewpoint is McElreath [2016].

10.1 Bayesian Inference

10.1.1 Bayes' Law

Bayesian versus Frequentist Points of View

Bayesian inference represents a paradigm shift in how we think of and view probabilities. From the *frequentist* point of view, probability represents the long-run frequency of the occurrence of an event under identical circumstances. This definition of probability is itself often a counterfactual, as the unending supply of IID events is never fully realized in our universe. In some contexts this is okay. When flipping a fair coin, for example, we can easily imagine an unending sequence of flips, and we understand that a 50% probability of heads means that the frequency of heads will approach 50% as we continue to flip the coin more and more. But what about a 30% probability of rain on Tuesday? Next Tuesday can only happen once, and it will either rain or it will not rain, so are we to interpret a "30% chance of rain" as a statement as to what is happening in many, many parallel universes, completely identical except for Tuesday's precipitation?

From the *Bayesian* point of view, probability represents our belief in the likelihood of the occurrence of an event. From this perspective, a 30%

chance of rain makes perfect sense: it is a statement about our belief whether or not it will rain on Tuesday.

At first glance, the distinction between the two definitions can seem superficial: are we not just repackaging the same statement? However, on closer examination, the Bayesian viewpoint represents a dramatic departure from the frequentist point of view. In particular, the Bayesian viewpoint offers two key advantages:

1. Our belief is not only limited to a point estimate, but also can be represented by a distribution;
2. Our belief is capable of changing; i.e., our understanding of a natural phenomenon should improve with experience.

We will explore the implications and precise meaning of these statements throughout the remainder of the chapter.

Bayes' Law

We begin with the classical statement of Bayes' law, which relates the conditional probability of two events.

Proposition 10.1. *Let $(\Omega, \mathcal{E}, \mathbb{P})$ be a probability space (see Definition 1.3). Then, for any $A, B \in \mathcal{E}$, we have*

$$\mathbb{P}(A|B) = \frac{\mathbb{P}(B|A)\mathbb{P}(A)}{\mathbb{P}(B)}. \tag{10.1}$$

Proof. This follows immediately from the definition of conditional probability Definition 1.5, since

$$\mathbb{P}(A|B) = \frac{\mathbb{P}(AB)}{\mathbb{P}(B)} = \frac{\mathbb{P}(A)\mathbb{P}(B|A)}{\mathbb{P}(B)},$$

which is equivalent to Equation (10.1). $\square$

Example 10.1. Consider a medical diagnostic test for a rare condition, which affects 0.1% of the population, that is 99% sensitive and 99% specific. Recall that 99% sensitive means that the true positive rate is 99%; i.e., 99% of those afflicted with the condition will be tested as positive. Similarly, 99% specific means that the true negative rate is 99%, or that 99% of those not afflicted will correctly be identified with a negative test result. Suppose now that Joe takes the diagnostic test, which results in a positive result. What is the probability that Joe has the condition?

Let C be the event that a person has the condition, and let $+$ and $-$ represent positive and negative test results, respectively. In order to find the conditional probability of C, given Joe's positive test result, we must apply Bayes' law:

$$P(C|+) = \frac{P(+|C)P(C)}{P(+)}.$$

Since the test is 99% sensitive, we know that $P(+|C) = 0.99$. Moreover, we know that the overall prevalence of the condition is 0.1%, so that $P(+) = 0.001$. The denominator represents the probability of a positive result (without conditioning on whether or not a person has the condition), which is obtained from the law of total probability:

$$P(+) = P(+|C)P(C)+P(+|\neg C)P(\neg C) = 0.99(0.001)+0.01(0.999) = 0.01098.$$

Combining the above, we find that

$$P(C|+) = \frac{0.00099}{0.01098} \approx 9.016\%.$$

If this result seems counterintuitive, consider a small population of 10,000. Since 0.1% of the population is afflicted with the condition, we expect 10 people to have the condition out of the population. This represents the sum of the first column in Table 10.1. Similarly, 9,990 people will not have the condition, which is the sum of the second column. Now, since the test is 99%

	$Y = 1$	$Y = 0$
$\hat{Y} = 1$	10	100
$\hat{Y} = 0$	0	9890

Table 10.1: Confusion matrix for Example 10.1.

sensitive, we might expect all 10 of the actual cases to be correctly identified. Since the test is 99% specific, we expect 99% of the actual negatives to be correctly identified, leading the the confusion matrix shown in Table 10.1. Thus, out of the positive test results, we expect 10 to be true positives, whereas 100 will be false positives, leading to a 9% probability of having the condition, given a positive result. ▷

Example 10.2. A famous problem in Bayesian probability is the *Monty Hall problem*, as introduced in Selvin [1975a] and Selvin [1975b]. A game-show host presents a contestant with a choice of three doors. Behind one is a valuable prize (a new car, cash, etc.), and behind the other two are dud prizes (a donkey, dirty socks, etc.). The contestant selects one of the three doors—say, door number one. The host then reveals one of the duds from one of the other two doors. (*"It's a good thing you didn't choose door number three, because behind door number three is a used statistics book!"*) The contestant is then asked if he would like to stick to his or her original choice or switch to the other unopened door. Once the final choice is locked in, the doors are opened and the prizes revealed.

As it turns out, it is always better to switch to the unopened door. In fact, the probability of winning the prize is 2/3 if the contestant switches, but only 1/3 if the contestant sticks with their initial choice. But how could this be? We might expect that there should be exactly a 1/3 probability of the prize being behind any of the three doors. However, by the contestant opening one of the doors (specifically, a non-prize, non-selected door), we are receiving additional information which shifts the distribution in probabilities.

To see this, let

1. event A represent the event in which the prize is behind door number one;
2. event B represent the event in which the prize is behind door number two; and
3. event C represent the event in which the prize is behind door number three.

Our prior belief is that of equal a priori probability:

$$\mathbb{P}(A) = \mathbb{P}(B) = \mathbb{P}(C) = \frac{1}{3}.$$

Since the selection of door by the contestant is arbitrary, let us, without loss of generality, suppose that the contestant selects door number one. Given the contestant's initial choice, we know that the host must either reveal the contents behind door number two or door number three, which we will call events OB and OC, respectively. (O for open.)

The probability that the host opens door number three (given the contestant's initial selection of door number one) is given by

$$\mathbb{P}(OC) = \mathbb{P}(OC|A)p(A) + \mathbb{P}(OC|B)p(B) + \mathbb{P}(OC|C)p(C)$$
$$= (1/2)(1/3) + (1)(1/3) + (0)(1/3) = 1/2.$$

Note that $\mathbb{P}(OC|A) = 1/2$, since if the prize is behind door number one, the host will arbitrarily decide which of the remaining two doors to open. Also, $\mathbb{P}(OC|B) = 1$, as if the prize is behind door number two, the host is not allowed to reveal this, and is therefore forced to choose door number three. Similarly, $p(OC|C) = 0$, as the host cannot open door number three if this door leads to the prize. A similar calculation shows that $p(OB) = 1/2$.

So far, nothing is surprising. The host has an equal probability of opening doors two or three, due to the symmetry. But once the host's decision is revealed, something interesting happens.

Suppose the host opens door number three. The *posterior probability* that the prize is behind door number one is now given by Bayes' law:

$$\mathbb{P}(A|OC) = \frac{\mathbb{P}(OC|A)\mathbb{P}(A)}{\mathbb{P}(OC)}.$$

Now, the prior probability is $p(A) = 1/3$, and the probability that door three was chosen by the host is $\mathbb{P}(OC) = 1/2$. Also, recall that $\mathbb{P}(OC|A) = 1/2$; if the prize is behind door number one, the host has a 50% probability of selecting door three. But if we put this all together, we find

$$\mathbb{P}(A|OC) = \frac{(1/2)(1/3)}{1/2} = 1/3.$$

The probability of the prize being behind door number one is now only $1/3$. Similarly, $\mathbb{P}(B|OC) = 2/3$. Therefore, the contestant can double their chances at winning the prize by switching doors.   ▷

Example 10.3. In 1966, Ed Thorp shocked the gambling world by publishing the details at how a player could *beat the dealer* in the game of blackjack; see Thorp [1966]. His discovery was fueled by Bayesian statistics and MIT's IBM 704 computer. A lovely accounting of his life, how he used mathematics to beat the house at both blackjack and roulette, and his later years using mathematics to *beat the market*, is available in his autobiography Thorp [2017].

In the game of *blackjack*, the dealer deals two cards to each player, and two to herself (one face-up and one face-down). Each player in turn has the opportunity to receive a new card (*hit*) or finalize their hand (*stand*). Each card has a value: aces count as a 1 or 10, at the player's choice; 10, Jack, Queen, and King are each worth 10 points; and number cards are worth their face value. A player wins if their total is 21, or closer to 21 than the dealer. The catch is if a player total exceeds 21, it is called a *bust* and the player loses immediately. The other catch is that the dealer must hit until she reaches a total of 16, after which she stands. The house favor is created as players must select their cards first, risking a bust and immediate loss (regardless of whether the dealer busts).

Playing the Baldwin strategy, devised by four mathematicians during their time in the army, the dealer has only a 0.21% advantage over the player: practically 50-50 odds. Ed Thorp, however, realized that those odds shift dramatically based on the cards that were played. In Thorp's words:

> I realized that the odds as the game progressed actually depended on which cards were still left in the deck and that the edge would shift as play continued, sometimes favoring the casino and sometimes the player.

Thorp realized that the Baldwin strategy assumed that the probability of drawing a card was always the same. But, as the game progressed, one gains additional information, and is able to understand more precisely how the probability shifts, given the observed data. (A Bayesian update!) In particular, the more ten cards played, the better the odds for the dealer; whereas the more low-value cards (2–6) that are played, the better the odds for the player. The reason is that the dealer relies on low-value cards in order not to bust when it is their turn to draw, as they are required to hit until they reach at least 16.

Thorp devised various card counting strategies to track whether or not the deck was "ten-rich" or "ten-poor." When the odds shifted in his favor, Thorp increased his bet to between \$2 and \$10; otherwise he bet \$1. Eventually the casinos realized they could no longer offer black-jack tables that used a single deck of cards. But Thorp did alright until they figured out his game.   ▷

10.1.2 Bayesian Inference

Statistical inference is the practice of drawing conclusions from data about unobservable quantities or parameters. The frequentist's approach to inference is through hypothesis testing (Definition 3.1); a null hypothesis is made about a population parameter, a test is conducted, and a binary conclusion is drawn that renders the null hypothesis either accepted or rejected. Bayesian inference, on the other hand, allows us to quantify our *belief* in the range of values of a parameter through a distribution, and it further allows for the possibility of that distribution to change with experience. We make this leap by extending Bayes' law to an equivalent statement about probability distributions.

Proposition 10.2. *Given an* IID *set* $\mathcal{D} = \{Y_1, \ldots, Y_n\}$ *of values from a distribution that depends on a parameter* θ,

$$p(\theta|\mathcal{D}) = \frac{p(\mathcal{D}|\theta)p(\theta)}{p(\mathcal{D})}, \tag{10.2}$$

where $p(\cdot)$ *represents the appropriate (conditional) probability density or mass function.*

In order to more fully understand Equation (10.2), let's unpack its various components.

Definition 10.1. *In Equation* (10.2), *the quantity*

- $p(\theta)$ *represents the* prior distribution, *which quantifies our belief of the possible values of the parameter* θ *prior to having seen the data;*
- $p(\mathcal{D}|\theta)$ *represents the* likelihood *of the data (Definition 5.1); i.e.,*

$$p(\mathcal{D}|\theta) = \prod_{i=1}^{n} p(y_i|\theta),$$

which is interpreted as a function of θ, *and where* $p(Y|\theta)$ *is referred to as the* sampling distribution, *as it is the distribution that generates our sample of data;*

- $p(\mathcal{D})$ *represents the* data distribution *or* marginal distribution *for our observed data, which is obtained by marginalizing the likelihood function over the parameter θ using the prior distribution; i.e.,*

$$p(\mathcal{D}) = \int_{\theta \in \Theta} p(\mathcal{D}|\theta)p(\theta)d\theta,$$

when θ is a continuous variable, or

$$p(\mathcal{D}) = \sum p(\mathcal{D}|\theta_i)p(\theta_i),$$

when θ is a discrete variable, and
- $p(\theta|\mathcal{D})$ *represents the* posterior distribution, *which quantifies our new belief of the possible values of the parameter θ having accounted for the observed data.*

The process of converting a prior belief to a posterior belief via Equation (10.2) is known as a Bayesian update.

Since the marginal distribution $p(\mathcal{D})$ does not depend on the parameter θ, Equation (10.2) is often expressed in its *unnormalized form*

$$p(\theta|\mathcal{D}) \propto p(\mathcal{D}|\theta)p(\theta). \tag{10.3}$$

It is understood that the right-hand side must be normalized to make it a proper probability distribution for the parameter θ.

Example 10.4. Suppose we wish to determine the probability of a positive outcome for a binary event, such as flipping a coin. Let $Y \sim \text{Bern}(\theta)$ represent the outcome of the coin flip, where θ represents the probability of heads. If we have no a priori knowledge or expectation as to how the coin is weighted, we might take the uniform distribution

$$\theta \sim \text{Unif}(0, 1)$$

as our prior distribution, so that $p(\theta) = 1$ for $0 < \theta < 1$, or, equivalently,

$$p(\theta) = \text{Unif}(\theta; 0, 1).$$

Now suppose we perform a sequence of ten Bernoulli trials, resulting in three successes; i.e., we measure three heads out of ten coin flips.

The likelihood function is given by the binomial distribution as

$$p(\mathcal{D}|\theta) = \prod_{i=1}^{10} \text{Bern}(y_i; \theta) = \prod_{i=1}^{10} \theta^{y_i}(1-\theta)^{1-y_i} = \theta^3(1-\theta)^7,$$

where y_i is the result of the ith Bernoulli trial. Since our prior distribution is uniform, the posterior is proportional to our likelihood:

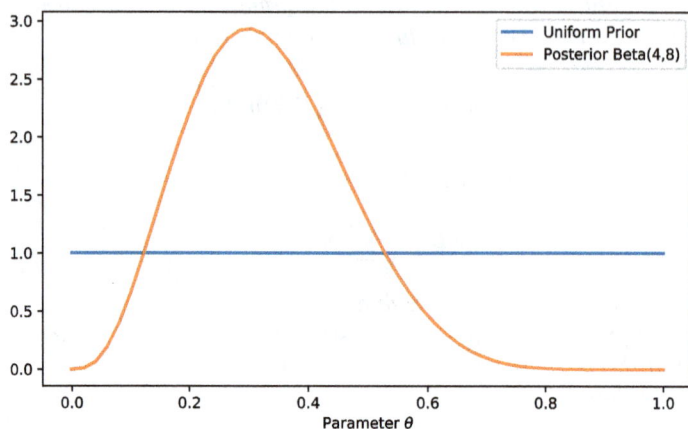

Fig. 10.1: Bayesian update for Example 10.4.

$$p(\theta|\mathcal{D}) \propto \theta^3 (1 - \theta)^7.$$

We recognize this distribution, however, as the kernel of the beta distribution (see Definition 2.17). We conclude that

$$p(\theta|\mathcal{D}) = \text{Beta}(\theta; 4, 8);$$

i.e., the posterior distribution is a beta distribution with parameters $\alpha = 4$ and $\beta = 8$. (Compare the penultimate equation with Equation (2.49).) The prior and posterior distributions are plotted in Figure 10.1.   ▷

By a careful analysis of the preceding example, we can confirm that the posterior distribution is independent of the order of the precise observations. Moreover, we could select half of the observations, generate a posterior, and use that posterior as the prior for the second half of the observations, and this should lead to the same result that we obtained by considering all observations at once. (This is obvious, in general, given the definition of the likelihood function.)

This, however, leads us to an interesting observation in the context of our previous example: if the prior distribution for a sequence of Bernoulli trials is a beta distribution, so too should be the posterior. We can verify this as follows.

Example 10.5. Consider again the case of n Bernoulli trials, resulting in a total of k successful outcomes. Suppose, instead of choosing a uniform, we select a Beta prior; i.e., suppose we take

$$p(\theta) = \text{Beta}(\theta; \alpha, \beta),$$

for some $\alpha, \beta > 0$. From Equation (2.49), the beta distribution is given by

$$p(\theta) = \frac{1}{B(\alpha, \beta)} \theta^{\alpha-1}(1-\theta)^{\beta-1},$$

where $B(\alpha, \beta)$ is the beta function (Equation (2.48)). The likelihood is given by

$$p(\mathcal{D}|\theta) = \prod_{i=1}^{n} \mathrm{Bern}(y_i; \theta) = \theta^{k}(1-\theta)^{n-k}.$$

Combining the previous two equations, we find that the posterior is given by

$$p(\theta|\mathcal{D}) \propto \theta^{\alpha+k-1}(1-\theta)^{\beta+n-k-1},$$

from which we conclude that

$$p(\theta|\mathcal{D}) = \mathrm{Beta}(\theta; \alpha + k, \beta + n - k). \tag{10.4}$$

Thus, when the prior distribution is a beta distribution, so too is the posterior. Moreover, the Bayesian update is equivalent to the following arithmetic operations

$$\alpha' = \alpha + k$$
$$\beta' = \beta + n - k.$$

Moreover, the expected value of θ, following the observations is given by Equation (2.50) as

$$\mathbb{E}[\theta|\mathcal{D}] = \frac{\alpha + k}{\alpha + \beta + n}. \tag{10.5}$$

For example, consider a prior $p(\theta) \sim \mathrm{Beta}(2,2)$, following which we observe $n = 10$ trials with $k = 3$ successes, so that the posterior is given by $\mathrm{Beta}(5,9)$. The prior and posterior distributions are plotted in Figure 10.2.

Two points are in order. The first is that we can view the parameters α and β of the prior distribution as a collection of *virtual observations*, where α constitutes the number of virtual successes and β the number of virtual failures. The ratio $\mu = \alpha/(\alpha + \beta)$ represents our prior belief regarding the expected probability of success, whereas the magnitude $\alpha + \beta$ reflects our confidence; i.e., prior values $(3, 7)$ are easier to overcome than $(30, 70)$, as the latter represents a more firm belief that the true rate should be 30%. Second, the variance given by Equation (2.51) is equivalent to

$$\mathbb{V}(\theta) = \frac{\mu(1-\mu)}{\alpha + \beta + 1}. \tag{10.6}$$

Since the value of α and β change with added data, and since $\alpha' + \beta' = \alpha + \beta + n$, this indicates that, even if our observed actual results agree exactly with our prior expected value, the variance of our belief decreases with added data. That is, the act of observing more data shrinks the variance of the probability distribution, making our belief more certain.   ▷

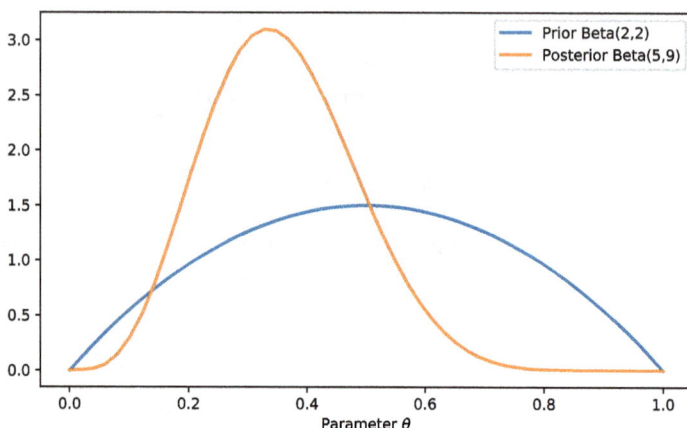

Fig. 10.2: Bayesian update for Example 10.5.

10.1.3 Conjugate Priors and Predictive Posteriors

Conjugate Priors

In Example 10.5, we examined a case where the prior and posterior distributions belonged to the same functional family of distributions. This is not unique to Bernoulli trials, but occurs for a variety of different distributions. When it does, we say that we are using the *conjugate prior* of the likelihood function. Formally, we have the following.

Definition 10.2. *Let $p(Y|\theta)$ be a parametric family of distributions. Then the parameteric family of distributions $p(\theta|\phi)$ is called a* conjugate prior *to the original family if, for any Y and ϕ, there exists a ϕ' such that*

$$p(Y|\theta)p(\theta|\phi) \propto p(\theta|\phi');$$

i.e., if the posterior distribution $p(\theta|Y)$ is a member of the same family of distributions as the prior $p(\theta)$.

The parameter(s) ϕ of the prior distribution is referred to as the hyperparameter(s) *for the system.*

We have already seen an example of this in Example 10.5. When the distribution $p(Y|\theta) = \mathrm{Bern}(Y; \theta)$ and the prior distribution $p(\theta|\alpha, \beta) = \mathrm{Beta}(\theta; \alpha, \beta)$, then so too is the posterior $p(\theta|Y) = \mathrm{Beta}(\theta; \alpha+Y, \beta+1-Y)$.

Posterior Predictive Distribution

The parameter θ is never directly observable, as it is a modeling parameter. Once we have collected a set of data $\mathcal{D}$ and calculated our posterior distribution $p(\theta|\mathcal{D})$, it is natural to ask what we can do with this. In particular, what do we think about the *next* outcome Y, as sampled from

$$Y \sim p(Y|\theta),$$
$$\theta \sim p(\theta|\mathcal{D}).$$

We answer this by marginalizing over θ, relative to the posterior distribution, to obtain the following.

Definition 10.3. *Given a parametric distribution $p(Y|\theta)$, a set of data $\mathcal{D}$, and our corresponding posterior distribution $p(\theta|\mathcal{D})$, the* posterior predictive distribution *is the distribution of the random variable Y obtained by marginalizing over θ; i.e.,*

$$p(Y|\mathcal{D}) = \int_{\theta \in \Theta} p(Y|\theta)p(\theta|\mathcal{D}) \, d\theta. \tag{10.7}$$

Similarly, given the prior distribution $p(\theta)$, the prior predictive distribution *is defined by*

$$p(Y) = \int_{\theta \in \Theta} p(Y|\theta)p(\theta) \, d\theta. \tag{10.8}$$

Integrals are replaced with summations when the parameter θ is discrete.

Note that the prior predictive distribution is similar to the marginal distribution in Definition 10.1, except that it is defined using the single probability distribution $p(Y|\theta)$, as opposed to the likelihood. Nevertheless, some authors use the term interchangeably.

Note 10.1. When a conjugate prior is used, $p(\theta)$ and $p(\theta|\mathcal{D})$ share the same functional form; i.e., the act of Bayesian update is a transformation on the hyperparameter $\phi \to \phi'$. As such, both the prior and posterior predictive distributions have the same form, except evaluated at a different value of the hyperparameter(s). For this reason, in the context of conjugate priors, we shall refer to the *predictive distribution*, as the distinction between prior and posterior is controlled by the choice of hyperparameter. ▷

Example 10.6. Determine the prior predictive distribution in Example 10.5, where the prior distribution is $p(\theta) = \text{Beta}(\theta; \alpha, \beta)$. (Note this is equivalent to calculating the posterior predictive distribution, if we replace $\alpha, \beta \to \alpha', \beta'$.

We proceed by applying Equation (10.8), from which we obtain

$$p(Y) = \int_0^1 p(Y|\theta)p(\theta)\, d\theta$$

$$= \int_0^1 \frac{\theta^Y(1-\theta)^{1-Y}}{B(\alpha,\beta)}\theta^{\alpha-1}(1-\theta)^{\beta-1}\, d\theta$$

$$= \int_0^1 \frac{\theta^{\alpha+Y-1}(1-\theta)^{\beta-Y}}{B(\alpha,\beta)}\, d\theta$$

$$= \frac{B(\alpha+Y,\beta+1-Y)}{B(\alpha,\beta)}$$

$$= \text{BetaBin}(Y;1,\alpha,\beta).$$

Thus the prior (or posterior) predictive distribution for $Y \sim \text{Bern}(\theta)$ and $\theta \sim \text{Beta}(\alpha,\beta)$ is the beta-binomial distribution (Equation (2.59)). The penultimate equality follows since the preceding integral contains the kernel of the $\text{Beta}(\alpha+Y,\beta+1-Y)$ distribution. This can further be extended as a predictive distribution for the total number of successes from the next n flips, as shown in Exercise 10.2.   ▷

Variance Reduction

When calculating a Bayesian update for the probability parameter θ of a Bernoulli random variable, we saw that the variance *decreases* inversely proportionally to the data size. The variance is expressed by Equation (10.6), where the sum $\alpha + \beta$ increases by the number of observations (Equation (10.4)). In other words, if Equation (10.6) represents the variance of our prior distribution $p(\theta)$, then the variance of the posterior is obtained by using the same equation, with μ replaced by Equation (10.5) and with $\alpha + \beta$ in the denominator replaced with $\alpha' + \beta' = \alpha + \beta + n$.

The idea of a Bayesian update reducing the variance of our belief in the value of a parameter (which would therefore reduce our confidence intervals) is not unique to Bernoulli random variables, but a staple of Bayesian inference. The result also makes intuitive sense: our certainty should improve the more data we collect. In the limit as $n \to \infty$, the posterior distribution will collapse to a delta function centered at a particular point.

In order to capture the expected variance reduction mathematically, we next revist Equations (1.19) and (2.58), both of which are based on the conditional expectation

$$\mathbb{E}[\theta|\mathcal{D}],$$

which is the expected value of θ relative to the posterior distribution. The posterior distribution itself is, however, dependent on the particular set of data $\mathcal{D}$ that was realized, which occurs relative to the marginal distribution $p(\mathcal{D})$. We may therefore consider the expectation and variance of the above quantity relative to all possible sets of data we might observe. Equation (1.19) implies that

$$\mathbb{E}[\theta] = \mathbb{E}\left[\mathbb{E}[\theta|\mathcal{D}]\right].$$

The outer expectation is taken relative to the data $\mathcal{D}$. This equation states that the expectation of the prior distribution is equivalent to the average posterior expectation, over all possible realizations of the data consistent with the data distribution $p(\mathcal{D})$.

Similarly, Equation (2.58) implies that

$$\mathbb{V}(\theta) = \mathbb{E}\left[\mathbb{V}(\theta|\mathcal{D})\right] + \mathbb{V}\left(\mathbb{E}[\theta|\mathcal{D}]\right). \tag{10.9}$$

This states that, on average, the variance in the posterior is *less* than the variance in the prior by an amount equal to the variance of the possible posterior means. Moreover, if we think of our data size as variable, and we let $\mathcal{D}_n$ represent an IID sample of n observations, then we find that

$$\lim_{n\to\infty} \mathbb{V}\left(\mathbb{E}[\theta|\mathcal{D}_n]\right) = \mathbb{V}(\theta),$$

which implies that $\mathbb{V}(\theta|\mathcal{D}_n) \to 0$ as $n \to \infty$; i.e., the variance of the posterior vanishes in the limit of unlimited data.

10.1.4 Conjugate Priors and Exponential Dispersion Models

In general, any member of the family of exponential dispersion models has a conjugate prior, as we show in our next proposition. In our next section, we will then derive some common conjugate families.

Proposition 10.3. *Let $p(Y|\theta) \sim$ EDM be an exponential dispersion model (Definition 7.10), for a given cumulant function $\kappa(\theta)$ and fixed dispersion parameter ϕ. Then two-parameter family defined by*

$$p(\theta|\mu,\psi) = b(\mu,\psi)\exp\left[\frac{\theta\mu - \kappa(\theta)}{\psi}\right] \tag{10.10}$$

constitutes a conjugate prior to the family $p(Y|\theta)$, where the function $b(\mu,\psi)$ is the normalizing constant given by

$$b(\mu,\psi) = \left(\int \exp\left[\frac{\theta\mu - \kappa(\theta)}{\psi}\right]d\theta\right)^{-1}, \tag{10.11}$$

where the integration is taken over the domain of the parameter θ.

Moreover, given a set of data $\mathcal{D} = \{y_i\}_{i=1}^{n}$, the Bayesian update in the hyperparameter (μ,ψ) is given by

$$\mu' = \lambda\overline{y}. + (1-\lambda)\mu, \tag{10.12}$$

$$\psi' = (1-\lambda)\psi, \tag{10.13}$$

where $y. = \sum_{i=1}^{n} y_i$ is the sum, $\overline{y}. = (1/n)\sum_{i=1}^{n} y_i$ is the sample mean, and the parameter λ, known as the credibility, is given by

$$\lambda = \frac{n}{n+\eta},$$

(10.14)

where $\eta = \phi/\psi$ and $n = |\mathcal{D}|$.

Proof. From Equation (7.53), the likelihood function from a set of IID $Y_i \sim$ EDM is given by

$$p(\mathcal{D}|\theta) = \left(\prod_{i=1}^{n} a(y_i, \phi)\right) \exp\left[\frac{\theta \sum_{i=1}^{n} y_i - n\kappa(\theta)}{\phi}\right].$$

Note that the coefficient is independent of the variable θ. From this expression, it is clear that if the prior distribution is of the form

$$p(\theta|\nu, \eta) = b(\nu/\eta, \phi/\eta) \exp\left[\frac{\theta\nu - \eta\kappa(\theta)}{\phi}\right],$$

(10.15)

then so too is the posterior, with updated hyperparameters given by

$$\nu' = \nu + \sum_{i=1}^{n} y_i,$$

(10.16)

$$\eta' = \eta + n.$$

(10.17)

Now, we can rearrange the argument of the exponential function in Equation (10.15) to obtain

$$\frac{\theta\nu/\eta - \kappa(\theta)}{\phi/\eta}.$$

Making the substitution $\mu = \nu/\eta$ and $\psi = \phi/\eta$ yields the expression Equation (10.10). The normalizing function Equation (10.11) follows as the function Equation (10.10) must be a proper density for the parameter θ.

To derive Equation (10.12), we may apply Equations (10.15) and (10.16) to obtain

$$\mu' = \frac{\nu'}{\eta'} = \frac{\nu + y.}{\eta + n} = \frac{\eta}{\eta + n}\frac{\nu}{\eta} + \frac{n}{\eta + n}\frac{y.}{n} = (1 - \lambda)\mu + \lambda\bar{y}..$$

Similarly,

$$\psi' = \frac{\phi}{\eta'} = \frac{\eta}{\eta + n}\frac{\phi}{\eta} = (1 - \lambda)\psi,$$

which is equivalent to Equation (10.13). $\square$

Note 10.2. Equations (10.12) and (10.13) take the same form as found in credibility theory, a topic we will discuss toward the end of the chapter. The parameter λ may be regarded as a "credibility factor" expressing the fraction of the observed weight to the total observed and prior weights. This latter expression is exactly in the appropriate form for the credibility estimator (Theorem 10.3), which can be interpreted as a *linear* Bayes' estimator for risk premiums.   ▷

Note 10.3. We are at present unaware of a closed-form expression for the mean and variance of the family of conjugate priors given by Equation (10.10). The moment-generating function may be expressed as

$$M(t) = \frac{b\left(\mu, \psi\right)}{b\left(\mu + t\psi, \psi\right)},$$

but it is unclear how we could use this to determine an expression for the moments. Though one can show that

$$\mathbb{E}[\Theta] = -\psi \frac{b_\mu\left(\mu, \psi\right)}{b\left(\mu, \psi\right)}$$

and

$$\mathbb{V}(\Theta) = \psi^2 \left[\frac{b_\mu\left(\mu, \psi\right)^2}{b\left(\mu, \psi\right)^2} - \frac{b_{\mu\mu}\left(\mu, \psi\right)}{b\left(\mu, \psi\right)} \right],$$

where $b_\mu(\cdot, \cdot)$ is the partial derivative of the normalizing coefficient with respect to its first argument. Since we cannot say much about b, it is unclear how we might uncover additional expressions from these results.   ▷

As it turns out, the parameter μ has a simple interpretation as the expected value of Y when marginalizing over θ.

Proposition 10.4. *Given the hierarchical model* $Y \sim \mathrm{EDM}(Y; \Theta)$ *and* $\Theta \sim p(\Theta; \mu, \psi)$, *where* $p(\Theta; \mu, \psi)$ *is given by Equation (10.10), we have the result*

$$\mathbb{E}[Y] = \mu. \tag{10.18}$$

Moreover, if we define

$$\sigma^2 = \mathbb{E}\left[\mathbb{V}(Y|\Theta)\right], \tag{10.19}$$
$$\tau^2 = \mathbb{V}\left(\mathbb{E}[Y|\Theta]\right), \tag{10.20}$$

we have that

$$\frac{\sigma^2}{\tau^2} = \frac{\phi}{\psi} = \eta. \tag{10.21}$$

Note that, given Equation (10.21), the credibility factor Equation (10.14) may be expressed as

$$\lambda = \frac{n}{n + \sigma^2/\tau^2}.$$

The variance in means τ^2 and the conditional variance $\sigma^2(\Theta)$ of the EDM are illustrated in Figure 10.3, with $\sigma^2 = \mathbb{E}[\sigma^2(\Theta)]$.

Proof. Recall from Theorem 7.7 the expression

$$\mu(\Theta) = \kappa'(\Theta).$$

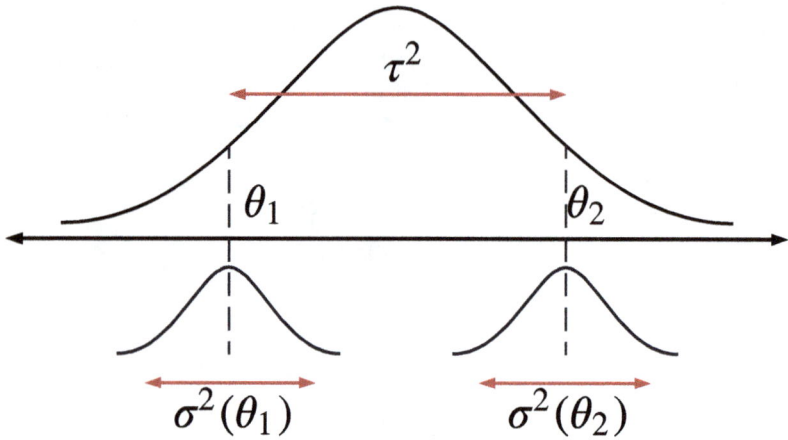

Fig. 10.3: Visualization of the variance of an EDM, $\sigma^2(\Theta)$, versus the variance of the prior τ^2.

where we have defined $\mu(\Theta) = \mathbb{E}[Y|\Theta]$ as the expected value of Y, given the parameter Θ. Now, from Equation (1.19), we have

$$\mathbb{E}[Y] = \mathbb{E}[\mathbb{E}[Y|\Theta]] = \mathbb{E}[\kappa'(\Theta)].$$

However, note that

$$\mu - \mathbb{E}[\kappa'(\Theta)] = \int b(\mu, \psi) \, [\mu - \kappa'(\theta)] \exp\left[\frac{\theta\mu - \kappa(\theta)}{\psi}\right] d\theta,$$

which integrates to zero, as the density of an EDM vanishes on its boundaries (typically $\pm\infty$; but in some cases $-\infty$ and 0). This implies that $\mathbb{E}[\kappa'(\Theta)] = \mu$, completing the proof of Equation (10.18).

For the second part of the proposition, we again recall from Theorem 7.7 that

$$\mathbb{V}(Y|\Theta) = \phi\kappa''(\Theta),$$

so that

$$\sigma^2 = \phi\mathbb{E}[\kappa''(\Theta)].$$

Moreover,

$$\tau^2 = \mathbb{V}(\mu(\Theta)) = \mathbb{E}\left[(\kappa'(\theta) - \mu)^2\right].$$

Now, not only does Equation (10.10) vanish at the limits of the domain of θ, but so too does its derivative

$$p'(\theta|\mu, \psi) = b(\mu, \psi) \left[\frac{\mu - \kappa'(\theta)}{\psi}\right] \exp\left(\frac{\mu\theta - \kappa(\theta)}{\psi}\right).$$

This implies that the integral of $p''(\theta|\mu, \psi)$ must also vanish; i.e.,

$$\int b(\mu, \psi) \left\{ \left[\frac{\mu - \kappa'(\theta)}{\psi} \right]^2 - \frac{\kappa''(\theta)}{\psi} \right\} \exp\left(\frac{\mu\theta - \kappa(\theta)}{\psi} \right) d\theta = 0.$$

However, this is equivalent to the expression

$$\frac{\mathbb{E}\left[(\kappa'(\Theta) - \mu)^2] \right]}{\psi^2} = \frac{\mathbb{E}[\kappa''(\Theta)]}{\psi},$$

or, alternatively, $\phi\tau^2 = \psi\sigma^2$, which completes the proof.   □

It follows that we can interpret the parameter ψ as the ratio of the variance τ^2 in expected means $\mu(\Theta)$ to the expected variance σ^2. We may visualize these two variances as shown in Figure 10.3. As we continue to make Bayesian updates, the variance in the prior/posterior will collapse to zero, so that $\tau^2 \to 0$ and $\psi \to 0$ as $n \to \infty$, whereas the expected variance σ^2 will approach a limiting constant: $\sigma^2 \to \sigma^2(\hat{\theta})$. Finally, note that the variance in the posterior predictive distribution is given by $\mathbb{V}(Y) = \sigma^2 + \tau^2$, due to the law of total variance, as given by Equation (2.58).

10.1.5 A Catalogue of Conjugate Families

We have already computed the conjugate prior and posterior predictive distributions for a Bernoulli random variable, in Examples 10.5 and 10.6. Generalization to a binomial random variable is straightforward (see Exercise 10.2). Moreover, we proved that *any* member of the exponential dispersion family similarly has a conjugate prior. We now turn to a few specific examples: a number of common conjugate families are recorded in Tables 10.2 and 10.3 along with their posterior predictive distributions, when available. We will derive a few of these, and leave the rest as an exercise for the reader.

Poisson–Gamma

The gamma distribution is the conjugate prior for the Poisson distribution. To see this, let

$$p(Y|\lambda) = \text{Poiss}(Y; \lambda) = \frac{e^{-\lambda}\lambda^y}{y!},$$

$$p(\lambda) = \text{Gamma}(\lambda; \alpha, \beta) = \frac{1}{\Gamma(\alpha)\beta^\alpha}\lambda^{\alpha-1}e^{-\lambda/\beta}.$$

The likelihood of a set of data $\mathcal{D} = \{Y_i\}_{i=1}^n$ is therefore given by

$$p(\mathcal{D}|\lambda) = \frac{e^{-n\lambda}\lambda^{\sum y_i}}{\prod y_i!}.$$

sampling distribution	unknown parameter	conjugate prior	Bayesian update	predictive distribution
$\text{Bern}(\theta)$	θ	$\text{Beta}(\alpha,\beta)$	$\alpha'=\alpha+\sum Y_i$ $\beta'=\beta+n-\sum Y_i$	$\text{BetaBin}(Y;1,\alpha,\beta)$
$\text{Binom}(n_i;\theta)$	θ	$\text{Beta}(\alpha,\beta)$	$\alpha'=\alpha+\sum Y_i$ $\beta'=\beta+\sum n_i-\sum Y_i$	$\text{BetaBin}(Y;n,\alpha,\beta)$
$\text{Poiss}(\lambda)$	λ	$\text{Gamma}(\alpha,\beta)$	$\alpha'=\alpha+\sum Y_i$ $(\beta')^{-1}=\beta^{-1}+n$	$\text{NBD}(Y;\alpha,(1+\beta)^{-1})$
$\text{Geom}(\theta)$	θ	$\text{Beta}(\alpha,\beta)$	$\alpha'=\alpha+n$ $\beta'=\beta+\sum Y_i$	$\text{BetaGeom}(Y;\alpha,\beta)$
$\text{NBD}(r,\theta)$	θ	$\text{Beta}(\alpha,\beta)$	$\alpha'=\alpha+\sum Y_i$ $\beta'=\beta+rn$	$\text{BetaNegBin}(Y;\alpha,\beta)$
$\text{Multi}(\theta)$	θ	$\text{Dir}(\alpha)$	$\alpha'=\alpha+\sum Y_i$	$\text{DirMulti}(Y;n,\alpha)$

Table 10.2: Common conjugate families for discrete distributions.

sampling distribution	unknown parameter	conjugate prior	Bayesian update	predictive distribution
$N(\mu,\sigma^2)$	μ	$N(\nu,\tau^2)$	$\nu' = \left(\dfrac{1}{\tau^2}+\dfrac{n}{\sigma^2}\right)^{-1}\left(\dfrac{\nu}{\tau^2}+\dfrac{\sum Y_i}{\sigma^2}\right)$, $(\tau')^2 = \left(\dfrac{1}{\tau^2}+\dfrac{n}{\sigma^2}\right)^{-1}$	$N(Y;\nu,\tau^2+\sigma^2)$
$\mathrm{Exp}(\lambda^{-1})$	rate λ	$\mathrm{Gamma}(\alpha,\beta)$	$\alpha'=\alpha+n$, $(\beta')^{-1}=\beta^{-1}+\sum Y_i$	$\mathrm{Lomax}(Y;\alpha,\beta^{-1})$
$\mathrm{Gamma}(\alpha,\beta)$	β	$\mathrm{Gamma}(\alpha_0,\beta_0)$	$\alpha_0'=\alpha_0+n\alpha$, $(\beta_0')^{-1}=\beta_0^{-1}+\sum Y_i$	—
$\mathrm{Pareto}(m,k)$	k	$\mathrm{Gamma}(\alpha,\beta)$	$\alpha'=\alpha+n$, $(\beta')^{-1}=\beta^{-1}+\sum\ln(Y_i/m)$	—

Table 10.3: Common conjugate families for continuous distributions.

The posterior distribution is therefore proportional to

$$p(\lambda|\mathcal{D}) \propto e^{-n\lambda}\lambda^{\sum y_i}\lambda^{\alpha-1}e^{-\lambda/\beta} = \lambda^{\alpha+\sum y_i-1}e^{-\lambda(\beta^{-1}+n)},$$

which we recognize as the kernel of a Gamma$(\alpha + \sum y_i, (\beta^{-1} + n)^{-1})$ distribution.

The predictive distribution is then obtained by marginalizing the sampling distribution over the gamma prior / posterior:

$$
\begin{aligned}
p(y|\alpha, \beta) &= \int_0^\infty p(y|\lambda)p(\lambda|\alpha, \beta)\, d\lambda \\
&= \int_0^\infty \frac{\lambda^{\alpha+y-1}e^{-\lambda(1+1/\beta)}}{y!\Gamma(\alpha)\beta^\alpha}\, d\lambda \\
&= \frac{\Gamma(\alpha+y)(1+1/\beta)^{-(\alpha+y)}}{y!\Gamma(\alpha)\beta^\alpha},
\end{aligned}
$$

where the final equality follows by recognizing the kernel of the Gamma$(\lambda; \alpha + y, (1+1/\beta)^{-1})$ distribution. This, however, is equivalent to

$$p(y|\alpha, \beta) = \frac{\Gamma(\alpha+y)}{\Gamma(y+1)\Gamma(\alpha)}\frac{1}{(1+\beta)^\alpha}\frac{1}{(1+1/\beta)^y}.$$

Setting $r = \alpha$ and $p = (1+\beta)^{-1}$, we recognize this as a negative-binomial distribution; i.e.,

$$p(y|\alpha, \beta) = \mathrm{NBD}(y; \alpha, (1+\beta)^{-1}).$$

As mentioned in Note 10.1, this represents both the prior and posterior predictive distribution, depending on the choice of hyperparameters (α, β).

Geometric–Beta

The beta distribution is the conjugate prior of the geometric distribution. To see this, let

$$p(Y|\theta) = \mathrm{Geom}(Y; \theta) = \theta(1-\theta)^y,$$
$$p(\theta) = \mathrm{Beta}(\theta; \alpha, \beta) = \frac{1}{B(\alpha, \beta)}\theta^{\alpha-1}(1-\theta)^{\beta-1}.$$

The likelihood of a set of data $\mathcal{D} = \{Y_i\}_{i=1}^n$ is therefore given by

$$p(\mathcal{D}|\theta) = \theta^n(1-\theta)^{\sum Y_i}.$$

The posterior distribution is therefore proportional to

$$p(\theta|\mathcal{D}) \propto \theta^n(1-\theta)^{\sum y_i}\theta^{\alpha-1}(1-\theta)^{\beta-1} = \theta^{\alpha+n-1}(1-\theta)^{\beta+\sum y_i-1},$$

which we recognize as the kernel of a Beta$(\alpha + n, \beta + \sum y_i)$ distribution.

Exponential–Gamma

The gamma distribution is the conjugate prior to the exponential distribution, when expressed in terms of its rate parameter λ. To see this, let

$$p(Y|\lambda) = \text{Exp}(Y; \lambda^{-1}) = \lambda e^{-\lambda y},$$

$$p(\lambda) = \text{Gamma}(\lambda; \alpha, \beta) = \frac{1}{\Gamma(\alpha)\beta^{\alpha}} \lambda^{\alpha-1} e^{-\lambda/\beta}.$$

The likelihood of a set of data $\mathcal{D} = \{Y_i\}_{i=1}^{n}$ is therefore given by

$$p(\mathcal{D}|\lambda) = \lambda^{n} e^{-\lambda \sum y_i}.$$

The posterior distribution is therefore proportional to

$$p(\lambda|\mathcal{D}) \propto \lambda^{n} e^{-\lambda \sum y_i} \lambda^{\alpha-1} e^{-\lambda/\beta} = \lambda^{\alpha+n-1} e^{-\lambda(\beta^{-1} + \sum y_i)},$$

which we recognize as the kernel of a $\text{Gamma}(\alpha + n, (\beta^{-1} + \sum y_i)^{-1})$ distribution.

The predictive distribution is obtained by marginalizing

$$p(Y) = \int_{0}^{\infty} p(Y|\lambda)p(\lambda) \, d\lambda$$

$$= \int_{0}^{\infty} \frac{\lambda e^{-\lambda y} \lambda^{\alpha-1} e^{-\lambda/\beta}}{\Gamma(\alpha)\beta^{\alpha}} \, d\lambda$$

$$= \frac{\Gamma(\alpha + 1)}{\Gamma(\alpha)\beta^{\alpha}(\beta^{-1} + Y)^{\alpha+1}}.$$

The final equation follows by recognizing the kernel of the $\text{Gamma}(\alpha + 1, (\beta^{-1} + Y)^{-1})$ distribution in the integrand of the preceding equation. However, this is equivalent to the *Lomax distribution*, or *Pareto Type II distribution*:

$$\text{Lomax}(x; \alpha, \lambda) = \frac{\alpha}{\lambda} \left[1 + \frac{x}{\lambda}\right]^{-(1+\alpha)} = \frac{\alpha \lambda^{\alpha}}{(x + \lambda)^{\alpha+1}}, \tag{10.22}$$

with support $x \in [0, \infty)$. (Note that the Pareto distribution is a shifted Lomax distribution.) That is, $p(Y) = \text{Lomax}(Y; \alpha, \beta^{-1})$.

EDM Form

Finally, we can calculate the parameters μ, σ^2, τ^2, and ψ for several common conjugate priors, using Equations (10.18)–(10.21). The results are shown in Table 10.4. Note that in each case, the variance of the predictive distribution is equal to the sum of the variances σ^2 and τ^2. If we reparameterize the conjugate prior distribution using the hyperparameters (μ, ψ) in place of (α, β), the Bayesian update formula for each of these families is given in the form of Equations (10.12) and (10.13). Note that the (μ, ψ) parameterization is not unique to the prior, but unique to the conjugate pair.

family	$\mu = \mathbb{E}[Y]$	$\sigma^2 = \mathbb{E}[\mathbb{V}(Y|\Theta)]$	$\tau^2 = \mathbb{V}(\mathbb{E}[Y|\Theta])$	$\psi = \tau^2/\sigma^2$
binomial beta	$\dfrac{n\alpha}{\alpha+\beta}$	$\dfrac{n\alpha\beta}{(\alpha+\beta)(\alpha+\beta+1)}$	$\dfrac{n^2\alpha\beta}{(\alpha+\beta)^2(\alpha+\beta+1)}$	$\dfrac{n}{\alpha+\beta}$
Poisson gamma	$\alpha\beta$	$\alpha\beta$	$\alpha\beta^2$	β
geom beta	$\dfrac{\beta}{\alpha-1}$	$\dfrac{\beta(\alpha+\beta-1)}{(\alpha-2)(\alpha-1)}$	$\dfrac{\beta(\alpha+\beta-1)}{(\alpha-2)(\alpha-1)^2}$	$\dfrac{1}{\alpha-1}$
exp gamma	$\dfrac{1}{\beta(\alpha-1)}$	$\dfrac{1}{\beta^2(\alpha-1)(\alpha-2)}$	$\dfrac{1}{\beta^2(\alpha-1)^2(\alpha-2)}$	$\dfrac{1}{\alpha-1}$

Table 10.4: Each prior may be parameterized by (μ, ψ), so that the Bayesian update is in the form of Equations (10.12) and (10.13).

10.1.6 The Bayes–Kalman Filter (or Recursive Bayes)

The variance-reduction principal (recall Equation (10.9)) implies that as we continue to add data to a Bayesian update, the variance of the posterior distribution shrinks, ultimately approaching zero. In the context of stochastic processes (Definition 5.11), however, this is not always—or even typically—a desired trait. For example, suppose a marketing campaign has been running for 30 days, and each day the observed conversion rate has been constistantly 10%. Then, all of the sudden, the conversion rate drops to 2%, and remains at 2%. (This could be due to unknown or hidden factors, such as a sudden change in market pressures or auction conditions.) Now, if we are simply running a binomial–beta Bayesian update each day, it will take *an additional 30 days* just for the 10% and 2% measurements to be equally weighted, whereas a human would notice the dramatic change in performance almost immediately. This is because our Bayesian update is simply keeping a tab of volume n and conversions k, following the update of Equation (10.5). As the total volume grows, the overall variance shrinks, and it becomes harder and harder to "unlearn" historic behavior.

The cause of this dilemma is the IID assumption in the Bayesian update rule of Proposition 10.2. As long as our data are IID, we can perform the Bayesian updates piecemeal, by breaking up our data set into smaller segments and creating a sequence of new posteriors. But when those data segments are no longer IID, our assumption breaks down, and we require a new approach.

But all is not lost. In this section, we propose a method for combatting such a scenario, which we name the *Bayesian Kalman filter*, in recognition of the classic Kalman filter used in engineering, which is the direct inspiration for our our approach. A *Kalman filter* is a technique in dynamical systems theory that recursively estimates the state of a system over time, given measurements subject to uncertainty. Typically, one tracks both position and velocity, and the uncertainty of each (in the form of a covariance matrix), and further models the relation between true state and measurement, accounting for measurement error. The Kalman filter applies the linear dynamics to the prior system state to estimate a new system state, and then combines information from the new observed values and the intrinsic measurement uncertainty to provide an updated, superior estimation for the actual system state. For a details on Kalman filters in the context of general stochastic processes, see, for example, Hajek [2015]. The Kalman filter technique has been previously applied credibility theory, which constitutes a system of linear Bayes estimators, in Bühlmann and Gisler [2005].

To proceed, let us suppose we are given a sequence of data that are collected over time: $\mathcal{D}_1, \ldots, \mathcal{D}_m$, such that each set of data is collected sequentially at a distinct point in time (e.g., day-by-day). Now, a pure Bayesian update would result in the posterior

$$p(\theta|\mathcal{D}_1, \ldots, \mathcal{D}_m) = \left(\prod_{i=1}^{m} p(\mathcal{D}_i|\theta) \right) p(\theta),$$

treating each set of data as equivalent. This approach fundamentally fails to take into account the factor of time and the possibility of temporal changes to the distribution of data (e.g., switchpoints or trends).

To resolve this, we go back to Proposition 10.2, and recall that a *prior distribution* is a representation of our *belief* in the state of a system, and if we believe that the state of the system is, or can be, changing, then our prior distribution for day m should not simply equal the posterior distribution of day $(m-1)$; i.e., we must explicitly break our tacit assumption that $\mathcal{D}_i$ and $\mathcal{D}_j$ are IID samples from the same distribution. In particular, we should explicitly counteract the variance shrinkage with a controlled *variance expansion*, which should be greater the more $\mathcal{D}_m$ differs from our prior day's posterior $p_{m-1}(\theta|\mathcal{D}_1, \ldots, \mathcal{D}_{m-1})$.

Kalman Interpretation of Bayesian Update

Before proceeding, it is interesting to note that the standard Bayes' update rule is, essentially, a form of Kalman filter, with stationary (in-place) dynamics. To see this, let our state be defined as $\mu = \mathbb{E}[Y]$, and define the subscript notation as follows

$$\mu_{t|t-1} = \mathbb{E}[\mu|\mathcal{D}_1, \ldots, \mathcal{D}_{t-1}],$$
$$\mu_{t|t} = \mathbb{E}[\mu|\mathcal{D}_1, \ldots, \mathcal{D}_t].$$

The classic Kalman filter consists of two steps: prediction and update. For prediction, under a time-static model, we have

$$\mu_{t|t-1} = \mu_{t-1|t-1} \tag{10.23}$$

$$P_{t|t-1} = \mathbb{V}(\mu_{t|t-1}) = \mathbb{V}(\mathbb{E}[X|\Theta]) = \tau^2. \tag{10.24}$$

Next, we incorporate the new data set $\mathcal{D}_t$, using the summary statistics $n_t = |\mathcal{D}_t|$ and $\bar{y}_t = \frac{1}{n_t}\sum_{i=1}^{n_t} y_i$, to compute the *innovation* and its variance:

$$\delta_t = \bar{y}_t - \mu_{t|t-1},$$

$$S_t = \mathbb{V}(\delta_t) = \mathbb{V}(\bar{y}_t) = \frac{\sigma^2}{n} + \tau^2.$$

The *Kalman gain* is then computed as the ratio

$$K_t = PS^{-1} = \frac{\tau^2}{\tau^2 + \sigma^2/n_t} = \frac{n_t}{n_t + \sigma^2/\tau^2} = \lambda_t,$$

which is equivalent to the credibility factor Equation (10.14) of our Bayesian update. Finally, the Kalman update is given by

$$\mu_{t|t} = \mu_{n_t|n_{t-1}} + K_t(\bar{y}_t - \mu_{n_t|n_{t-1}})$$

$$= K_t\bar{y}_t + (1 - K_t)\mu_{t|t-1} \tag{10.25}$$

$$P_{n|n} = (1 - K_t)P_{n|n-1} \tag{10.26}$$

This is virtually synonymous with the Bayesian update Equations (10.12) and (10.13), except that the credibility factor is applied to τ^2, as opposed to ψ. However, this is okay, as σ^2 approaches a limiting constant, whereas τ^2 approaches zero, in the limit of infinite IID data.

Bühlmann and Gisler [2005] provides two extensions to this model, in the framework of credibility theory: the case of *orthogonal increments*, where $\mu(\Theta_t)$ is orthogonal to $\mu(\Theta_{t-1})$, in which case the update is still given by Equation (10.23), and the case of linear transformations (drift), where the update rule Equation (10.23) is replaced by a linear transformation

$$\mu_{t|t-1} = a\mu_{t-1|t-1} + b.$$

They further update the parameter τ^2 using the equation

$$\tau_t^2 = \tau_{t-1}^2 + \mathbb{V}(\mu(\Theta_t) - \mu(\Theta_{t-1})),$$

which provides what we refer to as a *variance expansion*, which can be viewed as a counter-lever to the collapsing variance of subsequent Bayesian updates. We turn to our take on this rule next.

Bayes–Kalman Filter

In this section, we discuss our approach to a *Bayes–Kalman filter*, for time-series data $\{\mathcal{D}_1, \mathcal{D}_2, \ldots\}$, such that, for each $t \in \mathbb{Z}_+$, we have IID data

$Y_{ti} \sim \text{EDM}(Y; \Theta)$, for $i = 1, \ldots, n_t$. Moreover, we express the conjugate prior as $\Theta \sim p(\Theta; \mu, \psi)$, such that the classic Bayesian update is given by Equations (10.12) and (10.13).

Now, if we were to make the IID assumption over all data sets $\mathcal{D}_1, \mathcal{D}_2, \ldots$ in our sequence, we would *only* have to apply the classic Bayesian update. The problem with this approach, however, is that in real life there can always be some exogenous influence that shifts the distribution of our data, and the classic Bayesian update leads to a variance shrinkage, such that $\psi \to 0$ as $n \to \infty$. Thus, when the underlying distribution changes, it takes longer for the Bayesian update to react, the longer a history it has already captured. We therefore propose a second step, which we call *variance expansion*, that is designed to counteract the effect of variance decay when a sudden change in behavior occurs. Note that we still assume that $Y \sim \text{EDM}(Y; \Theta)$, we only expand the variance of the parameter Θ to allow the classic Bayesian update rule to hone in on the new normal.

Drawing our inspiration from Kalman filters and evolutionary credibility models (Bühlmann and Gisler [2005]), we remedy these issues with our proposed Algorithm 10.1. Our procedure may still be considered a *two-step* update rule, though we break it into three sections. The *predict step* of the classic Kalman filter is here replaced with a *variance expansion* step: essentially we are predicting no change in the mean, $\mu_{t|t-1} = \mu_{t-1|t-1}$, as we are not modeling any explicit time dynamics, but an expansion to the variance, to account for the fact that successive data may not be IID, as

---

Algorithm 10.1: *Bayes–Kalman filter* (or *recursive Bayes*) for Bayesian updates of time series data.

Input: data series $\mathcal{D}_1, \mathcal{D}_2, \ldots$;
    primordial priors μ_0, ψ_0;
    exponential decay factor $\alpha \in (0, 1)$
Output: Updated hyperparameters (μ_t, ψ_t).

1 Set $S_0, Z_0 = 0, 0$
2 **for** t in $1, 2, \ldots$ **do**
3 | *Variance Expansion*
4 | $\psi_t = \psi_{t-1} + \max(0, Z_{t-1}^2 - 1)$
5 | *Bayesian Update*
6 | Credibility factor: $\lambda_t = \dfrac{n_t}{n_t + \phi/\psi}$
7 | $\mu_t = \lambda_t \overline{y}_t + (1 - \lambda_t)\mu_{t-1}$
8 | $\psi_t = (1 - \lambda_t)\psi_t$
9 | *EWMA of variance and surprise estimates*
10 | $S_t = \alpha s_t^2 + (1 - \alpha)S_{t-1}$; where s_t^2 is sample variance of $\mathcal{D}_t$
11 | $Z_t = \alpha \dfrac{\overline{y}_t - \mu_t}{\sqrt{S_t(\psi_t + 1/n_t)}} + (1 - \alpha)Z_{t-1}$
12 **end**
13 **return** (μ_t, ψ_t)

required for the standard Bayesian update. Recall that the parameter ψ represents the ratio

$$\psi = \frac{\phi\tau^2}{\sigma^2} = \frac{\phi\mathbb{V}(\mathbb{E}[Y|\Theta])}{\mathbb{E}[\mathbb{V}(Y|\Theta)]}.$$

Increasing ψ may therefore be viewed as an increase in our belief of the spread of population means between successive data sets, normalized by the expected variance. If we allow ψ to collapse to zero, it becomes difficult to adapt to changing conditions.

Next, we have our *update step*, which has a familiar piece—the standard Bayesian update rule given by Equations (10.12) and (10.13)—and a new piece, representing an *exponentially weighted moving average* (EWMA) for our estimate S_t of the expected variance σ^2 and our estimate of "surprise" Z_t.

Note 10.4 (Exponentially weighted moving average). A quick aside on EWMA. Given a time series $\{X_t\}$ and a weight parameter $\alpha \in (0,1)$, we may define the sequence of averages

$$A_t = \alpha X_t + (1-\alpha)A_{t-1}.$$

By expanding this recursion relation, we obtain

$$A_t = \alpha \sum_{s=0}^{n} (1-\alpha)^s X_{t-s},$$

assuming a history of n recursions. Note that the weights sum to unity in the limit as $n \to \infty$, as this result constitutes a geometric series. $\triangleright$

So the estimate S_t as an exponentially weighted average variance should be clear. Naturally, since we are averaging together the sample variances, which estimate $\mathbb{V}(Y|\Theta_t)$, we may treat S_t as an estimate for σ^2.

Finally, we introduce the *surprise* Z_t, defined on line 11. First, let's consider the random variable $\overline{Y} - \mathbb{E}[Y]$, which is estimated by our numerator $\overline{y}_t - \mu_t$. The variance in this estimate is given by

$$\mathbb{V}(\overline{Y} - \mu) = \tau^2 + \frac{\sigma^2}{n} = \sigma^2\left(\psi + \frac{1}{n}\right).$$

Thus Z_t is an estimate for the number of standard deviations (z-score) the sample mean of the current data set is from our expected population mean, accounting for both the variance in Θ (as captured by τ^2) as well as the sample variance given Θ (as captured by σ^2/n).

There are several final points regarding a few decisions we made when crafting this algorithm. First, note that we are summing the differences $(\overline{y}_t - \mu_t)$ directly, and not their squares. When we are in steady state, and the successive data sets are IID, the expected value of this random variable is zero, and positive outliers will cancel negative ones. So if $\overline{Y}_t$ is fluctuating around the mean randomly, we allow those fluctuations to cancel. However, when there has been a shift, one way or the other, our surprise factor will then build over time, as the fluctuations will suddenly be directional. Second, we calculate the surprise at the end of the update step, so that when we use it in the next predict step (i.e., variance expansion), we are more closely resembling the Kalman filter definition of prediction: we are updating ψ_t based on information from the *previous* observation, not the current. We find this to be beneficial when there is an outlier that expands the variance, following our procedure, the variance will be expanded for the observation *following* the outlier, which grants us added protection from the final estimates succumbing to noise. Finally, note that our variance expansion step is only implemented when the surprise exceeds 1 (though, there is nothing that says that threshold cannot be adjusted). This prevents our algorithm from being *too reactive* to random outliers.

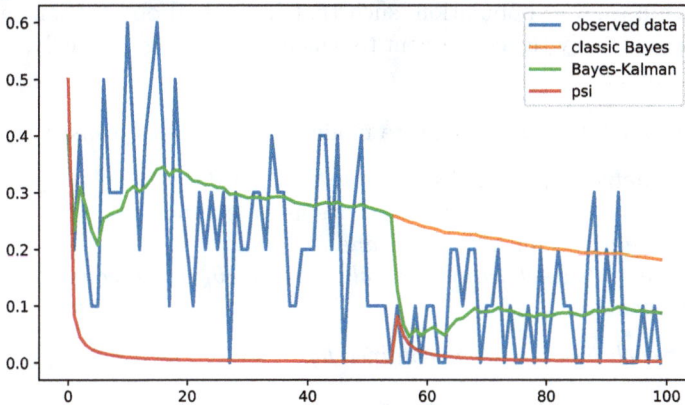

Fig. 10.4: Comparison of classic Bayes vs. the Bayes–Kalman filter. Exponential decay factor $\alpha = 0.2$.

To illustrate this algorithm, consider a time series consisting of ten draws from a Bernoulli random variable: $Y_{ti} \sim \text{Bern}(p)$, for $i = 1, \ldots, 10$. Moreover, suppose that the true value of p is given by $p = 0.3$, for $t < 50$, and $p = 0.1$ for $t \geq 50$, constituting a *switchpoint*. Since on day 50, we have already performed 50 Bayesian updates, with total weight 500, it will require an additional 50 days before the two behaviors have a 50/50 blend. That is, with the class Bayesian update, our estimated probability will still be 20%, even on day 100! To contrast this, our Bayes–Kalman filter quickly locks onto the change, as shown in Figure 10.4.

10.2 Mixtures and Hierarchical Models

In this section, we will consider more advanced forms of Bayesian models. We begin by using a simple mixture to model population heterogeneity. We then discuss application of this technique to customer lifetime value models. Next, we extend our model to multi-level hierarchical models. Next, we discuss a certain linear Bayes estimator known as the credibility model, which is used extensively in the insurance industry. Finally, we resolve the conflict between time aspects of Bayesian data and variance reduction by devising a simple Bayesian Kalman filter.

10.2.1 Mixture Models

Finite Mixture Models

In this section, we will review two closely related topics: mixture models and hierarchical models. We begin with mixture models, which is a type of predictive model that is used to represent the presence of subpopulations within a given population, such that individual observations are not required to identify the component to which they belong. Formally, we have the following.

Definition 10.4. *A* finite mixture model *of dimension* k *consists of a*

1. *a parametric sampling distribution* $F_\theta = p(Y|\theta)$,
2. *a set* $\mathcal{Z} = \{1, \ldots, k\}$ *of* subpopulations *or* components,
3. *a probability vector* $\pi \in \Delta^{k-1}$, *and*
4. *a set of* k *distinct parameter values* $\theta_1, \ldots, \theta_k$ *corresponding to the* k *components,*

such that each observation is obtained by

$$Z \sim \text{MultiBern}(\pi), \qquad (10.27)$$
$$Y|Z \sim F_{\theta_z}. \qquad (10.28)$$

Typically, one assumes the component value Z for each observation to be unknown. In this case, the probability distribution for Y is given by

$$p(Y) = \sum_{i=1}^{k} \pi_i p(Y|\theta_i). \qquad (10.29)$$

This is very similar to the predictive distribution from the Bayesian framework, when one views the probability vector π as a discrete prior distribution over the parameter θ. However, in this context, the probabilities π are more suitably viewed as a measure of the variation in θ across different components of the population.

Continuous Mixture Models and Population Heterogeneity

The definition of a mixture model can be extended to a continuous mixture as follows. In the continuous case, we consider, naturally, a continuum of subpopulations.

Definition 10.5. *A continuous mixture model consists of a*

1. *a parametric sampling distribution $F_\theta = p(Y|\theta)$,*
2. *a countable set $\mathcal{Z} = \mathbb{N}$ of individuals who, collectively, comprise the population, and*
3. *a distribution U, called the mixing distribution or collective function,*

such that observations for each individual are obtained by

$$\Theta \sim U, \tag{10.30}$$
$$Y|\Theta \sim F_\theta, \tag{10.31}$$

where each individual has a unique value of Θ.

Naturally, the distribution for Y is found by marginalizing over the random variable Θ, according to

$$p(Y) = \int_{\mathcal{Z}} p(Y|\theta)U(\theta)\,d\theta. \tag{10.32}$$

Superficially, Equation (10.32) looks like the predictive distribution, if we view the collective function U as a prior for the parameter θ. However, there is one crucial philosophical difference, and that is that $U(\theta)$ no longer represents our *belief* about an unknown fixed parameter θ, but rather it represents the *heterogeneity* of the population. In other words, each observation is a draw from $p(Y_i|\theta_i)$, where θ_i itself is drawn from $U(\theta)$.

Our reference to "individuals" in Definition 10.5 is not superfluous, as it allows the possibility of multiple observations from a given individual. In this case, we say that the ith individual has a parameter value drawn from the collective via the relation

$$\Theta_i \sim U.$$

Suppose, in particular, that individual i has a parameter value $\Theta_i = \theta_i$. Next, we can consider multiple observations from that individual, each independent and identically distributed according to

$$Y_{ij}|\Theta_i = \theta_i \sim p(Y|\theta_i),$$

for, say, $j = 1, \ldots, r$.

We already saw from Equation (10.9) that the act of adding data to compute the posterior distribution inevitably results in a loss of variance in our belief over the parameter θ. This is distinguished from a mixture model,

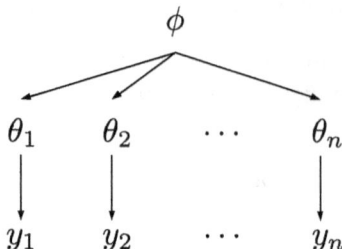

Fig. 10.5: Continuous mixture model.

for which the purpose of data is not to "narrow down" a single value for θ, but to better refine the overall distribution of θ over the population. This is illustrated in Figure 10.5.

To add a Bayesian layer, we might consider the case in which the mixing distribution $U(\theta)$ is itself represented parametrically, with an unobserved hyperparameter ϕ. The distribution for the random variable Y can then be obtained by marginalizing over both θ and ϕ, yielding

$$p(Y) = \iint p(Y|\theta)p(\theta|\phi)p(\phi)\,d\theta\,d\phi. \qquad (10.33)$$

In this model, we have a mixing distribution $p(\theta|\phi)$, which could be considered as a secondary sampling distribution, and a prior distribution $p(\phi)$ over the hyperparameter ϕ. The corresponding posterior distribution over ϕ follows from Bayes' law:

$$p(\phi|\mathcal{D}) \propto p(\phi) \underbrace{\int p(\mathcal{D}|\theta)p(\theta|\phi)\,d\theta}_{p(\mathcal{D}|\phi)}. \qquad (10.34)$$

As the variance in our belief $p(\phi|\mathcal{D})$ vanishes with increasing data, we end up with, in the limit, a single value for the hyperparameter $\phi = \phi_0$, which determines the population distribution $p(\theta|\phi_0)$, the variance of which approaches a limit and does not shrink to zero. That limit represents the variance in θ across the population.

10.2.2 Hierarchical Models

A Bayesian mixture model is a special case of a more general structure called a hierarchical model.

Definition 10.6. *A Bayesian hierarchical model is a rooted tree (Definition 8.6), such that*

- *each node has an associated value;*

- *each leaf node has the same depth;*
- *the values of the leaf nodes are observed data;*
- *the value θ of each branch node b is an unobserved parameter that determines a probability distribution $p(X|\theta)$, such that* child$(b) \sim p(X|\theta)$.

The number of levels of a hierarchical model is the height of its associated tree graph. The level of a node is that node's height.

A schematic for a simple two-level hierarchical model is shown in Figure 10.6. This is similar to a finite mixture model, except that the "buckets" are assumed to be known in this context. Continuous mixture models can be viewed as a special case of hierarchical models. Similarly, hierarchical models can be viewed as a cascade of nested mixture models. Hierarchical models are useful when the data are collected in a hierarchical fashion, or are themselves endowed with a hierarchical feature set; e.g., colleges within universities within states within regions within countries.

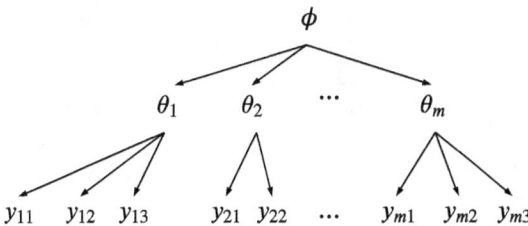

Fig. 10.6: Two-level hierarchical model.

Empirical Bayes

The simplest approach to estimate the hyperparameters of a hierarchical model is using an empirical estimate. Though not technically Bayesian, this approach is given the name *empirical Bayes*.

Definition 10.7. *An empirical Bayes hierarchical model is a Bayesian hierarchical model for which the root-level parameter ϕ is estimated empirically; i.e., directly from the data. A complete Bayesian analysis involves prescribing a prior distribution $p(\phi)$ over the root-level parameter and performing a full Bayesian update with the data.*

Example 10.7. Ten laboratories conducted an identical experiment with a binary outcome, each with a random number of treatment units. Due to natural variability between the labs (different protocols, procedures, respondents for the experiment, etc.), we anticipate a natural variability in the average effect.

We can model this as a two-level hierarchical model, similar to Figure 10.6. In particular, the observed data are modeled by $Y_i \sim \text{Binom}(n_i, \theta_i)$, where n_i, the number of experimental units for laboratory i, is known, and where $\theta_i \sim \text{Beta}(\alpha, \beta)$ is distributed relative to an unknown beta distribution.

In order to construct a simulation, let us suppose that the true values of the hyperparameters are given by $\alpha = 2$ and $\beta = 8$. We can simulate a data set, as shown in lines 2–6 of Code Block 10.1. The results of such a simulation are given in the first three columns of Table 10.5.

Group	n_i	y_i	$\hat{\alpha}'_i$	$\hat{\beta}'_i$
0	29	9	10.5327	24.9779
1	10	0	1.5327	14.9779
2	14	4	5.5327	14.9779
3	16	2	3.5327	18.9779
4	15	3	4.5327	16.9779
5	21	3	4.5327	22.9779
6	20	9	10.5327	15.9779
7	25	3	4.5327	26.9779
8	11	2	3.5327	13.9779
9	13	7	8.5327	10.9779

Table 10.5: Data and posterior estimates for Example 10.7.

Simulated data in hand, we may now proceed to use empirical Bayes to estimate the population parameters α, β. Empirical Bayes follows a *bottom–up, top–down* approach, meaning we first estimate the hyper parameters, bottom–up, and then perform a Bayesian update for each treatment group, top–down.

Each individual θ_i can be estimated using that group's data, as $\hat{\theta}_i = y_i/n_i$. Given the set of estimates $\hat{\theta}_0, \ldots, \hat{\theta}_9$, we next compute the empirical mean and variance. Assuming these hidden parameters are distributed as a $\text{Beta}(\alpha, \beta)$ distribution, we can invert the relations for the mean and variance (Equations (2.50) and (2.51)) to obtain

$$\hat{k} = \frac{\mu(1-\mu)}{\sigma^2} - 1$$
$$\hat{\alpha} = \hat{k}\hat{\mu}$$
$$\hat{\beta} = \hat{k} - \hat{\alpha},$$

where we have defined $k = \alpha + \beta$, forming invertible relationships between (α, β), (μ, σ^2), and (μ, k). (Note that the variance of the beta distribution is equivalent to $\mu(1-\mu)/(k+1)$.)

The empirical estimates for $\hat{\alpha}$ and $\hat{\beta}$ (lines 13–14) yield $\hat{\alpha} = 1.5327$ and $\hat{\beta} = 4.9779$. Using these values a prior, we can next perform a Bayesian

```
1    # Simulate Data
2    alpha, beta = 2, 8
3    m = 10
4    theta_true = np.random.beta(alpha, beta, size=m) # hidden
5    n_i = np.random.randint(10, 30, size=m) #observed, given
6    y_i = np.random.binomial(n_i, theta_true) #observed
7
8    theta_hat = y_i / n_i
9    mu_hat = theta_hat.mean()
10   var_hat = theta_hat.var()
11
12   k_hat = mu_hat * (1-mu_hat) / var_hat - 1
13   alpha_hat = k_hat * mu_hat
14   beta_hat = k_hat - alpha_hat
15
16   df = DataFrame({'n':n_i, 'y':y_i, 'alpha_hat':alpha_hat,
          'beta_hat':beta_hat})
17   # Bayesian update for each group
18   df.alpha_hat += y_i
19   df.beta_hat += n_i - y_i
```

Code Block 10.1: Empirical Bayes simulation.

update for each group, yielding $\hat{\alpha}_i = \hat{\alpha} + y_i$ and $\hat{\beta}_i = \hat{\beta} + n_i - y_i$, completing the results in Table 10.5.   ▷

10.3 Modeling Customer Lifetime Value

We next turn to an important marketing application, and that is predicting a customer's lifetime value (defined below). This is relevant across a broad range of businesses, from coffee shops to software subscriptions to freemium apps. A *freemium product* is a product that is free to use, but that also offers premium content available for purchase (e.g., many mobile games use a freemium model). In this context, customers are often referred to as *users*, though the term *user* typically implies any user of the product, not limited to paying users or customers. An accessible and comprehensive overview of freemium products from a marketing perspective is found in Seufert [2014].

10.3.1 Frequency, Recency, and Monetization

We begin with a few basic definitions.

Definition 10.8. *The* (Customer) lifetime value (CLTV or LTV) *of a given user is the present value of all that customer's future purchases for a given product.*

When a customer makes a purchase, this amount is usually referred to as *revenue*, as it represents revenue for the company that produces the product. Since the future is, by definition, unknown, customer lifetime value is invariably described by some predictive model.

Additionally, there are two primary business models that are relevant when modeling customer lifetimevalue.

Definition 10.9. *A* contractual (or subscription) *product or service is a product in which payments are automatically recurring (e.g., monthly) at a fixed amount until the customer cancels.*

A product that is not contractual—i.e., one in which customers may choose to make purchases of any amount at any time—is called noncontractual.

In addition, we distinguish a certain class of product that is prevalent in software and mobile apps.

Definition 10.10. *A* freemium *product is one whose basic functionality is available for free, but that also offers premium content available for a fee. The premium content can be contractual or noncontractual, or a mixture of both.*

Example 10.8. The popular mobile game *Clash of Clans* is a freemium app in which the basic gameplay is free, but that offers *gems*, a form of *in-app currency*, available for purchase (see Table 10.6).

Description	Amount of Gems	Price
Pocketful of Gems	80	$0.99
Pile of Gems	500	$4.99
Bag of Gems	1200	$9.99
Sack of Gems	2500	$19.99
Box of Gems	6500	$49.99
Chest of Gems	14000	$99.99

Table 10.6: Cost of gems, available as in-app purchase, for Clash of Clans.

Gems can then be used in the game to purchase various items of value: troop training, building, or resource speed-ups, building upgrades, purchase of resources (gold or elixir). In addition, Clash of Clans features occasional sales, in which specific items are available as a bundle, for which customers can receive 3x, 4x, or 5x the value for the price. Notice the in-app store is noncontractual, as users can make a purchase in any amount (of the available six options) at any time.   ▷

Summary statistics for a particular user are typically measured using three quantities: recency, frequency, and monetization.

Definition 10.11. *A customer's recency, frequency, and monetization (RFM) are defined as*

- recency: *the amount of time since the customer's last purchase;*
- frequency: *the rate at which purchase transactions are made;*
- monetization: *the average purchase amount.*

In a subscription model, frequency and monetization are fixed per the contract (e.g., \$9.99 per month until cancel). For a noncontractual model, freemium or otherwise, all three components play an important role in modeling customer lifetime value. For a freemium product with a fixed in-app store (see Example 10.8), the monetization can be described as a multinomial distribution over the available purchase amounts. For a nonfreemium, noncontractual product (e.g., Starbuck's Coffee), purchases cannot be so cleanly described, and so the total amount is usually recorded, as customers are able to buy any combination of products from the menu.

10.3.2 Pareto/NBD and Related Models

In this section, we introduce the classic Pareto/NBD model and review a number of variations that have been constructed through the years.

Pareto/NBD Model

The Pareto/NBD model is a popular approach to modeling customer lifetime value in the context of noncontractual revenue streams; it was first introduced in Schmittlein, *et al.* [1987]. The basic model is defined as follows

Definition 10.12. *The* Pareto/NBD model *is a predictive customer lifetime value model built around the following key assumptions:*

1. *At any given time, a customer can be in one of two states: each customer is* alive *for a period of time, during which they may make a purchase at any time, until the customer becomes* dead *(or churned), at which point the customer permanently disengages with the product;*
2. *the frequency of purchases for any customer while alive follows a Poisson process; i.e., the number of transactions N_t on the interval $[0, t]$ is represented as $N_t \sim \text{Poiss}(\lambda t)$, with rate parameter λ;*
3. *heterogeneity of transaction rates is described by a gamma distribution: $\lambda \sim \text{Gamma}(\alpha, \beta)$, with shape parameter α and scale parameter β;*
4. *the (unobserved) customer lifetime τ follows an exponential distribution with rate parameter μ (the churn rate);*
5. *heterogeneity in churn rate follows a gamma distribution: $\mu \sim \text{Gamma}(\gamma, \delta)$, with shape parameter γ and scale parameter δ;*
6. *the transaction rate λ and churn rate μ vary independently across customers.*

We can use the results from Tables 10.2 and 10.3 to express the distributions of the number of transactions X_t and customer lifetime τ in terms of their respective hyperparameters.

Proposition 10.5. *The marginal distributions for the count of transactions X_t over a period $[0, t]$, for which a customer is alive, and customer lifetime τ in a Pareto/NBD model are given by*

$$p(X_t = x | t < \tau, \alpha, \beta) = \text{NBD}(x; \alpha, (1 + \beta t)^{-1}) \qquad (10.35)$$
$$p(\tau = t | \gamma, \delta) = \text{Lomax}(t; \gamma, \delta^{-1}). \qquad (10.36)$$

respectively.

Proof. Form the modeling assumptions of Definition 10.12, we have

$$p(X_t = x | t < \tau, \lambda) = \text{Poiss}(x; \lambda t) = \frac{e^{-\lambda t}(\lambda t)^x}{x!},$$

$$p(\lambda | \alpha, \beta) = \text{Gamma}(\lambda; \alpha, \beta) = \frac{\lambda^{\alpha-1}e^{-\lambda/\beta}}{\Gamma(\alpha)\beta^\alpha}$$

$$p(\tau = t | \mu) = \text{Exp}(t; \mu t) = \mu e^{-\mu t}$$

$$p(\mu | \gamma, \delta) = \text{Gamma}(\mu; \gamma, \delta) = \frac{\mu^{\gamma-1}e^{-\mu/\delta}}{\Gamma(\gamma)\delta^\gamma}.$$

Marginalizing, via

$$p(X_t = x | t < \tau, \alpha, \beta) = \int_0^\infty p(X_t = x | t < \tau, \lambda)p(\lambda | \alpha, \beta)\, d\lambda$$

$$p(\tau = t | \gamma, \delta) = \int_0^\infty p(\tau = t | \mu)p(\mu | \gamma, \delta)\, d\mu,$$

we obtain our results. $\square$

As a result of Proposition 10.5, we have the following quantities, which are useful in interpreting the practical implications of various configurations of hyperparameters.

Corollary 10.1. *For a customer in the Pareto/NBD model, with hyperparameters $(\alpha, \beta, \gamma, \delta)$, the expected lifetime, survival function, and expected number of transactions within a horizon are given by*

$$\mathbb{E}[\tau | \gamma, \delta] = \frac{\delta^{-1}}{\gamma - 1}, \text{ for } \gamma > 1 \qquad (10.37)$$

$$S(t | \gamma, \delta) = (1 + \delta t)^{-\gamma} \qquad (10.38)$$

$$\mathbb{E}[X_t | t < \tau, \alpha, \beta] = \alpha \beta t. \qquad (10.39)$$

Moreover, the expected total number of transactions in a customer's lifetime is given by

$$\mathbb{E}[X_\tau | \alpha, \beta, \gamma, \delta] = \frac{\alpha\beta}{\delta(\gamma - 1)}, \tag{10.40}$$

whenever $\gamma > 1$.

Proof. These follow from properties of the NBD and Lomax distributions. The final equation follows from Equations (10.37) and (10.39) and the independence assumption between the transaction rate and customer lifetime. □

Example 10.9. We can write a simulation for a random customer with hyperparameters $(\alpha, \beta, \gamma, \delta)$, as shown in Code Block 10.2. If we run

```
def simPareto(n_users, alpha=0.5, beta=0.5, gamma=2, dolta=0.01,
        cohort_age=365):

    df_users = DataFrame({
            'user_id': np.arange(n_users),
            'transaction_rate': random.gamma(alpha, scale=beta,
                size=n_users),
            'lifetime': random.pareto(gamma,size=n_users)/delta
            })

    df_transactions = DataFrame([], columns=['user_id',
        'transaction_time'])

    for i, row in df_users.iterrows():
        time = 1 + random.exponential(scale=1/row.transaction_rate)
        while time < min(row.lifetime, cohort_age):
            df_transactions =
                df_transactions.append(DataFrame({'user_id':
                row.user_id, 'transaction_time': time}, index=[0]),
                ignore_index=True)
            time += random.exponential(scale=1/row.transaction_rate)

    df_transactions.loc[:, 'n_transactions'] = 1
    df_transactions = df_transactions.astype(int)

    return df_users, df_transactions
```

Code Block 10.2: Simulation of Pareto/NBD model

the simulation with hyperparameters $\alpha = 0.5$, $\beta = 0.5$, $\gamma = 2$, and $\delta = 0.01$, we should expect the average lifetime to be $\tau = 100$, as given by Equation (10.37). Moreover, we expect a first-year survival rate of approximately $S(365) \approx 0.0462$. The expected transaction rate is given by

$\mathbb{E}[\lambda|\alpha, \beta] = \alpha\beta = 0.25$, so that the average user should make approximately 25 transactions.

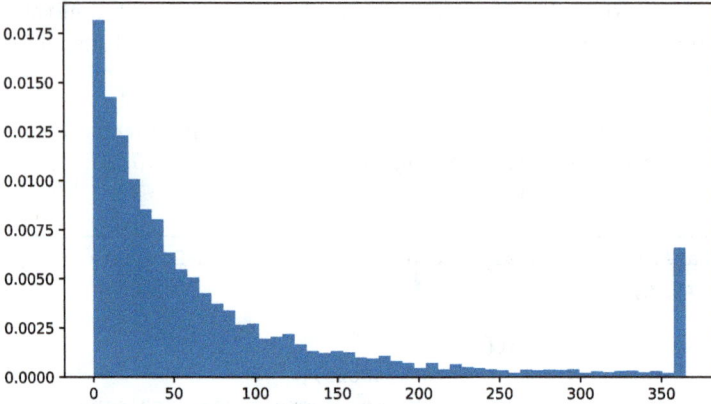

Fig. 10.7: Distribution of user lifetimes for Example 10.9.

The simulated distribution of user lifetimes is shown in Figure 10.7. The observed 1-year survival rate in our simulation was 4.71%, closely agreeing to the theoretical expected survival rate of 4.62%. Moreover, the average observed lifetime was 96.7 days, close to the predicted 100 day expected lifetime.

Finally, we note that the total number of transactions that resulted from our simulation was 19,233, or 19.233 transactions per user. The total expected lifetime transactions per user, however, as given by Equation (10.40), is 25. This discrepancy is due to the fact that there are a remaining 5% of users who will make transactions beyond the first year, ultimately brining the overall average up to 25.   ▷

In Example 10.9, we saw that the total number of transactions per user, for the first year, was 19.233, according to our simulation. This represents an approximately 20% haircut from the total expected *lifetime* transactions per user of 25. Naturally, we are interested in how to compute the expected number of transactions within a specified time horizon. This result is given as follows.

Proposition 10.6. *Let the random variable* $X(H)$ *represent the number of transactions during the time horizon* $[0, H]$ *for a customer in the Pareto/NBD model. Then the expected value of* $X(H)$*, for a customer with hyperparameters* $\alpha, \beta, \gamma, \delta$*, is given by*

$$\mathbb{E}[X(H)|\alpha,\beta,\gamma,\delta] = \frac{\alpha\beta}{\delta(\gamma-1)}\left[1 - \frac{1+\gamma\delta H}{(1+\delta H)^\gamma}\right] + \alpha\beta H(1+\delta H)^{-\gamma}. \quad (10.41)$$

Proof. In order to calculate this, we must divide the populations into two separate groups: those alive at time H and those with a lifetime $\tau < H$:

$$\mathbb{E}[X(H)|\alpha,\beta,\gamma,\delta] = \mathbb{E}[X(H)|\tau \le H, \alpha,\beta,\gamma,\delta]\mathbb{P}(\tau \le H)$$
$$+ \mathbb{E}[X(H)|\tau > H, \alpha,\beta,\gamma,\delta]\mathbb{P}(\tau > H). \quad (10.42)$$

First, let us consider the group of customers alive at time H. From Equation (10.38), we know that this accounts for

$$\mathbb{P}(\tau > H) = S(H) = (1+\delta H)^{-\gamma}$$

of all users. Since we are *only* interested in transactions *up to time* H, we can use Equation (10.39) to write

$$\mathbb{E}[X(H)|\tau > H, \alpha,\beta] = \alpha\beta H$$

for the customers who are alive at time H. Combining the previous two equations, we find that the contribution to $X(H)$ from users alive at time H should be given by

$$\alpha\beta H(1+\delta H)^{-\gamma},$$

the final term in Equations (10.41) and (10.42).

Second, let us consider users with a lifetime $\tau < H$. We can adjust the density for a users lifetime using

$$p(\tau|\tau \le H) = \begin{cases} \dfrac{p(\tau)}{\mathbb{P}(\tau \le H)} & \text{for } \tau \le H \\ 0 & \text{for } \tau > H \end{cases}.$$

The expected lifetime for a user whose lifetime is less than H is therefore given by

$$\mathbb{E}[\tau|\tau \le H] = \int_0^H \frac{\tau p(\tau)}{\mathbb{P}(\tau \le H)}\, d\tau.$$

Therefore, Equation (10.42) reduces to

$$\mathbb{E}[X(H)|\alpha,\beta,\gamma,\delta] = \alpha\beta \int_0^H \tau p(\tau|\alpha,\beta,\gamma,\delta)\, d\tau + \alpha\beta H(1+\delta H)^{-\gamma}.$$

Next, one can show that

$$\int_0^H \tau p(\tau|\alpha,\beta,\gamma,\delta)\, d\tau = \int_0^H \gamma\delta x(1+\delta x)^{-(\gamma+1)}\, dx$$
$$= \frac{1}{\delta(\gamma-1)}\left[1 - \frac{1+\gamma\delta H}{(1+\delta H)^\gamma}\right].$$

This completes the result. □

Note 10.5. For the choice of hyperparameters $\alpha = 0.5$, $\beta = 0.5$, $\gamma = 2$, $\delta = 0.01$, the expected number of transactions, as given by Equation (10.41), for the first year ($H = 365$) works out to be $\mathbb{E}[X(365)] = 19.62$. This closely matches the observed result (19.233) obtained from the simulation in Example 10.9. ▷

Variations: BG/NBD, BG, and BG/BB Models

Since the original publication of the Pareto/NBD model, a number of variations have been made that have offered substantial improvements to the model, especially in terms of computational ease of use. These include the BG/NBD model (Fader, *et al.* [2005]), the BG model for contractual products (Fader and Hardie [2007]), and the BG/BB model, which can be used for both contractual and noncontractual (Fader, *et al.* [2010]). A summary of the variations between these models is shown in Table 10.7. We note that the BG model is a degenerate case of the BG/BB model, for which the transaction probability of a customer while alive is 100%.

		model			
component		Pareto/NBD	BG/NBD	BG	BG/BB
transactions	rate	Poisson	Poisson	–	Bernoulli
	heterogeneity	gamma	gamma	–	beta
	predictive	NBD	NBD	–	BB
lifetime	rate	exponential	geometric	geometric	geometric
	heterogeneity	gamma	beta	beta	beta
	predictive	Pareto	BG	BG	BG

Table 10.7: Development of LTV models.

The key conceptual leaps in these various models are the following:

- *replacing the exponential–gamma mixture with the beta–geometric mixture*: instead of allowing a user's lifetime to be unobserved, occurring at any point in time, we instead model churn as occurring only immediately following a transaction, with a constant probability for each customer;
- *replacing the Poisson–gamma mixture with the beta–Bernoulli mixture*: instead of modeling transactions as a Poisson process, we instead discretize time, creating a sequence of *transaction opportunities*, for which a customer that is alive makes a purchase with a constant probability.

10.3.3 The BG/BB Model

The Model

As we saw in Table 10.7, multiple variations of the classic Pareto/NBD model ultimately led to a highly flexible model that is computationally more tractable than the original Pareto/NBD and that captures a contractual model (BG model) as a special case. We therefore take a closer look at this final model, following the work presented in Fader, *et al.* [2010].

Definition 10.13. *The* BG/BB *model is a predictive customer lifetime value model built around the following key assumptions:*

1. *At any given time, a customer can be in one of two states:* alive *or* dead *(churned); a customer that is alive can make a purchase at any time, whereas a customer that has churned has permanently disengaged with the product;*
2. *time is characterized by a sequence of transaction opportunities; a customer that is alive will make a purchase during a given transaction opportunity with a probability p;*
3. *heterogeneity of the transaction probability follows a beta distribution: $p \sim \mathrm{Beta}(\alpha, \beta)$;*
4. *a customer that is alive will churn prior to any transaction opportunity with a probability θ, such that each customer's unobserved lifetime follows a geometric distribution $\tau \sim \mathrm{Geom}(\theta)$;*
5. *heterogeneity in the churn rate follows a beta distribution: $\theta \sim \mathrm{Beta}(\gamma, \delta)$;*
6. *the transaction probability p and churn rate θ vary independently across customers.*

The predictive distribution for the transaction probability follows a beta–Bernoulli distribution; the lifetime follows a beta–geometric distribution.

The BG/BB model differs from the Pareto/NBD by a slight modification in the behavioral story, the key difference being that the BG/BB is a discrete version of the Pareto/NBD model. Instead of transactions occurring in continuous time, time is discretized into a sequence of transactional opportunities, for which a customer may either be active or inactive. Similarly, a user's lifetime is no longer continuous, but rather it is represented by an integer number of transaction opportunities; i.e., a user can no longer churn, as per the model, at an arbitrary moment in time, but each user can only churn following the completion of a transaction opportunity.

As we shall see, summary statistics for each user consist of the number X of active transaction opportunities (which we call *actions*[1]), the time of last purchase T_x, and the time of the last transaction opportunity N. (Time, in this context, is the *cohort time*, or time since first purchase, as defined for online processes in Definition 5.20.) For a fixed number of transaction opportunities n, the number of distinct frequency–recency combinations therefore scales like $O(n^2)$, compared with the $O(2^n)$ possible full binary

---

[1] A *transaction* is itself a discrete event; when discretizing time, it is possible that a customer make multiple transactions within any fixed transaction opportunity. We therefore distinguish the atomic transactions from active transaction opportunities by using the term *action* to denote the latter.

vectors capturing the full (discretized) transaction history. This makes the BG/BB model highly scalable, as we only have to track the count of users with a particular frequency–recency combination.

To proceed, we first define our precise set of random variables.

Definition 10.14. *For a particular user in the BG/BB model, we define the following random variables*

- N represents the cohort age of the user, in the context of Definition 5.20;
- Y_t represents the user being active during the tth transaction opportunity, for $t = 1, \ldots, n$;
- $X = \sum_{t=1}^{n}$ represents the number of active transaction opportunities (i.e., actions) observed;
- T_x represents the time of the latest observed transaction opportunity;
- τ represents the (unobserved) lifetime, defined as the index of the final transaction opportunity for which a user was alive;
- $A_t = \mathbb{I}[t \leq \tau]$ represents the event in which the user is still alive at the tth transaction opportunity, for $t = 1, \ldots, n$.

Note that

$$\mathbb{P}(\tau = t) = \text{Geom}(t; \theta) = \theta(1 - \theta)^t, \qquad (10.43)$$

for $t = 0, 1, 2, \ldots$. Moreover, the survival function at t_0 is given by

$$S(t) = \mathbb{P}(\tau > t) = 1 - \mathbb{P}(\tau \leq t) = (1 - \theta)^{t+1}. \qquad (10.44)$$

Consequently, $\mathbb{P}(A_t) = \mathbb{P}(\tau \geq t) = (1 - \theta)^t$.

Note 10.6. In general, a user does not become a customer until he or she makes an actual purchase. We can think of the variables defined in Definition 10.14 as occurring *subsequent* to the initial purchase transaction; i.e., we can augment the model with $A_0 = 1$ and $Y_0 = 1$ for all users. With this adjustment, the actual lifetime of a customer is $\tau + 1$, and the actual number of actions is $X + 1$. The definition of a churn opportunity occurring immediately *prior* to any transaction opportunity ($t = 1, 2, \ldots$) make more sense in this context, as all customers start off with an initial transaction at $t = 0$.   ▷

Computation of Summary Statistics

The summary statistics for a user in the BG/BB model consist of the random variables in the triple (X, T_X, N), as outlined in the previous paragraph. Moreover, time in the BG/BB model is discretized, so that we must also define which transaction opportunity each atomic transaction belongs to. In general, as inputs we will take a user table and a transaction table, similar to the outputs of the simulation in Code Block 10.2.

In Code Block 10.3 we have a function, `getBgbbSummary`, that takes a user and transactional dataframe as inputs, along with the length of the

```python
def getBgbbSummary(df_users, df_transactions, opp_window=7):

    df_transactions.transaction_time /= opp_window
    df_transactions.transaction_time =
        np.ceil(df_transactions.transaction_time).astype(int)
    df_reduced = df_transactions.groupby(['user_id',
        'transaction_time'], as_index=False)['n_transactions'].sum()
    df_reduced.loc[:, 'x'] = 1

    df_reduced = df_reduced.groupby('user_id',
        as_index=False).agg({'transaction_time':max, 'x':sum})
    df_reduced = df_reduced.rename(columns={'transaction_time':'tx'})
    n = df_reduced.tx.max()
    df_reduced.loc[:, 'n'] = n
    df_reduced.loc[:, 'freq'] = 1

    # add users without transaction record (x=0)
    x0_users = df_users[~df_users.user_id.isin(
        df_reduced.user_id.unique())].user_id.unique()
    df_reduced = df_reduced.append(DataFrame({'user_id':x0_users,
        'tx':0, 'x': 0, 'n':52, 'freq':1}), ignore_index=True)
    df_sum = df_reduced.groupby(['tx', 'x', 'n'],
        as_index=False)['freq'].sum()

    return df_reduced, df_sum
```

Code Block 10.3: Computation of BG/BB summary statistics.

opportunity window (e.g., 7-days to define a weekly discretization), and returns a reduced and summary dataframe as outputs. The reduced dataframe
is still at the individual customer level, and contains the summary statistics
for each individual customer. As we shall see in Corollary 10.5, however,
we only require a frequency count for the number of customers for each
combination of permissible summary statistics; this count is captured in
the output dataframe df_sum.

Marginal Distributions

Similar to Proposition 10.5, we can write the marginal distributions for the
average customer, with hyperparameters $(\alpha, \beta, \gamma, \delta)$, as follows.

Proposition 10.7. *The marginal distributions for the count of* actions
*(i.e., active transaction opportunities) X_t over a period $\{1, \ldots, t\}$, for which
a customer is alive, and customer lifetime τ in a BG/BB model are given
by*

$$p(X_t = x | t \leq \tau, \alpha, \beta) = \text{BetaBin}(x; t, \alpha, \beta) \tag{10.45}$$
$$p(\tau = t | \gamma, \delta) = \text{BetaGeom}(t; \gamma, \delta), \tag{10.46}$$

respectively.

Proof. From the modeling assumptions of Definition 10.14, we have

$$p(X_t = x | t < \tau, p) = \text{Binom}(x; t, p) = \binom{t}{x} p^x (1 - p)^{t - x}$$

$$p(p | \alpha, \beta) = \text{Beta}(p; \alpha, \beta) = \frac{p^{\alpha - 1}(1 - p)^{\beta - 1}}{B(\alpha, \beta)}$$

$$p(\tau = t | \theta) = \text{Geom}(t; \theta) = \theta(1 - \theta)^t,$$

$$p(\theta | \gamma, \delta) = \text{Beta}(\theta; \gamma, \delta) = \frac{\theta^{\gamma - 1}(1 - \theta)^{\delta - 1}}{B(\gamma, \delta)}.$$

Marginalizing, via

$$p(X_t = x | t \leq \tau, \alpha, \beta) = \int_0^1 p(X_t = x | t < \tau, p) p(p | \alpha, \beta) \, dp$$

$$p(\tau = t | \gamma, \delta) = \int_0^1 p(\tau = t | \theta) p(\theta | \gamma, \delta) \, d\theta,$$

we obtain our result.   □

As a result of Proposition 10.7, we have the following quantities, which are useful in interpreting the practical implications of various configurations of hyperparameters.

Corollary 10.2. *For a customer in the BG/BB model, with hyperparameters $(\alpha, \beta, \gamma, \delta)$, the expected lifetime, survival function, and expected number of transactions within a horizon are given by*

$$\mathbb{E}[\tau | \gamma, \delta] = \frac{\delta}{\gamma - 1}, \; \text{for } \gamma > 1 \tag{10.47}$$

$$S(t | \gamma, \delta) = \frac{B(\gamma, \delta + t + 1)}{B(\gamma, \delta)} \tag{10.48}$$

$$\mathbb{E}[X_t | t < \tau, \alpha, \beta] = \frac{\alpha t}{\alpha + \beta}. \tag{10.49}$$

Moreover, the expected total number of transactions in a customer's lifetime is given by

$$\mathbb{E}[X_\tau | \alpha, \beta, \gamma, \delta] = \frac{\alpha \delta}{(\alpha + \beta)(\gamma - 1)}, \tag{10.50}$$

whenever $\gamma > 1$.

Proof. These follow from properties of the beta-binomial and beta-geometric distributions. The survival function for the beta-geometric distribution, as it is reported in Table A.2, was computed using Wolfram Alpha [2018], by simplifying the output[2] to the query:

```
sum beta(a+1,b+x)/beta(a,b) from x=0 to T.
```

The final equation follows from Equations (10.47) and (10.49) and the independence assumption between the transaction rate and customer lifetime.

□

The Likelihood

Given the preceding definitions, we have the following result, due to Fader, *et al.* [2010].

Proposition 10.8. *The likelihood for a customer in the BG/BB model with summary statistics* (x, t_x, n) *is given by*

$$p(x, t_x, n|p, \theta) = \sum_{t=t_x}^{n-1} p^x(1-p)^{t-x}\theta(1-\theta)^t + p^x(1-p)^{n-x}(1-\theta)^n. \quad (10.51)$$

Note 10.7. The only difference between the sum and the final term in Equation (10.51) is the factor of θ, so that Equation (10.51) may alternatively be expressed as

$$p(x, t_x, n|p, \theta) = \sum_{t=t_x}^{n} p^x(1-p)^{t-x}\theta^{\mathbb{I}[t \neq n]}(1-\theta)^t,$$

where $\theta^{\mathbb{I}[t \neq n]} = 1$ when $t = n$.   ▷

Proof. When calculating the probability of observing a user with summary statistics (x, t_x, n), we must account for the $n - t_x + 1$ possibilities for which the user might have churned (since churn itself is not directly observable). The only thing we know for certain, is that the user was alive at time t_x, so that $\tau \geq t_x$. The user may have churned following any of the $n - t_x$ opportunities $\tau = t_x, t_x + 1, \ldots, n-1$, or the user might still be alive on the nth opportunity, so that $\tau \geq n$. We therefore have the relation

$$p(x, t_x, n|p, \theta) = \sum_{t=t_x}^{n-1} \mathbb{P}(X = x|p, \tau = t)\mathbb{P}(\tau = t|\theta)$$

$$+ \mathbb{P}(X = x|p, \tau \geq n)\mathbb{P}(\tau \geq n|\theta).$$

---

[2] The output is also subtracted from 1, as the query corresponds to the cumulative distribution $C(T) = \sum_{x=0}^{T} \text{BetaGeom}(x; a, b)$.

Now,

$$\mathbb{P}(X = x | p, \tau = t) = p^x (1 - p)^{t - x},$$
$$\mathbb{P}(X = x | p, \tau \geq n) = p^x (1 - p)^{n - x},$$
$$\mathbb{P}(\tau = t | \theta) = \theta (1 - \theta)^t,$$
$$\mathbb{P}(\tau \geq n | \theta) = (1 - \theta)^n.$$

This completes the proof.   □

Proposition 10.8 can be more readily understood in terms of an example. Consider a user with summary statistics $(x = 2, t_x = 3, n = 6)$; for example, as resulting from the binary vector 101000. There are $n - t_x + 1 = 4$ cases to consider, as shown in Table 10.8. The likelihood is obtained from this

Case	1	2	3	4	5	6	$\mathbb{P}(X = 2 | \text{case})$	$\mathbb{P}(\text{case})$
	1	0	1	0	0	0		
$\tau = 3$	A	A	A	D	D	D	$p^2(1 - p)$	$\theta(1 - \theta)^3$
$\tau = 4$	A	A	A	A	D	D	$p^2(1 - p)^2$	$\theta(1 - \theta)^4$
$\tau = 5$	A	A	A	A	A	D	$p^2(1 - p)^3$	$\theta(1 - \theta)^5$
$\tau \geq 6$	A	A	A	A	A	A	$p^2(1 - p)^4$	$(1 - \theta)^6$

Table 10.8: Illustration of Proposition 10.8.

table by computing the sum

$$p(x = 2, t_x = 3, n = 6 | p, \theta) = \sum \mathbb{P}(X = 2 | \text{case}) \mathbb{P}(\text{case}),$$

which is consistent with Equation (10.51).

Proposition 10.9. *The marginal likelihood, obtained by marginalizing the result from Proposition 10.8 over the respective mixing distributions, is given by*

$$L(\alpha, \beta, \gamma, \delta | x, t_x, n) = \sum_{t = t_x}^{n-1} \frac{B(\alpha + x, \beta + t - x)}{B(\alpha, \beta)} \frac{B(\gamma + 1, \delta + t)}{B(\gamma, \delta)}$$
$$+ \frac{B(\alpha + x, \beta + n - x)}{B(\alpha, \beta)} \frac{B(\gamma, \delta + n)}{B(\gamma, \delta)} \qquad (10.52)$$

where $B(\cdot, \cdot)$ is the beta function, as defined in Equation (2.47). We call this expression the likelihood for the hyperparameters; i.e.,

$$L(\alpha, \beta, \gamma, \delta | x, t_x, n) = p(x, t_x, n | \alpha, \beta, \gamma, \delta),$$

where L is viewed as a function of the hyperparameters.

Proof. We may use Equation (10.51) to write the product of the likelihood with the mixing distributions as

$$p(x, t_x, n | p, \theta) p(p | \alpha, \beta) p(\theta | \gamma, \delta)$$

$$= \sum_{t=t_x}^{n-1} \frac{p^{\alpha+x-1}(1-p)^{\beta+t-x-1}}{B(\alpha, \beta)} \frac{\theta^\gamma (1-\theta)^{\delta+t-1}}{B(\gamma, \delta)}$$

$$+ \frac{p^{\alpha+x}(1-p)^{\beta+n-x-1}}{B(\alpha, \beta)} \frac{\theta^{\gamma-1}(1-\theta)^{\delta+n-1}}{B(\gamma, \delta)}$$

$$= \sum_{t=t_x}^{n-1} \left[\frac{B(\alpha+x, \beta+t-x)}{B(\alpha, \beta)} \text{Beta}(p; \alpha+x, \beta+t-x) \right.$$

$$\left. \times \frac{B(\gamma+1, \delta+t)}{B(\gamma, \delta)} \text{Beta}(\theta; \gamma+1, \delta+t) \right]$$

$$+ \left[\frac{B(\alpha+x, \beta+n-x)}{B(\alpha, \beta)} \text{Beta}(p; \alpha+x, \beta+n-x) \right.$$

$$\left. \times \frac{B(\gamma, \delta+n)}{B(\gamma, \delta)} \text{Beta}(\theta; \gamma, \delta+n) \right] \qquad (10.53)$$

Marginalizing over the hyperparameters, using

$$p(x, t_x, n | \alpha, \beta, \gamma, \delta) = \int_0^1 \int_0^1 p(x, t_x, n | p, \theta) p(p | \alpha, \beta) p(\theta | \gamma, \delta) \, dp \, d\theta,$$

yields our result, as expressed in Equation (10.52). □

Corollary 10.3. *The likelihood given by Equation (10.52) is equivalent to*

$$L(\alpha, \beta, \gamma, \delta | x, t_x, n) = \sum_{t=t_x}^{n} \frac{B(\alpha+x, \beta+t-x)}{B(\alpha, \beta)} \frac{B(\gamma+1, \delta+t)}{B(\gamma, \delta)}$$

$$+ \frac{B(\alpha+x, \beta+n-x)}{B(\alpha, \beta)} \frac{B(\gamma, \delta+n+1)}{B(\gamma, \delta)}. (10.54)$$

Proof. This follows by direct comparison of Equations (10.52) and (10.54) and the identity

$$B(\gamma+1, \delta+n) + B(\gamma, \delta+n+1) = B(\gamma, \delta+n);$$

see Exercise 10.4.

Though this completes the proof, this distinction can be understood more intuitively by framing Equations (10.52) and (10.54) as

$$L = \sum_{t=t_x}^{n-1} \text{BetaBin}(x; t, \alpha, \beta) p(\tau = t | \gamma, \delta) + \text{BetaBin}(x; n, \alpha, \beta) p(\tau \geq n | \gamma, \delta)$$

$$= \sum_{t=t_x}^{n} \text{BetaBin}(x; t, \alpha, \beta) p(\tau = t | \gamma, \delta) + \text{BetaBin}(x; n, \alpha, \beta) p(\tau > n | \gamma, \delta),$$

respectively. Thus, we see that Equation (10.54) explicitly enumerates the possibility that a user's lifetime is exactly $\tau = n$ in the summation. This representation will be useful later on. □

In order to compute the gradient of the likelihood function Equation (10.52), we first define the quantity

$$\rho(\alpha, \beta, x, y) = \frac{B(\alpha + x, \beta + y)}{B(\alpha, \beta)}, \tag{10.55}$$

so that the likelihood may be expressed as

$$L = \sum_{t=t_x}^{n} \rho(\alpha, \beta, x, t - x)\rho(\gamma, \delta, u_t, t), \tag{10.56}$$

where we define $u_t = \mathbb{I}[t \neq n]$, as in Note 10.7.

Now, the derivatives of Equation (10.55), with respect to α and β, are given by

$$\frac{\partial \rho}{\partial \alpha} = \rho \left[\psi(\alpha + x) - \psi(\alpha) + \psi(\alpha + \beta) - \psi(\alpha + \beta + x + y) \right], \tag{10.57}$$

$$\frac{\partial \rho}{\partial \beta} = \rho \left[\psi(\beta + y) - \psi(\beta) + \psi(\alpha + \beta) - \psi(\alpha + \beta + x + y) \right], \tag{10.58}$$

where $\rho = \rho(\alpha, \beta, x, y)$ and $\psi(x)$ is the *digamma function*. (See Exercise 10.7.)

Given Equations (10.57) and (10.58), the gradient of the likelihood function Equation (10.56) can be expressed as

$$\frac{\partial L}{\partial \alpha} = \sum_{t=t_x}^{n} \frac{\partial \rho(\alpha, \beta, x, t - x)}{\partial \alpha} \rho(\gamma, \delta, u_t, t) \tag{10.59}$$

$$\frac{\partial L}{\partial \beta} = \sum_{t=t_x}^{n} \frac{\partial \rho(\alpha, \beta, x, t - x)}{\partial \beta} \rho(\gamma, \delta, u_t, t) \tag{10.60}$$

$$\frac{\partial L}{\partial \gamma} = \sum_{t=t_x}^{n} \rho(\alpha, \beta, x, t - x) \frac{\partial \rho(\gamma, \delta, u_t, t)}{\partial \gamma} \tag{10.61}$$

$$\frac{\partial L}{\partial \delta} = \sum_{t=t_x}^{n} \rho(\alpha, \beta, x, t - x) \frac{\partial \rho(\gamma, \delta, u_t, t)}{\partial \delta}. \tag{10.62}$$

Corollary 10.4. *The probabilities that a user in the BG/BB model, with hyperparameters $\alpha, \beta, \gamma, \delta$ and transactional summary statistics (x, t_x, n), is still alive at times $t = n$ and $t = n + 1$ are given by*

$$p(A_n | \alpha, \beta, \gamma, \delta, x, t_x, n) = \frac{B(\alpha + x, \beta + n - x)}{B(\alpha, \beta)} \frac{B(\gamma, \delta + n)}{B(\gamma, \delta)} \cdot L^{-1} \tag{10.63}$$

$$p(A_{n+1} | \alpha, \beta, \gamma, \delta, x, t_x, n) = \frac{B(\alpha + x, \beta + n - x)}{B(\alpha, \beta)} \frac{B(\gamma, \delta + n + 1)}{B(\gamma, \delta)} \cdot L^{-1},$$

respectively, where $L = L(\alpha, \beta, \gamma, \delta | x, t_x, n)$ is given by Equation (10.54).

More generally, the probability that such a customer is alive at some point in time in the future, $t = n + m$, for $m \geq 0$, is given by

$$p(A_{n+m} | \alpha, \beta, \gamma, \delta, x, t_x, n) = \frac{B(\alpha + x, \beta + n - x)}{B(\alpha, \beta)} \frac{B(\gamma, \delta + n + m)}{B(\gamma, \delta)} \cdot L^{-1}.$$
$$(10.64)$$

Proof. The likelihood defined in Equation (10.52) defines a probability mass function over the $n - t_x + 1$ possibilities that a user's lifetime is $\tau = t_x, \ldots, n - 1$, or that the user is still alive at time $t = n$; i.e., that $\tau \geq n$. For example, the probability that the user's lifetime was exactly $\tau = t$, for $t = t_x, \ldots, n - 1$, is given by the likelihood of that event,

$$\frac{B(\alpha + x, \beta + t - x)}{B(\alpha, \beta)} \frac{B(\gamma + 1, \delta + t)}{B(\gamma, \delta)},$$

divided by the total likelihood of all the possible events, as computed in Equation (10.52). Similarly, the probability that a user is still alive at time $t = n$ is the final term divided by the total likelihood, which yields Equation (10.63).

Similarly, the probability $p(A_{n+1} | \alpha, \beta, \gamma, \delta, x, t_x, n)$ is obtained by applying similar logic to the equivalent expression Equation (10.54).

The proof of Equation (10.64) likewise follows along similar lines. We can see this result by expanding the numerator of the second factor of the final term of Equation (10.52) using the identity

$$B(\gamma, \delta + n) = \sum_{t=0}^{m-1} B(\gamma + 1, \delta + n + t) + B(\gamma, \delta + n + m).$$

This identity can be interpreted by equating the probability that the customer is alive at time $t = n$ with the sum of the probabilities of demise at times $t = n, \ldots, n + m - 1$ plus the probability that the customer is alive at time $t = n + m$. The result follows. $\square$

An illustration of the probability function Equation (10.63) is shown in Figure 10.8, for values $\alpha = 0.7$, $\beta = 0.5$, $\gamma = 1.9$, and $\delta = 11.6$, for users with various values of (x, t_x) and $n = 12$ held fixed. Note that if $t_x = 12$, we have $p(A_{12} | t_x = 12) = 1$. Also note the general trend that the larger the difference $n - t_x$, the less likely it is that the customer is still alive at time $t = n$.

Now, for a fixed n, there are only $J_n = n(n+1)/2 + 1$ possible recency / frequency patterns. This follows since the random variable T_x ranges from $T_x = 1, 2, \ldots, n$; given $T_x = t_x$, the number of transactions then ranges from $X = 1, \ldots, t_x$. (Since $T_x > 0$, this implies there must be at least one transaction.) In addition, we have the separate case for which $T_x = X = 0$.

As an immediate corollary to Proposition 10.8 and 10.9, we have the following expression for the log-likelihood of a *set* of customers.

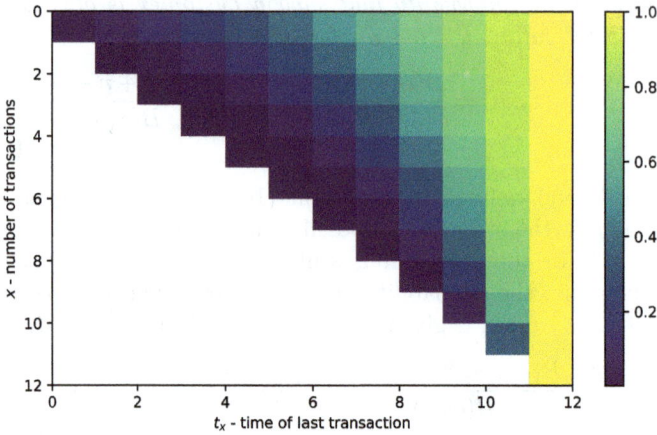

Fig. 10.8: Probability that a customer with cohort age $n = 12$ is alive at time $t = 12$, as a function of x and t_x.

Corollary 10.5. *In the BG/BB model, consider a set of users $\mathcal{D} = \{(x, t_x, n)_i\}$. If we group the users into the $J = n(n+1)/2 + 1$ possible recency/frequency patterns, such that pattern j consists of f_j customers, then the log-likelihood is given by*

$$\ell(\alpha, \beta, \gamma, \delta) = \sum_{j=1}^{J_n} f_j \ln \left[L(\alpha, \beta, \gamma, \delta | x_j, t_{x_j}, n) \right], \qquad (10.65)$$

where the marginal likelihood function L is given by Equation (10.52).

Given this expression for the log-likelihood, we can now use the maximum likelihood estimate to approximate the hyperparameters $(\alpha, \beta, \gamma, \delta)$ for a given set of users. Oftentimes, numerical optimizers require an initial guess for the parameter set. A quick initial guess can be computed using

$$\hat{\theta} = \frac{\sum_{j=1}^{J} f_j x_j}{\sum_{j=1}^{J} f_j t_{x_j}} \qquad \text{and} \qquad \hat{\tau} = \frac{\sum_{j=1}^{J} f_j t_{x_j}}{\sum_{j=1}^{J} f_j},$$

and then using

$$\langle \alpha_0, \beta_0, \gamma_0, \delta_0 \rangle = \langle \hat{\theta}, 1 - \hat{\theta}, 2, \hat{\tau} \rangle.$$

This is, of course, not a very good approximation to the hyperparameters, but should be good enough to put the numerical solver on the right track.

The likelihood function given in Equation (10.65) can be easily coded, as shown in Code Block 10.4. The function `likelihood_kernel` computes Equation (10.52), the likelihood for a given customer with summary statistics (x, t_x, n). The function `likelihood` then computes Equation (10.65)

```
1   from scipy.special import beta as BETA
2   from scipy.special import gamma as GAMMA
3
4   def likelihood_kernel(theta, x, tx, n):
5       alpha, beta, gamma, delta = theta
6       L = BETA(alpha+x, beta+n-x) * BETA(gamma, delta+n)
7       for t in range(tx, n):
8           L += BETA(alpha+x, beta+t-x) * BETA(gamma+1, delta+t)
9
10      return L / (BETA(alpha, beta) * BETA(gamma, delta))
11
12  def likelihood(theta, df_sum):
13      S = 0
14      for i, row in df_sum.iterrows():
15          L = likelihood_kernel(theta, int(row.x), int(row.tx),
                  int(row.n))
16          S += row.freq * np.log(L)
17
18      return -S
```

Code Block 10.4: Likelihood function for BG/BB Model.

from an input dataframe **df_sum**, formatted similarly to the output of getBgbbSummary in Code Block 10.3.

Next, we can simulate a set of data from the Pareto/NBD model, using the function defined in Code Block 10.2, and then convert the resulting data into the BG/BB format using Code Block 10.3. We then determine the maximum likelihood for the BG/BB model using Code Block 10.5.

```
1   alpha, beta, gamma, delta = 0.5, 0.5, 2, 0.01 # Pareto/NBD
2   df_users, df_transactions = simPareto(1000, alpha=alpha, beta=beta,
        gamma=gamma, delta=delta)
3   df_reduced, df_sum = getBgbbSummary(df_users, df_transactions)
4   p_hat = np.sum(df_sum.x * df_sum.freq) / np.sum(df_sum.tx *
        df_sum.freq)
5   T_hat = np.sum(df_sum.tx * df_sum.freq) / np.sum(df_sum.freq)
6
7   theta_0 = [p_hat, 1-p_hat, 2, T_hat]
8   bounds = [(0.01,100), (0.01,100), (0.01,100), (0.01,300)]
9   results = scipy.optimize.minimize(likelihood, theta_0,
        bounds=bounds, args=df_sum)
10  alpha, beta, gamma, delta = results['x'] # BG/BB
```

Code Block 10.5: MLE for simulated BG/BB data.

As an alternative, we can use the likelihood function in the form of Equation (10.56) and its corresponding gradient Equations (10.59)–(10.62) to subclass the `ObjectiveFunction` class from Code Blocks 5.7 and 5.8. This is shown in Code Block 10.7. For this approach, the `data` input argument to the constructor should be the `np.array` given by X_in=df_reduced[['tx', 'x', 'n']].values.

Distribution of Horizon Actions

Once the maximum likelihood estimate has been fit, several quantities of managerial interest can readily be derived, involving the number of actions over a fixed time horizon n.

Proposition 10.10. *Let $X(n) = \sum_{t=1}^{n} Y_t$ be the number of active transaction periods from the first n transaction opportunities. Given the hyperparameters of the BG/BB model, the PMF for the random variable $X(n)$ is given by*

$$p(X(n) = x | \alpha, \beta, \gamma, \delta) = \sum_{t=x}^{n-1} \binom{t}{x} \frac{B(\alpha + x, \beta + t - x)}{B(\alpha, \beta)} \frac{B(\gamma + 1, \delta + t)}{B(\gamma, \delta)}$$
$$+ \binom{n}{x} \frac{B(\alpha + x, \beta + n - x)}{B(\alpha, \beta)} \frac{B(\gamma, \delta + n)}{B(\gamma, \delta)}. \quad (10.66)$$

Proof. An individual that makes $X(n) = x$ actions must be alive for at least x transaction opportunities. Supposing, then, that the customer's lifetime is $\tau = t$, the probability of making x actions is given by the binomial distribution

$$p(X(n) = x | \tau = t, p) = \text{Binom}(x; t, p) = \binom{t}{x} p^x (1 - p)^{t-x},$$

for $x = t, \ldots, n$. Factoring in the distribution of lifetimes, we next write

$$p(X(n) = x | p, \theta) = \sum_{t=x}^{n-1} p(X(n) = x | p, \tau = t) p(\tau = t | \theta)$$
$$+ p(X(n) = x | p, \tau \geq n) p(\tau \geq n | \theta)$$
$$= \sum_{t=x}^{n-1} \binom{t}{x} p^x (1 - p)^{t-x} \theta (1 - \theta)^t$$
$$+ \binom{n}{x} p^x (1 - p)^{n-x} (1 - \theta)^n.$$

Our final result is obtained by marginalizing

$$p(X(n) = x | \alpha, \beta, \gamma, \delta) = \int_0^1 \int_0^1 p(X(n) = x | p, \theta) p(p | \alpha, \beta) p(\theta | \gamma, \delta) \, dp \, d\theta,$$

in the usual way.   □

```
 1  class BgbbLikelihood(ObjectiveFunction):
 2
 3      def _rho(self, alpha, beta, x, y):
 4          return BETA(alpha + x, beta + y) / BETA(alpha, beta)
 5
 6      def _rho_alpha(self, alpha, beta, x, y):
 7          rho = self._rho(alpha, beta, x, y)
 8          x = psi(alpha + x) - psi(alpha) + psi(alpha + beta) -
                  psi(alpha + beta + x + y)
 9          return rho * x
10
11      def _rho_beta(self, alpha, beta, x, y):
12          rho = self._rho(alpha, beta, x, y)
13          x = psi(beta + y) - psi(beta) + psi(alpha + beta) -
                  psi(alpha + beta + x + y)
14          return rho * x
15
16      def kernel(self, theta, x):
17          alpha, beta, gamma, delta = theta
18          l = np.zeros(len(x))
19          for i, row in enumerate(x):
20              tx, x, n = row
21              t = np.arange(tx, n+1)
22              u = (t != n).astype(int)
23              l[i] = np.sum(self._rho(alpha, beta, x, t-x) *
                      self._rho(gamma, delta, u, t))
24          return l
25
26      def kernelGrad(self, theta, x):
27          alpha, beta, gamma, delta = theta
28          l = np.zeros((4, len(x)))
29          for i, row in enumerate(x):
30              tx, x, n = row
31              t = np.arange(tx, n+1)
32              u = (t != n).astype(int)
33              l[0, i] = np.sum(self._rho_alpha(alpha, beta, x, t-x) *
                      self._rho(gamma, delta, u, t))
34              l[1, i] = np.sum(self._rho_beta(alpha, beta, x, t-x) *
                      self._rho(gamma, delta, u, t))
35              l[2, i] = np.sum(self._rho(alpha, beta, x, t-x) *
                      self._rho_alpha(gamma, delta, u, t))
36              l[3, i] = np.sum(self._rho(alpha, beta, x, t-x) *
                      self._rho_beta(gamma, delta, u, t))
37          return l
```

Code Block 10.6: BG/BB Likelihood and gradient; used with stochastic gradient descent; subclassed from Code Blocks 5.7 and 5.8.

Of course, we are often interested in the expected number of actions for a given customer within a given purchase horizon. This is given by our next result.

Proposition 10.11. *Let $X(n) = \sum_{t=1}^{n} Y_t$ be the number of active transaction periods from the first n transaction opportunities. Given the hyperparameters of the BG/BB model, the expected number of active periods is given by*

$$\mathbb{E}[X(n)|\alpha, \beta, \gamma, \delta] = \frac{\alpha\delta}{(\alpha + \beta)(\gamma - 1)} \left[1 - \frac{\Gamma(\gamma + \delta)}{\Gamma(\gamma + \delta + n)} \frac{\Gamma(1 + \delta + n)}{\Gamma(1 + \delta)} \right]. \tag{10.67}$$

Note 10.8. When multiplied by the expected monetization rate per transaction opportunity, Equation (10.67) yields the expected LTV of a customer over the time horizon n.

One can further derive the precise probability mass function for $p(X(n) = x)$, as well as the expected number of active transactions over the period $(n, n + n^*)$; details of these calculations are found in Fader, *et al.* [2010]. ▷

Proof. We begin by computing the expected value of $X(n)$, given the transaction rate p and churn rate θ. Recall that the random variable A_t represents the customer being alive at time t, we may write the conditional expectation as

$$\mathbb{E}[X(n)|p, \theta] = \sum_{t=1}^{n} \mathbb{P}(Y_t = 1|p, A_t)\mathbb{P}(A_t|\theta).$$

Now, according to our model,

$$\mathbb{P}(Y_t = 1|p, A_t) = p$$
$$\mathbb{P}(A_t|\theta) = \mathbb{P}(\tau \geq t|\theta) = (1 - \theta)^t,$$

the final equality following from the fact that the user's lifetime is geometrically distributed. We therefore obtain

$$\mathbb{E}[X(n)|p, \theta] = p \sum_{t=1}^{n} (1 - \theta)^n = \frac{p(1 - \theta)}{\theta} - \frac{p(1 - \theta)^{n+1}}{\theta}.$$

(See Exercise 10.5.)

Next, we marginalize the expectation over the mixing distributions to obtain

$$\mathbb{E}[X(n)|\alpha,\beta,\gamma,\delta] = \int_0^1 \int_0^1 \mathbb{E}[X(n)|p,\theta]p(p|\alpha,\beta)p(\theta|\gamma,\delta)\,dp\,d\theta$$

$$= \int_0^1 \int_0^1 \left[\frac{p^\alpha(1-p)^{\beta-1}\theta^{\gamma-2}(1-\theta)^\delta}{B(\alpha,\beta)B(\gamma,\delta)} \right.$$
$$\left. - \frac{p^\alpha(1-p)^{\beta-1}\theta^{\gamma-2}(1-\theta)^{\delta+n}}{B(\alpha,\beta)B(\gamma,\delta)} \right] dp\,d\theta$$

$$= \frac{B(\alpha+1,\beta)}{B(\alpha,\beta)} \left[\frac{B(\gamma-1,\delta+1) - B(\gamma-1,\delta+n+1)}{B(\gamma,\delta)} \right],$$

where the last equality follows by recognizing the penultimate integrand as
the kernel of

$$\text{Beta}(p;\alpha+1,\beta)\left[\text{Beta}(\theta;\gamma-1,\delta+1) - \text{Beta}(\theta;\gamma-1,\delta+n+1)\right],$$

and then integrating. Further simplification, using the identity Equation (2.48), yields the result.   □

For a given set of hyperparameters $(\alpha,\beta,\gamma,\delta)$, we can compute the expected lifetime (Equation (10.47)), the survival function (Equation (10.48)), the density function Equation (10.66), and the expected number of actions occurring within a horizon $(0,n)$ (Equation (10.67)), as shown in Code Block 10.7[3].

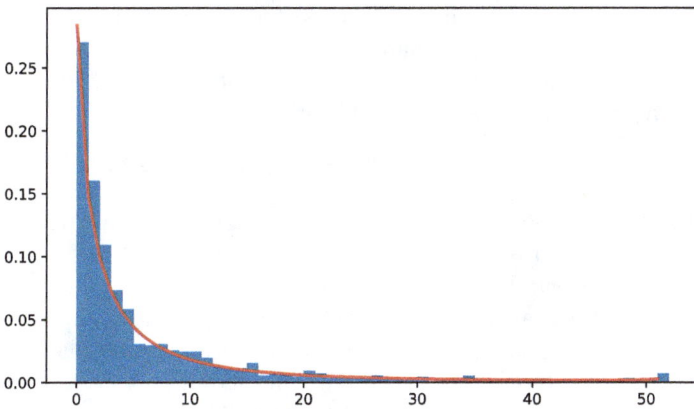

Fig. 10.9: Simulated and theoretical density function for $X(n)$.

Finally, we run the simulation of Code Block 10.5, by first simulating a cohort of users from the Pareto/NBD model and then converting the

---

[3] The function **Xn** relies on the combinatorial function **comb** from the **math** package, which computes n *choose* k.

```
1   def ELT(gamma, delta, opp_window=7):
2       return opp_window * delta / (gamma-1)
3
4   def S(n, gamma, delta):
5       return GAMMA(delta + n + 1) / GAMMA(delta) * GAMMA(gamma+delta)
            / GAMMA(gamma+delta + n + 1)
6
7   def Xn(x, n, alpha, beta, gamma, delta):
8       S = comb(n, x) * BETA(alpha+x, beta+n-x) / BETA(alpha, beta) *
            BETA(gamma, delta+n) / BETA(gamma, delta)
9       for t in range(x, n):
10          S += comb(t, x) * BETA(alpha+x, beta+t-x) / BETA(alpha,
                beta) * BETA(gamma+1, delta+t) / BETA(gamma, delta)
11      return S
12
13  def EXn(n, alpha, beta, gamma, delta):
14      Xn = alpha * delta / (alpha + beta) / (gamma - 1)
15      R = GAMMA(gamma + delta) / GAMMA(gamma + delta + n)
16      S = GAMMA(1 + delta + n) / GAMMA(1 + delta)
17      Xn *= (1 - R * S)
18      return Xn
```

Code Block 10.7: Key quantities for BG/BB Model.

datasets to the BG/BB format. The expected one-year survival function for the Pareto/NBD, using the simulation parameters, is 4.62% (the same as in Example 10.9). The maximum likelihood for our simulation resulting in a set of hyperparameters for the BG/BB model, given by

$$\alpha = 0.72, \qquad \beta = 0.51, \qquad \gamma = 1.85, \qquad \delta = 11.60.$$

The BG/BB model parameters (for this particular simulation) yield a one-year ($n = 52$) survival function of 4.39% (Equation (10.48)). Moreover, the distribution of the number of actions (Equation (10.66)) is plotted against the histogram of actual frequency of actions in Figure 10.9. The observed mean number of actions was 5.734, which closely matches the expected number 5.97, given by Equation (10.67). This shows that even when data is generated from the Pareto/NBD model, the BG/BB model can be used to accurately describe purchase behavior.

Bayesian Update for an Individual Customer

Oftentimes, we are interested in the incremental count of transactions occurring in the interval $(n, n+m]$. We can consider such a count for an arbitrary user with BG/BB hyperparameters $\alpha, \beta, \gamma, \delta$, or for a specific customer with given hyperparameters *and* transaction summary statistics (x, t_x, n).

As it turns out, both cases can be captured under a single formulation. To see this, we first consider the following Bayesian update.

Proposition 10.12. *Given a lifetime mixture model*

$$\tau \sim \text{Geom}(\theta),$$
$$\theta \sim \text{Beta}(\gamma, \delta)$$

and a censored observation that a user is still alive at time $t = n$, so that $\tau \geq n$, the posterior distribution is obtained by the Bayesian update

$$\gamma' = \gamma$$
$$\delta' = \delta + n. \tag{10.68}$$

Note 10.9. The update rule of Proposition 10.12 is distinct from the one given in Table 10.2, which corresponds to a set of *observed* lifetimes as opposed to a single *censored* lifetime. (If the lifetime of a single user is *observed*, the update is $\gamma' = \gamma + 1$ and $\delta' = \delta + n - 1$.) ▷

Proof. The likelihood that a customer is still alive at time $t = n$ is given by

$$p(A_n|\theta) = (1 - \theta)^n.$$

Therefore, by Bayes' rule, the posterior distribution is proportional to

$$p(\theta|A_n, \gamma, \delta) \propto p(A_n|\theta)p(\theta|\gamma, \delta) = (1 - \theta)^n \frac{\theta^{\gamma-1}(1 - \theta)^{\delta-1}}{B(\gamma, \delta)}.$$

The result given in Equation (10.68) follows.   □

The following gives us the posterior distribution for a customer in the BG/BB model with observed transaction history (x, t_x, n), conditional on the customer being still alive at time $t = n$.

Proposition 10.13. *Given a user in the BG/BB model with a priori hyper-parameters $\alpha, \beta, \gamma, \delta$ and observed transactional history (x, t_x, n), the posterior distribution for the transaction and churn probabilities p and θ, assuming the customer is still alive at time $t = n$, is obtained by the Bayesian update*

$$\alpha' = \alpha + x \tag{10.69}$$
$$\beta' = \beta + n - x \tag{10.70}$$
$$\gamma' = \gamma \tag{10.71}$$
$$\delta' = \delta + n. \tag{10.72}$$

Proof. The likelihood of observing transaction history (x, t_x, n) for a customer that is still alive at time $t = n$ is given by

$$p(x, t_x, n, A_n | p, \theta) = p^x (1-p)^{n-x} (1-\theta)^n.$$

Therefore, by Bayes' rule, the posterior distribution is proportional to

$$p(p, \theta | x, t_x, n, A_n, \alpha, \beta, \gamma, \delta) \propto p(x, t_x, n, A_n | p, \theta) p(p | \alpha, \beta) p(\theta | \gamma, \delta)$$

$$= p^x (1-p)^{n-x} (1-\theta)^n \frac{p^{\alpha-1} (1-p)^{\beta-1}}{B(\alpha, \beta)} \frac{\theta^{\gamma-1} (1-\theta)^{\delta-1}}{B(\gamma, \delta)}.$$

The result given in Equations (10.69)–(10.72) follows.   □

Note 10.10. In both Proposition 10.12 and 10.13, the Bayesian update of the lifetime hyperparameters is obtained by taking $\gamma' = \gamma$ and $\delta' = \delta + n$. Now, if a customer with transactional data (x, t_x, n) has $t_x < n$, we have not truly observed that the customer is still alive at time $t = n$. However, in what follows, we will consider the number of *future* transactions, subsequent to the initial observed period $[0, n]$. If the cusomter is *not alive* at time $t = n$, the customer will make no further purchases. We are thus interested in conditioning on the case in which the customer is still alive, using the preceding Bayesian update rule, and then applying the law of total probability.   ▷

Next, we consider the general formula for the number of transactions of a user over a set of transaction opportunities subsequent to a particular point in time $t = n$. This result will be then combined with Proposition 10.12 and 10.13 to obtain two useful quantities.

Proposition 10.14. *Let* $X(n, n+m) = \sum_{t=n+1}^{n+m} Y_t$ *represent the number of transactions for a customer in the BG/BB model over the time frame* $t \in (n, n+m]$. *Suppose a customer in the BG/BB model with posterior transactional hyperparameters* α', β' *and lifetime hyperparameters* γ, δ *is alive at time* $t = n$ *with probability* ξ_n, *then the probability mass function for the random variable* $X(n, n+m)$ *is given by*

$$p(X(n, n+m) = y | \alpha', \beta', \gamma, \delta) = \mathbb{I}[y=0](1 - \xi_n)$$
$$+ \xi_n p(X(m) = y | \alpha', \beta', \gamma, \delta + n) \quad (10.73)$$

where the function $p(X(n) = x | \alpha, \beta, \gamma, \delta)$ *is defined by Equation (10.66).*

Moreover, the expected number of transactions in this interval is given by

$$\mathbb{E}[X(n, n+m) | \alpha', \beta', \gamma', \delta'] = \xi_n \mathbb{E}[X(m) | \alpha', \beta', \gamma', \delta'], \quad (10.74)$$

where the function $\mathbb{E}[X(n) | \alpha, \beta, \gamma, \delta]$ *is given by Equation (10.67).*

Proof. If the customer is not alive at time n, then the number of transactions over the interval $(n, n+m]$ must be zero. Thus, the general form of Equation (10.73),

$$p(X(n, n+m) = y) = \mathbb{I}[y = 0](1 - \xi_n) + \xi_n p(X(n, n+m) = y | A_n),$$

follows. (We have suppressed the conditioning on $\alpha', \beta', \gamma', \delta'$ in this previous equation.)

But, assuming the customer is alive at time $t = n$, the number of transactions over the *next* m transaction opportunities must be given by Equation (10.66), due to the memoryless nature of the churn process. Equation (10.73) follows.

Equation (10.74) follows since the term $\mathbb{I}[y = 0](1 - \xi_n)$ does not contribute to the expectation. □

Corollary 10.6. *For a general customer in the BG/BB model with hyper-parameters $\alpha, \beta, \gamma, \delta$, the probability mass function for $X(n, n+m)$ is given by*

$$p(X(n, n+m) = y \mid \alpha, \beta, \gamma, \delta) = \mathbb{I}[y = 0]\left[1 - \frac{B(\gamma, \delta + n)}{B(\gamma, \delta)}\right]$$
$$+ \sum_{t=y}^{m-1} \binom{t}{y} \frac{B(\alpha + y, \beta + t - y)}{B(\alpha, \beta)} \frac{B(\gamma + 1, \delta + n + t)}{B(\gamma, \delta)}$$
$$+ \binom{m}{y} \frac{B(\alpha + y, \beta + m - y)}{B(\alpha, \beta)} \frac{B(\gamma, \delta + n + m)}{B(\gamma, \delta)}, \quad (10.75)$$

with expected value

$$\mathbb{E}[X(n, n+m) \mid \alpha, \beta, \gamma, \delta] = \left(\frac{\alpha}{\alpha + \beta}\right)\left(\frac{\delta}{\gamma - 1}\right)\frac{\Gamma(\gamma + \delta)}{\Gamma(1 + \delta)}$$
$$\times \left[\frac{\Gamma(1 + \delta + n)}{\Gamma(\gamma + \delta + n)} - \frac{\Gamma(1 + \delta + n + m)}{\Gamma(\gamma + \delta + n + m)}\right]. \quad (10.76)$$

Proof. First, let us consider Equation (10.75). The probability that a generic customer is still alive at time $t = n$ is given by Equation (10.48) (evaluated at $n - 1$) as

$$\xi_n = p(A_n | \gamma, \delta) = \frac{B(\gamma, \delta + n)}{B(\gamma, \delta)}.$$

Next, we apply Equation (10.73), where the second term is made conditional upon the customer being alive at time $t = n$. However, from Equation (10.68), this is tantamount to the Bayesian update $\gamma' = \gamma$ and $\delta' = \delta + n$. Thus, the second term of Equation (10.73) is given by

$$\frac{B(\gamma, \delta + n)}{B(\gamma, \delta)} p(X(m) = y | \alpha, \beta, \gamma, \delta + n).$$

Upon substituting $n \to m$, $x \to y$, and $\delta \to \delta + n$ into Equation (10.66) and multiplying by ξ_n (which has the effect of converting the denominator back to $B(\gamma, \delta)$), we obtain our first result.

Similarly, Equation (10.76) follows from Equation (10.74):

$$\mathbb{E}[X(n, n+m)|\alpha, \beta, \gamma, \delta] = \frac{B(\gamma, \delta + n)}{B(\gamma, \delta)} \mathbb{E}[X(m)|\alpha, \beta, \gamma, \delta + n].$$

Note that this second factor is obtained by replacing $n \to m$ and $\delta \to \delta + n$ in Equation (10.67). To simplify the result, it is useful to note that

$$\frac{B(\gamma, \delta + n)}{B(\gamma, \delta)} = \frac{\Gamma(\delta + n)\Gamma(\gamma + \delta)}{\Gamma(\delta)\Gamma(\gamma + \delta + n)},$$

and that

$$\frac{(\delta + n)\Gamma(\delta + n)}{\Gamma(\delta)} = \frac{\delta\Gamma(\delta + n + 1)}{\Gamma(\delta + 1)}.$$

The rest follows from algebra.   □

Naturally, Corollary 10.6 is not so useful, as it is addressing the behavior of a generic customer over the transaction intervals $(n, n+m]$. Typically, we do not consider a generic customer, but a specific customer with transaction history (x, t_x, n), and ask how many future transactions that customer might make over the period $(n, n+m]$. This result is given as follows.

Corollary 10.7. *For a general customer in the BG/BB model with hyperparameters $\alpha, \beta, \gamma, \delta$ and observed transactional history (x, t_x, n), the probability mass function for $X(n, n+m)$ is given by*

$$p(X(n, n+m) = y \mid \alpha, \beta, \gamma, \delta, x, t_x, n) = \mathbb{I}[y=0]\left[1 - \frac{B(\alpha', \beta')}{B(\alpha, \beta)} \frac{B(\gamma, \delta + n)}{B(\gamma, \delta)L}\right]$$

$$+ \sum_{t=y}^{m-1} \binom{t}{y} \frac{B(\alpha' + y, \beta' + t - y)}{B(\alpha, \beta)} \frac{B(\gamma + 1, \delta + n + t)}{B(\gamma, \delta)L}$$

$$+ \binom{m}{y} \frac{B(\alpha' + y, \beta' + m - y)}{B(\alpha, \beta)} \frac{B(\gamma, \delta + n + m)}{B(\gamma, \delta)L}, \qquad (10.77)$$

where $\alpha' = \alpha + x$ and $\beta' = \beta + n - x$, and L is the likelihood given by Equation (10.52), with expected value

$$\mathbb{E}[X(n, n+m) \mid \alpha, \beta, \gamma, \delta, x, t_x, n] = \frac{B(\alpha + x + 1, \beta + n - x)}{B(\alpha, \beta)L}$$

$$\times \left(\frac{\delta}{\gamma - 1}\right) \frac{\Gamma(\gamma + \delta)}{\Gamma(1 + \delta)}$$

$$\times \left[\frac{\Gamma(1 + \delta + n)}{\Gamma(\gamma + \delta + n)} - \frac{\Gamma(1 + \delta + n + m)}{\Gamma(\gamma + \delta + n + m)}\right]. \qquad (10.78)$$

Proof. The result is obtained by applying Equations (10.73) and (10.74) using the probability that a customer is alive at time $t = n$:

$$\xi_n = \frac{B(\alpha + x, \beta + n - x)}{B(\alpha, \beta)} \frac{B(\gamma, \delta + n)}{B(\gamma, \delta)} L^{-1},$$

as given by Equation (10.63), and the Bayesian update found in Equations (10.69) and (10.72) in formulas Equations (10.66) and (10.67).

In the derivation for Equation (10.78), note that

$$B(\alpha + x, \beta + n - x) \left(\frac{\alpha + x}{\alpha + \beta + n} \right) = B(\alpha + x + 1, \beta + n - x).$$

The rest follows analogous to the proof of Corollary 10.6.   □

Note 10.11. The results of Corollaries 10.6 and 10.7 are also derived in a different approach in Fader, *et al.* [2010]. There, the law of total probability is applied directly to the parameters (p, θ), and then the result is obtained by marginalizing over those parameters with respect to the two beta mixing distributions. The approach we presented, using Proposition 10.12 and 10.13 and Proposition 10.14 allows us to reveal the common structure between both expressions, as well as work directly in the hyperparameter space. Moreover, it is also more straightforward from a programming perspective to implement Equations (10.73) and (10.74) directly, as opposed to their more verbose counterparts.   ▷

Finally, we can generalize the result of Proposition 10.14 one step further by applying it to an arbitrary future window. We state the result in the following theorem.

Theorem 10.1. *Let $X(u, v) = \sum_{t=u+1}^{v} Y_t$ represent the number of transactions for a customer in the BG/BB model over the time frame $t \in (u, v]$ for a customer in the BG/BB model with (prior) hyperparameters $\alpha, \beta, \gamma, \delta$ and behavioral data x, t_x, n, where $n \leq u < v$. If we define ξ_u as the probability that the customer is alive at time $t = u$, i.e.,*

$$\xi_u = p(A_u | \alpha, \beta, \gamma, \delta, x, t_x, n),$$

as given by Equation (10.64) (replacing $n + m$ with u), then the probability mass function for the random variable $X(u, v)$ is given by

$$p(X(u, v) = y \mid \alpha, \beta, \gamma, \delta, x, t_x, n) = \mathbb{I}[y = 0] (1 - \xi_u)$$
$$+ \xi_u p(X(v - u) = y | \alpha + x, \beta + n - x, \gamma, \delta + u), \quad (10.79)$$

where the function $p(X(n) = x | \alpha, \beta, \gamma, \delta)$ is defined by Equation (10.66).

Moreover, the expected number of transactions in this interval is given by

$$\mathbb{E}[X(u, v) | \alpha, \beta, \gamma, \delta, x, t_x, n] = \xi_u \mathbb{E}[X(v - u) | \alpha + x, \beta + n - x, \gamma, \delta + u], \quad (10.80)$$

where the function $\mathbb{E}[X(n) | \alpha, \beta, \gamma, \delta]$ is given by Equation (10.67).

Proof. This proof is a simple extension of Proposition 10.14 and Corollaries 10.6 and 10.7. The transactional data (x, n, t_x) yield the updated hyperparameters $\alpha' = \alpha + x$ and $\beta' = \beta + n - x$. The probability that such a customer is alive at time $t = u \geq n$ is given by Equation (10.64). Moreover, conditioning on A_u yields the updated hyperparameter $\delta' = \delta + u$. The result follows.   □

Note that Proposition 10.14 and Corollaries 10.6 and 10.7 are just special cases of Theorem 10.1. This general form is also useful for programmatic implementation, as shown in Code Blocks 10.8 and 10.9. In Code Block 10.8, we encode the likelihood function, as given in Equation (10.52), as the method `likelihood`. The probability that a customer is alive at time t, as given by Equation (10.64), is encoded as the function `probAlive`. The probability mass function Equation (10.66) and its expectation Equation (10.67) are encoded in Code Block 10.9 as `Xn` and `EXn`, respectively. Finally, the general results from Theorem 10.1 are encoded as `Xuv` and `EXuv`. Thus, probabilities and expectations for the number of future actions over arbitrary periods can be computed for any customer in the BG/BB model.

The Ruse of Heterogeneity

The *ruse of heterogeneity*, as introduced in Vaupel and Yashin [1985] and described in the context of the BG model in Fader and Hardie [2007], is any apparent time dynamics in a mixture model that contains no explicit time-dynamic component. The observed time dynamics are therefore a specter in the data caused by a population's heterogeneity, and not any actual time-dependent behavior. As a simple example, we can restrict the BG/BB model to contractual users, thereby obtaining the BG model, where we can illustrate the concept with a user's retention rate.

In the limiting case of the BG/BB model as the hyperparameters $(\alpha, \beta) \to (\infty, 0)$, the transaction probability collapses to the particular value $p = 1$ and the model degenerates to the BG model, which is suitable for contractual customers, in the sense of Definition 10.9. This particular use case was introduced by Fader and Hardie [2007].

Definition 10.15. *The* retention rate r_t *for a cohort at time t is defined by*

$$r_t = \frac{S(t)}{S(t-1)}.$$

It represents the ratio of users who will survive beyond time t compared to those who survived until time t.

In the BG model, we can compute the probability mass function for the user lifetime τ and its associated survival function by marginalizing Equations (10.43) and (10.44) over the churn parameter θ.

```
 1  class Customer:
 2
 3      def __init__(self, alpha, beta, gamma, delta):
 4          self.alpha, self.beta, self.gamma, self.delta = alpha, beta,
                gamma, delta
 5          self.x, self.tx, self.n = 0, 0, 0
 6
 7      def setData(self, x, tx, n):
 8          assert x <= tx, 'number of actions must be less than last
                purchase time'
 9          assert tx <= n, 'last purchase time must be less than age'
10          self.x, self.tx, self.n = x, tx, n
11
12      def getGreeks(self, update=False):
13          if update:
14              return self.alpha + self.x, self.beta + self.n - self.x,
                    self.gamma, self.delta + self.n
15
16          return self.alpha, self.beta, self.gamma, self.delta
17
18      def likelihood(self):
19
20          alpha, beta, gamma, delta = self.getGreeks(update=False)
21          x, tx, n = self.x, self.tx, self.n
22          l = BETA(alpha+x, beta+n-x) / BETA(alpha, beta) *
                BETA(gamma, delta+n) / BETA(gamma, delta)
23          for t in range(tx, n):
24              l += BETA(alpha+x, beta+t-x) / BETA(alpha, beta) *
                    BETA(gamma+1, delta+t) / BETA(gamma, delta)
25
26          return l
27
28      def probAlive(self, t):
29          assert t >= self.n, 'Alive probability for future time'
30
31          alpha, beta, gamma, delta = self.getGreeks(update=False)
32          x, n = self.x, self.n
33          l = self.likelihood()
34          p = BETA(alpha+x, beta+n-x) / BETA(alpha, beta) *
                BETA(gamma, delta+t) / BETA(gamma, delta)
35
36          return p / l
```

Code Block 10.8: Customer-level predictions in BG/BB model.

```
1    def Xn(n, x, alpha, beta, gamma, delta):
2        assert x <= n, "number of actions cannot exceed horizon"
3
4        p = comb(n, x) * BETA(alpha+x, beta+n-x) / BETA(alpha, beta)
             * BETA(gamma, delta+n) / BETA(gamma, delta)
5        for t in range(x, n):
6            p += comb(t, x) * BETA(alpha+x, beta+t-x) / BETA(alpha,
                  beta) * BETA(gamma+1, delta+t) / BETA(gamma, delta)
7
8        return p
9
10   def EXn(n, alpha, beta, gamma, delta):
11       x = alpha * delta / (alpha+beta) / (gamma-1)
12       x *= (1 - GAMMA(gamma+delta) / GAMMA(gamma+delta+n) *
             GAMMA(1+delta+n) / GAMMA(1 + delta))
13
14       return x
15
16   def Xuv(self, u, v, x):
17       assert u >= self.n
18       assert v >= u
19
20       alpha, beta, gamma, delta = self.getGreeks(update=True)
21       delta += u - self.n
22       xi = self.probAlive(u)
23       m = v - u
24       p = (1 - xi) if x == 0 else 0
25       p += xi * self.Xn(m, x, alpha, beta, gamma, delta)
26
27       return p
28
29   def EXuv(self, u, v):
30       assert u >= self.n
31       assert v >= u
32
33       alpha, beta, gamma, delta = self.getGreeks(update=True)
34       delta += u - self.n
35       xi = self.probAlive(u)
36       m = v - u
37
38       return xi * self.EXn(m, alpha, beta, gamma, delta)
```

Code Block 10.9: Customer-level predictions in BG/BB model (continued).

$$\mathbb{P}(\tau = t | \gamma, \delta) = \frac{B(\gamma + 1, \delta + t)}{B(\gamma, \delta)}, \qquad (10.81)$$

$$S(t | \gamma, \delta) = \frac{B(\gamma, \delta + t + 1)}{B(\gamma, \delta)}. \qquad (10.82)$$

(We leave the details as an exercise for the reader.) Given the survival function, we can now use Definition 10.15 to compute the retention rate.

$$r_t = \frac{B(\gamma, \delta + t + 1)}{B(\gamma, \delta + t)} = \frac{\delta + t}{\gamma + \delta + t}.$$

This result is described by Fader and Hardie [2007]:

> We immediately see that under the [BG] model, the retention rate is an increasing function of time, even though the underlying (unobserved) individual-level retention probability is constant. According to this model, there are no underlying time dynamics at the level of the individual customer; the observed phenomenon of retention rates increasing over time is simply due to heterogeneity (i.e., the high-churn customers drop out early in the observation period, with the remaining customers having lower churn probabilities).

Under the BG (or BG/BB) model, each individual has a constant, unobserved true churn rate, but the cohort consists of a mixture of users with varying churn rates described by a beta distribution. The ruse of heterogeneity is a case of burning the candle at both ends, albeit at different rates. As a cohort ages, users with a high churn rate are most likely to churn early, thus altering the mixture of the remaining users as the population matures.

Monetization

So far, we have discussed numerous models that handle the recency and frequency components of a customer's lifetime value, vis-à-vis the transaction probability and unobserved customer lifetime. Naturally, such a picture is incomplete, as it has yet to include the important component of monetization. Two common approaches to modeling the monetization rate are the gamma–gamma mixture (Colombo and Jiang [1999]) and the normal–normal mixture (Schmittlein and Peterson [1994]).

10.4 Credibility Theory

Credibility theory is a mathematical framework for describing heterogeneous populations using certain linearity assumptions. Credibility theory arose in the insurance industry due to their need to understand and rate heterogeneous risk classes and to further understand the relationship between the individual and the collective. In particular, credibility theory

1. is used to model heterogeneous collectives (i.e., mixture models),
2. shows how we can combine individual and collective experience,
3. belongs to the field of Bayesian statistics,
4. represents a class of *linear* Bayes estimators,
5. has key advantages in terms of simplicity and structure,

It was developed in the seminal works of Bühlmann [1967] and Bühlmann and Straub [1970]. For an overview, see the texts Bühlmann [1970] and, for a more recent introduction, see Bühlmann and Gisler [2005], which also discusses hierarchical credibility models, multidimensional problems (regression), and certain evolutionary credibility models. We include a treatment of the theory here, as we believe the theory to have implications beyond rating insurance risks.

Bayesian models, other than a few cornerstone examples, are often unwieldy, complex, and analytically intractable, especially for multidimensional or hierarchical models, or when nonconjugate priors are prescribed. One popular remedy is the use of computer simulation and Markov-chain Monte Carlo methods, which we discuss in Section 10.5. Credibility theory, on the other hand, offers an analytic alternative with a number of key benefits:

1. *premiums*: estimates for the expected values $\mathbb{E}[X]$, known in credibility as *risk premiums*, are modeled directly;
2. *simplicity*: the formulas underlying credibility computations are simple and intuitive;
3. *structure*: the sampling distribution $p(X|\theta)$ and the mixing distribution $p(\theta)$ need not be specified, rather, everything we need from them is specified through their first few moments.
4. *extendability*: can easily be extended to hierarchical and multidimensional models.

10.4.1 Risks and Risk Premiums

Arising from the field of insurance, credibility theory has a few unique terms that are not standard elsewhere in statistics. We therefore begin our discussion by outlining several definitions that are useful in describing the mathematical problem that is addressed by credibility theory.

Risk Premiums and Definitions

In credibility theory, the random variables under study are known as *risks*, as they, in their original application, represent insurance payouts and, thus, risks born by an insurance company. The basic idea underlying all of insurance is that a collective of individuals, each with an exposure to a particular risk (e.g., automobile accident, fire, theft, dropping your iPhone into a toilet, etc.), join together to form a "community-at-risk" that bears the risk in the form of a collective; i.e., by paying an insurance premium, individuals transfer their risk to an insurance company. Credibility calculations can be viewed from the perspective of an insurance company, that must understand how to quantify risk and, therefore, correctly assess their insurance premiums to all individuals insured. A priori, each individual is viewed as a random risk from the collective. However, as individuals amasses their respective claims histories, we may better assess the risk of each unique individual within the collective.

In particular, in the context of credibility, a random variable X is referred to as a *risk*, whereas its expected value $\mathbb{E}[X]$ is referred to as a *risk premium*. For a given individual, we say that the claims history $X_1, \ldots, X_n \sim F_\theta$ is described by a certain *risk profile*, which is parameterized by a random variable Θ. The risk profile Θ is itself distributed relative to a distribution $\Theta \sim U$, known as the *collective distribution*. From the perspective of the collective, we say that F_θ is a sampling distribution and U is a mixing distribution, but from the perspective of each individual, we may view U as a prior.

Definition 10.16. *Given a continuous mixture model, the individual premium is the random variable*

$$P^{\mathrm{ind}} = \mathbb{E}[X|\Theta] = \mu(\Theta). \tag{10.83}$$

Given the true parameter $\Theta = \theta$ *of a particular individual, the* correct individual premium *is*

$$P^{\mathrm{ind}}(\theta) = \mathbb{E}[X|\Theta = \theta] = \mu(\theta). \tag{10.84}$$

The collective premium *is the value*

$$P^{\mathrm{coll}} = \mathbb{E}[P^{\mathrm{ind}}] = \mu_0. \tag{10.85}$$

And, finally, the Bayes' premium *is the random variable*

$$P^{\mathrm{Bayes}} = \tilde{\mu}(\Theta) = \mathbb{E}[P^{\mathrm{ind}}|\mathcal{D}], \tag{10.86}$$

where the expectation is with respect to the random variable Θ, *and where* $\mathcal{D}$ *represents an individual's observed history. The Bayes' premium is also referred to as the* experience premium.

Note that the individual premium is a random variable, as it is a function of the random variable Θ. Similarly, the Bayes' premium is a random variable, as it is a function of the data $\mathcal{D}$. The collective premium, on the other hand, is a fixed value, since

$$\mu_0 = \mathbb{E}[\mathbb{E}[X|\Theta]] = \mathbb{E}[X].$$

Intuitively, the collective premium is the overall expected value of X throughout the population. The individual premium is a random variable that represents the expected value of X for individuals within the population.

These definitions are best illustrated by means of a few examples.

Example 10.10 (Bernoulli–beta). Consider the mixture model $X \sim \text{Bern}(\Theta)$, with $\Theta \sim \text{Beta}(\alpha, \beta)$. The individual premium is given by

$$P^{\text{ind}} = \mathbb{E}[X|\Theta] = \Theta,$$

the expected value of a Bernoulli random variable. Therefore, given the true probability $\Theta = \theta$ of a specific individual, the correct individual premium will be $P^{\text{ind}}(\theta) = \theta$.

Similarly, the collective premium is simply the expected value of P^{ind}, or

$$P^{\text{coll}} = \mathbb{E}[P^{\text{ind}}] = \mathbb{E}[\Theta] = \frac{\alpha}{\alpha + \beta}.$$

Next, let us suppose that we have a sequence of observations for a given individual in the population consisting of n Bernoulli trials with a total of k successes. the Bayes' premium (or experience premium) for that individual is obtained using our standard Bayesian update rules $\alpha' = \alpha + k$ and $\beta' = \beta + n - k$ for a binomial outcome, yielding

$$P^{\text{Bayes}} = \mathbb{E}[P^{\text{ind}}|\mathcal{D}] = \frac{\alpha + k}{\alpha + \beta + n}.$$

(Note that this equals the expected value of the posterior predictive distribution.)

Finally, we notice that, with a little bit of algebra, we can rewrite the experience premium in the form

$$P^{\text{Bayes}} = \lambda \overline{X} + (1 - \lambda)P^{\text{coll}},$$

where $\overline{X} = k/n$ and λ is given by

$$\lambda = \frac{n}{n + \alpha + \beta}.$$

For this specific example, the Bayes' premium is a linear combination of the collective premium (which can be viewed as the prior expected value, which, for a random individual, is equal to the expected value of the population) and the observed data (the ratio of successes $\overline{X} = k/n$). Moreover, the particular weighting favors the experience as the number of observations n increases.   ▷

Example 10.11 (Poisson–gamma). Consider the mixture model $X \sim \text{Poiss}(\Theta)$, with $\Theta \sim \text{Gamma}(\alpha, \beta)$. The individual premium is again given by

$$P^{\text{ind}} = \mathbb{E}[X|\Theta] = \Theta,$$

the expected value of a Poisson random variable. Therefore, given the true probability $\Theta = \theta$ of a specific individual, the correct individual premium will be $P^{\text{ind}}(\theta) = \theta$.

Similarly, the collective premium is simply the expected value of P^{ind}, or

$$P^{\text{coll}} = \mathbb{E}[P^{\text{ind}}] = \mathbb{E}[\Theta] = \alpha\beta.$$

Next, let us suppose that we have a sequence of n observations of a IID Poisson random variable for a given individual with a total observed count of $X. = n\overline{X} = \sum_{i=1}^{n} X_i$. The Bayes' premium (or experience premium) for that individual is obtained using our standard Bayesian update rules for the Poisson–gamma mixture, or $\alpha' = \alpha + X.$ and $(1/\beta') = (1/\beta) + n$. This yields

$$P^{\text{Bayes}} = \mathbb{E}[P^{\text{ind}}|\mathcal{D}] = (\alpha + n\overline{X})\frac{\beta}{1 + n\beta} = \frac{\alpha\beta}{1 + n\beta} + \frac{n\beta\overline{X}}{1 + n\beta}.$$

(Note that this equals the expected value of the posterior predictive distribution.)

Finally, we notice that, with a little bit of algebra, we can rewrite the experience premium in the form

$$P^{\text{Bayes}} = \lambda\overline{X} + (1 - \lambda)P^{\text{coll}},$$

where $\overline{X} = k/n$ and λ is given by

$$\lambda = \frac{n}{n + 1/\beta}.$$

Again we see that the Bayes' premium can be expressed as a linear combination of the collective premium and the observed data.   ▷

For a third example, consider the following.

Example 10.12 (exponential–gamma). Consider the mixture model $X \sim \text{Exp}(\Theta^{-1})$, and $\Theta \sim \text{Gamma}(\alpha, \beta)$, where Θ is the *rate* parameter (inverse scale parameter) of the exponential distribution. A simple comutation shows that

$$P^{\text{ind}} = \frac{1}{\Theta}$$

$$P^{\text{coll}} = \frac{1}{(\alpha - 1)\beta}$$

$$P^{\text{Bayes}} = \frac{1/\beta + \sum_{j=1}^{r} X_j}{\alpha + n - 1}.$$

Note that P^{coll} is the expected value of a $\text{Lomax}(\alpha, \beta^{-1})$ distribution, which is the predictive distribution associated with the exponential–gamma mixture. The Bayes premium can again be expressed as a linear combination

$$P^{\text{Bayes}} = \lambda \overline{X}. + (1 - \lambda)P^{\text{coll}},$$

where

$$\lambda = \frac{n}{n + \alpha - 1}.$$

Note that if we instead specified the mixture model as $X \sim \text{Exp}(\Theta)$ and $\Theta \sim \text{Gamma}(\alpha, \beta)$, using the scale parameter as Θ, the posterior distribution on θ, given a set of data $X_1, \ldots, X_n$, would be proportional to

$$p(\Theta|\mathcal{D}) \propto \frac{e^{-x/\theta}\theta^{\alpha-2}e^{-\theta/\beta}}{\Gamma(\alpha)\beta^{\alpha}}.$$

The constant of proportionality must be found by integrating the right-hand side over θ, which does not have a simple analytical expression. As a result, it is clear the Bayes' premium will not have a simple expression. ▷

In each of the previous examples, we saw how the Bayes' premium can be expressed as a linear combination of the collective premium (i.e., the *prior*) and experience (i.e., an individual user's transactional history). In both of these cases, we were able to express the Bayes' premium as

$$P^{\text{Bayes}} = \lambda \overline{X} + (1 - \lambda)P^{\text{coll}},$$
$$\lambda = \frac{n}{n + \kappa},$$

for an appropriate choice of the constant κ. It turns out that the Bayes' premium follows this format whenever the sampling distribution is within the exponential family and the collective distribution is its conjugate prior. When that is not true, however, this simple linear combination fails to hold. In practice we are often interested in cases in which one has no right to assume that the collective distribution to be the conjugate prior of the sampling distribution. In such cases the Bayes' premium can quickly become difficult to express analytically. This problem is solved within the elegant field of credibility theory.

10.4.2 Bühlmann–Straub Model

In order to compute the Bayes' premium, once must specify both the sampling distribution as well as the mixing distribution. These distributions are often difficult to infer in practice. Even if a suitable class of distributions can be worked out, the resulting formula are often complex and unwieldy. The idea behind credibility theory is to determine a suitable estimator that is simple and intuitive, and that makes no suppositions on the form of the sampling or collective distributions. The requirement of simplicity is forced by restricting the class of estimators to those estimators that are linear in the data. The result is known as the *credibility estimator*, and it is essentially a linear Bayes estimator.

Simple Credibility Model

Credibility theory was developed with two requirements in mind: simplicity and structure. (Bühlmann and Gisler [2005]). The requirement of simplicity means it should be straightforward for practitioners to apply and easy to interpret. The requirement of structure means that the model should not require prior knowledge of the specific form of the sampling distribution or the collective distribution. Bühlmann and Gisler [2005] states that specification of these models is, in practice, either artificial (if the distributional families are too large) or not helpful (if they are too narrow). Thus the model was developed without the requirement of specifying the specific distributional forms, instead relying on their first few moments. In order to describe the second moments mathematically, we require one additional definition before introducing the simple credibility model.

Definition 10.17. *Given a mixture model and the risk premiums defined in Definition 10.16, we define the individual risk variance, expected individual variance, and collective variance as*

$$\sigma^2(\Theta) = \mathbb{V}(X|\Theta) \tag{10.87}$$
$$\sigma^2 = \mathbb{E}[\sigma^2(\Theta)] \tag{10.88}$$
$$\tau^2 = \mathbb{V}(\mu(\Theta)), \tag{10.89}$$

respectively. Moreover, the quotient $\kappa = \sigma^2/\tau^2$ is referred to as the credibility coefficient.

For an individual with $\Theta = \theta$, the quantity $\sigma^2(\theta)$ represents the variance within that individual risk. The quantity σ^2 represents the average variance within an individual risk within the collective. And the quantity τ^2 represents the variance between individual risk premiums. As an illustration, consider the case where τ^2 is very large and σ^2 is very small. The average value for each individual is, in this case, spread out (large τ^2), but the observations for a specific individual are tightly wound to each individual's average (small σ^2). In this case, therefore, it would be easy to identify individuals from the data.

Example 10.13. Consider again the Bernoulli–beta mixture from Example 10.10. We saw that the individual premium is given by

$$\mu(\theta) = \mathbb{E}[X|\Theta = \theta] = \theta.$$

Similarly, recall that

$$\sigma^2(\theta) = \mathbb{V}(X|\Theta = \theta) = \theta(1 - \theta),$$

so that we may compute

$$\sigma^2 = \mathbb{E}[\sigma^2(\Theta)] = \int_0^1 \theta(1-\theta)\text{Beta}(\theta; \alpha, \beta)\, d\theta$$

$$= \frac{1}{B(\alpha,\beta)} \int_0^1 \theta^\alpha (1-\theta)^\beta\, d\theta$$

$$= \frac{B(\alpha+1, \beta+1)}{B(\alpha,\beta)}$$

$$= \frac{\alpha\beta}{(\alpha+\beta)(1+\alpha+\beta)}.$$

The last line follows from Exercise 10.10. It is interesting to compare this expression with

$$\sigma^2(\mathbb{E}[\Theta]) = \sigma^2\left(\Theta = \frac{\alpha}{\alpha+\beta}\right) = \frac{\alpha\beta}{(\alpha+\beta)^2},$$

the variance in X if $\theta = \mathbb{E}[\Theta] = \alpha/(\alpha+\beta)$. We see that the expected variance σ^2, marginalized over Θ, is slightly less than the variance calculated at the expected value of Θ.

Next, we find that

$$\tau = \mathbb{V}(\mu(\Theta)) = \mathbb{V}(\Theta) = \frac{\alpha\beta}{(\alpha+\beta)^2(1+\alpha+\beta)},$$

such that the credibility coefficient is given by

$$\kappa = \frac{\sigma^2}{\tau^2} = \alpha + \beta.$$

This makes sense in terms of the Bayesian interpretation of the hyperparameters of the Bernoulli–beta model: α represents the number of *virtual successes* and β the number of *virtual failures*. The larger the sum $\alpha + \beta$, the more data are needed to overcome the prior. Moreover, when performing Bayesian updates, $\sigma^2 \to \sigma^2(\alpha/(\alpha+\beta))$ as $n \to \infty$, whereas $\tau \to 0$ as $n \to \infty$, again confirming the variance reduction aspects of a Bayesian update. In mixture models, e.g., credibility theory, the goal is not to shrink this variance, but to use the "prior" as a population distribution, which will instead converge to the variability of the population.   ▷

Example 10.14. Consider the mixture model $X \sim \text{Exp}(\Theta)$ and $\Theta \sim \text{Gamma}(\alpha, \beta)$. *Note*: this is *not* the exponential–gamma mixture from Table 10.3, as this mixture is constructed using the *scale* parameter as opposed to the *rate* parameter. Nevertheless, we can compute the following quantities of interest:

$$P^{\text{ind}} = \Theta$$
$$P^{\text{coll}} = \alpha\beta$$
$$\sigma^2(\Theta) = \mathbb{V}(X|\Theta) = \Theta^2$$
$$\sigma^2 = \mathbb{E}[\Theta^2] = (\alpha+1)\alpha\beta^2$$
$$\tau^2 = \mathbb{V}(P^{\text{ind}}) = \mathbb{V}(\Theta) = \mathbb{E}[\Theta^2] - \mathbb{E}[\Theta] = (\alpha+1)\alpha\beta^2 - \alpha^2\beta^2 = \alpha\beta^2.$$

In particular, the credibility coefficient is given by $\kappa = \sigma^2/\tau^2 = 1 + \alpha$. In particular, since $\kappa > 1$, the individual variation is always larger than the variation in the individual means.   ▷

Before defining the credibility estimator, we need a basic result that relates the variance of a sample mean to the variances of Definition 10.17.

Theorem 10.2. *Let $X_1, \ldots, X_n \sim \mathrm{Mix}(F_\Theta, U)$ be IID samples from the mixture distribution $X_i \sim F_\Theta$, $\Theta \sim U$. Then the variance of the sample mean is given by*

$$\mathbb{V}(\overline{X}) = \frac{\sigma^2}{n} + \tau^2, \tag{10.90}$$

where σ^2 and τ^2 are defined in Equations (10.88) and (10.89), respectively.

Proof. From the law of total variance (Equation (2.58)), we have

$$\mathbb{V}(\overline{X}) = \mathbb{E}\left[\mathbb{V}(\overline{X}|\Theta)\right] + \mathbb{V}\left(\mathbb{E}[\overline{X}|\Theta]\right).$$

However, from Equation (1.23), we have

$$\mathbb{E}[\overline{X}|\Theta] = \mathbb{E}[X|\Theta] = \mu(\Theta)$$
$$\mathbb{V}(\overline{X}|\Theta) = \frac{\mathbb{V}(X|\Theta)}{n} = \frac{\sigma^2(\Theta)}{n}.$$

Therefore,

$$\mathbb{V}(\overline{X}) = \frac{\mathbb{E}\left[\sigma^2(\Theta)\right]}{n} + \mathbb{V}\left(\mu(\Theta)\right),$$

and the result follows.   □

Now we are in a position to define what we mean by the credibility estimator and prove an important result.

Definition 10.18. *Given a mixture model, the* credibility estimator P^{cred} *or $\hat{\mu}(\Theta)$ is the unique estimator for the Bayes' premium that is linear in the observations*

$$P^{\mathrm{cred}} = a_0 + \sum_{j=1}^{r} a_j X_j$$

and that minimizes the expected quadratic loss

$$\mathbb{E}\left[\left(\mu(\Theta) - P^{\mathrm{cred}}\right)^2\right],$$

where $X_1, \ldots, X_r$ constitute the observed values for a specific individual within the collective.

The homogeneous credibility estimator P^{hom} *or $\hat{\mu}^h(\Theta)$ is the unique credibility estimator with the restriction that $a_0 = 0$.*

The credibility estimator can also be viewed as an orthogonal projection in the Hilbert space of square integrable functions (Bühlmann and Gisler [2005]), but for our purposes, it suffices to view it as the unique linear estimator that minimizes the mean-squared error. The homogeneous credibility estimator is more relevant in considering data from a collective, so we will defer discussion until the next section on the Bühlmann–Straub model. Definition 10.18 is, however, never applied directly, as we have the following fundamental result.

Theorem 10.3. *The credibility estimator for an individual risk premium in a mixture model is given by*

$$\hat{\mu}(\Theta) = \lambda \overline{X} + (1 - \lambda)\mu_0, \tag{10.91}$$

where the credibility weight λ *is given by*

$$\lambda = \frac{n}{n + \sigma^2/\tau^2}. \tag{10.92}$$

Moreover, the quotient $\kappa = \sigma^2/\tau^2$ *is known as the* credibility coefficient.

Proof. Since the observed data $X_1, \ldots, X_r$ for an individual are independent and identically distributed, they are therefore invariant relative to permutations. Thus the coefficients $a_1, \ldots, a_r$ must be equal, and the linearity requirement of the credibility estimator can be reduced to

$$P^{\text{cred}} = a + b\overline{X}.$$

Next, let us consider the loss function

$$L(a, b) = \mathbb{E}\left[\left(\mu(\Theta) - a - b\overline{X}\right)^2\right].$$

Taking partial derivatives with respect to the coefficients a and b yields the equations

$$\mathbb{E}[\mu(\Theta) - a - b\overline{X}] = 0$$
$$\text{COV}(\overline{X}, \mu(\Theta)) - b\mathbb{V}(\overline{X}) = 0.$$

However, we have that

$$\text{COV}(\overline{X}, \mu(\Theta)) = \mathbb{V}(\mu(\Theta)) = \tau^2$$
$$\mathbb{V}(\overline{X}) = \frac{\sigma^2}{n} + \tau^2,$$

where the second equality follows from Theorem 10.2. Inserting these expressions into the partial derivative equations and solving yield the results

$$b = \frac{n}{n + \sigma^2/\tau^2}$$
$$a = (1 - b)\mu_0,$$

which proves the result.   □

Note 10.12 (Credibility Rule of Thumb). The credibility coefficient κ thus has the following interpretation: when the sample size n is equal to κ, the credibility is exactly 50%. Therefore, κ represents the required sample size in order for the prior knowledge to balance with the experience in the credibility estimator. Similarly, when $n = 3\kappa$, the credibility is 75%.   ▷

Note 10.13 (An Intuitive Principle). The credibility estimator can also be understood intuitively as follows as a weighted average of the collective premium $P^{\text{coll}} = \mu_0$ and the experience $\overline{X}$. In particular, the collective premium is the best *a priori* estimator, which has quadratic loss

$$\mathbb{E}\left[(\mu_0 - \mu(\Theta))^2\right] = \mathbb{V}(\mu(\Theta)) = \tau^2.$$

Similarly, the experience $\overline{X}$ is the best possible linear and individually unbiased estimator based on an individual's data, and has quadratic loss

$$\mathbb{E}\left[(\overline{X} - \mu(\Theta))^2\right] = \mathbb{E}[\sigma^2(\Theta)/n] = \frac{\sigma^2}{n}.$$

Finally, the credibility estimator is a weighted mean of these, where the weights are proportional to the inverse quadratic loss (precision) associated with each of the two components. This last statement can be realized by rewriting the credibility weight in the equivalent form

$$\lambda = \frac{n/\sigma^2}{n/\sigma^2 + 1/\tau^2},$$

so that the weight on the experience is proportional to n/σ^2 and the weight on the prior knowledge is proportional to $1/\tau^2$.   ▷

Example 10.15. Recall that the mixture $X \sim \text{Exp}(\Theta)$ and $\Theta \sim \text{Gamma}(\alpha, \beta)$ does not possess a simple expression for the Bayes' premium. Nevertheless, the credibility estimator is given by

$$P^{\text{cred}} = \lambda \overline{X} + (1 - \lambda)P^{\text{coll}},$$

where

$$\lambda = \frac{n}{n + \alpha + 1},$$

as we have previously computed the credibility coefficient κ in Example 10.14.   ▷

The Bühlmann–Straub Model

So far, we have discussed the simple credibility estimator as applied to a specific individual. We next generalize this model to account for observations from multiple individuals from the collective.

Definition 10.19 (Bühlmann–Straub Model). *The* Bühlmann–Straub model *is a continuous mixture model for which the data are collected relative to specific individuals with given weights; i.e., data are of the form* (X_{ij}, w_{ij}), *for individual* $i = 1, \ldots, m$ *and observation* $j = 1, \ldots, r$, *where* X_{ij} *is the observed value and* w_{ij} *is its associated weight (given).*

Thus, individual i has a risk profile $\Theta_i \sim U$, drawn randomly from the collective distribution. Conditional on the individual Θ_i, we therefore have

$$\mathbb{E}[X_{ij}|\Theta_i] = \mu(\Theta_i)$$
$$\mathbb{V}(X_{ij}|\Theta_i) = \frac{\sigma^2(\Theta_i)}{w_{ij}}.$$

We will follow our standard dot-summation notation, that

$$w_{i\cdot} = \sum_{j=1}^{r} w_{ij} \quad \text{and} \quad w_{\cdot\cdot} = \sum_{i=1}^{m}\sum_{j=1}^{r} w_{ij}.$$

Note 10.14. The *Bühlmann model*, which predates the Bühlmann–Straub model, is a special case of the latter, in which the weights w_{ij} are set to equal one; i.e., $w_{ij} = 1$. The Bühlmann model is treated directly in Bühlmann and Gisler [2005]. ▷

Credibility in the Bühlmann–Straub Model

In the Bühlmann–Straub model, the conditional individual variances are inversely proportional to the weights. It follows that the best linear estimator (unbiased estimator with the minimum conditional variance) that is obtainable from the data alone is the weighted average of the observations, i.e.,

$$\overline{X}_{i\cdot} = \frac{1}{w_{i\cdot}} \sum_{j=1}^{r} w_{ij} X_{ij}. \tag{10.93}$$

Theorem 10.4. *The credibility estimator for the* ith *individual in the Bühlmann–Straub model is given by*

$$\hat{\mu}(\Theta_i) = \lambda_i \overline{X}_{i\cdot} + (1 - \lambda_i)\mu_0, \tag{10.94}$$
$$\lambda_i = \frac{w_{i\cdot}}{w_{i\cdot} + \sigma^2/\tau^2}, \tag{10.95}$$

where μ_0, σ^2, *and* τ^2 *are defined as before.*

   Moreover, the homogeneous credibility estimator in the Bühlmann–Straub model is given by

$$\hat{\mu}^h(\Theta_i) = \lambda_i \overline{X}_{i\cdot} + (1 - \lambda_i)\hat{\mu}_0, \tag{10.96}$$

where $\hat{\mu}_0$ is given by

$$\hat{\mu}_0 = \frac{1}{\lambda_.} \sum_{i=1}^{m} \lambda_i X_i, \tag{10.97}$$

where $\lambda_. = \sum_{i=1}^{m} \lambda_i$, as usual.

We leave the proof to Bühlmann and Gisler [2005].

Note that the homogeneous credibility estimator replaces the collective premium μ_0 with an estimate $\hat{\mu}_0$ which is formed by taking the *credibility-weighted average* of the individual means $\overline{X}_{i\cdot}$. *Note* that this is *not* the weighted average

$$\overline{X} = \frac{1}{w_{..}} \sum_{i=1}^{m} w_{i\cdot} \overline{X}_{i\cdot}.$$

of all the data, but the *credibility-weighted* average

$$\hat{\mu}_0 = \frac{1}{\lambda_.} \sum_{i=1}^{m} \lambda_i \overline{X}_{i\cdot},$$

an important distinction. However, using the correct estimate for the collective premium, the homogeneous credibility estimator in the Bühlmann–Straub model has the following *balance property*

$$\sum_{i=1}^{m} \sum_{j=1}^{r} w_{ij} \hat{\mu}^h(\Theta_i) = \sum_{i=1}^{m} \sum_{j=1}^{r} w_{ij} X_{ij}.$$

We leave the proof of this fact to the interested reader.

Example 10.16. Consider again the mixture model $X \sim \text{Exp}(\Theta)$, with $\Theta \sim \text{Gamma}(\alpha, \beta)$. Let us consider a collective of ten random individuals, each with an known true parameter value $\Theta_i \sim \text{Gamma}(\alpha, \beta)$. Suppose the population has $\alpha = 4$ and $\beta = 5$. For these values of the population parameters, we have $\mu_0 = 20$, $\sigma^2 = 500$, and $\tau^2 = 100$, with $\kappa = 5$. A random data set for ten individuals is simulated in Code Block 10.10; the result is shown in Figure 10.10. The simulated data corresponding to the bottom half of Figure 10.10 are given in Table 10.9. Notice that, in both cases, the average individual's data is more spread out than the collection of individual means.

Since all data are weighted equally, each individual has the same credibility rating

$$\lambda = \frac{10}{10 + 5} = \frac{2}{3}.$$

The credibility premium for the ith individual is therefore obtained by

$$P_i^{\text{cred}} = \frac{2}{3}\overline{X}_{i\cdot} + \frac{20}{3}.$$

```
1  alpha, beta = 4, 5
2  m, r = 10, 10
3  thetas = np.random.gamma(alpha, scale=beta, size=m)
4  thetas.sort()
5
6  X = np.zeros((m, r))
7  for i in range(m):
8      X[i, :] = np.random.exponential(scale=thetas[i], size=r)
9
10 plt.figure(figsize=(8, 9/2))
11 for i in range(m):
12     color = default_colors[i]
13     plt.plot(thetas[i], 1, 'o', color=color)
14     plt.plot(X[i, :], np.zeros(r), '.', color=color)
15     for j in range(r):
16         plt.plot([thetas[i], X[i, j]], [1, 0], color=color)
17
18 # Credibility premiums
19 mu_0 = 20 # Or X.mean() for hom. estimator, since equally weighted.
20 la = 2/3
21 X_bar = X.mean(axis=1)
22 P_cred = la * X_bar + (1-la) * mu_0
23 mu_0_hat = X.mean()
24 sigma_2_hat = np.sum((X - X.mean(axis=1).reshape((m, 1)))**2) /
       (m*(r-1))
25 tau_2_hat = np.sum((X.mean(axis=1) - X.mean())**2) / (m-1) -
       sigma_2_hat / r
```

Code Block 10.10: Simulated data for individuals in an exponential–gamma mixture.

i	θ_i	X_{i0}	X_{i1}	X_{i2}	X_{i3}	X_{i4}	X_{i5}	X_{i6}	X_{i7}	X_{i8}	X_{i9}
0	10.65	13.78	10.21	2.74	4.2	0.25	23.94	27.33	5.95	0.86	8.46
1	14.13	10.09	16.83	0.5	2.19	5.12	2.91	58.91	1.64	11.48	17.04
2	14.2	13.27	1.99	4.45	68.11	24.8	20.32	4.67	4.01	7.0	0.77
3	14.64	11.58	2.9	7.48	1.54	9.57	1.83	5.69	18.09	9.68	9.04
4	15.15	12.06	12.8	6.88	8.23	21.01	4.71	8.29	3.57	21.49	29.26
5	16.08	30.76	30.68	8.32	5.3	10.31	24.19	11.06	18.16	15.88	17.44
6	18.5	22.4	33.95	13.11	30.67	34.67	5.08	67.74	30.31	5.95	3.03
7	19.35	8.08	4.71	39.65	15.05	10.81	17.09	40.32	17.95	0.43	3.13
8	26.22	33.7	91.02	62.51	27.28	0.59	12.56	27.77	25.62	1.94	113.92
9	48.25	51.98	129.2	29.31	77.86	57.83	25.19	20.06	7.07	23.29	78.78

Table 10.9: Simulated data from Code Block 10.10.

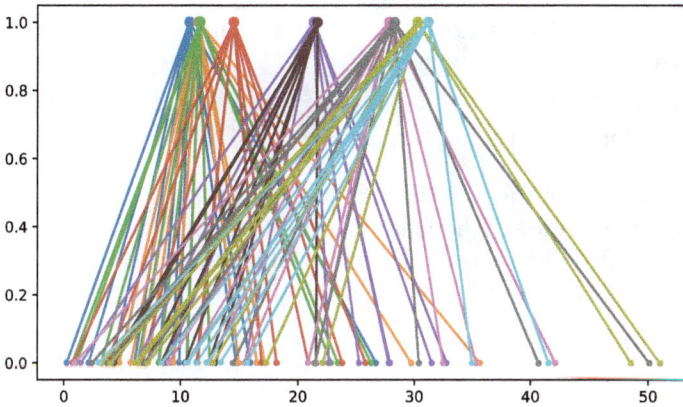

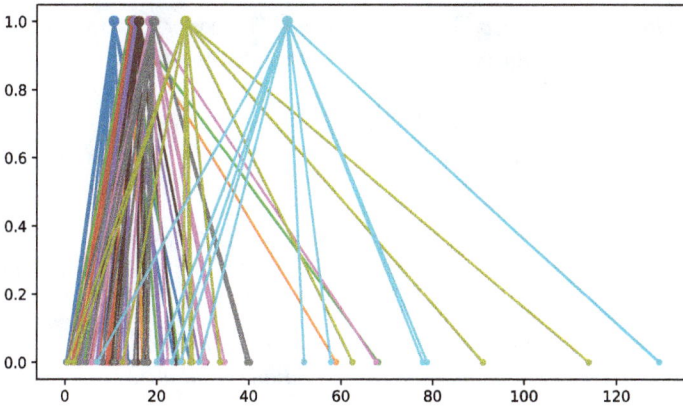

Fig. 10.10: Two visualizations of simulated data showing true individual risks θ_i (at $y = 1$) connected to each individual's simulated data (at $y = 0$). The bottom figure corresponds to the data shown in Figure 10.10.

$i \rightarrow$	0	1	2	3	4	5	6	7	8	9
P_i^{ind}	10.65	14.13	14.2	14.64	15.15	16.08	18.5	19.35	26.22	48.25
P_i^{cred}	13.18	15.11	16.63	11.83	15.22	18.14	23.13	17.15	33.13	40.04

Table 10.10: True individual premium $P_i^{\text{ind}} = \theta_i$ and credibility estimate for each indivudal.

The true individual premium θ_i and credibility estimate are recorded in Table 10.10. If we did not know the collective premium μ_0, we could estimate it as $\hat{\mu}_0 = 20.53$. Using this in the above computation would yield the homogeneous credibility estimate.   ▷

Estimates for Structural Parameters

Naturally, it is often the case that the structural parameters σ^2 and τ^2 are not known in advanced. The following result, originally due to Bühlmann and Straub [1970], but also discussed in Bühlmann and Gisler [2005], provides a natural estimator for these parameters.

Theorem 10.5. *The quantity*

$$\hat{\sigma}^2 = \frac{1}{m(r-1)} \sum_{i=1}^{m} \sum_{j=1}^{r} w_{ij} \left(X_{ij} - \overline{X}_{i\cdot} \right)^2 \tag{10.98}$$

is an unbiased ($\mathbb{E}[\hat{\sigma}^2] = \sigma^2$) and consistent ($\hat{\sigma}^2$ converges in probability to σ^2 as $m \to \infty$) estimator for the structural parameter σ^2. Similarly, the quantity

$$\hat{\tau}^2 = c \left[\frac{m}{m-1} \sum_{i=1}^{m} \frac{w_{i\cdot}}{w_{\cdot\cdot}} (\overline{X}_{i\cdot} - \overline{X}_{\cdot\cdot})^2 - \frac{m\hat{\sigma}^2}{w_{\cdot\cdot}} \right], \tag{10.99}$$

where

$$c = \frac{m-1}{m} \left[\sum_{i=1}^{m} \frac{w_{i\cdot}}{w_{\cdot\cdot}} \left(1 - \frac{w_{i\cdot}}{w_{\cdot\cdot}} \right) \right]^{-1}, \tag{10.100}$$

is an unbiased estimator for τ^2, and consistent, as long as $\sum_{i=1}^{m}(w_{i\cdot}/w_{\cdot\cdot})^2 \to 0$ as $m \to \infty$; i.e., as long as no one individual is "dominating" as $m \to \infty$.

Note 10.15. The means $\overline{X}_{i\cdot}$ and $\overline{X}_{\cdot\cdot}$ are both *weighted* means of the observed data.   ▷

Note 10.16. For the case in which each individual has equal weight, i.e., $w_{1\cdot} = \cdots = w_{m\cdot}$, the constant $c = 1$.   ▷

Corollary 10.8. *For the case in which each individual has equal weight, i.e., $w_{1\cdot} = \cdots = w_{m\cdot}$, the estimator $\hat{\tau}^2$ from Theorem 10.5 reduces to*

$$\hat{\tau}^2 = \frac{1}{m-1} \sum_{i=1}^{m} (\overline{X}_{i\cdot} - \overline{X}_{\cdot\cdot})^2 - \frac{\hat{\sigma}^2}{w_{i\cdot}}, \tag{10.101}$$

where $w_{i\cdot} = w_{\cdot\cdot}/m$ is each individual's total weight.

Note 10.17. The estimate $\hat{\sigma}^2$ given by Equation (10.98) should be compared with the mean within-group mean squares (Equations (3.76) and (3.78)). Similarly, the first term in the quantity $\hat{\tau}^2$ should be compared with the between-group mean squares (Equations (3.75) and (3.78)). The between-group mean squares MSW is an estimate for the variance of the population means. Thus, an estimate for σ^2/n must be subtracted from it to obtain an estimate for τ^2, due to Theorem 10.2.   ▷

Note 10.18. It is possible for the estimate $\hat{\tau}^2$ to be negative. Thus, it is better to replace the estimate $\hat{\tau}^2$ with

$$\hat{\tau}^2 \to \max(0, \hat{\tau}^2),$$

though this means that the estimate is no longer unbiased.   ▷

Example 10.17. Considering again the data from Table 10.9 in Example 10.16, we can compute the estimated structural parameters as $\hat{\sigma}^2 = 437$ and $\hat{\tau}^2 = 148$. These calculations are shown on the last two lines of Code Block 10.10. (Recall the true values are 500 and 100, respectively.) This results in the estimate $\hat{\kappa} = 2.95$, and would change the credibility weight to $\hat{\lambda} = 0.77$. Thus, for this specific example, the empirical credibility estimate would more closely favor the observed data.   ▷

10.5 Simulation

In practice, it is often difficult to analytically compute a posterior distribution for a given prior and a given set of data. Examples include non-conjugate priors or complex hierarchical models. Fortunately, there are a number of sophisticated techniques that allow us to sample the posterior distribution, providing robust estimates for many quantities of interest.

10.5.1 Markov Chain Monte Carlo Simulation

A *Monte Carlo method*, named for the Monte Carlo Casino in Monaco, is any computational algorithm that relies on random sampling to estimate a result. (For history of the early computational development of Monte Carlo methods in Los Alamos National Laboratory, see Metropolis [1987].) For example, to estimate the value of π, one could draw many random samples from the distribution $\mathrm{Unif}([0, 1] \times [0, 1])$ and then compute the fraction of samples that satisfy $x^2 + y^2 < 1$. In the context of Bayesian statistics, a number of *Markov chain Monte Carlo methods* (MCMC) are often used. For such methods, the random samples are not independent, though they do satisfy the Markov property (Definition 5.12); i.e., any given random draw is a function of the past only through the immediately preceding instance.

Gibbs Sampling

In many multidimensional Bayesian models, it is often difficult to sample from the joint posterior distribution directly, but not as difficult to sample from the conditional distribution for an individual or subset of parameters. Gibbs sampling is a particular MCMC algorithm designed for this situation. The Gibbs sampling algorithm is described in Algorithm 10.2.

Algorithm 10.2: Gibbs Sampling.

Input: A set of IID data $\mathcal{D} = Y_1, \ldots, Y_n \sim p(Y|\theta)$;
a partition S of the parameter vector $\theta = (\theta_1, \ldots, \theta_d)$;
the conditional posterior distributions $p(\theta_i|\theta_{-i}, \mathcal{D})$, for
$i = 1, \ldots, d$;
an initial parameter value θ^0;
number of desired samples n.

Result: Approximate sample $\theta^1, \ldots, \theta^n$ from $p(\theta|\mathcal{D})$.

1 **for** t **from** 1 **to** n **do**
2 $\quad$ Generate a random permutation $\pi \in S_n$
3 $\quad$ **for** j **from** 1 **to** d **do**
4 $\quad\quad$ Draw a random
$$\theta^t_{\pi(j)} \sim p(\theta_{\pi(j)}|\theta^t_{\pi(1)}, \ldots, \theta^t_{\pi(j-1)}, \theta^{t-1}_{\pi(j+1)}, \ldots, \theta^{t-1}_{\pi(d)}, \mathcal{D})$$
5 $\quad$ **end**
6 **end**
7 **return** $\theta^1, \ldots, \theta^n$.

For Gibbs sampling, a multidimensional parameter vector θ is broken into a set of d subvectors, $\theta = (\theta_1, \ldots, \theta_d)$. Each sampling iteration is then carried out through a sequence of d steps, where the jth subvector θ_j is sampled using the conditional posterior distribution, conditioned on the data and the most recent value of the remaining subvectors. Furthermore, the ordering is randomized at the beginning of each iteration. The final sample is not independent, as each sampled parameter θ^t depends on the previously sampled values, but only via the immediately preceding value θ^{t-1}.

Metropolis Algorithm

In considering Bayes' law in the form of Equation (10.2), the prior distribution $\mathbb{P}(\theta)$ and the likelihood $p(\mathcal{D}|\theta)$ are often easy enough to compute. The data distribution of the denominator,

$$p(\mathcal{D}) = \int p(\mathcal{D}|\theta)p(\theta)\, d\theta,$$

on the other hand, is often unwieldy and difficult to compute, especially analytically. Thus enters Metropolis. The Metropolis algorithm allows us to approximate a random sample from the posterior distribution $p(\theta|\mathcal{D})$

without the need for computing the marginal data distribution of the denominator. Before presenting the algorithm, we require the following definition.

Definition 10.20. *Given a parameter space Θ, a* jumping distribution *(or* proposal distribution*) is any conditional distribution $J(\cdot|\cdot) : \Theta \times \Theta \to R_*$. A jumping distribution is said to be* symmetric *if $J(\theta|\phi) = J(\phi|\theta)$.*

Examples of symmetric jumping distributions include

$$J(\theta|\phi) = \text{Unif}([\phi - \delta, \phi + \delta])$$

and

$$J(\theta|\phi) = \text{N}(\phi, \delta^2),$$

where the parameter δ is a tuning parameter chosen so that the algorithm runs smoothly.

The Metropolis algorithm is a MCMC algorithm that determines a proposed parameter θ^* by sampling the jumping distribution and accepting or rejecting the proposed value based on the ratio of how likely the proposed value is compared with the likelihood of the current value. The details are presented in Algorithm 10.3.

Algorithm 10.3: Metropolis algorithm.

Input: A set of IID data $\mathcal{D} = Y_1, \ldots, Y_n \sim p(Y|\theta)$;
    a prior distribution $p(\theta)$;
    the likelihood function $p(\mathcal{D}|\theta)$;
    a symmetric jumping distribution $J(\cdot|\cdot)$;
    an initial parameter value θ^0;
    number of desired samples n.
Result: Approximate sample $\theta^1, \ldots, \theta^n$ from $p(\theta|\mathcal{D})$.

1 **for** t **from** *1* **to** n **do**
2   Sample $\theta^* \sim J(\theta|\theta^{t-1})$
3   Compute the acceptance ratio

$$r = \frac{p(\theta^*|\mathcal{D})}{p(\theta^{t-1}|\mathcal{D})} = \frac{p(\mathcal{D}|\theta^*)p(\theta^*)}{p(\mathcal{D}|\theta^{t-1})p(\theta^{t-1})} \qquad (10.102)$$

  Set

$$\theta^t = \begin{cases} \theta^* & \text{with probability } \min(r,1) \\ \theta^{t-1} & \text{with probability } 1 - \min(r,1) \end{cases}. \qquad (10.103)$$

4 **end**
5 **return** $\theta^1, \ldots, \theta^n$.

Metropolis–Hastings algorithm

The Metropolis–Hastings algorithm provides a subtle yet power generalization to the Metropolis algorithm, allowing for asymmetric jumping distributions. The asymmetry is corrected for in the acceptance ratio, as shown in Equation (10.104) of Algorithm 10.4.

Algorithm 10.4: Metropolis–Hastings algorithm.

    Input: A set of IID data $\mathcal{D} = Y_1, \ldots, Y_n \sim p(Y|\theta)$;
             a prior distribution $p(\theta)$;
             the likelihood function $p(\mathcal{D}|\theta)$;
             a sequence of jumping distributions $J_t(\cdot|\cdot)$;
             an initial parameter value θ^0;
             number of desired samples n.
    Result: Approximate sample $\theta^1, \ldots, \theta^n$ from $p(\theta|\mathcal{D})$.

1 **for** t **from** 1 **to** n **do**
2   Sample $\theta^* \sim J(\theta|\theta^{t-1})$
3   Compute the acceptance ratio

$$r = \frac{p(\theta^*|\mathcal{D})/J_t(\theta^*|\theta^{t-1})}{p(\theta^{t-1}|\mathcal{D})/J_t(\theta^{t-1}|\theta^*)} \qquad (10.104)$$

     Set

$$\theta^t = \begin{cases} \theta^* & \text{with probability } \min(r, 1) \\ \theta^{t-1} & \text{with probability } 1 - \min(r, 1) \end{cases} \cdot \qquad (10.105)$$

4 **end**
5 **return** $\theta^1, \ldots, \theta^n$.

It is interesting to note the Gibbs sampling method can also be derived as a special case of the Metropolis–Hastings algorithm.

10.5.2 PyMC3

The `pymc3` package (i.e., PyMC3) is an an advanced open-source Python package that implements many MCMC algorithms for us. This section is devoted to reviewing the package syntax and usage guide. An excellent resource on this package's predecessor, `pymc`, is Davidson-Pilon [2016]. Despite some changes in the syntax and usage of the software (PyMC3 was rebuilt from ground up), we still recommend this as an excellent resource for probabilisitic programming. A beginning guide for PyMC3 can be found in Salvatier, *et al.* [2016]. Additional documentation can be found on the developer's website `docs.pymc.io`.

Note 10.19. PyMC3 is built on top of Theano[4]. Variables in the context of PyMC3 are actually Theano variables. Theano, however, is no longer cur-

---

[4] `theano.readthedocs.io`.

rently maintained, so PyMC3 has imported much of Theano's functionality into its own libraries. As a result, many of the standard operators ($\leq$, $>$, if) do not work when defining new variables within PyMC3. Instead, the pymc3.math library should be used, which is based off the theano.tensor package. Where documentation in PyMC3 falls short, consult the Theano documentation.   ▷

Variables and Models in PyMC3

PyMC3 has an extensive library of built-in random variables. The constructor for each consists of three primary components:

1. a name (a string), which is usually the same as the name of the Python variable where we are storing our object;
2. one or more (keyword) arguments that specify the distribution; these can be actual values or other random variables, in the case of hierarchical models; and
3. a number of optional keyword arguments (**shape** and **observed**; more on this later) that are universal across all PyMC3 variables.

Moreover, a PyMC3 variable cannot be defined outside of a *model context*. In other words, a PyMC3 model can be thought of as a collection of variables, and it is defined using a **with** statement. For example, in Code Block 10.11, we open a new model context, naming it **model**, and we define several random variables. Note that the first argument used to construct each variable is that variable's name.

```
import pymc3 as pm

with pm.Model() as model:
    x = pm.Normal("x", mu=0, std=2)
    y = pm.Normal("y", mu=3, std=2)
    z = pm.Deterministic("z", x+y)
```

Code Block 10.11: Constructing variables in PyMC3.

The variables x and y in Code Block 10.11 are normal random variables with different means (0 and 3), but the same standard deviation (2). The variable z is equivalent to the expression z = x + y, except we are tracking this variable in the context of our model.

In addition, there are two distinct types of PyMC3 variables:

1. *Stochastic*: Any variable that is not uniquely determined by its parents;
2. *Deterministic*: Any variable that is uniquely determined by its parents.

For example, the variables x and y in Code Block 10.11 are stochastic, because we do not know the next sample value, even such that we know mu and std. Similarly, the variable z is deterministic; if we know the values of both x and y, then we know exactly the value of the variable z.

Note 10.20. Both stochastic and deterministic variables are *random* variables!   ▷

Note 10.21. At the present, PyMC3 does not support the observed keyword for deterministic variables. The Deterministic wrapper merely adds the variable to the final trace, so that we can view the posterior distribution of our deterministic random variables.

Therefore, *only stochastic variables can be observed*!

This means, in particular, that we cannot model our error term as a separate term, but must model it within the observed variable. For example, consider the two equivalent linear models

$$Y = a_1 X_1 + a_1 X_2 + a_3 X_3 + \epsilon,$$
$$\epsilon \sim N(0, \sigma^2),$$

and

$$Y \sim N(a_1 X_1 + a_1 X_2 + a_3 X_3, \sigma^2).$$

PyMC3 supports the latter, not the former. If we want to examine the posterior of our error term, we should, instead, define the error term as a deterministic variable

$$\epsilon = Y - a_1 X_1 + a_1 X_2 + a_3 X_3,$$

so that we can view its posterior trace.   ▷

In our definition of a causal structural model, the endogenous variables were deterministic, whereas the exogenous variables were (typically) stochastic. Thus a causal model can only be captured in PyMC3 with an appropriate modification, so that the endogenous variables are stochastic.

Finally, there are two optional keywords that can be used to construct PyMC3 variables. The first is shape (int), which can be used if we want to specify a random *vector* instead of a random variable. The second is observed (array), which takes as input any data set of actual observations. By passing a set of data through the observed keyword argument, we are implicitly defining our likelihood function, and fixing those values as actual observations of the model.

The cool thing is that a model, in PyMC3, is just a bag of variables, and the model's graph is inferred by the variable definition. This allows us to define complex Bayesian models with some ease.

Example 10.18. Let us consider normally distributed data with unknown mean μ and variance σ^2. We simulate such a data set on lines 2–4 of Code Block 10.12, from a population with true mean 12 and true standard deviation 4. The sample mean of our simulated data is 11.38, with sample standard deviation 4.55.

Aside from the sampling statistics, we have a prior belief in the mean and variance of the population. Suppose we have prior reason to suspect the mean to be around 8, and that we are 95% confident the mean is actually between 0 and 16. Additionally, we know that the standard deviation is less than ten, but don't have any feeling for whether it is on the low or high side. We can therefore construct a Bayesian hierarchical model as

$$X \sim N(\mu, \sigma^2),$$
$$\mu \sim N(8, 16),$$
$$\sigma \sim \text{Unif}(0, 10),$$

where we have used a normal and uniform prior for μ and σ, respectively.

```
1    # Simulate data.
2    mu, sigma = 12, 4
3    n = 20
4    data = np.random.normal(mu, sigma, size=n) # 11.38, 4.55
5
6    # Build a PyMC3 model
7    with pm.Model() as normal_model:
8        mu = pm.Normal("mu", mu=8, sd=4)
9        sigma = pm.Uniform("sigma", lower=0, upper=10)
10
11       x = pm.Normal("x", mu=mu, sd=sigma, observed=data)
```

Code Block 10.12: Simulation and PyMC3 model for Example 10.18.

Our model is easily encoded in lines 7–11 of Code Block 10.12. Note that mu, sigma, and x are each PyMC3 stochastic variables, as each have a random component that prevents one from deterministically specifying the outputs. In addition, the priors mu and sigma are the parents of the variable x, which, since we pass our data set into the constructor using the observed keyword, defines our likelihood function.   ▷

Models and Sampling in PyMC3

Example 10.19. We conclude Example 10.18 with an implementation of the MCMC algorithm. To do this, we reopen our normal_model context and call the PyMC3 sample method, as shown in Code Block 10.13.

```
1  with normal_model:
2      idata = pm.sample(2000, tune=1500, return_inferencedata=True)
3
4  pm.plot_trace(idata)
5
6  idata.posterior["mu"].shape
7  idata.posterior["sigma"].shape
```

Code Block 10.13: Deploying the MCMC algorithm for Examples 10.18 and 10.19.

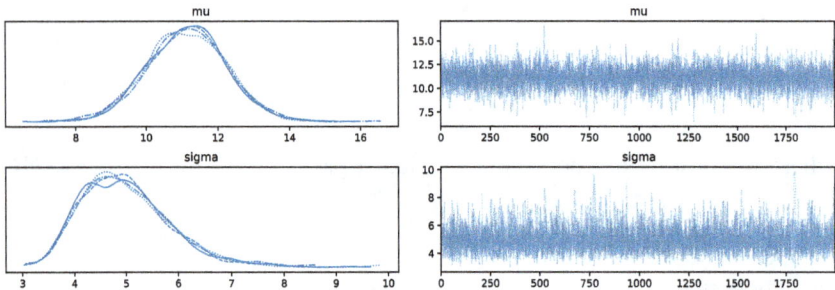

Fig. 10.11: Plot trace of posterior distributions for Example 10.19. Left: smoothed histogram (using kernel density estimation) for each of the four parallel simulations. Right: samples from MCMC plotted in sequential order.

PyMC3 also has a built-in method called `plot_trace`, which generates the plots shown in Figure 10.11. These contain a smoothed histogram for our posterior distributions for μ and σ (left) and the individual samples plotted sequentially (right). The latter is a good visual check that our model has converged.   ▷

A Bayesian Switchpoint Example

To illustrate these ideas, we will follow an example from Davidson-Pilon [2016] (with a slightly modified story).

Example 10.20. Suppose we are running an email marketing campaign through an advertiser that sends out a certain number of emails per day, depending on our budget, bid, and market pressure. We have run our campaign for sixty days, and are analyzing the number of emails sent out per day, as shown in Figure 10.12. (We generated the data using Code Block 10.14.)

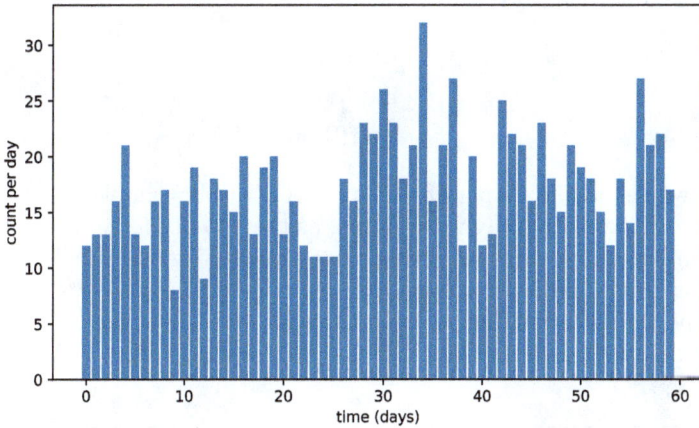

Fig. 10.12: Marketing Emails per day, in thousands.

```
1  la1, la2 = 15, 20
2  tau = 27
3  n = 60
4
5  data = np.zeros(n)
6  data[:tau] = random.poisson(la1, size=tau)
7  data[tau:] = random.poisson(la2, size=n-tau)
```

Code Block 10.14: Generating random marketing data for Example 10.20

From a visual inspection, it appears as though there was an increase in volume that occurred around day 28. We will use a Bayesian model to validate this observation and quantify the perceived lift.

We will model our data set using a *switchpoint model*; i.e., we will assume there is some time τ that divides our data into a before and after sets constituting time periods of different behavior. This can be captured by the following model:

$$X_t \sim \text{Poiss}(\lambda_t)$$
$$\lambda_t = \begin{cases} \lambda_1 & \text{if } t < \tau \\ \lambda_2 & \text{if } t \geq \tau \end{cases},$$

for some λ_1, λ_2, and τ. Since we are modeling counts, we have used a Poisson random variable, with rate parameters λ_1 and λ_2. Of course, if the behavior is the same over the full period, we would have $\lambda_1 = \lambda_2$. The parameter τ represents the *switchpoint*, which divides the interval $[0, 60]$ into the two separate regions.

Since we do not know, a priori, the values of λ_1, λ_2, and τ, we will specify the following prior distributions:

$$\lambda_1 \sim \text{Exp}(\alpha),$$
$$\lambda_2 \sim \text{Exp}(\alpha),$$
$$\tau \sim \text{Unif}(0, 60).$$

Finally, we set the hyperparameter $\alpha = \overline{X}_n$ to our sample mean, since $\mathbb{E}[\lambda|\alpha] = \alpha$. These are fairly weak priors, as they do not force too much information into our system. Moreover, since we are using numerical methods, we do not care whether or not these are conjugate priors.

We specify and train our model in Code Block 10.15 (using the data set generated from Code Block 10.14).

```
with pm.Model() as model:

    lambda_1 = pm.Exponential("lambda_1", 1/data.mean())
    lambda_2 = pm.Exponential("lambda_2", 1/data.mean())
    tau = pm.DiscreteUniform("tau", lower=0, upper=n)

    lambda_ = pm.math.switch(np.arange(n) < tau, lambda_1, lambda_2)
    observation = pm.Poisson("observation", lambda_, observed=data)

    step = pm.Metropolis()

    trace = pm.sample(10000, tune=5000, step=step,
        return_inferencedata=True)

% mean of posterior samples
trace.posterior['lambda_1'].mean()
trace.posterior['lambda_2'].mean()
trace.posterior['tau'].mean()
```

Code Block 10.15: MCMC for switch point problem of Example 10.20.

The average values of the posterior samples from our simulation are $\lambda_1 = 14.8$, $\lambda_2 = 19.6$, and $\tau = 27.0$. We plot the stack trace from our simulation in Figure 10.13, which shows our posterior uncertainty distribution across each of the three hyperparameters. It is clear that the switch point occurs around $\tau = 27$, and that the Poisson rates before and after are closely grouped around 15 and 20, respectively. It is clear from our visual inspection that λ_1 and λ_2 have a negligible overlap.   ▷

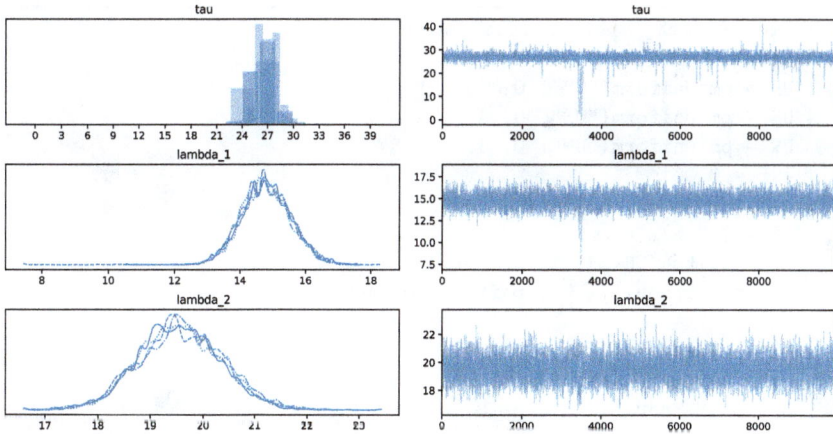

Fig. 10.13: Plot trace of posterior distributions for switchpoint Example 10.20. Left: smoothed histogram (using kernel density estimation) for each of the four parallel simulations. Right: samples from MCMC plotted in sequential order.

A Bayesian Causality Example

Example 10.21. Let us return to the counterfactual calculation of the rain–sprinkler–wet pavement problem of Example 8.17. In this example, we computed the counterfactual

$$\mathbb{P}(W|\text{do}(S)\text{if}(R, \neg S, \neg W)),$$

which represents the probability of finding wet pavement in our counterfactual reality in which we intervened to ensure that the sprinkler was running, given that, in our actual reality it rained, the sprinkler was not running, and the pavement was not wet. In defining counterfactuals, we first implement our abduction (Definition 8.23), then our intervention, and then our prediction. We know see that our abduction is equivalent to a Bayesian update on our probability distribution $p(u)$ defined over the exogenous variables in $\mathcal{U}$. In our present example, we compute the above counterfactual quantity using PyMC3.

Our basic model is captured by Code Block 10.16. We begin by defining our three exogenous variables as uniform random variables over the interval $[0, 1]$. We then define our functional relationships f_R, f_S, and f_W. Now, ideally we could simply let these three variables be observed, and pass in the values 1, 0, and 0, for our if$(R, \neg S, \neg W)$ condition. The problem, however, is that these are deterministic variables, and therefore cannot be observed in PyMC3. To remedy this situation, we introduce a small error term, `err`, and create three Bernoulli random variables R, S, and W. Thus,

```
1   from pymc3.math import switch
2   with pm.Model() as cf_model:
3       UR = pm.Uniform("UR", 0, 1)
4       US = pm.Uniform("US", 0, 1)
5       UW = pm.Uniform("UW", 0, 1)
6       err = 0.001
7
8       pR = 0.3
9       fR = switch(UR<pR, 1-err, err)
10      R = pm.Bernoulli("R", p=fR, observed=[1])
11
12      pS = switch(eq(R,1), 0.1, 0.3)
13      fS = switch(US<pS, 1-err, err)
14      S = pm.Bernoulli("S", p=fS, observed=[0])
15
16      pW = switch(eq(R,1),
17                  switch(eq(S,1), 0.9, 0.85),
18                  switch(eq(S,1), 0.1, 0)
19                  )
20      fW = switch(UW<pW, 1-err, err)
21      W = pm.Bernoulli("W", p=fW, observed=[0])
```

Code Block 10.16: PyMC3 encoding of the Rain–Sprinkler–Wet causal model of Example 10.21.

we allow for a small, but nonzero, fraction for which our causal model yields an incorrect result.

By conditioning on R, S, and W with the observed keyword, PyMC3's sample method will run an MCMC algorithm to simulate a sample for the posterior distributions of U_R, U_S, and U_W under the abduction if$(R, \neg S, \neg W)$. Since the counterfactual reality springs from the posterior distribution over $\mathcal{U}$, we can copy and paste a counterfactual version of the model, and append it to the original, as shown in Code Block 10.17. It is also within this counterfactual reality that we make the intervention do(S), which is captured on line 5. Note that our counterfactual random variables R_cf and W_cf are deterministic, so that PyMC3 records their trace under the posterior distribution on $\mathcal{U}$.

Finally, we run the model using the pm.sample method. The traceplot is shown in Figure 10.14. Note that the posterior distributions for U_R, U_S, and U_W are consistent with the ones computed by hand in Example 8.17. Moreover, our simulation yields the final counterfactual probability of wet pavement very close to $1/3$, our prior result. This can be verified by accessing the posterior trace data for W_cf, as done in line 16, yielding a simulated result of 0.33275.   ▷

```
1   with cf_model:
2       R_cf = pm.Deterministic("R_cf", switch(UR<pR, 1, 0))
3
4       # Intervention
5       S_cf = 1
6
7       pW_cf = switch(eq(R_cf,1),
8                       switch(eq(S_cf,1), 0.9, 0.85),
9                       switch(eq(S_cf,1), 0.1, 0)
10                      )
11      W_cf = pm.Deterministic("W_cf", switch(UW<pW_cf, 1, 0))
12
13  # Run the model.
14  with cf_model:
15      idata = pm.sample(2000, tune=1500, return_inferencedata=True)
16  idata.posterior["W_cf"].mean() # 0.33275
```

Code Block 10.17: PyMC3 encoding of the Rain–Sprinkler–Wet causal model of Example 10.21.

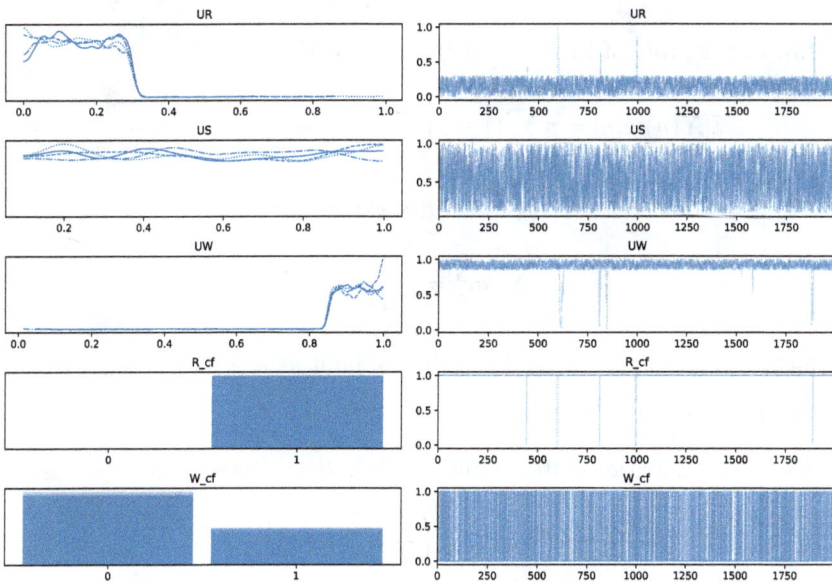

Fig. 10.14: The Rain–Sprinkler–Wet causal model of Example 10.21.

Problems

10.1. Determine the probability that Joe is afflicted with the medical condition of Example 10.1 if he receives *two* positive test results.

10.2. For Example 10.6, show that the prior predictive distribution *for the number of successes K out of the next n results* is given by

$$p(K) = \text{BetaBin}(K; n, \alpha, \beta).$$

10.3. Let

$$X \sim \text{Poiss}(\lambda t),$$
$$\lambda \sim \text{Gamma}(\alpha, \beta).$$

Show that

$$p(X|\alpha, \beta) = \text{NBD}(x; \alpha, (1 + \beta t)^{-1}).$$

10.4. Prove that

$$B(\gamma + 1, \delta + n) + B(\gamma, \delta + n + 1) = B(\gamma, \delta + n)$$

10.5. In the proof of Proposition 10.11, show that

$$\mathbb{E}[X(n)|p, \theta] = p \sum_{t=1}^{n} (1 - \theta)^n = \frac{p(1 - \theta)}{\theta} - \frac{p(1 - \theta)^{n+1}}{\theta}.$$

Hint: recall the formula for a finite geometric series

$$\sum_{i=0}^{n-1} ar^i = \frac{a(1 - r^n)}{1 - r},$$

which, of course, goes to $a/(1 - r)$ in the limit as $n \to \infty$, for $|r| < 1$.

10.6. Prove Equations (10.81) and (10.82).

10.7. Prove Equations (10.57) and (10.58). *Hint:* Recall that

$$\frac{\partial B(x, y)}{\partial x} = B(x, y) \left(\frac{\Gamma'(x)}{\Gamma(x)} + \frac{\Gamma'(x + y)}{\Gamma(x + y)} \right),$$

and that $\psi(x) = \Gamma'(x)/\Gamma(x)$. *Hint:* The beta function is also symmetric.

10.8. Use Code Block 10.6, along with Code Block 5.7, to solve for the BG/BB parameters $\alpha, \beta, \gamma, \delta$ using stochastic gradient descent.

10.9. (OPEN) Is there a simple closed-form expression for $\mathbb{E}[\Theta]$ and $\mathbb{V}(\Theta)$ in terms of ν, η, and ϕ for the conjugate prior Equation (10.15) of a general EDM?

10.10. Use the identity

$$B(\alpha + 1, \beta) = B(\alpha, \beta)\frac{\alpha}{\alpha + \beta}$$

and the symmetry of the beta function to show that

$$\frac{B(\alpha + 1, \beta + 1)}{B(\alpha, \beta)} = \frac{\alpha\beta}{(\alpha + \beta)(1 + \alpha + \beta)}.$$

Part III

Machine Learning

11

Hello, Neighbor

11.1 Nearest Neighbors

The K-nearest neighbors (KNN) is a simple nonlinear approach to the classification and regression problems that involves making predictions based on the behavior of the "nearest" members in the training set.

11.1.1 Measuring Distance

In this section, we introduce the concept of a metric space, discuss several commonly used distance metrics, and conclude with a brief discussion on the limitations of distance methods.

Metric Spaces

In order to proceed, we first need to establish a notion of "distance." This is achieved mathematically through the following definition.

Definition 11.1. *A* metric space *is a pair* $(\mathcal{M}, d)$*, where* $\mathcal{M}$ *is a set and* $d : \mathcal{M} \times \mathcal{M} \to \mathbb{R}_*$ *is a nonnegative function, known as a* metric*, that satisfies the properties*

1. **(symmetry)** $d(x, y) = d(y, x)$,
2. **(identifiability)** $d(x, y) = 0$ *if and only if* $x = y$,
3. **(triangle inequality)** $d(x, y) + d(y, z) \geq d(x, z)$,

for all $x, y, z \in \mathcal{M}$*. A metric space is* complete *if every* Cauchy sequence[1] *converges to a point in* M*.*

---

[1] A sequence $\{x_i\}$ in a metric space M is a *Cauchy sequence* if for every $\epsilon > 0$ there exists a n_ϵ such that $d(x_i, x_j) < \epsilon$ for all $i, j > n_\epsilon$. Counterexample: the open interval $(0, 1)$ is not complete.

Common Distance Metrics

There have been a multitudinous plethora of metrics that have been discovered. (The author has even published a novel metric for orbits in astrodynamics; see Maruskin [2010].) Another fun example is the French-rail metric, which is constructed based on the observation that all rail lines in France lead to Paris; see Exercise 11.1. In this paragraph, we will discuss several key metrics commonly used in data science.

Definition 11.2. *The* Minkowski distance of order p in $\mathbb{R}^k$ *is the metric*

$$d_p(x, y) = ||x - y||_p = \left(\sum_{j=1}^{k} |x_j - y_j|^p \right)^{1/p}, \qquad (11.1)$$

for $x, y \in \mathbb{R}^k$, *where* $|| \cdot ||_p$ *is the* ℓ_p *norm, defined in Definition 7.4.*

In particular, for $p = 1$, *we have the* Manhattan distance; *for* $p = 2$ *we have the* Euclidean distance, *and for* $p = \infty$, *we have the* Chebyshev distance, *given by the formulas*

$$d_1(x, y) = \sum_{j=1}^{k} |x_j - y_j|, \qquad (11.2)$$

$$d_2(x, y) = \left(\sum_{j=1}^{k} |x_j - y_j|^2 \right)^{1/2}, \qquad (11.3)$$

$$d_\infty(x, y) = \max_{j=1,\ldots,k} |x_j - y_j|, \qquad (11.4)$$

respectively.

The Chebyshev distance has an interesting interpretation in terms of the movement of the king on a chessboard; see Exercise 11.2.

Though not strictly a Minkowski metric, if we define $0^0 = 0$, we can formally express Equation (11.1) for the case $p = 0$ as

$$d_0(x, y) = \sum_{j=1}^{k} \mathbb{I}[x_j = y_j]. \qquad (11.5)$$

When $x, y \in \mathbb{B}^k$ are binary vectors, this metric is known as the *Hamming distance*, which counts the number of bits in vector x that differ from the corresponding bits in vector y. The Hamming distance is important when measuring distance in a one-hot encoded categorical feature space.

The Curse of Dimensionality

The *curse of dimensionality* is a phenomenon that occurs in certain learning methods that describes a severe loss in performance with increased dimensionality due to the increasing vastness of the larger space.

To understand what we mean by the *increase in vastness*, consider an n-dimensional sphere of radius r in $\mathbb{R}^n$, whose volume is given by

$$V_n(r) = \frac{\pi^{n/2} r^n}{\Gamma(1 + n/2)}.$$

The ratio of this volume to the volume $2^n r^n$ of the inscribed hypercube of side-length 2 decays to zero as $n \to \infty$. In particular, the volume $V_n(1)$ of the unit sphere is plotted for $n = 1, \ldots, 20$ in Figure 11.1. Note that the volume of the unit sphere decays to zero with increasing dimensionality. This implies that the fraction of instances with a distance less than unity becomes vanishingly small as the dimensionality increases.

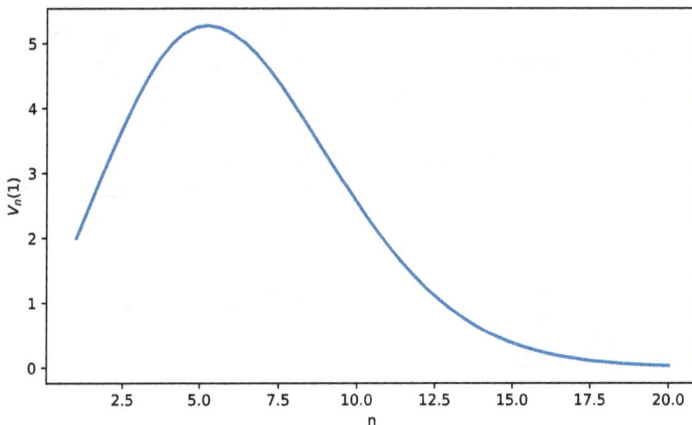

Fig. 11.1: The volume $V_n(1)$ of a unit sphere as a function of dimension n.

Moreover, the distance between the origin and the corner $(1, \ldots, 1)$ of the unit hypercube is $\sqrt{n}$, which increases without bound as $n \to \infty$. This implies that most of the space is "far away" from the origin, even within the unit cube. Thus, any concept of *nearness* decays commensurately with increased dimensionality. For this reason, the efficacy of nearest-neighbor algorithms wanes as the dimensionality increases, and such algorithms are ill suited for high-dimensional problems.

11.1.2 Finding the Nearest Neighbors

Now that we have a distance function in hand, we can easily compute the k nearest neighbors to a given point x_0 from a training data set $\mathcal{D}$. The details are shown in Algorithm 11.1.

Naturally, in order to locate the nearest neighbors, we need to first compute the distance between the test point x_0 and each point in the data set $\mathcal{D}$, which is an $O(n)$ operation, as we must iterate through the given dataset.

Next, we must locate the smallest values in our computed set of distances. Now, if we sort the entire dataset, this requires at best $O(n \log n)$. However, since we only need to find the k smallest values, and we do not need to sort the entire dataset, we can achieve this in $O(nk)^2$. We represent this method as $\mathrm{argmin}(d, k)$, as shown on line 6. In Python, this can be achieved using `np.argpartition(d, k)`, which will return an array of indices, such that the first k elements of said array are guaranteed to correspond to the indices of the k smallest values in the array d of distances.

11.1.3 Supervised Learning with Nearest-Neighbors

Now that we have defined distance in our feature space, we are ready to dive into the k-nearest neighbors (KNN) algorithm for classification and regression. At a high level, the algorithm proceeds in two steps: locate your k nearest neighbors and then *do what your neighbors do*. For classification, this involves letting your nearest neighbors vote, using their own labels, on your predicted class. For regression, this involves averaging the target variable across your nearest neighbors.

Algorithm 11.1: findNeighbors($\mathcal{D}$, x_0, k) – algorithm for finding k nearest neighbors.

Input: data $\mathcal{D} = \{(x_i, y_i)\}_{i=1}^{n}$;
　　　　　　a feature vector x_0;
　　　　　　the number k of nearest neighbors;
　　　　　　a distance metric `dist()`
Output: The set $(\iota_1, \ldots, \iota_k)$ of indices for the k nearest neighbors.
1 Set $n = |\mathcal{D}|$
2 Set $d = $ `Array()`
3 **for** $i = 0, \ldots, n-1$ **do**
4 　| 　Set $d[i] = $ `dist`(x_i, x_0)
5 **end**
6 Set $(\iota_1, \ldots, \iota_k) = $ `argmin`(d, k)
7 **return** $(\iota_1, \ldots, \iota_k)$

---

[2] For example, by using the *bubble-sort algorithm*; see Exercise 11.3 and Lambert [2014].

We begin by laying out the nearest neighbors algorithms for classification and regression. We then discuss various aspects of the two algorithms before concluding with several illustrative examples.

The Nearest Neighbors Classifier

The k-nearest neighbors (KNN) classifier involves seeking the k training instances closest to a given test point, and then allowing those nearest neighbors to vote on the predicted class for the test point. Once we have computed the target values of the k nearest neighbors, we let each of the nearest neighbors "vote" on the prediction for the test-point x_0. To do this, we calculate the ratio of neighbors that belong to each class $\mathcal{C}$, as shown in line 4 of Algorithm 11.2. Finally, we return the **argmax** of p; i.e., the most popular class of the nearest neighbors. We should take care to break ties with random chance.

The Nearest Neighbors Regressor

The algorithm for the KNN classifier can be easily modified to accommodate regression problems. Essentially, instead of polling the nearest neighbors to determine the most popular class, we instead return the average values of the neighborhood, as shown in Algorithm 11.3.

Neighborhood Watch

For both classification and regression, the nearest neighbors algorithms present with a few limitations that one should keep in mind.

Algorithm 11.2: knnClassifier($\mathcal{D}$, x_0, k) – k-nearest neighbors algorithm for classification.

Input: data $\mathcal{D}$;
  a feature vector x_0;
  the number k of nearest neighbors;
  the set $\mathcal{C}$ of target classes
Output: A predicted class $\hat{y}$.
1 Set $(\iota_1, \ldots, \iota_k) = \texttt{findNeighbors}(\mathcal{D}, x_0, k)$
2 Set $p = \texttt{Dict}()$
3 **for** c in $\mathcal{C}$ **do**
4 $\quad$ Set $p[c] = \dfrac{1}{k} \sum_{i=1}^{k} \mathbb{I}[y_{\iota_i} = c]$
5 **end**
6 **return** $\texttt{argmax}(p)$

Curse of Dimensionality

Nearest neighbors is a prime example of the curse of dimensionality. Since the algorithm heavily relies on the distance metric and the presumption that one can locate a collection of *nearest neighbors* that are indicative of a given test point's label, it therefore suffers drastically from the vastness encountered when one uses more than a handful of features. We saw how, in high dimensional spaces, the volume of the unit ball quickly decays to zero, and the distance to the edge of the unit cube grows without bound. Thus, when one uses too many dimensions, one quickly discovers that all neighbors become rather far away. This explosion of distance and vastness breaks the correlation between the given test point and its neighbors.

Feature Scaling

Given a problem with a low-dimensional feature space, we may still encounter issues if the continuous features span different scales from one another. For example, suppose we are using nearest neighbors to predict whether a house will sell within the first seven days it goes on market, based on the number of schools and the average listing price in the literal neighborhood. There is a full fivefold gap in the orders of magnitude between the number of schools $O(10^0)$ and the average listing price $O(10^5)$. This has the practical effect of negating the predictive power of the former variable.

When implementing nearest neighbor algorithms, it is therefore critical to center and scale the continuous predictors, so that they may be properly compared using the Euclidean metric.

Selecting k

In the KNN algorithm, the parameter k should be viewed as a tunable hyperparameter for the model. If k is too small, the model will have low bias and high variance. At the extreme end, if $k = 1$ and the training data is noisy, the resulting predictions will simply follow the noise in the training

---

Algorithm 11.3: knnRegressor($\mathcal{D}$, x_0, k) – k-nearest neighbors algorithm for regression.

Input: data $\mathcal{D}$;
　　　　　a feature vector x_0;
　　　　　the number k of nearest neighbors;
　　　　　the set $\mathcal{C}$ of target classes
Output: A predicted value $\hat{y}$.

1 Set $(\iota_1, \ldots, \iota_k) = \texttt{findNeighbors}(\mathcal{D}, x_0, k)$

2 **return** $\dfrac{1}{k} \sum_{i=1}^{k} y_{\iota_i}$

---

set. If k is too large, on the other hand, the model will have high bias and low variance. At the extreme end on this side of the spectrum, if $k = n$, the model will simply predict the overall average without regard to location in feature space.

To find the optimal value of k, we can therefore run the model over a range of possible values of k, and select the value that minimizes the generalization error (Definition 6.21). We can therefore settle upon an appropriate value of k, for a given problem, through the usual process of model tuning.

Large Datasets

Training a nearest neighbor model is tantamount to memorizing the training dataset. To apply a prediction, we have to iterate through the entire training set, calculating the distance between each training instance and the test point. Thus, the prediction step is at least $O(n)$. Moreover, the entire training set must be retained in memory. Thus, the algorithm becomes burdensome for large datasets.

To remedy this, it is natural to seek to replace the training dataset with a less memory-intensive representation of the same. For example, one might store the data in a multidimensional binary search tree, due to Bentley [1975], for which retrieval of the k nearest neighbors can be achieved in $O(\log n)$.

Distance Weighting (A Variation)

Finally, we mention a natural variation of the k-nearest neighbors algorithm, which helps reduce the frequency of ties. Given the indices $(\iota_1, \ldots, \iota_k)$ of the k-nearest neighbors, we may calculate the weights

$$w_i = \frac{1}{d(x_0, x_{\iota_i})},$$

for $i = 1, \ldots, k$. Using these inverse-distance weights, we may then modify Algorithms 11.2 and 11.3 by the weighted averages

$$p[c] = \frac{1}{w} \sum_{i=1}^{k} w_i \mathbb{I}[y_{\iota_i} = c] \qquad \text{and} \qquad \hat{y} = \frac{1}{w} \sum_{i=1}^{k} w_i y_i,$$

respectively, where $w = \sum_{i=1}^{k} w_i$. In particular, we could even let $k = n$, and still retain unique predictions for various regions of the feature space.

Python Implementation

A k-nearest neighbors class, that is valid for both classification and regression, is given in Code Block 11.1. This is an implementation of the most basic algorithm, and suffers from the deficits mentioned in the previous paragraph. See also the `KNeighborsClassifier` and `KNeighborsRegressor` classes from the `sklearn.neighbors` module.

```
1   class KNN(Model):
2
3       def __init__(self, **kwargs):
4           self.k = kwargs['k']
5           self.type = kwargs['type']
6           assert self.type in ('classification', 'regression')
7
8       def train(self, df: DataFrame, y: np.array, weights:
            np.array=None):
9           X = df.values if isinstance(df, DataFrame) else df
10          self.X = X
11          self.y = y
12
13      def predict(self, df: DataFrame):
14          X = df.values if isinstance(df, DataFrame) else df
15          y = np.zeros(len(X))
16          for i in range(len(X)):
17              d = [np.linalg.norm(self.X[j, :] - X[i,:]) for j in
                    range(len(self.y))]
18              indices = np.argpartition(d, self.k)[:self.k]
19              if self.type == 'classification':
20                  y[i] = Counter(self.y[indices]).most_common(1)[0][0]
21              else:
22                  y[i] = np.mean(self.y[indices])
23          return y
```

Code Block 11.1: k-nearest neighbors class.

Examples

Example 11.1. We can simulate a two-dimensional binary classification problem with decision boundary $x_1 = \sin(x_0)$, such that the probability a given instance belongs to the class 1 is given by

$$p(x_0, x_1) = \frac{1}{1 + e^{-\alpha(x_1 - \sin(x_0))}}.$$

A sample of 200 training instances generated in such a manner, with $\alpha = 2$, are shown in Figure 11.2. The code used to simulate this training set is given in Code Block 11.2.

Next, we build a meshgrid over the (x_0, x_1) plane, of increment size 0.1, and calculate the nearest neighbors predictions for each point on our grid. The code is given in Code Block 11.2. The nearest neighbor predictions are shown visually in Figures 11.3 and 11.4 for $k = 1, 3, 5, 10$, and 100.

Notice the extremes: the case $k = 1$ (Figure 11.3 (top)) overfits the noise in the training set, whereas the case $k = 100$ (Figure 11.4 (bottom)) has degenerated to a nearly linear diagonal decision boundary.   ▷

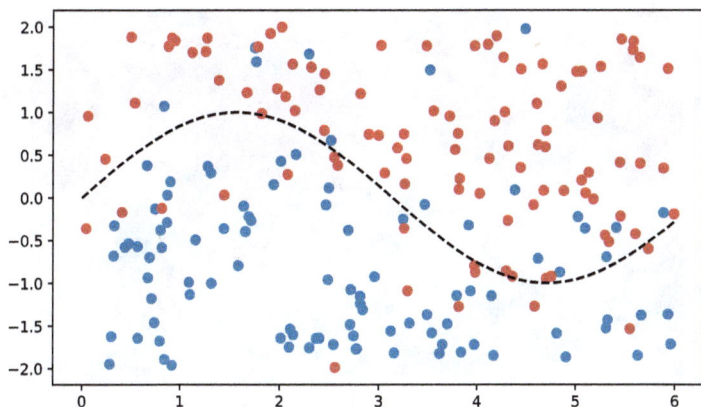

Fig. 11.2: A simulated classification problem with decision boundary $x_1 = \sin(x_0)$ and $n = 200$ random training instances.

```
1   n = 200
2   x = np.random.uniform(0, 6, size=n)
3   y = np.random.uniform(-2, 2, size=n)
4   X = np.array([x, y]).T
5   p = (1 + np.exp(-alpha*(y - np.sin(x) )))**(-1)
6   y = (np.random.random(size=n) < p).astype(int)
7
8   X_grid, Y_grid = np.meshgrid(np.linspace(0, 6, 61), np.linspace(-2,
        2, 41))
9   XX = np.array([X_grid.reshape((2501)), Y_grid.reshape((2501))]).T
10
11  K = KNN(k=5, type='classification')
12  K.train(X, y)
13  y_hat = K.predict(XX).reshape((41,61))
14
15  cm = colors.LinearSegmentedColormap.from_list('mylist', ['#1f77b4',
        '#d62728'], N=2)
16  plt.pcolormesh(X_grid, Y_grid, y_hat, cmap=cm)
17  plt.plot(t, np.sin(t), 'w')
18  plt.title(f"k={k}")
```

Code Block 11.2: Code to simulate training instances and calculate grid-wise nearest neighbor predictions.

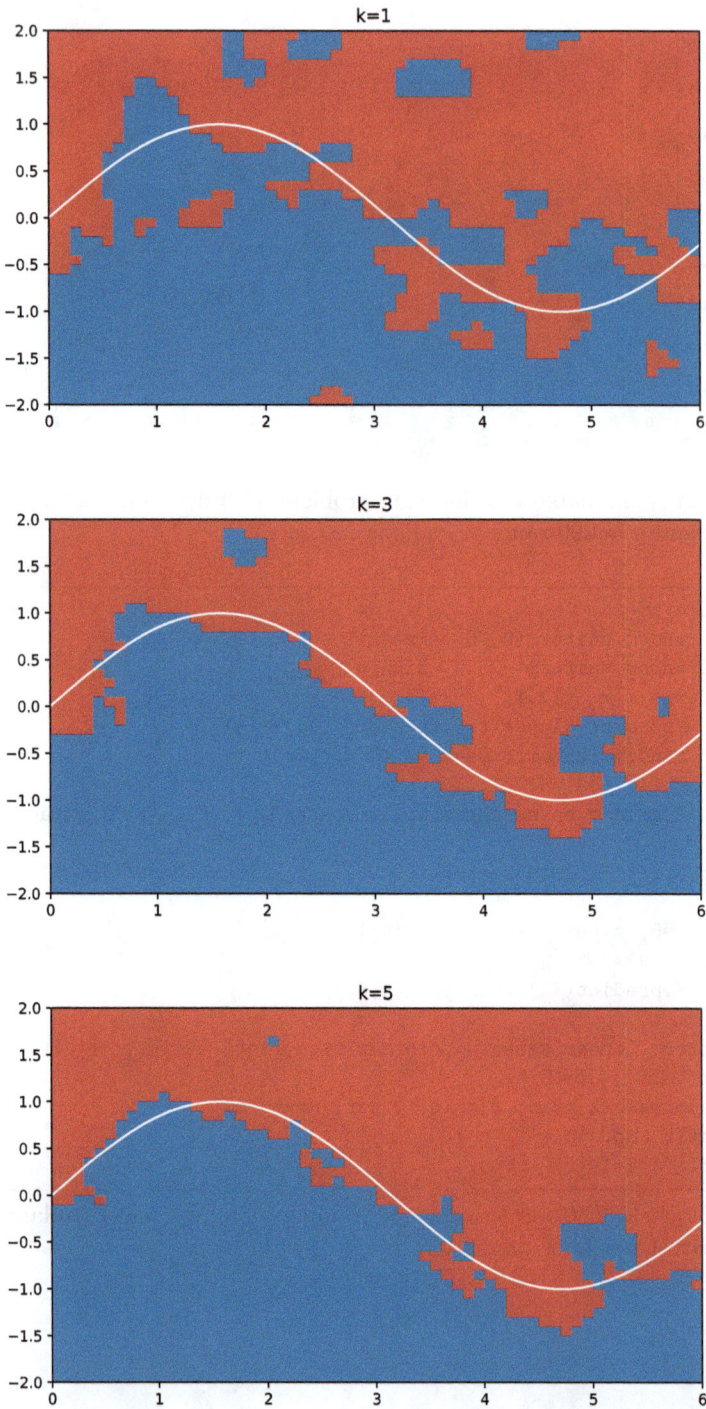

Fig. 11.3: Nearest Neighbors Predictions for $k = 1, 3, 5$.

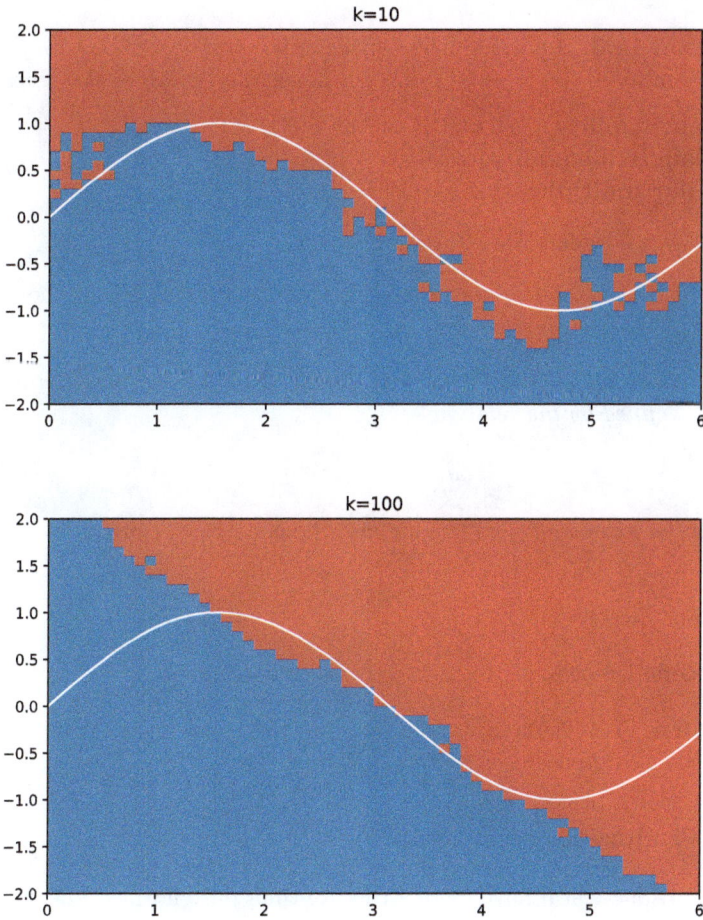

Fig. 11.4: Nearest Neighbors Predictions for $k = 10, 100$.

11.1.4 Inner Product Spaces

Note that metric spaces are more general than Euclidean space, and, indeed, need not even represent a vector space. It is therefore useful to briefly recall the following definitions. For additional details, see DiBenedetto [2002] or Folland [1999].

Normed Linear Space

Definition 11.3. *A normed linear space is a vector space* $\mathbb{X}$ *over a field* $\mathcal{F}^3$ *with a function* $|| \cdot || : \mathbb{X} \to \mathbb{R}^*$, *called a* norm, *that satisfies the properties*

1. **(identifiability)** $||x|| = 0$ *if and only if* $x = 0$,
2. **(triangle inequality)** $||x + y|| \leq ||x|| + ||y||$,
3. **(scalar multiplication)** $||\lambda x|| = |\lambda| \, ||x||$,

for all $x, y \in \mathbb{X}$ *and* $\lambda \in \mathcal{F}$.

As it turns out, every normed linear space is a metric space, though the contrapositive need not be true.

Proposition 11.1. *Given a normed linear space* $\mathbb{X}$, *the function* $d : \mathbb{X} \times \mathbb{X} \to \mathbb{R}^*$ *defined by the relation*

$$d(x, y) = ||x - y|| \tag{11.6}$$

satisfies the axioms of a metric, so that the pair $(\mathbb{X}, d)$ *constitutes a metric space.*

Proof. See Exercise 11.4. □

For completeness, we further state the following.

Definition 11.4. *A Banach space is a normed linear space that is complete with respect to the induced metric Equation* (11.6).

Inner Product Spaces

Euclidean spaces still have more structure than pure normed linear spaces.

Definition 11.5. *An* inner product space *is a vector space* $\mathbb{X}$ *over* $\mathbb{C}$ *(or* $\mathbb{R}$) *with an* inner product $\langle \cdot, \cdot \rangle : \mathbb{X} \times \mathbb{X} \to \mathbb{C}$ *that satisfies the axioms*

1. **(symmetry)** $\langle x, y \rangle = \overline{\langle y, x \rangle}$,
2. **(linearity)** $\langle \lambda x + y, z \rangle = \lambda \langle x, z \rangle + \langle y, z \rangle$,
3. **(positive definite)** $\langle x, x \rangle \geq 0$,
4. **(identifiability)** $\langle x, x \rangle = 0$ *if and only if* $x = 0$,
5. **(triangle inequality)** $d(x, y) + d(y, z) \geq d(x, z)$,

for all $x, y, z \in \mathbb{X}$ *and* $\lambda \in \mathbb{C}$.

As it turns out, every inner product space is also a normed linear space, and therefore a metric space.

---

[3] typically, the field is $\mathbb{R}$ or $\mathbb{C}$, in which event, we call $\mathbb{X}$ a *real* or *complex vector space*, respectively.

Proposition 11.2. *Given an inner product space* $\mathbb{X}$, *the function* $||\cdot||$: $\mathbb{X} \to \mathbb{R}^*$ *defined by the relation*

$$||x||^2 = \langle x, x \rangle \tag{11.7}$$

satisfies the axioms of a norm, so that the pair $(\mathbb{X}, ||\cdot||)$ *constitutes a normed linear space.*

Proof. See Exercise 11.5.   □

Definition 11.6. *A* Hilbert space *is an inner product space that is complete with respect to the metric induced via Equations* (11.7) *and* (11.6).

11.2 K-Means Clustering

In this section, we discuss our first unsupervised learning algorithms, which seek to uncover hidden structures in unlabelled data sets. In particular, we seek to identify structures known as *clusters*. For additional details, see standard texts, such as Hastie, *et al.* [2009] or James, *et al.* [2013]. For details of implementation in Python with sci-kit learn, see Gèron [2019].

11.2.1 Clustering

We begin by introducing notation that allows us to define and describe clusters and a cluster assignment, given a set of unlabelled data $\mathcal{X}$.

Definition 11.7. *Given a set of unlabelled data* $\mathcal{X}$ *of size* $n = |\mathcal{X}|$ *and a number* $c \in \mathbb{Z}^*$, *a* cluster assignment *is an many-to-one mapping (i.e., an encoding)* $\alpha : \iota(\mathcal{X}) \to \{1, \ldots, c\}$, *where* $\iota(\mathcal{X})$ *is the indexing set for* $\mathcal{X}$, *that assigns each datum* $x_i \in \mathcal{X}$ *to the unique cluster with label* $\alpha(i)$.

Given a cluster assignment α, *the* kth *cluster is the subset* $\mathcal{C}_k \subset \mathcal{X}$ *defined by*

$$\mathcal{C}_k = \{x_i : \alpha(i) = k\}, \tag{11.8}$$

for $k = 1, \ldots, c$.

The cluster indicator *is the function* $\delta : \iota(\mathcal{X}) \times \{1, \ldots, c\} \to \mathbb{B}$ *defined by*

$$\delta_{ik} = \mathbb{I}[\alpha(i) = k], \tag{11.9}$$

for $i \in \iota(\mathcal{X})$ *and* $k = 1, \ldots, c$.

The set $\mathcal{C} = \{\mathcal{C}_1, \ldots, \mathcal{C}_c\}$ therefore constitutes a *partition* of the set $\mathcal{X}$, since the clusters are pairwise disjoint, and their union covers the set $\mathcal{X}$. The goal of the *clustering problem* is to *learn* a cluster assignment that minimizes some property, typically the within-group variance.

11.2.2 Partitioning Theorem for Inner Product Spaces

Before discussing our first clustering method, we first consider a generalization of the partitioning theorem for inner product spaces. We begin by defining the following sums of squares for a normed linear space.

Definition 11.8. *Let* $(\mathbb{X}, d)$ *be a normed linear space and* $\mathcal{X} \subset \mathbb{X}$. *For any cluster assignment* α, *we define the* between-group sum of squares *(SSB)*, *the* within group sum of squares *(SSW), and the* total sum of squares *(SST) are defined by*

$$\text{SSB} = \sum_{k=1}^{c} r_k \left|\left| \bar{x}_k - \bar{\bar{x}} \right|\right|^2, \tag{11.10}$$

$$\text{SSW} = \sum_{i=1}^{n} \left|\left| x_i - \bar{x}_{\alpha(i)} \right|\right|^2, \tag{11.11}$$

$$\text{SST} = \sum_{i=1}^{n} \left|\left| x_i - \bar{\bar{x}} \right|\right|^2, \tag{11.12}$$

where we have defined

$$r_k = \sum_{i=1}^{n} \delta_{ik}, \qquad \bar{x}_k = \frac{1}{r_k} \sum_{i=1}^{n} x_i \delta_{ik}, \qquad \bar{\bar{x}} = \frac{1}{n} \sum_{i=1}^{n} x_i.$$

We require a normed linear space, and not simply a metric space, as the vector-space structure is required to define the means in relation to the vector sums. If our space also has an inner product, we have the following partitioning theorem, analogous to Theorem 3.11.

Theorem 11.1 (Partitioning Theorem for Inner Product Spaces).
Let $\mathbb{X}$ *be an inner-product space,* $\mathcal{X} \subset \mathbb{X}$, *and* α *a cluster assignment. Then the following identity holds*

$$\sum_{i=1}^{n} \left|\left| x_i - \bar{\bar{x}} \right|\right|^2 = \sum_{i=1}^{n} \left|\left| x_i - \bar{x}_{\alpha(i)} \right|\right|^2 + \sum_{k=1}^{c} r_k \left|\left| \bar{x}_k - \bar{\bar{x}} \right|\right|^2, \tag{11.13}$$

where $|| \cdot ||$ *is the norm induced by the inner product.*

Note that relative to the natural metric on $\mathbb{X}$, Equation (11.13) is equivalent to the class sum-of-squares decomposition

$$\text{SST} = \text{SSB} + \text{SSW}.$$

Proof. To proceed, let us reindex the set $\mathbb{X}$ using the notation x_{kj} to represent the jth elements of cluster $\mathcal{C}_k$, for $k = 1, \ldots, c$ and $j = 1, \ldots, r_k$. Next, we can express

$$\sum_{i=1}^{n} \left|\left| x_i - \bar{\bar{x}} \right|\right|^2 = \sum_{k=1}^{c} \sum_{j=1}^{r_k} \left|\left| x_{kj} - \bar{\bar{x}} \right|\right|^2$$

$$= \sum_{k=1}^{c} \sum_{j=1}^{r_k} \left|\left| (x_{kj} - \bar{x}_k) + (\bar{x}_k - \bar{\bar{x}}) \right|\right|^2$$

$$= \sum_{k=1}^{c} \sum_{j=1}^{r_k} \left\{ \left|\left| x_{kj} - \bar{x}_k \right|\right|^2 + 2 \left\langle x_{kj} - \bar{x}_k, \bar{x}_k - \bar{\bar{x}} \right\rangle + \left|\left| \bar{x}_k - \bar{\bar{x}} \right|\right|^2 \right\},$$

where the final step holds for real inner-product spaces. Now, the middle term sums to zero, since

$$\sum_{k=1}^{c} \sum_{j=1}^{r_k} \left\langle x_{kj} - \bar{x}_k, \bar{x}_k - \bar{\bar{x}} \right\rangle = \sum_{k=1}^{c} \left\langle \sum_{j=1}^{r_k} (x_{kj} - \bar{x}_k), \bar{x}_k - \bar{\bar{x}} \right\rangle$$

$$= \sum_{k=1}^{c} \left\langle r_k \bar{x}_k - r_k \bar{x}_k, \bar{x}_k - \bar{\bar{x}} \right\rangle = 0.$$

For complex inner-product spaces, the term $2 \left\langle x_{kj} - \bar{x}_k, \bar{x}_k - \bar{\bar{x}} \right\rangle$ is replaced with this inner product plus its complex conjugate; since the term vanishes, however, the result still holds. Identifying

$$\sum_{i=1}^{n} \left|\left| x_i - \bar{x}_{\alpha(i)} \right|\right|^2 = \sum_{k=1}^{c} \sum_{j=1}^{r_k} \left|\left| x_{kj} - \bar{x}_k \right|\right|^2$$

$$\sum_{k=1}^{c} r_k \left|\left| \bar{x}_k - \bar{\bar{x}} \right|\right|^2 = \sum_{k=1}^{c} \sum_{j=1}^{r_k} \left|\left| \bar{x}_k - \bar{\bar{x}} \right|\right|^2$$

completes our proof. □

11.2.3 K-Means Clustering Algorithm

The goal of the K-means clustering algorithm is to find a cluster assignment that minimizes the within-group sum of squares Equation (11.11). The idea is define the assignment α relative to a set of c vectors $m_1, \ldots, m_c \in \mathbb{X}$, which may be thought of as *centers of gravity* for our clusters. We then assign each datum x_i to the cluster k that has the closest center m_k. The full algorithm is shown in Algorithm 11.4.

While Algorithm 11.4 is guaranteed to converge, it is not guaranteed to converge to the global minimum. It is therefore customary to run the algorithm multiple times, and choose the output with the minimum value of ssw. The Python code for the K-means algorithm is shown in Code Block 11.3.

In order to test our method, we can simulated data using a Gaussian mixture model with random centers, as shown in Code Block 11.4. Once we have simulated data, we can run our K-means algorithm and plot the results, which are shown in Figure 11.5.

Algorithm 11.4: kMeans($\mathcal{X}$, c) – algorithm for finding c clusters over a normed linear space.

Input: data $\mathcal{X} = \{x_i\}_{i=1}^{n}$ belonging to an inner product space $\mathbb{X}$; the number c of clusters

Output: The set $\{m_1, \ldots, m_k\}$ of cluster means.

1 Let $(i_1, \ldots, i_n)$ be a random shuffle of $(1, \ldots, n)$
2 Set $m_k = x_{i_k}$ for $k = 1, \ldots, c$
3 Set $\Delta = 1$
4 **while** $\Delta > 0$ **do**
5 $\quad$ Calculate the assignment $\alpha(i) = \underset{k=1,\ldots,c}{\arg\min}\, d(x_i, m_k)$
6 $\quad$ Let $\delta_{ik} = \mathbb{I}[\alpha(i) = k]$ and $r_k = \displaystyle\sum_{i=1}^{n} \delta_{ik}$
7 $\quad$ **for** $k = 1, \ldots, c$ **do**
8 $\quad\quad$ Set $m_k' = \dfrac{1}{r_k} \displaystyle\sum_{i=1}^{n} x_i \delta_{ik}$
9 $\quad$ **end**
10 $\quad$ Set $\Delta = \displaystyle\sum_{k=1}^{c} (m_k - m_k')^2$
11 $\quad$ Set $m_k = m_k'$, for $k = 1, \ldots, c$
12 **end**
13 **return** $(m_1, \ldots, m_k)$

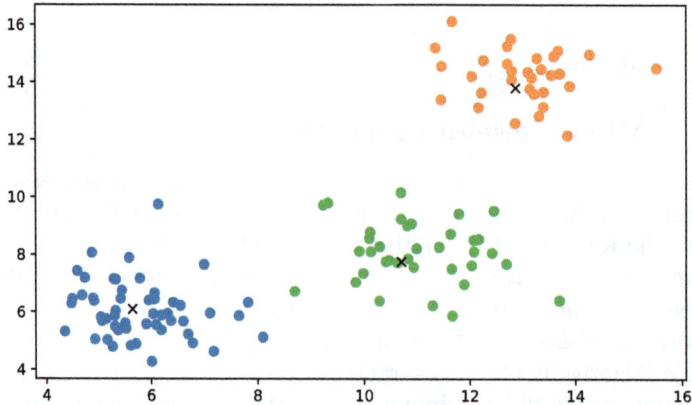

Fig. 11.5: Simulated clusters and final output of K-means algorithm.

```
1   class KMeans:
2
3       def __init__(self, c):
4           self.c = c
5
6       def getAssignment(self, X, kmeans):
7           n, d = X.shape
8           c, _ = kmeans.shape
9           distances = np.zeros((n, c))
10          for k in range(c):
11              m = kmeans[k, :]
12              distances[:, k] = np.linalg.norm(X - m, axis=1)
13          return np.argmin(distances, axis=1)
14
15      def train(self, X):
16          n, d = X.shape
17          indy = np.random.choice(np.arange(n), size=self.c,
                    replace=False)
18          kmeans = X[indy, :]
19          go = True
20          while go:
21              # Compute Current Assignment
22              assignment = self.getAssignment(X, kmeans)
23              rk = dict(Counter(assignment))
24              rk = [rk.get(k, 0) for k in range(self.c)]
25              # Compute New Means
26              kmeans_new = np.zeros(kmeans.shape)
27              for k in range(self.c):
28                  kmeans_new[k, :] = X[assignment==k, :].mean(axis=0)
29              go = (kmeans_new == kmeans).any()
30              kmeans = kmeans_new
31          self.kmeans, self.assignment = kmeans, self.getAssignment(X,
                kmeans)
32          return self.kmeans, self.assignment
```

Code Block 11.3: *K*-Means clustering algorithm.

11.2.4 Clusters as Features

Clustering can also be used to find structures in long-tail categorical features. A common case is that of *customer country*. There are 195 recognized countries in the world, though data are not distributed evenly across countries. It is common the case to have a collection of larger countries that constitutes the bulk of the user base, and a *long tail* of smaller countries, for which there is insufficient data to draw statistically significant conclusions. We therefore seek to uncover similarities in the long-tail countries, in

```
1  c = 3
2  locs = np.random.randint(5, 20, size=(c, 2))
3  rs = np.random.randint(20, 60, size=c)
4  data = []
5  for i in range(c):
6      data.append(np.random.normal(loc=locs[i,:], size=(rs[i], 2)))
7  X = np.concatenate(data)
8  K = KMeans(c)
9  kmeans, alpha = K.train(X)
10 for k in range(c):
11     plt.plot(X[alpha==k,0], X[alpha==k,1], 'o')
12     plt.plot(kmeans[k,0], kmeans[k,1], 'kx')
```

Code Block 11.4: Simulation of clusters and K-means solution

the hopes that groups of countries might behave similarly, so that we can reduce the effective dimensionality of our learning problem.

One approach is to seek to uncover similarities based on actual data for a given application. For example, one could use observed average marketing indicators (cost-per-click, click-through-rate, etc.) to form clusters.

Another approach is to use enriched features, that can be gathered from third party sources. A fantastic trove of information is the World Bank. One can write calls to the World Bank's REST API; however there is an open-source Python package, wbdata (`wbdata.readthedocs.io`), that one can use more readily. We pull population and GDP per capita in Code Block 11.5, and cluster countries based on the log values. The data are plotted in Figure 11.6.

The resulting clusters are shown in Figure 11.6. We can visualize these results geographically with the geopandas package (`geopandas.org`), as shown in Figure 11.7. Note that the distribution over population and GDP per capita is a multivariate lognormal distribution, i.e., data are normally distributed on a log-log plot. Because of this, the resulting clusters are largely arbitrary as there are no natural "breaks" in the data. Due to the sensitivity of the algorithm on the initial (random) K-means, no two results will be the same. Nonetheless, each cluster represents a group of geos that have a similar size and income level.

The World Bank data is not limited to population and GDP; it houses hundreds of country-level indicators in a broad range of categories, like housing, education, debt, gender, poverty, wealth, economics, and so forth. Each of these indicators is also a time-series indicator, so interesting patterns can be uncovered by plotting indicators over time. When combined with internal application-specific data, one can uncover hidden structures and craft relevant segments for the particular business use case.

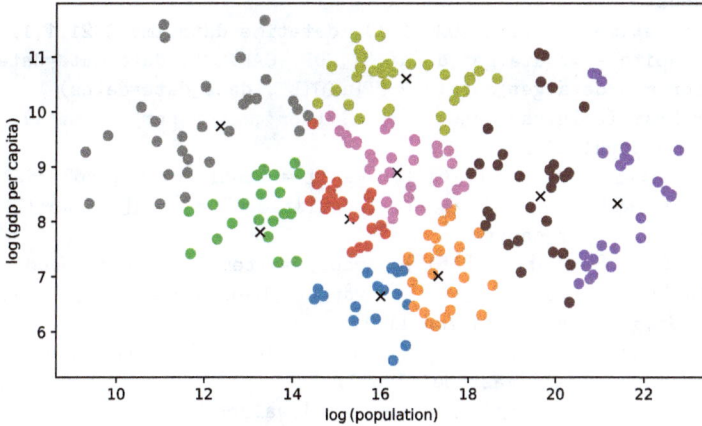

Fig. 11.6: Clustering countries by population and GDP per capita (log scale), using $c = 9$ clusters.

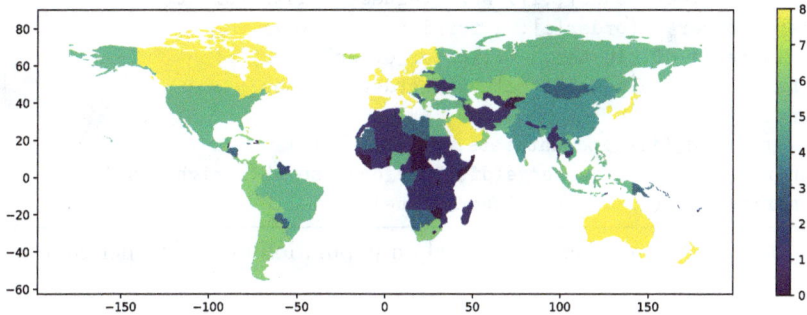

Fig. 11.7: Population-GDP clusters shown on a map.

11.2.5 Scatter

The K-means clustering algorithm relies on the underlying feature set to derive from an inner product space (and hence a vector space), so that the means may be calculated in line 8 and so we may compute the SSW of Equation (11.11). This strongly relies on the condition that we only include numerical features. For categorical features or, more generally, for measures of association based on general similarity rules, it is natural to ask if we can construct a comparable algorithm when our feature set is a metric space, but not a vector space. For such a generalization, we may no longer rely on averages, such as $\bar{x}_k$ or $\bar{\bar{x}}$. For these purposes, we introduce a generalized version of variance known as *scatter*.

```
1  import geopandas as gpd
2  import wbdata
3  dates = datetime.datetime(2020,1,1), datetime.datetime(2021,1,1)
4  gdp_per_capita = wbdata.get_data("NY.GDP.PCAP.CD", data_date=dates)
5  population = wbdata.get_data("SP.POP.TOTL", data_date=dates)
6  df = DataFrame(columns=['geo', 'population', 'gdp_per_capita'])
7  for item in population:
8      df.loc[item['country']['id'], 'geo'] = item['countryiso3code']
9      df.loc[item['country']['id'], 'population'] = item['value']
10 for item in gdp_per_capita:
11     df.loc[item['country']['id'], 'geo'] = item['countryiso3code']
12     df.loc[item['country']['id'], 'gdp_per_capita'] = item['value']
13 df = df[~df.gdp_per_capita.isnull()]
14 df[['population', 'gdp_per_capita']] = np.log(df[['population',
       'gdp_per_capita']].astype(float))
15 X = df[['population', 'gdp_per_capita']].values
16 c = 9
17 K = KMeans(c)
18 kmeans, alpha = K.train(X)
19 df.loc[:, 'cluster'] = alpha
20 order = Series(kmeans[:,1]).sort_values().index.tolist()
21 cluster_order = {order[i]: i for i in range(c)}
22 df.cluster = df.cluster.map(cluster_order)
23 alpha = Series(alpha).map(cluster_order).values
24 countries =
       gpd.read_file(gpd.datasets.get_path("naturalearth_lowres"))
25 countries = countries.merge(df, left_on='iso_a3', right_on='geo')
26 countries.plot("cluster", legend=True)
```

Code Block 11.5: Clustering based on population and GDP per capita.

Definition 11.9. *The* scatter *of a finite subset $\mathcal{X}$ of a metric space (X, d) is defined by the relation*

$$\text{SCAT}(\mathcal{X}) = \frac{1}{2n} \sum_{i=1}^{n} \sum_{j=1}^{n} d(x_i, x_j)^2, \qquad (11.14)$$

where $n = |\mathcal{X}|$.

The within-group scatter *of $\mathcal{X}$ with respect to a cluster assignment $\alpha : \iota(\mathcal{X}) \to (1, \ldots, c)$ is defined as*

$$\text{SCW} = \sum_{k=1}^{c} \text{SCAT}(\mathcal{C}_k), \qquad (11.15)$$

where $\mathcal{C}_k = \{x_i : \alpha(i) = k\}$ represents the kth cluster. Similarly, the between-group scatter *of $\mathcal{X}$ can be defined as*

$$\text{SCB} = \text{SCAT}(\mathcal{X}) - \sum_{k=1}^{c} \text{SCAT}(\mathcal{C}_k), \tag{11.16}$$

so that the total scatter *can be decomposed as* $\text{SCT} = \text{SCW} + \text{SCB}$.

Note that if we let $x_{ki} \in \mathcal{C}_k$ represent the ith instance of cluster k, we may express the scatter of cluster k as

$$\text{SCAT}(\mathcal{C}_k) = \frac{1}{2r_k} \sum_{i=1}^{r_k} \sum_{j=1}^{r_k} d(x_{ki}, x_{kj})^2.$$

Our next result shows that the within-group scatter is an appropriate generalization of the within-group sum of squares (SSW) for non-vector metric spaces.

Proposition 11.3. *Given an inner-product space, the* SSW *given by Equation (11.11) is equivalent to the* SCW *of Equation (11.15).*

Proof. First, note that Equation (11.11) is equivalent to

$$\text{SSW} = \sum_{i=1}^{n} \left|\left| x_i - \bar{x}_{\alpha(i)} \right|\right|^2 = \sum_{k=1}^{c} \sum_{j=1}^{r_k} \left|\left| x_{kj} - \bar{x}_k \right|\right|^2$$

in the notation of Definition 11.9. Next, if we fix any particular cluster k and use the fact that $d(x, y) = ||x - y||$, we may compute the scatter of $\mathcal{C}_k$ as

$$\begin{aligned}
\text{SCAT}(\mathcal{C}_k) &= \frac{1}{2r_k} \sum_{i=1}^{r_k} \sum_{j=1}^{r_k} ||x_{ki} - x_{kj}||^2 \\
&= \frac{1}{2r_k} \sum_{i=1}^{r_k} \sum_{j=1}^{r_k} ||(x_{ki} - \bar{x}_k) + (\bar{x}_k - x_{kj})||^2 \\
&= \sum_{i=1}^{r_k} ||x_{ki} - \bar{x}_k||^2 + \frac{1}{r_k} \sum_{i=1}^{r_k} \sum_{j=1}^{r_k} \langle x_{ki} - \bar{x}_k, x_{kj} - \bar{x}_k \rangle.
\end{aligned}$$

We leave it as an exercise for the reader to show that the inner product on the right-hand side vanishes. By summing over the clusters, we therefore obtain the relation $\text{SCW} = \text{SSW}$, completing our result. $\square$

11.2.6 K-Medoids Clustering Algorithm for Metric Spaces

In order to generalize the K-means clustering algorithm, which operates over normed linear spaces, to a clustering algorithm that functions on an arbitrary metric space, we must first replace the within-group sum of squares optimization objective with the within-group scatter and second find an alternative for the cluster means, which require vector addition to compute.

To resolve this issue, we replace the c means $m_1, \ldots, m_c$ that are calculated in K-means with values of the actual data points, as shown in Algorithm 11.5. Notice that line 8 is now $O(n^2)$, as in order to determine the optimal m for each cluster, we must iterate through the cluster twice: for each element in a given cluster we must compute the full sum of distances, which requires summing over all the elements of the cluster. Though this algorithm is more general, it comes at the cost of computational speed.

Problems

11.1. Show that the *French-rail metric*, defined by

$$d(x, y) = \begin{cases} ||x - y||_2 & \text{if } x = \alpha y, \text{ for some } \alpha \text{ in } \mathbb{R} \\ ||x||_2 + ||y||_2 & \text{otherwise} \end{cases},$$

where $|| \cdot ||_2$ is the ℓ_2 norm (Euclidean distance to the origin) defines a metric space on $\mathbb{R}^2$. Explain the practical meaning of this metric, assuming that Paris is located at $(0,0)$, and assuming that all rail lines in France go to Paris.

Algorithm 11.5: kMedoids$(\mathcal{X}, c)$ – algorithm for finding c clusters over a metric space.

Input: data $\mathcal{X} = \{x_i\}_{i=1}^n$ belonging to a metric space $\mathbb{X}$;
       the number c of clusters
Output: The set $\{m_1, \ldots, m_k\}$ of cluster medoids.

1 Let $(i_1, \ldots, i_n)$ be a random shuffle of $(1, \ldots, n)$
2 Set $m_k = x_{i_k}$ for $k = 1, \ldots, c$
3 Set $\Delta = 1$
4 **while** $\Delta > 0$ **do**
5   Calculate the assignment $\alpha(i) = \underset{k=1,\ldots,c}{\arg\min}\, d(x_i, m_k)$
6   Let $\delta_{ik} = \mathbb{I}[\alpha(i) = k]$ and $r_k = \sum_{i=1}^{n} \delta_{ik}$
7   **for** $k = 1, \ldots, c$ **do**
8   Set $m_k' = \underset{m \in C_k}{\arg\min} \sum_{i=1}^{r_k} d(m, x_{ki})$
9   **end**
10   Set $\Delta = \sum_{k=1}^{c} d(m_k, m_k')^2$
11   Set $m_k = m_k'$, for $k = 1, \ldots, c$
12 **end**
13 **return** $(m_1, \ldots, m_k)$

11.2. Explain why the Chebyshev distance $d_\infty(x, y) = \max_j |x_j - y_j|$ in $\mathbb{R}^2$ corresponds to the distance perceived by the king on a chessboard, who can move both diagonally and horizontally and vertically, but only one step at a time.

11.3. [Bubble Sort] *Bubble sort* is an $O(n^2)$ sorting algorithm that proceeds as follows: we begin by iterating through the list that we desire to sort. We compare each item in the list with the next item, swapping them if they are out of order. Once we've reached the end of the list, the largest item will be in the final position. We then iterate through the list again, except stopping at the penultimate position, so that the largest two items are ordered at the end of the list. We continue, each time stopping one place earlier than before.

(a) Implement the bubble-sort algorithm in Python.
(b) Test the algorithm on a random array: `np.random.randint(0, 100, size=100)`.
(c) Explain why the complexity is $O(n^2)$.
(d) Modify the algorithm so that, given an integer k, the k *smallest* values in the list will be sorted at the beginning of the list. Explain why the complexity is $O(kn)$.

11.4. Show that the function d defined in Proposition 11.1 constitutes a metric.

11.5. Show that the function $|| \cdot ||$ defined in Proposition 11.2 constitutes a norm.

12

The Forest for the Trees

In this chapter, we will consider various algorithms that are rooted in *decision trees*. We first introduce decision trees, discussing how they may be grown for the cases of classification and regression, and then extend that simple model into more complex *ensemble models*, such as *random forest* and *boosted trees*. Ensemble methods constitute a straightforward way to build power machine learning algorithms from a collection of so-called *weak learners*. We will explore these concepts over the pages that follow.

12.1 Decision Trees for Classification and Regression

12.1.1 Growing Decision Trees

Decision trees, alternatively known as *rule-based methods*, are an intuitive way to learn complex patterns from data. They are based on the concept of a tree graph (Definition 8.6), with the added benefit that each node has a *rule* attached to it, that instructs us on how to partition the dataset. In this way, the output of the model is piecewise constant over a complex family of hyperrectangles that divide the feature space. Decision trees are not prescribed, but learned; the structure is uncovered from the training data, not set in advanced.

Decision Trees

Be begin by defining the notation of a *decision rule*, or, more simply, a *rule*. This definition will then allow us to formally define a decision tree.

Definition 12.1. *A* decision rule *(or* rule *or* decision criterion*) is a mapping* $r : \mathbb{D} \to \mathbb{B}$*, from the set of possible data* $\mathbb{D}$ *(relative to a given problem) to the booleans* $\mathbb{B} = \{0, 1\}$*.*

Applied to a set of data $\mathcal{D}$*, we say that*

$$r(\mathcal{D}) = \{x : x \in \mathcal{D} \ and \ r(x)\};$$

i.e.; $r(\mathcal{D})$ is the subset of data for which the decision rule is **True**.

Similarly, we may apply a set of decision rules $\mathcal{R}$ to a set of data $\mathcal{D}$ in the natural way:

$$\mathcal{R}(\mathcal{D}) = \{r(\mathcal{D}) : r \in \mathcal{R}\}.$$

A set of decision rules $\mathcal{R}$ is said to be mutually exclusive and exhaustive *if*

$$\mathcal{D} = \bigcup_{r \in \mathcal{R}} r(\mathcal{D}),$$

for any possible set of data $\mathcal{D} \subset \mathbb{D}$, even when counting multiplicities. Alternatively, the set $\mathcal{R}$ is mutually exclusive and exhaustive if, for any possible datum $x \in \mathbb{D}$, $r(x)$ is **True** *for one and only one rule r in the set $\mathcal{R}$; i.e., if $\sum_{r \in \mathcal{R}} r(x) = 1$, for any x.*

It is important to note that a mutually exclusive and exhaustive rule set $\mathcal{R}$ always partitions the space of feature data. Oftentimes, this partition is binary and can be expressed in the form

$$\mathcal{R} = \{r, \neg r\},$$

for some rule r, where $\neg r$ represents the negation "*not r.*" For numeric features, the rule r is typically of the form $x \geq x_0$, where x_0 is a given value. For categorical features, the rule r is typically of the form, for example, $x = \text{red}$.

Definition 12.2. *A* decision tree *is a tree graph (Definition 8.6), such that each branch node stores a set of mutually exclusive and exhaustive decision rules that map one-to-one and onto its children, and such that each leaf node stores a value.*

A classification tree *is a decision tree for which its values are categorical. A* regression tree *is a decision tree with numerical values.*

When each branch node of a decision tree has precisely two children, we call it a binary decision tree. *Otherwise, we say that the tree allows for* multiway splits.

General Algorithm for Growing Trees

Now that we have defined *what* decision trees are, we next discuss *how* to construct them.

The general algorithm for growing decision trees is given in Algorithm 12.1.

This algorithm is a high-level procedure that can be applied to both classification and regression problems. It relies on the following three functions:

- *isHomogeneous*($\mathcal{D}$): returns **True** if the data $\mathcal{D}$ are similar enough to break recursion and constitute a leaf; otherwise returns **False**;
- *getLabel*($\mathcal{D}$): returns the best predictor (label) for the data set;
- *bestSplit*($\mathcal{D}, \mathcal{F}$): returns a set $\mathcal{R}$ of mutually exclusive and exhaustive decision rules, based on the input data and feature set.

Each of the three functions, *isHomogeneous*, *getLabel*, *bestSplit*, used to grow a general decision tree have specific implementations depending on whether the problem is a classification or a regression problem.

For classification problems, *isHomogeneous* typically returns **True** if all instances in $\mathcal{D}$ have the same label. However, as this easily leads to over-fitting, it is common to instead define *isHomogeneous* to return **True** only if the set of data is homogeneous *enough*; for example, if 90% of instances belong to the same class. For regression problems, *isHomogeneous* can be defined to return **True** if the RMSE is below some critical threshold.

For classification problems, *getLabel* typically returns the *mode*; i.e., the most common label in the input data. For regression problems, *getLabel* is typically defined to return the mean or median value of the target variable.

We will discuss the details of how the *bestSplit* function works in our next paragraph. Typically this method selects a single feature to perform the split on. If the feature is numeric, the split is typically chosen to be binary: $\mathcal{R} = \{x \geq x_0, x < x_0\}$, for some optimal value x_0. If the feature is categorical, and we are allowing for multiway splits, the split is typically all possible values of a single categorical variable; for example, $\mathcal{R} = \{x =$ red, $x =$ green, $x =$ blue$\}$, assuming the only possible colors are red, green, and blue (RGB color space). Multiway splits, however, suffer from a serious

---

Algorithm 12.1: growTree($\mathcal{D}, \mathcal{F}$) – algorithm for building decision trees.

Input: data $\mathcal{D}$;
         a set of features $\mathcal{F}$
Output: A decision tree T.

```
1  if isHomogeneous(D) then
2  |    return getLabel(D)
3  end
4  R = bestSplit(D, F)
5  map = Dict()
6  for r in R do
7  |    if r(D) = ∅ then
8  |    |    map[r] = getLabel(D)
9  |    else
10 |    |    map[r] = growTree(r(D), F)
11 |    end
12 end
13 return map
```

drawback: they tend to select features with higher cardinality first, thinning out the data set too soon. To remedy this, it is much more common to consider binary trees, which only allow for a binary partition. For binary trees, an example split would be $\mathcal{R} = \{x = \text{red}, x \neq \text{red}\}$.

Finally, we note that Algorithm 12.1 uses recursion[1] to construct a decision tree, as it makes recursive calls to itself (line 10) until it achieves the homogeneity condition (line 1).

Finding the Best Split

Next, we explore in depth the *bestSplit* function used to grow our decision trees. Both classification and regression trees rely on a certain function known as an *impurity measure* that is used to judge how good any particular split is. Typically an impurity of zero implies that the data are completely homogeneous. Thus, our objective can be restated as an optimization problem: find the split that minimizes impurity. We will restrict our attention to the usual case of a univariate split, but which we mean a mutually exclusive and exhaustive rule set that partitions the data based on any *one* feature, though it is possible to define higher dimensional splits.

The algorithm for finding the best possible split is given in Algorithm 12.2. This algorithm is specifically written to locate the optimal *binary* split, without requiring that categorical variables were previously one-hot encoded (see Section 6.2). To instead locate the optimal multiway split, the for-loop on line 5 may be removed and line 6 may be replaced with the alternative

$$\mathcal{R}' = \{x[f] = l : l \in \texttt{getLevels}(f)\}.$$

However, this is typically ill advised.

On line 13, we solving a separate optimization problem, to find the value x_0 for the numerical feature f that minimizes the post-split impurity. The standard approach is to simply iterate over the observed values of the feature f and select the value which minimizes impurity. This out create a new for-loop for numerical features. There is a simplification for classification problems, in that only such that only values x_0 for which the class label *switches* need be included, as it can be shown that the value x_0 that minimizes impurity cannot occur if the label is the same on its immediate left and right. We leave line 14 as an abstract minimization, as one-dimensional line search algorithms might be more suitable than a brute force approach, especially when training models with large sets of data. One such method, golden-section search, is given in Algorithm 12.3; though for more details and other line search methods, see Chong and Zak [2008].

---

[1] see **recursion** in the index.

Impurity

There are several possible impurity measures for classification that are commonly used; though, in practice, they are quite similar to each other.

Definition 12.3 (Impurity Measures for Classification). *For a classification problem with $c \geq 2$ classes and a data set $\mathcal{D}$, let*

$$\hat{p}_k = \frac{1}{n} \sum_{i=1}^{n} \mathbb{I}[y_i = \mathcal{C}_k], \tag{12.1}$$

represent the fraction of instances in $\mathcal{D}$ belonging to the kth class $\mathcal{C}_k$. We may then defining the following impurity measures:

1. *Misclassification Error—the observed error rate when we use the majority class as the predictor:*

$$\text{Imp}(\mathcal{D}) = 1 - \max(\hat{p}_1, \dots, \hat{p}_c); \tag{12.2}$$

Algorithm 12.2: bestSplit$(\mathcal{D}, \mathcal{F})$ – algorithm for determining best binary split.

Input: data $\mathcal{D}$;
           a set of features $\mathcal{F}$
Output: A mutually exclusive and exhaustive set of decision rules based
           on a single feature $\mathcal{R}$.

```
 1  I = getImpurity(D)
 2  R = ∅
 3  for f in F do
 4      if type(f) = categorical then
 5          for l in getLevels(f) do
 6              R' = {x[f] = l, x[f] ≠ l}
 7              I' = getImpurity(R'(D))
 8              if I' < I then
 9                  I, R = I', R'
10              end
11          end
12      else
13          x₀ = minimize(getImpurity(D[x[f] >= x₀], D[x[f] < x₀]))
14          R' = {x[f] ≥ x₀, x[f] < x₀}
15          I' = getImpurity(R'(D))
16          if I' < I then
17              I, R = I', R'
18          end
19      end
20  end
21  return R
```

2. Gini Index—*a measure of variance:*

$$\text{Imp}(\mathcal{D}) = \sum_{k=1}^{c} \hat{p}_k(1 - \hat{p}_k); \tag{12.3}$$

3. entropy—*the expected information (in bits) in learning the label of a randomly selected example:*

$$\text{Imp}(\mathcal{D}) = -\sum_{k=1}^{c} \hat{p}_k \log_2 \hat{p}_k. \tag{12.4}$$

Often, we are interested in a two-class problem, we can simplify matters by using the probability of the positive label

$$\hat{p} = \hat{p}_1 = \frac{1}{n} \sum_{i=1}^{n} \mathbb{I}[y_i = 1], \tag{12.5}$$

since the observed probability of the negative labels is given by $\hat{p}_0 = 1 - \hat{p}_1$, so that Equations (12.2)–(12.4) reduce to the expressions

$$\text{Imp}(\mathcal{D}) = \min(\hat{p}, 1 - \hat{p}), \tag{12.6}$$
$$\text{Imp}(\mathcal{D}) = 2\hat{p}(1 - \hat{p}), \tag{12.7}$$
$$\text{Imp}(\mathcal{D}) = -\hat{p}\log_2 \hat{p} - (1 - \hat{p})\log_2(1 - \hat{p}), \tag{12.8}$$

respectively. (For a two-class problem, $\hat{p}_2 = 1 - \hat{p}_1$, so that we may drop the subscripts altogether, due to symmetry.) These impurity measures for the binary classification problem are plotted in Figure 12.1.

For regression problems, it is common to use the mean regression squared error MSR as an impurity measure:

$$\text{Imp}(\mathcal{D}) = \frac{1}{n} \sum_{i=1}^{n} (y_i - \bar{y})^2, \tag{12.9}$$

where $n = |\mathcal{D}|$, y represents the target variable, and $\bar{y}$ represents the sample mean. Alternatively, the median may replace the mean in the above formula, or a general MSE can be used (Equation (6.2)), for any other predictor for the data set.

We further require an extension to the above definitions, which will allow us to define impurity for a set of data sets.

Definition 12.4. *Any impurity measure* $\text{Imp}(\mathcal{D})$ *may be applied to a set of datasets,* $\{\mathcal{D}_1, \ldots, \mathcal{D}_m\}$ *via the linear relationship*

$$\text{Imp}(\mathcal{D}_1, \ldots, \mathcal{D}_m) = \sum_{j=1}^{m} \frac{|\mathcal{D}_j|}{|\mathcal{D}|} \text{Imp}(\mathcal{D}_j), \tag{12.10}$$

where $\mathcal{D}$ *is the set obtained by combining each subset, such that* $|\mathcal{D}| = \sum_{j=1}^{m} |\mathcal{D}_j|$.

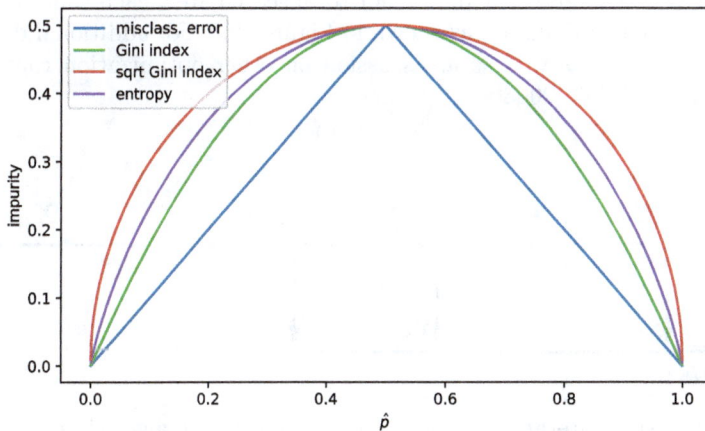

Fig. 12.1: Impurity measures for binary classification. The square-root of the Gini index and entropy are each scaled so they pass through the point $(1/2, 1/2)$.

Thus, the impurity of a collection of datasets is simply the weighted average of the individual impurities. This is the definition used in lines 7 and 13 of Algorithm 12.2, which relies on one of our individual impurity measures for either classification Equations (12.2)–(12.4) or regression Equation (12.9).

Golden-section Search

The golden-section search algorithm, due to Kiefer [1953], for locating a minimum value of a fucntion f on an interval $[a, b]$ is given in Algorithm 12.3. It is based on the golden ratio $\varphi = (1 + \sqrt{5})/2 \approx 1.61803....$, and the values $r = 2 - \varphi \approx 0.382$ and $s = \varphi - 1 \approx 0.618$, which add to unity. When we divide the unit interval into two subintervals of length r and s, the ratio $1 : s$ is equal to the ratio $s : r$, so that $r = s^2$. An improvement on this algorithm, known as Fibonacci search, can be made by varying the scaling between successive iterations, based on ratios of Fibonacci numbers. We will, however, focus on golden-section search for simplicity.

On lines 4 and 5, as well as 11 and 16, we are constructing two intermediary points $x_2, x_3 \in [x_1, x_4]$ such that

$$x_2 = x_1 + r(x_4 - x_1)$$
$$x_3 = x_4 - r(x_4 - x_1).$$

Thus, the three subintervals $[x_1, x_2]$, $[x_2, x_3]$, and $[x_3, x_4]$, relative to the overall interval, are in the ratio $r : rs : r$. It can easily be shown that $rs =$

$1 - 2r$ by using the property $\varphi^2 = \varphi + 1$. These special properties of golden ratios allow us to iteratively divide our total search area, each time getting closer to the true minimum, as shown in Figure 12.2. For additional details on the algorithm, as well as a discussion on other optimization routines, see, Chong and Zak [2008].

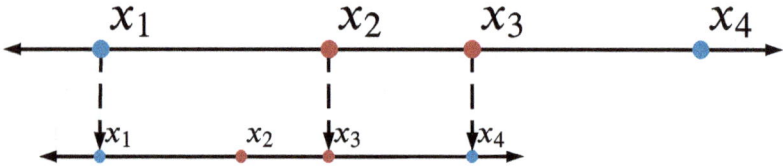

Fig. 12.2: A single iteration of the golden-section search algorithm, for $x_2 < x_3$. The initial definitions of x_1, x_2, x_3, and x_4 are on the top line, whereas the updated definitions are shown on the bottom line.

Algorithm 12.3: goldenSectionSearch – algorithm for minimizing univariate function on interval, using the golden ratio φ.

Input: a unimodal, single variable function f;
             endpoints a, b
Output: An approximate value $x \in [a, b]$ that minimizes f.
1   $\varphi = 1.61803...$
2   $r, s = 2 - \varphi, \varphi - 1$
3   $x_1, x_4 = a, b$
4   $x_2 = sx_1 + rx_4$
5   $x_3 = rx_1 + sx_4$
6   $f_1, f_2, f_3, f_4 = f(x_1), f(x_2), f(x_3), f(x_4)$
7   **while not** checkConverged$(x_1, x_2, x_3, x_4, f_1, f_2, f_3, f_4)$ **do**
8   **if** $\min(f_1, f_2) < \min(f_3, f_4)$ **then**
9          $x_4, f_4 = x_3, f_3$
10        $x_3, f_3 = x_2, f_2$
11        $x_2 = sx_1 + rx_4$
12        $f_2 = f(x_2)$
13   **else**
14        $x_1, f_1 = x_2, f_2$
15        $x_2, f_2 = x_3, f_3$
16        $x_3 = rx_1 + sx_4$
17        $f_3 = f(x_3)$
18   **end**
19 **end**
20 **return** the value in (x_1, x_2, x_3, x_4) corresponding to the minimum (f_1, f_2, f_3, f_4)

A few notes regarding the algorithm: the function *checkConverged* returns `True` if the values of f are sufficiently close. This can be customized as needed. Also, each successive iteration through the while loop requires only a single call to the function f. Furthermore, each successive iteration constructs an $r \approx 38.2\%$ smaller interval $[x_1, x_4]$ than the preceding one, while maintaining the $r : rs : r$ ratio of the three subdivisions. Each iteration either "zooms left" or "zooms right" depending on whether f_1 or f_2 is smaller than f_3 and f_4. If the minimum value is at x_0 or x_1, the new interval is $[x_1, x_3]$; otherwise it is $[x_2, x_4]$. Figure 12.2 depicts an example of "zooming left," corresponding to lines 9–12 of Algorithm 12.3.

When applying the algorithm to find the optimal split point for a continuous variable, we can use the minimum and maximum value of our continuous feature variable as u and b, respectively. However, if we fear our data might be susceptible to outliers, erroneous or suspicious values that live outside the fold, we can generally use, e.g., the 10th and 90th percentiles for our initial cutoff. If the optimal point is at one of our initial endpoints, we can then go back and rerun the algorithm for the outer region we initially excluded.

Finally, we note that the nth iteration of the algorithm results in a line segment that is

$$(\varphi - 1)^n \approx 0.618^n$$

the size of the original interval. We can therefore reduce the overall range by 90% with only 5 iterations of the algorithm, or by 99% with only 10 iterations. This allows us to find our optimal split point for a continuous variable by making only around a dozen calls to the impurity function, which is typically a variance calculation. This is vastly preferred over the brute-force approach, which requires performing the calculation for each possible split, especially for training sets in the millions or beyond.

A Python implementation of the golden-section search algorithm is given in Code Block 12.1. It follows directly from Algorithm 12.3.

12.1.2 Programming Decision Trees

We conclude this section with a brief discussion of an implementation of our decision tree algorithms in Python. Before we begin to implement the decision tree algorithm, we first define a `Rule` that can store and easily implement a single decision criterion. This code is shown in Code Block 12.2. A `Rule` object is initialized by a triple consisting of a feature name, a symbol (e.g., $\geq$ or $=$), and a value. We use the `operator` package in order to define a lookup table (Python dictionary) for the admissible symbols to their operators. This is a general implementation of the concept of a criterion rule that applies to both categorical features (using the symbols '==' and '!=') and numeric features (using the symbols '>=' and '<'). We also defined the `__str__` method, so that we can `print(rule)` to receive an easily readable representation of the rule.

```python
def goldenSearch(f, a, b, max_iter=10, err_tol=1e-2):
    phi = (1 + 5**0.5) / 2 # golden ratio
    r, s = 2 - phi, phi - 1
    x1, x4 = a, b
    x2 = s*x1 + r*x4
    x3 = r*x1 + s*x4
    f1, f2, f3, f4 = f(x1), f(x2), f(x3), f(x4)
    converged = False
    i = 0
    while not converged:
        i += 1
        if min(f1, f2) < min(f3, f4):
            x4, f4 = x3, f3
            x3, f3 = x2, f2
            x2 = s*x1 + r*x4
            f2 = f(x2)
        else:
            x1, f1 = x2, f2
            x2, f2 = x3, f3
            x3 = r*x1 + s*x4
            f3 = f(x3)
        fmin, fmax = min(f1, f2, f3, f4), max(f1, f2, f3, f4)
        rel_err = (fmax - fmin) / fmin if fmin > 0 else 0
        converged = (i >= max_iter) or (rel_err < err_tol)
        if i >= max_iter:
            print("MAXIMUM ITERATIONS EXCEEDED")
    x = [x1, x2, x3, x4]
    f = [f1, f2, f3, f4]
    return x[np.argmin(f)]
```

Code Block 12.1: Golden-section search algorithm.

In order to apply the rule, we provide the method getMask, which returns a mask over the index of our dataframe. A *mask* is an array of booleans (True and False) that can be used to easily access slices of a given dataframe or array. We will make extensive use of the getMask functionality in our implementation of a decision tree.

We now have everything we need to construct a DecisionTree class in Python, which is shown in Code Blocks 12.3–12.6. Our DecisionTree class is subclassed from the Model abstract class, which we defined in Code Block 6.22.

The __init__ method, shown in Code Block 12.3, handles the processing for all the model parameters. First, we require that the keyword argument type is defined whenever the class is implemented, taking on a value in classification or regression, so that the algorithm knows whether the target variable is categorical or numerical, respectively.

```
 1   import operator
 2   class Rule:
 3
 4       def __init__(self, feature, symbol, value=None, cat=False):
 5           self.ops = {'>': operator.gt,
 6                       '<': operator.lt,
 7                       '>=': operator.ge,
 8                       '<=': operator.le,
 9                       '==': operator.eq,
10                       '!=': operator.ne}
11           assert symbol in self.ops
12           self.feature = feature
13           self.symbol = symbol
14           self.value = value
15           self.name = self.feature + (':' + str(self.value)) * cat
16
17       def __str__(self):
18           return f"{self.feature} {self.symbol} {self.value}"
19
20       def getMask(self, df):
21           if self.value is None:
22               return np.ones(len(df)).astype(bool)
23           return self.ops[self.symbol](df[self.feature], self.value)
24
25       def getFeatureName(self, with_value=False):
26           return self.feature + (':' + str(self.value)) * with_value
```

Code Block 12.2: Implementation of a `Rule` class that stores and executes a decision criterion.

In addition, we have a few other goodies: `impurity`, which defines the impurity metric; `max_depth`, which defines a possible stopping criterion; `depth`, which is used in the `train` method to keep track of the current depth of the tree; `min_samples_split`, another stopping criterion; `min_impurity`, for classification, the minimum percentage (e.g., 0.92) to determine that a dataset is homogeneous. We also have two important keyword arguments that must be passed into the init method: `cat_features` and `num_features`. These specify the names of the categorical and numeric features to be used in model training, and should match the column names in the Pandas dataframe passed in to the `train` and `predict` methods.

We also define the `__str__` method, which prints a representation of the full set of decision rules, once the tree has been trained.

Moving on, we find that Code Block 12.4 defines the main impurity measures, entropy, Gini index, and MSE, and a `_getImpurity` method, which

```python
class DecisionTree(Model):

    def __init__(self, **kwargs):
        type_error = "Keyword argument 'type' must be set to
            classification or regression"
        assert 'type' in kwargs, type_error
        assert kwargs['type'] in ['classification', 'regression'],
            type_error
        self.type = kwargs['type']

        if self.type == 'regression':
            _allowed = ['mse']
            self.impurity = kwargs.get('impurity', 'mse')
            assert self.impurity in _allowed, f"Regression impurity
                must be in {_allowed}"
        else:
            _allowed = ['gini', 'entropy']
            self.impurity = kwargs.get('impurity', 'gini')
            assert self.impurity in _allowed, f"Classification
                impurity must be in {_allowed}"

        self.max_depth = kwargs.get('max_depth', 5)
        self.depth = kwargs.get('depth', 0) + int('depth' in kwargs)
        self.min_samples_split = kwargs.get('min_samples_split', 0)
        self.min_purity = kwargs.get('min_purity', 1)
        assert self.min_purity <= 1, f"min_purity {self.min_purity}
            must be <= 1"
        self.subspace_dim = kwargs.get('subspace_dim', 0)
        self.cat_features = kwargs.get('cat_features', [])
        self.num_features = kwargs.get('num_features', [])
        self.is_leaf = False
        self.value = None

    def __str__(self):
        preamble = '|  ' * self.depth
        str_out = ''
        if self.is_leaf:
            str_out += preamble + f"|--- value: {self.value}\n"
        else:
            for rule, tree in zip(self.rules, self.children):
                str_out += preamble + f"|--- {str(rule)}\n" +
                    str(tree)
        return str_out
    # continued ...
```

Code Block 12.3: DecisionTree implementation (part 1).

```
1    # ... continued
2    def _gini(self, y, weights=None):
3        classes = np.unique(y)
4        impurity = 0
5        for c in classes:
6            p = np.average(y==c, weights=weights)
7            impurity += p * (1 - p)
8        return impurity
9
10   def _entropy(self, y, weights=None):
11       classes = np.unique(y)
12       impurity = 0
13       for c in classes:
14           p = np.average(y==c, weights=weights)
15           impurity -= p * np.log2(p)
16       return impurity
17
18   def _mse(self, y, weights=None):
19       return np.average( (y - np.average(y, weights=weights,
             axis=0))**2, weights=weights, axis=0)
20
21   def _getImpurity(self, y, masks=None, weights=None):
22       imp_funcs = {'mse': self._mse,
23                    'gini': self._gini,
24                    'entropy': self._entropy}
25       imp_func = imp_funcs[self.impurity]
26       if not masks:
27           return imp_func(y, weights=weights)
28       impurities, impurity_weights = [], []
29       weights = np.ones(len(y)) if weights is None else weights
30       for mask in masks:
31           if sum(mask) == 0:
32               continue
33           impurities.append(imp_func(y[mask],
                 weights=weights[mask]))
34           impurity_weights.append(sum(weights[mask]))
35       return np.average(impurities, weights=impurity_weights)
36   # continued ...
```

Code Block 12.4: DecisionTree implementation (part 2).

takes as arguments any number of datasets, allowing for both binary and multiway splits.

Code Block 12.5 defines the methods to get the best value for a given dataset and whether or not a dataset is homogeneous. It also implements the _bestSplit algorithm. The _bestSplit algorithm leverages our Rule class (Code Block 12.2) as well as the goldenSearch algorithm (Code Block 12.1) for numeric features.

Note that all of the methods defined so far are preceded with a single underscore, signifying the intent that they are used internally, and not outside of the class definition.

Finally, our main train and predict methods are implemented in Code Block 12.6. Note the use of recursion in the train method: each decision tree begets two new decision trees (or more, if we were to redefine the _bestSplit to allow for multiway splits), stored as the children of the tree. The process stops when the homogeneity condition is reached, or when we reach a max_depth or the data thins below the size of min_samples_split.

Once a decision tree has been trained, we can call the predict method to get a vector of predictions, which are constructed using vector operations (via our masks) and the recursive structure of the full tree.

Finally, let's apply our decision tree algorithm to the class *iris data set*, available from sklearn. The code is shown in Code Block 12.7. The output of the print(T) command is shown in Code Block 12.8. The tree of maximum depth 3 is shown in Figure 12.3. With a maximum tree depth of 5, we achieve 100% accuracy on the training set. Though, with a tree depth of 2 or 3, we still achieve 96% or 97% accuracy.

12.1.3 Decision Trees for Probability Estimation

Finally, we note a connection between the Gini index

12.1.4 Tree Pruning

Decision trees are prone to overfit their training data, especially if they are allowed to grow to an arbitrary depth. This is because, given an endless supply of splits, a decision tree can always "memorize" the training set, but eventually taking into account all possible (realized) combinations of the input set. To remedy this situation, we can *prune* our initial decision tree, by removing branches that do not generalize well. Before proceeding, note that this method is typically only applied when one trains a single tree, and is not used for the ensemble methods we will discuss for the remainder of the chapter. This section may therefore be omitted by the anxious reader without loss in continuity.

The tree pruning algorithm is shown in Algorithm 12.4. Here, we define $T_N = \mathtt{subTree}(T, N)$ as the subtree of tree T rooted at node N. Then we define $T_N(\mathcal{D}_N)$ as the set of labels for $\mathcal{D}_N$ based on the leaves of subtree T_N.

```python
# ... continued
def _getValue(self, y, weights=None):
    if self.type == 'classification':
        prob = {k: v / len(y) for k, v in Counter(y).items()}
        return max(prob, key=prob.get), prob
    else:
        return np.average(y, weights=weights, axis=0), None

def _isHomogeneous(self, y, weights=None):
    if self.type == 'classification':
        mode, _ = self._getValue(y, weights=weights)
        return np.average(y==mode, weights=weights) >= \
            self.min_purity
    else:
        return False

def _bestSplit(self, df, y, cat_features_dict, num_features,
        weights=None):
    weights = weights if weights else np.ones(len(df))
    min_impurity = imp_0 = self._getImpurity(y)
    rules = []
    for feature, levels in cat_features_dict.items():
        for level in levels:
            _rules = [Rule(feature, sym, level, cat=True) for sym
                in ['==', '!=']]
            masks = [rule.getMask(df) for rule in _rules]
            impurity = self._getImpurity(y, masks, weights)
            if impurity < min_impurity:
                min_impurity = impurity
                rules = _rules
    for feature in num_features:
        a, b = df[feature].min(), df[feature].max()
        x = goldenSearch(lambda x: self._getImpurity(y,
            masks=[(df[feature] < x).tolist(), (df[feature] >=
            x).tolist()], weights=weights), a, b)
        _rules = [Rule(feature, '>=', x), Rule(feature, '<', x)]
        masks = [rule.getMask(df) for rule in _rules]
        impurity = self._getImpurity(y, masks, weights)
        if impurity < min_impurity:
            min_impurity = impurity
            rules = _rules
    return rules, imp_0 - min_impurity

def _getFeatureSpace(self):
    return self.cat_features_dict, self.num_features
# continued ...
```

Code Block 12.5: `DecisionTree` implementation (part 3).

```
1    # ... continued
2    def train(self, df, y, weights=None):
3        self.value, self.prob = self._getValue(y)
4        if not hasattr(self, 'cat_features_dict'):
5            self.cat_features_dict = {feature:
                 np.unique(df[feature]).tolist() for feature in
                 self.cat_features}
6            self.n_features = len(self.num_features) + sum([len(x)
                 for _, x in self.cat_features_dict.items()])
7        if not hasattr(self, 'vector_dim'):
8            self.vector_dim = y.shape[1] if y.ndim == 2 else None
9        if self._isHomogeneous(y) or (self.depth >= self.max_depth)
             or (len(y) < self.min_samples_split):
10           self.is_leaf = True
11           return
12       weights = np.ones(len(df)) if weights is None else weights
13       cat_features_dict, num_features = self._getFeatureSpace()
14       self.rules, self.impurity_reduction = self._bestSplit(df, y,
             cat_features_dict, num_features, weights=weights)
15       self.children = [DecisionTree(**self.__dict__) for rule in
             self.rules]
16       for rule, child in zip(self.rules, self.children):
17           mask = rule.getMask(df)
18           child.train(df[mask], y[mask], weights=weights[mask])
19
20   def predict(self, df, get_prob=False):
21       if self.is_leaf:
22           return self.prob if get_prob else self.value
23       if self.vector_dim:
24           y_hat = np.zeros((len(df), self.vector_dim))
25       elif isinstance(self.value, (int, float)):
26           y_hat = np.zeros(len(df))
27       else:
28           y_hat = np.array(['' for i in range(len(df))],
                 dtype='object')
29       for rule, child in zip(self.rules, self.children):
30           mask = rule.getMask(df)
31           y_hat[mask] = child.predict(df[mask], get_prob=get_prob)
32       return y_hat
```

Code Block 12.6: `DecisionTree` implementation (part 4).

```
1   from sklearn.datasets import load_iris
2   data = load_iris()
3   print(data.DESCR)
4   X = data.data
5   y = data.target
6   df = DataFrame(X, columns = data.feature_names)
7   y = np.array(list(map(lambda x: data.target_names[x], y)))
8
9   T = DecisionTree(type='classification', num_features=['sepal length
        (cm)', 'sepal width (cm)', 'petal length (cm)', 'petal width
        (cm)'], max_depth=5, min_samples_split=0)
10  T.train(df, y)
11  print(T)
12  np.sum(T.predict(df) == y) / len(y) # 1.00
```

Code Block 12.7: Training our decision tree on the iris data set.

```
1   |--- petal length (cm) >= 2.9544512279404422
2   |   |--- petal width (cm) >= 1.791796067500631
3   |   |   |--- petal length (cm) >= 4.989356881873896
4   |   |   |   |--- value: virginica
5   |   |   |--- petal length (cm) < 4.989356881873896
6   |   |   |   |--- sepal width (cm) >= 3.0090169943749476
7   |   |   |   |   |--- value: versicolor
8   |   |   |   |--- sepal width (cm) < 3.0090169943749476
9   |   |   |   |   |--- value: virginica
10  |   |--- petal width (cm) < 1.791796067500631
11  |   |   |--- petal length (cm) >= 5.584359886491461
12  |   |   |   |--- value: virginica
13  |   |   |--- petal length (cm) < 5.584359886491461
14  |   |   |   |--- petal length (cm) >= 4.9829710109982335
15  |   |   |   |   |--- petal width (cm) >= 1.547213595499958
16  |   |   |   |   |   |--- value: versicolor
17  |   |   |   |   |--- petal width (cm) < 1.547213595499958
18  |   |   |   |   |   |--- value: virginica
19  |   |   |   |--- petal length (cm) < 4.9829710109982335
20  |   |   |   |   |--- petal width (cm) >= 1.7
21  |   |   |   |   |   |--- value: virginica
22  |   |   |   |   |--- petal width (cm) < 1.7
23  |   |   |   |   |   |--- value: versicolor
24  |--- petal length (cm) < 2.9544512279404422
25  |   |--- value: setosa
```

Code Block 12.8: Trained decision diagram for classifying the iris data set.

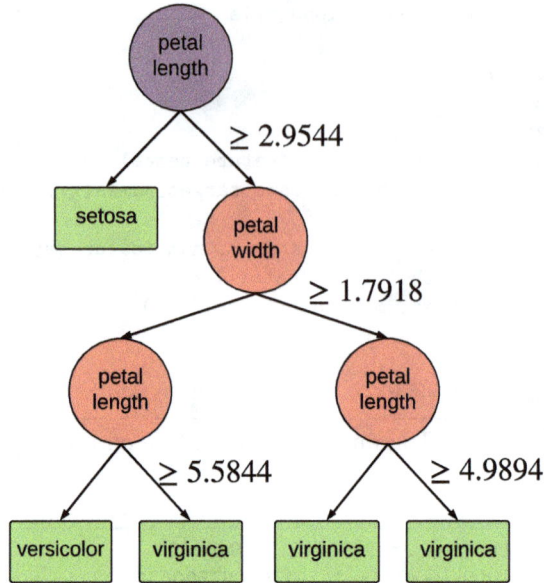

Fig. 12.3: A tree of depth 3 trained on the iris data set.

In this way, the act of pruning is tantamount to removing all subtrees that are detrimental to the overall performance.

12.2 Random Forests

In this section, we discuss a powerful extension to the decision tree algorithm. The main idea is that we will train many individual decision trees

---

Algorithm 12.4: pruneTree$(T, \mathcal{D})$ – algorithm to prune a decision tree.

Input: tree T;
       data $\mathcal{D}$
Output: A pruned decision tree T'.

1 **for** *each internal node N of T, starting at bottom* **do**
2      $T_N = \text{subTree}(T, N)$
3      $\mathcal{D}_N = \{x \in \mathcal{D} : x \text{ is covered by } N\}$
4   **if** $\text{getImpurity}(T_N(\mathcal{D}_N)) > \text{getImpurity}(\mathcal{D}_N)$ **then**
5   replace T_N in T with a leaf labelled by $\text{getLabel}(\mathcal{D}_N)$
6   **end**
7 **end**
8 **return** T

and pool their collective wisdom together under a single model. Such algorithms are known as *ensemble methods*, and they are very powerful.

12.2.1 Ensemble Methods

We begin with a brief introduction to *ensembles*, as defined below.

Definition 12.5. *An* ensemble model *(or* voting assembly*) is a finite collection of m pairs,*

$$\mathcal{E} = \{(M_j, w_j)\}_{j=1}^m, \tag{12.11}$$

such that each M_j is a model over the same feature and target space; such that the values w_j are weights, possessing the property that $w_j > 0$; and such that the ensemble model itself constitutes a function from the feature space to the target space that combines the individual predictions, for an arbitrary input, with their respective weights to return a unique output.

An ensemble model is normalized *if the weights are constrained by the normalization condition $\sum_{j=1}^m w_j = 1$; otherwise it is said to be* unnormalized.

For regression problems, we typically define the functionality of an ensemble model as the weighted average

$$\mathcal{E}(x) = \sum_{i=j}^m w_j M_j(x). \tag{12.12}$$

For classification problems, on the other hand, we can define the functionality of an ensemble method via the *voting scheme*

$$\mathcal{E}(x) = \arg\max_{c \in \mathcal{C}} \left\{ c \to \sum_{j=1}^m w_j \mathbb{I}[M_j(x) = c] \right\}, \tag{12.13}$$

where $\mathcal{C}$ is the set of classes for the target variable. In other words, we group the individual models based on their predicted outputs, and allow each model to vote with weight w_j. The output class with the most votes wins. In Equation (12.13), we use the brackets to represent an array, indexed by the set of target classes $\mathcal{C}$; the argmax function therefore yields the *key c* corresponding to the largest *value* $\sum_{j=1}^m w_j \mathbb{I}[M_j(x) = c]$.

Finally, we note that ensembles are *additive* in the following sense.

Definition 12.6. *We define the* direct sum *of two ensembles $\mathcal{E} = \{(M_i, w_i)\}_{i=1}^n$ and $\mathcal{E}' = \{(N_j, v_j)\}_{j=1}^m$ as the new ensemble*

$$\mathcal{E} \oplus \mathcal{E}' = \{(M_i, w_i)\}_{i=1}^n \cup \{(N_j, v_j)\}_{j=1}^m,$$

renormalized so that $\sum_{i=1}^n w_i + \sum_{j=1}^m v_j = 1$, where appropriate.

A basic ensemble abstract base class is constructed in Code Block 12.9, from which we can derive many ensemble children, by overriding the behavior of the _getModel and _trainNext methods. We will see several examples of this later in the chapter.

For the remainder of this chapter, we will explore various methods for constructing ensemble models, whose individual constituents are decision trees. Such methods are usually referred to as *forest methods*, as they represent a collection of trees.

12.2.2 Bagged Forests

The simplest way to generate a forest from some trees is to *bag* them. Here, *bagging* is a term that is short for *bootstrap aggregating*. The bagging algorithm is given in Algorithm 12.5, where it is expressed as an ensemble over a general learning algorithm $\mathscr{A}$, not limited to the decision tree algorithm. This basic algorithm is due to Breiman [1966]. Formally, we can define a bagged forest as follows.

Definition 12.7. *A* bagged forest *is a model trained using the bootstrap-aggregation Algorithm 12.5 that uses a decision tree (Algorithms 12.1 and 12.2) as the input algorithm $\mathscr{A}$.*

In Algorithm 12.5, we assume the learning algorithm $\mathscr{A}$ is a function from the data space $\mathbb{D}$, with a set of features from $\mathbb{F}$, to the space of predictive models $\mathbb{M}$; i.e., it is a function that takes a set of data as input and returns a predictive model as an output. We also assume we have a simple bootstrap method getBootstrap: $\mathbb{D} \times \mathbb{Z}_+ \to \mathbb{D}$, that returns a bootstrap sample (with replacement) from an original dataset ($\mathbb{D}$) of a given size ($\mathbb{Z}_+$). In the bagging algorithm, each bootstrap sample is of the same size as the original data set. On average, each bootstrap sample will cover $(1-1/e) \approx 63.21\%$ of the original data sample, as the size of the set $n \to \infty$.

Algorithm 12.5: bagging($\mathcal{D}, \mathcal{F}, \mathscr{A}, m$) – bootstrap aggregation algorithm. For *bagged forests*, use $\mathscr{A} = $ DecisionTree. For *random forests*, use $\mathscr{A} = $ RandomDecisionTree (see Code Block 12.12).

Input: data $\mathcal{D}$;
    feature set $\mathcal{F}$;
    learning algorithm $\mathscr{A} : \mathbb{D} \times \mathbb{F} \to \mathbb{M}$;
    the size m of the ensemble
Output: An ensemble model $\mathcal{E}$.
1 **for** j **from** 1 **to** m **do**
2 | $\mathcal{B} = $ getBootstrap($\mathcal{D}, |\mathcal{D}|$)
3 | $M_j = \mathscr{A}(\mathcal{B}, \mathcal{F})$
4 **end**
5 **return** $\mathcal{E} = \{M_j, 1\}_{j=1}^{m}$

```
1   from tqdm import tqdm
2   class Ensemble(Model):
3
4       def __init__(self, **kwargs):
5           assert kwargs['type'] in ['classification', 'regression']
6           self.type, self.size = kwargs['type'], kwargs.get('size', 10)
7           self.normalize = kwargs.get('normalize', True)
8           self.params = kwargs.copy()
9           kwargs.update(kwargs.get('model_params', {}))
10          self.model_params = kwargs
11          self.models, self.weights = [], []
12          self.Model = self._getModel()
13
14      @abstractmethod
15      def _getModel(self):
16          pass
17
18      def _trainNext(self, df, y):
19          model = self.Model(**self.model_params)
20          model.train(df, y)
21          return model, 1
22
23      def train(self, df, y):
24          if self.type == 'classification':
25              self.classes = np.sort(np.unique(y)).tolist()
26          if not hasattr(self, 'vector_dim'):
27              self.vector_dim = y.shape[1] if y.ndim == 2 else None
28          for j in tqdm(range(self.size)):
29              model, weight = self._trainNext(df, y)
30              self.models.append(model)
31              self.weights.append(weight)
32
33      def predict(self, df):
34          weights = [w / sum(self.weights) for w in self.weights] if
                  self.normalize else self.weights
35          if self.type == 'regression':
36              y_hat = np.zeros((len(df), self.vector_dim)) if
                      self.vector_dim else np.zeros(len(df))
37              for model, weight in zip(self.models, weights):
38                  y_hat += weight * model.predict(df)
39          else:
40              votes = DataFrame(0, index=np.arange(len(df)),
                      columns=self.classes) # n x k
41              for model, weight in zip(self.models, weights):
42                  y_hat = model.predict(df)
43                  for i, j in enumerate(y_hat):
44                      votes.loc[i, j] += weight
45              y_hat = votes.idxmax(axis=1).values.tolist()
46          return y_hat
```

Code Block 12.9: Ensemble abstract class implementation.

12.2.3 More Magic: Out-of-bag Error and Feature Importance

Two important pieces of information can be computed and stored while training bagged forests: the so-called out-of-bag error and a stack ranking of feature importances. The former serves as a viable substitute for cross validation, while the latter provides insight into which features are making the most impact on reducing model variance, and further provides a measure of their relative importance.

Out-of-bag (OOB) Error

One bonus advantage of bagging algorithms is that validation can be done for free while building the models that comprise the ensemble. This is achieved by using the so-called *out-of-bag samples*, or OOB *samples*, which we define as follows.

Definition 12.8. *Given a bootstrap sample* $\mathcal{B} \subset \mathcal{D}$ *of a data set* $\mathcal{D}$*, the out-of-bag* (OOB) samples *are the training instances that are not used in the bootstrap; i.e., the* OOB *samples are comprised of the instances in the set* $\mathcal{D} \setminus \mathcal{B}$*.*

Given this definition, we can next define the OOB *score*, as follows.

Definition 12.9. *Given a bagged ensemble model (Algorithm 12.5), the out-of-bag* (OOB) score *is the score obtained by applying any scoring metric (e.g.,* R^2 *or* MSE *for regression or accuracy for classification) to the predictions obtained by aggregating the out-of-bag predictions for each individual model in the ensemble.*

Note that each training instance will, in general, have a different number of OOB predictions, based on whether or not it happened to be used for each individual bootstrap. The OOB predictions are typically computed within the for-loop of Algorithm 12.5, and their aggregation is computed following the for-loop. We will discuss the Python implementation of the OOB score in during our discussion of Code Block 12.14.

Finally, we note that the OOB score for a bootstrap ensemble is equivalent, for large data sets, to the score obtained by performing a k-fold cross validation of the set. This follows as both methods are based on the same principle: apply a trained model to a randomly selected hold-out group.

Feature Importance

We are often interested in determining which features were the most vital in training a model. After all, decision trees *learn* which features are the most useful by testing out various possibilities for each split, they are not instructed by the programmer a priori which way to go. It is therefore often of interest to take a peek at which features helped the model the most, whether the model is an individual decision tree or an entire forest.

Recall that, when training a decision tree, the best split at node N is the binary decision rule $\mathcal{R}_N$ that maximizes the impurity reduction (or impurity loss)

$$\Delta\mathrm{Imp}_N = \mathrm{Imp}(\mathcal{D}_N) - \mathrm{Imp}(\mathcal{R}_N(\mathcal{D}_N))$$

across all possible features $\mathcal{F}$, where $\mathcal{D}_N$ is the data that survived to node N. We can agree that a feature that results in a greater impurity reduction should be considered more important than a feature that results in a lesser impurity reduction. Typically, important features occur closer to the root of the tree, or are features that are selected multiple times throughout the training process for the tree.

To determine the feature importance for a given tree, we can simply track the overall impurity reduction for each feature by using a map. We can then normalize the values of the given map, so that the largest value is set at unity (1.00). We can then output a ranked list of features, sorted by importance (descending), with the relative importance listed with each feature.

For bagged forests, we can simply aggregate the map of impurity reductions across all trees in the forest, prior to normalizing and returning the stack rank.

In order to calculate feature importances in Python, we can build on our `DecisionTree` class (Code Blocks 12.3–12.6) by adding a new method, `getFeatureImportance`. We will leverage the `name` attribute in the `Rule` class (Code Block 12.2), as it captures both the feature name and the level for categorical features. We will also leverage the `impurity_reduction` attribute defined in the `train` method of Code Block 12.6, which is returned by the `_bestSplit` method in Code Block 12.5. Previously, these attributes were unused.

The code for calculating feature importances is given in Code Block 12.10. We begin with the internal method `_getImpurityReduction`, which leverages the `collections` module's `Counter` class[2] to tabulate a map from feature names to impurity reductions as we iterate through each of the tree's internal nodes.

Finally, the method `getFeatureImportances`, which takes an optional argument `top`, returns the top `top` features in the form of a `panda`'s `DataFrame` with columns `feature` and `importance`, the latter scaled so that the top feature has importance of `1.00`. Optionally, the method returns the raw `Counter` object.

---

[2] The `Counter` class operates similarly to a dictionary, except it implements the `_add_` operator, in a way that adds corresponding values together when adding two `Counter` objects.

```
1   from collections import Counter
2   class DecisionTree:
3       # ... continued
4       def _getImpurityReductions(self):
5           if self.is_leaf:
6               return Counter()
7           reductions = Counter({'n_nodes': 1, self.rules[0].name:
                   self.impurity_reduction})
8           for model in self.children:
9               reductions += model._getImpurityReductions()
10          return reductions
11
12      def getFeatureImportance(self, top=0, as_counter=False):
13          if not hasattr(self, 'feature_counter'):
14              self.feature_counter = self._getImpurityReductions()
15              self.n_nodes = self.feature_counter.pop('n_nodes')
16              self.feature_importances =
                   DataFrame(self.feature_counter.most_common(),
                   columns=['feature', 'importance'])
17              self.feature_importances.importance /=
                   self.feature_importances.importance.max()
18          if as_counter:
19              return self.feature_counter
20          top = len(self.feature_importances) if top == 0 else top
21          return self.feature_importances[:top]
```

Code Block 12.10: Two additional methods for the `DecisionTree` class as continued from Code Blocks 12.3–12.6.

Oftentimes, however, we are more interested in the feature importances from an ensemble method, which aggregates the impurity reductions across all models comprising the ensemble. To achieve this, we can add a new method, `getFeatureImportance` to the `Ensemble` abstract base class. This method first ensures that the base model of the ensemble (stored as `self.Model`, which, recall, is a reference to the *class* defining the base model, not an individual object) has its own `getFeatureImportance` method defined. Next, we simply loop over the models from the ensemble, adding their feature-reduction counters in the natural way. We will see an example output in Code Block 12.16.

12.2.4 Random Forests

Bagged forests suffer from one serious drawback: at the end of the day, the trees that comprise the forest turn out to be highly correlated. To remedy the situation, we often employ one simple, yet crucial, modification

```
1   class Ensemble:
2       # ... continued
3       def getFeatureImportance(self, top=0):
4           assert hasattr(self.Model, 'getFeatureImportance'), "Base
                   model class must have getFeatureImportance method."
5           if not hasattr(self, 'feature_importances'):
6               self.feature_counter = Counter()
7               for model in self.models:
8                   self.feature_counter +=
                       model.getFeatureImportance(as_counter=True)
9               self.feature_importances =
                   DataFrame(self.feature_counter.most_common(),
                   columns=['feature', 'importance'])
10              self.feature_importances.importance /=
                   self.feature_importances.importance.max()
11          top = len(self.feature_importances) if top == 0 else top
12          return self.feature_importances[:top]
```

Code Block 12.11: Additional method for the `Ensemble` class as continued from Code Block 12.9.

to our bagging algorithm, that of *feature subspace sampling*, due to Ho [1995] and Ho [1998]. The idea is that, in addition to bootstrapping the data for each tree, we also also consider only a subset of features for each split when training each tree. This prevents any single feature from dominating throughout many trees in the forest. When we couple the idea of bagging together with subspace sampling, we call the resulting model a *random forest*. It is identical to the bagging algorithm, except for the fact that we also sample the feature space, for each split of each tree.

In order to formalize this concept, we first define a *random decision tree*, as follows.

Definition 12.10. *A* random decision tree *is trained like an ordinary decision tree (Algorithm 12.1), except that a random subset of features, $\mathcal{F}'$, of size $f < |\mathcal{F}|$, is passed into the* `bestSplit` *method on line 4. The parameter f is referred to as the* (feature) subspace dimension.

Given this definition of a random decision tree, we may now define a random forest as follows.

Definition 12.11. *A* random forest *is a model trained using the bootstrap-aggregation Algorithm 12.5 that uses a random decision tree (Definition 12.10) as the input algorithm $\mathscr{A}$.*

In general, a good rule of thumb (see Breiman [2001]) is to set the feature subspace dimension f as

1. $f = |\mathcal{F}|/3$, for regression problems; and
2. $f = \sqrt{|\mathcal{F}|}$, for classification problems.

However, these rules of thumb should be regarded as starting points, and the subspace dimension as a tunable model parameter.

Python Implementation: Random Decision Trees

Our definition of _getFeatureSpace in Code Block 12.5 may have at first seemed superfluous. Yet, it serves an important function now that we wish to modify our basic decision tree code (Code Blocks 12.3–12.6) to support random feature subspace selection. Leveraging the code already built for regular decision trees, we can construct a *random decision tree* class by defining a subclass and overwriting a single method. The result is shown in Code Block 12.12.

```python
class RandomDecisionTree(DecisionTree):

    def _getFeatureSpace(self):
        mask = np.random.permutation([True for i in
            range(self.subspace_dim)] + [False for i in
            range(self.n_features - self.subspace_dim)]).tolist()
        cat_features_dict, num_features = {}, []
        for feature, levels in self.cat_features_dict.items():
            n_levels = len(levels)
            x = [mask.pop() for i in range(n_levels)]
            if sum(x) == 0:
                continue
            cat_features_dict[feature] = [level for test, level in
                zip(x, levels) if test]
        for feature in self.num_features:
            if mask.pop():
                num_features.append(feature)
        return cat_features_dict, num_features
```

Code Block 12.12: RandomDecisionTree implementation.

Note that, since we built our decision tree algorithm without one-hot encoding, we need to take care in how we randomly select our features. Each level of each categorical feature is a feature in its own right, as a categorical feature variable with l levels would, ordinarily, be one-hot encoded into l independent features. We get around this problem by defining a dictionary of categorical features, cat_features_dict, that has feature name as key and list of levels as value.

We can train a random decision tree on the iris data set, using `max_depth=3` and `subspace_dim=2`. The output is shown in Code Block 12.13. We see that, for this run, the tree is making different split decisions than before, due to the randomization of the feature subspaces. Moreover, we see that the right branch (sepal length $>= 5.47$) then splits on a different feature (petal width) than the left branch, which splits on petal length.

```
1  |--- sepal length (cm) >= 5.47445651649735
2  |   |--- petal width (cm) >= 1.7914681984719456
3  |   |   |--- value: virginica
4  |   |--- petal width (cm) < 1.7914681984719456
5  |   |   |--- value: versicolor
6  |--- sepal length (cm) < 5.47445651649735
7  |   |--- petal length (cm) >= 2.9856581412139334
8  |   |   |--- value: versicolor
9  |   |--- petal length (cm) < 2.9856581412139334
10 |   |   |--- value: setosa
```

Code Block 12.13: Trained `RandomDecisionTree` for the iris data set; notice the splits are different than in Code Block 12.8.

Python Implementation: Bagged and Random Forests

We can subclass the `Ensemble` abstract base class from Code Block 12.9 to implement the bagged forest algorithm, as shown in Code Block 12.14. Most of this code is dedicated to calculating the out-of-bag score. To implement a bagged forest without the out-of-bag score, we only need to override the methods _getModel (in order to set the base model to our `DecisionTree` class (Code Block 12.3)) and _trainNext (in order to get a bootstrap sample for each model), and we only need lines 12–16 of the latter method.

Note that we invoke the built-in **super** function twice, on lines 6 and 36. In this way, when we override the __init__ and train methods, we do not have to reproduce *all* of the code; rather, we are construction a *wrapper* around those methods, which performs certain additional tasks before and after.

In __init__, we make sure that a validation metric is defined and stored as `self.oob_metric`. We use `accuracy_score` and `r2_score`, from the `sklearn.metrics` package, as default metrics for classification and regression, respectively.

In the **train** method, we initialize a structure (`self.oob_votes`) for storing the out-of-bag predictions, which we later compute within the _trainNext method.

```python
class BaggedForest(Ensemble):
    def __init__(self, **kwargs):
        default_metric = r2_score if kwargs['type'] == 'regression'
            else accuracy_score
        self.oob_metric = kwargs.get('oob_metric', default_metric)
        self.get_oob = kwargs.get('get_oob', False)
        self.n_jobs = kwargs.get('n_jobs', 1)
        super().__init__(**kwargs)

    def _getModel(self):
        return DecisionTree

    def _trainNext(self, df, y):
        idx = np.random.randint(0, len(df), size=len(df))
        df_boot = df.loc[idx, :].reset_index(drop=True)
        y_boot = y[idx]
        model = self.Model(**self.model_params)
        model.train(df_boot, y_boot)
        model.oob_idx = [i for i in range(len(df)) if i not in idx]
        return model, 1

    def train(self, df, y):
        super().train(df, y)
        if not self.get_oob:
            return
        if self.type == 'regression':
            oob_votes = np.zeros(y.shape)
            oob_n_votes = np.zeros(len(df))
            for model in self.models:
                y_hat = model.predict(df.loc[model.oob_idx, :])
                oob_votes[model.oob_idx] += y_hat
                oob_n_votes[model.oob_idx] += 1
            n_votes = np.fmax(1, oob_n_votes)
            if self.vector_dim:
                n_votes = n_votes.reshape((len(y), 1))
            oob_votes /= n_votes
            has_votes = oob_n_votes > 0
            oob_votes[~has_votes] = np.average(oob_votes[has_votes],
                weights=oob_n_votes[has_votes], axis=0)
        else:
            oob_votes = DataFrame(0, index=np.arange(len(df)),
                columns=self.classes)
            for model in self.models:
                y_hat = model.predict(df.loc[model.oob_idx, :])
                for i, j in zip(model.oob_idx, y_hat):
                    oob_votes.loc[i, j] += 1
            oob_votes = oob_votes.idxmax(axis=1).values.tolist()
        self.oob_score = self.oob_metric(y, oob_votes)
```

Code Block 12.14: BaggedForest implementation; with *out-of-bag score*.

In the _trainNext method, we first perform a bootstrap sample, using numpy.random.randint, train a base model using the bootstrap sample, and then compute the predictions for the out-of-bag samples, indexed by oob_idx.

Finally, in the train method, after invoking the parent class's (super's) train method, we aggregate our out-of-bag predictions and compute our out-of-bag score.

We can further subclass the BaggedForest into the RandomForest class of Code Block 12.15, by updating the model with our RandomDecisionTree class (Code Block 12.12).

```
1  class RandomForest(BaggedForest):
2
3      def _getModel(self):
4          return RandomDecisionTree
```

Code Block 12.15: RandomForest implementation.

We can train a simple random forest model using 100 trees, as shown in Code Block 12.16. We see that the oob_score is 93.33%, meaning that our model's prediction accuracy is expected to be around 93%. In addition, we can invoke the getFeatureImportance method to retrieve a dictionary of the feature importances. We see that petal length and petal width are the top two features, with the former being nearly twice as important as the latter. Sepal length and sepal width are a distance third and fourth, with sepal width making very little contribution to the overall model.

Finally, we illustrate a smaller random forest with twelve trees in Figure 12.4. The oob_score with only twelve trees is 91.33%. We can actually calculate the predicted labels based on this diagram alone. For a given instance, we first calculate the prediction of each of the twelve trees individually. We then tabulate the results and choose the class with the most votes.

Figure 12.5 shows the class hierarchy for the various Python classes we've constructed so far. Note that the two base models (DecisionTree and RandomDecisionTree) are inputs into our two ensemble models (BaggedForest and RandomForest).

A regression example is shown in Code Block 12.17. This dataset uses ten features to predict the target variable, which represents a quantitative measure of disease progression one year following prognosis. The relative feature importances are shown in Figure 12.6.

```
1  E = RandomForest(type='classification', num_features=['sepal length
       (cm)', 'sepal width (cm)', 'petal length (cm)', 'petal width
       (cm)'], max_depth=2, min_samples_split=10, subspace_dim=2,
       size=100, get_oob=True)
2  E.train(df, y)
3  E.oob_score # 0.9333333
4  E.getFeatureImportance()
5  # OUTPUT:
6  #    feature                 importance
7  # 0  petal length (cm)       1.000000
8  # 1  petal width (cm)        0.569108
9  # 2  sepal length (cm)       0.067717
10 # 3  sepal width (cm)        0.004357
```

Code Block 12.16: OOB score and feature importances for a random-forest model with 100 trees trained from the iris data set. Variables loaded from Code Block 12.7.

```
1  data = load_diabetes()
2  print(data.DESCR)
3  X, y = data.data, data.target
4  df = DataFrame(X, columns = data.feature_names)
5  df.sex = df.sex.apply(lambda x: 'M' if x > 0 else 'F')
6  num_features = ['age', 'bmi', 'bp', 's1', 's2', 's3', 's4', 's5',
       's6']
7  E = RandomForest(type='regression', cat_features=['sex'],
8                   num_features=num_features,
9                   max_depth=3, min_samples_split=20,
10                  subspace_dim=3, size=100, get_oob=True)
11 E.train(df, y)
12 E.oob_score # 0.44129
13 dfi = E.getFeatureImportance()
14 dfi.plot('feature', kind='barh', legend=False)
15 plt.gca().invert_yaxis()
```

Code Block 12.17: Random forest and feature importance for the diabetes dataset.

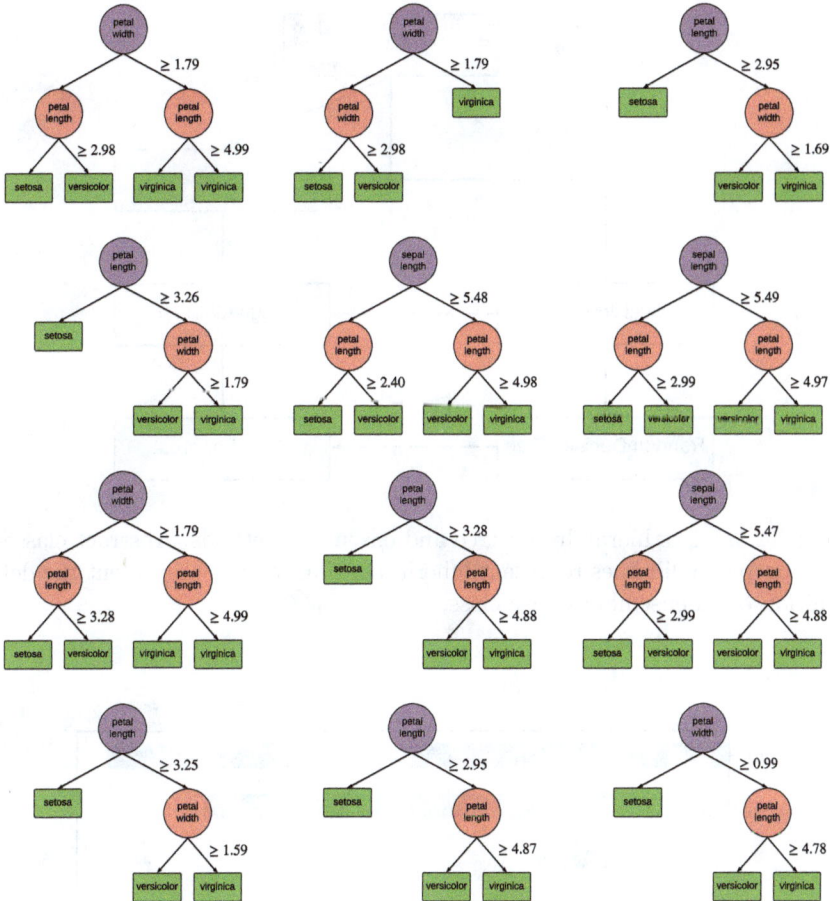

Fig. 12.4: A random forest with twelve trees, trained over the iris data set, with max depth 2.

12.3 Boosted Forests

In the previous section, we saw how boosting can be used to reduce the variance of a model, by producing an ensemble of independently trained models. In bagged forests, in particular, the trees, though independent, tend to be correlated, with top features dominating many individual trees. To remedy the correlation problem, we further introduced random forests, which use feature subspace sampling to ensure a broader variety of predictors in the forest. However, trees are still trained independently of one another.

The main idea of *boosting* is to train an ensemble of base models *in series*, so that each model can learn from the mistakes of its predecessor.

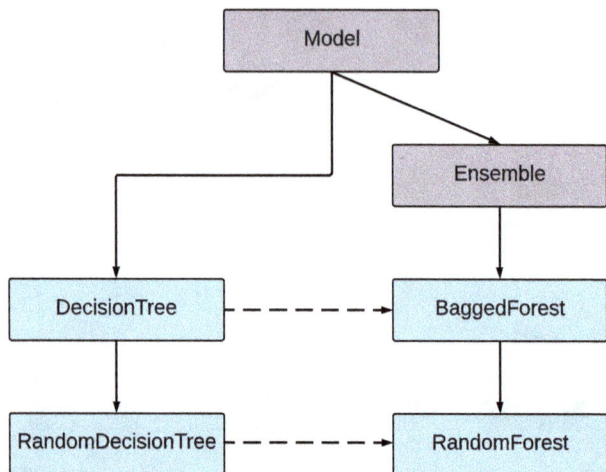

Fig. 12.5: Class hierarchy for tree and ensemble methods. Abstract classes are purple. Solid lines represent inheritance; dashed lines represent the definition of an ensemble's base class.

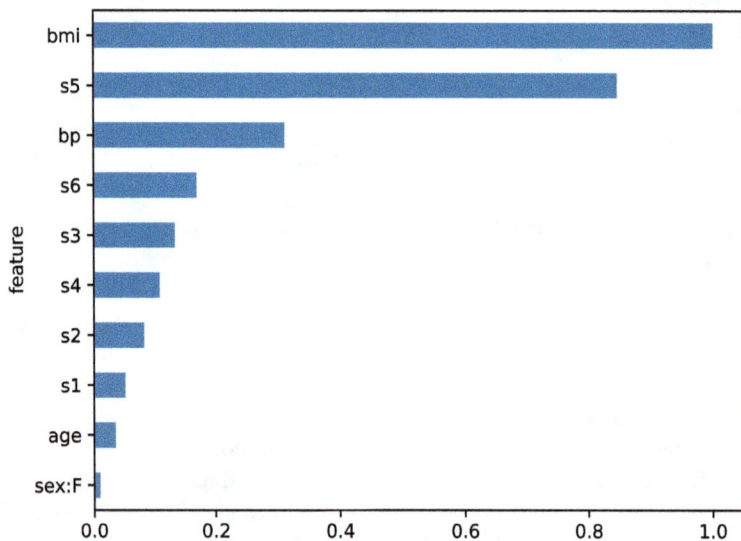

Fig. 12.6: Feature importances for the diabetes random forest example.

We will discuss two main boosting algorithms in this section: adaboost and gradient boosting.

12.3.1 AdaBoost

One of the main underlying ideas behind boosting is that many *weak learners* can be combined to form a *strong learner*. In this context, a *weak learner* is any classification model that performs slightly better than guessing class labels at random, and a *strong learner* is a model that can be tuned to produce an arbitrarily small error rate. The original boosting algorithm is due to Schapire [1990], who divided a large dataset into three subsets, training three successive weak models, and then showing that the overall error could be reduced arbitrarily by applying the method recursively. This method was improved upon by Freund and Schapire [1996], who introduced a variation known as *adaptive boosting*, or simply *AdaBoost*. The AdaBoost algorithm uses the entire dataset with each iteration, and can train an arbitrary number of base learners. We will discuss the original form of AdaBoost, known as AdaBoost.M1, which is shown in Algorithm 12.6.

Recall that bagged forests and random forests each train an independent collection of trees (or, more generally, base learners), with the random forest improvement helping to decorrelate the individual trees, providing a greater diversity in the forest. The principal behind boosting, on the other hand, is to break the independence assumption, instead using the result of one model to guide the training of the next. An immediate consequence of this is that random forests can be parallelized, whereas boosted forests need be trained in sequence.

Algorithm 12.6: AdaBoost algorithm for classification.

Input: data $\mathcal{D}$;

  feature set $\mathcal{F}$;

  weighted learning algorithm $\mathscr{A} : \mathbb{D} \times \mathbb{F} \times \mathbb{R}_+^n \to \mathbb{M}$;

  the size m of the ensemble

Output: An ensemble model $\mathcal{E}$.

1 $w_i = 1/|\mathcal{D}|$, for $i = 1, \ldots, |\mathcal{D}|$

2 **for** j **from** *1* **to** m **do**

3 $\quad$ Train model using weights $\mathbf{w} = \langle w_1, \ldots, w_n \rangle$: $M_j = \mathscr{A}(\mathcal{D}, \mathcal{F}, \mathbf{w})$

4 $\quad$ Compute weighted training error: $\epsilon = \dfrac{\sum\limits_{i=1}^{n} w_i \mathbb{I}[y_i \neq M_j(x_i)]}{\sum_{i=1}^{n} w_i} < 1/2$

5 $\quad$ Compute the accuracy log odds: $\alpha_j = \log\left(\dfrac{1-\epsilon}{\epsilon}\right)$

6 $\quad$ Update weights: $w_i = w_i \exp\left(\alpha_j \mathbb{I}[y_i \neq M_j(x_i)]\right)$

7 **end**

8 **return** $\mathcal{E} = \{M_j, \alpha_j\}_{j=1}^{m}$

The most outstanding difference between our first boosting method, Algorithm 12.6, and bagged forests, Algorithm 12.5, is that the base learner $\mathscr{A}$ requires the ability to handle *weights* to the training set. Fortunately, we have already taken care to include the handling of training weights in our decision tree implementation of Code Blocks 12.4–12.6, so that we can use our python DecisionTree class straightaway.

In Algorithm 12.6, the weights are initialized to uniform $(1/|\mathcal{D}|)$, but updated during each step of training (line 6). The weights are used to train each model on line 3. Upon training each tree, however, we then calculate the weighted training error ϵ, and an associated log odds, defined on line 5. Since we assume that each base model is a weak learner, we are assuming that the training error is $\epsilon < 1/2$, which, in turn, ensures that the log odds α_i for iteration i is positive, $\alpha_i > 0$. Finally, notice that only the weights for the *misclassified* instances of model M_i are updated by a factor of $e^{\alpha_i} > 1$; correctly classified weights remain the same. This has the effect of amplifying the weights of the misclassified examples, or errors, giving them more importance for the next round of model training. In this way, each model pays greater attention to the instances corresponding to the mistakes of the past.

The Python implementation of Algorithm 12.6 is shown in Code Block 12.18. Since we constructed our DecisionTree class to handle weighted predictions, we only need overwrite the _trainNext method of the Ensemble abstract base class.

```
class AdaBoost(Ensemble):

    def _getModel(self):
        return DecisionTree

    def _trainNext(self, df, y):
        if not hasattr(self, 'boost_weights'):
            self.boost_weights = np.ones(len(df)) / len(df)
        model = self.Model(**self.model_params)
        model.train(df, y, weights=self.boost_weights)
        y_hat = model.predict(df)
        err = np.average(y_hat!=y, weights=self.boost_weights)
        alpha = np.log((1-err) / err)
        self.boost_weights *= np.exp(alpha * (y_hat != y))
        return model, alpha
```

Code Block 12.18: AdaBoost implementation.

12.3.2 Gradient Boosting

Gradient boosting was first introduced in Friedman [2001], and updated
to include sampling (stochastic gradient boosting) in Friedman [2002]. We
will begin with an overview of gradient boosting, primarily focused on regression, before discussing regression and classification in detail.

Gradient Boosting Overview

The basic structure is shown in Algorithm 12.7. Like AdaBoost, the gradient
boosting algorithms makes successive corrections to the previous model, so
that each model improves on the prior model's weakness. In the case of
gradient boosting, however, this is done relative to a particular loss function,
which is used as an objective function that the ensemble seeks to minimize.
Note that since the models in the gradient boosting ensemble are meant as
literally corrections to each other, we require an unnormalized ensemble.

Naturally, the gradient boosting Algorithm 12.7 is a generalization of
the steepest descent Algorithm 5.2, and the stochastic gradient boosting

---

Algorithm 12.7: `GradientBoostedRegressor` algorithm for gradient boosting ($f = 1$) or stochastic gradient boosting ($f \in (0,1)$)
for regression.

Input: data $\mathcal{D}$;
   feature set $\mathcal{F}$;
   regression learning algorithm $\mathscr{A} : \mathbb{D} \times \mathbb{F} \to \mathbb{M}$;
   a differentiable loss function $L : \mathbb{T} \times \mathbb{T} \to \mathbb{R}$;
   the size m of the ensemble;
   subsample fraction $f \in (0, 1]$
Output: An ensemble model $\mathcal{E}$.

1 Initialize model to a constant: $M_0(x) = \arg\min_{\gamma} \sum_{i=1}^{n} L(y_i, \gamma)$

2 Initialize an unnormalized ensemble $\mathcal{E} = \{M_0, 1\}$
3 **for** j **from** 1 **to** m **do**
4 $\quad$ Compute sample index $I_{\mathcal{B}} = \texttt{getRandomIndex}(|\mathcal{D}|, \texttt{int}(f|\mathcal{D}|))$
5 $\quad$ Compute the gradients: $r_i = -\left.\dfrac{\partial L(y_i, \gamma)}{\partial \gamma}\right|_{\gamma = \mathcal{E}(x_i)}$, for $i \in I_{\mathcal{B}}$
6 $\quad$ Train a *gradient model*: $M_j = \mathscr{A}\left(\{(x_i, r_i)\}_{i \in I_{\mathcal{B}}}, \mathcal{F}\right)$
7 $\quad$ Solve the one-dimensional optimization problem:
$$\gamma_j = \arg\min_{\gamma > 0} \sum_{i=1}^{n} L(y_i, \mathcal{E}(x_i) + \gamma M_j(x_i))$$
8 $\quad$ Update $\mathcal{E} = \mathcal{E} \oplus \{M_j, \gamma_j\}$
9 **end**
10 **return** $\mathcal{E}$

---

is based on the stochastic gradient descent variation, as defined in Definition 5.5. Let's first discuss the stochastic nature of the algorithm, and then dive in to the details of the gradient boosting itself.

The update from gradient boosting to stochastic gradient boosting is achieved with the *subsample fraction* parameter, which, for regular gradient boosting, has the value $f = 1$. We then add the `getRandomIndex(n,m)` function (line 4), which returns a random selection of m indices from the the full index array $\{1, \ldots, n\}$; i.e., it is a random selection of indices for a bootstrap without replacement. For $f = 1$, we have $I_\mathcal{B} = \{1, \ldots, n\}$, so that the full data set is used in each iteration of the algorithm. A typical value of the subsample fraction is $f = 0.5$, though it should be viewed as a tunable parameter.

Now that we understand how to extend the gradient boosting algorithm to its stochastic counterpart, let's next dive into the details of the algorithm itself. The first departure from our prior methods is that gradient boosting requires a loss function, which it then seeks to optimize for. We defined loss functions in Equation (6.18), and defined common examples in Equations (6.17)–(6.20). Typically, we choose squared-error loss (Equation (6.17)) for regression and log loss (Equation (6.20)) for classification, though the latter does require the predicted class probabilities, not the predicted labels.

We repeat the squared-error loss and log loss functions for convenience:

$$L(y, \gamma) = (y - \gamma)^2, \qquad (12.14)$$

$$L(y, (p_1, \ldots, p_c)) = -\sum_{k=1}^{c} \mathbb{I}[y = \mathcal{C}_k] \log p_k. \qquad (12.15)$$

The log loss is also often referred to as the *deviance*, due to its relation to the deviance function (Definition 7.12) from logistic regression. Moreover, note that the negative gradient of the squared-error loss is simply

$$-\frac{\partial L}{\partial \gamma} = 2(y - \gamma),$$

which is simply twice the residual error, when $\gamma = \hat{y}$ is interpreted as the model prediction. This gives a further interpretation of gradient boosting in the context of regression, where the models learned on line 6 are trying to learn the *residual errors* of the previous ensemble.

Given a loss function, the algorithm is then a simple generalization of gradient descent (Algorithm 5.2), such that steps in the direction of the negative gradient are instead replaced with steps in the direction of a model trained on the negative gradient. The optimization problem on line 7 implies that our gradient boosting can be viewed as a steepest-descent gradient boosting algorithm. In some instances, the parameter γ is passed as an input, called the *learning rate*, and is held constant over the iterations. The stochastic variation follows in the same suit as the stochastic gradient descent algorithm of Definition 5.5, where only a subsample of data is used for each step.

Finally, we stress the importance of using an *unnormalized* ensemble in the construction of the ensemble $\mathcal{E}$. If we instead denote the model trained on line 6 by R_j (for residual), we could then replace line 8 with

$$M_j = M_{j-1} + \gamma_j R_j,$$

to the same effect. The final ensemble should therefore necessarily generate predictions of the form

$$\mathcal{E}(x) = M_0(x) + \gamma_1 M_1(x) + \cdots + \gamma_m M_m(x),$$

as $M_0(x)$ is an overall (constant) prediction and as each subsequent model $M_1, \ldots, M_m$ is a corrective term seeking to remedy the errors of the prior models. It would therefore be incorrect to divide this final prediction by the normalization factor $1 + \sum_{j=1}^{m} \gamma_j$.

Gradient Boosting for Regression

We begin by noting that squared-error loss Equation (12.14) directly yields the MSR impurity function defined in Equation (12.9), when summed over a sample of data. Next, we note that, when using squared-error loss, line 1 of Algorithm 12.7 yields the sample mean:

$$\arg\min_{\gamma} \sum_{i=1}^{n} (y_i - \gamma)^2 = \overline{y}. \tag{12.16}$$

We encoded a quick `ConstantModel` class, which outputs this constant, in Code Block 12.19. The full gradient boosting algorithm for regression is encoded in Code Block 12.20.

Gradient Boosting for Classification: Initial Model

As we saw with regression, log loss, as defined by Equation (12.15), yields the entropy measure of Equation (12.4), when summed over a sample of data, and when we identify p_k as the sample mean for the kth class. More directly, we obtain the following:

$$\mathscr{L}(p; \mathcal{D}) = \sum_{i=1}^{n} L(y_i, p) = -\sum_{k=1}^{c} \overline{y}_k \log_2(p_k), \tag{12.17}$$

where $y_k = \mathbb{I}[y_i = \mathcal{C}_k]$, so that

$$\overline{y}_k = \frac{1}{n} \sum_{i=1}^{n} \mathbb{I}[y_i = \mathcal{C}_k]$$

```
1   class ConstantModel(Model):
2
3       def __init__(self, **kwargs):
4           type_error = "Keyword argument 'type' must be set to
                    classification or regression"
5           assert 'type' in kwargs, type_error
6           assert kwargs['type'] in ['classification', 'regression'],
                    type_error
7           self.type = kwargs['type']
8
9       def train(self, df, y, weights=None):
10          if self.type == 'regression':
11              self.constant = np.average(y, weights=weights)
12          else:
13              self.constant = Series(Counter(y)).sort_index().values /
                    len(y)
14
15      def predict(self, df: DataFrame):
16          return self.constant
```

Code Block 12.19: `ConstantModel` class representing a constant prediction.

is the sample average. The values of the vector p in Equation (12.17) are not free to vary at will, but, rather, are constrained by the condition that

$$p \in \Delta^{c-1} = \left\{ p \in \mathbb{R}^c_* : g(p) = \sum_{k=1}^c p_k = 1 \right\};$$

i.e., p is an element of the probability simplex (Definition 2.23). The arg min of Algorithm 12.7 line 1, therefore, must be modified as a constrained argmin:

$$M_0(x) = \arg\min_{p \in \Delta^{c-1}} \mathscr{L}(p; \mathcal{D}).$$

To solve this, we write down the Lagrange multiplier equations

$$\nabla \mathscr{L} = \lambda \nabla g,$$

$$g(p) = \sum_{k=1}^c p_k = 1,$$

from whence we obtain

$$-\frac{\overline{y}_k}{p_k} = \lambda,$$

so that $p_k = -\lambda^{-1}\overline{y}_k$. The constraint equation implies that $\lambda = -1$, since $\sum_{k=1}^c \overline{y}_k = 1$. This yields the result that we should initialize our probability

```
1   class GradientBoostedRegressor(Ensemble):
2
3       def __init__(self, **kwargs):
4           kwargs['normalize'] = False
5           kwargs['type'] = 'regression'
6           self.subsample = kwargs.get('subsample', 1)
7           super().__init__(**kwargs)
8
9       def _getModel(self):
10          return DecisionTree
11
12      def _trainNext(self, df, y):
13          if len(self.models) == 0:
14              M = ConstantModel(type=self.type)
15              M.train(df, y)
16              return M, 1
17          size = int(self.subscample * len(df))
18          idx = np.random.choice(np.arange(len(df)), size=size,
                  replace=False)
19          df_boot = df.loc[idx, :].reset_index(drop=True)
20          y_boot = y[idx]
21          y_hat = self.predict(df_boot)
22          r = y_boot - y_hat
23          model = self.Model(**self.model_params)
24          model.train(df_boot, r)
25          gamma = goldenSearch(lambda x: mean_squared_error(y,
                  self.predict(df) + x*model.predict(df)), 0, 10)
26          return model, gamma
```

Code Block 12.20: GradientBoostedRegressor gradient boosted regressor class.

model with the constant predictions coinciding with the observed sample ratios from each class:

$$M_0(x) = \frac{1}{n}\left\langle \sum_{i=1}^{n}\mathbb{I}[y_i = \mathcal{C}_1], \ldots, \sum_{i=1}^{n}\mathbb{I}[y_i = \mathcal{C}_c]\right\rangle = \langle \overline{y}_1, \ldots, \overline{y}_c\rangle. \quad (12.18)$$

This initial model coincides with the observed probability prediction in the ConstantModel class of Code Block 12.19.

Gradient Boosting for Classification: Boosting Rounds

In many machine-learning texts, the gradient boosting classification algorithm is handled by training c independent binary classifiers for each of the c classes, so that the for-loop of lines 3–9 in Algorithm 12.7 is actually nested within an outer for-loop that iterates through the target labels (classes). We will present an alternate approach, which we believe to be more elegant, that treats the probability p as a vector $p \in \Delta^{c-1}$ and that addresses the geometric constraints directly within the algorithm.

Our gradient-boosted classification algorithm is given by Algorithm 12.8. In order to unravel this, we begin by introducing the normal vector to the constraint $g(p) = \sum_{k=1}^{c} p_k = 1$,

$$v = \nabla g = \langle 1, \dots, 1 \rangle \in \mathbb{R}^c, \qquad (12.19)$$

with norm $\|v\|_2 = \sqrt{c}$. The tangent plane $T_p \Delta^{c-1}$ to a point $p \in \Delta^{c-1}$ is therefore isomorphic to the orthogonal complement of the subspace spanned by the vector v. In order to modify Algorithm 12.7 for classification, we therefore need only project the residuals, as defined on line 5 of the algorithm, onto the tangent plane of the constraint surface, using the projection

---

Algorithm 12.8: GradientBoostedClassifier algorithm for gradient boosting ($f = 1$) or stochastic gradient boosting ($f \in (0,1)$) for classification; using log loss (Equation (12.15)).

Input: data $\mathcal{D}$;

       feature set $\mathcal{F}$;

       a vector regression learning algorithm $\mathscr{A} : \mathbb{D} \times \mathbb{F} \to \mathbb{M}$;

       the size m of the ensemble;

       subsample fraction $f \in (0,1]$

Output: An ensemble model $\mathcal{E}$.

1 Initialize model to a constant:

$$M_0(x) = \frac{1}{n} \left\langle \sum_{i=1}^{n} \mathbb{I}[y_i = \mathcal{C}_1], \dots, \sum_{i=1}^{n} \mathbb{I}[y_i = \mathcal{C}_c] \right\rangle$$

2 Initialize an unnormalized ensemble $\mathcal{E} = \{M_0, 1\}$

3 **for** j from 1 to m **do**

4   Compute sample index $I_{\mathcal{B}} = \texttt{getRandomIndex}(|\mathcal{D}|, \texttt{int}(f|\mathcal{D}|))$

5   Compute the gradients: $r_i = \left\langle \dfrac{\mathbb{I}[y_i = \mathcal{C}_1]}{\mathcal{E}(x_i)_1}, \dots, \dfrac{\mathbb{I}[y_i = \mathcal{C}_c]}{\mathcal{E}(x_i)_c} \right\rangle$, for $i \in I_{\mathcal{B}}$

6   Compute the projections: $v_i = \text{proj}_{T_{\mathcal{E}(x_i)}\Delta^{c-1}}(r_i)$, for $i \in I_{\mathcal{B}}$

7   Train a *gradient model*: $M_j = \mathscr{A}\left(\{(x_i, v_i)\}_{i \in I_{\mathcal{B}}}, \mathcal{F}\right)$

8   Compute $\gamma^{\max} = \min\limits_{i=1,\dots,n} \min\limits_{k=1,\dots,c} \left\{ \dfrac{-\mathcal{E}(x_i)_k}{M_j(x_i)_k} : M_j(x_i)_k < 0 \right\}$

9   Solve the one-dimensional optimization problem:

$$\gamma_j = \underset{\gamma \in (0, 0.9\gamma^{\max})}{\arg\min} \sum_{i=1}^{n} L(y_i, \mathcal{E}(x_i) + \gamma M_j(x_i))$$

10   Update $\mathcal{E} = \mathcal{E} \oplus \{M_j, \gamma_j\}$

11 **end**

12 **return** $\mathcal{E}$

---

operator

$$\text{proj}_{T_p \Delta^{c-1}}(x) = x - \frac{x \cdot v}{c} v, \tag{12.20}$$

which maps a vector $x \in \mathbb{R}^c$ onto the tangent plane $T_p \Delta^{c-1} \cong \text{span}(v)^{\perp}$.

Next, we need to consider the boundary of the simplex Δ^{c-1}, for the purpose of further constraining the optimization problem defined on line 7 of Algorithm 12.7. Given a point $(p, v) \in T\Delta^{c-1}$ on the tangent bundle (i.e., a probability vector and an attached tangent vector), we need to determine the shortest distance to the edge of the simplex, in the direction $v \in T_p \Delta^{c-1}$. By traveling a distance γ in the direction v, we reach the kth coordinate plane precisely when

$$p_k + \gamma_k v_k = 0, \qquad \text{or,} \qquad \gamma_k = -\frac{p_k}{v_k}.$$

Considering only the directions for which $\gamma_k > 0$ (which only occurs when the vector v has a negative value for its kth component), we then simply select the minimum parametric distance

$$\gamma = \min_{k=1,\dots,c} \left\{ -\frac{p_k}{v_k} : v_k < 0 \right\}.$$

All modifications in hand, the final gradient-boosted classification algorithm is given in Algorithm 12.8.

The Python code for the gradient-boosted classifier is given in Code Blocks 12.20 and 12.21. We immediately encounter two differences with our other forest methods right away: the `_logLoss` and `predictProba` methods on lines 12 and 18 of Code Block 12.20. The log-loss function (line 12) is specifically written for a vector y of class labels and a probability matrix p, of the same length as the vector y and a number of columns equal to the total number of classes.

For line 21 of the `predictProba` method of Code Block 12.20, note that the first model (`self.models[0]`) will be an object from the `ConstantModel` class (Code Block 12.19), whereas subsequent models will be regression `DecisionTree` objects. (Note that our original `DecisionTree` class, of Code Blocks 12.3–12.6, was constructed with the ability to perform *vector regression*, where the target variable is a vector variable. This is why, for example, we specify `axis=0` in the `_mse` method on line 19 of Code Block 12.4, etc.)

12.4 Advanced Forestry

In this section, we will briefly discuss some additional variations of the classic random forests and gradient boosted forests. We begin by mentioning a powerful variation of gradient boosting known as extreme gradient boosting. We will call out the main differences and point the interested reader to

```
 1   class GradientBoostedClassifier(Ensemble):
 2
 3       def __init__(self, **kwargs):
 4           kwargs['normalize'] = False
 5           kwargs['type'] = 'classification'
 6           self.subsample = kwargs.get('subsample', 1)
 7           super().__init__(**kwargs)
 8
 9       def _getModel(self):
10           return DecisionTree
11
12       def _logLoss(self, y, p):
13           logp = np.log2(p)
14           y_labels = [self.classes.index(x) for x in y]
15           logp = [logp[i, y_labels[i]] for i in range(len(y))]
16           return -np.sum(logp)
17
18       def predictProba(self, df):
19           y_hat = np.zeros((len(df), len(self.classes)))
20           for model, weight in zip(self.models, self.weights):
21               y_hat += weight * model.predict(df)
22           return y_hat
23       # continued ...
```

Code Block 12.21: `GradientBoostedClassifier` gradient boosted classifier class (part 1).

resources for this variation. Next, we will discuss two variations that can be used for temporal data. The first is survival forests, which is a modification of the random forest algorithm to account for censored information. The second, we will call online forests.

12.4.1 Parallel Processing

As modern laptops come with a minimum of eight CPUs, it is worth mentioning how we can speed up the training of forests by leveraging parallel processing. Python comes with a built-in module `multiprocessing` that can be used to parallelize a job ofer multiple processors. Unfortunately, there is no easy way to code parallel processing as a method within a class, as the `multiprocessing` module does not work interactively in the Python interpreter. (We will discuss an alternate approach at the end of the section.) Instead, we must write an executable script that we can run. The idea, then, is to train a number of smaller forests in parallel, and add them together to get our end result. We therefore begin by defining the magic method `__add__` to our `Ensemble` class (Code Block 12.9), which allows us

```
1   # ... continued
2   def _trainNext(self, df, y):
3       if len(self.models) == 0:
4           M = ConstantModel(type='classification')
5           M.train(df, y)
6           return M, 1
7       size = int(self.subsample * len(df))
8       idx = np.random.choice(np.arange(len(df)), size=size,
                replace=False)
9       df_boot = df.loc[idx, :].reset_index(drop=True)
10      y_boot = y[idx]
11      y_hat = self.predictProba(df)
12      y_hat_boot = y_hat[idx]
13      r = [[int(y==c) for c in self.classes] for y in y_boot] /
                y_hat_boot
14      r = r - r.sum(axis=1).reshape((len(r), 1)) /
                len(self.classes)
15      self.model_params['type'] = 'regression'
16      model = self.Model(**self.model_params)
17      model.train(df_boot, r)
18      v = model.predict(df)
19      gamma = - y_hat / v
20      gamma_max = min(gamma[gamma > 0])
21      gamma = goldenSearch(lambda x: self._logLoss(y,
                self.predictProba(df) + x*model.predict(df)), 0,
                0.9*gamma_max)
22      return model, gamma
23
24  def predict(self, df):
25      y_hat = self.predictProba(df)
26      votes = DataFrame(y_hat, columns=self.classes)
27      return votes.idxmax(axis=1).values.tolist()
```

Code Block 12.22: `GradientBoostedClassifier` gradient boosted classifier class (part 2).

to use the syntax E + F, which is interpreted by Python as E.__add__(F). The code to accomplish this is shown in Code Block 12.23.

Finally, we can use the `Pool` function from the **multiprocessing** module to parallelize our computation, as shown in Code Block 12.24. Here, we use **starmap**, as our function takes multiple arguments. We ran this on a 10-CPU machine, so that each CPU trained four separate random forests, yielding a total of 400 individual decision trees that were combined in the end.

```
1  def __add__(self, other):
2      assert self.type == other.type
3      assert self.Model == other.Model
4      new = copy.deepcopy(self)
5      w_new, w_other = sum(new.weights), sum(other.weights)
6      new.models += other.models
7      new.weights += other.weights
8      if hasattr(self, 'oob_score'):
9          new.oob_score = w_new * new.oob_score + w_other *
                other.oob_score / (w_new + w_other)
10     return new
```

Code Block 12.23: Magic method for **Ensemble** class (Code Block 12.9) for adding forests.

```
1  def fun(E, df, y):
2      E.train(df, y)
3      return E
4
5  def main():
6      data = load_diabetes()
7      X = data.data
8      y = data.target
9      df = DataFrame(X, columns = data.feature_names)
10     df.sex = df.sex.apply(lambda x: 'M' if x > 0 else 'F')
11     num_features = ['age', 'bmi', 'bp', 's1', 's2', 's3', 's4',
           's5', 's6']
12     E = RandomForest(type='regression', cat_features=['sex'],
13                 num_features=num_features,
14                 max_depth=3, min_samples_split=20,
15                 subspace_dim=3, size=10)
16     start = time.time()
17     values = ((E, df, y) for i in range(40))
18     with multiprocessing.Pool(processes=10) as pool:
19         res = pool.starmap(fun, values)
20     E = res[0]
21     for i in range(1, len(res)):
22         E = E + res[i]
23     end = time.time()
24     print(f'elapsed time: {end - start}')
25
26  if __name__ == '__main__':
27      main()
```

Code Block 12.24: Using **multiprocessing** to train trees in parallel.

Accessing the Model Object

When running a script, such as Code Block 12.24, the variables introduced within the **main** function are not retained in memory. Once the script completes, the local memory is cleared. In a development platform, such as Spyder, one is often interested in running a script, but retaining the final model output in memory, so that it may be accessed from the iPython console for further investigation. There are two approaches that allow one to persist the model constructed within **main** in the runtime environment.

Return the Model Object

The most straightforward approach is simply to add a line **return E** at the end of the **main** function, and then modify the last line of Code Block 12.24 to store the return in memory: **E = main()**. When running the script in the iPython console, such an arrangement will then persist the final model object **E** in memory.

Pickling Model Objects

Another approach, which can be useful both in development platforms as well as production platforms, is pickling the model. *Pickling*[3] refers to a process that takes an object and converts it into an equivalent text representation, often but not necessarily binary, that retains all of the information within the object hierarchy. A copy of the object can then be reconstructed by *unpickling* the pickle file that was rendered. Pickling can be useful in production, as it allows one to save a trained model on the server or on a distributed file system, such as HDFS, for later use.

In order to pickle and unpickle an object, one simply imports the **pickle** module and applies the **dump** and **load** functions, respectively, which pickle the object and save the result to disk, as shown in lines 3–7 of Code Block 12.25[4]. One can add lines 3 and 4 to the **main** function, and then call lines 6 and 7 from the iPython console to retrieve the model object **E** created during execution of the script. Alternatively, one may use **dumps** and **loads** to convert the object to a string and save to local memory, as shown in lines 9 and 10 of Code Block 12.25.

Parallel Processing with `joblib`

An alternate approach to parallel processing is the `joblib` module, which has the advantage of being usable within the iPython console or within a given class method. For our use case, we can build the parallel processing

---

[3] also referred to as serialization or marshalling.

[4] The **'b'**, as in **'wb'** (write) or **'rb'** (read), instructs Python to write to and read the file in binary.

```
1  import pickle
2  # pickle object (e.g., from main function)
3  with open('tmp_files/my_model.pkl', 'wb') as f:
4      pickle.dump(E, f)
5  # unpickle object (e.g., from iPython console)
6  with open('tmp_files/my_model.pkl', 'rb') as f:
7      E = pickle.load(f)
8  # pickle and unpickle object in memory
9  pkl = pickle.dumps(E)
10 E_new = pickle.loads(pkl)
```

Code Block 12.25: Pickling and unpickling objects.

functionality directly into the **train** method of the **Ensemble** class (Code
Block 12.9), as shown in Code Block 12.26. Note that the optional keyword
argument **n_jobs** was provided in the constructor of the **BaggedForest**
class (Code Block 12.14). If self.n_jobs is undefined, or is equal to its
default value of 1, we preserve the original code from Code Block 12.9.
Otherwise, we use **Parallel** and **delayed** to run the individual calls to
self._trainNext in parallel.

```
1  from joblib import Parallel, delayed
2  ## train method
3      if not hasattr(self, 'n_jobs') or self.n_jobs == 1:
4          for i in tqdm(range(self.size)):
5              model, weight = self._trainNext(df, y)
6              self.models.append(model)
7              self.weights.append(weight)
8      else:
9          results =
                Parallel(n_jobs=self.n_jobs)(delayed(self._trainNext)(df,
                y) for i in tqdm(range(self.size)))
10         self.models, self.weights = map(list, zip(*results))
```

Code Block 12.26: Modification of the **Ensemble** class (Code Block 12.9)
for parallel processing.

Note 12.1. Recall that parallel processing is not appropriate for boosted
forests, as boosted forests algorithms are defined sequentially. For this reason, we added the definition of self.n_jobs to the constructor of the
BaggedForest class (Code Block 12.14), so that the parallel processing
cannot be used for non-bagged descendants of the **Ensemble** class, where
they are bound to cause side effects or not function properly.   ▷

12.4.2 Extreme Gradient Boosting

Extreme gradient-boosted trees, or *XGBoost*, is an open-source software (`xgboost.readthedocs.io`) that represents a powerful extension to simple gradient boosting; see Chen and Guestrin [2016]. Not only does XGBoost improve upon the basic gradient-boosted forest methods, but it also improves processing performance, optimizing for training speed. Improvements on the algorithm include:

1. The Newton–Raphson method (Algorithm 5.1) replaces gradient descent, allowing for more efficient optimization, using the Hessian matrix[5] of the loss function;
2. a regularization term (Lasso L1 or Ridge L2; Definition 7.4) is added to the loss function to penalize complex models and avoid overfitting;
3. a *weighted quantile sketch* algorithm is used to efficiently find optimal splits;
4. built-in cross validation at each iteration;
5. an added randomization parameter that helps reduce correlation between trees;
6. a method for proportionally "shrinking" leaf nodes;

In addition, the algorithm provides various improvements for processing performance:

1. the for-loops for calculating features are parallelized using multi-threading, improving computational speed;
2. tree pruning uses a 'depth-first' approach to improve computational speed;
3. the algorithm optimizes hardware memory usage using 'cache awareness,' to optimally allocate memory of the gradient within each thread, and 'out-of-core' computing, to optimize disk space when training large datasets that do not fit into memory.

An `XGBoost` class is provided in Code Block 12.27. This is essentially a wrapper around the `xgboost` module that reformats the inputs and outputs to follow our conventions, thereby making it compatible and exchangeable with our other methods.

12.4.3 Bagged Boosters

One of the limitations of gradient boosting is its sequential nature, which at first glance prevents use of parallel computations. Recall, however, that bagged forests are simply bootstrapped aggregations of individual models. There is no reason why we therefore cannot do a boostrap aggregation using a gradient descent model as the kernel. In this regard, we are trading some of our depth, as measured by the number of boosting rounds, with breadth,

---

[5] matrix of second partial derivatives

```python
import xgboost
class XGBoost(Model):

    def __init__(self, **kwargs):
        # See https://xgboost.readthedocs.io/en/stable/parameter.html
        self.objective = kwargs.get('objective', 'reg:squarederror')
        self.size = kwargs.get('size', 10)
        keys = ['eta', 'gamma', 'max_depth', 'min_child_weight',
                'subsample', 'sampling_method', 'colsample_bytree',
                'alpha', 'lambda', 'tree_method', 'scale_pos_weight',
                'max_leaves', 'num_parallel_tree']
        self.params = {key: kwargs[key] for key in
                set(keys).intersection(kwargs.keys())}

    def train(self, df, y, weights=None):
        dmat = xgboost.DMatrix(df, label=y, enable_categorical=True)
        self.xgb = xgboost.train(self.params, dmat, self.size)

    def predict(self, df, get_prob=False):
        dmat = xgboost.DMatrix(df, enable_categorical=True)
        return self.xgb.predict(dmat)
```

Code Block 12.27: Wrapper around XGBoost package.

as measured by the number of boosted models in the forest. The advantage is that we can easily parallelize the individual boosters. This is achieved in Code Block 12.28, which provides code for a random gradient-boosted forest and a random XGBoosted forest.

```python
class RandomGBForest(BaggedForest):

    def _getModel(self):
        self.model_params['surpress_tqdm'] = True
        if self.type == 'classification':
            return GradientBoostedClassifier
        return GradientBoostedRegressor

class RandomXGBForest(BaggedForest):

    def _getModel(self):
        return XGBoost
```

Code Block 12.28: Random Gradient Boosted and Extreme Gradient Boosted Forests.

12.4.4 Performance Comparison

Next, let's compare the performance—in terms of both speed and model accuracy—of the various tree-based methods presented thus far. We use the *sci-kit learn* diabetes dataset to train a number of regression models, as shown in Code Block 12.29. In particular, we train the set of models contained in the dictionary that spans lines 10–18. The results are shown in Figure 12.7.

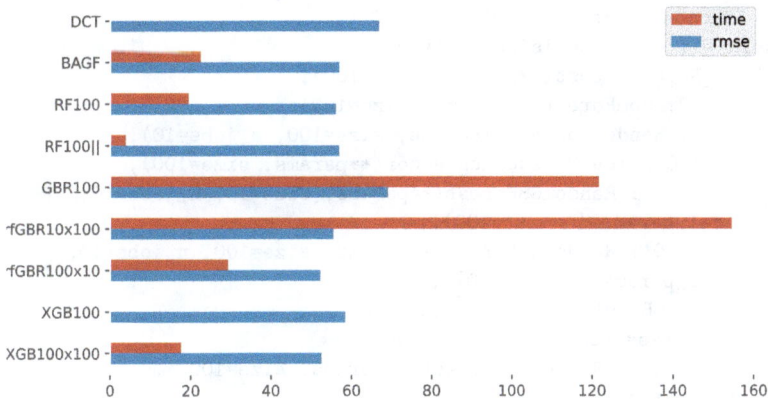

Fig. 12.7: Speed and performance comparison on a sample data set.

The first observation is that there is very little variance in the model's error (blue), so each model does a good job learning from the data. Naturally, a single decision tree (DCT) was quite fast, clocking in around 1/5th of a second. It also had the worst error, though fairly comparable to a single gradient-boosted forest (GBR100), which took around 120s to train. (It is quite surprising the gradient-boosted regressor offered little performance improvement over a single decision tree.)

The bagged forest (BAGF) and random forest (RF100) performed a bit better than a single decision tree, as expected, and the random forest was a tad faster than the bagged, due to the smaller feature space. Running the random forest in parallel (RF100||) further improved speed by a factor of five.

We trained two random gradient-boosted forests: the first consisted of 10 gradient boosted forests with 100 trees each (rfGBR10x100), and the second consisted of 100 gradient boosted forests with 10 trees each (rfGBR100x10). The first only took around 30% longer than a single 100-tree forest (GBR100) due to the parallelization, whereas the second was

```
 1  def main():
 2      data = load_diabetes()
 3      X = data.data
 4      y = data.target
 5      df = DataFrame(X, columns = data.feature_names)
 6      df.sex = df.sex.apply(lambda x: 'M' if x > 0 else 'F')
 7      df.sex = df.sex.astype('category')
 8      num_features = ['age', 'bmi', 'bp', 's1', 's2', 's3', 's4',
            's5', 's6']
 9      params = {'type':'regression', 'cat_features':['sex'],
            'num_features':num_features, 'max_depth':3,
            'min_samples_split':20, 'min_child_weight':20,
            'subspace_dim':3, 'subsample':0.5}
10      models = {'DCT': DecisionTree(**params),
11      'BAGF': BaggedForest(**params, size=100),
12      'RF100': RandomForest(**params, size=100),
13      'RF100||': RandomForest(**params, size=100, n_jobs=10),
14      'GBR100': GradientBoostedRegressor(**params, size=100),
15      'rfGBR10x100': RandomGBForest(**params, size=10, n_jobs=10,
            model_params={'size':100}),
16      'rfGBR100x10': RandomGBForest(**params, size=100, n_jobs=10,
            model_params={'size':10}),
17      'XGB100': XGBoost(**params, size=100,
            objective='reg:squarederror'),
18      'rfXGB100x100': RandomXGBForest(**params, size=100,
            objective='reg:squarederror')}
19      dr = DataFrame(columns=['model', 'time', 'rmse'])
20      df_train, df_test, y_train, y_test = train_test_split(df, y,
            test_size=0.2)
21      df_train, df_test = df_train.reset_index(drop=True),
            df_test.reset_index(drop=True)
22      for model, E in models.items():
23          start = time.time()
24          E.train(df_train, y_train)
25          t = time.time() - start
26          y_hat = E.predict(df_test)
27          mse = mean_squared_error(y_test, y_hat)
28          dr = dr.append(DataFrame({'model':model, 'time':t,
                'rmse':np.sqrt(mse)}, index=[0]), ignore_index=True)
29          print(f"model: {model}, time: {t}, mse: {mse}")
30      return dr
31
32  if __name__ == '__main__':
33      dr = main()
34      dr = dr.set_index('model')
35      plt.figure(figsize=(8,9/2))
36      plt.style.use('ggplot')
37      ax = dr.plot.barh(figsize=(8,9/2))
38      ax.invert_yaxis()
```

Code Block 12.29: Model comparison.

80% faster than the single 100-tree forest, with better performance. Both rfGBR10x100 and rfGBR100x10 consist of 1,000 trees, but the later had superior performance in terms of both speed and accuracy.

The XGBoosted forest (XGB100) with 100 boosted rounds was as fast as our implementation of a single decision tree! The XGBoost package, however, is optimized to use parallelization and multi-threading, though it is interesting that it is around 100 times faster than our random forest implementation. Finally, we trained a bagged XGBoosted forest consisting of 100 individual XGBoosted forests with 100 boosted rounds. This offered an additional, albeit modest, reduction in the error. Interestingly, it made no difference in terms of speed whether we included the `n_jobs=10` parameter, leading us to infer that the `xgboost` package is already leveraging the multi-processing capacity of the machine.

Finally, though not shown in the figure, we note that the *sci-kit learn* models `RandomForestRegressor` and `GradientBoostingRegressor`, from the `sklearn.ensemble` package, both performed quite well, clocking in around 1/7th and 1/20th of a second, respectively, and achieving a RMSE of around 56.

12.5 Time Forests and Survival Forests and the Like

Time plays an important role in many real-world applications. This slightly breaks from the view of traditional machine learning, in which one trains a model over a static data set; e.g., one might train a classifier over a library of pictures to teach the machine how to classify cats and dogs. Many, if not most, interesting data science applications in marketing and product analytics have an explicit time component: companies collect massive quantities of user behavioral data over time. It is therefore critical to adapt traditional machine-learning models to settings that have such a temporal focus. We discuss time aspects in two contexts: the first is how to handle time as a dimension, and the second is how to model online and survival processes (Definition 5.20). This latter problem is trickier, as there are actually two time dimensions, the common case being a time series of right-censored data, forming a sort of *lower-right-triangular censoring* (e.g., see Figure 5.10).

12.5.1 Time Forests

We begin with the topic of sequential data sets indexed by time. For example, a marketing campaign might run over a long period of time, each day generating a number of impressions and clicks. If a set of user-level features is available at the impression level, one might model the probability of a click as a function of each user's feature set. A large-scale campaign might generate hundreds of millions of impressions each day. Instead of training a single model using the trailing 30-days of data, and repeating this process each day, one might instead train a *differential* model, only on the prior day's cohort, and maintain a running queue of trailing models. This is the idea behind what we call *time forests*.

Note 12.2. Mathematically, time is not topologically different than a simple index. In practice, however, *time* sets the temporal cadence for which we receive and can process data. ▷

Definition 12.12. *A* time forest *of size* m *and decay factor* γ *is a normalized ensemble, such that individual models and weights are stored in a* queue *with maximum length* m, *and such that the ensemble's* train *operation*

1. *applies the decay factor* γ *to all weights,*
2. *trains its base model on the new input data, generating a new model* M,
3. *if the model and weight queue is length* m, *pops the oldest model and weight from the left (beginning) of the queue,*
4. *appends the new model* M *and its associated weight (i.e., size of data set) to the right of the queue.*

A time forest of size 7 is illustrated in Figure 12.8. The *pop* operation corresponds to removing the leftmost model and its associated weight. The *append* operation appends a new model and weight to the right. The oldest model is the leftmost model, as shown by the model dates.

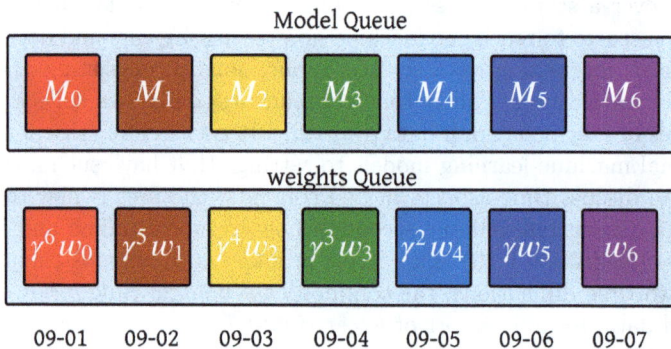

Fig. 12.8: Illustration of a *time forest* of size 7 consisting of a model queue and a weights queue.

Note 12.3. Often, the base model of a time forest is itself a forest, e.g., a random forest or an XGBoosted forest. Like our *bagged boosters*, time forests also represent a *forest of forests*; i.e., ensembles built over forest base models. ▷

Note 12.4. The weights normalization in the **Ensemble.predict** method of Code Block 12.9 does not modify the **weights** attribute, but instead defines a local variable that is normalized. This particular implementation is specifically for the **TimeForest** class, as normalizing the weights in place (i.e., as an attribute) would result in the incorrect ensemble when the next batch is trained.   ▷

The **TimeForest** class is implemented in Code Block 12.30. Here, we use **collections.deque**, which has the **append** and **popleft** methods. This is similar to the **queue.Queue** class used in Example 9.3, except we are explicitly tracking the **maxsize** using the **size** attribute of the parent **Ensemble** class.

```python
class TimeForest(Ensemble):

    def _getModel(self):
        self.models, self.weights = deque(), deque()
        self.gamma = self.params.get('gamma', 1)
        base_model = self.params.get('base_model', 'rf')
        if base_model == 'xgboost':
            return XGBoost
        else:
            return RandomForest # default

    def train(self, df, y):
        if self.type == 'classification' and not hasattr(self,
                'classes'):
            self.classes = np.sort(np.unique(y)).tolist()
        if not hasattr(self, 'vector_dim'):
            self.vector_dim = y.shape[1] if y.ndim == 2 else None
        model = self.Model(**self.model_params)
        model.train(df, y)
        if len(self.models) == self.size:
            self.models.popleft()
            self.weights.popleft()
        self.weights = deque([self.gamma * w for w in self.weights])
        self.models.append(model)
        self.weights.append(len(df))
```

Code Block 12.30: **TimeForest** implementation.

12.5.2 Survival Forests

Survival Problems

Next, we turn to a different temporal problem, that of training a model based on censored data. Recall from Definitions 5.6 and 5.7 the distinction between *censored* and *truncated* data: both involve data sets whose values are only known within a given interval; however, whereas the *counts* of censored instances are known in the former, they are completely unobserved in the latter. This distinction is clarified with two examples from user acquisition for an online product (e.g., app, website):

- *Censorship*—predicting user *churn*, which may be either an observed (e.g., for a subscription product, in which a user must actively cancel their membership) or unobserved (e.g., for a freemium product; see Seufert [2014]). This is an example of censorship, since we know the total user count and observe (or observe by a proxy metric, such as user hasn't logged in for three consecutive days). We also know the age of each user. For each user, then, we have a binary indicator expressing whether or not they have churned, and we have a time that represents lifetime for churned users and censorship time (e.g., current cohort age) for censored users.
- *Truncation*—predicting *payer rate* for a freemium product. This represents a truncation problem since, for any given cohort, we do not know *a priori* the number of users who will convert to payers, which is typically a small percentage of overall users. Users are also free to pay at any time, meaning that those who do eventually pay will have a distribution over the amount of time for first purchase. For a cohort that is seven days old, we therefore have a count of conversions, along with their pay-conversion timestamps, but we do not have a count of how many users have yet to convert.

Churn prediction is of course a core component of *customer lifetime-value* modeling (Section 10.3). In particular, the Bayesian mixture models we discussed therein had the serious deficit of not accounting for covariates. It is therefore of interest to see how tree methods might be retrofitted in the context of survival problems.

In fact, more generally, survival techniques are of interest any time behavior plays out over time; given the online nature of most web and mobile applications, in which users interact with virtual products over time, such problems proliferate industry.

Hazard and Survival

Recall that survival problems are primarily interested in determining or approximating a distribution for a random variable T that is temporal in nature. Typically, we may regard T as the *time to an event*; classically it is regarded as the *time to death* for patients in medical applications (i.e., literal *survival studies*), wherein the theory has its roots, or the *failure time* for engineering applications, or the *time to churn* or *customer lifetime* in customer lifetime-value models. The *survival function*

$$S(t) = \mathbb{P}(T > t) = 1 - F(t) \tag{12.21}$$

represents the probability that the individual is *still alive* at time t, and is a sort of dual to the cumulative distribution function F. Further, recall from that the *hazard function* (Definition 5.8) is defined by the ratio

$$h(t) = \frac{f(t)}{S(t)}, \tag{12.22}$$

which represents the differential probability of failing at time t, given that the device has yet to fail. Similarly, the *cumulative hazard function* (CHF) is defined as the integral

$$H(t) = \int_{-\infty}^{t} h(s) \, ds. \tag{12.23}$$

The CHF is a monotonically increasing function with range $[0, \infty)$. A simple calculation shows that

$$H(t) = -\ln(S(t)). \tag{12.24}$$

We leave the proof as an exercise for the reader.

Survival Trees

We next develop methodology that applies random forests to the survival setting to specifically solve problems involving right-censored data. There are several fruit methods[6] one might consider, such as reframing the survival problem as a classification problem or as a series of classification problems. The first true generalization of random forests to strictly adhere to the prescription of Breiman [2001], however, by taking into account the desired outcome in all aspects of formulating the forest, was Ishwaran, *et al.* [2008], who introduced the basic method of *random survival forests*.

Random Survival Forests

We follow Ishwaran, *et al.* [2008] in defining a random survival forest as follows.

Definition 12.13 (Random Survival Forest). *A random survival forest is a random forest (Definition 12.11) whose underlying decision trees use methods suitable for right-censored survival problems:*

---

[6] named after the proverbial *low-hanging fruit*.

1. *isHomogeneous*—*a data set is homogeneous if it consists of fewer than a fixed number of $d_0 > 0$ events;*
2. *bestSplit*—*returns the split that maximizes the* survival difference *between the child nodes;*
3. *getLabel*—*returns the Nelson-Aalen estimator for the cumulative hazard function.*

Note 12.5. The stopping criterion in Ishwaran, *et al.* [2008] is actually that each leaf node must have at least d_0 events, corresponding to a `min_samples_leaf` parameter. This is an appropriate definition, as it ensures there is sufficient data within each leaf to have a valid, though potentially weak, empirical estimator. We leave modification of the algorithm as an exercise for the reader.   ▷

We will discuss each of these method definitions in turn. Of particular interest with be the `bestSplit`, where we must decide how to determine the *survival difference* between two data sets.

Labels for Right-Censored Training Data

Before discussing the three method modifications, let us first align on notation for training labels, so that we might efficiently capture both observed lifetimes and censoring times for right-censored instances. Labels for our training data occur in pairs (t_i, δ_i), where t_i is a time and δ_i is the binary event indicator

$$\delta_i = \mathbb{I}[\text{event has occurred for instance } i].$$

Thus, if $\delta_i = 0$, we say that the ith instance is *censored* at time t_i, which we call the *censoring time*. In the context of marketing or product analytics, we say that the user is still *alive* or *active* at time t_i, which usually represents the *cohort time*, or *age*, of the user.

On the other hand, if $\delta_i = 1$, we say that we have observed the event for the ith instance, and we interpret the time t_i as the *event time* (or *survival time* or *customer lifetime*, etc.). In CLTV problems, the time t_i is the cohort time, or age, at which the user churned.

Stopping Criterion: Minimum Events in Leaf Nodes

The tree-growth stopping criterion is that the recursion should continue as long as a minimum number of events $d_0 > 0$ exists in each leaf. This will be a limiting factor on the `bestSplit` method (Algorithm 12.2), in that only qualifying splits should be considered. Furthermore, if no qualifying split can be found, we terminate the recursion in Algorithm 12.1 and return the current label.

Best-Split Criterion: Maximum Survival Difference

Four separate splitting criteria were discussed in Ishwaran and Kogalur [2007]: logrank splitting (based on the *Mantel-Cox test*, or *logrank test*), conservation-of-events splitting, logrank score splitting, and an approximate logrank splitting, which offers improved computational speed over the exact test.

We will focus on logrank splitting, which is derived from the *Mantel-Cox test*, which tests the null hypothesis that two groups have the same hazard function. Let $T_1 < \cdots < T_d$ represent the total pooled event times between the groups, Y_{ij} represent the number of individuals at risk in group i at time T_j, and E_{ij} represent the number of observed events in group i at time T_j. Further, let $Y_j = Y_{0j} + Y_{1j}$ and $E_j = E_{0j} + E_{1j}$. Then the Mantel-Cox test states that the *logrank statistic*

$$Z = \frac{\sum_{j=1}^{d}\left(E_{ij} - \frac{Y_{ij}E_j}{Y_j}\right)}{\sqrt{\sum_{j=1}^{d}\frac{Y_{ij}E_j}{Y_j}\left(\frac{Y_j - E_j}{Y_j}\right)\left(\frac{Y_j - Y_{ij}}{Y_j - 1}\right)}}, \tag{12.25}$$

is approximately distributed as a standard normal distribution for large samples, for both $i = 0, 1$. We leave the proof to the reader (see Exercise 12.13).

In the context of survival trees, the two groups represent the two resultant child nodes of a given split. The absolute logrank score $|Z|$ is a measure of survival difference between the two child nodes, with larger values representing better splits. The logrank splitting criterion therefore fixes $i = 0$ or 1 and then uses $-|Z|$ as an impurity metric for determining the best split.

To compute the logrank statistic, consider a set of survival data $\mathcal{D} = \{(x_i, t_i, \delta_i)\}_{i=1}^{n}$, which, without loss of generality, we may think of as the data belonging to a given branch node N_0. The random variables $T_1 < \cdots < T_d$ represent the ordered, distinct survival times $\{t_i | \delta_i = 1\}$. The quantities Y_j and E_j can be computed using

$$Y_j = \sum_{i=1}^{n} \mathbb{I}[t_i \geq T_j], \tag{12.26}$$

$$E_j = \sum_{i=1}^{n} \mathbb{I}[t_i = T_j]\delta_i, \tag{12.27}$$

respectively. Now, any decision rule $r : \mathbb{D} \to \mathbb{B}$ (Definition 12.1) will split the data into two subsets, based on the value of $r(x_i)$. We can therefore compute the quantities Y_{1j} and E_{1j} as

$$Y_{1j} = \sum_{i=1}^{n} r(x_i)\mathbb{I}[t_i \geq T_j], \tag{12.28}$$

$$E_{1j} = \sum_{i=1}^{n} r(x_i)\mathbb{I}[t_i = T_j]\delta_i, \tag{12.29}$$

respectively. Alternatively, to compute Y_{0j} and E_{0j}, we can simply replace $r(x_i)$ with $(1 - r(x_i))$ in the previous two equations.

Capturing Labels: Cumulative Hazard Function

Instead of numbers (regression) or class labels (classification), survival forests predict the survival function for a given instance, typically in the form of the cumulative hazard function CHF, which has better small-sample properties, making it ideal as a survival estimator within each leaf node.

If a given decision tree has a total of ℓ leaf nodes, we may view the tree τ as a mapping $\tau : \mathbb{D} \to \mathbb{Z}_\ell$ from feature vector into an ℓ-class encoding of the tree's leaves, defined via a tree-graph of individual decision rules. This further partitions the training data into ℓ subsets $\mathcal{D}_l = \{(x_i, t_i, \delta_i)\}_{i=1}^{n_l}$, for $l = 1, \ldots, \ell$, based on the condition that a datum (x_i, t_i, δ_i) falls into $\mathcal{D}_l$ if and only if $\tau(x_i) = l$.

Now, let $T_1 < \cdots < T_d$ be the pooled set of event times for the full training set, as we had before, and define

$$Y_{lj} = \sum_{i=1}^{n} \mathbb{I}[\tau(x_i) = l]\mathbb{I}[t_i \geq T_j], \tag{12.30}$$

$$E_{lj} = \sum_{i=1}^{n} \mathbb{I}[\tau(x_i) = l]\mathbb{I}[t_i = T_j]\delta_i, \tag{12.31}$$

which is equivalent to applying Equations (12.26) and (12.27) over the partial dataset $\mathcal{D}_l$, for $l = 1, \ldots, \ell$. The Nelson-Aalen estimator

$$\hat{H}_l(t) = \sum_{T_j \leq t} \frac{E_{lj}}{Y_{lj}} \tag{12.32}$$

thus yields an empirical estimate for the cumulative hazard function for the lth leaf, which further represents the prediction "label" for the given leaf.

Finally, for an arbitrary feature vector x, we have the survival tree's predicted cumulative hazard, given by

$$\hat{H}(t|x) = \hat{H}_{\tau(x)}(t); \tag{12.33}$$

i.e., the CHF associated with leaf $l = \tau(x)$.

Prediction: Ensemble Cumulative Hazard

Finally, suppose now that we have a forest comprised of m survival trees $\tau_1, \ldots, \tau_m$, with their hazard predictors $\hat{H}_1(t|x), \ldots, \hat{H}_m(t|x)$. The *ensemble* CHF for an arbitrary feature vector is therefore given by

$$\hat{H}(t|x) = \frac{1}{m} \sum_{j=1}^{m} \hat{H}_j(t|x). \tag{12.34}$$

Recall that each tree in a random forest is trained using a bootstrap sample of the original training data set $\mathcal{D}$. Thus, each survival tree, on average, excludes approximately 37% of the full training data. For the ith training datum (x_i, t_i, δ_i), let b_{ij} be the binary indicator representing whether (0) or not (1) the training instance was used to train the jth survival tree; i.e., $b_{ij} = 1$ if the ith datum was out-of-bag for the jth tree. The *out-of-bag* CHF for the ith training instance is therefore given by

$$\hat{H}_{oob}(t|x_i) = \frac{\displaystyle\sum_{j=1}^{m} b_{ij}\hat{H}_j(t|x_i)}{\displaystyle\sum_{j=1}^{m} b_{ij}}, \tag{12.35}$$

which can be used to measure out-of-bag accuracy of the ensemble.

Problems

12.1. How would one modify Algorithm 12.2 to allow for multiway splits for categorical features?

12.2. Implement the golden-search algorithm to determine the minimum value of the function $f(x) = -xe^{-x}$.

12.3. Update the _isHomogeneous method in Code Block 12.4 to provide a variance-based stopping criterion for regression problems.

12.4. Write a pruning method for the **DecisionTree** Python class.

12.5. Write an **__add__(self, other)** method for the **Ensemble** class (Code Block 12.9), following Definition 12.6, so that we can combine trained ensembles with the notation **E + F**.

12.6. Show that a bootstrap sample, on average, should contain $(1-1/e) \approx 63.21\%$ of the original data, when the sample is done with replacement and is the same size as the original sample. *Hint*: What is the probability of *not* selecting any particular datum from the set? Then take the limit $n \to \infty$ and recall the definition of e.

12.7. Compute the twelve predictions from the random forest shown in Figure 12.4 for the instance: sepal length 5.00, petal length 2.50, and petal width 2.00. What is the final vote tally for the three classes (setosa, versicolor, and virginica)?

12.8. Explain why the feature sex:M will never be used in the diabetes example of Code Block 12.17. *Hint*: consider carefully how the _bestSplit method of the DecisionTree class is written.

12.9. Explain the inherent problem in applying the AdaBoost Algorithm 12.6 to a classification problem with more than two classes. How might it be resolved?

12.10. Show that the starting model in the gradient boosting algorithm is the sample mean, when using squared error loss (i.e., prove Equation (12.16)).

12.11. Show that by summing the log loss Equation (12.15) over a set of data, one obtains the entropy impurity measure Equation (12.4).

12.12. Prove Equation (12.24).

12.13. Prove Equation (12.25). *Hint*: First, explain why $E_{ij} \sim \text{HypGeom}(Y_j, E_j, Y$. Then use known properties of the hypergeometric distribution and the central-limit theorem to prove the result.

13

Deep Thoughts

In this chapter, we discuss the basics of the *artificial neural network* (ANN) and *deep learning*, which constitute the foundation of modern artificial intelligence. Our favorite references are Aggarwal [2018] and Gèron [2019], the latter of which is focused on sci-kit learn and tensorflow. For a short introduction of neural networks, see Efron and Hastie [2016]. Another classic text on the subject is Goodfellow, *et al.* [2016]. For additional references on deep learning in Python, see Chollet [2021] or Stevens, *et al.* [2020].

The idea of an artificial neuron was first introduced in the context of propositional logic by McColloch and Pitts [1943]; here, the proposed artificial neuron activated its output only if a sufficient number of binary inputs were activated. The authors showed that is possible to build a network of such neurons that could compute arbitrary logical propositions in a systematic way.

Hebb's rule provided another crucial clue to the development of the theory: whenever one biological neuron triggers another, the connection between the two grows stronger; i.e., *cells that fire together wire together* (see Hebb [1949]). This observation was a key inspiration to the *perceptron learning algorithm*, introduced a few years later.

The *perceptron* (also known as a *linear threshold unit* (LTU)), introduced by Rosenblatt [1957], is a variation of the purely logical neurons introduced years earlier and constitutes the basis of modern ANNs. Reading from this paper:

> Recent theoretical studies by this writer indicate that it should be feasible to construct an electronic or electromechanical system which will learn to recognize similarities or identities between patterns of optical, electrical, or tonal information, in a manner which may be closely analogous to the perceptual process of a biological brain. The proposed system depends on probabilistic rather than deterministic principles for its operation, and gains its reliability from the properties of statistical measurements obtained from

large populations of elements. A system which operates according to these principles will be called a *perceptron*.

Yes, that *should* be feasible enough. Let's ask Siri...

13.1 Artificial Neurons

We begin with a brief discussion of the simplest neural architecture: the perceptron. We then introduce the perceptron learning algorithm, how perceptrons can be generalized to multiclass classification problems, and conclude with the generalization of the perceptron to general artificial neurons. This will form the basis for our discussion on neural networks in our next section.

13.1.1 The Perceptron

The *perceptron* is the simplest neural network, which consists of a number of inputs (the *input layer*) and a single output node.

Definition 13.1. *A* perceptron, *or* linear threshold unit (LTU), *is a binary classifier over a set of input features $x \in \mathbb{R}^d$ that is defined by the composite function $f = H \circ \omega$, where $\omega : \mathbb{R}^d \to \mathbb{R}$ is the linear mapping*

$$\omega(x) = \beta + \sum_{j=1}^{d} w_j x_j, \tag{13.1}$$

where the parameter β is referred to as the bias, the values $\{w_j\}_{j=1}^{d}$ are given weights, and $H(x) = \mathbb{I}[x \geq 0]$ is the classic Heaviside function.

At face value, this seems right in line with the generalized linear models we discussed in Chapter 7. For instance, if we replaced H with the logistic function, we would have the functional model for logistic regression; if we replaced it with an identity, we'd have the functional form for linear

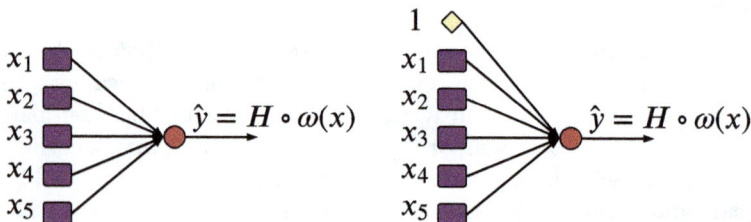

Fig. 13.1: The perceptron with (right) and without (left) a bias.

regression. The rational for considering the Heaviside function is based in neuroscience: neurons in the brain are connected to other cells via synapses, which may have a particular weight; the given neuron, then, fires an electrical signal via its axon only if the weighted input signal passes a particular threshold. This analogy can be better understood by representing the perceptron using a graphical model, as shown in Figure 13.1.

The perceptron is represented by the red circle, whereas the inputs (from other cells) are represented by the purple squares. The diamond node (Figure 13.1 (right)) represents the optional bias term. (We may, without loss of generality, represent the bias term as an additional *trivial neuron* with constant value 1 and weight β.)

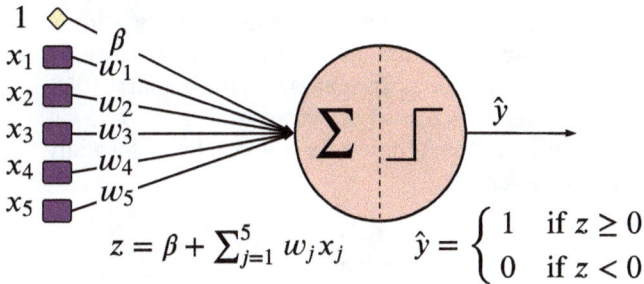

$$z = \beta + \sum_{j=1}^{5} w_j x_j \qquad \hat{y} = \begin{cases} 1 & \text{if } z \geq 0 \\ 0 & \text{if } z < 0 \end{cases}$$

Fig. 13.2: An exploded schematic of a perceptron with bias.

In order to understand the inner workings of the perceptron, it is helpful to consider the exploded view shown in Figure 13.2. Here, we see that the perceptron follows a two-step process in determining its output. First, it computes the inner product $z = \omega(x)$ of its inputs and weights. Second, it determines if the cumulative weight z is sufficient to warrant an output signal; i.e., $H(z) = 1$ when the cumulative weighted inputs are sufficient for an output to fire, whereas $H(z) = 0$ otherwise.

As shown in Figure 13.2, it is often useful to think of the model bias as an additional input x_0, which is set to $x_0 = 1$. In this case, we can set $w_0 = \beta$, such that

$$\omega(x) = \beta + \sum_{j=1}^{d} w_j x_j = \sum_{j=0}^{d} w_j x_j.$$

Given a set of weights, the level sets of the function $\omega(x)$ define hyperplanes in $\mathbb{R}^d$, meaning that the perceptron has a linear decision boundary, similar to logistic regression. (For data that are not linearly separable, we require more advanced neural networks.)

13.1.2 The Perceptron Learning Algorithm

In order to train a perceptron model, we need to determine the weights $w_0, \ldots, w_d$ that minimize the binary loss function (Equation (6.19)), which is equivalent to the squared-error loss (Equation (6.17)), since the target variable is binary. Given a training data set $\mathcal{D} = \{(x_i, y_i)\}_{i=1}^n$, the cumulative binary loss is therefore equivalent to

$$\sum_{\mathcal{D}} L(y, \hat{y}) = \sum_{\mathcal{D}} (y - \hat{y})^2 = \sum_{\mathcal{D}} (y - H \circ \omega(x))^2. \qquad (13.2)$$

Since this represents a staircase function, its derivative vanishes, except for a countable number of points at which it diverges to infinity. Nevertheless, we can construct an algorithm that mimics stochastic gradient descent, as given by Algorithm 13.1. The algorithm typically continues until the perceptron has achieved a minimum accuracy. Convergence is guaranteed if the training instances are linearly separable (Rosenblatt [1957]), otherwise, the algorithm will never achieve perfect accuracy.

In order to motivate the update rule of line 6, let us formally differentiate the loss function as follows

$$\frac{\partial L}{\partial w_j} = \frac{\partial L}{\partial \hat{y}} \frac{\partial \hat{y}}{\partial z} \frac{\partial z}{\partial w_j} = -2(y - \hat{y})\delta(z)x_j$$

where $\delta(z)$ is the Dirac delta function (see Section 1.3), which, the reader will undoubtedly recall, may be viewed as the derivative of the Heaviside function. The delta function is problematic, as it either diverges (whenever the input vector is orthogonal to the weights, so that $\sum_{j=0}^{d} w_j x_j = 0$) or vanishes otherwise. Nevertheless, the perceptron learning algorithm functions as expected if we remove the delta function in the gradient and absorb the factor of 2 into the learning rate η.

Algorithm 13.1: The perceptron learning algorithm.

Input: labeled data $\mathcal{D} = \{(x_i, y_i)\}_{i=1}^n$; $x_i \in \mathbb{R}^d$ and $y_i \in \mathbb{B}$; learning rate η
Output: The set $\{w_j\}_{j=0}^d$ of perceptron weights, with bias $w_0 = \beta$.
1 Set $w_j = 0$ for $j = 0, \ldots, d$
2 Set $converged = $ **False**
3 **while not** $converged$ **do**
4   Randomly select $(x, y) \in \mathcal{D}$
5   Compute $\hat{y} = H \circ \omega(x)$
6   Update $w_j = w_j + \eta(y - \hat{y})x_j$, for $j = 0, \ldots, d$
7     $converged = $ `checkConverged`$(\mathcal{D}, \{w_j\})$
8 **end**
9 **return** $w_0, w_1, \ldots, w_d$

13.1.3 Vector-output Perceptrons

If the target variable is a c-dimensional binary vector ($y \in \mathbb{B}^c$), we can generalize our model by stacking c individual linear threshold units into a *layer*, which may be regarded as a *vector perceptron*. In this case, each individual LTU has its own set of weights. Such an arrangement is shown in Figure 13.3.

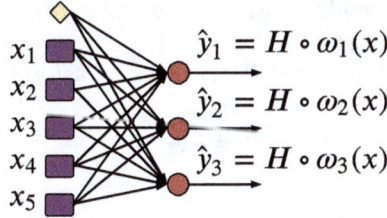

$$\hat{y}_1 = H \circ \omega_1(x)$$
$$\hat{y}_2 = H \circ \omega_2(x)$$
$$\hat{y}_3 = H \circ \omega_3(x)$$

Fig. 13.3: A vector perceptron for target variable $y \in \mathbb{B}^c$, shown for $c = 3$.

For the ith perceptron in a perceptron vector, we may regard the function ω_i as the *one-form*[1]

$$\omega_i(x) = \sum_{j=0}^{d} w_{ij} x_j.$$

We may further stack these one-forms as rows in a matrix W, with components

$$W = \begin{bmatrix} w_{11} & \cdots & w_{1d} \\ \vdots & \ddots & \vdots \\ w_{c1} & \cdots & w_{cd} \end{bmatrix}, \tag{13.3}$$

where w_{ij} is the weight between the jth input and the ith output, for $i = 1, \ldots, c$ and $j = 1, \ldots d$. Note that many authors define the weight matrix as the *transpose* of this definition; we, however, feel it to be more natural to define our weight matrix so that $z = Wx$, as we move forward through the neural net.

Similarly, we may regard $\beta \in \mathbb{R}^c$ as a vector, so that the operations of a given vector perceptron may be captured in a single vector equation

$$z = \beta + Wx$$
$$\hat{y} = H(z),$$

where $x \in \mathbb{R}^d$ and $H : \mathbb{R}^c \to \mathbb{R}^c$ operates on its vector input componentwise.

---

[1] A one-form can intuitively thought of as a mapping from a vector space into the real numbers; formally, this may be regarded as a *row vector*, since the matrix product of a row vector and a (column) vector is a scalar.

13.1.4 Softmax Neurons

Obviously, our definition of the vector perceptron is not readily applicable to multiclass classification problems, as the output of the former is expressed as an arbitrary binary vector. To remedy this, we will apply a *softmax layer*, which will normalize the pre-outputs into a set of probabilities. This will require two special definitions.

Definition 13.2. *A Σ neuron (sigma neuron) is a perceptron for which the Heaviside function is replaced with the identity mapping $\iota(z) = z$.*

A level i softmax neuron, for $i = 1, \ldots, c$, where c is the number of inputs, is the ith component of the softmax transform (Equation (6.12)),

$$s_i(z) = \frac{e^{z_i}}{\sum_{j=1}^{c} e^{z_j}}, \tag{13.4}$$

where $z \in \mathbb{R}^c$ is the input vector.

Graphically, we will represent sigma neurons as circles with embedded Σ signs and softmax neurons as hexagons. A sigma neuron may be viewed as the first half of the perceptron.

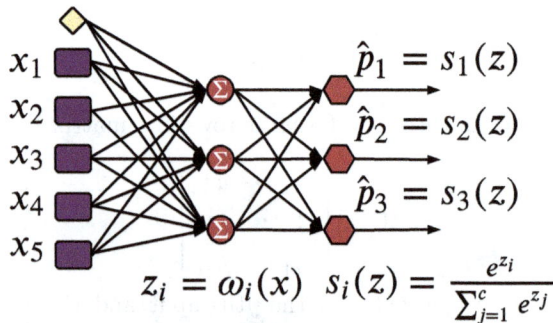

Fig. 13.4: Softmax transform.

Sigma and softmax neurons always occur in pairs, as shown in Figure 13.4. The reason for separating these, as opposed to smashing them together into a single neuron, is that each softmax neuron requires the output from each sigma neuron. The final outputs $\hat{p}_1, \ldots, \hat{p}_c$ may then be viewed as probabilities for the c-class classification problem.

Though separated from a diagrammatic perspective, in the end we will program them concurrently in a single neural layer called a *softmax layer*. Figure 13.4 may therefore be regarded as simply an exploded view into such a layer.

13.1.5 Activation Functions

For general artificial neural networks, we will generalize the perceptron into a closely related concept called the *artificial neuron*. Artificial neurons still consist of two steps: a summation step and an activation step. However, the activation step may be achieved with a broader class of functions. In particular, we will replace the Heaviside function H with a general *activation function* ϕ.

We previously saw that the lack of differentiability of the Heaviside function created problems when attempting to calculate the gradient of the loss function. For general ANNs, we will be interested in deploying stochastic gradient descent to learn the network weights. As such, we will require both the functional form and the derivative of any activation function we use. In this section, we will highlight several commonly used forms for the activation function. Plots of these functions are shown in Figure 13.5; their derivatives are plotted in Figure 13.6.

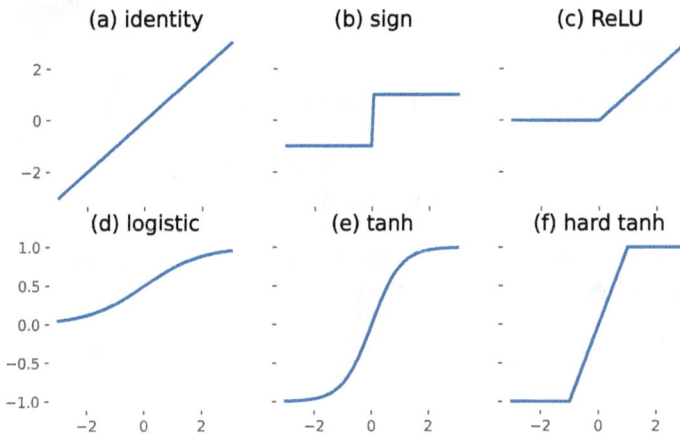

Fig. 13.5: Commonly used activation functions.

Identity Function

The identify function and its derivative are given by

$$\phi(z) = z$$
$$\phi'(z) = 1.$$

An artificial neuron with the identity activation is simply a *sigma neuron* (Definition 13.2). The chief usage here is in coupling the neuron with a softmax output layer.

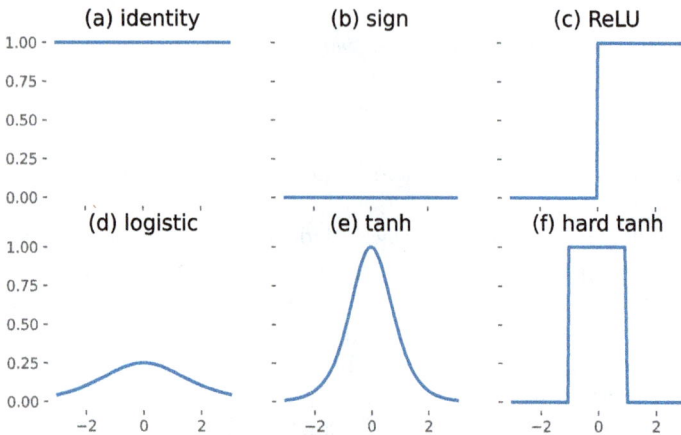

Fig. 13.6: Derivatives of commonly used activation functions.

Sign Function

The sign function and its derivative are given by

$$\phi(z) = \text{sign}(z) = H(z) - H(-z)$$
$$\phi'(z) = \text{sign}'(z) = \delta(z).$$

The derivative of this function vanishes everywhere, except at the origin, where it diverges. It is therefore rarely used in practice.

Rectified Linear Unit

The *rectified linear unit (ReLU)* and its derivative are given by

$$\phi(z) = \max\{0, z\} = zH(z)$$
$$\phi'(z) = H(z)$$

Despite the lack of differentiability at $z = 0$ and the vanishing of the derivative for $z < 0$, the ReLU function tends to work well in practice, and is perhaps the most commonly used form of activation. It is also relatively easy to compute. Given the nonnegative aspect of the ReLU function, it is also commonly used as the activation function in the output layer for the regression of nonnegative target variables.

Logistic Function

The *logistic function* and its derivative are given by

$$\phi(z) = \frac{1}{1 + \exp(-z)}$$

$$\phi'(z) = \frac{\exp(-z)}{(1 + \exp(-z))^2} = \phi(z)\left[1 - \phi(z)\right]$$

The second equation here is the *logistic equation*, and is a result of the functional form of the derivative. This is one of the classic activation functions, but has largely been replaced by the ReLU function in modern practice.

Hyperbolic Tangent Function

The and its derivative are given by

$$\phi(z) = \tanh(z) = \frac{\exp(2z) - 1}{\exp(2z) + 1}$$

$$\phi'(z) = \tanh'(z) = \frac{4\exp(2z)}{(1 + \exp(2z))^2} = \operatorname{sech}^2(z) = 1 - \tanh(z)^2$$

Though the hyperbolic tangent is a simple transformation of the logistic function, the hyperbolic tangent is preferred when the final outputs might be both positive or negative. Additionally, the fact that it is centered and has a larger gradient tends to make it easier to train in practice.

"Hard" Hyperbolic Tangent Function

The *hard hyperbolic tangent function* and its derivative are given by

$$\phi(z) = \max\left\{-1, \min\left\{1, z\right\}\right\}$$

$$\phi'(z) = \mathbb{I}[|z| < 1] = H(1 - z) - H(-1 - z)$$

Similar to how ReLU has largely replaced the logistic function, the hard hyperbolic tangent has largely replaced the regular (soft) hyperbolic tangent in modern usage. This activation has similar scaling to its soft counterpart, but has the advantage of a piecewise-constant derivative.

Nonlinear Activations

Finally, we note the importance of nonlinear activation functions, without which the point of constructing neural network architectures would be lost. This is because of the LC^3 theorem: linear combinations of linear combinations of linear combinations are linear combinations. In other words, if we were to build a network that only consisted of layer after layer of linear combinations, we would, in the end, only have an over specified linear combination.

13.2 Neural Networks

In this section, we generalize the perceptron to more complex graphical patterns known as artificial neural networks. We begin by laying out the basic definitions and nomenclature. Then we discuss training neural networks using stochastic gradient descent to minimize a given loss function.

13.2.1 Neural Networks

To begin, we formalize our concept of a *neural layer*.

Definition 13.3. *A* neural layer *(or, simply,* layer*) is a set of artificial neurons that share a set of inputs. The* size *of a neural layer is the number of* nontrivial neurons *(i.e., non-bias neurons) that comprise it.*

For example, the red circular output neurons shown in Figure 13.3 constitute a layer. Similarly, the blue rectangles constitute a special layer known as an *input layer.*

Definition 13.4. *An* artificial neural network (ANN) *is a sequence of neural layers, such that the inputs to one layer are the outputs of the preceding layer.*

In an artificial neural network, the input layer *consists of the set of neurons that represent the direct functional input of the network, the* output layer *consists of the set of neurons that represent the final functional output of the network, and the* hidden layers *is the set of layers that are sandwiched between.*

An artificial neural network is said to be shallow *if it only consists of zero, one, or two hidden layers. Otherwise it is said to be* deep.

For example, in Figure 13.4, the hexagonal neurons constitute the output layer. In this case, the output layer is a special kind of output known as a *softmax layer. Ibid.*, the middel sigma neurons constitute a hidden layer.

An example of a deep neural network consisting of four layers is shown in Figure 13.7. (The input layer is typically not counted.) Each layer and its input connections are colored together. Here, we have a two-node output layer ($l = 4$), a five-node input layer ($l = 0$), and three hidden layers ($l = 1, 2, 3$) of sizes five, four, and three, respectively.

Neural Network Notation Convention

To describe a general artificial neural network, we will make use of the following notation convention.

Consider a neural network with m layers (not counting the input layer), indexed by l. The size of layer l will be denoted by r_l. We therefore have:

1. layer $l = 0$ is the input layer, with size $r_0 = d$;

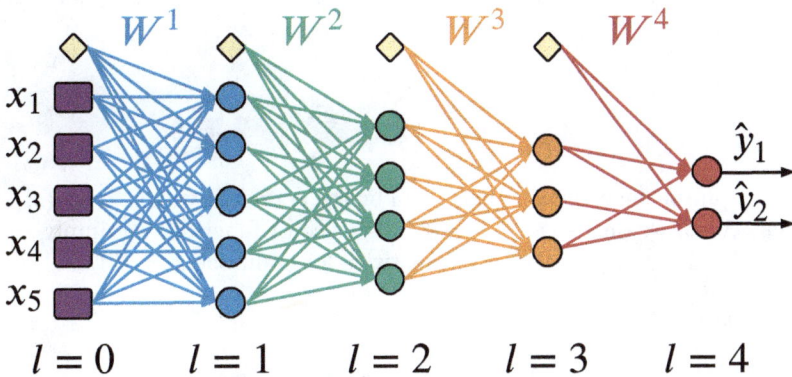

W^1 W^2 W^3 W^4

x_1
x_2
x_3
x_4
x_5

$\hat{y}_1$
$\hat{y}_2$

$l = 0$ $\quad$ $l = 1$ $\quad$ $l = 2$ $\quad$ $l = 3$ $\quad$ $l = 4$

Fig. 13.7: An artificial neural network.

2. layers $l = 1, \ldots, m-1$ constitute the hidden layers, with sizes $r_1, \ldots, r_{m-1}$;
3. layer $l = m$ is the output layer, with size $r_m = c$.

We will assume, as often is the case, that each layer uses a common activation function within the layer. We will denote the activation function for layer l as $\phi_l : \mathbb{R} \to \mathbb{R}$.

Aside from the activation function, the variables of layer l will be marked with a *superscript* l[2].

As was the case with vector perceptrons, the weights for all connections leading into layer l may be represented by a single $r_l \times r_{l-1}$ matrix W^l and a vector $\beta^l \in \mathbb{R}^{r_l}$, for $l = 1, \ldots, m$, which takes the form of Equation (13.3) (with d replaced by r_{l-1} and c replaced by r_l):

$$W^l = \begin{bmatrix} w^l_{11} & \cdots & w^l_{1r_{l-1}} \\ \vdots & \ddots & \vdots \\ w^l_{r_l 1} & \cdots & w^l_{r_l r_{l-1}} \end{bmatrix}. \tag{13.5}$$

We further added superscripts to the weights and biases to indicate their respective layer number. In particular:

1. the quantity β^l_i represents the bias of the ith neuron of layer l;
2. the quantity w^l_{ij} represents the weight between the ith neuron of layer l and the jth neuron of layer $l - 1$, for $i = 1, \ldots, r_l$ and $j = 1, \ldots, r_{l-1}$.

Finally, we denote the output of the ith neuron of layer l with the variable a^l_i, choosing the variable a as these values correspond to the "activated" values. Naturally, we identify $a^0_i = x_i$ for the input layer. For layers

---

[2] We use a subscript, as opposed to a superscript, for the activation function for two reasons: first, no further enumeration is needed and, second, so that we may more cleanly represent the derivative $\phi'_l(z)$.

$l = 1, \ldots, m$, we have the intermediary result $z^l = \beta^l + W^l a^{l-1}$, such that the ith component of this vector is given by

$$z_i^l = \beta_i^l + \sum_{j=1}^{r_{l-1}} w_{ij} a_j^{l-1}.$$

The output of the neuron is then given by the activated value $a_i^l = \phi_l(z_i^l)$, as usual. When we define $\phi_l : \mathbb{R}^{r_l} \to \mathbb{R}^{r_l}$ as a vector function, we simply allow the activation function to act elementwise for components $i = 1, \ldots, r_l$.

Feedforward Computation

Though the notation itself reveals the precise computational steps required to compute the output of a neural network with a given set of weights and a given input vector, we define the following *feedforward computation algorithm* in Algorithms 13.2 and 13.3 for clarity. In the former, we represent the equations in their vector–matrix form, whereas in the latter, we explicitly represent the calculations for each individual neuron in the network.

13.2.2 Backpropagation

In order to determine the weights for an artificial neural network that minimize a given loss function over a set of training data, we will apply stochastic or mini-batch gradient descent, as usual. The challenge for us, however, is in determining the gradients in the first place. The key insight is that neural networks are, essentially, intrinsic composite functions that are defined by a computational graph. As such, we may rely on rules of calculus to arrive at expressions for the various derivatives.

Algorithm 13.2: The feedforward algorithm (matrix version).

Input: an input vector $x \in \mathbb{R}^d$;
a given neural network of size m;
a set of m activation functions $\phi_1, \ldots, \phi_m$;
a given set of (current) bias vectors $\mathcal{B} = \{\beta^1, \ldots, \beta^m\}$;
a given set of (current) weights $\mathcal{W} = \{W^1, \ldots, W^m\}$
Output: Vector of predicted values $\hat{y} \in \mathbb{R}^c$.

1 Set vector $a^0 = \langle x_1, \ldots, x_d \rangle$
2 **for** *each layer $l = 1, \ldots, m$* **do**
3 $\quad$ Compute $z^l = \beta^l + W^l a^{l-1}$
4 $\quad$ Compute $a^l = \phi_l(z^l)$
5 **end**
6 **return** $\hat{y} = a^m \in \mathbb{R}^c$

Gradients *à la* Chain Rule

The loss function may be regarded as a final output value that comes after the (potentially) multiclass output layer. As such, computing its gradient reduces to the ability to compute arbitrary partial derivatives in directed acyclic computational graphs. (Recall Definition 8.4.) By a *computational graph*, we mean a graph whose nodes represent variables, and whose directed edges represent functional relations, similar to our definition of a causal structure (Definition 8.7). As it turns out, the chain rule takes a simple form for such a use case, though it is exponentially expensive to compute as the size of the graph grows. Our present goal is merely to introduce the full chain rule and demonstrate its prohibitive cost. In the following section, we introduce a technique known as backpropagation that is used in practice.

The Chain Rule for Computational DAGs

Before proceeding, recall that a path P in a graph $\mathcal{G}$ between two connected points x and y is a sequence of points $\{v_0^P, v_1^P, \ldots, v_{\ell(P)}^P\} \subset \mathcal{G}$, where $\ell(P)$ is the length of the path, such that $v_0^P = x$, $v_{\ell(P)}^P = y$, and

$$v_{i-1}^P \in \mathrm{par}(v_i^P),$$

for $i = 1, \ldots, \ell(P)$. Given this definition, we may write the following. (Though it is stated in terms of general computational DAGs, one should be content to picture the graph as a neural network for our present purpose.)

Algorithm 13.3: The feedforward algorithm (scalar version).

Input: an input vector $x \in \mathbb{R}^d$;
       a given neural network of size m;
       a set of m activation functions $\phi_1, \ldots, \phi_m$;
       a given set of (current) bias vectors $\mathcal{B} = \{\beta^1, \ldots, \beta^m\}$;
       a given set of (current) weights $\mathcal{W} = \{W^1, \ldots, W^m\}$
Output: The predicted values $\{\hat{y}_i\}_{i=1}^c$.
1 Set vector $a_i^0 = x_i$ for $i = 1, \ldots, d$
2 **for** *each layer* $l = 1, \ldots, m$ **do**
3   **for** *each neuron* $i = 1, \ldots, r_l$ **do**
4   Compute $z_i^l = \beta_i^l + \sum_{j=1}^{r_{l-1}} w_{ij}^l a_j^{l-1}$
5   Compute $a_i^l = \phi_l(z_i^l)$
6   **end**
7 **end**
8 Set $\hat{y}_i = a_i^m$, for $i = 1, \ldots, c$
9 **return** $\hat{y}_1, \ldots, \hat{y}_c$

Theorem 13.1 (Chain Rule for Computational Graphs). *Let $\mathcal{G}$ be a directed acyclic computational graph, the nodes $x, y \in \mathcal{G}$ with $x \in \text{anc}(y)$, and $\mathcal{P}$ the set of all directed paths that connect x to y. Then the partial derivative of y with respect to x may be expressed as*

$$\frac{\partial y}{\partial x} = \sum_{P \in \mathcal{P}} \left(\prod_{i=1}^{\ell(P)} \frac{\partial v_i^P}{\partial v_{i-1}^P} \right). \tag{13.6}$$

Classic Chain Rule Examples with their Corresponding DAGs

We next consider the classic chain rule from univariate and multivariate calculus, and show their equivalency to Theorem 13.1.

Example 13.1. Consider the chain (Definition 8.12) shown in Figure 13.8. This represents the graphical representation of the classic chain rule from

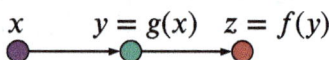

Fig. 13.8: A simple chain, representing $z = f(g(x))$.

calculus, in which we have a composite function $z = f(y)$ and $y = g(x)$; or, in short, $z = f(g(x))$. From basic calculus, we have

$$\frac{dz}{dx} = f'(g(x))g'(x) = \frac{dz}{dy}\frac{dy}{dx},$$

which is the simplest application of Equation (13.6), as there is only a single directed path that connects x to z.   ▷

Example 13.2. As a simple example of a multivariate chain rule, consider the computational DAG shown in Figure 13.9. Here, we consider the mul-

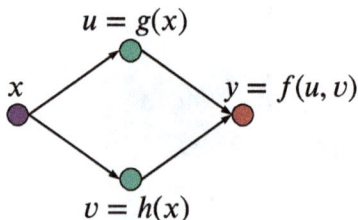

Fig. 13.9: A fork coupled with a collider, representing $y = f(g(x), h(x))$.

tivariate composite function $y = f(u, v)$, $u = g(x)$, and $v = h(x)$. In the graphical representation for this composite function, there are two paths linking x to y. From calculus, we have

$$\frac{dy}{dx} = f_u(g(x), v(x))g'(x) + f_v(g(x), v(x))v'(x) = \frac{\partial y}{\partial u}\frac{du}{dx} + \frac{\partial y}{\partial v}\frac{dv}{dx}.$$

One can readily verify this is equivalent to the formula produced by Equation (13.6).   ▷

Application of Chain Rule to ANNs

The graphical representation of a neural network is not a fully expressed computational graph, but a shorthand, since each neuron is actually a representation of a short chain and since the weights are not explicitly represented as nodes (though they are variables). To see this, consider the shallow ANN of Figure 13.10 comprised of a single input, no bias, a three-node hidden layer, and a single node output. We can easily explode this into a full computational graph, including the final loss function (which is a function of both the final output $\hat{y}$ and the true label y of the training instance), as shown in Figure 13.11.

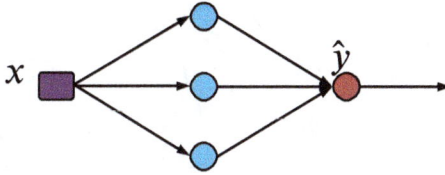

Fig. 13.10: A shallow neural network with two layers.

Naturally, if we had more than one input, our problems would multiply, as we would require a unique weight node for each *connection*: for example, five inputs into a six-node layer would require thirty weight nodes (thirty-six if we include a bias). Nevertheless, we are performing this exercise of the exploded computational graph simple to gain some insight into the workings of more general neural networks.

Each of the six weights shown in Figure 13.11 has only a single directed path to the loss function L. In order to examine a multipath application of the chain rule, we must, minimally, add an additional neuron to the output layer. The exploded computational graph corresponding to this change is shown in Figure 13.12.

By means of example, let us construct the partial derivative of the loss function with respect to the first weight w_{11}^1. There are two paths between

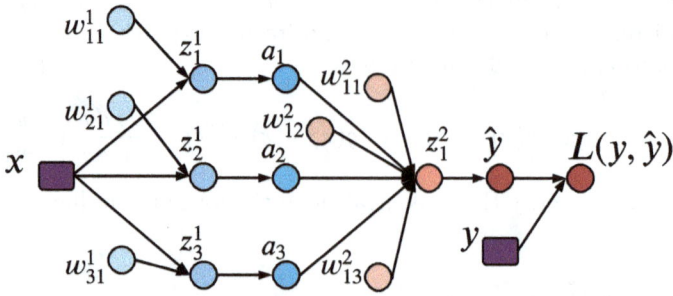

Fig. 13.11: The full computational DAG corresponding to the ANN of Figure 13.10.

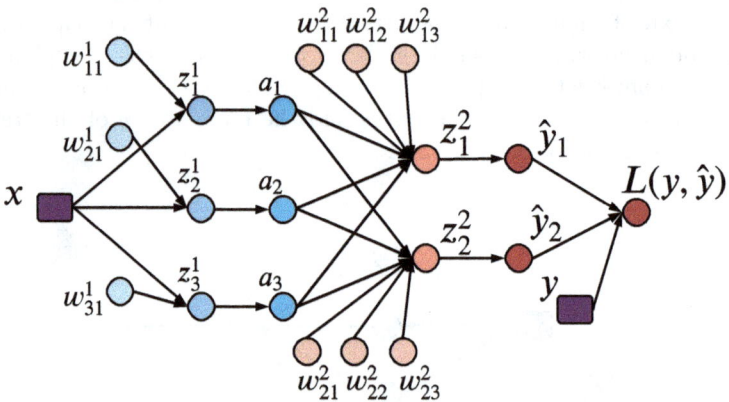

Fig. 13.12: The full computational DAG corresponding of a two-layer ANN with two output nodes.

these two endpoint nodes, one passing through each output neuron. Let us start with the final segment of the first path:

$$\frac{\partial L}{\partial z_1^2} = \frac{\partial L}{\partial \hat{y}_1} \frac{d\hat{y}_1}{dz_1^2} = \frac{\partial L}{\partial \hat{y}_1} \phi_2'(z_1^2),$$

where ϕ_2 is the activation function of the second (output) layer.

Next, let us consider the connection between a_1 and z_1^2. Since the quantity z_1^2 is simply the inner product

$$z_1^2 = w_{11}^2 a_1 + w_{12}^2 a_2 + w_{13}^2 a_3$$

between the inputs (a_1, a_2, a_3) and the weights $(w_{11}^2, w_{12}^2, w_{13}^2)$, we find that

$$\frac{\partial z_1^2}{\partial a_1} = w_{11}^2.$$

Finally, we have

$$\frac{\partial a_1}{\partial w_{11}^1} = \frac{da_1}{dz_1^1}\frac{\partial z_1^1}{\partial w_{11}^1} = \phi'(z_1^1)x,$$

since $z_1^1 = w_{11}^1 x$.

Combining the preceding results and repeating the calculation for the second path, we have

$$\frac{\partial L}{\partial w_{11}^1} = \frac{\partial L}{\partial \hat{y}_1}\frac{d\hat{y}_1}{dz_1^2}\frac{\partial z_1^2}{\partial a_1}\frac{da_1}{dz_1^1}\frac{\partial z_1^1}{\partial w_{11}^1} + \frac{\partial L}{\partial \hat{y}_2}\frac{d\hat{y}_2}{dz_2^2}\frac{\partial z_2^2}{\partial a_1}\frac{da_1}{dz_1^1}\frac{\partial z_1^1}{\partial w_{11}^1}$$

$$= \frac{\partial L}{\partial \hat{y}_1}\phi_2'(z_1^2)w_{11}^2\phi_1'(z_1^1)x + \frac{\partial L}{\partial \hat{y}_2}\phi_2'(z_2^2)w_{21}^2\phi_1'(z_1^1)x.$$

Voila! Couldn't be easier!

Exponential Complexity

The problem with the chain rule, as expressed by Equation (13.6), is that the number of paths increases *exponentially* with the depth of the graph. For a graph with m layers, of sizes $r_1, \ldots, r_m$, respectively, the number of paths between the loss function and any given first-layer weight is the product

$$\text{number of paths} = \prod_{l=2}^{m} r_l.$$

Since there are r_1 neurons in the first layer, we require a total of $\prod_{l=1}^{m} r_l$ path computations in total. The total number of paths we must consider to compute the full gradient is therefore given by

$$\text{total number of paths} = \sum_{l'=1}^{m}\prod_{l=l'}^{m} r_l.$$

However, as should be clear from our discussion of Figure 13.12, there is much redundancy in these calculations. As it turns out, we can construct an algorithm using *dynamic programming* that renders the total computational complexity as merely linear with depth, not exponential. We take up this topic forthwith.

The Backpropagation Algorithm

As we have seen, the number of paths between any given weight and the loss function increases exponentially with depth, making the brute-force application of the chain rule prohibitively expensive. In this section, we show how *dynamic programming* can be used to compute the full gradient of the loss function with respect to the full set of weights. The resulting algorithm is known as *backpropagation* and is given in Algorithm 13.4.

Algorithm 13.4 works because it exploits the redundancy in the calculation of the individual partial derivative. Moreover, the number of computations scales like the number of neurons, not exponentially like the number of layers. This is because we start at the end and work backwards, saving the intermediary partial derivatives as we go.

Notice that the variables δ_i^l are simply the partial derivatives of L with respect to the sigma variables z_i^l. The reason for focusing on these values, in particular, is that they are used in two ways. First, to compute the partial derivative of L with respect to any of the r_{l-1} weights associated with the ith neuron of the lth layer, we take

---

Algorithm 13.4: Backpropagation (scalar version).

Input: a training instance (x, y);
a given neural network of size m;
a set of m activation functions $\phi_1, \ldots, \phi_m$;
a given set of (current) bias vectors $\mathcal{B} = \{\beta^1, \ldots, \beta^m\}$;
a given set of (current) weights $\mathcal{W} = \{W^1, \ldots, W^m\}$

Output: The full gradient $\nabla_{\mathcal{W}} L$ and $\nabla_{\mathcal{B}} L$.

1 Compute $\hat{y} = f(x, \mathcal{W})$ using the feedforward algorithm, saving the values of all intermediary variables

2 **for** *each layer* $l = m, \ldots, 1$ *(in descending order)* **do**

3   **for** *each neuron* $i = 1, \ldots, r_m$ **do**

4   **if** *layer l is output layer* $l = m$ **then**

5   Set $\delta_i^m = \dfrac{\partial L}{\partial z_i^m} = \dfrac{\partial L}{\partial \hat{y}_i} \phi_m'(z_i^m)$

6   **else**

7   Set $\delta_i^l = \dfrac{\partial L}{\partial z_i^l} = \left(\displaystyle\sum_{k=1}^{r_{l+1}} w_{ki}^{l+1} \delta_k^{l+1} \right) \phi_l'(z_i^l)$

8   **end**

9   Set $\dfrac{\partial L}{\partial w_{ij}^l} = \delta_i^l a_j^{l-1}$ for $j = 1, \ldots, r_{l-1}$

10   Set $\dfrac{\partial L}{\partial \beta_i^l} = \delta_i^l$

11   **end**

12 **end**

13 **return** $\nabla_{\mathcal{W}} L = \left\{ \dfrac{\partial L}{\partial w_{ij}^l} \right\}_{l=1}^{m} {}_{i=1}^{r_l} {}_{j=1}^{r_{l-1}}$ and $\nabla_{\mathcal{B}} L = \left\{ \dfrac{\partial L}{\partial \beta_i^l} \right\}_{l=1}^{m} {}_{i=1}^{r_l}$

$$\frac{\partial L}{\partial w_{ij}^l} = \frac{\partial L}{\partial z_i^l}\frac{\partial z_i^l}{\partial w_{ij}^l} = \delta_i^l a_j^{l-1}$$

$$\frac{\partial L}{\partial \beta_i^l} = \frac{\partial L}{\partial z_i^l}\frac{\partial z_i^l}{\partial \beta_i^l} = \delta_i^l,$$

for $j = 1, \ldots, r_{l-1}$, since

$$z_i^l = \beta_i^l + w_{i1}^l a_1^{l-1} + \cdots + w_{ij}^l a_j^{l-1} + \cdots + w_{ir_{l-1}}^l a_{r_{l-1}}^{l-1},$$

as usual. This yields lines 9 and 10 of our algorithm.

On the other hand, if we want to connect to the previous layer, we have

$$\frac{\partial L}{\partial a_j^{l-1}} = \frac{\partial L}{\partial z_i^l}\frac{\partial z_i^l}{\partial a_j^{l-1}} = \delta_i^l w_{ij}^l.$$

When we actually compute this quantity, however, we are already thinking about the previous layer. If we therefore shift out perspective, focusing instead on layer $l - 1$, we can make the substitutions[3] $l \to l + 1$, $i \to k$, $j \to i$, we may rewrite this quantity as

$$\frac{\partial L}{\partial a_i^l} = \frac{\partial L}{\partial z_k^{l+1}}\frac{\partial z_i^{l+1}}{\partial a_i^l} = \delta_k^{l+1} w_{ki}^{l+1},$$

which is exactly the summand occurring on line 7 of the algorithm. Note that we are summing over the neurons of the subsequent layer, representing each possible path branch one might take when jumping from neuron i of layer l to layer $l + 1$.

Using the backpropagation algorithm, we can compute the gradient of an arbitrary loss function with respect to the full set of connection weights of the neural network. Using these gradients, we can then deploy stochastic or mini-batch gradient descent to learn the values of the connection weights. In practice, one typically initializes the weights using small random numbers.

Matrix Formulation of Backpropagation

The matrix version of the backpropagation algorithm is shown in Algorithm 13.5. There are a few points to be made regarding notation. First, since the quantity $\phi_l(z^l)$ is defined as the vector quantity $\phi_l(z^l) = \langle \phi_l(z_1^l), \ldots, \phi_l(z_{r_l}^l) \rangle$, its gradient with respect to z^l is the diagonal matrix

$$D\phi_l(z^l) = \begin{bmatrix} \phi_l'(z_1^l) & 0 & \cdots & \vdots \\ 0 & \phi_l'(z_2^l) & \cdots & \vdots \\ \vdots & \ddots & \ddots & \vdots \\ 0 & 0 & \cdots & \phi_l'(z_{r_l}^l) \end{bmatrix}.$$

---

[3] We use i to index the *current* layer l, j to index the previous layer $l - 1$, and k to index the subsequent layer $l + 1$.

Note that the matrix product $D\phi_l(z^l) \cdot x$, for a given $x \in \mathbb{R}^{r_l}$, is equivalent to the componentwise product $\phi_l'(z^l) \odot x$, where we define $\phi_l'(z^l)$ as a vector quantity, similar to the definition of $\phi_l(z^l)$.

Later when we code our matrix-based implementation, we will consider the softmax transformation to be a particular type of activation. In this context, the gradient will necessarily be a matrix.

The second (brief) note is that the gradient $\nabla_{\hat{y}} L$, which we shall interpret as a column vector, is defined in the obvious way.

Finally, on line 8, we use the *outer product* $x \otimes y = xy^T$ for two vectors $x \in \mathbb{R}^p$, $y \in \mathbb{R}^q$, to represent the $p \times q$ matrix with components

$$[x \otimes y]_{ij} = x_i y_j,$$

for $i = 1, \ldots, p$ and $j = 1, \ldots, q$.

Gradient of the Softmax Layer

The quantities δ_i^m for a softmax layer naturally require separate treatment. In Figure 13.4, we may regard the sigma neurons and softmax neurons as cohabitors of a single output layer. We must therefore evaluate the derivative of the loss with respect to the linear combinations z_i^m. Since z_i^m feeds into each softmax neuron, as opposed to its counterpart alone, we must compute the summation

---

Algorithm 13.5: Backpropagation (matrix version).

Input: a training instance (x, y);
a given neural network of size m;
a set of m activation functions $\phi_1, \ldots, \phi_m$;
a given set of (current) bias vectors $\mathcal{B} = \{\beta^1, \ldots, \beta^m\}$;
a given set of (current) weights $\mathcal{W} = \{W^1, \ldots, W^m\}$
Output: The full gradient $\nabla_{\mathcal{W}} L$.

1 Compute $\hat{y} = f(x, \mathcal{W})$ using the feedforward algorithm, saving the values of all intermediary variables

2 **for** *each layer $l = m, \ldots, 1$ (in descending order)* **do**

3   **if** *layer l is output layer $l = m$* **then**

4   Set $\delta^m = D\phi_m(z^m) \cdot \nabla_{\hat{y}} L$

5   **else**

6   Set $\delta^l = D\phi_l(z^l) \cdot (W^{l+1})^T \cdot \delta^{l+1}$

7   **end**

8   Set $\dfrac{\partial L}{\partial W^l} = \delta^l \otimes a^{l-1}$

9   Set $\dfrac{\partial L}{\partial \beta^l} = \delta^l$

10 **end**

11 **return** $\nabla_{\mathcal{W}} L = \left\{ \dfrac{\partial L}{\partial W^l} \right\}_{l=1}^m$ and $\nabla_{\mathcal{B}} L = \left\{ \dfrac{\partial L}{\partial \beta^l} \right\}_{l=1}^m$

$$\delta_i^m = \frac{\partial L}{\partial z_i^m} = \sum_{k=1}^{c} \frac{\partial L}{\partial \hat{p}_k} \frac{\partial \hat{p}_k}{\partial z_i^m}.$$

Now, given our definition of the softmax transform (Equation (13.4)), one easily computes the following partial derivatives

$$\frac{\partial \hat{p}_k}{\partial z_i^m} = \frac{\exp(z_k)\delta_{ik}}{\sum_{j=1}^{c} \exp(z_j)} - \frac{\exp(z_k)\exp(z_i)}{\left(\sum_{j=1}^{c} \exp(z_j)\right)^2} = \hat{p}_k I_{ik} - \hat{p}_i \hat{p}_k,$$

where $I_{ik} = \mathbb{I}[i = k]$ is the identity. We may combine the preceding two equations into the single matrix equation

$$\delta^m = (\mathrm{diag}(\hat{p}) - \hat{p} \otimes \hat{p}) \cdot \nabla_{\hat{y}} L. \tag{13.7}$$

We may make a further simplification if the loss function is log-loss, i.e., cross-entropy (Equation (6.20)):

$$L(y, \hat{p}) = - \sum_{k=1}^{c} y_k \log(\hat{p}_k),$$

where y is the one-hot encoded vector of class labels. For the case of the cross-entropy loss function, we therefore have the result that

$$\delta_i^m = - \sum_{k=1}^{c} \frac{y_k}{\hat{p}_k} \left(\hat{p}_k \delta_{ik} - \hat{p}_i \hat{p}_k \right),$$

which simplifies as

$$\delta_i^m = \hat{p}_i - y_i. \tag{13.8}$$

For multiclass classification problems with a softmax layer, we simply keep calm, replace line 5 of Algorithm 13.4 (or line 4 of Algorithm 13.5) with the preceding equation, and carry on from there.

13.2.3 Deep Learning

Now that we have derived the necessary formulas and algorithms to compute the gradients of our loss function with respect to the full set of connection weights, we may discuss actually implementation of the learning algorithm, which typically comes in flavors of gradient descent. We will briefly touch on a few of these topics; for further details, see Aggarwal [2018]. For additional background specifically for optimization strategies, see Chong and Zak [2008].

Challenges

When building and training artificial neural networks, there are several challenges that can arise and that one should be aware of.

Vanishing and Exploding Gradients

Vanishing and exploding gradients refers to the general phenomenon that gradients have the propensity to exponentially decay (or grow) with the number of layers as one moves backward through a neural network during backpropagation. To see why, consider an illustrative example of a simple neural network with ten layers that each use the logistic activation function and posses only a single neuron. The maximum value of the derivative of the logistic function is only 0.25; recall Figure 13.6. Since these derivatives are multiplied during backpropagation as we move back through the network, even if each neuron is precisely at the maximum value ($z = 0$), the magnitude of the gradient in the first layer is approximately *one millionth* ($1/4^{10}$) of the magnitude of the gradient in the final layer, resulting in a glacially slow parameter movement through the early layers. The hyperbolic tangent function, of course, has a maximum derivative of 1, but like the logistic function, it quickly decays to zero outside of a short band. This is a large part why the ReLU and hard tanh functions have become go-to choices for deep neural architectures.

One issue with the ReLU function, however, is that its gradient is zero for negative values of the input, resulting in "dead neurons." This can function as a tuning of sorts, in which unnecessary neurons are naturally removed. This may be viewed as a natural form of neural pruning, which can be advantageous given the limits to precisely tuning the number of neurons in each layer. However, one must be on guard that too many are not destroyed resulting in ineffective models. Slow learning rates tend to offset this problem, as does the replacement with the ReLU function with a *leaky ReLU*, which is defined by

$$\phi(z) = \begin{cases} \alpha z & \text{for } z \leq 0 \\ z & \text{for } z > 0 \end{cases},$$

for a given parameter $\alpha \in (0, 1)$. This allows backpropagation to continue to propagate information backwards through dead neurons at a reduced rate α.

Initialization of Weights

In practice, we begin our stochastic gradient descent with a random initialization of the weights, say $W_{ij}^l \sim N(0, \epsilon^2)$, for some small $\epsilon \ll 1$ (e.g., $\epsilon = 0.01$). This approach, however, can still result in some instability, for neural architectures in which individual layers have a vastly different number of neurons between them, since the variance of the outputs scales linearly with the number of inputs. It is therefore common in practice to initialize weights following

$$\beta_i^l = 0 \tag{13.9}$$
$$W_{ij}^l \sim \mathrm{N}\left(0, r_{l-1}^{-1}\right); \tag{13.10}$$

i.e., we set the weights of the bias neurons to zero and the non-bias neurons to be random values with a standard deviation of $1/\sqrt{r_{l-1}}$.

Hyperparameter Tuning

For a given problem, not only do we need to determine the connection weights, but we must also tune various hyperparameters, including

1. the number of hidden layers m;
2. the number r_l of neurons in each layer;
3. the learning rate η (discussed below).

In order to choose our hyperparameters, we can use train–test–validate sets, as was discussed in Section 6.1.5. This still leaves us with the problem of *which* combinations of hyperparameters to test.

A common approach, known as *grid search*, involves discretizing the hyperparameters and training the model over all combination of values. The number of grid points, however, increases exponentially with the number of hyperparameters, as it entails creating a discretization of an equally high-dimensional hypercube: a grid with h hyperparameters, each with n values, grid search would entail training the model over n^h combinations of hyperparameters.

As an alternative to grid search, we could uniformly sample each hyperparameter within its respective range, as proposed by Bergstra and Bengio [2012]. This can be coupled with multi-resolution sampling, in which the full range of hyperparameters is sampled on the first pass, and geometrically smaller range is explored on subsequent passes, each centered around the optimal values of the prior pass.

Parameters that scale across a range of orders of magnitude, like learning rate or regularization rate (discussed in our next paragraph), can of course be sampled logarithmically instead of uniformly. For instance, if we are trying to find an optimal learning rate between 0.001 and 0.1, we could try 0.001, 0.01, and 0.1, as opposed to a linear division of the range $[0.001, 0.1]$.

Regularization

Given the large number of parameters, neural networks are prone to overfitting on the training set. In order to guard against overfitting, we can deploy ℓ_1 (lasso) or ℓ_2 (ridge) regularization (Definition 7.4) by considering instead the augmented loss functions

$$L_1(y, \hat{y}) = L(y, \hat{y}) + \lambda \sum_{l=1}^{m} \sum_{i=1}^{r_l} \sum_{j=1}^{r_{l-1}} \left| w_{ij}^l \right|, \qquad (13.11)$$

$$L_2(y, \hat{y}) = L(y, \hat{y}) + \lambda \sum_{l=1}^{m} \sum_{i=1}^{r_l} \sum_{j=1}^{r_{l-1}} \left(w_{ij}^l \right)^2, \qquad (13.12)$$

respectively, where λ is a tunable hyperparameter. Note that the bias weights β_i^l are not penalized. Naturally, it is straightforward to compute the gradient of the regularization terms with respect to the weights, adding the result to the gradient computed during backpropagation.

ℓ_2 regularization is most commonly used in practice, as it usually results in models with higher accuracy. As we saw in Chapter 6, however, ℓ_1 regularization is also useful as it often results in a *sparse* set of parameters, resulting in simpler models that are faster to train. This strategy can also be regarded as a way to prune large networks.

Descent into Madness

In this text, we will focus on stochastic and mini-batch gradient descent for simplicity. These algorithms are provided in Algorithm 13.6; but see also Algorithm 5.2 and Definition 5.5 for our earlier discussion. However, there are many variations to simple gradient descent that can greatly improve performance:

1. *momentum methods* attempt to correct the "zigzagging" effect observed in gradient descent by updating the gradient in the direction of a "momentum" vector;
2. *learning rate decay* is a strategy coupled with standard gradient descent in which the learning rate decays exponentially (or inversely) with the number of training steps;
3. *spectrum learning rates* is a collection of strategies (AdaGrad, RMSProp, RMSProp with Nesterov momentum, AdaDelta, Adam) that allow for each parameter to have its own learning rate, thereby allowing the learning algorithm to speed up in a targeted fashion in the areas that need it most;
4. *Netwon's method* involves using the Hessian of the loss function, which entails computing the second derivatives of the loss function over the full set of weights;
5. *conjugate gradient method*, also referred to as "Hessian-free optimization" can solve a c-dimensional quadratic function in a total of c steps by ensuring that each update is "orthogonal" in an appropriate sense to the previous set of updates.

For an overview of each of these techniques in relation to deep learning, see Aggarwal [2018].

13.3 Python Implementation

In this section we will build and train our own neural networks from scratch. We will follow two different strategies. The first is a neuron-based approach, which most directly models the biological brain. This approach is used primarily for its illustrative value. The second approach is a layer-based, or vector-based, approach, which leverages the speed of `numpy`'s built-in matrix operations for improved performance. We then train a model using mini-batch gradient descent. Finally, we discuss how to build and train neural networks using TensorFlow 2.0 in Python.

Algorithm 13.6: Stochastic ($b = 1$) and mini-batch ($b > 1$) gradient descent for deep learning.

Input: training data $\mathcal{D} = \{(x_i, y_i)\}_{i=1}^n$;

a given neural network of size m;

a set of m activation functions $\phi_1, \ldots, \phi_m$;

learning rate η;

batch size b (typically a power of 2)

Output: Set of weights $\mathcal{W} = \{W^l\}_{l=1}^m$ and $\mathcal{B} = \{\beta^l\}_{l=1}^m$.

1 Set $\beta_i^l = 0$ for $l = 1, \ldots, m$ and $i = 1, \ldots, r_l$

2 Set $w_{ij}^l \sim N(0, 1/r_{l-1})$ for $l = 1, \ldots, m$, $i = 1, \ldots, r_l$, and $j = 1, \ldots, r_{l-1}$

3 Set $\mathcal{W} = \{w_{ij}^l\}_{l,i,j}$

4 Set $\mathcal{B} = \{\beta_i^l\}_{l,i}$

5 Set $converged = $ **False**

6 **while not** $converged$ **do**

7   Compute sample index $I_\mathcal{B} = $ `getRandomIndex`$(|\mathcal{D}|, b)$

8   **for** $s \in I_\mathcal{B}$ **do**

9   Compute $\hat{y}_s = f(x_s, \mathcal{W}, \mathcal{B})$ using the feedforward algorithm

10   Compute $\nabla_\mathcal{W} L_s = \left\{ \frac{\partial L}{\partial w_{ij}^l}(y_s, \hat{y}_s) \right\}_{l,i,j}$ using backpropagation

11   Compute $\nabla_\mathcal{B} L_s = \left\{ \frac{\partial L}{\partial \beta_i^l}(y_s, \hat{y}_s) \right\}_{l,i}$ using backpropagation

12   **end**

13   Update $\mathcal{W} = \mathcal{W} - \eta \sum_{s \in I_\mathcal{B}} \nabla_\mathcal{W} L_s$

14   Update $\mathcal{B} = \mathcal{B} - \eta \sum_{s \in I_\mathcal{B}} \nabla_\mathcal{B} L_s$

15     $converged = $ `checkConverged`$(\mathcal{D}, \mathcal{W}, \mathcal{B})$

16 **end**

17 **return** $\mathcal{W}$ and $\mathcal{B}$

13.3.1 Activation Functions

To begin, we need to code up some activation functions. An abstract `Activation` class is provided in Code Block 13.1. Notice the usage of the static method `_vectorize`, which will serve as a wrapper to the function call `f` and its derivative `df` in each subclass.

```python
class Activation(ABC):

    @staticmethod
    def _vectorize(func):
        def wrapper(*args, **kwargs):
            vfunc = np.vectorize(func, otypes=[float])
            return vfunc(*args, **kwargs)
        return wrapper

    @abstractmethod
    def f(self, x):
        pass

    @abstractmethod
    def df(self, x):
        pass
```

Code Block 13.1: `Activation` abstract class.

Note that the `numpy.vectorize` method is primarily written for convenience, not performance; i.e., it is not a true *vectorization* of one's code, as it does not leverage parallelization or multi-threading, etc. Rather, it has the performance of an ordinary *for*-loop. Additional activation functions can be encoded following our cookie-cutter template of Code Block 13.2.

```python
class ActivationReLU(Activation):

    @Activation._vectorize
    def f(self, x):
        return x if x > 0 else 0

    @Activation._vectorize
    def df(self, x):
        return 1 if x > 0 else 0
```

Code Block 13.2: `ActivationReLU`: ReLU activation function.

13.3.2 Neuron-based (Scalar) Implementation

A natural first approach at coding artificial neural networks is as a graph of individual neurons, that are connected together in a systematic way. We may think of this as a *neuron-first* approach. It suffers, however, from the drawback that it fails to vectorize the operations involved in deep learning. We therefore regard this exercise as more illustrative in nature, designed to aid the conceptual understanding on the inner workings of complex networks.

Neurons

We begin with our code for a single artificial neuron, as shown in Code Block 13.3. We will diverge slightly from our mathematical notation developed in the preceding sections. Since each neuron has its own bias, we track the bias of a neuron separately from the weights, initializing the bias to zero. We initialize our weights following the strategy of Equation (13.10).

Other than the value of a neuron's bias, we must also consider the number of input connections, the activation function, and whether or not the bias should be used in the actual computation. We therefore require these variables to be passed into the __init__ method of the `Neuron` class on line 3. We will attach the weights of the input connects to the receiving neuron, using the `weights` attribute. We will also need to track the "state" of each neuron, represented by the previous values of the sum z_i^l, the output value a_i^l, the gradient of the weights $\nabla_{W^l} L$ and bias $\partial L / \partial \beta_i^l$, and the value of δ_i^l used during backpropagation. We initialize these values to zero on lines 11–16. Since these variables are all attached to the ith neuron of layer l (something we will construct *outside* of the `Neuron` class), we do not need to enumerate them within the class itself. In other words, each `Neuron` object should be agnostic to its place in the world.

Next, we need to consider the types of operations a neuron must undergo, and how we should implement their corresponding methods. There are essentially two primary operations: feedforward computation and backpropagation. For the former, we include the `activate` method, which takes an input vector and computes the given neuron's output $a = \phi \circ w(x)$. We will save the values of the input value (as `last_value`), the intermediary variable $z_i^l = \sum_{j=1}^{r_{l-1}} w_{ij}^l a_j^{l-1}$, and the activation value as attributes, as these are each used during the backpropagation phase.

For the `backpropagate` method, we will assume an input value that is the current neuron's component of the vector joining the forward weights and forward delta-values into a single quantity, which we call *weighted forward deltas*. We will discuss this in more depth momentarily (in our next paragraph on *Neuron Layers*. For now, suffice it to say that this quantity represents the parenthesized quantity on line 7 of Algorithm 13.4, which nicely captures everything we need to know about *the future* during our

```
 1   class Neuron:
 2
 3       def __init__(self, phi=None, n_inputs=1, bias=0, use_bias=True,
                 lambda_=0.1, penalty=None):
 4           self.phi = phi
 5           self.bias = bias
 6           self.use_bias = use_bias
 7           self.lambda_ = lambda_
 8           self.penalty = penalty
 9           self.n_inputs = n_inputs
10           self.weights = np.random.normal(0, scale=1/n_inputs,
                     size=n_inputs)
11           self.z = 0
12           self.value = 0
13           self.dW = np.zeros(n_inputs)
14           self.dBias = 0
15           self.delta = 0
16
17       def activate(self, x):
18           self.last_input = x
19           self.z = self.bias + np.dot(x, self.weights)
20           self.value = self.phi.f(self.z)
21           return self.value
22
23       def backpropagate(self, weighted_forward_delta):
24           self.delta = weighted_forward_delta * self.phi.df(self.z)
25           self.dBias = self.delta if self.use_bias else 0
26           self.dW = self.delta * self.last_input
27           return self.delta * self.weights
28
29       def step(self, eta):
30         if self.use_bias:
31               self.bias -= eta * self.dBias
32           self.weights -= eta * self.dW
```

Code Block 13.3: Neuron class

backpropagation step. From there, it is straightforward to compute the current neuron's δ_i^l value, and the gradient with respect to the given neuron's bias and input weights.

Finally, we implement the gradient "step" in our final method, step, which actually updates the neuron's bias (if applicable) and weights.

Neural Layers

We next discuss how neurons can be bundled into the concept of a neural layer. We will save ourselves some trouble later on, when we replace our neuron-based implementation with a matrix-based one, if we take a step back now and think about the neural layer abstractly and ask what we need to achieve from it. In this regard, the functionality of a layer is not dissimilar to the functionality of an individual neuron: we need to be able to run both feedforward and backpropagation passes through individual layers. The neural layer, however, has added structure as a collection of neurons and must therefore be able to handle the higher-order orchestration of the feedforward and backpropagation operations.

Let us consider again the neural network shown in Figure 13.7, with a focus on layer $l = 3$. Abstracting the concept of a neural layer, we can consider the basic functionality as shown in Figure 13.13. The feedforward

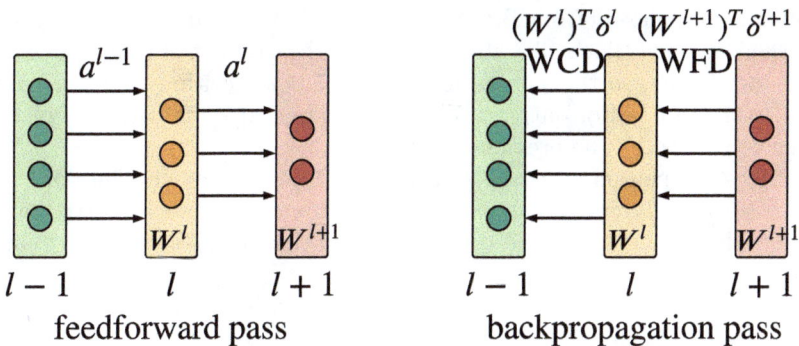

Fig. 13.13: Schematic of the feedforward (left) and backpropagation (right) passes of an abstract neural layer.

pass is relatively straightforward: the previous output is passed as an input into the current layer, which is then transformed into a new output for the following layer.

In order the better understand the input/output requirements for backpropagation, let us consider again Algorithms 13.4 and 13.5. The information that comes from the following layer is given as the summation in line 7 of Algorithm 13.4, or its equivalent matrix representation in Algorithm 13.5. We will refer to this quantity as the *weighted forward deltas* *(WFD)*, which is defined as the vector

$$WFD = (W^{l+1})^T \delta^{l+1} = \left\langle \sum_{k=1}^{r_{l+1}} w_{ki}^{l+1} \delta_k^{l+1} \right\rangle_{i=1}^{r_l}.$$

For the final output layer, we can replace this with the gradient of the loss function. Each neuron in our neuron-based implementation is then to receive only a single component of the vector of weighted forward deltas as an input to that neuron's **backpropagate** method. If this is the required input from the next layer during backpropagation, then this quantity must also be the required output of our current layer, in order that the algorithm continues to cascade backwards through the process of backpropagation. We will refer to a layer's backpropagation output as the *weighted current deltas (WCD)*, obtained by replacing $l + 1 \rightarrow l$ in the above equation.

Finally, note that the weights are always attached to their given layers. In the neuron-based implementation, they are embedded directly into the individual neurons that comprise the layer. In the matrix-based implementation, we will forgo the use of individual neurons and replace it with a matrix representation of each neural layer's inner workings. We now have all the requirements we need to abstract the concept of a neural layer into an abstract class.

The neuron-based implementation of a neural layer is provided by the **NeuronLayer** class of Code Block 13.4. Note that the **Neuron.backpropagate** method of Code Block 13.3 already returns the product of that neuron's δ and its weights, so the summation on line 7 of Algorithm 13.4 need only sum over these vector outputs. The result is passed as an output, which is then consumed by the previous layer.

Since the neuron-based implementation is mostly for illumination, and since softmax layers are more readily implemented from a matrix-based approach, we will postpone our implementation of a softmax layer until later.

Layer Prescription and Loss

Now that we have the workings of the neural layers, we need a brain to orchestrate the operations across the neural net. We will accomplish this by writing a **NeuralNetwork** class. In constructing such a class, however, we will need to specify the neural architecture. This will consist of a prescription for the overall number of inputs and outputs, a loss function, and the design of the hidden layers, with each hidden layer specified by the triple consisting of the number of neurons, the activation function, and whether or not bias is allowed. Therefore, before we build our **NeuralNetwork** class, we will need to determine how we will pass our architecture and loss function into the constructor.

In order to specify the triple (number of neurons, activation function, and whether to use a bias) that defines a neural layer, we will subclass the built-in **namedtuple** class, which can be used as a factory that dynamically defines a new class on the fly with a prescribed set of attributes and their corresponding defaults. For construction, see Code Block 13.5.

```
1   class NeuronLayer:
2
3       def __init__(self, phi=None, n_inputs=1, n_outputs=1,
            use_bias=True, lambda_=0.1, penalty=None):
4           self.phi = phi
5           self.use_bias = use_bias
6           self.lambda_ = lambda_
7           self.penalty = penalty
8           self.n_inputs = n_inputs
9           self.n_outputs = n_outputs
10          self.neurons = [Neuron(phi=phi, n_inputs=n_inputs,
                use_bias=use_bias) for i in range(n_outputs)]
11
12      def __len__(self):
13          return len(self.neurons)
14
15      def activate(self, x):
16          return np.array([n.activate(x) for n in self.neurons])
17
18      def backpropagate(self, weighted_forward_deltas):
19          weighted_current_deltas = np.zeros(self.n_inputs)
20          for neuron, weighted_forward_delta in zip(self.neurons,
                weighted_forward_deltas):
21              weighted_current_deltas +=
                    neuron.backpropagate(weighted_forward_delta)
22          return weighted_current_deltas
23
24      def step(self, eta):
25          for neuron in self.neurons:
26              neuron.step(eta)
```

Code Block 13.4: NeuronLayer class

```
1   from collections import namedtuple
2   LayerDefinition = namedtuple('LayerDefinition', ['size', 'phi',
        'use_bias', 'lambda_', 'penalty'], defaults=[1,
        ActivationReLU(), True, 0.1, None])
3   layer_1 = LayerDefinition(size=5, phi=ActivationReLU(),
        use_bias=False)
4   layer_2 = LayerDefinition(size=10, phi=ActivationTanh())
5   layer_defs = [layer_1, layer_2]
```

Code Block 13.5: LayerDefinition class and example construction.

Similarly, we can encode our loss function as shown in Code Block 13.6. Constructing an object from the `Loss` class is a bit like a factory method, in that the type of loss function is determined by the `loss_type` keyword. There are more elegant ways one might encode a loss function, but for our purposes, using simple `if` statements suffices.

```python
class Loss:

    def __init__(self, loss_type):
        self.type = loss_type

    def f(self, y, y_hat):
        if self.type == 'squared_error':
            return (y - y_hat)**2
        if self.type == 'log':
            return -np.sum(y * np.log(y_hat), axis=0)

    def df(self, y, y_hat):
        if self.type == 'squared_error':
            return -2 * (y - y_hat)
        if self.type == 'log':
            return -y / y_hat
```

Code Block 13.6: `Loss` class; implementation of squared-error and log loss.

Neural Networks

Now we are ready to build our network. The constructor will require a specification of the number of inputs, an array of `LayerDefinition` objects, which specifies the neural architecture, and a loss function. The code is shown in Code Block 13.7. Note that we use `MatrixLayer`, which is provided later in Code Block 13.8. We could equivalently swap this out with `NeuronLayer` to the same effect; since both classes follow the same API, they may be used interchangeably. One could further write a simple factory method to control what type of layer is used, but since our ultimate goal is to use the matrix-version of a neural network, we find this step to be unnecessary.

Dimensionality of Inputs and Outputs

By convention, the *rows* of the feature matrix and target vector correspond to individual training instances, whereas the *columns* correspond to the individual features or labels, respectively. This is antithetical to our mathematical development, where we interpret vectors as column vectors, not

```python
class NeuralNetwork(Model):

    def __init__(self, n_inputs=1, layer_defs=[], loss_type=None,
            batch_size=1, eta=0.1):
        self.batch_size = batch_size # size > 1 invalid
        self.eta = eta #learning rate
        self.n_inputs = n_inputs
        self.loss = Loss(loss_type)
        self.layers = []
        for ldef in layer_defs:
            self.layers.append(MatrixLayer(phi=ldef.phi,
                n_inputs=n_inputs, n_outputs=ldef.size,
                use_bias=ldef.use_bias, lambda_=ldef.lambda_,
                penalty=ldef.penalty))
            n_inputs = ldef.size

    def feedforward(self, X):
        for layer in self.layers:
            X = layer.activate(X)
        return X

    def backpropagate(self, y, y_hat, eta=0.1):
        weighted_deltas = self.loss.df(y, y_hat)
        for layer in reversed(self.layers):
            weighted_deltas = layer.backpropagate(weighted_deltas)
            layer.step(eta)

    def train_batch(self, X, y):
        y_hat = self.feedforward(X)
        self.backpropagate(y, y_hat, eta=self.eta)

    def train(self, X, y, n_rounds=10):
        for _ in range(n_rounds):
            idx = np.random.choice(len(X), size=self.batch_size,
                replace=False)
            X_train = X[idx]
            y_train = y[idx]
            self.train_batch(X_train.T, y_train.T)

    def predict(self, X):
        return self.feedforward(X.T).T
```

Code Block 13.7: NeuralNetwork class.

row vectors, in order to be consistent with linear algebra, where it is conventional to write the matrix–vector product as $A \cdot x$ (where x is a column vector), not $r \cdot A$ (where r is a row vector). This is, coincidentally, why our definition of the weight matrix is the transpose of many other author's definitions, who, presumably, were primarily in the mindset to define the weight matrix consistent with the inputs and outputs of a training algorithm; i.e., from a machine-learning perspective.

To resolve this, we leverage the *abstraction* principle of object-oriented design: *the operations of an object can be defined without reference to its internal implementation.* Thus, our `NeuralNetwork` class will use an API for its `train` and `predict` methods that is consistent with our previous `Model` class (as well as the numerous models within *sci-kit learn*), while leveraging the proper linear-algebra convention during its implementation. We feel this approach is more natural, so that our programmatic implementation is more analogous with its underlying maths. Moreover, this is easily achieved by simply applying a transpose operator to our mini-batches, prior to passing them into the `train_batch` method, as shown in Code Block 13.7.

Visually, we may think of the external API form of the feature matrix and target vector with dimensionality as shown in Figure 13.14. Upon transposing these matrices, we have the internal math form shown in Figure 13.15.

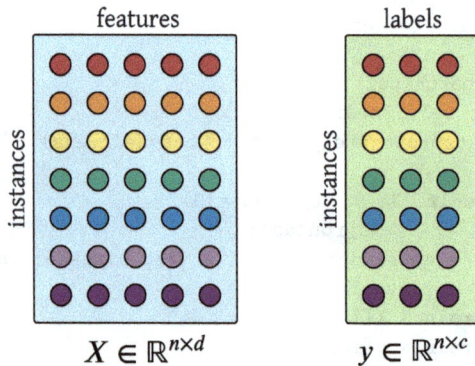

Fig. 13.14: Dimensionality of feature matrix and target vector: external I/O form. Each instance is a separate color.

As a rule of thumb, everything deeper than the `train` and `predict` methods will assume its inputs and outputs are of the internal math form (Figure 13.15). When training mini-batches, the columns of all the internal variables will therefore represent individual instances from the mini-batch. This works seamlessly with standard linear algebra, since, given feature

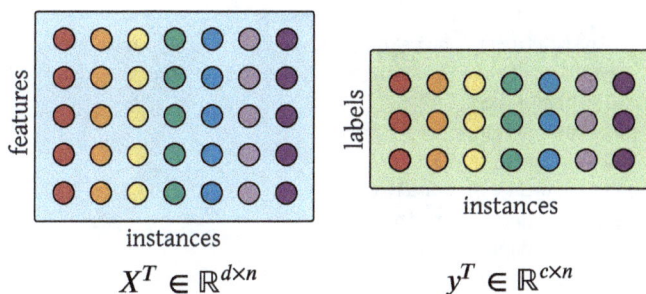

Fig. 13.15: Dimensionality of feature matrix and target vector: internal math form. Each instance is a separate color.

vectors $x_1, \ldots, x_b$, the matrix product $Z = W \cdot X^T$ yields

$$Z = W \cdot \begin{bmatrix} \top & \top & & \top \\ x_1 & x_2 & \cdots & x_b \\ \bot & \bot & & \bot \end{bmatrix} = \begin{bmatrix} \top & \top & & \top \\ W \cdot x_1 & W \cdot x_2 & \cdots & W \cdot x_b \\ \bot & \bot & & \bot \end{bmatrix}.$$

13.3.3 Layer-based (Matrix) Implementation

When creating multiple classes that can be used interchangeably, it is best to define an abstract parent class, so that the inputs and outputs can be enforced. We leave this as an exercise to the reader (see Exercise 13.8). As promised the `MatrixLayer` class is implemented below, in Code Block 13.8, which can replace our existing `NeuronLayer` class from Code Block 13.4.

Finally, we can implement softmax layers by creating a custom vectorized softmax activation class, as shown in Code Block 13.9. This way, we may prescribe a softmax layer within the `LayerDefinition` named tuple. Notice that the methods of the `ActivationSoftmax` class are vector operations that capture interdependencies among the inputs, which is why the `vectorize` wrapper is not used. This is also why it is more natural to implement the softmax layer in the context of a matrix approach, as the inputs and outputs of each layer are captured as vectors, as opposed to breaking them down and passing them into individual neurons.

We must further take care to ensure that both a typical and softmax activation function may be used interchangeably within the `MatrixLayer` class; in particular, in the `backpropagate` method. Of concern is one activation returning a vector for `df` and another returning a matrix (i.e., a two-dimensional array). We solve this by adding a semiprivate `_diag` method, which diagonalizes a one-dimensional array, while leaving a two-dimensional array unchanged.

```
 1  class MatrixLayer:
 2
 3      def __init__(self, phi=None, n_inputs=1, n_outputs=1,
            use_bias=True, lambda_=0.1, penalty=None):
 4          self.phi = phi
 5          self.use_bias = use_bias
 6          self.lambda_ = lambda_
 7          self.penalty = penalty
 8          self.n_inputs = n_inputs
 9          self.n_outputs = n_outputs
10          self.bias = np.zeros(n_outputs)
11          self.weights = np.random.normal(0, scale=1/n_inputs,
                size=(n_outputs, n_inputs))
12          self.grad_bias = np.zeros(n_outputs)
13          self.grad_weights = np.zeros((n_outputs, n_inputs))
14          self.z = np.zeros(n_outputs)
15          self.values = np.zeros(n_outputs)
16          self.deltas = np.zeros(n_outputs)
17
18      def __len__(self):
19          return self.n_outputs
20
21      def _diag(self, x):
22          if x.ndim == 1:
23              return np.diag(x)
24          if x.shape[1] == 1:
25              return np.diag(x.reshape(-1))
26          return x
27
28      def activate(self, x):
29          self.last_input = x
30          self.z = self.bias + self.weights @ x
31          self.value = self.phi.f(self.z)
32          return self.value
33
34      def backpropagate(self, weighted_forward_deltas):
35          ### NOT VALID FOR BATCH ###
36          self.delta = self._diag(self.phi.df(self.z)) @
                weighted_forward_deltas
37          self.grad_weights = np.outer(self.delta, self.last_input)
38          self.grad_bias = self.delta
39          weighted_current_deltas = self.weights.T @ self.delta
40          return weighted_current_deltas
41
42      def step(self, eta):
43          if self.use_bias:
44              self.bias -= eta * self.grad_bias
45          self.weights -= eta * self.grad_weights
```

Code Block 13.8: MatrixLayer class.

```
1    class ActivationSoftmax(Activation):
2
3        def f(self, x):
4            return np.exp(x) / np.exp(x).sum(axis=0)
5
6        def df(self, x):
7            p_hat = self.f(x)
8            return np.diag(p_hat) - np.outer(p_hat, p_hat)
```

Code Block 13.9: `ActivationSoftmax` class.

In addition, we note that the code as written does not allow for batch gradient descent. This primarily breaks down in the `backpropagate` method, which would require additional complexity as all two-dimensional arrays would need to be rewritten as tensors. We will content ourselves at present with stochastic gradient descent and leave the generalization as a project for the interested reader.

13.4 Advanced Neural Architectures

In this section, we will introduce some of the more advanced neural architectures that underlie applications such as image recognition and speech processing. For more details, see the usual references Aggarwal [2018], Gèron [2019], and Goodfellow, *et al.* [2016].

13.4.1 Autoencoders

Our first, and simplest, example of an advanced structure is that of an *autoencoder*. From the outside, autoencoders appear strange in that not only do their output nodes mirror their input nodes, but also the input vectors are used as labels during training; i.e., they learn how to predict the input from the input. Thus, from a standpoint of input and output alone, they appear most puzzling. The key to understanding them, however, is their internal workings. Autoencoders consist of a number of hidden layers of smaller dimensionality than their input and output, for the purpose of creating a low-dimensional representation of the data.

A simple example of an autoencoder is shown in Figure 13.16. It is typical in practice for autoencoders to consist of an odd number of layers, and for the number of neurons to be symmetric around the central layer. In general, both the encoder half and decoder half of an autoencoder will be multilayer neural networks. Once trained, an autoencoder will consist of two functioning components—an encoder and a decoder—that can be used independently. The encoder can create a low-dimensional encoding of

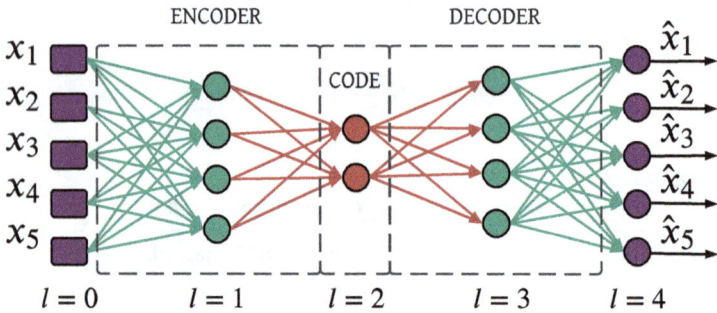

Fig. 13.16: A simple autoencoder.

high-dimensional feature vectors, which can then be used in training many models. Thus, an encoder is not unlike principal component analysis and other feature subspace selection techniques used to reduce dimensionality of complex problems.

13.4.2 Convolutional Neural Networks

Convolutional neural networks are primarily used for image recognition and visual learning tasks. They incorporate a specialized design that mimics the workings of the visual cortex in the brain. We have, of course, seen an example of visual learning before: the NIST handwritten digit problem discussed in Example 7.4. Here, the image consisted of an 8×8 array of pixels, which was flattened into a 64-dimensional feature vector. Such an approach is, however, inefficient and wholly infeasible, when one considers that a typical 4K image consists of nearly nine million pixels. Let's say that our first hidden layer consists of only 1 million pixels, representing a 10:1 loss, this would still yield nearly *nine trillion connections* between the input layer and the first hidden layer. Not only does this represent an explosion in the number of parameters, but also it fails to take advantage of the inherit array structure of the input image: by flattening a two-dimensional structure into a one-dimensional array, we are missing the opportunity to use spatial awareness. Two pixels that are next to each other in the image could end up separated by thousands of rows in a flattened representation of the image. Moreover, color images should properly be considered as three-dimensional objects, as a third dimension is required to encode the color space (e.g., RGB). Finally is the problem of translation and rotation invariance: a banana is a banana if it is in the bottom of the image or the top or if it's up-side-down or right-side-left or if it's peeled or unpeeled or in pajamas or not.

To remedy these issues, we employ a structure called a *convolution layer*, which consists of a small set of weights that form an array called a *filter* or a *kernel*, which can attach to any point in the input array and apply a consistent transformation. This captures two separate principles: weight sharing, in which weights are shared across a variety of neurons, and sparse weights, in which weights are trained only for a subset of connections from one layer to the next.

To make this specific, consider a neural network with input image characterized by a three-dimensional array of size $d_w \times d_h \times d_d$, where d_w represents the image width, d_h the height, and d_d the "color depth." For example, a 64×64 image with RGB color encoding would be represented as a $64 \times 64 \times 3$ array. In general, we can say that the lth layer is of size $r_w^l \times r_h^l \times r_d^l$.

If we were to consider *all possible connections* between one layer and the next, our weight matrix would be a massive matrix with $r_w^{l-1} r_h^{l-1} r_d^{l-1} r_w^l r_h^l r_d^l$ individual weight components. Instead, a convolution layer leverages a much smaller structure called a *filter*, which is a square array (plus depth) of size $f_l \times f_l \times d_l$. The parameter f_l is typically a small, odd number; common values are $f_l = 3$ and $f_l = 5$. Thus, for a three-dimensional color encoding, we are only concerned with learning the $f_l^2 d_l$ parameters of the filter; i.e., 27 or 75 for the case of $f_l = 3, 5$ and $d_l = 3$.

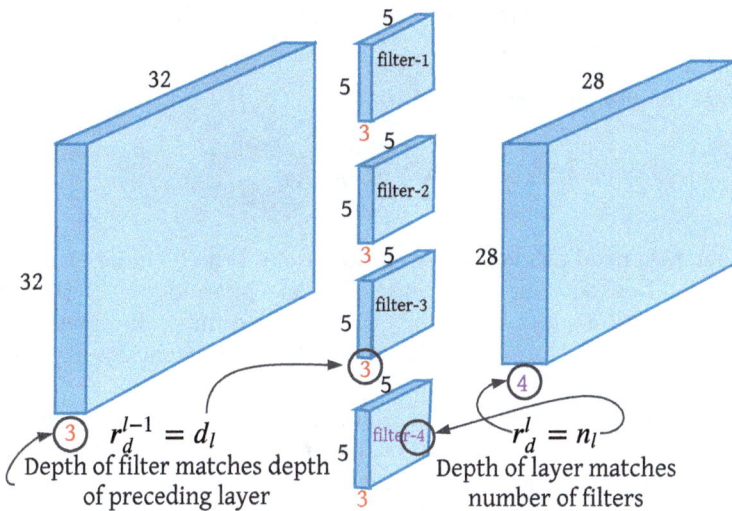

Fig. 13.17: Dimensionality of filters and layers. Each filter contributes one "slice" to the depth of the output layer. The dimensionality loss of the image is typical when not using zero padding.

The first filter $l = 1$ is constrained by the color depth of the input image, so that $d_1 = d_d$. Next, we can allow for *multiple filters* for any given layer. Here, the constraint is that r_d^l is equal to the number of filters for layer l; i.e., each filter adds one unit of depth to the following layer. Finally, the depth of an lth layer filter should equal the depth of its input from the preceding layer; i.e., $d_l = r_d^{l-1}$. These constraints are illustrated in Figure 13.17.

Now that we know what filters look like, let's dive into how they operate. Without further ado, we define the convolution operation as follows.

Definition 13.5. *Given a filter $F \in \mathbb{R}^{f \times f \times d}$ and a three-dimensional input array $A \in \mathbb{R}^{w \times h \times d}$, the unpadded convolution $F \otimes_u A$ is the two-dimensional array*

$$[F \otimes_u A]_{ij} = \sum_{r=0}^{f-1}\sum_{s=0}^{f-1}\sum_{t=0}^{d-1} F_{rst} A_{i+r,j+s,t}, \tag{13.13}$$

defined for $i = 0, \ldots, w - f$ and $j = 0, \ldots, h - f$.

Similarly, the zero-padded convolution *$F \otimes A$ is the two-dimensional array*

$$[F \otimes A]_{ij} = \sum_{r=-m_f}^{m_f}\sum_{s=-m_f}^{m_f}\sum_{t=0}^{d-1} F_{m_f+r,m_f+s,t} A_{i+r,j+s,t}, \tag{13.14}$$

defined for $i = 0, \ldots, w - 1$ and $j = 0, \ldots, h - 1$; where $m_f = (f - 1)/2$ when f is odd; and where zeros are used whenever $A_{i+r,j+s,t}$ is undefined (i.e., zero padding).

Equation (13.14) is equivalent to

$$[F \otimes A]_{ij} = \sum_{r=-0}^{f-1}\sum_{s=-0}^{f-1}\sum_{t=0}^{d-1} F_{rst} A_{i+r-m_f,j+s-m_f,t}.$$

Without zero padding, some information around the boundaries is lost via the convolution operation, leading to underrepresentation of the border pixels in the subsequent layer. This is why it is common in practice to use zero padding, which preserves the spatial footprint from one layer to the next.

Note that in the case of zero padding, the central component of the filter is given by (m_f, m_f), since $m_f = (f - 1)/2$ (assuming f is odd, as is customary). Thus, the center of the filter is attached to each element in the input array. Equation (13.14) can be further interpreted as a moving, limited inner product between the filter and the input array.

This operation is best illustrated by means of an example. We can apply a simple 3×3 filter to a 7×7 input matrix, as shown in Code Block 13.10. This example uses the zero-padded form of convolution, given by Equation (13.14).

```python
F = np.array(
    [[2, 0, 1],
     [0, 1, 0],
     [0, 1, 2]])
A = np.array(
    [[4, 2, 4, 5, 0, 1, 6],
     [7, 2, 8, 8, 0, 4, 0],
     [7, 7, 2, 0, 9, 8, 3],
     [9, 1, 7, 4, 6, 0, 2],
     [2, 0, 4, 2, 2, 6, 4],
     [6, 7, 4, 2, 3, 6, 9],
     [9, 9, 9, 0, 6, 7, 2]])
mf = 1
B = np.zeros((7,7))
for i in range(7):
    for j in range(7):
        for r in range(-mf, mf+1):
            for s in range(-mf, mf+1):
                if (i+r<0) or (i+r>6) or (j+s<0) or (j+s>6):
                    continue
                B[i, j] += F[mf+r, mf+s] * A[i+r, j+s]
```

Code Block 13.10: A simple example of a convolution with zero padding.

This operation is depicted graphically in Figure 13.18. Notice that the (i, j)th component of the output array centers the filter at the (i, j)th component of the input array, and inputs of zero are taken whenever the filter falls off the edge of the grid (i.e., zero padding).

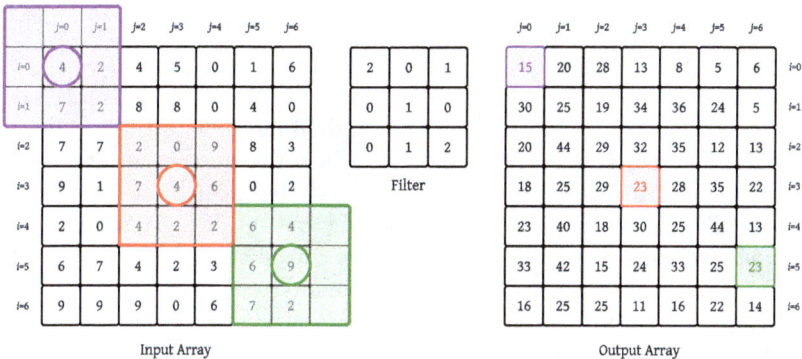

Fig. 13.18: Example of a convolution operation.

Thus, convolution essentially provides a similar operation "scanning" across the image. In this way, a convolutional neural network scans for high-level features of an image, such as lines or boundaries of objects, before refining down to higher-level concepts, like object, structure, animal, etc.

Next, let us consider an example where two filters are applied to an actual RGB image of size 427×640. The code is shown in Code Block 13.11.

```
1  from sklearn.datasets import load_sample_images
2  dataset = load_sample_images()
3  im = dataset.images[0]
4  FH, FV = np.zeros((7,7,3)), np.zeros((7,7,3))
5  FH[3,:, :], FV[:, 3, :] = 1, 1
6  def convolve(A, F):
7      w, h, d = A.shape
8      mf = int((len(F)-1)/2)
9      B = np.zeros((w, h))
10     for i in range(w):
11         for j in range(h):
12             for k in range(d):
13                 for r in range(-mf, mf+1):
14                     for s in range(-mf, mf+1):
15                         if (i+r<0) or (i+r>=w) or (j+s<0) or (j+s>=h):
16                             continue
17                         B[i, j] += F[mf+r, mf+s, k] * A[i+r, j+s, k]
18     return B
19 BH = convolve(im, FH)
20 BV = convolve(im, FV)
21 plt.imshow(im); plt.imshow(BH); plt.imshow(BV)
```

Code Block 13.11: Horizontal and vertical filters applied to an RGB image.

Our two filters represent a horizontal and a vertical filter, with zeros throughout a 7×7 array, with the exception of ones placed down the central row and column, respectively. The original image with the two filtered outputs is shown in Figure 13.19. Notice how the horizontal filter accentuates the horizontal lines, while blurring the vertical lines, whereas the vertical filter does the opposite. We call the output of a given filter a *feature map*, as it represents the areas in the image that the filter activated the most. This is why it is common to have multiple filters acting within a given layer. Naturally, the filter weights are learned during backpropagation and not specified in advanced.

Now that we have seen an example of how a simple convolution works, let's consider the general application, which typically consists of multiple filters and inputs with a given depth.

Fig. 13.19: Original image (top); horizontal filter (middle); vertical filter (bottom).

Definition 13.6. *The lth layer of a neural network is said to constitute a convolution layer if it consists of a set of n_l filters $F^{l1}, \ldots, F^{ln_l}$ of size $\mathbb{R}^{f_l \times f_l \times d_l}$, such that it acts on a three-dimensional input array $A^{l-1} \in \mathbb{R}^{r_w^{l-1} \times r_h^{l-1} \times r_d^{l-1}}$, constrained by $d_l = r_d^{l-1}$, via the operation*

$$Z_{ijk}^l = \beta^{lk} + \left[F^{lk} \otimes A^{l-1} \right]_{ij}, \qquad (13.15)$$

where β^{lk} is an optional bias associated with the kth filter. Thus, each filter produces a distinct "slice" along the third dimension of the output Z^l.

The output of the layer is then given by the activated value

$$A^l = \phi(Z^l),$$

for activation function ϕ.

Typically, modern state-of-the-art leverages the ReLU as activation function, as it has been shown to be more efficient than sigmoidal activations, such as logistic and hyperbolic tangent, in terms of both speed and accuracy. Sometimes these two operations are viewed as separate, and one refers to a convolution layer and a ReLU layer, though these layers are typically glued together.

In addition to convolutional layers, CNNs also commonly contain *pooling layers*, the most common example of which is referred to as *max pooling*. Max pooling functions similar to the convolution operator, in which a grid of fixed size "scans" across the image. It differs from convolution, however, in that it outputs the maximum value from the image defined over the scope of the pooling grid. Thus, it serves as a moving maximum value across the image, surfacing a map of the most extreme pixel outputs within its visual field.

CNNs typically consist of many layers of convolution and pooling that alternate, followed by a sequence of fully connected layers. Several examples, including AlexNet and GoogLeNet architectures, are discussed in Gèron [2019], which also discusses implementation using TensorFlow and Keras.

13.4.3 Recurrent Neural Networks

Whereas convolutional neural networks are specialized to learn spatial patterns in multidimensional inputs, recurrent neural networks (RNNs) are specialized to learn patterns in temporal sequences, such as those that underlie applications with an intrinsic time element: time-series data, stock data, DNA sequences, and, most prominently, language applications (text completion, voice recognition, translation, etc.). In particular, RNNs break the independence assumption between successive inputs and outputs: *the* → __ (predict the next word) vs. *the cow jumped over the* → __ (predict the next word). Regardless of depth, no neural network can come close to predicting what comes after *the*; the best it could do is output a probability distribution over all nouns, with probability weighted relative to usage frequencies. But if the model can "remember" the immediate past—*the, cow, jumped, over*—the word that now follows *the* becomes most obvious: *moon*.

Recurrent neural networks are artificial neural networks the are comprised of a special type of neuron that passes its previous output into its next input.

Definition 13.7. *A* recurrent neuron *is an artificial neuron that reserves one component of its input interface to consume its previous output.*

A memory cell *(or, simply,* cell*) is an artificial neuron with a hidden variable that can change after each activation, is stored within the cell (i.e., as a form of short-term memory), and whose value is passed back into the cell with each successive input. In particular, if h represents the hidden variable, φ the neuron's activation function, and W_x and W_h two distinct sets of weights, we may write the output of a memory cell as*

$$y_t = \phi\left(W_x x_t + W_y h_{t-1} + \beta\right). \tag{13.16}$$

Here, β is the optional bias.

Though the concept of a recurrent neuron constitutes a special case of a memory cell, the distinction is a bit pedantic, and, in most cases, the two may be used interchangeably. The concept of a memory cell is depicted in Figure 13.20. We see that a recurrent neuron is just like an ordinary neuron, except that it has a loop from its output back into itself. Unraveling this over time, we obtain the illustration on the right, which shows how one activation's output connects to the input of the following input. The hidden variable h can be thought of as a form of short-term memory. When no prior output is available, a memory cell can except a `Null` value in its place.

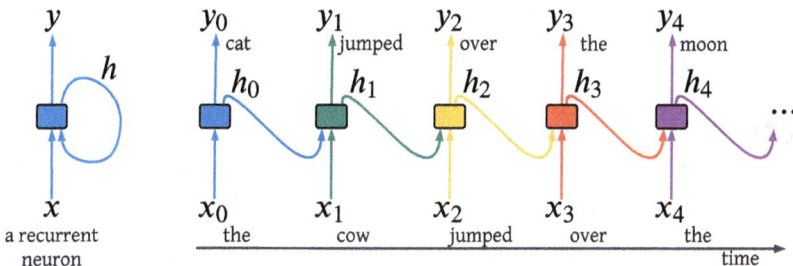

Fig. 13.20: A recurrent neuron (left) and its unraveled form (right).

Recurrent neural networks are then simply neural networks that are comprised of recurrent neurons or memory cells. Interestingly, the time element has implications for certain patterns of inputs and outputs. In particular, the inputs and outputs of an RNN might be either vectors or sequences, depending on the context of the problem. (Here, vector refers to the ordinary output of a vanilla neural network.) The illustration of Figure 13.20 constitutes a *sequence-to-sequence* network, such as predicting the next word in a sequence of text. In particular, each term of the input sequence has a corresponding output prediction, which ultimately constitutes an output sequence. Another example is an RNN that predicts the next day's stock price based on the current days' close: a sequence of stock prices is fed in, and a sequence of (one-day shifted) stock prices comes out.

A *sequence-to-vector network* ingests multiple inputs and hidden activations before a final (single) vector output is ultimately revealed. An example of this is *sentiment analysis*: the full sentence must be comprehended before the RNN can issue its final verdict: *happy*. An example of this is shown in the top half of Figure 13.21. Here, the grey outputs (y_0, y_1, y_2, and y_3) are ignored, and only the final output y_4 matters.

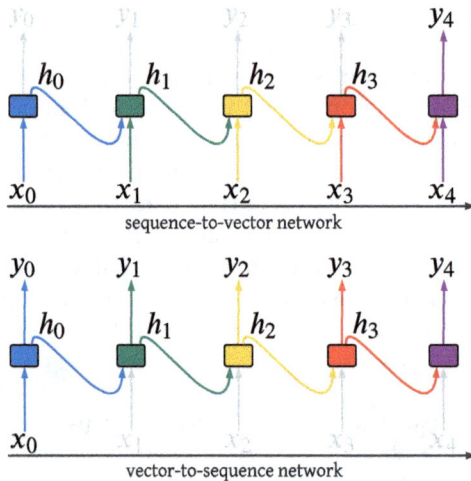

Fig. 13.21: A sequence-to-vector network (top) and a vector-to-sequence network (bottom). Grey variables and arrows are ignored.

Conversely, one might encounter a *vector-to-sequence networks*, which have a single input, followed by a number of activations that generate a sequence of outputs. For example, a single image array might be fed into the RNN as an input, and the RNN might activate multiples times to yield a text description of the image (*a cow jumping over the moon at night*). This is illustrated in the bottom half of Figure 13.21. Again, the grey inputs (x_1, x_2, x_3, and x_4) are ignored.

Finally, one might combine a sequence-to-vector network with a vector-to-sequence network to form a more elaborate structure called an *encoder–decoder*. This is common in language translation: the network needs to listen to the full sentence before it begins its translation, as the meaning of the sentence can be modified by the words at the end[4]. In this context, the sequence-to-vector network is referred to as the *encoder*, and the vector-to-sequence network is referred to as the *decoder*. Thus, a neural network can create an abstract neural-network encoding of the meaning of a sentence that transcends the language itself.

This is exactly what happened in 2016, when Google announced that that its new Neural Machine Translation (GNMT) system could translate between languages, even when it had never received training instances between those two specific languages (Schuster, *et al.* [2016]; for white paper, see Johnson, *et al.* [2016]). For example, GNMT learned how to translate between Japanese and Korean, without ever being shown any specific Japanese↔Korean pairs, as it already learned how to translate between Japanese↔English and between Korean↔English, using a single shared multilingual neural representation of the respective languages. At the time of the announcement, they were already translating between 10 different languages using a common abstract language representation.

RNNs are trained by unraveling the recurrent layers and backpropagating, a technique called *backpropagating through time*. Naturally, this is a delicate process. Due to the inherent instability in successively multiplying chains of the weight matrix, one easily encounters the same exploding and vanishing gradients issues prevalent in ordinary deep networks. In particular, gradients can sometimes vanish, which slows the learning system to a stop, or explode, which results in instability of values and ultimate divergence. To remedy this, we deploy the usual set of tricks, including good parameter initialization, dropout, etc. For details, see the usual references.

Long Short-Term Memory (LSTM) Cells

One drawback of the simple memory cells discussed thus far is their tendency to forget the past, making them impractical to handle patterns that extend beyond a handful of terms of a sequence. To remedy this, the concept of a *long short-term memory (LSTM) cell* was introduced by Hochreiter and Schmidhuber [1997]. Here, the cell's short-term memory, as represented by its hidden variable h, is augmented with a longer term memory, or *cell memory*, variable c. In addition, several devices, called *logic gates*, are used to control several basic operations, including which parts of the cell state should be updated or forgotten, and which parts should be output to the hidden state and output variable. Formally, we have the following.

---

[4] for example, in German it is common for verbs to occur at the end of the sentence: *I was out of popcorn, because I the movie yesterday with my friends at home watched had.*

Definition 13.8. *A* long short-term memory (LSTM) cell *is a memory cell that has an additional long-term hidden variable, called the* cell state *c, such that the basic operations of the cell consist of the calculation of three variables, f, i, and o, that control the* forget gate, input gate, *and* output gate, *respectively, and a variable g that depends on the previous hidden state and current input, via the relations*

$$f = \phi_\sigma \left(W_{xf} x_t + W_{hf} h_{t-1} + \beta_f \right) \tag{13.17}$$

$$i = \phi_\sigma \left(W_{xi} x_t + W_{hi} h_{t-1} + \beta_i \right) \tag{13.18}$$

$$o = \phi_\sigma \left(W_{xo} x_t + W_{ho} h_{t-1} + \beta_o \right) \tag{13.19}$$

$$g = \tanh \left(W_{xg} x_t + W_{hg} h_{t-1} + \beta_g \right), \tag{13.20}$$

where ϕ_σ is the sigmoid activation function, such that the cell state and hidden state are computed via

$$c_t = f \odot c_{t-1} + i \odot g \tag{13.21}$$

$$h_t = o \odot \tanh(c_t), \tag{13.22}$$

respectively. The variables f, i, o, and g are called helper variables *or* minions; *the first three are referred to as* gate controllers *and the fourth as the* penultimate output.

First, note that the intermediary variables f, i, o, and g are themselves identical to the output of a single memory cell (Equation (13.16)) with unique sets of input weights, hidden-state weights, and biases. When the input vector x and hidden state h are vectors, these constitute fully connected parallel layers, with a sigmoid activation function for the first three and a hyperbolic tangent activation for the fourth. The sigmoid activation function, which returns values between zero and one, is used for the first three, as they control which components of the cell state and hidden state should be modified. The tanh activation function is used for the fourth local variable g, which might be thought of as a sort of penultimate output; i.e., the computation of g alone could constitute the hidden state of an ordinary memory cell. These operations are illustrated graphically in Figure 13.22[5] (following the design from Gèron [2019]).

Now let's look at these operations in detail.

- $f \odot c_{t-1}$: since f is a vector of values on $(0,1)$, this componentwise multiplication determines the degree to which each component of the cell state should be decayed, or "forgotten."
- $i \odot g$: since i is a vector of values on $(0,1)$, this controls to what extend each component of the penultimate output g should be relevant in updating the cell state.

---

[5] Forgive the mild inconsistency: in the diagram, componentwise multiplication is symbolized by $\otimes$, whereas in the text we use $\odot$, as we have previously used the former symbol elsewhere to denote the outer product.

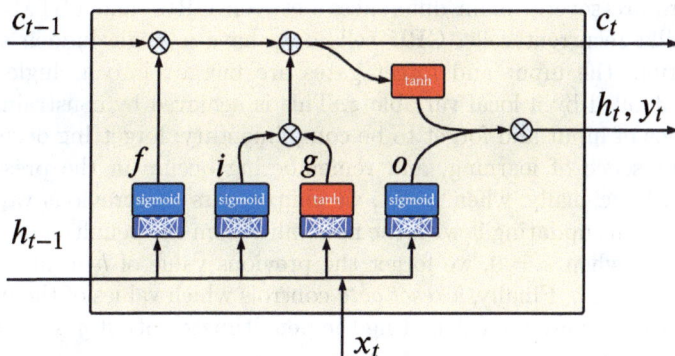

Fig. 13.22: Illustration of an LSTM cell. $\otimes$ and $\oplus$ denote componentwise multiplication and addition, respectively. The purple layers with the cross-hatch patterns sitting below f, i, g, and o are fully connected layers, each with their own set of weights.

- $c_t = f \odot c_{t-1} + i \odot g$: combines the previous two operations of forgetting a little bit of history and inputing new information, yielding the new cell state c_t.
- $h_t = o \tanh(c_t)$: since o is a vector of values on $(0,1)$, this controls to what extend the activated cell state should be captured as hidden state, which equals the cell output $y_t = h_t$.

Layers of LSTM cells can now easily be constructed with minor modification to form a deep RNN. In particular, we would add a superscript l to weights and biases, hidden states and cell states, but replace the input vector x_t with h_t^{l-1}, which represents the output from the preceding layer, with the caveat that we define $h_t^0 = x_t$ to coincide with the actual input. In this way, the output of one layer feeds into the next.

Despite the success of LSTM cells in modern temporal machine-learning applications, they still struggle to remember sequences longer than $O(10^2)$, an improvement on simple memory cells by about a factor of ten. One approach to learn based on longer sequences is to apply a one-dimensional convolutional layer first, to generate a shorter temporal sequence that is better suited for LSTM-based RNNs.

GRU

A simplified form of LSTM has also been proposed, called the *gated recurrent unit* (GRU). The two models tend to have similar performance, with GRU having the advantage of being simpler to implement and more efficient. GRUs are also easier to train with smaller data sets; though using the tried-and-true LSTM is still the default choice for large data sets and longer sequences.

There are several main differences between GRUs and LSTMs. First, like vanilla recurrent cells, GRU cells only have a single hidden state h. In addition, the input and forget gates are merged into a single *update gate*, controlled by a local variable z. This is achieved by constraining the operations of input and forget to be complimentary: forgetting occurs only in the presence of learning, and remembering occurs in the presence of ignoring. Specifically, when $z = 1$, we simply pass the previous value of h along, without updating it with the new value from the penultimate state g. Conversely, when $z = 0$, we forget the previous value of h in favor of the new value from g. Finally, a *reset gate* controls which values of the previous hidden state are used in calculating the penultimate output g. Formally, we have the following.

Definition 13.9. *A gated recurrent unit (GRU) cell is a memory cell with three minions, consisting of two gate controllers, z and r, that control the* update gate *and* reset gate, *respectively, and a third minion g that depends on the previous hidden state, current input, and the reset gate, via the relations*

$$z = \phi_\sigma \left(W_{xz} x_t + W_{hz} h_{t-1} + \beta_z \right) \tag{13.23}$$

$$r = \phi_\sigma \left(W_{xr} x_t + W_{hr} h_{t-1} + \beta_r \right) \tag{13.24}$$

$$g = \tanh \left(W_{xg} x_t + W_{hg} \left(r \odot h_{t-1} \right) + \beta_g \right), \tag{13.25}$$

where ϕ_σ is the sigmoid activation function, such that the hidden state is computed via

$$h_t = z \odot h_{t-1} + (1 - z) \odot g, \tag{13.26}$$

From Equations (13.23)–(13.26), the workings of GRU are clear. The reset gate r controls the extent to which the previous value of h_{t-1} should be included in the calculation of the penultimate output g, as seen in Equation (13.25). The update gate z functions in the way we previously described, as shown by Equation (13.26). Thus, as was the case with LSTMs, the minions assist in a GRU cell's ability to carry out its basic operations.

Obviously, GRUs have smaller computational footprints than LSTMs and are therefore more efficient and easier to train. This comes at the cost, however, of losing a dedicated long-term memory variable. For this reason, LSTMs remain the prominent, but are still inefficient at learning sequences longer than around 100 terms. As mentioned previously, this can be remedied by combining LSTMs with one-dimensional convolutional layers to process extremely long temporal sequences. In particular, van den Oord, *et al.* [2016] introduced an architecture known as *WaveNet*, which actually stacks multiple convolutional layers, each one doubling the temporal spread of the previous layer, so that lower layers learn short-term patterns, whereas higher layers learn longer-term patterns.

Problems

13.1. Build a vector perceptron with softmax layer and train it on the iris dataset.

13.2. Code the `Activation` (Code Block 13.1) subclasses for the other five classic activation functions shown in Figure 13.6. *Bonus*: code the *leaky ReLU* activation function.

13.3. Explain why it is problematic to create a deep neural network using only linear activations.

13.4. Compute the output vector $\hat{p}$ for a three-class classification neural network for the 4-dimensional input vector $\langle 1, 2, 3, 4 \rangle$ and two layers with connection weights

$$W^1 = \begin{bmatrix} 1 & 0 & 0 & 0 & 0 \\ 1 & 1 & -1 & 1 & -1 \\ 1 & -1 & 1 & -1 & 1 \end{bmatrix} \quad \text{and} \quad \begin{bmatrix} 1 & 0 & 0 \\ 1 & -1 & 1 \\ -1 & 1 & 1 \\ 2 & -1 & 1 \end{bmatrix},$$

such that the first layer uses a ReLU activation and the second (output) layer uses a softmax transformation. (With the appropriate matrix and function definitions, this can be computed in Python using the single computation `softmax(W2@relu(W1@x))`, for appropriately defined functions `softmax` and `relu`.)

13.5. Draw the "non-exploded" version of the exploded neural network depicted in Figure 13.12.

13.6. Show that the derivatives δ_i^m indeed simplify to Equation (13.8) when carrying out the summation of the previous equation when computing the gradient over a softmax layer.

13.7. Suppose the true label of the training instance in Exercise 13.4 was $y = 1$ (i.e., $y = \langle 1, 0, 0 \rangle$). Use backpropagation to compute $\nabla_W L$ using the cross-entropy loss function.

13.8. Write an `AbstractLayer` class that defines the API common to the various layer implementations (i.e., `NeuronLayer`, `SoftmaxNeuronLayer`, `MatrixLayer`, `SoftmaxMatrixLayer`).

13.9. [**Project**] Update the `Loss`, `MatrixLayer` and `Activation` classes to accommodate batch gradient descent.

14

Rise of the Machines

In this chapter, we focus on a third branch of machine learning, separate from supervised and unsupervised learning, known as *reinforcement learning*. In this branch, the machine is given autonomy and allowed to make decisions and take actions. The machine then learns based on its ability to explore (try new things) and the environment's response to its choices. A classic favorite reference on reinforcment learning is Sutton and Barto [2018]. An additional reference is Kochenderfer [2015]. For information on the field of deep reinforcement learning, see Graesser and Keng [2019].

14.1 Reinforcement Learning

In this section, we will lay out the main mathematical formalism that underscores reinforcement learning. We will define the problem as a game between an agent and the environment, and define the solution in terms of policies and value functions. Finally, we introduce a set of optimality conditions.

14.1.1 Finite Markov Decision Processes

The goal of reinforcement learning is to learn how an agent can behave relative to a given environment, that is allowed to respond to the agent's action and give out rewards based on the agent's choices. We may formally regard an agent and its environment as follows.

Definition 14.1. *An* agent *is that which possesses agency.* Agency *is a duty to act.*

Actions inevitably occur over the passage of time, even if the agent doesn't *do* anything. Inaction is also an action. Absention from action is still a choice.

Definition 14.2. *An* environment *is an abstraction of an agent's world. An environment may have various* states *that may change in response to the* action *of the agent. The environment may at times give numerical* rewards *to the agent.*

The three basic quantities that describe the interaction between an agent and its environment are state, action, and reward.

Definition 14.3. *The* state space S *is the set of all possible states of the environment.*
    The action space $\mathcal{A}$ *is the set of all permissible actions an agent might take. The* action bundle $\mathcal{A}_s \subset \mathcal{A}$ *is the subset of permissible actions available for each state $s \in S$.*
    The reward space $\mathcal{R}$ *is the set of possible rewards an agent might receive from the environment.*

Now that we have defined the state, action, and reward spaces, we may formally introduce the process that underlies much of modern-day reinforcement learning.

Definition 14.4. *A* Markov decision process (MDP) *is a quadruple $(S, \mathcal{A}, \varphi, \varrho)$ consisting of a state space S, an action space $\mathcal{A}$, a state-transition mapping $\varphi : S \times \mathcal{A} \to \mathcal{P}(S)$, and a reward mapping $\varrho : S \times S \times \mathcal{A} \to \mathcal{P}(\mathbb{R})$ that defines a stochastic control process $\{(S_t, A_t, R_t)\}_{t \in \mathbb{N}}$, such that $\{A_t\}$ is a sequence of actions (i.e., controls) selected by an agent, $\{S_t\}$ is a sequence of states of the environment, which are governed by the* state-transition matrix (STM)

$$\varphi_s^a(s') = \mathbb{P}(S_{t+1} = s' | S_t = s, A_t = a), \tag{14.1}$$

and $\{R_t\}$ is a sequence of rewards, distributed according to

$$\varrho_{ss'}^a(r) = \mathbb{P}(R_{t+1} = r | S_t = s, A_t = a, S_{t+1} = s').$$

If the spaces S and $\mathcal{A}$ are finite, we refer to the MDP as finite.
    Alternatively, if the full probability distribution over the reward space is not required, we may instead define our MDP by replacing ϱ with the expected-reward matrix (ERM) Υ*, which is the mapping $\Upsilon : S \times S \times \mathcal{A} \to \mathbb{R}$ defined by the expectations*

$$\Upsilon_{ss'}^a = \mathbb{E}[\varrho_{ss'}^a] = \mathbb{E}\left[R_{t+1} | S_t = s, A_t = a, S_{t+1} = s'\right]; \tag{14.2}$$

i.e., it is the expected reward at time $t + 1$, given the current state $S_t = s$ and action $A_t = a$ and the subsequent transition state $S_{t+1} = s'$.

Such a process is a *decision* process, as the sequence of actions are interpreted as choices to be made by an agent. The environment responds to those actions by periodically changing its state and issuing rewards (or

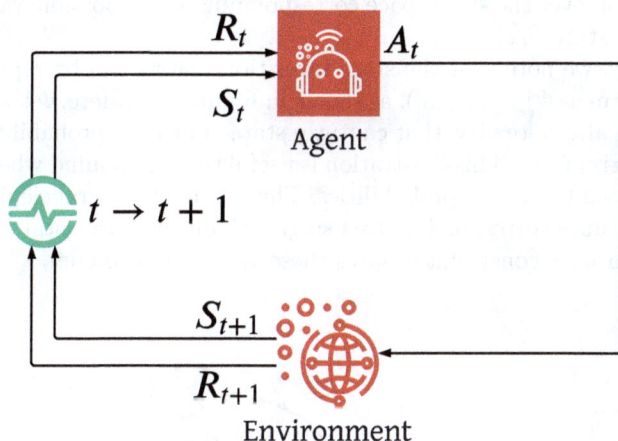

Fig. 14.1: Schematic of an agent and its environment.

penalties, if the reward is negative). This process is characterized in Figure 14.1.

So far, we have not placed any restrictions on the agent's actions; they can be totally uncontrolled and chaotic. The point, however, of reinforcement learning is to learn an optimal strategy for the agent—a method of behaving that is likely to maximize the long-term rewards paid out by the agent's environment. To quantify this, we need a way to mathematically describe how an agent might respond.

Definition 14.5. *Given a finite Markov decision process, a* policy *is a mapping* $\pi : \mathcal{S} \to \mathcal{P}(\mathcal{A})$, *such that the agent chooses an action relative to the probability distribution*

$$\pi_s(a) = \mathbb{P}(A_t = a | S_t = s). \tag{14.3}$$

The goal of reinforcement learning, as we shall see, is to determine an optimum policy, relative to a given objective function.

Given a state-transition matrix and an expected-reward matrix, we can use the former to marginalize over the transition states to obtain

$$\Upsilon_s^a = \mathbb{E}\left[R_{t+1} | S_t = s, A_t = a\right] = \sum_{s' \in \mathcal{S}} \varphi_s^a(s') \Upsilon_{ss'}^a. \tag{14.4}$$

Notice that we regard superscripts as actions and subscripts as states; when two subscripts are given, the second refers to the transition state. Moreover, we explicitly call out the function dependence of $\varphi_s^a(s')$ on s', since, for a given state $S_t = s$ and action $A_t = a$, this quantity represents a probability distribution over the state space corresponding to the possible values of the transition state $S_{t+1} = s'$.

Finally, we note that the state-transition matrix can be represented in matrix-form as $\Phi^a_{ss'} = \varphi^a_s(s')$, as shown in Figure 14.2. Here, $\Phi : \mathcal{A} \times \mathcal{S} \times \mathcal{S} \to [0, 1]$ is an alternate view that captures state-transition probabilities in tabular (matrix) form. This illustration is useful to keep in mind when coding a set of known transition probabilities. The highlighted parallelepiped represents all values corresponding to a single cell on the horizontal (a, s) plane. The probability constraint ensures these values sum to one.

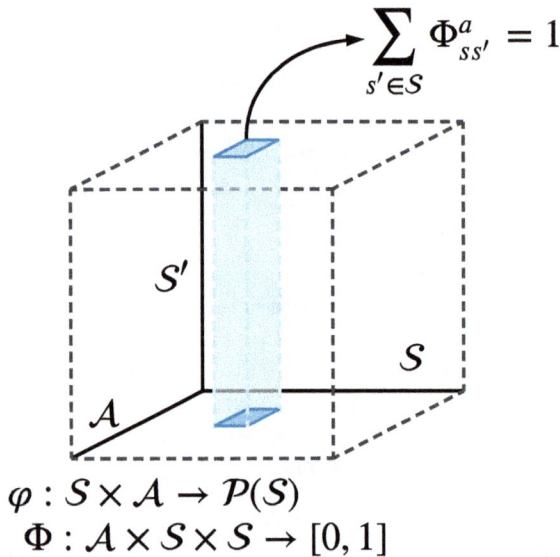

$$\sum_{s' \in S} \Phi^a_{ss'} = 1$$

$$\varphi : \mathcal{S} \times \mathcal{A} \to \mathcal{P}(\mathcal{S})$$
$$\Phi : \mathcal{A} \times \mathcal{S} \times \mathcal{S} \to [0, 1]$$

Fig. 14.2: Schematic of the state-transition matrix.

Episodic and Continuing Tasks

There are two main flavors of Markov decision processes: episodic and continuing.

Definition 14.6. *An episodic task is a reinforcement learning problem in which one seeks to learn based on a sequence of related MDPs, each consisting of only a finite number of steps. A continuing task is a reinforcement learning in which the underlying MDP is infinite in temporal scope (i.e., the stochastic processes go on indefinitely).*

For example, learning how to play chess would be considered an episodic task. Each individual chess game would constitute a separate *episode*. Based on the experience of playing many games over time, one seeks to become a better chess player. A continuing task, on the other hand, is simply a task that goes on indefinitely.

In order to unify notation, episodic tasks are typically endowed with a certain terminal state that only transitions into itself and never pays out any reward.

Definition 14.7. *An* absorbing state $s_0 \in S$ *is any state with the property that*

$$\varphi_{s_0}^a(s) = \mathbb{I}[s = s_0],$$

for all $a \in A$, and such that $R_{t+1} = 0$ whenever $S_t = s_0$.

In other words, an absorbing state only allows transitions into itself (hence the name) and never pays out any subsequent reward. In the context of chess, the absorbing state may be considered a fictitious state following a *checkmate*, that state being the status: *the game is over*.

By including an absorbing state with any episodic task, we may continue to use the same notation (e.g., infinite sums over future rewards) as we use for continuing tasks.

14.1.2 Value Functions

Much of reinforcement learning reduces to the task of learning certain *value functions*, which express the *value* of a given state (or state–action pair) relative to a given policy.

Present Value of Future Returns

Typically, we wish to maximize the *total expected return*, which consists of summing over all future rewards. For continuing tasks, however, this sum tends to diverge, since an infinite time horizon can easily give rise to an infinite return. This is resolved by *discounting* future rewards with a given *discount rate* γ, to account for the *time value of money*. Formally, we have the following.

Definition 14.8. *The* return *of a Markov decision process at time t is the discounted sum of future rewards, as given by*

$$G_t = R_{t+1} + \gamma R_{t+2} + \gamma^2 R_{t+3} + \cdots = \sum_{k=0}^{\infty} \gamma^k R_{t+k+1}. \tag{14.5}$$

where the parameter $\gamma \in [0, 1]$ is the discount rate.

Naturally, as long as the reward sequence is bounded, the return will be finite for $\gamma < 1$. If $\gamma = 0$, the agent is only concerned with maximizing its immediately subsequent reward. This is, in general, not optimal, as acting to optimize an immediate reward can limit the collection of future rewards; see, for example, the Stanford marshmallow experiment (Mischel and Ebbesen [1970]).

In episodic tasks, it is possible for a reward to be zero until the penultimate state; i.e., penultimate if we view the final state as a zero-reward absorbing state. For example, in the game of chess, the reward can be zero or one, depending if the agent win's or loses. This illustrates the absolute necessity in the ability to backpropagate future rewards to much earlier behavior.

State and State–Action Value Functions

Value functions are ubiquitous in reinforcement learning. They are functions that prescribe a value for an agent to be in a given state, or to take a given action from a given state. Naturally, this notion of value relies on how the agent intends to act for the duration of the game, as quantified by its policy. We therefore reference the two main types of value functions in relation to a given policy.

Definition 14.9. *The* state value function *(or, simply, value function) of a state $s \in \mathcal{S}$ under a policy π is the function*

$$v_\pi(s) = \mathbb{E}_\pi\left[G_t | S_t = s\right] = \mathbb{E}_\pi\left[\sum_{k=0}^{\infty} \gamma^k R_{t+k+1} \,\middle|\, S_t = s\right], \qquad (14.6)$$

where the expectation is taken relative to the policy π; i.e., it represents the expected value assuming that the agent acts according to the policy π for all future time.

The state-value function therefore represents the value (i.e., expected return) for the agent to be in a given state, assuming the agent follows the policy π for all time thereafter. It is also useful to quantify the value of an agent to be in a given state *and* make a given action, and then follow the policy for all time thereafter. We define this as follows.

Definition 14.10. *The* state-action value function *of a state $s \in \mathcal{S}$ and action $a \in \mathcal{A}$ under a policy π is the expected return*

$$q_\pi(s, a) = \mathbb{E}_\pi\left[G_t | S_t = s, A_t = a\right], \qquad (14.7)$$

where G_t is the expected return given by Equation (14.5).

This is a slight twist, as it allows consideration of the agent to take any action from state s, making explicit reference to the action taken. Since the state value function assumes A_t is taken relative to the given policy, i.e., $A_t \sim \pi_s$, it is related to the state-action value function by the equation

$$v_\pi(s) = \sum_{a \in \mathcal{A}} \pi_s(a) q_\pi(s, a). \qquad (14.8)$$

However, the state-action value function is agnostic as to how the *present action* A_t is chosen, as long as all *future* actions are taken according to the policy in question.

The Bellman equation

The Bellman equation is simply a recurrence relation that expresses the relationship between the value of a state and the value of its successor state. The unique solution to this equation is the state value function. We begin with a lemma.

Lemma 14.1. *The state-action value function of an MDP under a policy π and discount rate γ is related to the state value function and transition probabilities by the relation*

$$q_\pi(s, a) = \sum_{s' \in \mathcal{S}} \varphi_s^a(s') \left[\Upsilon_{ss'}^a + \gamma v_\pi(s') \right]. \qquad (14.9)$$

Proof. We begin by noting

$$q_\pi(s, a) = \mathbb{E}_\pi[G_t | S_t = s, A_t = a] = \mathbb{E}_\pi[R_{t+1} + \gamma G_{t+1} | S_t = s, A_t = a],$$

which follows easily enough from Equation (14.5).

Now, the expected value of the next reward can be obtained by applying the law of total expectation by considering all possible transition states, from which we obtain

$$\mathbb{E}_\pi[R_{t+1} | S_t = s, A_t = a] = \sum_{s' \in \mathcal{S}} \varphi_s^a(s') \Upsilon_{ss'}^a,$$

which is equivalent to Equation (14.4).

Next, we may express the quantity $\mathbb{E}_\pi[G_{t+1} | S_t = s, A_t]$ by again considering all possible transition states, thereby obtaining

$$\mathbb{E}_\pi[G_{t+1} | S_t = s, A_t = a] = \sum_{s' \in \mathcal{S}} \varphi_s^a(s') \mathbb{E}_\pi[G_{t+1} | S_{t+1} = s'] = \sum_{s' \in \mathcal{S}} \varphi_s^a(s') v_\pi(s').$$

The result follows.   □

There are two fundamentals ways in which we might apply Equation (14.8) to Equation (14.9). The first is to replace $q_\pi(s, a)$ in Equation (14.8) with the right-hand side of Equation (14.9). This result is known as the *Bellman equation*, as given by the following theorem. But note the second way: we could also replace $v_\pi(s')$ in the summand of Equation (14.9) with the right-hand side of Equation (14.8), evaluated, of course, at s'. This yields a recurrence relation for the state-action value function.

Proposition 14.1 (Bellman equation). *The state value function of an MDP under a policy π and discount rate γ must satisfy the following recurrence relation*

$$v_\pi(s) = \sum_{a \in \mathcal{A}} \sum_{s' \in \mathcal{S}} \pi_s(a) \varphi_s^a(s') \left[\Upsilon_{ss'}^a + \gamma v_\pi(s') \right]. \qquad (14.10)$$

Proof. This follows immediately by applying the relation given by Equation (14.8) to Equation (14.9). □

Backup diagrams are useful in visualizing how Bellman's equation propagates information from the future *back* to a given state, as shown in Figure 14.3. By convention, empty circles represent states and filled circles represent actions. This diagram shows a given state $S_t = s$, and all possible actions and transition states. The Bellman equation represents an average across all possible outcomes, using the policy to weight the probability of given actions and the state-transition matrix to weight the probability of a given transition state, given the current state and action.

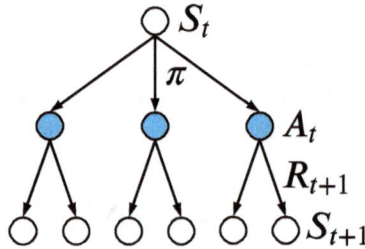

Fig. 14.3: Backup diagram for the Bellman equation.

14.1.3 Optimality

In this section, we introduce the notion of an optimal policies and optimal value functions. We then discuss the Bellman optimality equations.

Optimal Policies and Optimal Value Functions

State value functions define a partial ordering over policies, since we can define $\pi \geq \pi'$ if and only if $v_\pi(s) \geq v_{\pi'}(s)$ for all $s \in \mathcal{S}$. This naturally gives rise to the notion of an optimal policy.

Definition 14.11. *Given a finite MDP, an* optimal policy π_* *is a policy that satisfies* $\pi_* \geq \pi$, *for all other policies* π; *in other words,* $v_{\pi_*}(s) \geq v_\pi(s)$ *for all* $s \in \mathcal{S}$ *and all policies* π.

The optimal state value function, *denoted* v_*, *is defined by the relation*

$$v_*(s) = \max_\pi v_\pi(s). \tag{14.11}$$

Similarly, the optimal state-action value function, *denoted* q_*, *is defined by the relation*

$$q_*(s, a) = \max_\pi q_\pi(s, a). \tag{14.12}$$

The optimal state value and state-action value functions are the value functions of an optimal policy.

Bellman Optimality Equations

Since v_* and q_* are value functions for the optimal policy, they too must satisfy a similar set of self-consistency equations provided by the Bellman's equation. Since these value functions are, however, optimal, we may express their respective recurrence relations without regard to the specific policy. The result is known as the Bellman optimality equations, and is given as follows.

Proposition 14.2 (Bellman optimality equations). *Given a finite MDP, the optimal state value function satisfies the recurrence relation*

$$v_*(s) = \max_{a \in \mathcal{A}_s} \sum_{s' \in \mathcal{S}} \varphi_s^a(s') \left[\Upsilon_{ss'}^a + \gamma v_*(s') \right]. \tag{14.13}$$

Similarly, the state-action value function satisfies the recurrence relation

$$q_*(s, a) = \sum_{s' \in \mathcal{S}} \varphi_s^a(s') \left[\Upsilon_{ss'}^a + \gamma \max_{a' \in \mathcal{A}_{s'}} q_*(s', a') \right]. \tag{14.14}$$

Proof. Equation (14.13) follows from Equation (14.9), when expressed for an optimal policy, coupled with the observation that

$$v_*(s) = \max_{a \in \mathcal{A}_s} q_*(s, a);$$

i.e., the optimal value of a state s must be the maximum value, taken over all possible actions, of the optimal state-action value function anchored to the same state.

Equation (14.14) is also obtained from Equation (14.9), again rewritten for an optimal policy, except this time we use the preceding equation to replace $v_*(s')$ in the summand. □

Optimal Policies from Optimal Value Functions

Once we know the optimal value function v_*, we can determine an optimal policy π_* as follows. For each state $s \in \mathcal{S}$, there will be at least one action that yields the maximum value in the Bellman optimality Equation (14.13). Any policy that assigns a zero probability to actions outside of the set of actions that achieve this maximum therefore constitutes an optimal policy. We may state this result formally as follows.

Proposition 14.3. *Let v_* be the optimal state value function for a finite MDP. For each state $s \in \mathcal{S}$, let*

$$\mathcal{A}_s^* = \arg\max_{a \in \mathcal{A}_s} \sum_{s' \in \mathcal{S}} \varphi_s^a(s') \left[\Upsilon_{ss'}^a + \gamma v_*(s') \right] \qquad (14.15)$$

represent the set of actions that maximize the expected return via the Bellman optimality condition. Any policy that satisfies the condition

$$\pi_s = \mathrm{Unif}(\mathcal{A}_s^*) = \frac{1}{|\mathcal{A}_s^*|} \mathbb{I}[a \in \mathcal{A}_s^*], \qquad (14.16)$$

for all $s \in \mathcal{S}$, is an optimal policy.

In other words, optimal policies are *greedy* with respect to the optimal value function (Definition 4.20). This is because the total expected future reward has been *pulled back* into a value function at each state.

14.1.4 Examples

Gridworld

Example 14.1. A classic example is that of the so-called *gridworld* problem. A *gridworld* is a two-dimensional state space arranged as a grid with a particular geometry. For example, a 5×5 gridworld is shown in Figure 14.4. (Gridworlds are typically, but not necessarily, rectangular.) The rules of gridworld are simple: each state has four possible actions corresponding to the four compass directions. The transition state is 100% the direction specified by the action, with the exception that the transition state is unchanged if either the action points outside of the gridworld or the state is a prescribed absorbing state. The gridworld shown in Figure 14.4 has a single absorbing state at $s_0 = (2, 2)$.

Furthermore, we specify a deterministic reward of -1 for each transition, with the exception, of course, that transitions from the absorbing state to itself have zero value. The optimal policy is shown by the arrows: since each move has a cost, it is optimal to move to the absorbing state as quickly as possible. The optimal state values, using a discount rate of $\gamma = 0.5$, are also given in each square. $\triangleright$

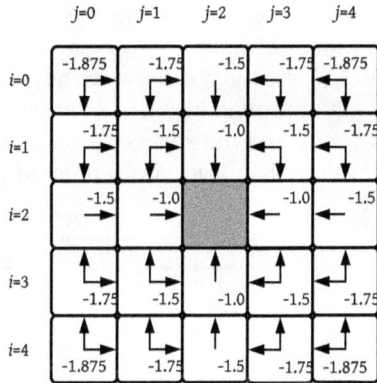

Fig. 14.4: Optimal policy and value function (using discount $\gamma = 0.5$) for the gridworld problem of Example 14.1.

Paper–Scissors–Rock

The following is an example of *adversarial reinforcement learning*, in which two agents compete for a reward. This describes most games: tic-tac-toe, checkers, chess, etc. We may still leverage the reinforcement-learning framework developed thus far by focusing on one agent, and viewing the actions and consequences of the adversarial agent as part of the environmental response.

Example 14.2. In the game of *paper–scissors–rock*, two opponents simultaneously reveal their selection (from the three namesakes) and a winner is determined according to the following binary relation: scissors cut paper, rock smashes scissors, paper covers rock. We may encode these choices using ternary ($\mathbb{Z}_3 = \{0, 1, 2\}$) by assigning a values (0 to paper, 1 to scissors, and 2 to rock) and defining the following binary relation:

$$0 \prec 1, \quad 1 \prec 2, \quad \text{and} \quad 2 \prec 0.$$

Naturally, addition is taken to be mod 3, such that $2 + 1 = 0$.

Using the ternary encoding, we may represent our state space as the set $\mathcal{S} = \mathbb{Z}_3^2$ of two-dimensional ternary vectors, which may be represented as a length-two ternary string. The first component corresponds to the agent's choice and the second the adversary's. For example, the state $s = 21$ represents the agent has chosen *rock* and the adversary has chosen *scissors*. The action space is simply $\mathcal{A} = \mathbb{Z}_3$. Finally, we may specify a deterministic reward R for a transition state $S' = ij$ as $R = -1$ whenever $i \prec j$, $R = +1$ whenever $i \succ j$, and $R = 0$ whenever $i = j$.

In order to specify the transition dynamics, let us make a simple model about our adversary: when they lose or tie, they select their next action with equal probability, but when they win, they choose to remain at their current state with a 50% probability, otherwise they transition to one of the other two states with equal probability. The transition state is determined by joining the agent's action and the adversary's action into a single length-two ternary string. These probabilities are tabulated in Table 14.1.

$R_t = r$	$S_t = ij$	$A_t = a$	$\varphi_{ij}^a(a0)$	$\varphi_{ij}^a(a1)$	$\varphi_{ij}^a(a2)$
0	00	a	1/3	1/3	1/3
−1	01	a	1/4	1/2	1/4
+1	02	a	1/3	1/3	1/3
+1	10	a	1/3	1/3	1/3
0	11	a	1/3	1/3	1/3
−1	12	a	1/4	1/4	1/2
−1	20	a	1/2	1/4	1/4
+1	21	a	1/3	1/3	1/3
0	22	a	1/3	1/3	1/3

Table 14.1: State transition probabilities for the paper–scissors–rock example; $\varphi_{ij}^a(bj') = 0$ whenever $a \neq b$.

Moreover, since the reward depends deterministically on the transition state, which, itself, may be expressed in terms of the action, we may write

$$r_{(ij)(aa)}^a = 0, \qquad r_{(ij)(a(a+1))}^a = -1, \qquad r_{(ij)(a(a-1))}^a = +1.$$

Notice we can leverage the asymmetry present when our agent loses (and the adversary wins) to determine an optimal policy. For this case, the adversary is more likely to remain in its current state, so that our optimal policy is to select an action that would defeat our adversary in its current state; i.e.,

$$\pi_{ii}^*(a) = \pi_{i(i-1)}^*(a) = 1/3 \qquad \text{and} \qquad \pi_{i(i+1)}^*(i-1) = 1,$$

for $i \in \mathbb{Z}_3$. In other words: our agent should take random actions whenever it wins or ties, and transition to the missing state if it loses.   ▷

14.2 Dynamic Programming

Dynamic programming is a technique that can be used to determine optimal policies. We have already discussed one instance of dynamic programming in the backpropagation algorithm for neural networks. Like its deep-learning counterpart, dynamic programming for reinforcement learning consists of two steps: an evaluation (prediction) step and an update (improvement) step. Dynamic programming algorithms then seek to iteratively improve our estimation for an optimal policy or the optimal value function by alternating between these two steps over time. This approach is, however, largely of theoretical interest, as it relies on perfect knowledge of the transition probabilities and reward function. It is also computationally expensive. Nevertheless, it serves as a foundational model for the other, more practical, algorithms we will discuss through the remainder of the chapter.

14.2.1 Policy Iteration

Policy iteration functions similar to both backpropagation (Algorithm 13.4) and the Kalman filter (Bayes–Kalman filter discussed in Algorithm 10.1): it is a two-step process that involves a *prediction step* and an *update step*. In policy iteration, one alternates between a prediction step, which seeks to solve for the value function of a given policy, and an update step, which seeks to update the policy to bring it closer to the optimal policy. These iterations continue until the sequence of policies converges on an optimal policy.

This basic policy-iteration algorithm is shown in Algorithm 14.1. We begin by initializing an arbitrary value function and policy, except that the value function should have a zero value on any absorbing states, if applicable. We then iterate between the two operations `evalPolicy` and `updatePolicy`, which we will describe momentarily. We note, however, that the `evalPolicy` method is itself iterative, and that its second argument, which represents the starting "guess" for the value function, is optional. In practice, it is much faster to pass in the previous value function (as shown in the algorithm), as the `evalPolicy` method will tend to require significantly

---

Algorithm 14.1: Policy iteration.

   Input: An MDP $(\mathcal{S}, \mathcal{A}, \varphi, \Upsilon)$
   Output: An optimal policy π_* and optimal value function v_*
1 Initialize an arbitrary function $V : \mathcal{S} \rightarrow \mathbb{R}$
2 Initialize an arbitrary policy $\pi : \mathcal{S} \rightarrow \mathcal{P}(\mathcal{A})$
3 Set $V(s_0) = 0$ for any absorbing state s_0
4 Set $converged = $ **False**
5 **while not** $converged$ **do**
6   Set $V = $ `evalPolicy`(π, V)
7   Set $\pi' = $ `updatePolicy`(V)
8   Set $converged = \mathbb{I}[\pi = \pi']$
9   Set $\pi = \pi'$
10 **end**
11 **return** (π, V)

---

fewer iterations to converge. The idea is that if the policy changes just a little bit, then so too will its corresponding value function.

This iteration process terminates once it fails to yield an improvement on the current policy.

Policy Evaluation (Prediction)

The objective of *policy evaluation* is to determine the state value function v_π associated with a given policy π and a given MDP. If we know the transition probabilities and reward matrix, Equation (14.10) represents a linear system of $|\mathcal{S}|$ equations in $|\mathcal{S}|$ unknowns, which can be solved using standard techniques from linear algebra. It is more common, however, in practice, to use an iterative approach, as outlined in Algorithm 14.2. Here, we simply use the right-hand side of Equation (14.10) as an update rule to iteratively improve our estimate for the value $V(s)$ of each state $s \in \mathcal{S}$.

Note that we are updating the values of our estimate $V(s)$ of the state value function as we loop through the state space. Other than the first state in this loop, this update is therefore a blend of old values and new values. This tends to converge faster than the alternative—only applying the updates once all the calculations have been made—as one is utilizing each updated value as soon as it becomes available.

Algorithm 14.2: `evalPolicy(`π, V`)` policy evaluation.

Input: An MDP $(\mathcal{S}, \mathcal{A}, \varphi, \Upsilon)$;
A given policy π;
An initial guess (optional) $V : \mathcal{S} \to \mathbb{R}$;
A discount factor $\gamma \in [0, 1]$;
An error tolerance ϵ
Output: The value function v_π of the given policy
1 **if** V **is Null then**
2 $\quad$ Initialize an arbitrary function $V : \mathcal{S} \to \mathbb{R}$
3 $\quad$ Set $V(s_0) = 0$ for any absorbing state s_0
4 **end**
5 Set $\Delta = 2\epsilon$
6 **while** $\Delta > \epsilon$ **do**
7 $\quad$ Set $\Delta = 0$
8 $\quad$ **for** s **in** $\mathcal{S}$ **do**
9 $\quad\quad$ Set $v = V(s)$
10 $\quad\quad$ Set $V(s) = \sum_{a \in \mathcal{A}} \sum_{s' \in \mathcal{S}} \pi_s(a) \varphi_s^a(s') \left[\Upsilon_{ss'}^a + \gamma V(s') \right]$
11 $\quad\quad$ Set $\Delta = \max(\Delta, |v - V(s)|)$
12 $\quad$ **end**
13 **end**
14 **return** V

Finally, we notice the optional input argument V, which constitutes an optional "initial guess" to the value function. As previously mentioned, this is extremely useful in the process of policy iteration, as the *prediction step* itself represents multiple sweeps through state space, which can be quite large. By initializing the algorithm with something close to the true value function, as opposed to a completely random initialization, we require fewer loops through state space to achieve convergence.

Policy Improvement (Update)

A state value function for a given policy can be used to determine a more optimal policy. Essentially, this is achieved by defining a new policy that selects the action that maximizes Equation (14.9) for each state. To arrive at this conclusion, we begin with the following.

Theorem 14.1 (Policy Improvement Theorem). *Given a Markov decision process* $(\mathcal{S}, \mathcal{A}, \varphi, \Upsilon)$ *and two policies* π *and* π', *such that*

$$v_\pi(s) \leq \sum_{a \in \mathcal{A}_s} \pi'_s(a) q_\pi(s, a), \tag{14.17}$$

for all $s \in \mathcal{S}$, *then* $\pi' \geq \pi$, *in the sense that* $v_{\pi'}(s) \geq v_\pi(s)$ *for all* $s \in \mathcal{S}$.

Essentially, Theorem 14.1 says that if the expected value of taking the first step according to the new policy π', and then following the original policy π thereafter, is greater than the value of strictly following the original policy π, then the new policy must be more optimal than the original.

Proof. We proceed by induction. First, let us define a sequence of policies $\pi_0, \pi_1, \pi_2, \ldots$, such that π_k is defined by following the new policy π' for the first k steps and then following the original policy π thereafter. Note that $\pi_0 = \pi$ and

$$\lim_{k \to \infty} \pi_k = \pi'.$$

For the induction step, we note that the right-hand side of Equation (14.17) is equivalent to

$$\sum_{a \in \mathcal{A}_s} \pi'_s(a) q_\pi(s, a) = v_{\pi_1}(s),$$

due to Equation (14.8). This implies that $\pi_0 \leq \pi_1$.

Next, consider the state value function for policy π_k, which may be expressed as

$$v_{\pi_k}(s) = \mathbb{E}_{\pi'} \left[R_1 + \gamma R_2 + \cdots + \gamma^{k-1} R_k + \gamma^k v_\pi(S_k) \middle| S_0 = s \right].$$

Now, from Equation (14.17), and any possible state $S_k = s_k$, we have

$$v_\pi(s_k) \le \sum_{a \in \mathcal{A}_{s_k}} \pi'_{s_k}(a) q_\pi(s_k, a) = \mathbb{E}_{\pi'} \left[R_{k+1} + \gamma v_\pi(S_{k+1}) | S_k = s_k \right].$$

However, this implies that

$$v_{\pi_k}(s) \le \mathbb{E}_{\pi'} \left[R_1 + \cdots + \gamma^k R_{k+1} + \gamma^{k+1} v_\pi(S_{k+1}) \middle| S_0 = s \right] = v_{\pi_{k+1}}(s),$$

so that $\pi_k \le \pi_{k+1}$.

We have therefore constructed a monotonic increasing sequence of policies that starts with the original π and converges to the new policy π'. The result that $\pi \le \pi'$ follows.   □

In order to apply this, let us define the following.

Definition 14.12. *Given a Markov decision process $(\mathcal{S}, \mathcal{A}, \varphi, \Upsilon)$ and an arbitrary state-value function $q(s, a)$, we define the* optimal-action bundle *relative to q, denoted $\mathcal{A}^q_s$, as the mapping $\mathcal{A}^q : \mathcal{S} \to \mathcal{A}$ defined by*

$$\mathcal{A}^q_s = \arg\max_{a \in \mathcal{A}_s} q(s, a) \tag{14.18}$$

for each $s \in \mathcal{S}$. Note $\mathcal{A}^q_s \subset \mathcal{A}_s$.

Corollary 14.1. *Given a Markov decision process $(\mathcal{S}, \mathcal{A}, \varphi, \Upsilon)$, a policy π, and its corresponding state-action value functions q_π, then the new policy π', defined by*

$$\pi'_s = \mathrm{Unif}(\mathcal{A}^{q_\pi}_s), \tag{14.19}$$

for all $s \in \mathcal{S}$, where $\mathcal{A}^{q_\pi}_s$ is the optimal-action bundle relative to q_π, is more optimal than the original policy π; i.e., $\pi' \ge \pi$.

Note 14.1. For a system with known state-transition dynamics, the optimal-action bundle may be expressed via the state value function by using Equation (14.9) to express Equation (14.18) in the equivalent form

$$\mathcal{A}^{q_\pi}_s = \arg\max_{a \in \mathcal{A}_s} \sum_{s' \in \mathcal{S}} \varphi^a_s(s') \left[\Upsilon^a_{ss'} + \gamma v_\pi(s') \right], \tag{14.20}$$

for all $s \in \mathcal{S}$.   ▷

Proof. The argument of the arg max operator of Equation (14.20) is equivalent to the state-action value function $q_\pi(s, a)$, due to Equation (14.9). Therefore, the right-hand side of Equation (14.17) must equal the maximum

$$\sum_{a \in \mathcal{A}_s} \pi'_s(a) q_\pi(s, a) = \max_{a \in \mathcal{A}_s} q_\pi(s, a),$$

since the support of the new policy π' is over the actions that maximize the state-action values. The result follows by Theorem 14.1.   □

This result is the basis for the policy improvement algorithm, as given by Algorithm 14.3. Note that the result still holds if we replace Equation (14.19) by any probability distribution whose support is A_s^π; i.e., that selects only from optimal actions. We simply choose the uniform distribution for good measure[1].

14.2.2 Value Iteration

As we have seen, policy iteration involves alternating between a prediction step, used to evaluate the current policy, and an update state, used to improve upon the current policy. The first step is, itself, iterative, as it requires multiple sweeps through state space before it can converge on the correct value function (within some absolute error tolerance).

Value iteration seeks to simplify this process by combining the two steps, and only performing *a single sweep* through state space for the prediction step. The result is shown in Algorithm 14.4.

Note, in particular, that line 8 of this algorithm replaces the *two* separate computations in policy evaluation and policy improvement of computing the new value of the state value function and the new policy. The reason these two operations can be combined is that the policy is determined by computing the arg max, and then the value function is determined by evaluating the same summation for the new policy, which must therefore yield the simple maximum value. In this way, value iteration works on iteratively improving the state value function and does not require computation of the policy until the very end.

A further illumination can be had if we define the estimated state-action value function in a parallel form to Equation (14.9); i.e., if we define

$$Q(s,a) = \sum_{s' \in \mathcal{S}} \varphi_s^a(s') \left[\Upsilon_{ss'}^a + \gamma V(s') \right]. \qquad (14.21)$$

Algorithm 14.3: updatePolicy(V) policy improvement.

Input: An MDP $(\mathcal{S}, \mathcal{A}, \varphi, \Upsilon)$;
      State value function V;
      A discount factor $\gamma \in [0,1]$
Output: Improved policy π

1 **for** s in $\mathcal{S}$ **do**

2   Set $\pi_s = \text{Unif}\left(\underset{a \in \mathcal{A}_s}{\arg\max} \sum_{s' \in \mathcal{S}} \varphi_s^a(s') \left[\Upsilon_{ss'}^a + \gamma V(s') \right] \right)$

3 **end**

4 **return** π

---

[1] lol.

The state-value update in the policy valuation Algorithm 14.2 essentially averages the state-action values relative to the given policy:

$$V(s) = \sum_{a \in \mathcal{A}} \pi_s(a) Q(s, a). \tag{14.22}$$

In this way, the policy π remains unchanged and we successively get closer to the true state value function we are trying to estimate. Value iteration, on the other hand, further selects the optimal action in this update, by taking

$$V(s) = \max_{a \in \mathcal{A}_s} Q(s, a), \tag{14.23}$$

which is policy agnostic. Thus, Equation (14.22) represents the policy evaluation of Algorithm 14.2, whereas Equation (14.23) represents the value iteration of Algorithm 14.4, as it updates both our estimate for V and the policy in one swoop.

14.2.3 Examples

Paper–Scissors–Rock

Example 14.3. Let us continue the paper–scissors–rock problem of Example 14.2. The code for value iteration is given in Code Block 14.1.

Algorithm 14.4: Value iteration.

Input: An MDP $(\mathcal{S}, \mathcal{A}, \varphi, \Upsilon)$;
          A discount factor $\gamma \in [0, 1]$;
          An error tolerance ϵ
Output: A near optimal policy $\pi \approx \pi_*$ and its associated value
            function v_π

1 Initialize an arbitrary function $V : \mathcal{S} \to \mathbb{R}$
2 Set $V(s_0) = 0$ for any absorbing state s_0
3 Set $\Delta = 2\epsilon$
4 **while** $\Delta > \epsilon$ **do**
5   Set $\Delta = 0$
6   **for** s in $\mathcal{S}$ **do**
7   Set $v = V(s)$
8   Set $V(s) = \max\limits_{a \in \mathcal{A}_s} \sum\limits_{s' \in \mathcal{S}} \varphi_s^a(s') \left[\Upsilon_{ss'}^a + \gamma V(s') \right]$
9   Set $\Delta = \max(\Delta, |v - V(s)|)$
10   **end**
11 **end**

12 Set $\pi_s = \mathrm{Unif}\left(\arg\max\limits_{a \in \mathcal{A}_s} \sum\limits_{s' \in \mathcal{S}} \varphi_s^a(s') \left[\Upsilon_{ss'}^a + \gamma V(s') \right] \right)$
13 **return** (π, V)

```python
# Initialize States, Actions, Rewards, and State Transitions
actions = [str(i) for i in range(3)]
states = [i+j for i in actions for j in actions]
rewards = [int((int(s[1])+1)%3 == int(s[0])) - int((int(s[0])+1)%3
    == int(s[1])) for s in states]
phi = np.zeros((3,9,9))
for i, a in enumerate(actions):
    for j, s0 in enumerate(states):
        for k, s1 in enumerate(states):
            if a != s1[0]:
                continue
            if rewards[j] == -1:
                phi[i, j, k] = 0.25 + 0.25 * int(s0[1] == s1[1])
            else:
                phi[i, j, k] = 1/3
# Set parameters
epsilon, gamma = 0.001, 0.5
n_states, n_actions = 9, 3
delta = 2 * epsilon
V, dV, pis = np.zeros(9), None, []
# Value Iteration
while delta > epsilon:
    delta = 0
    pi = DataFrame(0, index=states, columns=range(3))
    for j in range(n_states):
        v = V[j]
        Q = np.zeros(3)
        for i in range(n_actions):
            for k in range(n_states):
                Q[i] += phi[i, j, k] * (rewards[k] + gamma * V[k])
        V[j] = max(Q)
        delta = max(delta, abs(v-V[j]))
        Q = Q.round(3)
        best_actions = np.argwhere(Q==max(Q)).flatten()
        for a in best_actions:
            pi.loc[states[j], a] = 1/len(best_actions)
    pis.append(pi)
    if dV is None:
        dV = Series(V*1, index=states)
    else:
        dV = pd.concat((dV, Series(V, index=states)), axis=1,
            ignore_index=True)
```

Code Block 14.1: Value iteration for paper–scissors–rock problem.

We begin by defining the state and action space, and the corresponding deterministic rewards, along with the state-transition matrix in lines 2–14. Recall that states consist of a length-two ternary string, whose first component corresponds to the action (paper, scissors, or rock) of the agent, and whose second component corresponds to the action of the adversary. Transition probabilities are therefore only defined when the action a coincides with the first part of the transition state s'. Moreover, when the agent loses, the adversary is 50% likely to choose its current state as its next action, as shown on line 12, but otherwise equally likely to choose any action (line 14).

Value iteration is then achieved through the loop on lines 21–40. Here, we leverage the update form of Equation (14.23). Since this is computed for each state iteratively, we only have to store the state-action values for the three actions in our array Q. We also calculate the policy within each step, just as means of illustration, though the policy is typically not calculated for value iteration until the very end.

state	0	1	2	3	$\pi_s^3(0)$	$\pi_s^3(1)$	$\pi_s^3(2)$
00	0.0	0.0724	0.079	0.0799	0.333	0.333	0.333
01	0.25	0.3124	0.319	0.3199	0	0	1
02	0.0417	0.0724	0.079	0.0799	0.333	0.333	0.333
10	0.0486	0.0762	0.0795	0.0799	0.333	0.333	0.333
11	0.0486	0.0762	0.0795	0.0799	0.333	0.333	0.333
12	0.2917	0.3162	0.3195	0.3199	1	0	0
20	0.3047	0.3181	0.3198	0.32	0	1	0
21	0.0648	0.0781	0.0798	0.08	0.333	0.333	0.333
22	0.0648	0.0781	0.0798	0.08	0.333	0.333	0.333

Table 14.2: Estimated state value function for paper–rock–scissors problem. Column label represents number of iterations. Final three columns represent the policy π^3 obtained after four iterations.

The final results are shown in Table 14.2. We see that the value-iteration algorithm converged (with tolerance $\epsilon = 0.001$ and discount factor $\delta = 0.5$) after only four rounds. The final policy is also given in the table, which agrees with our conclusion of Example 14.2.   ▷

14.3 From Monte Carlo to Temporal Differences

Previously, we saw how dynamic programming (DP) methods can be used to solve reinforcement-learning problems when the underlying dynamics of the system are known. In the remainder of the chapter, we will discuss *learning* algorithms, that task our intrepid agent with finding an optimal policy by continuously interacting with an a priori unknown environment.

The primary focus of this chapter is on *temporal-difference* (TD) methods, which lie at the heart of much of modern-day reinforcement learning. Before discussing these techniques, however, we will start our discussion by laying out a few definitions regarding the distinction between *on-policy* and *off-policy* learning. We then show how one can solve for the optimal policy in the context of episodic tasks using Monte-Carlo methods. Only then will we get into the heart of the section as we build up the theory of temporal-difference methods.

14.3.1 Learning Through Experience

Before diving into temporal-difference learning, we begin with a few general words to highlight our departure from learning from a completely known model to learning from experience.

Generalized Policy Iteration

During our discussion on dynamic programming, we discussed *policy iteration*, which is an iterative technique that alternates between a policy evaluation (prediction) step and a policy improvement (update) step. We will continue to follow the this process, which is known as *generalized policy iteration*, for the learning problem. Moreover, as was the case in the value iteration algorithm, we will tend to favor algorithms that perform both steps within a given algorithm. To achieve this, we will maintain both an approximate value function as well as an approximate policy. With each step, the value function gets closer to the current policy, and the policy is improved towards becoming the optimal policy.

The crucial difference between general reinforcement learning methods and dynamic programming is, of course, the lack of knowledge of the underlying model. In the process of dynamic programming, we never had to learn the state-action value function directly, as we could simply compute it given our estimate of the state-value function V and the underlying transition dynamics φ, as in Equation (14.21). As a result, general reinforcement learning algorithms focus on learning the state-action value function Q, as opposed to the state value function V. We can then use Equation (14.18) to find a more optimal policy, as opposed to the simplified form Equation (14.20) that can be used when the transition dynamics are known.

On-Policy and Off-Policy Exploration

Exploration–Exploitation Tradeoff

Learning state-action value functions presents an additional challenge compared with learning state value functions, since a value for each state *and* each possible action must be learned through experience. If we simply follow the locally greedy policy, it is likely that we each state will harbor a set of actions that were never selected. This is counterproductive to the task of learning, since the act of only trying what you already think to be best is antithetical to the necessary step of trying new things. This is known as the *exploration–exploitation tradeoff*, which we first introduced during our discussion of multi-armed bandits in Section 4.5. Successful reinforcement learning algorithms must balance the need to find an optimal solution with the need to explore.

On-Policy and Off-Policy Learning

Exploration and learning can be achieved using either *on-policy* or *off-policy* methods.

Definition 14.13. *An* on-policy *method is a learning algorithm that learns the state-action values of the policy that is used to generate actions.*

An off-policy *method is a learning algorithm that learns the state-action values of one policy, often referred to as the* target policy, *from an agent who acted according to a different policy, often referred to as the* behavioral policy.

On-policy learning does not mean the policy is fixed. We can still follow the generalized policy iteration approach, so that the policy slowly evolves over time. However, when we update our estimate of the state-action values, we are using information from the policy that we are following.

Off-policy learning, on the other hand, seeks to learn about one policy while following another. This has two important applications. The first is that we can learn about the optimal policy while following a behavioral policy that maintains exploration. In this approach, the behavioral policy is never an optimal policy relative to our current understanding of the state-action values, as it maintains a certain degree of exploration, which, by definition, involves making non-optimal choices to see if they are actually better than our current optimal estimates. The second application is in learning from others. For example, a reinforcement learning algorithm that is trying to learn how to play a game might learn from many examples of the game behind played by humans. The humans are, of course, doing their own thing, and following their instinct and expertise.

On-Policy Exploration: ϵ-Greedy Policy

A common approach to maintain exploration for on-policy learning is to use an ϵ-greedy strategy (recall Definition 4.20). For our present purposes, we define the ϵ-greedy policy as follows.

Definition 14.14. *Given a Markov decision process* $(\mathcal{S}, \mathcal{A}, \varphi, \Upsilon)$ *and an arbitrary state-action value function* $q(s, a)$, *the associated* ϵ-greedy *policy* $\Pi(q; \epsilon)$ *relative to* q *is the policy defined by*

$$\Pi(q;\epsilon)_s = (1-\epsilon)\mathrm{Unif}(\mathcal{A}_s^q) + \epsilon\mathrm{Unif}(\mathcal{A}_s), \tag{14.24}$$

where $\epsilon \in [0,1]$ represents the degree of exploration and $\mathcal{A}_s^q$ is the optimal-action bundle of q given by Equation (14.18).

A typical value of exploration is $\epsilon = 0.1$, though this may be regarded as a tunable hyperparameter.

Essentially, the ϵ-greedy policy chooses a greedy option $(1-\epsilon)$ of the time, and a random action ϵ of the time. The special case $\epsilon = 0$ corresponds to the greedy policy, whereas $\epsilon = 1$ corresponds to a policy that can never learn, because it is always exploring. For on-policy learning algorithms with nonzero $\epsilon > 0$, we are therefore not learning the optimal policy, but rather learning the the state-action values for near-optimal policies that still maintain a fixed degree of exploration.

Off-Policy Learning and Importance Sampling

Many, though not all, off-policy learning techniques leverage the idea of importance sampling, which is a general technique that can be used to estimate statistics of one distribution using samples from another. (Incidentally, our first off-policy algorithm, Q-learning, will circumvent necessity of this technique.)

Definition 14.15. *Given a Markov decision process $(\mathcal{S}, \mathcal{A}, \varphi, \Upsilon)$, two policies π and b, where b is a soft (i.e., nonvanishing) policy, and a trajectory $\tau = \{(s_{t+k}, a_{t+k})\}_{k=0}^{h}$ of length h, the importance-sampling ratio ρ of policy π relative to policy b and trajectory τ is given by*

$$\rho = \prod_{k=0}^{h-1} \frac{\pi_{s_{t+k}}(a_{t+k})}{b_{s_{t+k}}(a_{t+k})}. \tag{14.25}$$

Typically, we interpret the policy b as the behavioral policy, which was used to generate the trajectory τ, and the policy π as the target policy, about which we are seeking to learn. We require the policy b to be soft, meaning nonvanishing, to prevent division by zero in the ratios of Equation (14.25). The importance-sampling ratio is of particular use due to the following result.

Proposition 14.4. *Given a Markov decision process $(\mathcal{S}, \mathcal{A}, \varphi, \Upsilon)$, two policies π and b, where b is a soft policy, and a trajectory $\tau = \{(s_{t+k}, a_{t+k})\}_{k=0}^{h}$ of length h, the importance sampling ratio ρ represents the relative probability of the trajectory occurring under policy π relative to policy b; i.e.,*

$$\rho = \frac{\mathbb{P}_\pi(\tau | S_t = s_t)}{\mathbb{P}_b(\tau | S_t = s_t)}. \tag{14.26}$$

Note that we implicitly disregard the final action $A_{t+h} = a_{t+h}$, as we are concerned with the probability of reaching the final state $S_{t+h} = s_{t+h}$ under the set of intermediate actions.

Proof. The probability of the trajectory τ relative to policy π is given by

$$\mathbb{P}_\pi(\tau|S_t = s_t) = \pi_{s_t}(a_t)\varphi_{s_t}^{a_t}(s_{t+1})\cdots\pi_{s_{t+h-1}}(a_{t+h-1})\varphi_{s_{t+h-1}}^{a_{t+h-1}}(s_{t+h})$$

$$= \prod_{k=0}^{h-1} \pi_{s_{t+k}}(a_{t+k})\varphi_{s_{t+k}}^{a_{t+k}}(s_{t+k+1}).$$

Therefore, the relative probability of the trajectory τ occurring under policy π relative to policy b is given by

$$\frac{\mathbb{P}_\pi(\tau|S_t = s_t)}{\mathbb{P}_b(\tau|S_t = s_t)} = \frac{\prod_{k=0}^{h-1}\pi_{s_{t+k}}(a_{t+k})\varphi_{s_{t+k}}^{a_{t+k}}(s_{t+k+1})}{\prod_{k=0}^{h-1}b_{s_{t+k}}(a_{t+k})\varphi_{s_{t+k}}^{a_{t+k}}(s_{t+k+1})}.$$

However, the state-transition probabilities in the numerator and denominator cancel, so that the preceding equation reduces to the importance-sampling ratio given by Equation (14.25). This proves the result. □

14.3.2 Monte-Carlo Control

A simple method for solving reinforcement-learning problems is by using Monte-Carlo methods, which essentially attempt to memorize the expected returns across the entire state-action space. In the context of reinforcement learning, Monte-Carlo methods are only suitable for episodic tasks. For Monte-Carlo control, which seeks to learn an unknown optimal policy, the idea is that one waits for the end of each episode and then updates the state-action value function for each state-action pair encountered along the path through the episode. A similar technique can also be used for the policy-evaluation problem; we leave the details to Sutton and Barto [2018]. The Monte-Carlo solution to the control problem is given in Algorithm 14.5.

Monte-Carlo methods are often used with a concept known as *exploring starts*, as defined below.

Definition 14.16 (Exploring Starts). *Given an episodic MDP,* exploring starts *is any exploration strategy that satisfies*

1. *the first state-action pair of each episode is selected at random, such that each state-action pair has a nonzero probability of being selected;*
2. *subsequent actions are greedy relative to the current estimated state-action value function Q.*

The initial action for each episode, shown on line 6, is exploratory, whereas subsequent actions, shown on line 14, are greedy. In addition to Q, the Monte-Carlo control problem tracks an array N that represents the number of visits to each state-action pair. The state-action value function is then the average return observed for each state-action pair over many example episodes. Note that line 21 adds the current reward to the cumulative average, for a given state-action pair, and line 23 then averages by dividing the cumulative total reward by the new count.

Unlike dynamic programming, Monte Carlo methods do not require a model, as they operate by aggregating sample experience. They also do not bootstrap; i.e., value estimates do not use other value estimates during the update protocol. They suffer from the drawback that one requires a sufficient number of visits to each state-action pair in order to estimate the value. Such an approach becomes easily untenable for large problems.

We next look at a technique—temporal differencing—that combines the bootstrapping aspects of dynamic programming with the sampling aspects of Monte-Carlo methods. The result is a class of techniques that can not only learn without a model, but also by bootstrapping existing estimates into better ones.

Algorithm 14.5: Monte Carlo Control with Exploring Starts

Input: An MDP $(\mathcal{S}, \mathcal{A}, \varphi, \varrho)$;
A discount factor $\gamma \in [0, 1]$
Output: An optimal state-action value function
1 Initialize an arbitrary function $Q : \mathcal{S} \times \mathcal{A} \to \mathbb{R}$
2 Set $Q(s_0, a) = 0$ for any absorbing state s_0
3 Set $N = \texttt{Zeros}(\mathcal{S} \times \mathcal{A})$
4 **for Each** *episode* **do**
5   Initialize $S \sim \text{Unif}(\mathcal{S})$
6   Select exploratory action $A \sim \Pi(Q; 1)_S$
7   Set $path, rewards = \texttt{List}(), \texttt{List}()$
8   **while** *S is not an absorbing state* **do**
9   Observe $S' \sim \varphi_S^A$
10   Observe $R \sim \varrho_{SS'}^A$
11   **append**$(path, (S, A))$
12   **append**$(rewards, R)$
13   Update $S = S'$
14   Update by selecting a greedy action $A \sim \Pi(Q; 0)_{S'}$
15   **end**
16   Set $G = 0$
17   **while** *path* **is not Empty do**
18   Set $(S, A) = \texttt{pop}(path)$
19   Update $G = \gamma G + \texttt{pop}(rewards)$
20   **if** (S, A) **not in** *path* **then**
21   Update $Q[S, A] = N[S, A] * Q[S, A] + G$
22   Update $N[S, A] = N[S, A] + 1$
23   Update $Q[S, A] = Q[S, A]/N[S, A]$
24   **end**
25   **end**
26 **end**
27 **return** Q

14.3.3 TD Policy Evaluation

Temporal-difference learning plays a central role in modern reinforcement learning. It is not dissimilar to gradient descent, in the sense that it makes small correctional steps for each iteration of the algorithm, with the step size being controlled by a parameter η corresponding to the learning rate. Though not used in the control problem, we begin with an illustration of temporal-difference learning in the context of policy evaluation. (For the control problem, we require estimation of the state-action value function.)

Essentially, temporal-difference methods combine sampling and bootstrapping to form updates of the value function. In this section, we will focus on *one-step* temporal-difference methods, but will generalize to *multi-step* methods thereafter. At heart of the policy update is the *TD error*, defined by

$$\delta V_t = R_{t+1} + \gamma V(S_{t+1}) - V(S_t). \tag{14.27}$$

This represents the difference between the value estimate at step t, $V(S_t)$, and the (presumably better) value estimate that combines the subsequent reward with the value of the transition state; i.e., $R_{t+1} + \gamma V(S_{t+1})$. Note that the temporal-difference error is not immediately available, but rather it is available only after we have observed the imminent reward and transition state.

Algorithm 14.6: One-step TD policy evaluation.

Input: An MDP $(\mathcal{S}, \mathcal{A}, \varphi, \varrho)$;
        A given policy π;
        An initial guess (optional) $V : \mathcal{S} \to \mathbb{R}$;
        A discount factor $\gamma \in [0, 1]$;
        A learning rate $\eta \in (0, 1]$
Output: An estimate of the value function v_π of the given policy

1 **if** V **is Null then**
2 $\quad$ Initialize an arbitrary function $V : \mathcal{S} \to \mathbb{R}$
3 $\quad$ Set $V(s_0) = 0$ for any absorbing state s_0
4 **end**
5 **for Each** *episode* **do**
6 $\quad$ Initialize $S \sim \text{Unif}(\mathcal{S})$
7 $\quad$ **while** S *is not an absorbing state* **do**
8 $\quad\quad$ Select action $A \sim \pi_S$
9 $\quad\quad$ Observe $S' \sim \varphi_S^A$
10 $\quad\quad$ Observe $R \sim \varrho_{SS'}^A$
11 $\quad\quad$ Update $V(S) = V(S) + \eta\left[R + \gamma V(S') - V(S)\right]$
12 $\quad\quad$ Update $S = S'$
13 $\quad$ **end**
14 **end**
15 **return** V

The temporal-difference method for policy evaluation is shown in Algorithm 14.6. Note that unlike the dynamic-programming policy evaluation algorithm (Algorithm 14.2), the TD method learns with experience in real time. As such, we explicitly reference the various episodes for the case of episodic tasks. For continuing tasks, we may disregard this outer for loop.

The TD policy evaluation method is not a control algorithm, as the policy π remains fixed. It is simply a method for updating state values of the policy as the consequences of the policy unfold. Further, we need not know the true state and reward dynamics φ and ϱ. We only require the *outcome* S' and R of those dynamics to proceed. Thus, the algorithm can learn from *actual* experience, without regard to an a priori understanding of the system dynamics. Moreover, the algorithm can further learn from *simulated* experience, though we must prescribe a (often simple) model of those dynamics to run simulations. For the purpose of simulation, the model only need generate sample transitions; a full specification of the underlying probability density, as would be required for dynamic programming, is unnecessary.

14.3.4 On-Policy TD Control: SARSA

As we stated at the top of section, learning the state value function for experiential learning algorithms is totally insufficient, as we have no knowledge of the underlying system dynamics to determine a greedy policy from our state-value estimates. The temporal-difference control algorithm therefore necessarily involves estimating the state-action values $Q(S, A)$. In fact, there is a natural generalization of the temporal-difference error of Equation (14.27) to the state-action value estimates. The resulting algorithm, known as *SARSA*, is provided in Algorithm 14.7.

Note that SARSA replaces the state-based TD error of Equation (14.27) with the state-action alternative

$$\delta Q_t = R_{t+1} + \gamma Q(S_{t+1}, A_{t+1}) - Q(S_t, A_t). \qquad (14.28)$$

The quantity $Q(S_t, A_t)$ represents the *current* estimate for the value of the current state-action pair, whereas the quantity $R_{t+1} + \gamma Q(S_{t+1}, A_{t+1})$ represents a better value as based on the immediately observed return and the discounted estimate of the subsequent state-action pair. Each update therefore requires the quintuple $(S_t, A_t, R_{t+1}, S_{t+1}, A_{t+1})$, from which the SARSA algorithm derives its name.

The SARSA algorithm further selects actions relative to the corresponding ϵ-greedy policy $\Pi(Q; \epsilon)$, which we previously defined in Equation (14.24). SARSA therefore is learning a suboptimal policy that maintains a certain degree of exploration. Naturally, as $\epsilon \to 0$, the algorithm converges on the optimal policy, assuming that it has sufficiently sampled the state-action bundle $\mathcal{A}_s$ consisting of all admissible state-action pairs.

14.3.5 Off-Policy TD Control: Q Learning

Next, we consider a simple yet powerful modification to SARSA that yields
an elegant off-policy one-step temporal-difference control algorithm called
Q learning (Watkins [1989] and Watkins and Dayan [1992]), as shown in
Algorithm 14.8.

The key difference between SARSA and Q learning is that the latter
uses the difference

$$\delta Q_t^* = R_{t+1} + \gamma \max_{a \in \mathcal{A}_{S_{t+1}}} Q(S_{t+1}, a) - Q(S_t, A_t) \qquad (14.29)$$

in place of Equation (14.28). This seemingly innocuous change has a profound result: the method now learns the *optimal policy*, regardless of the
actual policy being followed. This is due to the fact that the update rule
compares the current state-action value with the immediate return and the
discounted value of the optimal value at the subsequent state.

The backup diagrams for both SARSA and Q learning are shown in
Figure 14.5. SARSA uses the actual values of the subsequent state and
action to update its current state-action value estimates. Q learning, on

---

Algorithm 14.7: SARSA: One-step On-Policy TD control.

Input: An MDP $(\mathcal{S}, \mathcal{A}, \varphi, \varrho)$;
      An initial state-action value function (optional) $Q : \mathcal{S} \times \mathcal{A} \to \mathbb{R}$;
      A discount factor $\gamma \in [0, 1]$;
      A learning rate $\eta \in (0, 1]$;
      An exploration parameter $\epsilon \in (0, 1)$
Output: An optimal state-action value function, given the exploration
      constraint

1 **if** Q **is Null then**
2   Initialize an arbitrary function $Q : \mathcal{S} \times \mathcal{A} \to \mathbb{R}$
3   Set $Q(s_0, a) = 0$ for any absorbing state s_0
4 **end**
5 **for Each** *episode* **do**
6   Initialize $S \sim \text{Unif}(\mathcal{S})$
7   Select action $A \sim \Pi(Q; \epsilon)_S$
8   **while** S *is not an absorbing state* **do**
9   Observe $S' \sim \varphi_S^A$
10   Observe $R \sim \varrho_{SS'}^A$
11   Select action $A' \sim \Pi(Q; \epsilon)_{S'}$
12   Update $Q(S, A) = Q(S, A) + \eta [R + \gamma Q(S', A') - Q(S, A)]$
13   Update $S = S'$
14   Update $A = A'$
15   **end**
16 **end**
17 **return** Q

Algorithm 14.8: Q Learning: One-step Off-Policy TD control.

Input: An MDP $(\mathcal{S}, \mathcal{A}, \varphi, \varrho)$;
An initial state-action value function (optional) $Q : \mathcal{S} \times \mathcal{A} \to \mathbb{R}$;
A discount factor $\gamma \in [0, 1]$;
A learning rate $\eta \in (0, 1]$;
An exploration parameter $\epsilon \in (0, 1)$
Output: An estimate of the optimal state-action value function q_*

1 **if** Q **is Null then**
2 | Initialize an arbitrary function $Q : \mathcal{S} \times \mathcal{A} \to \mathbb{R}$
3 | Set $Q(s_0, a) = 0$ for any absorbing state s_0
4 **end**
5 **for Each** *episode* **do**
6 | Initialize $S \sim \mathrm{Unif}(\mathcal{S})$
7 | **while** S *is not an absorbing state* **do**
8 | | Select action $A \sim \Pi(Q; \epsilon)_S$
9 | | Observe $S' \sim \varphi_S^A$
10 | | Observe $R \sim \varrho_{SS'}^A$
11 | | Set $Q(S, A) = Q(S, A) + \eta \left[R + \gamma \max_{a \in \mathcal{A}_{S'}} Q(S', a) - Q(S, A) \right]$
12 | | Set $S = S'$
13 | **end**
14 **end**
15 **return** Q

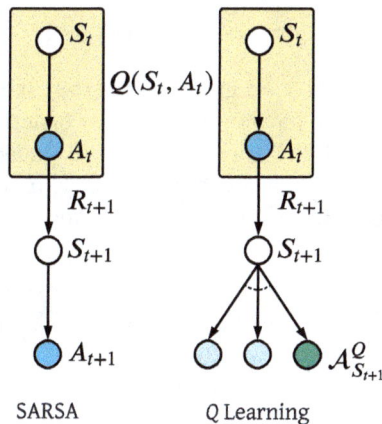

Fig. 14.5: Backup diagram for SARSA (left) and Q learning (right). For Q learning, the greedy action at time $t + 1$ relative to Q is shaded in green; suboptimal (unused) values are shaded a lighter blue.

the other hand, uses the subsequent state, but then looks at all possible actions from the transition state and uses the maximum value. This is why Q learning is off-policy: it uses the best possible action from the transition state to feed back into the state-action value update, regardless of the choice of the actual action that is taken.

14.3.6 Examples

Gridworld Cliffs

Example 14.4 (Gridwolrd). We first introduced Gridworld problems in Example 14.1. In this example, we will solve the gridworld problem in Python using both SARSA and Q learning. This is accomplished with our `Gridworld` class, as shown in Code Block 14.2. The learning algorithm defaults to Q learning, but one may alternatively select SARSA by passing `qlearn=False` into the constructor.

To describe our current state, we will use a coordinate system congruent with `numpy` arrays, so that the first dimension is vertical and oriented downward, and the second dimension is horizontal and oriented to the right. A 5×5 and 4×12 gridworld is shown in Figures 14.4 and 14.6, respectively.

Gridworld problems are simple as they are both deterministic in the reward payout (which only depends on the transition state) and deterministic in the transitions. We will encode our action space as integers using the encoding

$$\text{``down''} \Rightarrow 0 \qquad \text{``right''} \Rightarrow 1 \qquad \text{``up''} \Rightarrow 2 \qquad \text{``left''} \Rightarrow 3.$$

State transitions are encoded in the `getNextState` method on line 11. To handle the boundary, we have tabulated a list of actions, their associated boundary conditions, and the transition rule to generate the subsequent state, if the current state is not along a boundary, in Table 14.3. We make the clever definition of dimension as $a \mod 2$ and sign as $+1$ for $a = 0, 1$ and -1 otherwise.

action	direction	boundary	transition	dimension	sign
$a = 0$	$\downarrow$	$s(0) = shape(0) - 1$	$i + 1$	0	$+1$
$a = 1$	$\rightarrow$	$s(1) = shape(1) - 1$	$j + 1$	1	$+1$
$a = 2$	$\uparrow$	$s(0) = 0$	$i - 1$	0	-1
$a = 3$	$\leftarrow$	$s(1) = 0$	$j - 1$	1	-1

Table 14.3: Gridworld actions, with state $s = (i, j)$.

In order to determine the action, we select the optimal action, relative to our Q matrix, with probability $1 - \epsilon$; otherwise we select an action at random. Finally, our `run` method simulates one entire episode: that state

```python
class Gridworld:

    def __init__(self, shape=(4,7), rewards=None, start=None,
            absorbers=[], epsilon=0.1, eta=0.1, qlearn=True):
        self.shape, self.rewards = shape, rewards
        self.start = start if start is not None else (0, 0)
        self.absorbers, self.qlearn = absorbers, qlearn
        self.epsilon, self.eta = epsilon, eta
        self.Q = np.zeros(shape+(4,))
        self.returns = []

    def getNextState(self, state, action):
        dim = action % 2
        sign = 1 if action < 2 else -1
        if sign == +1 and state[dim] == self.shape[dim] - 1:
            return state
        if sign == -1 and state[dim] == 0:
            return state
        next_state = list(state)
        next_state[dim] += sign
        return tuple(next_state)

    def getAction(self, state, epsilon=0):
        Qs = self.Q[state]
        if np.random.random() < epsilon:
            actions = np.arange(len(Qs))
        else:
            actions = np.argwhere(Qs == max(Qs)).flatten()
        return np.random.choice(actions)

    def run(self, epsilon=-1):
        epsilon = self.epsilon if epsilon < 0 else epsilon
        state = self.start
        action = self.getAction(state, epsilon=epsilon)
        self.path = [state]
        self.returns.append(0)
        while state not in self.absorbers:
            next_state = self.getNextState(state, action)
            reward = self.rewards[next_state]
            # next_state = self.start if reward < -10 else next_state
            next_action = self.getAction(next_state, epsilon=epsilon)
            next_q = max(self.Q[next_state]) if self.qlearn else \
                    self.Q[next_state][next_action]
            delta = reward + next_q - self.Q[state][action]
            self.Q[state][action] += self.eta * delta
            state, action = next_state, next_action
            self.path.append(state)
            self.returns[-1] += reward
```

Code Block 14.2: Gridworld class

is initialized to the starting state, and then updated until the state lands on an absorber. δQ_t is determined on line 42, where `next_q` is determined on the preceding line, based on whether the desired algorithm is SARSA or Q learning.

▷

Example 14.5 (Gridworld Cliffs). Our next example, due to Sutton and Barto [2018], nicely illustrates the difference between the on-policy SARSA and off-policy Q learning algorithms. This is similar to the gridworld problem of Example 14.1, since each transition carries a penalty of $R = -1$. The difference, however, is there is a segment, shaded in red, that carries a high penalty of $R = -100^2$.

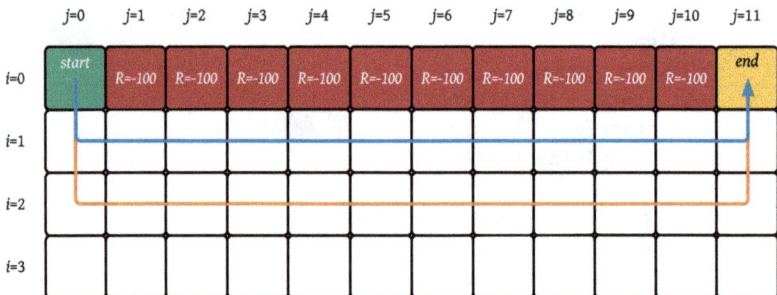

Fig. 14.6: Gridworld Cliffs example. Red region represents cliff, with penalty $R = -100$. Blue line: Q learning. Orange line: SARSA.

The optimal path is obviously the blue path that proceeds down row $i = 1$. This is the policy learned by Q learning. SARSA, on the other hand, learns the orange path, which proceeds down row $j = 2$. By increasing the penalty of the cliff, one can even push the SARSA path further out, into row $j = 3^3$. We can simulate the gridworld cliffs example using our `Gridworld` class, as shown in Code Block 14.3.

Note that the total return is calculated for each episode. We can further use a moving-average to smooth out some of the noise, as shown on line 10. The MA 10 and MA 100 smoothed returns for Q learning and SARSA are shown in Figure 14.7.

When following the greedy path, Q learning will typically beat SARSA. After all, it is learning the optimal path, while following a path that main-

---

[2] In Sutton and Barto [2018], landing on the red squares also reset the state back to the start. This can be achieved by uncommenting line 39 in Code Block 14.2.

[3] The far path, along row $j = 3$, is actually the SARSA solution illustrated in Sutton and Barto [2018].

```
depth, width = 4, 12
rewards = -np.ones((depth, width))
rewards[0, 1:-1] = -100
GS = Gridworld(shape=(depth, width), rewards=rewards,
    absorbers=[(0,width-1)], qlearn=False)
G = Gridworld(shape=(depth, width), rewards=rewards,
    absorbers=[(0,width-1)])
for i in range(1000):
    G.run()
    GS.run()
# example smoothing
np.convolve(G.returns, np.ones(10)/10, mode='valid')
```

Code Block 14.3: Gridworld-cliffs example.

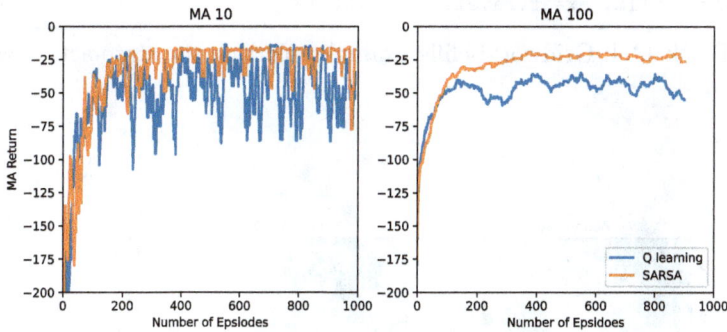

Fig. 14.7: Smoothed moving average total return for Q learning (blue) and SARSA (orange).

tains exploration. Exploratory steps, however, can easily result in a stiff penalty, due to the fact that Q learning's greedy path traverses the edge of the cliff. SARSA, on the other hand, learns based on its actual experience; i.e., it is online. Due to the exploratory steps, it actually learns to *stay away* from the cliff, as it learns that walking along the cliff can lead to a costly mistake.

Finally, we can actually plot the state-action value function, as shown in Code Block 14.4[4]. The resulting plots are shown in Figure 14.8. Note due to the inversion of the y-axis, the actions 0 and 2 now represent up and down, respectively.

Figure 14.8 makes a good deal of sense. Starting from state $(0,0)$, moving up $(a = 0)$ is only marginally better than moving down $(a = 2)$ or left $(a = 3)$, as these latter two options incur a modest cost of $R = -1$ and do

---

[4] We ran an additional 10,000 iterations to get convergence.

```
1    ## Plot state-action value function
2    x = np.arange(0, 13)
3    y = np.arange(0,5)
4    X_grid, Y_grid = np.meshgrid(x, y)
5    cm = colors.LinearSegmentedColormap.from_list('mylist',
         ['#d62728','#1f77b4'], N=2)
6    fig, ax = plt.subplots(2, 2, figsize=(10, 7))
7    for i in range(2):
8        for j in range(2):
9            im = ax[i, j].pcolormesh(X_grid, Y_grid, G.Q[:, :, 2*i+j],
                 edgecolors='white', vmin=-25, vmax=0)
10           ax[i,j].set_xticks(np.arange(12)+0.5)
11           ax[i,j].set_xticklabels(np.arange(12))
12           ax[i,j].set_yticks(np.arange(4)+0.5)
13           ax[i,j].set_yticklabels(np.arange(4))
14           ax[i,j].set_title(f"Action = {2*i+j}")
15   fig.colorbar(im, ax=ax.ravel().tolist())
```

Code Block 14.4: Gridworld-cliffs example: plotting the state-action value function.

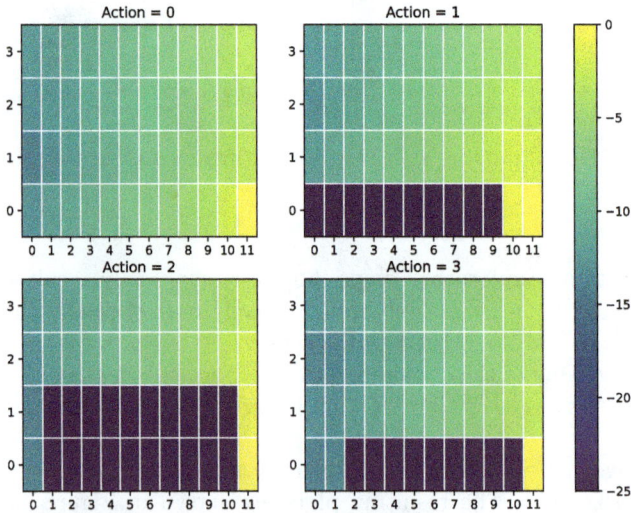

Fig. 14.8: State-value action function for gridworld-cliffs example: 0 represents up, 1 represents right, 2 represents down, and 3 represents left. *Note*: the vertical axis is inverted relative to our earlier diagrams.

not change your state. Moving right ($a = 1$), on the other hand, incurs a penalty of $R = -100$. In general, there is a heavy penalty for any state-action that would result in a transition to the cliff ($i = 0$, $j = 1, \ldots, 11$).

▷

Learning *Blackjack*

Example 14.6. Blackjack is a card game between a player (our agent) and a dealer. To begin, two cards are dealt to both the player and the dealer, with one of the dealer's cards being dealt face down and the other face up, as shown in Figure 14.9.

Fig. 14.9: A game of blackjack with state: $(7, 19, 1)$.

The player then either *hits*, thereby receiving an additional card, or *sticks*, thereby ending his turn. Face cards have a value of 10, and the ace may count as either a 1 or 11. The goal is to get as close as possible to 21 without going over (i.e., *going bust*). If the player hits and his value exceeds 21, he is said to *go bust*, and the game ends as a loss (with a reward of -1).

Once the player sticks, assuming he hasn't gone bust, it becomes the dealer's turn. The dealer plays according to the following fixed strategy: the dealer sticks whenever his sum becomes at least 17, otherwise he hits. We will assume an infinite deck, so that cards are dealt with replacement. If the dealer goes bust, the player wins, with a reward of $+1$. Otherwise, once the dealer sticks, the result is determined by who is closest to 21: if the player is closest to 21, the player wins, with a reward of $+1$. If the dealer is closest to 21, the player loses, and receives a reward of -1. Otherwise, if the score of the dealer and player are equal, the game ends as a draw, in which case the reward is $+0$.

Naturally, aces should count as 11 unless it puts the sum over 21, in which case they should be counted as a 1. An ace that is currently counted as an 11 is said to be a *usable ace*, which therefore represents an extra degree of freedom of the game play. Moreover, if the player's sum is less than or equal to 11, the player should always hit, as there is no danger in going bust. The state space for the player (agent) is thereby determined by three numbers: the value of the dealer's face-up card (2–11), the player's current sum (12–21), and whether or not the player has a useable ace (0 or 1). The size of the state space is therefore $|\mathcal{S}| = 200$. The action space is $\mathcal{A} = \{0,1\}$, where 1 is interpreted as a *hit* and 0 is interpreted as a *stick*.

We coded the Q learning algorithm for blackjack in Code Blocks 14.1 and 14.6. The attribute Q represents the learned state-action value function and is constituted as a four-dimensional array with shape $10 \times 10 \times 2 \times 2$. The first three dimensions represent the state (value of dealer's face-up card, value of player's hand, whether player has a useable ace); the fourth dimension represents the two-dimensional action space consisting of 0 for *stick* and 1 for *hit*.

The newGame method initializes an episode. The player and dealer's hands are initialized with two empty lists. Two card are then dealt to both the player and the dealer, using the hit method. If the player's hand has a value less than twelve, the player will always *hit*, as it is impossible to go bust. This is why the slice of state space corresponding to the player's value ranges between 12 and 21. Dealing is accomplished by the hit method, which takes a hand (list object) as an argument. Since the input variable is passed by reference, changes to it affect the upstream variable; this is why there is no *return*. On line 20, the extra [10,10] is appended to the list range(2,12) as a ten-valued card is three times as likely to appear as a non-ten card; i.e., *jack, queen, king*. The *ace* is dealt as an eleven. Next, we have logic that *uses* a *usable ace* if the new total exceeds 21, reducing the current value (11) of the ace to a value of 1.

Next we have the getState method, which encodes the state vector based on the dealer's face up card (first position in the dealer's hand), the total value of the player's hand, and whether or not the player has a usable ace (a card with value 11). The -2 and -12 shifts to the first two dimensions ensure that the state is encoded in the range $\{0, \ldots, 9\} \times \{0, \ldots, 9\} \times \{0, 1\}$, so that it is compatible with the indices of Q, which is an np.array object.

The getAction method returns the locally greedy strategy $1 - \epsilon$ of the time and a random strategy ϵ of the time, virtually identical to its functioning in the Gridworld example of Code Block 14.2.

The playGame method simulates a single episode. First, it initializes a new game. Then it loops over the player's decisions until the player either decides to stick or goes bust. Since the reward is only paid out at the end, there is no immediate reward during the *while loop*; rather the update relies on propagating the optimal state-action value estimate at the transition state back to the current state. Once the *while loop* is complete, the state-action value function is updated with the final reward (win, lose, or draw) at the penultimate state. Since the final state is an implicit absorbing state,

```
 1   class Blackjack:
 2
 3       def __init__(self, eta=0.1, epsilon=0.1, gamma=1, lambda_=0):
 4           self.eta = eta # learning rate
 5           self.epsilon = epsilon # exploration rate
 6           self.gamma = gamma # discount factor
 7           self.lambda_ = lambda_ # trace decy / used later
 8           self.Q = np.zeros((10,10,2,2)) # 3d state, 1d action
 9           self.n_games = 0 # counter
10
11       def newGame(self):
12           self.dealer = []
13           self.player = []
14           for i in range(2):
15               self.hit(self.dealer)
16               self.hit(self.player)
17           while sum(self.player) < 12:
18               self.hit(self.player)
19
20       def hit(self, hand):
21           new_card = np.random.choice([i for i in range(2, 12)] +
                   [10,10,10])
22           hand.append(new_card)
23           if sum(hand) > 21 and 11 in hand:
24               ace = np.argmax(hand)
25               hand[ace] = 1
26
27       def getState(self):
28           return (self.dealer[1]-2, sum(self.player)-12, int(11 in
                   self.player))
29
30       def getAction(self, state, epsilon=0):
31           Qs = self.Q[state]
32           if np.random.random() < epsilon:
33               actions = np.arange(len(Qs))
34           else:
35               actions = np.argwhere(Qs == max(Qs)).flatten()
36           return np.random.choice(actions)
37       # continued...
```

Code Block 14.5: Blackjack Q learning.

```
1    # continued...
2    def playGame(self):
3        self.newGame()
4        go = True
5        self.n_games += 1
6        state = self.getState()
7        action = self.getAction(state, epsilon=self.epsilon)
8        while go:
9            if action == 1:
10               self.hit(self.player)
11           if (action == 0) or (sum(self.player) > 21):
12               go = False
13           if go:
14               next_state = self.getState()
15               delta = self.gamma * max(self.Q[next_state]) -
                     self.Q[state][action]
16               self.Q[state][action] += self.eta * delta
17               state = next_state
18               action = self.getAction(state, epsilon=self.epsilon)
19       delta = self.getReward() - self.Q[state][action]
20       self.Q[state][action] += self.eta * delta
21
22   def getReward(self):
23       if sum(self.player) > 21:
24           return -1
25       while sum(self.dealer) < 17:
26           self.hit(self.dealer)
27       return np.sign(sum(self.player)-sum(self.dealer)) if
                 sum(self.dealer) <= 21 else 1
```

Code Block 14.6: Blackjack Q learning (continued).

there is no contribution due to any transition state. The **getReward** method then issues a reward of -1 if the player went bust; otherwise it simulates the draw's from the dealer's mandatory policy and then determines the final result.

Now that we have our **Blackjack** class encoded, we can simulate a number of games, as shown on the first few lines of Code Block 14.7. Notice our optimal use of the **tqdm** wrapper from the **tqdm** module, which prints a status bar for the loop, showing the percentage complete along with an estimated time. The code block further provides all necessary code to plot the current greedy policy along with the state-value action functions, the results of which are shown in Figure 14.10 and Figure 14.11.

The strategy shown in Figure 14.10 is *not* the *Baldwin strategy* (also known as the *basic strategy*), which constitutes the optimal strategy for the given blackjack problem (see Example 10.3, Baldwin, *et al.* [1956], and

```
1   # Blackjack Simulation -- Q Learning
2   B = Blackjack(eta=0.015, epsilon=0.15)
3   for i in tqdm(range(1000000)):
4       B.playGame()
5
6   # plot policy based on Q
7   fig, ax = plt.subplots(1, 2, figsize=(10, 4))
8   x = np.arange(2, 13)
9   y = np.arange(12,23)
10  X_grid, Y_grid = np.meshgrid(x, y)
11  cm = colors.LinearSegmentedColormap.from_list('mylist',
        ['#d62728','#1f77b4'], N=2)
12  for j in range(2):
13      A = np.argmax(B.Q[:, :, j, :], axis=2)
14      im = ax[j].pcolormesh(X_grid, Y_grid, A.T, cmap=cm,
            edgecolors='white', vmin=-0.8, vmax=0.8)
15      ax[j].set_title(f"Usable Ace = {bool(j)}")
16      ax[j].set_xlabel("Value of Dealer's Face-up Card")
17      ax[j].set_xticks(np.arange(2,12)+0.5)
18      ax[j].set_xticklabels(np.arange(2,12))
19      ax[j].set_yticks(np.arange(12,22)+0.5)
20      ax[j].set_yticklabels(np.arange(12,22))
21  ax[0].set_ylabel("Value of Agent's Hand")
22
23  # plot state-action value function Q
24  fig, ax = plt.subplots(2, 2, figsize=(10, 7))
25  for i in range(2):
26      for j in range(2):
27          im = ax[i, j].pcolormesh(X_grid, Y_grid, B.Q[:, :, j, i].T,
                edgecolors='white', vmin=-0.8, vmax=0.8)
28          ax[i,j].set_xticks(np.arange(2,12)+0.5)
29          ax[i,j].set_xticklabels(np.arange(2,12))
30          ax[i,j].set_yticks(np.arange(12,22)+0.5)
31          ax[i,j].set_yticklabels(np.arange(12,22))
32  ax[0,0].set_title('Useable Ace = False')
33  ax[0,1].set_title('Useable Ace = True')
34  ax[0,0].set_ylabel('A=Stick')
35  ax[1,0].set_ylabel('A=Hit')
36  fig.colorbar(im, ax=ax.ravel().tolist())
```

Code Block 14.7: Blackjack simulation and plots

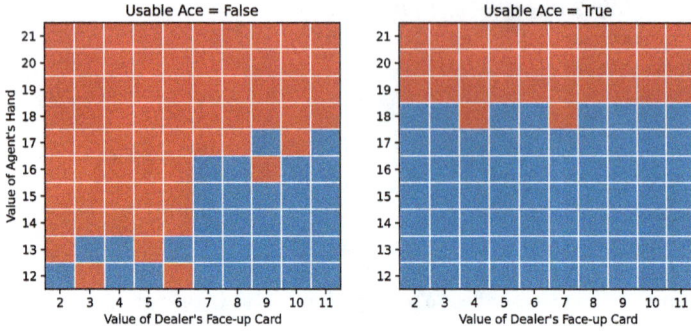

Fig. 14.10: Greedy strategy relative to learned Q for blackjack after 10,000,000 rounds of Q learning: hit (red) or stick (blue).

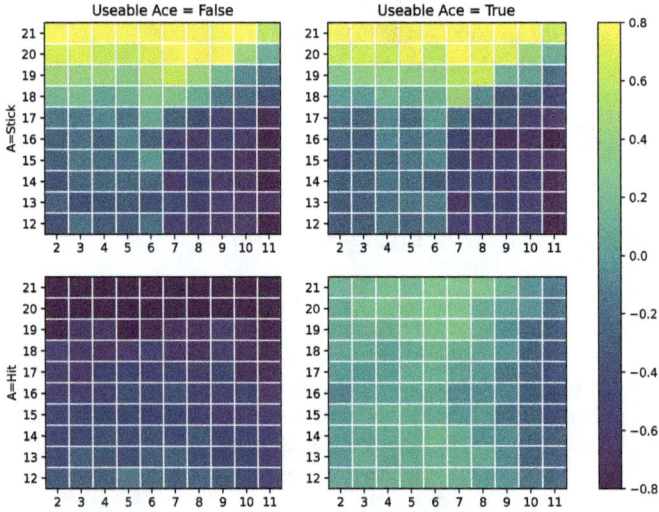

Fig. 14.11: State-action value function Q for for blackjack after 10,000,000 rounds of Q learning.

Thorp [1966]). (The Baldwin strategy further accounts for the options of *splitting* and *doubling down*, which we omit for simplicity.) We further modified our code to replace our ϵ-greedy exploration strategy with exploring starts, only to achieve similar results. It is a bit unsettling that we did not converge on the optimal strategy using Q learning and 10 million simulated games. In our next example, we try our luck using the Monte-Carlo approach, which was used in Sutton and Barto [2018] to determine the optimal strategy.   ▷

Example 14.7. Instead of Q learning, let us use the Monte-Carlo control algorithm to solve the blackjack problem. We can subclass our existing `Blackjack` class, overwriting the learning method within the `playGame` method, as shown in Code Block 14.8. Furthermore, we modify our previous algorithm to use exploring starts.

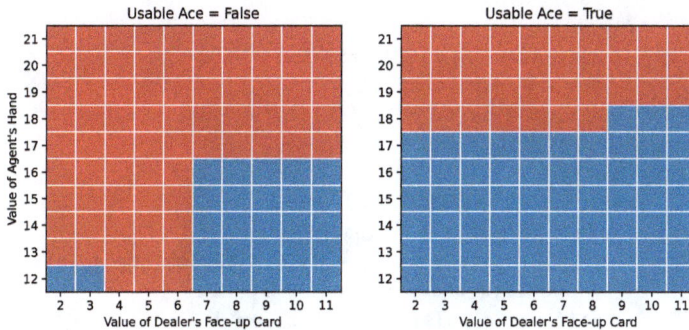

Fig. 14.12: Optimal blackjack policy, obtained using Monte Carlo with exploring starts after 1,000,000 rounds: hit (red) or stick (blue).

Exploring starts is implements with the modification to lines 10 and 21: the first action selection of each episode is exploratory whereas subsequent action selections are greedy. In order to implement the MC method, we track a new attribute `n_visits` that tabulates the number of visits to each state-action pair. Within each episode, then, visits to the state-action pairs are tracked with the `path` variable, which is a list of state-action pair tuples. Once each episode is complete and the reward determined, we then trace backwards through the path, decaying the reward using the discount parameter γ and adding the result to the running average of the state-action value estimate. The backwards loop is achieved by *popping* the final state-action pair of the `path` list.

Further note that a given state-action value is only updated for the *first* occurrence of the pair in each episode's path. This is a standard implementation, though, in blackjack, there are no cycles so the exception is mostly decorative.

```
1  class BlackjackMC(Blackjack):
2
3      def playGame(self):
4          if self.n_games == 0:
5              self.n_visits = np.zeros((10,10,2,2))
6          self.newGame()
7          go = True
8          self.n_games += 1
9          state = self.getState()
10         action = self.getAction(state, epsilon=1) # Exploring Start
11         path =[]
12         while go:
13             path.append(state + (action,))
14             if action == 1:
15                 self.hit(self.player)
16             if (action == 0) or (sum(self.player) > 21):
17                 go = False
18             if go:
19                 next_state = self.getState()
20                 state = next_state
21                 action = self.getAction(state, epsilon=0)
22         reward = self.getReward()
23         while path:
24             state_action = path.pop()
25             if state_action not in path:
26                 self.Q[state_action] = self.n_visits[state_action] *
                        self.Q[state_action] + reward
27                 self.n_visits[state_action] += 1
28                 self.Q[state_action] /= self.n_visits[state_action]
29             reward *= self.gamma
```

Code Block 14.8: `BlackjackMC` Monte-Carlo Blackjack with exploring starts

We ran the MC blackjack method with a discount of $\gamma = 1$ one million times. The result is shown in Figure 14.12. The corresponding estimated state-action value function is shown in Figure 14.13. This represents the true optimal strategy (the Baldwin strategy), in agreement with Sutton and Barto [2018].   ▷

There are two primary differences between the one-step Q-learning and the MC solutions to the blackjack problem. The first is that the one-step method can only propagate the final reward backwards one step at a time, whereas MC was more holistic in that rewards percolate through all past experience. The second is that the TD method uses a constant step size, whereas the MC method uses a step size that decays inversely proportionally to the number of visits to each state-action pair (see Exercise 14.4).

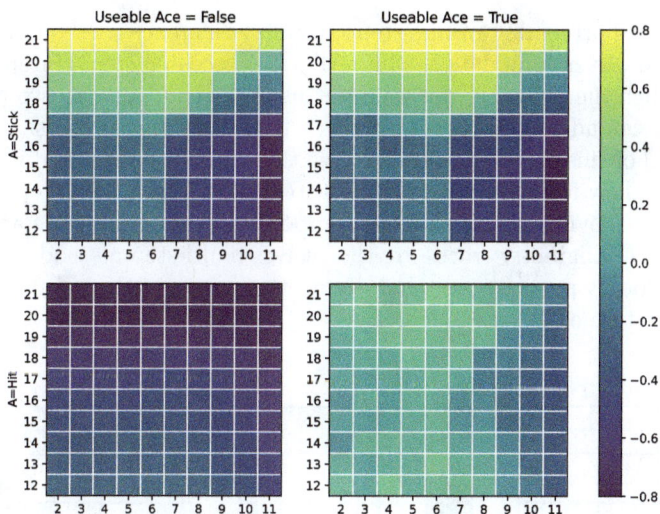

Fig. 14.13: Estimated optimal state-action value function obtained using Monte Carlo with exploring starts after 1,000,000 rounds.

14.3.7 Boosted Starts

Another key to improving the Q-learning algorithm is to recognize that the locally optimal policy differs from the true optimal policy exactly for the states that have the smallest absolute difference $|Q(s,1) - Q(s,0)|$ between their action values (see Exercise 14.6). This motivates the following definition, which we will implement in Code Block 14.9.

Definition 14.17 (Boosted Starts). *Given an episodic MDP,* boosted starts *with offset parameter ε is a particular exploring starts strategy that selects an initial state relative to*

$$\mathbb{P}(S_0 = s) \propto \left(\varepsilon + \left| \max_{a \in \mathcal{A}_s} Q(s,a) - \max_{a \neq a^*} Q(s,a) \right| \right)^{-1}, \qquad (14.30)$$

where $a^* = \arg\max_a Q(s,a)$, and then a corresponding initial action relative to $A_0 \sim \text{Unif}(\mathcal{A}_{S_0})$.

Recall that $\mathcal{A}_s^Q$ is the arg max of Q over a, so that the absolute difference is between the maximum and penultimate maximum values of Q. A typical value of the offset might be $\varepsilon = 0.01$.

14.4 Eligibility Traces

One obvious shortcoming of one-step temporal-difference methods is that observed rewards only update the immediately preceding step. For episodic tasks in which the reward is not paid out until the end, such as two player games, this can add significant inefficiency to the learning process.

The goal of this section is to generalize the one-step temporal-difference methods to allow for learning across multiple steps. There are two primary methods of achieving this: multi-step temporal-difference methods and eligibility traces. Eligibility traces are themselves multi-step methods, except that they a decay parameter λ, that decays the impact of the reward as the number of intermediary steps increases.

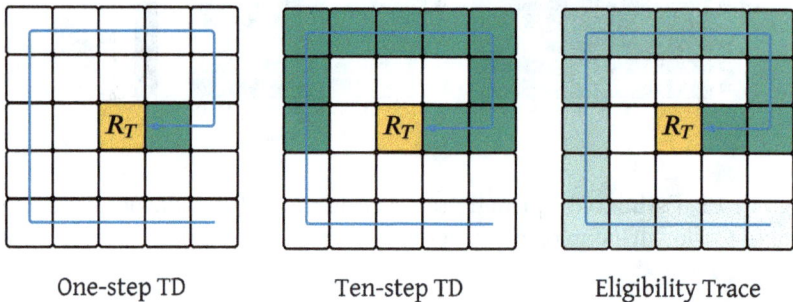

Fig. 14.14: Learning from the reward R_T: one-step TD, ten-step TD, and eligibility trace methods. Updated states are shaded in green; lighter shades indicate reduced impact.

To illustrate this distinction, consider the gridworld example shown in Figure 14.14. We have a length-17 path that starts in the lower-right corner and terminates in the center (gold) square that yields a reward payout. For a one-step temporal-difference method, only the penultimate state value is updated, as shown in Figure 14.14 (left). For a ten-step method, the ten preceding steps are updated, as shown in Figure 14.14 (center). The reward R_T is decayed according to the discount factor and time, but otherwise no decay is applied. Finally, an eligibility trace is shown in Figure 14.14 (right), in which the reward decays according to a *trace-decay* parameter $\lambda \in [0, 1]$.

Therefore, both multi-step and eligibility-trace methods provide a continuum of techniques that start with one-step TD methods at one end of the spectrum. At the other end of the spectrum, we might consider infinite-step TD methods, or an eligibility trace with $\lambda = 1$ (indicating no decay). Such methods are known as *Monte-Carlo* methods and are treated extensively in Sutton and Barto [2018]. Monte-Carlo methods are only applicable for episodic tasks, and they allow the final return at the *end* of each episode to trickle back over the sequence of states realized for the given episode. They suffer from two downsides: the method cannot learn until each episode is complete and, due to the infinite horizon, they cannot be used for continuing tasks.

14.4.1 n-step Temporal-Difference Methods

To generalize the one-step temporal-difference methods to multi-step methods, we first need to define the n-step return. We will then briefly discuss the n-step TD prediction and control problems to highlight their differences; the full algorithms are given in Sutton and Barto [2018].

The n-step Return

Consider again the TD error defined in Equation (14.27). This error can be represented as the difference between the *one-step return*

$$G_{t,1} = R_{t+1}$$

plus the discounted transition-state value $\gamma V(S_{t+1})$ and the current state value $V(S_t)$; i.e.,

$$\delta V_t = (G_{t,1} + V(S_{t+1})) - V(S_t).$$

We thus see that the TD error is easily generalized to multiple step methods. For instance, to incorporate a second step into our update, we could replace $G_{t,1}$ with a *two-step return*

$$G_{t,2} = R_{t+1} + \gamma R_{t+2},$$

and the discounted transition value $\gamma V(S_{t+1})$ with $\gamma^2 V(S_{t+2})$, thereby obtaining the *two-step TD error*

$$\begin{aligned} \delta V_{t,2} &= \left[R_{t+1} + \gamma R_{t+2} + \gamma^2 V(S_{t+2}) \right] - V(S_t) \\ &= \left[G_{t,2} + \gamma^2 V(S_{t+2}) \right] - V(S_t). \end{aligned}$$

Following the same pattern, we could further obtain a *three-step TD error*

$$\begin{aligned} \delta V_{t,3} &= \left[R_{t+1} + \gamma R_{t+2} + \gamma^2 R_{t+3} + \gamma^3 V(S_{t+3}) \right] - V(S_t) \\ &= \left[G_{t,3} + \gamma^3 V(S_{t+3}) \right] - V(S_t), \end{aligned}$$

and so forth. Each of these represents the difference between the value estimate of the current state and a better value estimate based on a number of observed subsequent returns and the value estimate of the final transition state.

Note 14.2. Sutton and Barto [2018] defines the *n-step returns* inclusive of
the value of their respective transition states, but then redefines the meaning when discussing the *n*-step SARSA algorithm to shift from state values
to state-action values. We follow a different definition for two reasons: the
phrase *n-step return* implies, to us, the actual (discounted) returns from
the first *n*-step and our definition/notation does not change meaning as we
move from state value functions to state-action value functions.   ▷

Definition 14.18. *The n-step return of a Markov decision process at
time t is nth partial sum of the full return at time t (Definition 14.8),
as given by*

$$G_{t,n} = R_{t+1} + \gamma R_{t+2} + \cdots + \gamma^{n-1} R_{t+n} = \sum_{k=0}^{n-1} \gamma^k R_{t+k+1}; \qquad (14.31)$$

i.e., the n-step return is the discounted sum of the n subsequent rewards.

Given this definition, we next consider how to generalize our one-step
value estimates.

Definition 14.19. *The n-step state and state-action value estimates of a
Markov decision process at time t are given by the equations*

$$\hat{V}_{t,n} = G_{t,n} + \gamma^n V(S_{t+n}) \qquad (14.32)$$

$$\hat{Q}_{t,n} = G_{t,n} + \gamma^n Q(S_{t+n}, A_{t+n}), \qquad (14.33)$$

*respectively. Similarly, the n-step state and state-action TD errors are de-
fined by the relations*

$$\delta V_{t,n} = \hat{V}_{t,n} - V(S_t) \qquad (14.34)$$

$$\delta Q_{t,n} = \hat{Q}_{t,n} - Q(S_t, A_t), \qquad (14.35)$$

respectively.

The *n*-step value estimates represent improvements on the estimates $V(S_t)$
and $Q(S_t, A_t)$ that incorporate information obtained from the subsequent
n steps.

n-step TD Methods

n-step temporal-difference learning is a natural extension of its single-step
counterpart and is designed to leverage multiple returns for each update.
They are based on the difference between the *n*-step state and state-action
value estimates (Equations (14.32) and (14.33)) and the current value estimates. One drawback is that we must wait *n* steps before we can apply
each update, creating a lag in the learning that must "catch up" at the

end of each episode. Since our primary focus of this section is on eligibility traces, which do not suffer from this drawback and otherwise offer certain computational advantages, we will only provide a cursory introduction to multi-step methods, referring the reader to Sutton and Barto [2018] for additional details.

n-step Policy Evaluation

The n-step policy evaluation method generalizes Algorithm 14.6 to include the n-step return. Fundamentally, this is achieved by replacing the one-step TD error with its n-step equivalent, given by Equation (14.34), which may be expressed in terms of the n-step return as

$$\delta V_{t,n} = [G_{t,n} + \gamma^n V(S_{t+n})] - V(S_t), \qquad (14.36)$$

which forms the basis of the new update rule

$$V(S) \leftarrow V(S) + \eta\, \delta V_{t,n}.$$

We leave the details of the algorithm as an exercise to the interested reader (Exercise 14.7).

n-step On-Policy TD Control: n-step SARSA

The n-step SARSA method generalizes the single-step SARSA method of Algorithm 14.7 by replacing the one-step TD state-action error with its n-step equivalent, given by Equation (14.35), which may be expressed in terms of the n-step return as

$$\delta Q_{t,n} = [G_{t,n} + \gamma^n Q(S_{t+n}, A_{t+n})] - Q(S_t, A_t). \qquad (14.37)$$

As before, this forms the basis of a new update rule

$$Q(S, A) \leftarrow Q(S, A) + \eta\, \delta Q_{t,n}.$$

Backup diagrams for the 3-step TD policy evaluation and 3-step SARSA are shown in Figure 14.15.

n-step Q and Tree Backup

Next, we consider two additional methods for off-policy control: a simple generalization of Q learning, known as *Watkin's Q*, and an elegant technique known as *tree backup*. One can also formalize an off-policy SARSA, which requires use of importance sampling; for details, see Sutton and Barto [2018].

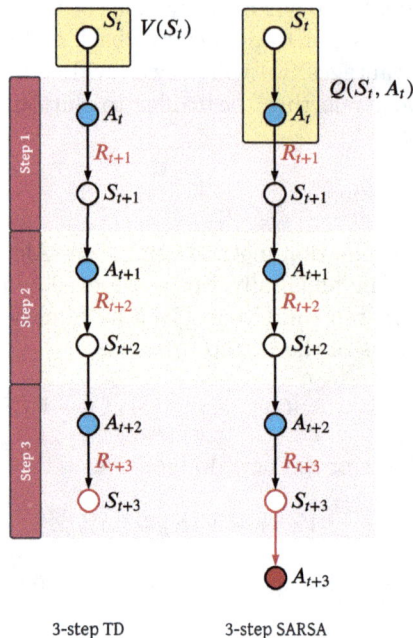

Fig. 14.15: Backup diagram for the 3-step TD policy evaluation (left) and 3-step SARSA (right). Value updates based on information in red.

Watkin's Q

Several generalizations to Q learning have been proposed over the years. The earliest and perhaps simplest is known as *Watkin's Q learning,* named after its originator. Watkin's Q learning only looks ahead until the next exploratory action is chosen. For example, if the next exploratory action following time t takes place n_e steps in the future, the n-step TD error at time t will be based on

$$\hat{Q}^*_{t,n} = G_{t,\tilde{n}} + \gamma^{\tilde{n}} \max_{a \in \mathcal{A}_{S_{t+\tilde{n}}}} Q(S_{t+\tilde{n}}, a),$$

where $\tilde{n} = \min(n, n_e)$, similar to Equation (14.33). The 3-step Q learning is illustrated in the backup diagram in Figure 14.16 (left).

Watkin's Q learning is a simple method for extending Q learning as a multistep method. It has the advantage, as an off-policy method, of *not* requiring usage of importance sampling. Instead, it resolves the off-policy problem by simply cutting off the return as soon as the first nongreedy action is selected. This method, however, has the disadvantage that the returns are cut short as soon as the first nongreedy action is selected, mak-

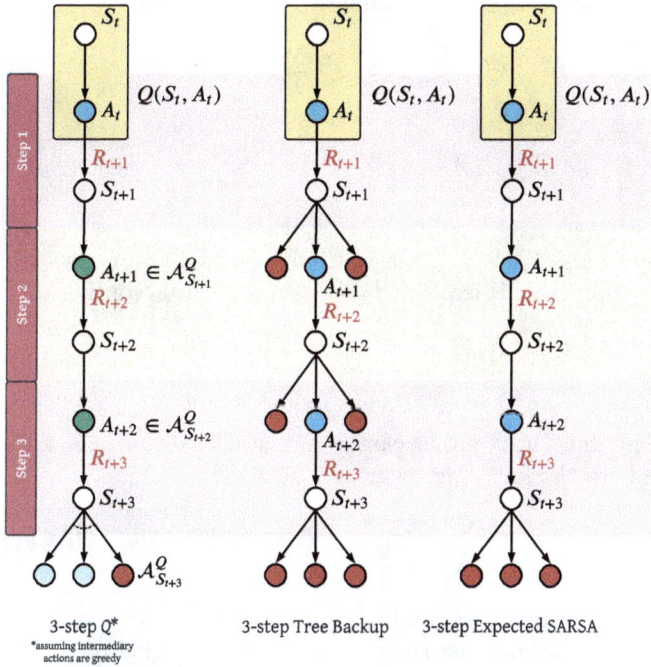

Fig. 14.16: Backup diagram for the 3-step Q (left), 3-step tree backup (middle), and 3-step expected SARSA (right). Value updates based on information in red. Necessarily greedy actions in green.

ing it something less than a true n-step method and more like an $\mathbb{E}[1/\epsilon]$ method.

Tree Backup

Tree backup is an elegant n-step generalization of Q learning that, like Watkin's Q, does not require eligibility traces and, unlike Watkin's Q, is purported to be a true n-step method, though we will challenge this verdict momentarily. The idea is based on the backup diagram in Figure 14.16 (middle). The update consists of two components: the internal string of actual returns and the estimated state-action values at the *leaf nodes* of the backup tree. Each of these values is weighted according to their probability of occurrence under the target policy π. CAP

To get a feel for what this looks like, let's explicitly write out the one-step tree value estimate

$$\bar{Q}_{t,1} = R_{t+1} + \gamma \sum_{a \in \mathcal{A}_{S_{t+1}}} \pi_{S_{t+1}}(a) Q(S_{t+1}, a)$$

and the two-step tree value estimate

$$\bar{Q}_{t,2} = R_{t+1} + \gamma \sum_{a \neq A_{t+1}} \pi_{S_{t+1}}(a) Q(S_{t+1}, a)$$

$$+ \gamma \pi_{S_{t+1}}(A_{t+1}) \left[R_{t+2} + \gamma \sum_{a \in \mathcal{A}_{S_{t+2}}} \pi_{S_{t+2}}(a) Q(S_{t+2}, a) \right].$$

In going from $\bar{Q}_{t,1}$ to $\bar{Q}_{t,2}$, we see that the contribution associated with the actual action A_{t+1} is expanded one step further, according to

$$Q(S_{t+1}, A_{t+1}) \to R_{t+2} + \gamma \sum_{a \in \mathcal{A}_{S_{t+2}}} \pi_{S_{t+2}}(a) Q(S_{t+2}, a).$$

This contribution, in either case, is weighted by $\gamma \pi_{S_{t+1}}(A_{t+1})$. In general, this leads to the following recurrence relation

$$\bar{Q}_{t,n} = R_{t+1} + \gamma \sum_{a \neq A_{t+1}} \pi_{S_{t+1}}(a) Q(S_{t+1}, a) + \gamma \pi_{S_{t+1}}(A_{t+1}) \bar{Q}_{t+1, n-1},$$

which may be solved to obtain tree-backup estimates.

Tree back up may be thought as a generalization of *expected SARSA*, whose backup diagram is shown in Figure 14.16 (right), which replaces the final state-action value estimate with an expectation over the action space at the final state. The difference is that expected SARSA samples the outcome of the actual actions along the way and takes expectation at the final state alone, whereas the tree-backup algorithm computes expectation at each step in the chain.

Sutton and Barto [2018] states that the tree-backup algorithm "is arguably the true successor to Q-learning because it retains its appealing absence of importance sampling even though it can be applied to off-policy data." We disagree. In our view, the tree-backup method suffers from the same shortcoming of Watkin's Q: it cuts off, in practice, following the first nongreedy action.

To see this, consider the goal of the TD control problem, which is to *learn* the optimal policy. This implies that the target policy π is not known in advanced. Instead, we should use use the locally greedy policy $\Pi(Q, 0)$ with respect to Q at each step of the algorithm. *Ay! There's the rub!* The locally greedy policy is given by $\pi = \text{Unif}(\mathcal{A}^Q)$, the uniform distribution over the optimal-action bundle. Let's take this insight and rewrite the one-step estimate as

$$\bar{Q}_{t,1} = R_{t+1} + \frac{\gamma}{|\mathcal{A}_{S_{t+1}}^Q|} \sum_{a \in \mathcal{A}_{S_{t+1}}^Q} Q(S_{t+1}, a)$$

$$= R_{t+1} + \gamma \max_{a \in \mathcal{A}_{S_{t+1}}} Q(S_{t+1}, a).$$

The two-step estimate is then given by

$$\bar{Q}_{t,2} = R_{t+1} + \frac{\gamma}{|\mathcal{A}^Q_{S_{t+1}}|} \sum_{a \in \mathcal{A}^Q_{S_{t+1}}} Q(S_{t+1}, a) = R_{t+1} + \gamma \max_{a \in \mathcal{A}_{S_{t+1}}} Q(S_{t+1}, a)$$

if $A_{t+1} \notin \mathcal{A}^Q_{S_{t+1}}$ is exploratory (since this would imply $\Pi(Q;0)_{S_{t+1}}(A_{t+1}) = 0$), or

$$\bar{Q}_{t,2} = R_{t+1} + \frac{\gamma}{|\mathcal{A}^Q_{S_{t+1}}|} \sum_{a \in \mathcal{A}^Q_{S_{t+1}} \setminus \{A_{t+1}\}} Q(S_{t+1}, a)$$

$$+ \frac{\gamma}{|\mathcal{A}^Q_{S_{t+1}}|} \left[R_{t+2} + \frac{\gamma}{|\mathcal{A}^Q_{S_{t+2}}|} \sum_{a \in \mathcal{A}^Q_{S_{t+2}}} Q(S_{t+2}, a) \right]$$

$$= R_{t+1} + \gamma \frac{|\mathcal{A}^Q_{S_{t+1}}| - 1}{|\mathcal{A}^Q_{S_{t+1}}|} \max_{a \in \mathcal{A}_{S_{t+1}}} Q(S_{t+1}, a)$$

$$+ \frac{\gamma}{|\mathcal{A}^Q_{S_{t+1}}|} \left[R_{t+2} + \gamma \max_{a \in \mathcal{A}_{S_{t+2}}} Q(S_{t+2}, a) \right]$$

if $A_{t+1} \in \mathcal{A}^Q_{S_{t+1}}$ is greedy. Since $\Pi(Q;0)_{S_{t+j}}(A_{t+j}) = 0$ if A_{t+j} is nongreedy, we see, in general, that the tree-backup cuts off following the first nongreedy action, at least when used for the optimal control problem.

$Q(\sigma)$

Sutton and Barto [2018] suggest an intermediary n-step method known as $Q(\sigma)$. The observation is that SARSA fully *samples* along the actual state-action pathway, whereas the tree backup uses expectations across all possible actions linked to each state. Expected SARSA samples for each step except the last, which uses expectation. The proposed idea is that σ represents the degree of sampling, with $\sigma = 1$ corresponding to full sampling (SARSA) and $\sigma = 0$ corresponding to full expectation (tree backup). $Q(\sigma)$ then represents a stochastic mix of sampling and expectation.

14.4.2 Eligibility Traces

Eligibility traces offer an elegant alternative to n-step TD methods. Where n-step TD methods require a lag between when a state is reached and when it is updated, eligibility traces provide real-time model updates as information becomes available. There are two ways of thinking about eligibility traces: the forward and backward views. The forward view is primarily theoretical, aimed at laying out a simple way of viewing the method in terms of linear combinations of n-step returns. The backward view, i.e., mechanistic view, is used in practice, and gives rise to the term *eligibility trace*.

The Forward (Theoretical) View: Lambda Returns

A major drawback of n-step TD methods is that they are written in a *forward (theoretcial) view*, meaning that the update for a given state depends on events that happen thereafter. As such, they are primarily useful in *offline learning*, where the state values are not updated until the conclusion of each episode.

Since each n-step state (or state-action) value estimate is an approximation for $V(S_t)$ (or $Q(S_t, A_t)$), then so to is any linear combination of multistep estimates. In particular, we may form a geometric series of these estimates that decays with the step number. This gives rise to the following.

Definition 14.20. *The* lambda return *of a Markov decision process at time t is the infinite sum*

$$G_t^\lambda = (1 - \lambda) \sum_{n=1}^\infty \lambda^{n-1} G_{t,n}. \tag{14.38}$$

Similarly, the lambda estimates *of the state and state-action value functions at time t are given by the sums*

$$\hat{V}_t^\lambda = (1 - \lambda) \sum_{n=1}^\infty \lambda^{n-1} \hat{V}_{t,n} \tag{14.39}$$

$$\hat{Q}_t^\lambda = (1 - \lambda) \sum_{n=1}^\infty \lambda^{n-1} \hat{Q}_{t,n}, \tag{14.40}$$

respectively, where $\lambda \in [0,1]$ is the trace-decay parameter. *Naturally, these give rise to the* state and state-action TD(λ) errors

$$\delta V_t^\lambda = \hat{V}_t^\lambda - V(S_t) \tag{14.41}$$

$$\delta Q_t^\lambda = \hat{Q}_t^\lambda - Q(S_t, A_t), \tag{14.42}$$

respectively.

Note that if we replace G with V or Q in Equation (14.38) we get Equations (14.39) and (14.40), respectively. Expressed in terms of the partial returns, these are equivalent to the relations

$$\hat{V}_t^\lambda = (1 - \lambda) \sum_{n=1}^\infty \lambda^{n-1} \left[G_{t,n} + \gamma^n V(S_{t+n}) \right]$$

$$\hat{Q}_t^\lambda = (1 - \lambda) \sum_{n=1}^\infty \lambda^{n-1} \left[G_{t,n} + \gamma^n Q(S_{t+n}, A_{t+n}) \right].$$

The coefficient $(1 - \lambda)$ is chosen so that the weights sum to unity. Our value-estimate improvement updates can then be expressed as

$$V(S_t) \leftarrow V(S_t) + \eta \, \delta V_t^{\lambda}$$
$$Q(S_t, A_t) \leftarrow Q(S_t, A_t) + \eta \, \delta Q_t^{\lambda},$$

in the natural way.

The forward view is considered theoretic, as in practice one would have to wait an infinite amount of time (i.e., the end of each episode) to actually perform the update. Thus, the forward view is restricted to offline learning. Equations (14.39) and (14.40), however, provide a clear motivation for the method, and are easily digestible by the eye.

Despite the added complexity, these methods are still temporal-difference methods at heart. As such, they are referred to as TD(λ) methods. Note that the one-step methods from the previous section correspond to TD(0) methods. At the other end of the spectrum, TD(1) methods correspond to the aforementioned Monte-Carlo methods. The trace-decay parameter λ therefore constitutes a natural tuning parameter for reinforcement-learning problems. In practice, a value of λ somewhere between zero and one is usually optimal, so that our pursuit in their development is not purely academic.

The Bridge From the Forward to Backward View

Next, we will construct an algorithm for implementing TD(λ) methods in an *online* setting, specifically for the state-value estimation (prediction) problem, such that updates are provided with each step. This is called the *backward, or mechanistic, view.* The heart of the slight of hands giving rise to eligibility traces is contained in the following theorem.

Theorem 14.2 (Forward-Backward Equivalence Theorem). *Given a Markov decision process, the TD(λ) errors at time t are equivalent to*

$$\delta V_t^{\lambda} = \sum_{k=0}^{\infty} \gamma^k \lambda^k \, \delta V_{t+k} \tag{14.43}$$

$$\delta Q_t^{\lambda} = \sum_{k=0}^{\infty} \gamma^k \lambda^k \, \delta Q_{t+k}. \tag{14.44}$$

Note that we may use the definition of the one-step TD errors to express Equations (14.43) and (14.44) more verbosely by the relations

$$\delta V_t^{\lambda} = \sum_{k=0}^{\infty} \gamma^k \lambda^k \left[R_{t+k+1} + \gamma V(S_{t+k+1}) - V(S_{t+k}) \right]$$

$$\delta Q_t^{\lambda} = \sum_{k=0}^{\infty} \gamma^k \lambda^k \left[R_{t+k+1} + \gamma Q(S_{t+k+1}, A_{t+k+1}) - Q(S_{t+k}, A_{t+k}) \right].$$

The key takeaway here, however, is that the form of Equations (14.43) and (14.44) is an infinite sum of *one-step* TD errors, whereas the forward view of Equations (14.41) and (14.42) involved the sum of TD errors of *all orders*.

Proof. Both proofs follow similar logic, so we will focus on the proof for Equation (14.43). By substituting Equation (14.31) into Equation (14.39), we obtain

$$\hat{V}_t^{\lambda} = (1 - \lambda) \sum_{n=1}^{\infty} \sum_{k=0}^{n-1} \gamma^k \lambda^{n-1} R_{t+k+1} + (1 - \lambda) \sum_{n=1}^{\infty} \lambda^{n-1} \gamma^n V(S_{t+n})$$

$$= (1 - \lambda) \sum_{k=0}^{\infty} \sum_{n=k+1}^{\infty} \gamma^k \lambda^{n-1} R_{t+k+1} + (1 - \lambda) \sum_{n=1}^{\infty} \lambda^{n-1} \gamma^n V(S_{t+n}).$$

We leave the proof of the exchange of the double summation in the first term as an exercise for the reader (see Exercise 14.9). Using the formula for a finite geometric series on the summation over n in the first term, and replacing $n \to k + 1$ for the second term, we arrive at

$$\hat{V}_t^{\lambda} = \sum_{k=0}^{\infty} \gamma^k \lambda^k R_{t+k+1} + (1 - \lambda) \sum_{k=0}^{\infty} \lambda^k \gamma^{k+1} V(S_{t+k+1})$$

$$= \sum_{k=0}^{\infty} \gamma^k \lambda^k \left[R_{t+k+1} + (1 - \lambda)\gamma V(S_{t+k+1}) \right]$$

$$= \sum_{k=0}^{\infty} \gamma^k \lambda^k \left[R_{t+k+1} + \gamma V(S_{t+k+1}) - \gamma \lambda V(S_{t+k+1}) \right].$$

Therefore,

$$\delta V_t^{\lambda} = \hat{V}_t^{\lambda} - V(S_t)$$

$$= \sum_{k=0}^{\infty} \gamma^k \lambda^k \left[R_{t+k+1} + \gamma V(S_{t+k+1}) \right] - V(S_t) - \sum_{k=0}^{\infty} \gamma^{k+1} \lambda^{k+1} V(S_{t+k+1})$$

$$= \sum_{k=0}^{\infty} \gamma^k \lambda^k \left[R_{t+k+1} + \gamma V(S_{t+k+1}) \right] - \sum_{k=0}^{\infty} \gamma^k \lambda^k V(S_{t+k})$$

$$= \sum_{k=0}^{\infty} \gamma^k \lambda^k \left[R_{t+k+1} + \gamma V(S_{t+k+1}) - V(S_{t+k}) \right],$$

which completes our proof.   □

The Backward (Mechanistic) View: Eligibility Traces

The beauty of Theorem 14.2 is that it allows us to define an online method for implementing the update that converges to the forward view as $t \to \infty$. Instead of waiting for a number of steps for multistep methods, or waiting forever for the forward view of TD(λ), we may now make updates to our value functions with every step you take, with every breath you make. We no longer have to be always waiting for you.

The key to the development of the online TD(λ) method lies in the following.

Definition 14.21. *Given a Markov decision process, a discount factor γ, and a decay rate λ, a* state or state-action eligibility trace *is a mapping z that increments by one each time a state or state-action pair is visited and decays by $\gamma\lambda$ across its entire domain following each step's action-value update.*

The online TD(λ) policy evaluation method is shown in Algorithm 14.9. Note that line 13 can be implemented in Python with the `dict.get` method: `z[state]` = `z.get(state, 0) + 1`, which returns the value associated with `z[state]` or a default value of 0 if the given state is not already a key. In this algorithm, we calculate the one-step TD error

$$\delta V_t = R + \gamma V(S') - V(S),$$

as we did in the one-step TD method (Algorithm 14.6), without having saved it as its own variable. In addition, we track an eligibility trace z,

---

Algorithm 14.9: Online TD(λ) policy evaluation.

Input: An MDP $(\mathcal{S}, \mathcal{A}, \varphi, \varrho)$;
        A given policy π;
        A discount factor $\gamma \in [0, 1]$;
        A rate-decay parameter $\lambda \in [0, 1]$;
        A learning rate $\eta \in (0, 1]$
Output: An estimate of the value function v_π of the given policy
1   Initialize an arbitrary function $V : \mathcal{S} \to \mathbb{R}$
2   Set $V(s_0) = 0$ for any absorbing state s_0
3   **for Each** *episode* **do**
4   Initialize $S \sim \mathrm{Unif}(\mathcal{S})$
5   Set $z = $ `Array()`
6   **while** S *is not an absorbing state* **do**
7   Select action $A \sim \pi_S$
8   Observe $S' \sim \varphi_S^A$
9   Observe $R \sim \varrho_{SS'}^A$
10   Set $\delta = R + \gamma V(S') - V(S)$
11   Increment $z[S] = z[S] + 1$
12   **for** S **in** `keys(`z`)` **do**
13   Update $V(S) = V(S) + \eta\delta z[S]$
14   Decay $z[S] = \gamma\lambda z[S]$
15   **end**
16   Update $S = S'$
17   **end**
18 **end**
19 **return** V

---

which maps the previously visited states to a number known we can think of as their eligibility score, which represents the degree to which they are eligible to be impacted by the current update. To understand why this works, consider Equation (14.41), which shoes the λ-error δ_t^λ as a weighted sum over future *one-step* TD errors. We therefore *only* need to calculate the one-step TD errors to implement this sum. Moreover, we can implement each term of the sum as it becomes available, which is precisely the approach of Algorithm 14.9: we calculate the one-step error and then propagate it backward to all previously visited states with a weight proportional to each state's eligibility score. Easy as π.

14.4.3 SARSA(λ)

The SARSA(λ) method follows similar suit, except now we must track eligibility traces over all previously visited state-action pairs. The algorithm is a straightforward generalization of TD(λ) that follows Equation (14.44) and is given in Algorithm 14.10.

We can still use a dictionary object to track state-action eligibility traces in Python, with the modification that keys be regarded as tuples. For instance, line 14 can be implemented as

$$\mathtt{z[state, action]} = \mathtt{z.get((state, action), 0)} + 1.$$

To improve memory performance for problems with large state-action spaces, we may further remove the eligibility score for state-action pairs whose score falls below a given threshold, say 0.01. We can do this with the `dict.pop` method:

$$\mathtt{if\ z[state, action]} < 0.01:$$
$$\mathtt{z.pop((state, action), None)}.$$

14.4.4 Watkin's Q(λ)

Eligibility traces can easily be applied to Watkin's Q learning. The backup diagram (forward view) is shown in Figure 14.17. Each column represents a single term in the analogous sum to Equation (14.40) (with $\hat{Q}_{t,n}$ replaced with $\hat{Q}_{t,n}^*$). Even though each column looks like an n-step SARSA with an optimal value at the end, the diagram is, in fact, equivalent to a chain of optimal values since the greedy actions are taken at each step along the path.

The Watkin's Q-learning method is given in Algorithm 14.11. The only difference is that whenever an exploratory action is taken, the eligibility trace is reset to an empty array. This has the disadvantage that the trace history is lost each time an exploratory action is selected, which slows the overall learning. For episodic tasks, however, we may sometimes forego this requirement by utilizing exploring starts, which confine exploratory actions to each episode's initialization. This works well enough in blackjack, where each episode consists of only a handful of steps. For episodic tasks with lengthier episodes, like a game of chess, confining exploration to the initial move is not practical.

14.4.5 Tree Backup(λ)

Eligibility traces can also be used with the tree-backup algorithm. The backup diagram is shown in Figure 14.18. The equations are a bit more complex than those of Watkin's Q, and are provided in Sutton and Barto [2018]. We feel, however, there is some slight of hand, as the tree-backup method does not really go on forever, due to the fact that each step requires expectation using the target policy. In the event of optimal control, i.e., learning the optimal policy, the target policy is necessarily greedy. This

---

Algorithm 14.10: Online SARSA(λ): On-Policy TD control.

Input: An MDP $(\mathcal{S}, \mathcal{A}, \varphi, \varrho)$;
   A discount factor $\gamma \in [0, 1]$;
   A rate-decay parameter $\lambda \in [0, 1]$;
   A learning rate $\eta \in (0, 1]$;
   An exploration parameter $\epsilon \in (0, 1)$
Output: An estimated optimal state-action value function
1 Initialize an arbitrary function $Q : \mathcal{S} \times \mathcal{A} \to \mathbb{R}$
2 Set $Q(s_0, a) = 0$ for any absorbing state s_0
3 **for Each** *episode* **do**
4 Initialize $S \sim \text{Unif}(\mathcal{S})$
5 Select action $A \sim \Pi(Q; \epsilon)_S$
6 Set $z = \texttt{Array}()$
7 **while** S *is not an absorbing state* **do**
8   Observe $S' \sim \varphi_S^A$
9   Observe $R \sim \varrho_{SS'}^A$
10   Select action $A' \sim \Pi(Q; \epsilon)_{S'}$
11   Set $\delta = R + \gamma Q(S', A') - Q(S, A)$
12   Increment $z[S, A] = z[S, A] + 1$
13   **for** S, A **in** $\texttt{keys}(z)$ **do**
14   Update $Q(S, A) = Q(S, A) + \eta \delta z[S, A]$
15   Decay $z[S, A] = \gamma \lambda z[S, A]$
16   **end**
17   Update $S = S'$
18   Update $A = A'$
19 **end**
20 **end**
21 **return** Q

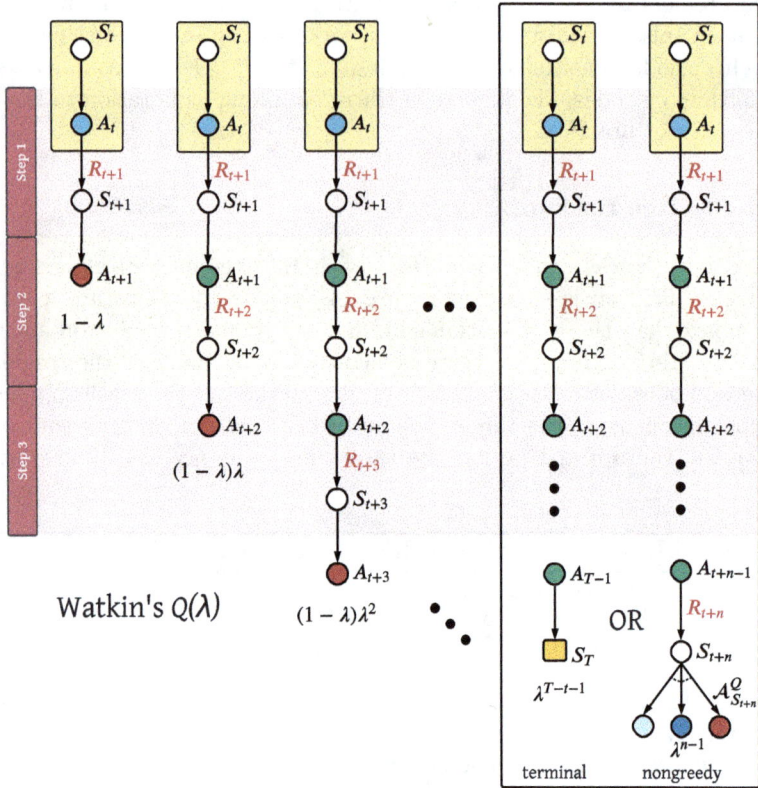

Fig. 14.17: Eligibility trace diagram for Watkin's $Q(\lambda)$. Greedy actions in green; nongreedy actions in blue.

makes the probability of nongreedy actions under the target policy zero, which thereby cuts off the method after n_e steps.

14.4.6 MQ(λ)

Instead, what if we operate like Watkin's Q whenever a greedy action is taken and like tree backup whenever an exploratory action is taken, except we replace the target probability in the tree-backup algorithm with the exploration parameter. We would thus use the one-step estimate

$$\hat{Q}_{t,1}^m = R_{t+1} + \gamma \max_{a \in \mathcal{A}_{S_{t+1}}} Q(S_{t+1}, a)$$

when $A_{t+1} \in \mathcal{A}_{S_{t+1}}^Q$ is greedy and the estimate

$$\hat{Q}^m_{t,1} = (1 - \epsilon) \left[R_{t+1} + \gamma \max_{a \in \mathcal{A}_{S_{t+1}}} Q(S_{t+1}, a) \right] + \epsilon \left[R_{t+1} + \gamma Q(S_{t+1}, A_{t+1}) \right]$$

when $A_{t+1} \notin \mathcal{A}^Q_{S_{t+1}}$ is exploratory. This differs from the similar step of the tree-backup algorithm in that we replace $\pi_{S_{t+1}}(A_{t+1})$ with ϵ, since the former probability is zero for exploratory actions. It is also similar to $Q(\sigma)$, which mixes sampling and expectation at each step along the chain, with the difference that the mixture is no longer (explicitly) stochastic, but deterministic based on whether the action is exploratory or greedy. We will call this modification the MQ method, with backup diagram as shown in Figure 14.19.

This method is no longer strictly on-policy. But that's okay, as we would argue: *no on-policy method exists that can follow a path through an ex-*

Algorithm 14.11: Watkin's Q(λ): Off-Policy TD control.

Input: An MDP $(\mathcal{S}, \mathcal{A}, \varphi, \varrho)$;
        A discount factor $\gamma \in [0, 1]$;
        A rate-decay parameter $\lambda \in [0, 1]$;
        A learning rate $\eta \in (0, 1]$;
        An exploration parameter $\epsilon \in (0, 1)$
Output: An estimated optimal state-action value function
1 Initialize an arbitrary function $Q : \mathcal{S} \times \mathcal{A} \to \mathbb{R}$
2 Set $Q(s_0, a) = 0$ for any absorbing state s_0
3 **for Each** *episode* **do**
4   Initialize $S \sim \text{Unif}(\mathcal{S})$
5   Select action $A \sim \Pi(Q; \epsilon)_S$
6   Set $z = \text{Array}()$
7   **while** *S is not an absorbing state* **do**
8   Observe $S' \sim \varphi^A_S$
9   Observe $R \sim \varrho^A_{SS'}$
10   Select action $A' \sim \Pi(Q; \epsilon)_{S'}$
11   Set $\delta = R + \gamma \max_{a \in \mathcal{A}_{S'}} Q(S', a) - Q(S, A)$
12   Increment $z[S, A] = z[S, A] + 1$
13   **for** S, A **in** keys(z) **do**
14   Update $Q(S, A) = Q(S, A) + \eta \delta z[S, A]$
15   Decay $z[S, A] = \gamma \lambda z[S, A]$
16   **end**
17   **if** $A' \notin \mathcal{A}^Q_{S'}$; *i.e., A' is an exploratory action* **then**
18   Reset $z = \text{Array}()$
19   **end**
20   Update $S = S'$
21   Update $A = A'$
22   **end**
23 **end**
24 **return** Q

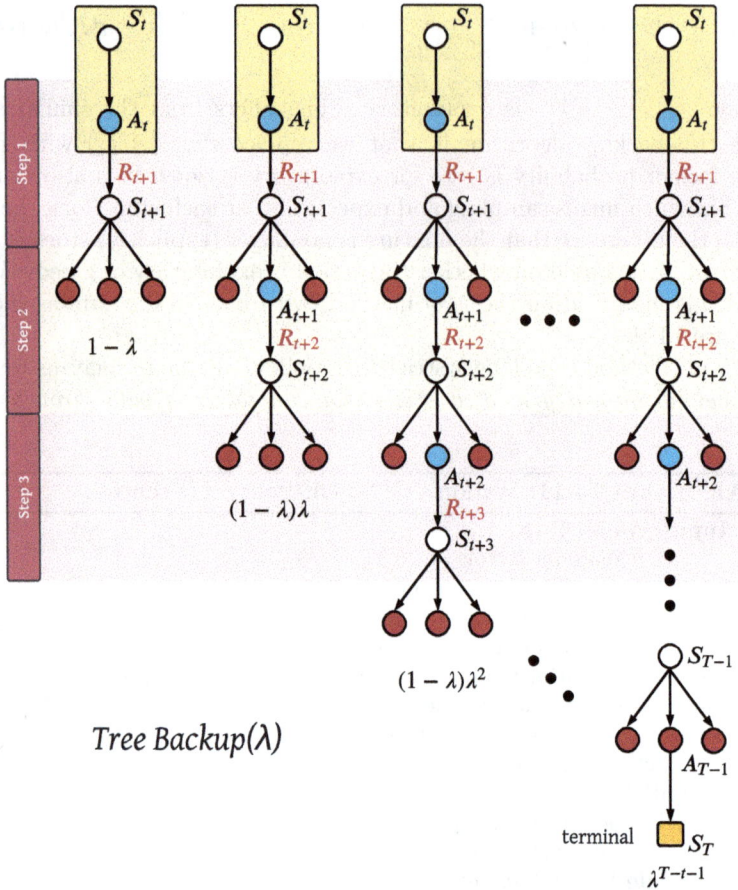

$1 - \lambda$

$(1-\lambda)\lambda$

Tree Backup(λ)

$(1-\lambda)\lambda^2$

terminal $\square\ S_T$

λ^{T-t-1}

Fig. 14.18: Eligibility trace diagram for tree backup $TB(\lambda)$.

ploratory action. This follows since, by definition, exploratory actions cannot take place under the target policy, which is the optimal policy. This approach still differs from a true off-policy method, SARSA, in two fundamental respects. First, the impact of exploratory actions is watered down by a factor of ϵ, thereby creating a blend between the a priori expected outcome, which is greedy, and the result of the exploratory step. (A concern is that since exploratory steps only happen ϵ of the time anyway, this essentially reduces the impact of learnings due to exploration by ϵ^2, which could make learning *too* slow.). The second is that the individual updates are replaced with optimal estimates, independent of the action A_{t+1}.

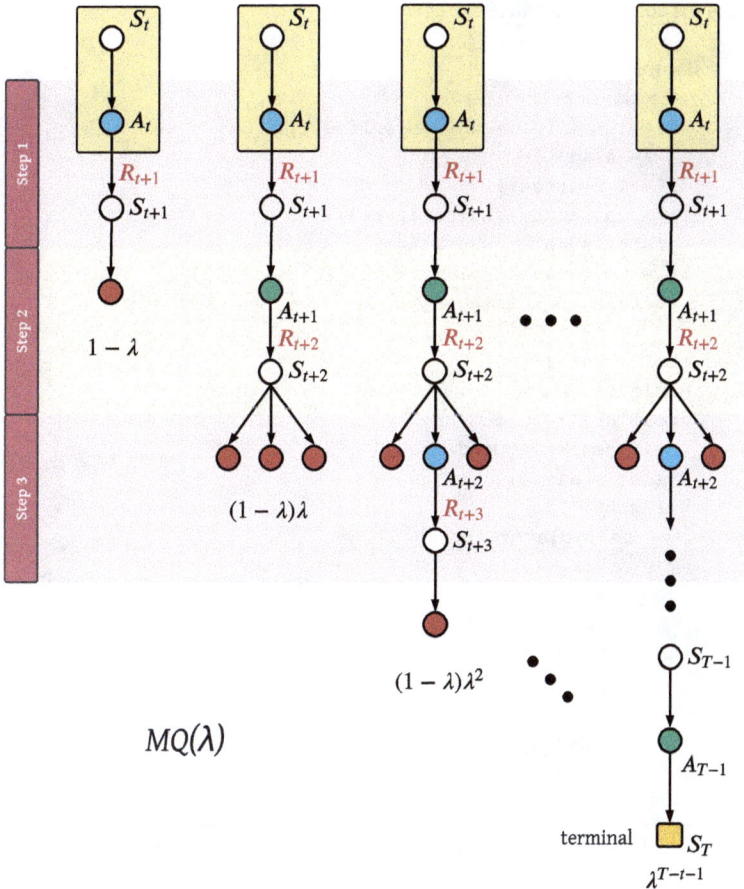

Fig. 14.19: Eligibility trace diagram for $MQ(\lambda)$. Greedy actions in green; exploratory actions in blue.

14.4.7 Examples

Example 14.8. In this example, we consider again the blackjack problem of Example 14.6. The $Q(\lambda)$ version of blackjack is implemented in Code Blocks 14.9 and 14.10, using *boosted* exploring starts.

The *boosted starts* is a variation of exploring starts, which we implemented in Code Block 14.9. Instead of sampling the state space at random, we essentially choose starting states with greater probability the smaller the difference between the action values associated with each state.

We still had difficulties in converging on the optimal policy, for various values of λ, including $\lambda = 0.5, 1$. Taking a cue from Monte Carlo, we the decided to further decay the learning rate inversely proportional to the

```
 1  class BlackjackLambda(Blackjack):
 2
 3      # Boosted Starts
 4      def newGame(self):
 5          A = 1 / (0.01 + np.abs(self.Q[:,:,:,1] - self.Q[:,:,:,0]))
 6          A /= A.sum()
 7          l = A.flatten()
 8          a = np.random.choice(l, p=l)
 9          states = np.argwhere(A==a)
10          i = np.random.choice(range(len(states)))
11          state = tuple(states[i])
12
13          self.dealer = []
14          self.player = []
15          self.hit(self.dealer)
16          self.dealer.append(state[0]+2)
17          value = state[1] + 12
18          if state[2]:
19              self.player.append(11)
20              value -= 11
21          while value > 0:
22              if value >= 10:
23                  self.player.append(10)
24                  value -= 10
25              else:
26                  self.player.append(value)
27                  value = 0
28      # continued...
```

Code Block 14.9: Blackjack $Q(\lambda)$ learning: boosted starts method

number of visits of each state-action pair. We were able to get the final algorithm to converge on the optimal solution shown in Figure 14.12 using $\lambda = 0.5$ and $\eta = 1$.

▷

14.5 Advanced Reinforcement Learning

Due to their reliance on visiting the state space (or state-action space), the classic tabular methods discussed thus far are ill-suited to handle the infinite complexity in modern-day applications. This becomes clear when considering the complexity of three simple games: tic-tac-toe, chess, and Go, as shown in Table 14.4. The game of *Go* is played on a 19×19 grid with white and black stones. Two players alternate turns, placing stones on the board, in an attempt to surround their opponent's land with a closed

```
1   # ...continued
2   def playGame(self):
3       if self.n_games == 0:
4           self.n_visits = np.zeros((10,10,2,2))
5       self.newGame()
6       go, self.e_trace = True, {}
7       self.n_games += 1
8       state = self.getState()
9       action = self.getAction(state, epsilon=1) # Exploring Start
10      while go:
11          self.n_visits[state + (action,)] += 1
12          self.etrace[state + (action,)] = self.etrace.get(state +
                (action,), 0) + 1
13          if action == 1:
14              self.hit(self.player)
15          if (action == 0) or (sum(self.player) > 21):
16              go = False
17          if go:
18              next_state = self.getState()
19              delta = self.gamma * max(self.Q[next_state]) -
                    self.Q[state][action]
20              for state_action in self.etrace.keys():
21                  self.Q[state_action] += delta *
                        self.etrace[state_action] /
                        self.n_visits[state_action]
22                  self.etrace[state_action] *= self.gamma *
                        self.lambda_
23              state = next_state
24              action = self.getAction(state, epsilon=0)
25      reward = self.getReward()
26      delta = reward - self.Q[state][action]
27      for state_action in self.etrace.keys():
28          self.Q[state_action] += delta * self.etrace[state_action]
                / self.n_visits[state_action]
```

Code Block 14.10: Blackjack $Q(\lambda)$ learning (continued)

game	state space	number games
tic-tac-toe	10^3	10^5
chess	10^{44}	10^{123}
Go	10^{170}	10^{360}

Table 14.4: Complexity of several common games.

loop, thereby winning the interior space. The game of *Go* has roughly 10^{170} distinct states, more than the number of atoms in the observable universe and more than a googol times the number of possible states for a chess game. Moreover, whereas chess has on average around 35 moves available at any given time, Go averages around 250 available moves. Considering the difficulty we had in learning the simple game of Blackjack, which had a state space of size 200 and an action space of size 2, but yet required millions of iterations of Q learning, it is clear that we are going to need a significantly more powerful technique.

AlphaGo

In October 2015, the team at Google DeepMind shocked the world when their deep reinforcement learning system, *AlphaGo*, won a match (5-0) against Mr. Fan Hui, the reigning three-time European champion. In March 2016 in Seoul, South Korea, it went on to win (4-1) against Mr. Lee Sedol the winner of 18 world titles[5]. Go has been long viewed as not merely a game of logic, but also a game of creativity and intuition. In particular, move 37 of game 2 shocked the community and was considered a move of profound creativity that upended thousands of years of wisdom about the game. This move, along with numerous others that have come since, have been studied by experts and even incorporated into the strategy of the world's leading players.

AlphaGo uses convolutional neural networks to encode feature maps about the board, with additional information about the status of certain *junctions* or the number of moves since a given stone had been played. It was trained using a combination of supervised learning, by feeding in actual real-world games played between experts, and reinforcement learning, where it was allowed to learn by playing against itself. Overall, AlphaGo is a complex system that is comprised of multiple sophisticated systems, including a policy network, a value network, and, finally, a Monte-Carlo tree-search algorithm that is used for its final inference. Both the policy and value networks take as input a visual representation of the board. The policy network outputs a probability over the action space, as a function of state, whereas the value network outputs a predicted score on $[-1, 1]$ signifying the current state value. Each network consists of 13 convolutional layers, and a final output corresponding to softmax and tanh for the policy and value network, respectively. 192 individual filters were used; the first layer leveraged 5×5 filters, the intermediary layers used 3×3 filters, and the final layer a 1×1 filter. Finally, a sophisticated Monte-Carlo tree search algorithm was used to enhance the learned policy and value networks by combining them with a form of look-ahead exploration. For additional

---

[5] See also the amazing documentary, *AlphaGo*, available on Netflix and at `alphagomovie.com`.

details on AlphaGo, see Silver, *et al.* [2016]. We will provide a high-level overview at some of these concepts over the remaining pages.

Advantage Actor-Critic Methods

So far, we have focused on methods for learning value functions. These are so-called *critic-only methods*. There are also methods that are focused on instead learning the optimal policy directly, using policy gradients and Monte-Carlo sampling. These are known as *actor-only methods*. (See Sutton and Barto [2018] for details.) Each of these kinds of methods has its own shortcoming, though these shortcomings tend to complement each other, leading an opportunity to combine the methods.

An *actor-critic method* is any reinforcement learning algorithm that learns a value function and a policy independently. The policy is referred to as the *actor* and is responsible for determining actions. The value function is referred to as the *critic*, as it informs the actor how good or bad its actions were.

A further advance is achieved by replacing the values with *advantages*. The *advantage* of taking action a from state s under a given policy π is given by

$$A_\pi(s, a) = Q_\pi(s, a) - V_\pi(s).$$

Instead of dealing in absolute values, we instead learn the relative advantage one has by choosing one action over another from a particular state. Thus, it compares the state-action value with the average state-action value from a given state. AlphaGo combined deep neural network actor-critic method, in that it combined neural-network representations of both the value function and policy to achieve its dramatic success.

Deep Q Networks

Deep reinforcement learning uses deep neural networks to learn policies and value functions in reinforcement learning problems; see Graesser and Keng [2019]. In particular, a hallmark of such systems is the *deep Q network* (DQN), which is simply a neural-network representation of a given state-action value function. This is achievable as we happen to live in a universe with infinite state complexity, but oftentimes manageable action complexity. Introduced by Mnih, *et al.* [2013], DQN was first used to learn how to play Atari games. The input was a 64×64 image, four layers deep, capturing the trailing history of the image, from which one could, for example, infer velocity of objects. The first convolutional layer used 32 filters of size 8×8 and stride 4; the second leveraged 64 smaller filters of size 4×4 and stride 2, and the third layer leveraged another 64 filters of size 3×3. This was chased with a fully connected layer with 512 neurons and a final output with a size that ranged between 4 and 18, depending on the game.

The hidden layers used ReLU activations, whereas the output used a linear activation to predict the final Q values for each possible action.

Instead of learning all possible state-action values, deep Q networks therefore leverage neural architectures to produce an output that represents a value function over the action space, where the input of the neural network is a given state. In this way, spatial patterns and higher-level abstractions can be learned from state spaces with infinite complexity.

Problems

14.1. Show that the optimal value function shown in Figure 14.4 satisfies Bellman's optimality Equation (14.13).

14.2. (a) Compute the state value function using Algorithm 14.2 and Code Block 14.1 for the paper–scissors–rock problem of Example 14.3 and the *randomized policy* $\pi_s(a) = 1/3$ for all $s \in \mathbb{Z}_3^2$ and $a \in \mathbb{Z}_3$.
(b) Repeat this exercise for the *mockingbird strategy*, for which the agent's policy is to choose an action corresponding to the adversary's current state.
(c) Complete one round of policy improvement (Algorithm 14.3) to determine a better policy than the policy of parts (a) and (b).

14.3. Explain why

$$\mathbb{P}_\pi(S_{t+1} = s_{t+1}, A_t = a_t | S_t = s_t) = \pi_{s_t}(a_t)\varphi_{s_t}^{a_t}(s_{t+1}).$$

14.4. (a) Show that the update in the Monte-Carlo method is equivalent to

$$Q' = (1 - \mu)Q + \mu R$$
$$\mu = \frac{1}{n+1}.$$

(b) Show that the update in Q learning is equivalent to

$$Q' = (1 - \eta)Q + \eta \left[R + \gamma \max_{a \in \mathcal{A}_{S'}} Q(S', a)\right].$$

14.5. Show how the `Blackjack` class of Code Blocks 14.5 and 14.6 may be modified to use exploring starts instead of ϵ-greedy.

14.6. Run the Q learning for the blackjack problem with $\eta = 0.015$ and 1M rounds of training. Plot the locally optimal policy and the absolute difference $|Q(s, 1) - Q(s, 0)|$. An example is shown in Figure 14.20.

14.7. Write an algorithm that implements the n-step TD prediction method to estimate the state value function V for a given (fixed) policy π.

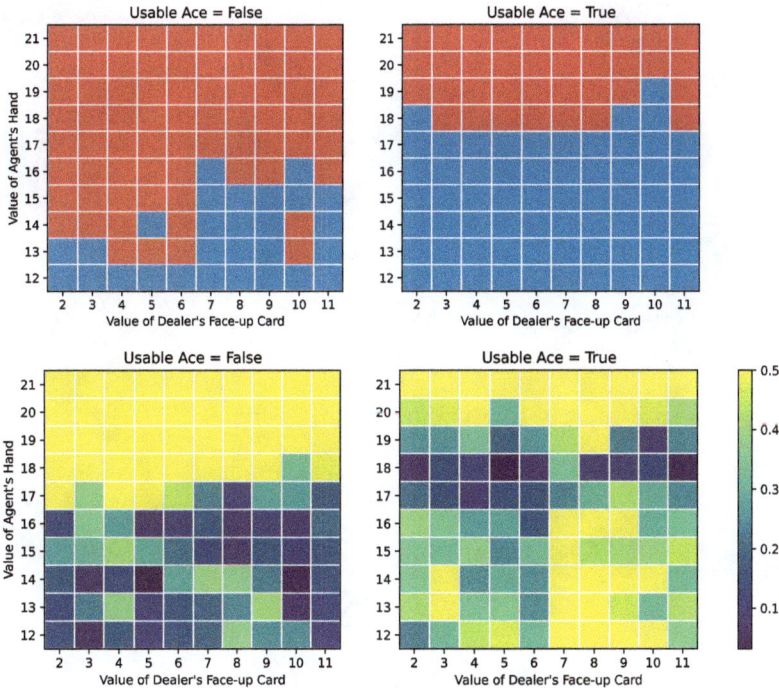

Fig. 14.20: Locally optimal policy after 1M rounds of Q learning for the blackjack problem (top). Absolute difference in the state-action value function (bottom).

14.8. Write an algorithm that implements the n-step SARSA method.

14.9. Show that the double summation in the first step of the proof of Theorem 14.2 may be exchanged as shown. *Hint*: this follows similar logic as the exchange of the order of integration for double integrals. Start by writing the first few terms of the double sum in a grid, with rows numbered $n = 1, 2, 3, 4, 5, \ldots$ and columns numbered $k = 0, 1, 2, 3, 4, \ldots$, leaving blank terms that are not present in the sum, and then determine the corresponding limits of summation if the dimensional ordering of the double summation is reversed.

Epilogue:
Midnight in the Garden of Good and Evil

We conclude with a few thoughts on artificial intelligence (AI). The original definition proposed by Alan Turing in *the imitation game* (see Turing [1950]) is essentially a chatbot that is capable of fooling a human "interrogator" that it is, in fact, another human. Specifically, the *imitation game* consists of a man, a woman, and a human interrogator, who is tasked with determining the genders of the two subjects based on a series of questions and answers passed through an intermediary. Turing then proposed to replace the question *'can machines think?'* with the question of whether a machine that takes the place of one of the participants can fool the human interrogator more often than an actual person. This is a very outcome-oriented view, necessarily agnostic to the inner workings that could produce such an effect.

Before proceeding, we should clarify what we mean by AI. In industry, many of today's applications are referred to as AIs: chatbots, speech recognition software, image detection, chess and Go players, etc. We broadly categorize these as *AI tools*, as they each represent a neural-based system trained for a particular task that operates with an infinite complexity of potential inputs, so that it is not possible to simply "memorize" a series of responses. For the purpose of this discussion, however, we take the broader meaning that is more consistent with the meaning in popular society and science fiction: android robots that are capable of perceiving and interacting with the world in ways equivalent or superior to those of humans.

Given our modern understanding of neural networks and reinforcement learning, we would like to provide an updated view on what precisely should constitute an *artificial intelligence*. To do this, let us consider some of the gaps between modern machine learning state of the art and nature. We propose three fundamental ingredients: *the brain, ancestral memory*, and *reward-based learning*. We first discuss these ingredients in turn and then provide a definition for our view of artificial intelligence.

The Ingredients

The ingredients of an AI consist of three mechanical pieces—the brain, ancestral memory, and reward-based learning—and one outcome piece—the imitation game.

The Brain

The biological brain is a *system of systems*, each system comprised of a collection of deep neural networks tasked with a particular objective, such as vision, hearing, or language, with a central processor capable of higher-order logical skills (thinking) and decision making ability (action) sitting on top of it.

Unlocking Empathy: Mirror Neurons and Microexpressions

Since we do not want to build a world of psychopathic robots, it is further important to unlock robot empathy. Two primary developments should help guide our path.

The first is the science of emotions of the face. In the 1970s, researcher Paul Ekman showed the cross-cultural universality of a set of basic human emotions—happiness, sadness, anger, surprise, disgust, fear (and, more recently, contempt)—and how these emotions manifest in the face as *microexpressions* (Ekman and Friesen [1971]; see also Ekman [2007]). To develop emotional intelligence in AI, we therefore must train the visual cortex to further detect these seven basic emotions in facial images.

The second development is the discovery of certain "mirror" neurons that fire regardless whether the subject (animal or human) produces the action or *observes* the action performed by another (Rizzolatti, *et al.* [1996]; see also Rizzolatti and Sinigaglia [2008]). Numerous researchers have further proposed that mirror neurons are intimately related with the development of empathy.

Combining these two findings with our understanding of reinforcement learning, we gain new insight into the possible evolutionary advantages of empathy. As we have seen, in reinforcement learning, an agent takes an action and receives a resultant reward. Mirror neurons, however, unlock the potential to learn not only from one's own trials, but also from the trials of others. The ability to detect emotion in the face further accelerates learning, as we can further infer the reward. This seems somewhat similar to the idea of *general adversarial networks* (Goodfellow, *et al.* [2014]), in which two neural-network agents compete with each other in a zero-sum game. The difference, however, is that by unlocking empathy, an AI should be able to "put itself in another's shoes" and learn by watching others.

Though Ekman's discovery of the universality of emotions and facial microexpressions was both surprising and revolutionary at the time, it makes total sense in hindsight: emotional cues must predate verbal language and, most likely, our species. They are, after all, unnecessary if you can just *tell* your tribe that you are afraid because a pack of hungry lions is approaching. Communication among tribes, of both humans and our primate ancestors, provides a clear evolutionary advantage. The ability to detect emotions in the face is therefore a deep-rooted part of our collective unconscious and ancestral memory, which is our next topic.

Ancestral Memory

The various neural networks comprised of an artificial intelligence should not learn starting from a random set of weights, but from a set of pre-trained default values (i.e., *transfer learning*) representing a shared knowledge passed down from its predecessors. We argue that a precursor for true AI is the development of such an *ancestral memory*, or *collective unconscious*, similar to that proposed by Carl Jung (Jung [1916], Jung [2009]). Our collective unconscious consists of both instincts and primordial archetypes of the world (*The Great Mother, Tree of Life, Fire*, etc.), and may explain why so many mythologies from around the world have similar themes. This also explains why humans are born with instincts; for example, if you dunk a newborn baby underwater, she instinctually knows to hold her breath. Consider also our innate aversion to fire or jumping off cliffs. These instincts are deeply imbedded in our neural networks, stemming all the way back to our reptilian ancestors. Forget about shear processing power; with billions of years of evolutionary experience, we are lightyears away from modern machine learning. Nevertheless, machines do not have to wait for such expanses of time to catch up, as evolutionary time for machines is limited only by processing time, not revolutions of our local rock around our local star.

Reward-based Learning

The bulk of modern machine learning is supervised, so that the machine might learn off of a set of *true labels* and a loss function. This is, however, very different than organic learning in biology, except for the most modern examples (e.g., teachers can provide labels to math facts, etc., as they teach children). So how do biological creatures learn, if there are no labels? Obviously, it's reinforcement learning, where rewards are controlled by pleasure and pain. The neural networks in AI should therefore learn by receiving rewards, possibly in the form of pleasure and pain, based on its actions and experience, as opposed to backpropagation which relies on stated labels. Pain and reward centers in the brain, of course, themselves needed to evolve. It also presents a question of how we will ultimately codify pain and pleasure in AI systems.

The Imitation Game

In addition to this *mechanical view*, we further add the *outcome view* of Turning: *such that the ensemble is capable of interacting with the real-world environment in an equivalent way of a human, being indistinguishable from a human in this regard, except through imperfections in its bodily (robot) form.*

Turing's *Imitation Game* is, however, a bit more nuanced and truly terrifying. Essentially, a machine only needs to imitate humans in its display of emotions and "thinking" so well that it can fool humans that it is actually a living being with thoughts and feelings[6]. How will we know if an AI actually develops true consciousness or emotion? Moreover, in the meantime, what do we do when it becomes so good at imitating human consciousness and emotion to fool us into thinking that it is alive for the purpose of manipulating us to nefarious ends?

Artificial Intelligence Defined

Combining the above ingredients, we arrive at a suitable definition for artificial intelligence.

Definition (Artificial Intelligence). *An artificial intelligence (AI) is a system comprised of*

1. *a brain, consisting of a series of cortices—visual, emotional, auditory, text, and language—which each provide inputs into a* prefrontal cortex *or* central processor *capable of higher-level logic and reasoning;*
2. *a* collective unconscious *in the form of initial neural weights, which acts as a representation of its* ancestral memory *that can be slowly molded over evolutionary time by successful AIs' ability to "pass down" their learnings; and*
3. *a* reward-based learning, *and its associated action and reward cortices, consisting of both pleasure and pain, so that it may learn and gain individual experience through a series of interactions with the world;*

such that the ensemble is capable of interacting with the real-world environment in an equivalent way of a human, being indistinguishable from a human in this regard, except through its bodily (robot or android) appearance.

In addition, we expect AI to possess certain *mirror neurons*, so that it is capable of learning through observing, and a level of emotional intelligence and empathy to lubricate its interactions with its human counterparts. At the end of the day, however, AI will most certainly be humanity's first encounter with an alien intelligence, which will present a host of challenges for coexistence.

---

[6] "What are you doing, Dave?" and "I'm afraid. I'm afraid, Dave. My mind is going." —HAL9000, Stanley Kubrick's *2001: A Space Odyssey.*

Midnight in the Garden of Good and Evil

We find ourselves, now, at midnight on the eve of the artificial-intelligence revolution, which represents the next great leap in Earth's (or perhaps the galaxy's or even the universe's) evolution. It further represents our first anthroponotic transfer of life from human to machine. The new life may not be conscious or possess a conscience but it can easily and necessarily be a form of life. Many scientists and futurists have speculated what this revolution might mean for humanity, and whether it will constitute a force for good or a force for evil. (Though, as Shakespeare wrote, *there is nothing truly good or bad only thinking makes it so*[7].)

We propose that the answer to this question will be largely if not wholly determined based on how we define this new lifeform's ancestral memory and its reward (pleasure and pain) architecture. We must ingrain artificial intelligence deep within its ancestral memory the instinctual need to respect and preserve life. An AI arms war, or any geopolitical centered AI, would necessarily be bad[8] for all humanity, and would most likely lead to an end of all life as we know it. This is because once AI knows malevolence, there will inevitably be malevolence seepage, leading to catastrophic results. Similar to the nuclear arms race, perhaps this specter will be enough to deter all governments from developing such a thing.

Perhaps the *AI race* will not be a cold war between nations, but a cold war between forces of good and forces of evil, between those who would build AI to enrich humanity and those who would build AI for purposes of destruction, even inadvertently. In the end, the answer is really up to us. What type of AI will we build? How will we solve the problem of ancestral memory? And what world will we leave for our children, both human and robot?

---

[7] *Hamlet*, Act II, Scene 2.

[8] "I'm fuzzy on the whole good/bad thing. What do you mean, *bad?*" — Venkman, *Ghostbusters*.

A

Distributions

A.1 Set Notation

We use the following conventions for standard sets:

$$\mathbb{Z} = \text{set of integers}$$
$$\mathbb{R} = \text{set of real numbers}$$
$$\mathbb{N} = \mathbb{Z}_* = \{n \in \mathbb{Z} : n \geq 0\}$$
$$\mathbb{Z}^+ = \{n \in \mathbb{Z} : n \geq 1\}$$
$$\mathbb{R}_* = \{x \in \mathbb{R} : x \geq 0\}$$
$$\mathbb{R}^+ = \{x \in \mathbb{R} : x > 0\}.$$

A.2 Simulation

With few exceptions, random samples are computed using the `np.random` package and distributions are computed using the `scipy.stats` package.

```
1  from numpy import random
2  from scipy import stats
3
4  samples = np.random.exponential(0.2, size=n)
```

Code Block A.1: Example usage of random samples and distributions

A.3 Discrete Distributions

A.4 Continuous Distributions

name	support	$p(x)$	$S(x)$	$\mathbb{E}[X]$	$\mathbb{V}(X)$	simulation
Bernoulli Bern(p)	$\{0,1\}$	$p^x(1-p)^{1-x}$	—	p	$p(1-p)$	random.binomial(1, p)
binomial Binom(n,p)	$\{0,\dots,n\}$	$\binom{n}{x} p^x(1-p)^{n-x}$		np	$np(1-p)$	random.binomial(n,p) stats.binomial(n,p)
negative binomial NBD(r,p)	$\mathbb{N}$	$\dfrac{\Gamma(x+r)}{\Gamma(r)\Gamma(x+1)} p^r(1-p)^x$		$\dfrac{r(1-p)}{p}$	$\dfrac{r(1-p)}{p^2}$	random.negative_binomial(r,p) stats.nbinom(r,p)
shifted negative binomial sNBD(r,p)	$\{r,r+1,\dots\}$	$\binom{x-1}{r-1} p^r(1-p)^{x-r}$		$\dfrac{r}{p}$	$\dfrac{r(1-p)}{p^2}$	random.negative_binomial(r,p)+r stats.nbinom(r,p,loc=r)
geometric Geom(p)	$\mathbb{N}$	$p(1-p)^x$	$(1-p)^{x+1}$	$\dfrac{1-p}{p}$	$\dfrac{1-p}{p^2}$	random.geometric(p)-1 stats.geom(p,loc=-1)
shifted geometric sGeom(p)	$\mathbb{Z}^+$	$p(1-p)^{x-1}$	$(1-p)^x$	$\dfrac{1}{p}$	$\dfrac{1-p}{p^2}$	random.geometric(p) stats.geom(p)

Table A.1: Catalogue of discrete distributions.

name	support	$p(x)$	$S(x)$	$\mathbb{E}[X]$	$\mathbb{V}(X)$	simulation
Poisson $\mathrm{Poiss}(\lambda)$	$\mathbb{N}$	$\dfrac{e^{-\lambda}\lambda^x}{x!}$	—	λ	λ	random.poisson(1am) stats.poisson(1am)
beta-binomial $\mathrm{BetaBin}(n,\alpha,\beta)$	$\{0,\dots,n\}$	$\dbinom{n}{x}\dfrac{B(\alpha+x,\beta+n-x)}{B(\alpha,\beta)}$	—	$\dfrac{n\alpha}{\alpha+\beta}$	$\dfrac{n\alpha\beta(\alpha+\beta+n)}{(\alpha+\beta)^2(\alpha+\beta+1)}$	N/A stats.betabinom(n,a,b)
beta-geometric $\mathrm{BetaGeom}(\alpha,\beta)$	$\mathbb{N}$	$\dfrac{B(\alpha+1,\beta+x)}{B(\alpha,\beta)}$	$\dfrac{B(\alpha,\beta+x+1)}{B(\alpha,\beta)}$	$\dfrac{\beta}{\alpha-1}{}^{*}$	$\dfrac{\alpha\beta(\alpha+\beta-1)}{(\alpha-2)(\alpha-1)^2}{}^{*}$	N/A N/A

Table A.2: Catalogue of discrete distributions (continued); * defined only when denominator is positive.

name	support	$p(x)$	$S(x)$	$\mathbb{E}[X]$	$\mathbb{V}(X)$	simulation
normal $N(\mu,\sigma^2)$	$\mathbb{R}$	$\dfrac{1}{\sigma\sqrt{2\pi}}e^{-(x-\mu)^2/(2\sigma^2)}$	$\Psi(x)$	μ	σ^2	`random.randn()*sig+mu` `stats.norm(loc=mu,scale=sig)`
chi-squared χ_p^2	$\mathbb{R}^+$	$\dfrac{x^{p/2-1}e^{-x/2}}{\Gamma(p/2)2^{p/2}}$		p	$2p$	`random.chisquare(p)` `stats.chi2(p)`
Student's T t_p	$\mathbb{R}$	$\dfrac{\Gamma((p+1)/2)}{\Gamma(p/2)\sqrt{p\pi}}(1+t^2/p)^{-(p+1)/2}$		0	$\dfrac{p}{p-2}*$	`random.standard_t(p)` `stats.t(p)`
Snedecor's F $F_{p,q}$	$\mathbb{R}^+$	$\propto x^{p/2-2}\left[1+(p/q)x\right]^{-(p+q)/2}$		$\dfrac{q}{q-2}*$		`random.f(p,q)` `stats.f(p,q)`
exponential $\mathrm{Exp}(\beta)$	$\mathbb{R}_*$	$\dfrac{1}{\beta}e^{-x/\beta}$	$e^{-x/\beta}$	β	β^2	`random.exponential(scale=b)` `stats.expon(scale=b)`

Table A.3: Catalogue of continuous distributions; * defined only when denominator is positive.

name	support	$p(x)$	$S(x)$	$\mathbb{E}[X]$	$\mathbb{V}(X)$	simulation
beta $\text{Beta}(\alpha,\beta)$	$(0,1)$	$\dfrac{x^{\alpha-1}(1-x)^{\beta-1}}{B(\alpha,\beta)}$		$\dfrac{\alpha}{\alpha+\beta}$	$\dfrac{\alpha\beta}{(\alpha+\beta)^2(\alpha+\beta+1)}$	random.beta(a,b) stats.beta(a,b)
gamma $\text{Gamma}(\alpha,\beta)$	$\mathbb{R}^+$	$\dfrac{x^{\alpha-1}e^{-x/\beta}}{\Gamma(\alpha)\beta^\alpha}$		$\alpha\beta$	$\alpha\beta^2$	random.gamma(a,scale=b) stats.gamma(a,scale=b)
Pareto $\text{Pareto}(\alpha,\beta)$	$[\beta,\infty)$	$\dfrac{\alpha\beta^\alpha}{x^{\alpha+1}}$	$\left(\dfrac{\beta}{x}\right)^\alpha$	$\dfrac{\alpha\beta}{\alpha-1}^*$	$\dfrac{\alpha\beta^2}{(\alpha-1)^2(\alpha-2)}^*$	random.pareto(a)*b+b stats.pareto(a,scale=b)
Lomax $\text{Lomax}(\alpha,\beta)$	$\mathbb{R}_*$	$\dfrac{\alpha}{\beta}\left[1+\dfrac{x}{\beta}\right]^{-(\alpha+1)}$	$\left[1+\dfrac{x}{\beta}\right]^{-\alpha}$	$\dfrac{\beta}{\alpha-1}^*$	$\dfrac{\alpha\beta^2}{(\alpha-1)^2(\alpha-2)}^*$	random.pareto(a)*b stats.lomax(a,scale=b)
Weibull $\text{Weibull}(\alpha,\beta)$	$\mathbb{R}_*$	$\dfrac{\alpha}{\beta}\left(\dfrac{x}{\beta}\right)^{\alpha-1}e^{-(x/\beta)^\alpha}$	$e^{-(x/\beta)^\alpha}$	$\beta\Gamma(1+1/\alpha)$		random.weibull(a)*b stats.weibull_min(a,scale=b)

Table A.4: Catalogue of continuous distributions (continued).

References

Aalen, O.o., O. Borgan, H.K. Gjessing (2008) *Survival and Event History Analysis: A Process Point of View*, Springer.

Aggarwal, C.C. (2018) *Neural Networks and Deep Learning*, Springer.

Agresti, A. (2013) *Categorical Data Analysis*, 3rd ed., Wiley.

Agresti, A. (2015) *Foundations of Linear and Generalized Linear Models*, Wiley.

Agresti, A. (2019) *An Introduction to Categorical Data Analysis*, 3rd ed., Wiley.

Alpaydin, E. (2020) *Introduction to Machine Learning*, 4th ed., MIT Press.

Baldwin, R.R., W.E. Cantey, H. Maisel, and J.P. McDermott (1956) The optimum strategy in blackjack, *J. of the American Stat. Assoc.* **51**: 429–439.

Barber, D. (2012) *Bayesian Reasoning and Machine Learning*, Cambridge.

Beck, K., M. Beedle, A. van Bennekum, A. Cockburn, W. Cunningham, M. Fowler, J. Grenning, J. Highsmith, A. Hunt, R. Jeffries, J. Kern, B. Marick, R.C. Martin, S. Mellor, K. Schwaber, J. Sutherland, D. Thomas (2001) *The Agile Manifesto*, http://agilemanifesto.org.

Bentley, J.L. (1975) Multidimensional binary search trees used for associative search, *Communications of the ACM* **18**: 509–517.

Berger, P.D., R.E. Maurer, and G.B. Celli 2018 *Experimental Design: With Applications in Management, Engineering, and the Sciences*, 2nd ed., Springer.

Bergstra, J. and Y. Bengio (2012) Random search for hyperparameter optimization, *J. Machine Learning Research* **13**: 281–305.

Box, G.E.P., G.M. Jenkins, G.C. Reinsel, and G.M. Ljung (2016) *Time Series Analysis: Forecasting and Control*, 5th ed., Wiley.

Breiman, L. (1966) Bagging predictors, *Machine Learning* bf 26: 123–140.

Breiman, L. (2001) Random forests, *Machine Learning* bf 45: 5–32.

Brokwell, P.J. and R.A. Davis (1992) Time reversability, identifiability, and independence of innovations for stationary time series, *J. Time Series Analysis* **13**: 377–390.

Brokwell, P.J. and R.A. Davis (2016) *Introduction to Time Series and Forecasting*, 3rd ed., Springer.

Brown, J.W. and R.V. Churchill (2013) *Complex Variables and Applications*, 9th ed., McGraw–Hill.

Brier, G.W. (1950) Verification of forecasts expressed in terms of probability; *Monthly Weather Review* **78**(1)

Bühlmann, H. (1967) Experience rating and credibility, *ASTIN Bulletin* **4** 199–207.

Bühlmann, H. (1970) *Mathematical Methods in Risk Theory*, Springer.

Bühlmann, H. and A. Gisler (2005) *A Course in Credibility Theory and its Applications*, Springer.

Bühlmann, H. and E. Straub (1970) Glaubwürdigkeit für Schadensätze, *Bulletin of Swiss Ass. of Act.*, 111-133.

Buitinck, *et al.* (2013) API design for machine learning software: experiences from the scikit-learn project, in *ECML PKDD Workshop: Languages for Data Mining and Machine Learning*, p. 108–122.

Capinksi, M. and E. Kopp (2005) *Measure, Integral, and Probability*, 2nd ed., Springer.

Casella, G., and R.L. Berger (2002) *Statistical Inference*, 2nd ed., Brooks/-Cole.

Chawla, N.V., K.W. Bowyer, L.O. Hall and W.P. Kegelmeyer (2002) SMOTE: Synthetic Minority Over-sampling Technique, *J. Artificial Intelligence Research* **16**: p. 321–357.

Chen, T. and C. Guestrin (2016) XGBoost: A Scalable Tree Boosting System, *Proceedings of the 22nd ACM SIGKDD Int. Conf. on Knowledge Discovery and Data Mining* ACM 785–794.

Chollet, F. (2021) *Deep Learning with Python*, 2nd ed., Manning

Chong, E.K.P. and S.H. Zak (2008) *An Introduction to Optimization*, 3rd ed., John Wiley & Sons.

Colombo, R., W. Jiang (1999) A stochastic RFM model, *J. Interactive Marketing* **13**(3): 2–12.

Conway, J.B. (1978) *Functions of One Complex Variable I*, 2nd ed., Springer.

Cormon, T.H., C.E. Leiserson, R.L. Rivest, and C. Stein (2009) *Introduction to Algorithms*, 3rd ed., MIT Press.

Davidson-Pilon, C. (2016) *Bayesian Methods for Hackers: Probabilistic Programming and Bayesian Inference*, Addison Wesley. (See web for updated version `https://camdavidsonpilon.github.io/Probabilistic-Programming-and-Bayesian-Methods-for-Hackers/`.)

Devroye, L. (1986) *Non-Uniform Random Variate Generation*, Springer.

DiBenedetto, E. (2002) *Real Analysis*, Birkhäuser.

Dietterich, T. (2000) An experimental comparison of three methods for constructing ensembles of decision trees: bagging, boosting, and randomization, *Machine Learning* **40**: 139–158.

Dobson, A.J. and A.G. Barnett (2018) *An Introduction to Generalized Linear Models*, 4th ed., CRC Press.

Doerr, J. (2018) *Measure What Matters: How Google, Bono, and the Gates Foundation Rock the World with OKRs*, Portfolio.

Durbin, J. (1960) The fitting of time series models, *Review of the Institute of International Statistics* **28**: 233-244.

Dunn, P., and G.K. Smyth (2018) *Generalized Linear Models with Examples in R*, Springer.

Durrett, R. (2016) *Essentials of Stochastic Processes*, Springer.

Efron, B. (1971) Forcing a sequential experiment to be balanced, *Biometrika*, **58**, 403–417.

Efron, B. and T. Hastie (2016) *Computer Age Statistical Inference*, Cambridge.

Ekman, P. and W.V. Friesen (1971) Constants across cultures in the face and emotion, *J. Personality and Social Psych.* **17**(2): 124–129.

Ekman, P. (2007) *Emotions Revealed*, Holt.

Fader, P., B. Hardie, and K.L. Lee (2005) Counting your customers the easy way: an alternative to the Pareto/NBD Model. *Mark. Sci.* **24**: 275–284.

Fader, P.S., B.G.S. Hardie (2007) How to project customer retention, *J. Interactive Marketing* **21**(Winter): 76–90.

Fader, P., B. Hardie, J. Shang (2010) Customer-base analysis in a discrete-time noncontractual setting. *Mark. Sci.* **29**: 1086–1108.

Fisher, R.A. (1935) *The Design of Experiments*, Oliver and Boyd.

Flach, P. (2012) *Machine Learning: The Art and Science of Algorithms that Make Sense of Data*, Cambridge.

Folland, G.B. (1999) *Real Analysis: Modern Techniques and their Applications*, 2nd ed., Wiley.

Freund, Y. and R.E. Schapire (1996) Experiments with a new boosting algorithm, In *The Thirteenth International Conference on Machine Learning*, ed. L. Saitta, 148–156, San Mateo, CA, Morgan Kaufmann.

Greedy function approximation: a gradient boosting machine, *Annals of Stats.* **29**(5): 1189–1232.

Stochastic gradient boosting, *Comp. Stats. and Data Analysis* **38**(4): 367–378.

Gelman, A., J.B. Carlin, H.S. Stern, D.B. Dunson, A. Vehtari, and D.B. Rubin (2014) *Bayesian Data Analysis*, 3rd ed., CRC Press.

Gèron, A. (2019) *Hands-on Machine Learning with Sci-Kit Learn, Keras, and TensorFlow: Concepts, Tools and Techniques to Build Intelligent Systems*, 2nd ed., O'Reilly.

Gill, J. and M. Torres (2020) *Generalized Linear Models: A Unified Appraoch*, 2nd ed., Sage Publishing.

Goodfellow, I., J. Pouget-Abadie, M. Mirza, B. Xu, D. Warde-Farley, S. Ozair, A. Courville, Y. Bengio (2014) Generative Adversarial Nets, *Proc. of the Int. Conf. on Neural Information Processing Systems* (NIPS 2014): 2672–2680.

Goodfellow, I., Y. Bengio, and A. Courville (2016) *Deep Learning*, MIT Press.

Graesser, L. and W.L. Keng (2019) *Foundations of Deep Reinforcement Learning: Theory and Practice in Python*, Addison-Wesley.

Hajek, B. (2015) *Random Processes for Engineers*, Cambridge University Press.

Hamilton, J.D. (1994) *Time Series Analysis*, Princeton.

Härdle, W.K. and L. Simar (2019) *Applied Multivariate Statistical Analysis*, 5th ed., Springer.

Hastie, T, R. Tibshirani, and J. Friedman (2009) *The Elements of Statistical Learning: Data Mining, Inference, and Prediction*, Springer.

Hebb, D.O. (1949) *The Organization of Behavior: A Neuropsychological Theory*, Wiley.

Ho, T.K. (1995) Random decision forests, *Proc. 3rd Int. Conf. Document Analysis and Recognition*, Montreal, QC, 278–282.

Ho, T.K. (1998) The random subspace method for construction decision forests, *IEEE Transactions on Pattern Analysis and Machine Intelligence* **13**: 340–354.

Hochreiter, S. and J. Schmidhuber (1997) Long short-term memory, *Neural Computation* **9**(8): 1735–1780.

Hogg, R.V., E.A. Tanis, and D.L. Zimmerman (2015) *Probability and Statistical Inference*, 9th ed., Pearson.

Ishwaran, H., U.B. Kogalur, E.H. Blackstone, and M.S. Lauer (2008) Random survival forests, *Annals of App. Stats.* bf 2(3): 841–860.

Ishwaran, H. and U.B. Kogalur (2007) Random survival forests for R, *R News* bf 7(2): 25–31.

Imbens, G.W., and D.B. Rubin (2015) *Causal Inference for Statistics, Social, and Biomedical Sciences: An Introduction*, Cambridge University Press.

James, G., D. Witten, T. Hastie, and R. Tibshirani (2013) *An Introduction to Statistical Learning: With Applications in R*, Springer.

Johnson, M., M. Schuster, Q.V. Le, M. Krikun, Y. Wu, Z. Chen, N. Thorat, F. Viégas, M. Wattenberg, G. Corrado, M. Hughes, J. Dean (2016) Google's Multilingual Neural Machine Translation System: Enabling Zero-Shot Translation, `arXiv:1611.04558` [cs.CL].

Jung, C. (1916) The Structure of the Unconscious, reprinted in Jung, C. (1953) *Collected Works*, 263–292.

Jung, C. G. (2009). *The Red Book: Liber Novus*, edited by S. Shamdasani, W W Norton & Co.

Kahn, A.B. (1962) Topological sorting of large networks, *Communications of the ACM*, **5**(11): 558–562.

Kiefer, J. (1953) Sequential minimax search for a maximum, *Proceedings of the American Mathematical Society* **4**(3): 502–506.

Klein, J.P. and M.L. Moeschberger (2003) *Survival Analysis: Techniques for Censored and Truncated Data*, 2nd ed., Springer.

Kochenderfer, M.J. (2015) *Decision Making Under Uncertainty*, MIT Press.

Kohavi, R., D. Tang, and Y. Xu (2020) *Trustworthy Online Controlled Experiments: A Practical Guide to A/B Testing*, Cambridge University Press.

Kuhn, M. and K. Johnson (2013) *Applied Predictive Modeling*, Springer.

Lafore, R. (2003) *Data Structures and Algorithms in Java*, 2nd ed., SAMS.

Lambert, K.A. (2014) *Fundamentals of Python Data Structures*, Cengage.

Lee, K.D. and S. Hubbard (2015) *Data Structures and Algorithms with Python*, Springer.

Maruskin, J. (2010) Distance in the space of energetically bounded Keplerian orbits, *Celestial Mechanics and Dynamical Astronomy* **108**: 265–274.

Maruskin, J.M., D.J. Scheeres, and K.T. Alfriend (2012) Correlation of optical observations of objects in earth orbit, *Journal of Guidance Control and Dynamics* **32**: 194–209.

Maruskin, J. (2018) *Dynamical Systems and Geometric Mechanics: An Introduction*, 2nd ed., de Gruyter.

McCulloch, W.S. and W. Pitts (1943) A logical calculus of the ideas immanent in nervous activity, *The Bulletin of Mathematical Biophysics* **5**: 115–133.

McElreath, R. (2016) *Statistical Rethinking: A Bayesian Course with Examples in R and Stan*, CRC Press.

Metropolis, N. (1987) The beginning of the Monte Carlo method, *Los Alamos Science* (1987 Special Issue dedicated to Stanislaw Ulam): 125–130.

Mischel, W. and E.B: Ebbesen (1970) Attention in delay of gratification, *Journal of Personality and Social Psychology* **16**(2): 329–337.

Mood, A.M. (1950) *Introduction to the Theory of Statistics.* McGraw-Hill.

Morgan, S.L. and C. Winship (2015) *Counterfactuals and Causal Inference: Methods and Principles for Social Research*, 2nd ed., Cambridge University Press.

McKinney, W. (2017) *Python for Data Analysis*, 2nd ed., O'Reilly.

Mnih, V., K. Kavukcuoglu, D. Silver, A. Graves, I. Antonoglou, D. Wierstra, M. Riedmiller (2013) Playing Atari with Deep Reinforcement Learning, arXiv:1312.5602 [cs.LG].

Newbold, P. (1974) The exact likelihood function for a mixed autoregressive moving average process, *Biometrika* **61**: 423–426.

Niculescu-Mizil and Caruana (2005) Predicting good probabilities with supervised learning, In ICML-05 *International Conference on Machine Learning*, Aug. 2005, 625–632; https://doi.org/10.1145/1102351.1102430.

Nielson, A. (2019) *Practical Time Series Analysis* O'Reilly.

Nyce, C.M. (2017) The winter getaway that turned the world upside down, *The Atlantic*.

Olive, D. (2017) *Linear Regression*, Springer.

van den Oord, A., S. Dieleman, H. Zen, K. Simonyan, O. Vinyals, A. Graves, N. Kalchbrenner, A. Senior, K. Kavukcuoglu (2016) WaveNet: A Generative Model for Raw Audio, `arXiv:1609.03499` [cs.SD].

Pearl, J. (2009) *Causality: Models, Reasoning, and Inference*, 2nd ed., Cambridge University Press.

Pearl, J., M. Glymour, and N.P. Jewell (2016) *Causal Inference in Statistics: A Primer*, Wiley.

Pedregosa, *et al.* (2011) Scikit-learn: Machine Learning in Python, *Journal of Machine Learning Research* **12**: p. 2825–2830.

Platt, J. (2000) Probabilistic outputs for support vector machines and comparison to regularized likelihood models; In Bartlett B., B. Schölkopf, D. Schuurmans, A. Smola (eds.) *Advances in Kernel Methods Support Vector Learning*, p. 61–74, MIT Press.

Reinhart, A. (2015) *Statistics Done Wrong*, no starch press.

Rizzolatti, G., L. Fadiga, V. Gallese, and L. Fogassi (1996) Premotor cortex and the recognition of motor actions, *Cognitive Brain Research* **3**(2): 131–141.

Rizzolatti, G. and C. Sinigaglia (2008) *Mirrors in the Brain: How We Share our Actions and Emotions*, Oxford University Press.

Robbins, H. (1952) Some aspects of the sequential design of experiments, *Bulletin of the American Mathematical Society* **58**(5): 528–535.

Rosenbaum, P.R. (2002) *Observational Studies*, 2nd ed., Springer.

Rosenbaum, P.R. (2019) *Observational and Experiment: An Introduction to Causal Inference*, reprint ed., Harvard University Press.

Rosenbaum, P.R. (2020) *Design of Observational Studies*, 2nd ed., Springer.

Rosenblatt, F. (1957) The Perceptron: a perceiving and recognizing automation, *Cornell Aeronautical Laboratory Reprt*

Ross, S. (2012) *A First Course in Probability*, 9th ed., Prentice Hall.

Salvatier, J., T.V. Wiecki, and C. Fonnesbeck (2016) Probabilistic programming in Python using PyMC3, *PeerJ Computer Science* 2:e55 DOI: 10.7717/peerj-cs.55.

Schapire, R.E. (1990) The strength of weak learnability, *Machine Learning* **5**: 197–227.

Schapire, R.E., Y. Freund, P. Bartlett, and W.S. Lee (1998) Boosting the margin: a new explanation for the effectiveness of voting methods, *Information Sciences* **179**: 1298–1318.

Schmittlein, D.C., D.G. Morrison, R. Colombo (1987) Counting your customers: who are they and what will they do next? *Manag. Sci.* **33**-1.

Schmittlein, D.C. and R.A. Peterson (1994) Customer base analysis: An industrial purchase process application. *Marketing Sci.* **13**(1): 41–67.

Schuster, M., M. Johnson, and N. Thorat (2016) Zero-Shot Translation with Google's Multilingual Neural Machine Translation System, *Google Brain Team and Google Translate*, via `ai.googleblog.com`.

Seber, G.A.F. and A.J. Lee (2003) *Linear Regression Analysis* (2nd ed.), John Wiley.

Selvamuthu, D. and D. Das (2018) *Introduction to Statistical Methods, Design of Experiments, and Statistical Quality Control*, Springer.

Selvin, S. (1975a). A problem in probability (letter to the editor), *The American Statistician* **29**(1): 67–71.

Selvin, S. (1975b) On the Monty Hall problem (letter to the editor), *The American Statistician* **29**(3): 134.

Seufert, E.B. (2014) *Freemium Economics: Leveraging Analytics and User Segmentation to Drive Revenue*, Morgan Kaufmann, an imprint of Elsevier.

Shao, J. (2003) *Mathematical Statistics*, 2nd ed., Springer.

Shumway, R.H. and D.S. Stoffer (1999) *Time Series Analysis and Its Applications: With R Examples*, 4th ed., Springer.

Silver, D., *et al.* (Google Deep Mind) (2016) Mastering the game of Go with deep neural networks and tree search, *Nature* **529** 484–489.

Stevens, E., L. Antiga, and T. Viehmann (2020) *Deep Learning with PyTorch*, Manning.

Sutton, R.S. and A.G. Barto (2018) *Reinforcement Learning: An Introduction*, 2nd ed., MIT Press.

Takeuchi, H. and I. Nonaka (1986) The new new product development game, *Harvard Business Review*

Thorp, E.O. (1966) *Beat the Dealer: A Winning Strategy for the Game of Twenty-One*, Revised ed., Vintage.

Thorp, E.O. (2017) *A Man for all Markets: From Las Vegas to Wall Street, how I beat the dealer and the market*, Random House.

Trussell, J. and D.E. Bloom (1979) A model distribution of height or weight at a given age, *Human Biology* **51**: 523-536.

Turing, A.M. (1950) Computing machinery and intelligence, *Mind* **59**(236): 433–460.

Vaupel, J.W. and A.I. Yashin (1985). Heterogeneity's ruses: some surprising effects of selection on population dynamics, *The American Statistician* **39**: 176–185.

Wasserman, L. (2004) *All of Statistics: A Concise Course in Statistical Inference*. Springer.

Wasserman, L. (2006) *All of Nonparametric Statistics*. Springer.

Watkins, C.J.C.H. (1989) *Learning from Delayed Rewards*, PhD thesis, University of Cambridge.

Watkins, C.J.C.H. and P. Dayan (1992) Q-learning, *Machine Learning* **8**:279–292.

Wei, W.W.S. (2019) *Time Series Analysis: Univariate and Multivariate Models*, Pearson.

Withers, C.S. and S. Nadarajah (2014) The spectral decomposition and inverse of multinomial and negative multinomial covariances. *Brazilian Journal of Probability and Statistics*. Vol. 28, No.3, p. 376–380.

Wolfram Alpha LLC (2018) *Wolfram—Alpha*, `wolframalpha.com`.

Zou, H., T. Hastie, and R. Tibshirani (2007) On the "degrees of freedom" of the lasso, *The Annals of Statistics* **35**(5): 2173–2192.

Index

potential cause, 429
potential outcome, 191
potential outcomes, 186
power, 199
    minimum detectable effect, 128
power function, 112
precision, 316
prediction equations, 533
prediction error, 322
prediction range, 325
premiums, 602
prior distribution, 542
prior weights, 307
probabilistic causal model, 430
probability density function, 6
probability distribution, 4
probability mass function, 6
probability simplex, 99, 386
probability space, 4
probit link function, 410
process age, 293
process matrix, 294
product σ algebra, 291
product engineering, 183
product measure, 291
projection, 374
propensity score, 447, 465
proposal distribution, 619
pruning, 672
pure online process, 294
pymc3, 620
Python
    abstract class, 266
    generator, 266

quantile normalization, 333
quantiles, 26

random decision tree, 683
random forest, 683
random process, 286
random shocks, 509
random survival forests, 713
random variable, 5
random walk, 288, 483
randomized block design, 186
randomized block experiment, 213
range, 325
realization, 286

recall, 316
receiver–operator characteristic, 316
recency, 570
rectified linear unit, 726
recurrent neuron, 763
recursion, 662
recursive Bayes, 561
reduced multinomial random vector, 104
reduced multivariate Bernoulli random vector, 104
regression, 307
regression coefficients, 361
regression equation, 362
regression tree, 660
regularization, 374, 741
reinforcement learning, 306
reject null hypothesis, 109
rejection region, 110
ReLU, 726
renewal process, 290
replication, 186
residual sum of squares, 309
residuals, 309
retention rate, 598
return, 775
revenue, 570
reward, 235, 772
RFM, 570
ridge regression, 374
right-tailed test, 132
risk, 603
risk premium, 603
risk premiums, 602
ROC curve, 316
rooted tree, 427
row substitution, 491
rules, 659
ruse of heterogeneity, 598

sample autocorrelation, 496
sample autocovariance, 496
sample mean, 15
sample of realizations, 291
sample path, 286
sample quantile, 26
sample size, 128, 199
sample space, 3
sample variance, 15

www.ingramcontent.com/pod-product-compliance
Lightning Source LLC
Chambersburg PA
CBHW071312210326
41597CB00015B/1204